17th International Conference on Aluminium Alloys 2020 (ICAA17)

Held online due to COVID-19

MATEC Web of Conferences Volume 326 (2020)

Grenoble, France
26-29 October 2020

ISBN: 978-1-7138-2099-4

Printed from e-media with permission by:

Curran Associates, Inc.
57 Morehouse Lane
Red Hook, NY 12571

Some format issues inherent in the e-media version may also appear in this print version.

For additional information, please contact EDP Sciences – Web of Conferences
at the address below.

EDP Sciences – Web of Conferences
17, Avenue du Hoggar
Parc d'Activité de Courtabœuf
BP 112
F-91944 Les Ulis Cedex A
France

Phone: +33 (0) 1 69 18 75 75
Fax: +33 (0) 1 69 28 84 91

contact-edps@webofconferences.org

Additional copies of this publication are available from:

Curran Associates, Inc.
57 Morehouse Lane
Red Hook, NY 12571 USA
Phone: 845-758-0400
Fax: 845-758-2633
Email: curran@proceedings.com
Web: www.proceedings.com

TABLE OF CONTENTS

PLENARY LECTURE & ECR AWARD RECIPIENTS

PHASE TRANSFORMATIONS

STRENGTH, PLASTIC DEFORMATION & FORMABILITY

DURABILITY: FATIGUE, FRACTURE, CORROSION & SURFACE TREATMENTS

THERMOMECHANICAL PROCESSING, TEXTURE & RECRYSTALLIZATION

CASTING & SOLIDIFICATION, RECYCLING, FUNDAMENTALS OF ADDITIVE MANUFACTURING

NEW DIRECTIONS IN ALLOY AND PROCESS DEVELOPMENT I: ADDITIVE MANUFACTURING

NEW DIRECTIONS IN ALLOY AND PROCESS DEVELOPMENT II: JOINING, SEVERE PLASTIC DEFORMATION, EMERGING PROCESSES

ADVANCED CHARACTERIZATION

Statement of Peer review

In submitting conference proceedings to *Web of Conferences*, the editors of the proceedings certify to the Publisher that

1. They adhere to its **Policy on Publishing Integrity** in order to safeguard good scientific practice in publishing.
2. All articles have been subjected to peer review administered by the proceedings editors.
3. Reviews have been conducted by expert referees, who have been requested to provide unbiased and constructive comments aimed, whenever possible, at improving the work.
4. Proceedings editors have taken all reasonable steps to ensure the quality of the materials they publish and their decision to accept or reject a paper for publication has been based only on the merits of the work and the relevance to the journal.

Title, date and place of the conference

ICAA17 The 17th International Conference on Aluminium Alloys 2020
October 26-29, 2020
Grenoble, France
(Virtual conference)

Proceedings editor(s):

Frédéric De Geuser, Alexis Deschamps, Jean-Christophe Ehrström,
Philippe Jarry, Georges Salloum-Abou-Jaoude, Luc Salvo, Christophe Sigli

Date and editor's signature

October, 20th, 2020

17th International Conference on Aluminium Alloys – ICAA17

Frédéric De Geuser[1], *Alexis* Deschamps[1], *Jean-Christophe* Ehrström[2], *Philippe* Jarry[2], *Georges* Salloum-Abou-Jaoude[2], *Luc* Salvo[1], *Christophe* Sigli[2]

[1]Univ. Grenoble Alpes, CNRS, Grenoble INP, SIMAP, 38000 Grenoble, France,
[2]Constellium Technology Center, CS 10027, 38341 Voreppe Cedex, France

Our world is changing rapidly. Climate change, the exhaustion of resources, are strong drivers for modifying profoundly the way we travel, the way we heat and cool our buildings, the way we consume products and services. Among the industrial materials, aluminium has a key role to play in these transitions. Although the energy required to reduce alumina into aluminium is very large, it is then almost infinitely recyclable and makes it possible to build lighter and stronger structures, among many applications. The current research on aluminium alloys is reflecting these evolutions, where new alloys, new processing methods, are invented to address the growing needs of vehicle weight reduction, recyclability, and durability. The 17th International Conference on Aluminium Alloys (ICAA), for the first time in a virtual format, has gathered more than 350 scientists from all continents to exchange on the most recent advances on all aspects of processing, microstructure and properties of Aluminium alloys. These proceedings comprise the published version of a subset of the 220 presentations given at this conference, in full open access.

The papers in the proceedings are organized in plenary lectures, and 8 symposia listed below. Among the plenary lectures, six were devoted to Early Career Research Award recipients, who are PhD students and post-doctoral fellows selected among applicants, based on the excellence of their research and their publication, which appears in these proceedings.

- Plenary lecture & ECR award recipients
- Symposium 1: Phase transformations
- Symposium 2: Strength, plastic deformation & formability
- Symposium 3: Durability: fatigue, fracture, corrosion & surface treatments
- Symposium 4: Thermomechanical processing, texture & recrystallization
- Symposium 5: Casting & solidification, recycling, fundamentals of additive manufacturing
- Symposium 6: New directions in alloy and process development I: additive manufacturing
- Symposium 7: New directions in alloy and process development II: joining, severe plastic deformation, emerging processes
- Symposium 8 : Advanced characterization

Exploring the hidden world of solute atoms, clusters and vacancies in aluminium alloys

John Banhart[1*], *Zi* Yang[1], *Meng* Liu[1], *Mazen* Madanat[1], *Xingpu* Zhang[1], *Qianning* Guo[1], *Yong* Yan[1], *Andreas* Röhsler[1], *Konrad* Fricke[1], *Zeqin* Liang[2], *David* Leyvraz[2], *Armin* Hoell[1], *Eike* Gericke[1], *Robert* Wendt[1], *Chunhui* Liu[3]

[1]Helmholtz-Centre Berlin for Materials and Energy, Hahn-Meitner-Platz 1, 14109 Berlin, Germany
[2]Novelis R&T Centre, Sierre, Route des Laminoirs 15, 3960 Sierre, Switzerland
[3]State Key Laboratory of High-Performance Complex Manufacturing, Central South University, Changsha 410083, China

Abstract. Precipitation hardening involves solutionising, quenching and annealing steps, the latter often at various temperatures. The phenomena observed in Al-Mg-Si alloys are very complicated and partially not well understood. During and after quenching, solute atoms diffuse through the lattice assisted by vacancies and form atom clusters that gradually grow. These act back onto vacancies, which complicates the situation. We apply positron annihilation techniques in addition to traditional hardness, resistivity and thermal measurements to clarify what happens in various stages of thermal treatment: The quenching process can be divided into a stage of vacancy loss and of precipitation. Very short artificial ageing treatments after heating at different rates show that there is a competition between vacancy losses and cluster formation as the temperature increases. The difference between natural ageing and artificial ageing can be defined based on the importance of excess vacancies. Based on such results the behaviour of "invisible" objects such as vacancies and small clusters can be better understood but some open question remain such as the kinetics of secondary ageing or the details of the negative effect of natural ageing on artificial ageing.

1 Introduction

Precipitation strengthening in aluminium alloys involves dissolving solute atoms in a matrix at high temperatures, preserving the solid solution by quenching, and ageing the alloy after, during which solute atoms diffuse through the matrix and form *clusters*, *zones* and later *precipitates* – terms that are not always defined in the same way in the literature – and increase strength (or hardness). The high vacancy fraction at the solutionising temperature is partially quenched-in and enables diffusion. Knowing vacancy properties is essential for understanding the kinetics of ageing. Thermo-dynamically, age hardening is a non-equilibrium process during which both the vacancy fraction and solute supersaturation change continuously. Various species of atoms may interact and the temperature course can be complex. A variety of different precipitates may form and interact with moving dislocations once there is a mechanical load and thus increase strength.

How complex the scenario is can be seen by considering rate equations for the relative precipitated volume α:

$$\frac{d\alpha(t)}{dt} = \varphi(1-\alpha(t))\, \mathrm{K(T)}\, x_v(t,T), \tag{1}$$

where the function φ expresses the driving force given by the decreasing solute supersaturation, K the temperature dependence of precipitation and x_v the non-equilibrium vacancy fraction. In thermal equilibrium, x_v is a known constant and Eq. (1) can be solved analytically after assuming a reaction function φ, e.g. the Johnson-Mehl-Avrami-Kolgomorov (JMAK model [1]. In non-equilibrium, $x_v(t,T)$ does not depend on temperature and time in a simple way, but also on α, since emerging precipitates interact with vacancies, e.g. trap them. As the interactions between solute atoms, clusters and vacancies are in general not known, Eq. (1) cannot be easily solved.

As precipitation starts with the smallest possible objects, solute-vacancy and solute-solute pairs, knowledge of the earliest stages is crucial to explain subsequent stages in which the final hardened microstructure is formed. However, both vacancies and atomic clusters in such early ageing stages are elusive and experiments meant to gain information are difficult. Positron annihilation spectroscopy (PAS) provides direct access to vacancies, while atom probe tomography (APT) has been used to investigate atomic clusters, but both methods can provide ambiguous results.

In this paper, some of the problems associated with early stages during and shortly after quenching from the solutionising temperature and their influence on subsequent heat treatment are addressed. We start with an overview of the available methods with an emphasis on PAS and then discuss a problem associated with age hardening of Al-Mg-Si alloys. Two attempts to extend currently available methods and two problems associated to Al-Mg-Si alloys, which are waiting to be solved, are presented.

* Corresponding author: banhart@helmholtz-berlin.de

2 Available methods

2.1 General overview

We give a short overview of methods suitable for investigating the small atomic clusters that form during natural ageing (NA) or short artificial ageing (AA). In an earlier paper, some of the methods have already been reviewed [2].

Clustering during NA normally results in changes of *mechanical properties*. Ageing curves are measured ex-situ in various stages of ageing. The main problem is that the exact relationship between cluster parameters and hardness or strength are not known. Fig. 1 illustrates this as it shows TEM images of two samples aged to the same hardness at different temperatures. In (a), fine particles give rise to very little contrast, possibly tiny clusters without an ordered crystal structure. In (b), in contrast, fine particles of monoclinic structure appear, sometimes without even one complete unit cell. This demonstrates that there is no one-to-one relationship between hardness and particle properties.

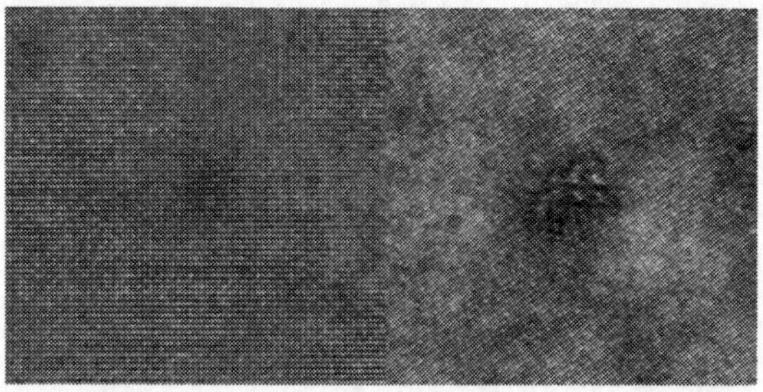

Fig. 1. HR-TEM images of alloy 6014 hardened to 70 HBW 1/10 in two different ways directly after solutionising and quenching, (a) 40 h at 80 °C, (b) 20 min at 160 °C. Images recorded without objective aperture display phase contrast.

Electrical resistivity is easy to apply ex-situ [3] or in-situ [4]. Sensitivity is high but as for hardness, it is not known precisely how cluster parameters influence resistivity. Vacancies also contribute but it is hard to separate their contribution from that of clusters and solute atoms.

Hardness and resistivity do not vary in the same way during ageing as can be seen by plotting one as a function of the other. It has been argued that an increase of the number of small clusters rather increases resistivity [5], while hardness is influenced more by cluster growth [6].

Thermoelectric power measurements have been used to determine the solute content in the matrix and the influence of precipitates both during NA [7] and AA [8], but to isolate the cluster contribution remains difficult.

DSC clearly shows clustering reactions during isochronal heating [9-11], but interpretations are difficult once various reactions (exothermal, endothermal) overlap. Isothermal DSC has been applied to AA in Al-Mg-Si alloys [12] and also to NA in Al-Cu-Mg alloys [13], but deconvolution of signals is equally difficult.

Dilatometry has only recently been developed to a method sensitive enough to yield precise information on precipitation in Al-Mg-Si alloys and even the effect of vacancy annihilation during NA has been captured [14].

Small-angle scattering is the classical method to detect nm-sized objects. Neutrons (SANS) yield sufficient contrast to monitor precipitation phenomena in Al-Mg-Si alloys even in-situ [15] during AA but it has not been possible to resolve clusters formed during NA. X-rays (SAXS) have not resolved NA in Al-Mg-Si alloys and as shown below insufficient contrast between clusters and the matrix prevents SAXS from being used successfully.

Table 1. Methods to study early precipitation phenomena in aluminium alloys. Column 'Vac' specifies sensitivity to (mono)vacancies, 'In-situ' whether phenomena can be followed in the same sample during ageing, 'Examples' provides some references for Al-Mg-Si alloys.

Observable / Method	Vac ?	in-situ?	Examples
Hardness	no	yes	Sec. 4.1
Strength	no	no	Fig. 4
Electrical resistivity	yes	yes	[3]
Thermoelectric power	no	yes	[7]
Calorimetry	no	yes	[9], Fig. 6,7
Dilatometry	yes	yes	[14]
Small-angle scattering	no	yes	Sec. 4.2
X-ray absorption spectr.	no	yes	(AA: [16])
Nuclear magnetic resonance	no	yes	[17]
Magnetisation	no	yes	[18, 19]
Muon spectroscopy	yes	?	[20]
Positron Annihilation	yes	yes	[17, 21]
Atom probe tomography	no	no	[22-24]
Electron microscopy (TEM)	no	no	[25, 26]

X-ray absorption spectroscopy has been applied to NA in alloys such as Al-Cu [27], while for Al-Mg-Si alloys only AA has been investigated [16].

Nuclear magnetic resonance spectroscopy has been used to follow NA in Al-Mg-Si. Only the ^{25}Mg resonance is usable there but an estimate of the Mg content in NA clusters as a function of NA time could be obtained [17].

Magnetisation has recently been found to vary during NA. The effects are small (≤0.5% variation) but measurable and show a similar pattern as positron lifetimes. Si-rich and Mg-rich clusters are thought to give rise to different magnetisation contributions but the exact mechanisms are not yet known [18, 19].

Muon spectroscopy relies on measuring the depolarisation rate of the magnetic moments of muons trapped in point defects. As for positron spectroscopy (see Sec. 2.2) it allows one to detect vacancies in alloys [20].

Atom probe tomography provides a direct visualisation of NA clusters and allows for a quantitative analysis. Unfortunately, the results published in the literature are not consistent. The size of clusters hardly seems to evolve during NA [28] and the evolution of their composition has been presented differently. One issue might be the way cluster search algorithms are applied but there might also be a fundamental limitation linked to the trajectories of atoms that prevent very small clusters to be analysed properly [29].

While high-resolution TEM is a remarkable success story when it comes to revealing the smallest details of precipitates in Al-Mg-Si alloys, clusters formed during NA remain elusive and no generally accepted images of NA clusters have been presented.

https://doi.org/10.1051/matecconf/202032601001

2.2 Positron annihilation lifetime spectroscopy (PALS)

PALS has been successfully applied to characterise ageing phenomena in various 2xxx [30], 7xxx [31] and 6xxx [21] alloys. Its basic principle is simple: Positrons generated during the radioactive decay of ^{22}Na are injected into a sample where they thermalise and start a random diffusional walk. Once they encounter an electron they annihilate. The emitted radiation allows one to measure the lifetime of the positron (PLT). In complex structures such as Al-Mg-Si alloys containing various different objects, positrons can be trapped by vacancies or precipitates and have a typical lifetime within, or they remain in the bulk Al lattice and decay there. The individual lifetimes have to be extracted from the lifetime spectrum containing signals from many events. The most simple way to view the situation is given in Fig. 2a.

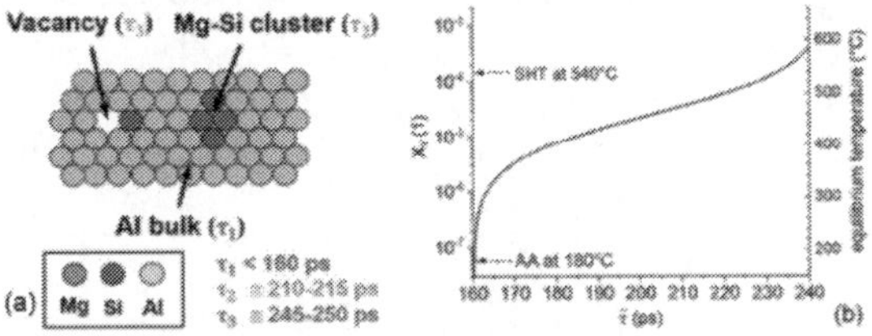

Fig. 2. (a) Schematic of positron annihilation in Al-Mg-Si alloys after NA or short AA including the three typical occurring PLTs. (b) relationship between average PLT and vacancy site fraction in aluminium containing vacancies only and corresponding calculated temperature.

The three associated lifetimes (τ_1, τ_2, τ_3) are known with only small uncertainty and are given in Fig. 2a. From experimental lifetime spectra, the different contributions have to be extracted by fitting, which is a challenge due to the general problem of de-convoluting decaying exponentials [32], but also because additional corrections such as for background, spectrometer resolution and for annihilation in the positron source itself are necessary [33, 34]. The result of a fit includes the relative intensities of each of the 3 contributions in Fig. 2a and defines an average lifetime:

$$\bar{\tau} = I_1\tau_1 + I_2\tau_2 + I_3\tau_3; \qquad I_1+I_2+I_3 = 1.$$

Such decompositions work only in few favourable cases and require high-count spectra, e.g. Ref. [35]. One can try to treat the traps as one average component characterised by I_{2+3} and τ_{2+3}, thereby reducing the number of fitting parameters or to fit just a single exponential and obtain the one-component lifetime τ_{1C}, which, however, can deviate from the average lifetime $\bar{\tau}$ if two or more lifetime components are present.

Another problem associated with lifetime decompositions is that more than 2 different traps might co-exist: It is plausible that τ_2 varies with cluster composition and therefore there is a range of PLTs. Moreover, the precipitates occurring during AA have different values of τ_2 too, e.g. τ_2~210 ps for GP zones and β" precipitates, but a much higher τ_2~235 ps for β' precipitates [21]. Finally, solutes attached to vacancies can modify the associated lifetime τ_3 [21] and give rise to a distribution of PLTs as well.

If the only trap are vacancies, the two-state trapping model can be used to calculate their site fraction as shown in Fig. 2b. We find: (i) for very low $\bar{\tau}$ just above 160 ps the slope is very large and therefore the vacancy fraction x_v cannot be determines, for example, the equilibrium site fraction at 180 °C. (ii) x_v values between 5×10^{-7} and 5×10^{-4} can be measured. (iii) A specific trapping rate for vacancies has to be used in calculating vacancy fractions. It is known within a margin of about a factor of 2. Unfortunately, the specific trapping rate for other objects such as clusters is not known, so that a calculation in analogy to Fig. 2b for the cluster component τ_2 cannot be carried out. Clusters in Al-Mg-Si alloys tend to trap positrons efficiently so that τ_2 often dominates the positron lifetime spectrum as almost all positrons annihilate within clusters ('saturated trapping').

With these and other facts in mind we can summarise the strengths and weaknesses of PALS in studies of vacancies and clusters in Al-Mg-Si alloys:

+ Sample preparation and measurement is easy,
+ Reproducibility is very good,
+ Highest sensitivity to changes of vacancy/cluster configuration if clustering is not too strong, i.e. average PLT is not between ~200 and ~220 ps,
+ In-situ experiments allow for the detection of small changes even if many clusters are present,
– Lifetime decompositions often not reliable,
– Vacancy and cluster contributions hard to separate,
– Ad-hoc assumptions concerning the nature of positron traps sometimes necessary: lack of calculations of PLTs and specific trapping rates.

In the following, PALS will be applied to gain knowledge about vacancy and cluster evolution in Al-Mg-Si alloys.

3 Role of excess vacancies

The terms 'natural ageing' and 'artificial ageing' refer to ageing carried out at 'room temperature' (mostly without defining what this is) and at 'elevated temperature' (usually specified and mostly between 160 °C and 190 °C for 6xxx alloys), but it is not clear what the fundamental difference between the two ageing regimes is.

It has been claimed that there is a transition temperature above which the structure of the precipitates formed during ageing differs from that below and researchers have provided additional evidence [36]. The transition temperature is seen at around 70 °C and the differences include that clusters formed during NA:

- are richer in Si [37, 38]
- contain vacancies [39, 40],
- further precipitation during subsequent AA can only take place after cluster dissolution [41, 42].

Fig. 3 shows how hardness evolves at different temperatures. A common feature is the increase in hardness as ageing proceeds. Overageing is seen above 160 °C and might eventually also occur for lower temperatures, but for NA overageing has never been observed. Notably, there is a range, where ageing at lower temperature leads to higher hardness than ageing at higher temperature for the same time, namely for a few hours.

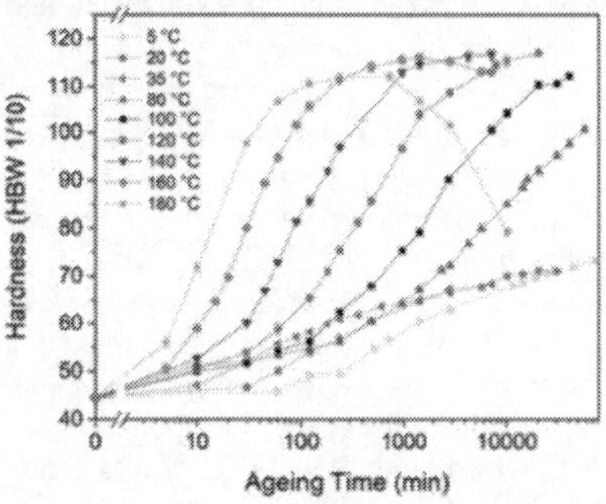

Fig. 3. Isothermal hardening curves for alloy 6014 [43, 44].

This crossover has been used to define a division between NA and AA in terms of strengthening kinetics [45]. To show this clearer the strength reached after an ageing treatment is given as a function of temperature for a fixed ageing time in Fig. 4. The data show two distinct peaks and a minimum in between.

According to the authors the high-temperature peak corresponds to artificial ageing. The more time is available for ageing the higher the achievable strength. Regime III expresses that for a given time higher temperatures give rise to faster ageing and higher strength, but in regime IV this eventually leads to over-ageing. The same applies to the low-temperature peak (regimes I and II). What is remarkable is the higher strength that can be achieved for ageing at a lower temperature for a given time. For example, ageing for 3 h at 35 °C leads to a higher strength than 3 h at 75 °C (purple line)

In thermodynamic equilibrium, a reaction at a lower temperature should always be slower than at a higher. A possible explanation would be that at low temperature the type of precipitate formed is different than at higher and that this type is more efficient in increasing strength or that more nuclei are formed at lower temperature. However, as hardening requires non-shearable precipitates of a certain size it is unlikely that this is the only reason. An alternative explanation is that the vacancy fraction has not yet reached thermal equilibrium during NA and that ageing is accelerated by the still available excess vacancies. It has been known for a long time that at 'room temperature' NA can proceed only so fast because the fraction of excess vacancies is many orders of magnitude above the equilibrium value directly after quenching [46]. We therefore claim:

During ***natural ageing*** *the rate of the strengthening precipitation process is largely determined by excess vacancies, whereas* ***artificial ageing*** *is characterised by precipitation at equilibrium vacancy site fractions.*

The 4 regimes in Fig. 4 are then due to:

I. During ageing, excess vacancies anneal out as they go to vacancy sinks but the loss is slow enough to allow for the clustering of a significant amount of solutes. A given vacancy diffuses so slowly that it encounters a sink only after assisting solute atoms in clustering many times, i.e. vacancy loss is not a rate determining factor. At higher temperatures, clustering accelerates, hence the increase in stage I for every given ageing time.

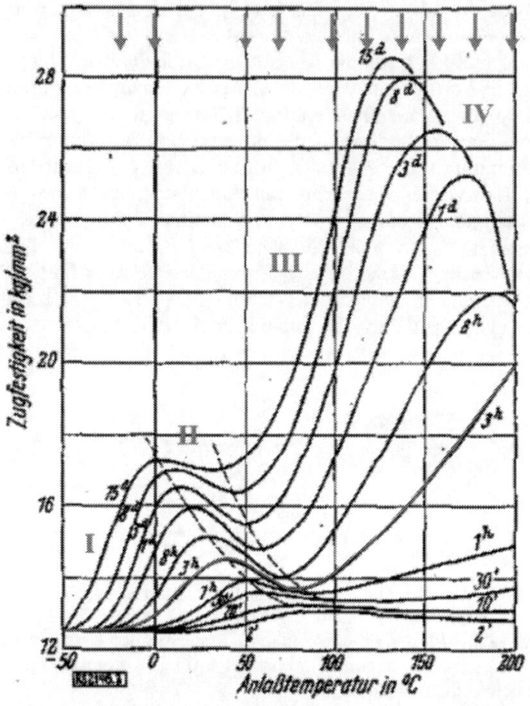

Fig. 4. Tensile strength (1 kg/mm² = 9.81 MPa) as a function of ageing temperature for 10 different ageing times ranging from 2 min (2') to 15 days (15^d). No individual data points are given, just interpolated lines based on data at the measurement temperatures marked by vertical arrows at the top. Alloy Al-0.4 Mg-0.55 Si-0.2 Fe (probably wt.%) was solutionised for 1 h at 530 °C and water quenched after. After immersion into oil or a salt bath the final ageing temperature was reached in less than 1 min [45].

II. After exceeding a certain temperature, vacancy losses during a given period of ageing start influencing ageing kinetics. When ageing for 8 h, for example, the vacancies are still largely preserved at 20 °C, but when moving to 80 °C, vacancy concentration drops so quickly during ageing that clustering kinetics is retarded despite the in general much faster solute diffusion.

III. Once vacancy concentration has dropped to an equilibrium value shortly after the onset of artificial ageing, kinetics are mainly determined by temperature. For higher temperatures, diffusion under equilibrium conditions produces more precipitates in a given time.

IV. For very high temperatures, fewer precipitates form and over-ageing sets in.

Before presenting experimental evidence for this concept we apply a simple model. First, the diffusion of a Al or solute atom is represented by a product of vacancy fraction and a Al/solute migration enthalpy H^m related term

$$D(T,t) \sim x_v(T,t)e^{-H^m/kT} \qquad (2)$$

Unlike in the definition of an equilibrium diffusion coefficient, x_v is a non-equilibrium value here and depends on time and thermal history. Its equilibrium value after a long time is $x_v(T) = \exp(S/k - H_v^f/kT)$. In equilibrium, the activation energy of self diffusion is $Q_{\rm Al}^d = H_v^f + H_{\rm Al}^m$, where the vacancy formation enthalpy H_v^f and the migration enthalpy of Al, H_{Al}^m can be measured or calculated. H_{Al}^m also describes the migration of vacancies. We use values, S/k =0.7, H_v^f=0.665 eV [47, 48], and H_{Al}^m=0.655 eV [48]. Published diffusion data for impurities in Al show a large variance and none are available for 'room temperature'. Especially the breakdown of diffusion values into pre-factor D_0 and activation energy Q is notoriously unreliable. Therefore, we partially rely on calculated data here. H_{Si}^m for Si is 0.48 eV as calculated from $Q_{\rm Si}^d$ = 1.15 eV [49]. For Mg, this value is higher by 0.12 eV, but note that Ref. [48] quotes almost swapped values for Mg and Si.

Using a model of vacancy annihilation allows one to calculate $x_v(t,T)$ after quenching from 540 °C and during subsequent isothermal ageing [50] and to apply Eq. (2). Diffusing solute atoms reach other solutes and form clusters. The diffusion range of a solute atom in time t is

$$\bar{r} = \sqrt{6D_s(T,t)\,t}. \qquad (3)$$

We assume that in a given time interval Δt a solute leads to a precipitation activity $\Delta\alpha$ proportional to the distance $\Delta\bar{r}$ covered and obtain for the precipitation rate

$$\frac{d\alpha}{dt} = A\frac{d\bar{r}}{dt} = \frac{A}{2\sqrt{t}}\sqrt{6D(T,t)}, \qquad (4)$$

with A a constant. As precipitation proceeds, solute atoms are consumed and the driving force decreases. $\frac{d\alpha}{dt}$ in Eq. (4) has to be multiplied by a solute supersaturation factor φ decreasing with α to prevent α from increasing indefinitely (in analogy to Eq. (1) in Ref. [4]). As we do not know the functional relationship between φ, the driving force, and solute concentration we write

$$\varphi(\alpha) = (1-\alpha), \qquad (5)$$

Eq. (4) multiplied with Eq. (5) and using Eq. (2) is now equivalent to the general rate equation Eq. (1) and ensures $\alpha \in [0,1]$. Eq. (4) is just a special case for $n = 1$ of the generalised JMAK equation [1]:

$$\varphi(\alpha) = n(1-\alpha)[-\ln(1-\alpha)]^{1-\frac{1}{n}}. \qquad (5a)$$

We now numerically integrate $\frac{d\alpha}{dt}\varphi(\alpha)$ using the calculated $x_v(T,t)$ and obtain $\alpha(t)$. It is experimentally known that the precipitated volume is proportional to hardness or strength during NA [44, 51] and we extend this to AA as we are only interested in trends, i.e. write $\Delta S = \alpha$, with ΔS the relative strength increase. The adjustable constant A in Eq. 3 is chosen such that ΔS during NA is about 1/3 of ΔS during AA, which matches reality, see Fig. 3.

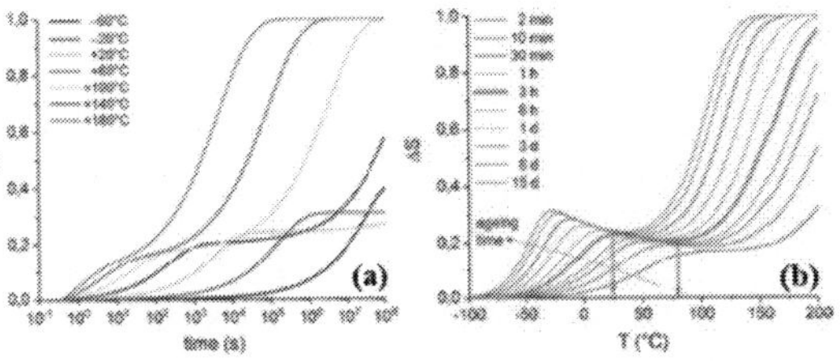

Fig. 5. Calculated strengthening as a function of a) time for fixed T or b) of T for fixed times (same as in Fig. 4). Model parameters: dislocation density 3×10[12] m[-2], jog spacing 0.02, grain diameter 50 µm, H_v^f=0.665 eV, S/k =0.7, H_{Al}^m=0.655 eV, H_{Al}^m=0.48 eV (see Ref. [50] for description of model)

Fig. 5a features a clear two-stage ageing behaviour for the higher temperatures. Fig. 5b represents the experimental hardening curves in Fig. 4 except for the over-ageing regime IV that is not included in the model. As an example, ageing for 3 h (purple curve) leads to a maximum of precipitation for 25 °C and a subsequent minimum at 80 °C, which is what the experiment shows as well, see purple curve in Fig. 4.

Choosing values $n \neq 1$ in Eq. (5a) or the function proposed in Eq. (4) of Ref. [4] does not have a pronounced influence on the principle course in Fig. 4. This is because the existence of the decrease in stage II is due to the fact that $H_{Si}^m < H_{Al}^m$ has been adopted for Si atoms (see comments above). A lower ageing temperature can lead to more precipitation due to an excess vacancy concentration that exceeds that at a higher temperature by a larger factor than the solute diffusion rate is lower. In other words, the H_{Al}^{fm}=0.656 eV, which determines the kinetics of vacancy migration to sinks, penalizes ageing at higher temperature more than the gain in solute migration (and precipitation) rate given by H_{Si}^m=0.48 eV. Thus, this simple kinetic model that ignores details of precipitate nucleation and type explains the behaviour of ageing at different temperatures surprisingly well and underlines that excess vacancies cause strong ageing at natural ageing temperatures. Thus, a main feature distinguishing NA and AA is the role of vacancies and the transition temperature is not fixed but depends on ageing time.

We now refer to 2 experiments to show that at the usual artificial ageing temperature of 180 °C ageing kinetics are determined by the equilibrium vacancy concentration only.

In the first, DSC is used to visualise the exothermic clustering peak that is observed when 6xxx alloys are heated [10]. Fig. 6a shows the part of the DSC trace in which this clustering peak occurs, namely from 30 °C to 130 °C. If samples are heat treated at 180 °C for up to 30 s before DSC is run, this peak disappears as expressed by the relative area of the clustering peak in Fig. 6b.

There are two possible explanations for the disappearance of the peak in such a short time: Due to ageing before DSC (i) solutes precipitate within seconds, so that during DSC no more clusters can be formed due to a low solute level; (ii) the vacancy concentration drops to such a low value so that during ensuing DSC kinetics are slowed down so much that no more precipitation takes place in the 10 minutes of heating from 30 °C to 130 °C.

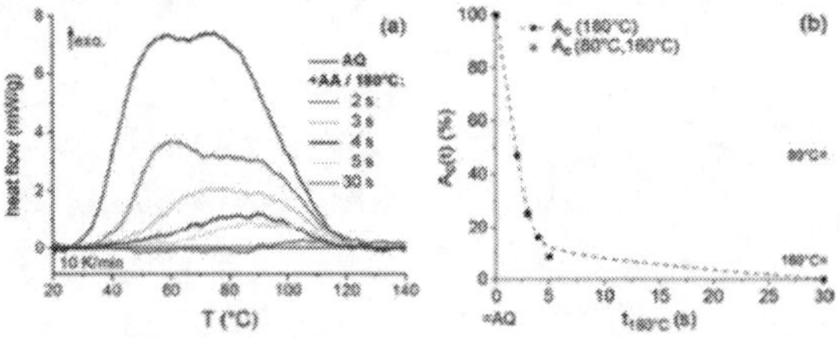

Fig. 6. (a) Clustering peak observed by DSC directly after quenching (AQ) and after an additional short treatment at various temperatures in an oil bath. (b) Areas A_c of clustering peak with respect to the AQ state as a function of time. Alloy Al-0.6Mg-0.8Si (wt.%). Data taken from Ref. [52].

As heating for a few seconds at 180 °C does not lead to any measurable hardening (just 2 HBW in 1 min, see Fig. 9) and not to more than a slight change of resistivity, which can be negative or positive [53], option (i) is unlikely. This is different for NA performed before DSC which also brings down the area of the DSC peak to about 50% after 1 d, see Fig. 7, but the associated hardness increase from 44 to 65 HV is notable (Fig. 3), which shows that in this case (i) is significant. Moreover, the low-temperature part of the clustering part is selectively eliminated here, whereas short AA decreases the entire profile of the DSC clustering peak equally as vacancies are removed and kinetics are slowed down globally while the solute configuration remains largely unchanged.

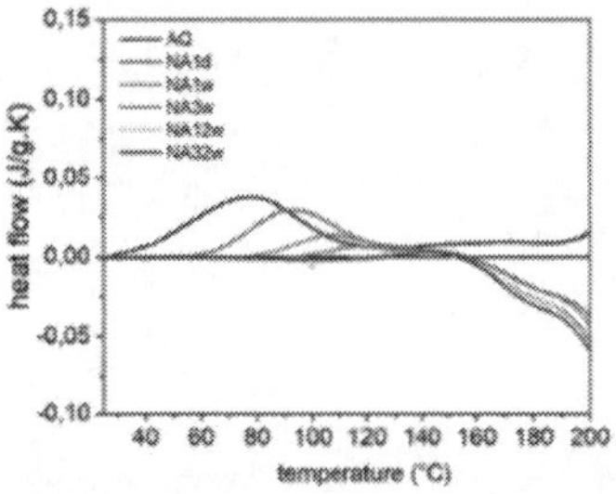

Fig. 7. DSC traces of alloy 6014 directly after quenching and after NA at 20 °C for different times.

A positron lifetime experiment provides further insights. Al-Mg-Si alloys were subjected to ageing treatments directly after quenching as short as 1 s and positron lifetime (PLT) measured directly after.

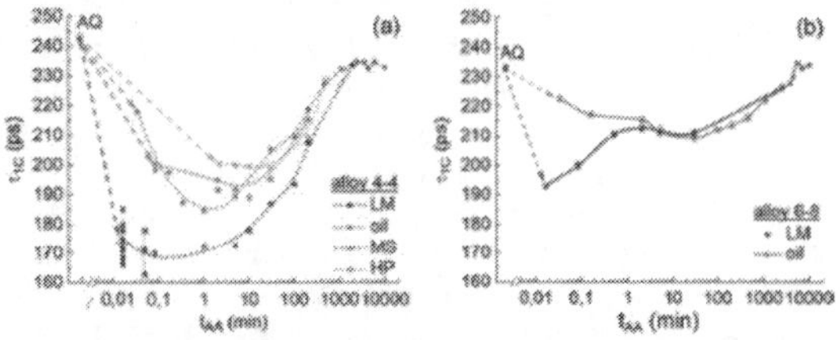

Fig. 8. Positron lifetime measured after solutionising, quenching and artificial ageing at 180 °C of a) alloy Al-0.4Mg-0.4Mg and b) Al-0.6Mg-0.8Si (wt.%) in different media [53].

In Fig. 8a, the initially high positron lifetime of a lean alloy directly after quenching ('AQ') is associated to a high vacancy fraction. It is reduced to a very low level by AA just for 1 s. Values close to 160 ps – defect free Al – imply that most of the vacancies have been eliminated by such a short treatment at 180 °C, see Fig. 2b. The degree of lifetime reduction depends on the heat bath used. It is most pronounced for liquid Bi-Sn alloy where the final temperature is reached almost instantaneously. We can use the lowest value observed in Fig. 8 to estimate the vacancy fraction after short AA by applying the two-state trapping model and assuming the absence of any clusters and obtain x_v=2.5×10^{-7} [53]. This is just 3 times the equilibrium vacancy fraction of 180 °C and possibly the real x_v was even lower if some clusters were present.

When ageing in molten salt ('MS'), oil or on a heating plate ('HP', where temperature is ramped up at 3 K/s) PLT reduction by ageing at 180 °C is less pronounced. This is because slower heating allows the alloy to be exposed more to the low-temperature range where clusters form and add a lifetime component of ~215 ps to the spectrum.

In the alloy richer in solute, see Fig. 8b, the positron lifetime reductions are less pronounced with 190 ps the lowest value measured in LM and slower heating leads to an even shallower minimum than in the leaner alloy. This is either due to the presence of more clusters or more vacancies than in Fig. 8a. Clusters may have formed already during quenching and further grown in the short time passing through low-temperature regimes when heating up the alloy at a reduced rate. The mechanism behind this could be that clusters formed in large numbers assisted by excess vacancies help preventing annihilation of these vacancies for a little while during heating, which in turn supports further clustering [53]. For very long ageing times the PLT increases to high values up to 235 ps. This is due to a high PLT in the β' precipitates that form in late stage of (over-)ageing.

There is another argument against the existence of excess vacancies during artificial ageing for an extended period of, say, minutes. Hardening directly after solutionising and quenching for up to 10 min would be initially very fast and then level off in this case. In reality this is not observed, see Fig. 9 and the hardening rate is rather constant.

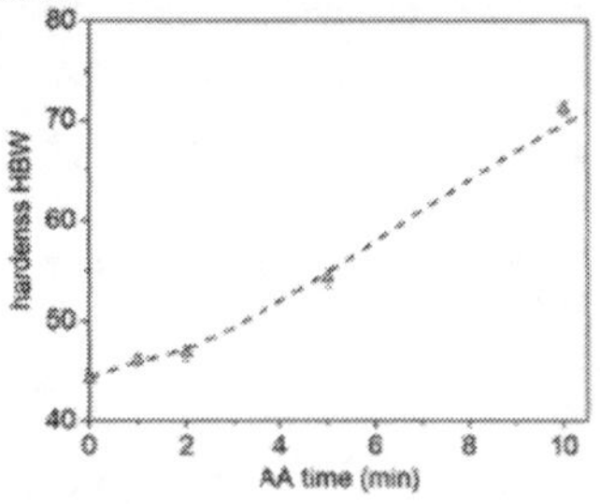

Fig. 9. Hardness of alloy 6014 after short artificial ageing at 180°C.

Positron lifetime spectroscopy is most useful in states where lifetimes are very high or very low. This is illustrated in Fig. 10, where information about the vacancy and cluster evolution during slow quenching is obtained by allowing samples to cool to a given temperature after solutionising and then interrupting by a fast quench. The positron lifetimes clearly change in 2 stages. (i) First, a continuous drop down to values around

170 ps for both cooling rates. Such values point at very low fractions of both vacancies and clusters, namely vacancy fractions $\leq 2\times10^{-6}$ in the absence of clusters, see Fig. 2b, and even lower fractions if clusters form during cooling. (ii) If slow cooling is interrupted at temperatures lower than 200 °C, the resulting PLT is higher again, reaching ~215 ps for uninterrupted slow quenching. As the vacancy site fraction cannot increase from its minimum achieved at 200 °C, the PLT increase below 200 °C must be entirely due to clustering. Therefore, the quenching process can be divided to 2 stages: predominantly vacancy losses down to 200 °C and clustering after.

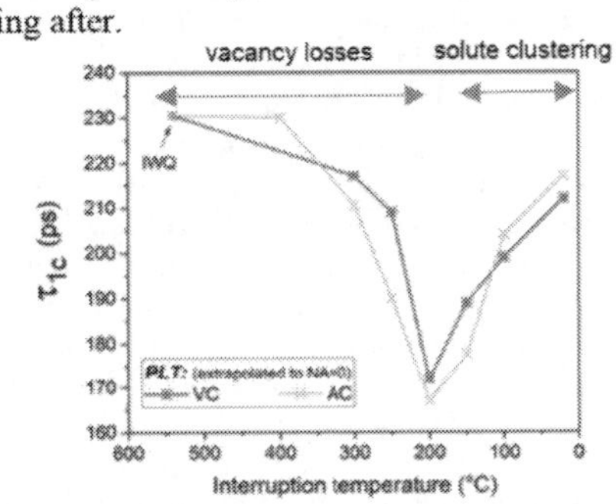

Fig. 10. Positron lifetime in alloy 6014 samples solutionised and cooled at 27 K/s (VC) and 5 K/s (AC) down to a target temperature, after which fast quenching to 20 °C in ice water largely preserves the resulting microstructure. The first point corresponds to uninterrupted fast quenching, the last to uninterrupted slow quenching.

Fig. 11 shows another experiment with an interrupted temperature course. Here, as-quenched samples were heated to a target temperature at three different rates, after which heating was interrupted by fast quenching. Directly after quenching, PLT is high, indicating predominance of vacancies. After heating to 60 °C, the PLT is almost at the level of NA for 1 h, i.e. clusters dominate the signal. A weak vacancy-related signal survives up to 100 °C, for the highest heating rate. From the previous discussion and Fig. 6a, we know that during linear heating clusters occur and that this is driven by excess vacancies. In this case, however, PALS cannot provide clear information about the fraction of vacancies due to the predominance of clusters. After heating from 140 °C to 220 °C, a minimum is found that might be due to the decrease of cluster site fraction and the development of a coarser distribution of precipitates. The final increase corresponds to the increase in Fig. 8 for long ageing times caused by formation of β' precipitates.

We conclude this section by noting that non-equilibrium vacancies have an important influence on ageing and that a combination of PALS with DSC and hardness measurements allows one to elucidate the co-evolutions of clusters and precipitates both during isothermal ageing and linear heating. However, ambiguities still exist, which is why new experimental options should be evaluated. This will be briefly review in the following section.

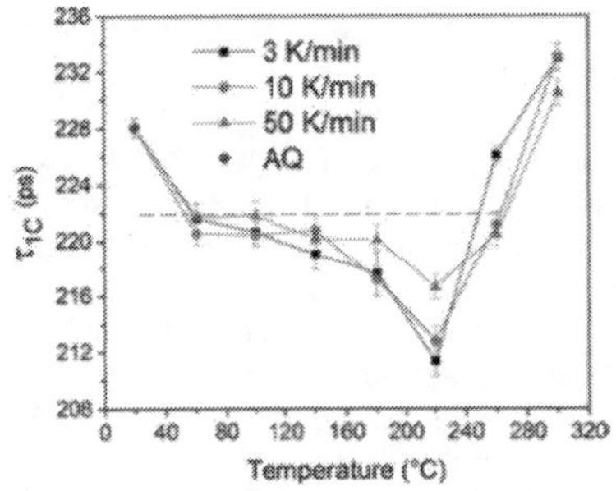

Fig. 11. Positron lifetimes after solutionising and quenching and subsequent linear heating at 3 different heating rates from 20 °C to a selected end temperature. Horizontal line marks approximate PLT level after 1 h of NA. Alloy 6014, [44].

4. Extension of characterisation methods

4.1 In-situ NA hardness measurements

Hardness measurements provide an indirect measure for clustering or precipitation. Although their interpretation is not always straight-forward they are often applied as a first orientation. The basic problem is the labour intensive procedure when carried out manually so that often ageing curves consist only of a small number of measurements, for example 14 NA times and 10 hardness measurements for each time of the NA curve in Ref. [2] where two or three distinct linear stages on the logarithmic time scale were postulated with a transition time between 50 and 80 min. By performing automated micro-hardness measurements one can largely increase the number of measurements and obtain a clearer indication whether hardness evolves in stages, as seen in Fig. 12, where the existence of two stages with a transition around 1 h of NA is seen much more clearly than in Ref. [2].

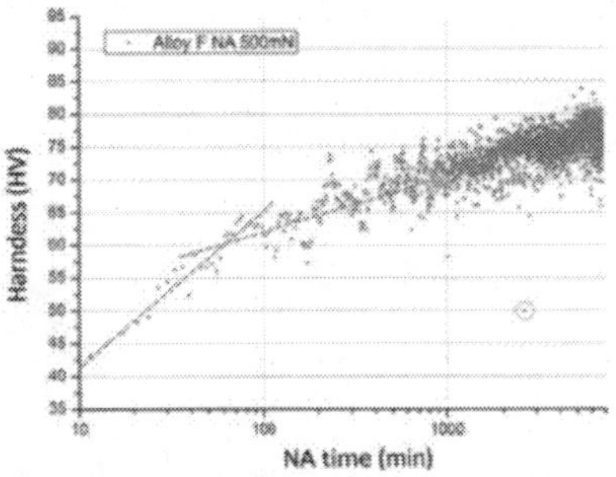

Fig. 12. In-situ micro-Vickers hardness measurement on alloy Al-0.6Mg-0.8Si (wt.%) carried out with a Fisher 'Picodentor HM500' during NA following solutionising and quenching. One hardness indentation every 3 minutes was automatically obtained and in total 2195 data points acquired over a period of 110 h. Force was 0.5 N. The broken green lines are linear fits to the data for <100 min and >100 min NA time. The broken circle is an example for irregular results removed from the data set.

4.2 In-situ SAXS measurements

SAXS is expected to yield only small signals from Mg and Si rich clusters in Al-Mg-Si alloys due to the small difference in scattering length between Al, Mg and Si that is proportional to the number of electrons in an atomic species. The scattering length of almost equal quantities of Mg and Si in clusters nearly cancels out compared to Al [2]. Thus, the contrast between the clusters and the matrix is only caused by differences of the mass density of both phases. Still, we attempted to overcome the difficulties by applying an in-situ experiment on a solutionised and quenched sample in which all conditions remained constant so that any change during NA could be ascribed to clustering phenomena.

Alloy Al-0.6Mg-0.8Si (wt.%) was polished to a 60 μm-thick platelet, solutionised at 540 °C between two protective titanium plates and water quenched. Within 90 s the sample was transferred to a transport box filled with liquid nitrogen. SAXS was measured using the HZB Anomalous SAXS instrument [54] at the four-crystal monochromator (FCM) beamline of the PTB at the BESSY-II storage ring using an in-vacuum Pilatus 1M detector [55] at a X-ray energy of (8980.5±2.0) eV. The sample was mounted onto a holder in liquid nitrogen and rapidly transferred to the sample chamber. After reaching 4×10^{-4} mbar during evacuation, the measurement was started, initially at a temperature of 3 °C that eventually rose to 25 °C during the SAXS experiment (see Fig. 13c, blue line).

SAXS scattering images were measured at five different spots on the alloy platelet. The X-ray transmission and sample thickness (calculated from the transmission coefficient) varied slightly between the spots on the platelet but were found to be constant over time. Acquisition times of initially 30 seconds and later 120 and 300 seconds were used per scattering image. In further data treatment, all images were azimuthally averaged and normalized to absolute scattering cross-sections. The instrumental scattering background was subtracted.

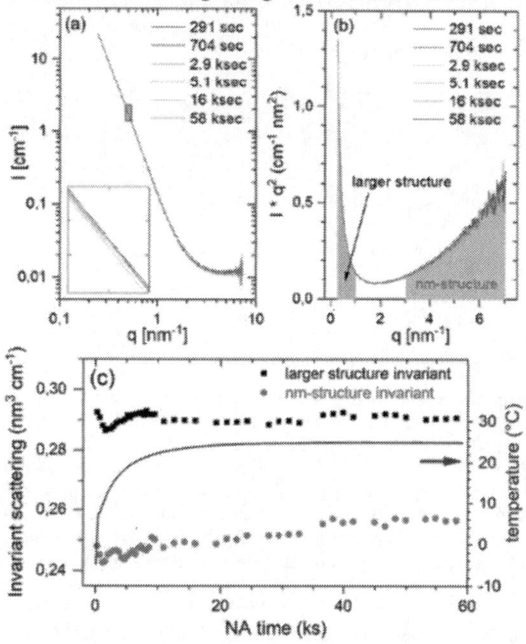

Fig. 13. SAXS results on a logarithmic scale (a) and as a Kratky plot (b). NA time dependence of the two integrals marked by a grey and a red area in the Kratky plot are shown in (c).

The SAXS curves change only slightly over time, see Fig. 13a. The inset provides a close up of the SAXS curves in the q-range (0.48–0.50) nm^{-1} and indicate a nearly non-recognizable change during the in-situ experiment. The scattering intensity is plotted against the value of momentum transfer (q). The q-value is related to the scattering angle (2θ) by $q = 4\pi / \lambda \sin(\theta)$, with λ the X-ray wavelength). In Fig. 13b, the Kratky plot ($I\cdot q^2$ vs. q) of the SAXS curves is shown. The integral of $I\cdot q^2$ over q is the so-called invariant of scattering and allows for the calculation of a particle volume fraction in two-phase systems. The integral between q-values of 3 and 7 nm^{-1} depends on very small nano-structures (0.9 to 2 nm size) (Fig. 13b, red area). In contrast, the integral between the smaller q-values of 0.26 and 1 nm^{-1} depends on structures larger than 6 nm (Fig. 13b, grey area). In Fig. 13c, both integrals are plotted against aging time as black squares (larger-structural invariant) and red circles (nano-structural invariant). Only the latter shows a change with NA time, which is a weak signature of clustering.

This experiment confirms that SAXS, although it captured a faint signature of clustering is not capable of yielding detailed data in pure ternary Al-Mg-Si alloys. The situation would be different in quaternary systems containing elements that scatter stronger, for example Cu. By carrying out SAXS at various energies slightly below the absorption edges of the elements involved here (e.g. Si and Cu) one could obtain element specific scattering curves. Such ASAXS experiments could reveal the structure of Si or Cu containing clusters when carried out near the Si K-edge (1.84 keV) or Cu K-edge (8.98 keV).

5 Open questions

5.1 Natural secondary ageing (NSA)

Unlike (primary) NA that is carried out after solutionising and quenching, natural secondary ageing (NSA) takes place after an underageing treatment (UA) at AA temperatures short enough to still allow for ageing phenomena to occur during subsequent NA. Such treatments have been investigated with an application background in mind (called T6I4 and T6I6 [56]) but the phenomena during NSA are also worth being studied because UA represents a state of high solute super-saturation but low vacancy fraction and NSA provides indirect information about ageing phenomena during UA.

The kinetics of NSA show some peculiarities that are not easy to understand. Fig. 14 shows the rate of electrical resistivity change during NA/NSA after different UA treatments (including no UA). As expected, the highest rate is found directly after solutionising and quenching as the vacancy and solute supersaturation are the highest. During NA the rate continuously decreases. This is also the case after UA with the notable exception of alloys containing tin. In these alloys, the rate increases by one order of magnitude after 1 week of NSA and peaks, see arrows. Such an increase is not explainable in terms of the picture of a rate proportional to excess vacancy fraction as given by Eq. (1).

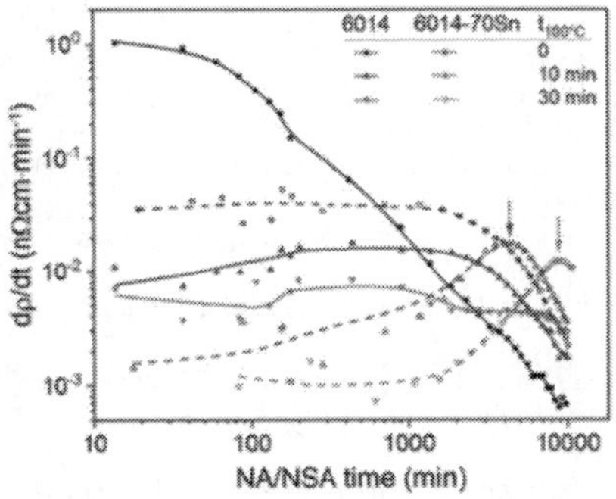

Fig. 14. Rate of electrical resistivity increase after solutionising and quenching as a function of NA/NSA time in 6014 alloys without and with 70 ppm Sn addition [57].

Hardening during NSA after short UA at 180 °C is also seen to yield unexpected results, see Fig. 15. Here, NSA directly after quenching leads to hardening from 42 to 70 HV within one year. After UA for up to 5 min, the total NSA course is only slightly changed, most notably in the first 100 min. One would expect that removal of most of the vacancies from the matrix by short exposure to 180 °C, see discussion above and Fig. 6, would notably slow down NSA. However, there is a mechanism that seems to equalise conditions, for example by trapping most excess vacancies in clusters, so that after 100 min of NSA the slope of hardness increase is very similar for AA times up to 5 min. Only for much longer UA times the hardness increase during UA has set in and the hardness during NSA is largely constant.

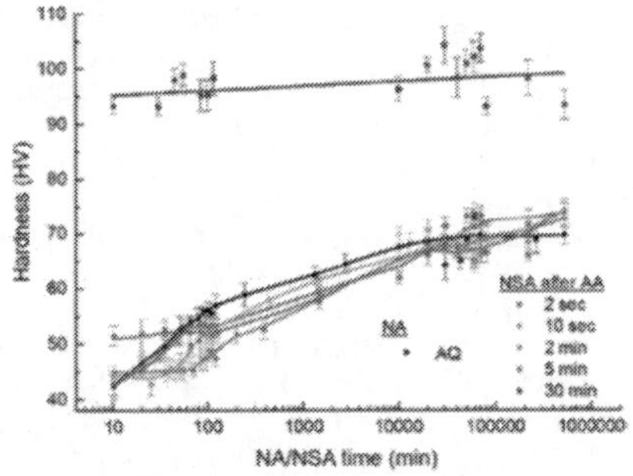

Fig. 15. Hardness evolution during NA/NSA at 20 °C after UA for 2 s up to 30 min.

5.2 Reversal of negative effect of natural ageing

In solute-rich Al-Mg-Si alloys, artificial ageing is compromised by prior natural ageing. Solute and vacancies are locked up in clusters or clusters might have to dissolve before strengthening β" particles can be formed because direct conversion is not possible. The relative importance for each mechanism is not known but an even bigger surprise is that the negative effect does not continuously build up with longer NA time but sometimes is reversed. Fig. 16 demonstrates this for the alloy Al-0.6Mg-0,8Si (wt.%). For intermediate AA times (2 h, 4 h, 8 h), NA up to 1 day increasingy reduces AA response (arrow 1), but for very long times the negative influence is partially reversed (arrow 2), after which it starts to be restored again (arrow 3). For very short AA (2, 5 min), short NA can increase AA response (arrow 4), but for very long NA, AA up to 30 min decreases hardness (arrow 5) [52]. Two papers have also discussed these effect [58, 59]. The clusters formed in different stages of NA possibly have different stabilities and the ones corresponding to the largest negative effect are more difficult to dissolve, but the structural reasons for this behaviour are not clear. It has been observed that cluster dissolution, which is a pre-requisite for fast AA, depends on the alloy composition [42], which could be a key for a consistent explanation, but further studies are needed before one can claim to understand and eventually control the effect. Table 2 lists some references of work containing data showing some of the effects of Fig. 16.

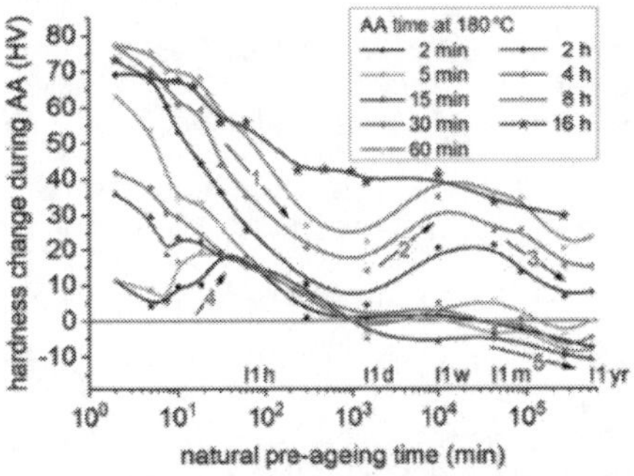

Fig. 16. Hardening response of alloy Al-0.6Mg-0.8Si (wt.%) during AA at 180 °C for different times after natural pre-ageing (calculated from data of Refs. [52, 60]). Arrows: effects discussed, 1: establishment of negative effect of NA, 2: partial reversion of negative effect, partial restoration of negative effect, 4: positive effect of short NA for short AA treatments, 5: cluster dissolution effect.

Table 2. Studies reporting reversal and restoration of the negative effect of NA on AA.

Phenomenon observed	This work (from [52, 60])	References
Partial reversal (arrow 2)	yes	[58, 59, 61-66]
Partial restoration (arrow 3)	yes	[59, 61, 66]
Initial increase (arrow 4)	only for short AA	[58, 66, 67]

6 Conclusions

Vacancies play a key role in determining the kinetics of precipitation in aluminium alloys. A high excess vacancy fraction during ageing at low temperatures is crucial for the fast hardening observed there. Positron annihilation lifetime spectroscopy provides direct evidence and quantitative estimates of the evolution of vacancy fraction in Al-Mg-Si alloys during quenching and short artificial ageing. Due to the limitations of the currently known methods, new developments are needed for deeper insights into the subtle problems of early ageing phenomena.

Acknowledgements: We acknowledge help in the analysis of SAXS data by our student Sarah Flöter.

References

[1] M.J. Starink, Int Mater Rev **49**, 191 (2004)
[2] J. Banhart, C.S.T. Chang, Z.Q. Liang, et al., Adv Eng Mater **12**, 559 (2010)
[3] C. Panseri, T. Federighi, J I Met **94**, 99 (1966)
[4] M. Madanat, M. Liu, J. Banhart, Acta Mater **159**, 163 (2018)
[5] S. Esmaeili, D. Vaumousse, M.W. Zandbergen, et al., Philos Mag **87**, (2007)
[6] S.N. Kim, J.H. Kim, H. Tezuka, et al., Mater Trans **54**, 297 (2013)
[7] D. Dafir, G. Guichon, R. Borrelly, et al., Mater Sci Engnr **A 144**, 311 (1991)
[8] V. Massardier, T. Epicier, P. Merle, Acta Mater **48**, 2911 (2000)
[9] A.K. Gupta, D.J. Lloyd, Metall Mater Trans A **30**, 879 (1999)
[10] I. Dutta, S.M. Allen, J Mater Sci Lett **10**, 323 (1991)
[11] C.S.T. Chang, J. Banhart, Metall Mater Trans A **42A**, 1960 (2011)
[12] S. Esmaeili, D.J. Lloyd, W.J. Poole, Acta Mater **51**, 3467 (2003)
[13] A.M. Zahra, C.Y. Zahra, C. Alfonso, et al., Scripta Mater **39**, (1998)
[14] M. Luckabauer, E. Hengge, G. Klinser, et al., in: Magnesium Technology 2017, K.N. Solanki (Ed.) The Minerals, Metals, & Materials Society, 2017, pp. 669-674
[15] C.S.T. Chang, F. De Geuser, J. Banhart, J Appl Crystallogr **48**, 455 (2015)
[16] H. Adachi, H. Nakanishi, M. Asano, J. Japan Inst. Light Met. **65**, 411 (2015) [in Japanese]
[17] M.D.H. Lay, H.S. Zurob, C.R. Hutchinson, et al., Metall Mater Trans A **43A**, 4507 (2012)
[18] K. Nishimura, K. Matsuda, T. Namiki, et al., Mater Trans **56**, 1307 (2015)
[19] K. Nishimura, K. Matsuda, Q.K. Lei, et al., Mater Trans **57**, 627 (2016)
[20] S. Wenner, C.D. Marioara, K. Nishimura, et al., Metall Mater Trans A **50A**, 3446 (2019)
[21] J. Banhart, M.D.H. Lay, C.S.T. Chang, et al., Phys Rev B **83**, 014101 (2011)
[22] G.A. Edwards, K. Stiller, G.L. Dunlop, et al., Acta Mater **46**, 3893 (1998)
[23] M. Murayama, K. Hono, Acta Mater **47**, 1537 (1999)
[24] A. Serizawa, S. Hirosawa, T. Sato, Metall Mater Trans A **39A**, 243 (2008)
[25] V. Fallah, A. Korinek, N. Ofori-Opoku, et al., **82**, (2015)
[26] C. Marioara, S.J. Andersen, J. Friis, et al., in: International Conference on Aluminium Alloys (ICAA16), M. Wells, M. Brochu (Eds.) Canadian Institute of Mining, Metallurgy & Petroleum, Montréal, 2018, p. 403435
[27] D. Petschke, F. Lotter, E. Bläss, et al., J Appl Cryst **51**, 1339 (2018)
[28] M.W. Zandbergen, Q. Xu, A. Cerezo, et al., Acta Mater **101**, 136 (2015)
[29] F. de Geuser, B. Gault, Acta Mater **188**, 406 (2020)
[30] A. Somoza, A. Dupasquier, I.J. Polmear, et al., Phys Rev B **61**, 14464 (2000)
[31] R. Ferragut, A. Somoza, A. Dupasquier, J Phys-Condens Mat **10**, 3903 (1998)
[32] A.A. Istratov, O.F. Vyvenko, Rev Sci Instrum. **70**, 1233 (1999)
[33] B. Somieski, T.E.M. Staab, R. Krause-Rehberg, Nucl Instrum Meth A **381**, 128 (1996)
[34] T.E.M. Staab, B. Somieski, R. Krause-Rehberg, Nucl Instrum Meth A **381**, 141 (1996)
[35] M. Liu, J. Čížek, C.S.T. Chang, et al., Acta Mater **91**, 355 (2015)
[36] M. Saga, Y. Sasaki, M. Kikuchi, et al., Mater Sci Forum **217**, 821 (1996)
[37] I. Kovács, J. Lendvai, E. Nagy, Acta Metall **20**, 975 (1972)
[38] A. Serizawa, T. Sato, in: 11th International Conference on Aluminium Alloys (ICAA11), J. Hirsch, B. Skrotzki, G. Gottstein (Eds.) Wiley VCH, Aachen, Germany, 2008, pp. 915-921
[39] S. Pogatscher, H. Antrekowitsch, H. Leitner, et al., Acta Mater **60**, 4496 (2012)
[40] J.D. Bryant, Metall Mater Trans A **30A**, 1999 (1999)
[41] K. Yamada, T. Sato, A. Kamio, Mater Sci Forum **331-337**, 669 (2000)
[42] A. Poznak, R.K.W. Marceau, P.G. Sanders, Mat Sci Eng A **721**, 47 (2018)
[43] Z. Yang, Z.Q. Liang, D. Leyvraz, et al., Materialia **7**, 100413 (2019)
[44] Z. Yang, PhD thesis, Technische Universität Berlin, Berlin, (2019)
[45] C. Haase, H. Wurst, Z Metallkde **33**, 399 (1941) [in German]
[46] F. Federighi, Acta Mater **6**, 379 (1958)
[47] R.W. Siegel, J Nucl Mat **69-70**, 117 (1978)
[48] E. Povoden-Karadeniz, Therodynamic and diffusion databases of program Matcalc (mc_al_v2.032.tdb mc_al_v2.004.ddb), Vienna (2014).
[49] M. Mantina, Y. Wang, L.Q. Chen, et al., Acta Mater **57**, 4102 (2009)
[50] F.D. Fischer, J. Svoboda, F. Appel, et al., Acta Mater **59**, 3463 (2011)
[51] M.J. Starink, L.F. Cao, P.A. Rometsch, Acta Mater **60**, 4194 (2012)
[52] Y. Yan, PhD thesis, Technische Universität Berlin, Berlin, (2014) [doi: 10.14279/depositonce-4020]
[53] M. Madanat, M. Liu, X.-P. Zhang, et al., Phys Rev Mater **4**, 063608 (2020)
[54] A. Hoell, I. Zizak, H. Bieder, et al., Patent DE 10 2006 029 449, 2006.
[55] J. Wernecke, C. Gollwitzer, P. Müller, et al., J. Synchrotron Rad., 529 (2014)
[56] R.N. Lumley, I.J. Polmear, A.J. Morton, Mater Sci Tech-Lond **21**, 1025 (2005)
[57] X.P. Zhang, PhD thesis, Technische Universität Berlin, Berlin, (2019) [doi: 10.14279/depositonce-9247]
[58] F.A. Martinsen, F.J.H. Ehlers, M. Torsæter, et al., Acta Mater **60**, 6091 (2012)
[59] J.H. Kim, E. Kobayashi, T. Sato, Mater Trans **56**, 1771 (2015)
[60] A. Röhsler, Diploma thesis, Bergakademie Freiberg, Freiberg, (2015)
[61] P.E. Fortin, Can Metall Quart **2**, 143 (1963)
[62] S. Muromachi, T. Mae, J Jap Inst Metals Mater **38**, 130 (1974) [in Japanese]
[63] K. Yamada, A. Sato, A. Kamio, J Jap Inst Light Met **51**, 215 (2001) [in Japanese]
[64] J. Røyset, T. Stene, J.A. Sæter, et al., Mater Sci Forum **519-521**, 239 (2006)
[65] L.Z. He, H.-T. Zhang, J.-H. Cui, J Mater Sci Technol **26**, 141 (2010)
[66] S. Wenner, C.D. Marioara, S.J. Andersen, et al., Int J Mater Res **103**, 948 (2012)
[67] H. Borchers, M. Kainz, Z Metallkde **52**, 685 (1961) [in German]

https://doi.org/10.1051/matecconf/202032601002

Twin relationship in between the variant-pair of η precipitates in the Al-Zn-Mg-Cu aluminium alloy

Tsai-Fu Chung[1], Yo-Lun Yang[2], Chien-Nan Hsiao[3], Wei-Chih Li[4], Jer-Ren Yang[1,]*
[1] Department of Materials Science and Engineering, National Taiwan University, Taipei, Taiwan.
[2] Department of Mechanical Engineering, Imperial College London, London SW7 2AZ, UK.
[3] Taiwan Instrument Research Institute, Hsinchu, Taiwan.
[4] E.A. Fischione Instruments, Inc., 9003 Corporate Circle, Export, PA 15632, U.S.A.

Abstract. For Al-Zn-Mg-Cu aluminium alloys, 15 types of η precipitates would possess the symmetrically variants distributed on the closed planes of the Al matrix, parallel to the $(0001)\eta$, $(2\bar{1}\bar{1}0)\eta$, or $(10\bar{1}0)\eta$ interfaces of precipitates. The η_2 precipitates, possessing the crystallographic orientation of $(1\bar{1}\bar{1})_{Al}$ // $(0001)\eta_2$ and $[110]_{Al}$ // $[10\bar{1}0]\eta_2$, would exhibit four equivalent variants, i.e., $\eta_2^{(1)}$ to $\eta_2^{(4)}$, on the $(1\bar{1}\bar{1})_{Al}$ habit planes. In the present work, along the zone axis of $[110]_{Al}$ // $[10\bar{1}0]\eta$, the edge-on configurations showing the twin-like atomic arrays would occur as the growth/coalescence of two η_2 variants grown on $(1\bar{1}\bar{1})_{Al}$ and $(1\bar{1}1)_{Al}$ planes, respectively. The twin relationship can be revealed in term of the crystallographic relationship at 70.5° with respect to their habit planes. Alternatively, on the $(\bar{1}\bar{1}0)_{Al}$ image, it also indicates the nearly-twinning configuration in between the variant-pair of η_2.

1. Introduction

The Al-Zn-Mg-Cu aluminium alloys, exhibiting the high strength mechanical property, are well known to be applied in the automotive and aerospace applications. The precipitation hardening of the AA7050 aluminium alloys [1-9] begins from the fully or nearly coherent GP zones (GPI and GPII zones), and subsequently transforms into the semi-coherent η' precipitates (4 variants), and then the incoherent η precipitates (15 types, η_1-η_{14}, and η_4'). In the recent work, The nucleation and transformation mechanisms of GPII zones → η' precipitates and η' precipitates → η_2 precipitates have been recognized as the separated and in-situ nucleations, which are conducted by the in-situ frames of micrographs and the serial Fast-Fourier-Transform (FFT) diffractograms of high-resolution transmission electron microscopy (HRTEM), respectively [1].

According to the in-situ transformation mechanism of η' precipitates → η_2 precipitates and their identical crystallographic orientations with respect to the Al matrix : $[10\bar{1}0]\eta'$ // $[110]_{Al}$ and $(0001)\eta'$ // $(1\bar{1}\bar{1})_{Al}$ of η' precipitates, and $[10\bar{1}0]\eta_2$ // $[110]_{Al}$ and $(0001)\eta_2$ // $(1\bar{1}\bar{1})_{Al}$ of η_2 precipitates [1, 10], it would be observed four equivalent variants of η_2 (i.e., $\eta_2^{(1)}$ precipitates to $\eta_2^{(4)}$ precipitates), grown on the respective $(1\bar{1}\bar{1})_{Al}$ habit planes, as portrayed in Fig. 1. Afterwards, employed by Cs-corrected high-angle annular-dark-field (HAADF) scanning-transmission electron microscopy (STEM) [2], the edge-on configurations of four η_2 along the zone axis of $[10\bar{1}0]\eta_2$ // $[110]_{Al}$ are characterized as the sandwiched stacking arrangements, in which the regular variations of Zn and Mg layers subsequently stacks on the $(0001)\eta_2$ // $(1\bar{1}\bar{1})_{Al}$ habit planes.

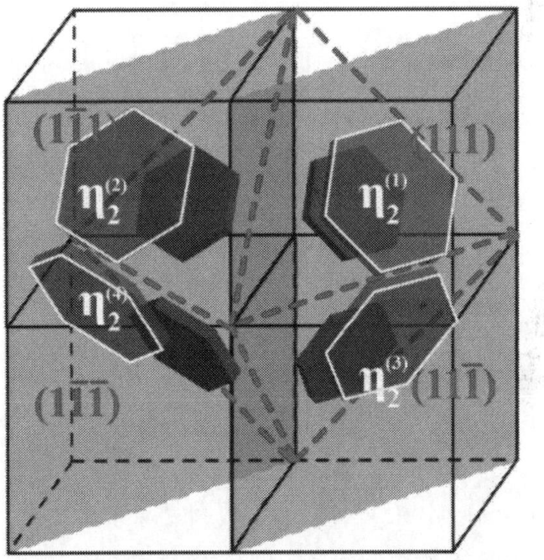

Fig. 1 The schematic diagram of four variants of η_2 precipitates (i.e., $\eta_2^{(1)}$ to $\eta_2^{(4)}$) grow on $\{111\}_{Al}$ habit planes. As the precipitate growing, the coarsened η_2 are indicated by yellow frames.

It has been reported that the distributions in between the orientational variants of precipitates would be easily misunderstood from the 2-dimensional projections of the 3-dimensional phenomena [11, 12]. For example, in the Al-Cu-(Li) aluminium alloys [11], the impingements are inevitably caused by the overlap of two edge-on variants of θ'-Al_2Cu precipitates, in which their habit planes are parallel to the $(100)_{Al}$ and $(010)_{Al}$ planes of the Al matrix, respectively. On the other hand, in the recent work employed by HAADF-STEM [12], the joint of two edge-on variants of θ' incisively revealed the distributed

* Corresponding author: author@e-mail.org

arrangements, implying that the possibility of the interaction between the orientational variants of precipitates. However, so far there is no direct experimental to support the crystallographic relationship such as twinning relationship in between two variants of θ'. Tacitly, in the Al-Zn-Mg-Cu aluminium alloys, whether the crystallographic distribution of the 15 typed η precipitates (i.e., η_1-η_{14}, and η_4') and their corresponding variants [1, 2] would also exist the crystallographic relationship in between themselves has yet to be elucidated.

In the present work, employed by Cs-corrected HAADF STEM, we especially focused on the crystallographic relationships of η_2 and their variant pairs in Al-Zn-Mg-Cu aluminium alloys. Along the zone axis of $[110]_{Al}$, the detailed atomic configurations of η_2 has been revealed. Additionally, the crystallographic relationship in between variant-pair of η_2 has been confirmed under the observation of $[110]_{Al}$ and $[\bar{1}\bar{1}0]_{Al}$ zone axes

2. Materials and methods

The AA7050 aluminium alloy (Al-6.25Zn-2.14Mg-2.23Cu, wt.%) was used in this study. After solution treatment (at 475 °C for 1 h with water quenching), the samples were treated with a three-step ageing treatment: (1) ageing at 120 °C for 8 h, (2) ageing at 165 °C for 1 h, and subsequently (3) ageing at 174 °C for 8 h with an applied constant tensile stress of 162.5 MPa at the same time (named as creep-age forming samples, CAF samples). The TEM specimens were prepared by cutting discs from the selected samples for observing the pancaked structures and thinning the discs to the thickness of 65 - 70 μm before they were twin-jet electropolished in a mixture of 0.33 HNO_3 and 0.67 CH_3OH at -20 °C with a working voltage of 10-11 V, as shown in Fig. 2.

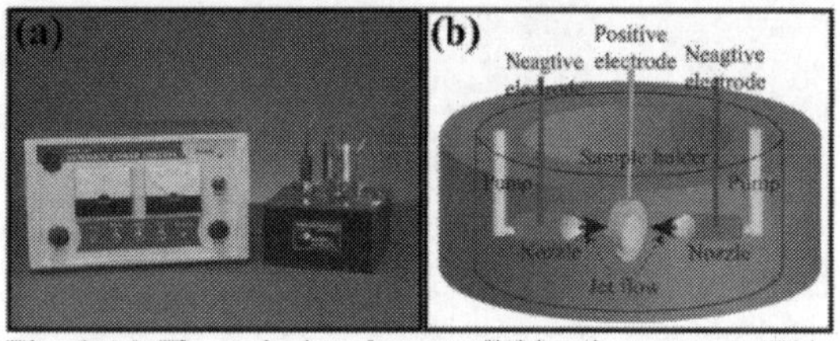

Fig. 2 (a) The twin-jet electropolishing instrument (E.A. Fischione Instrument), and (b) the schematic inward of twin-jet system.

The oxidation layers of specimens was removed and the corresponding thickness was reduced with an M1040 NanoMill system (E.A. Fischione Instrument), as shown in Fig. 3a and then the hydrocarbon contamination was cleaned by an M1070 Nanoclean plasma (E.A. Fischione Instrument), as shown in Fig. 3b. Afterwards, the TEM specimens were observed with an FEI Titan Chemi-STEM equipped with a spherical aberration corrector (Cs-corrector), under the 135 nm camera length and with a collecting angle range of ~8.24 mrad (inner angle) to ~143.6 mrad (outer angle) of a HAADF detector.

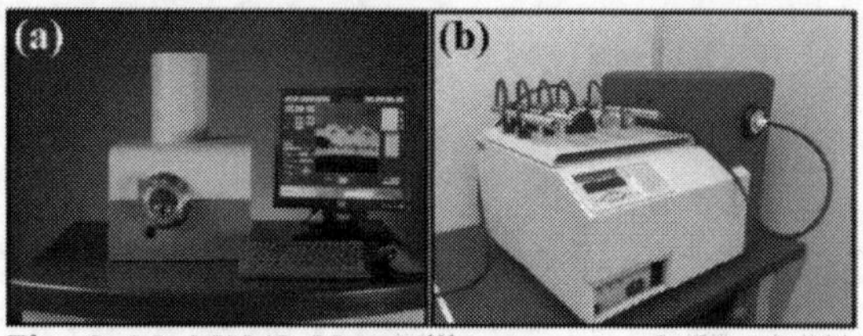

Fig. 3 (a) M1040 NanoMill system, and (b) M1070 Nanoclean plasma system.

3. Results & Discussion

Fig. 4a shows the edge-on configuration of η_2 grown on the $(\bar{1}11)_{Al}$ habit plane, and its atomic arrangement along the $[110]_{Al}$ zone axis. The alternative brightness-darkness Z-contrast can be explained by the stimulated atomic image based on first principle calculation, as shown in Fig. 4b. The interweaving Zn and Mg atomic-layers are consistent with the brighter Z-contrast representing Zn-rich layers and the lower Z-contrast, the Mg/Zn layers. This result is similar to the sandwiched stacking of η_2, repored in our previpus work [2]. Moreover, in Fig. 4c, the FFT diffractogram, this paeticle can be recognized as $\eta_2^{(4)}$ crystal exhibiting the crystallographic relationship with respect to the Al matrix: $[10\bar{1}0]\eta_2^{(4)}$ // $[110]_{Al}$ and $(0001)\eta_2^{(4)}$ // $(\bar{1}11)_{Al}$.

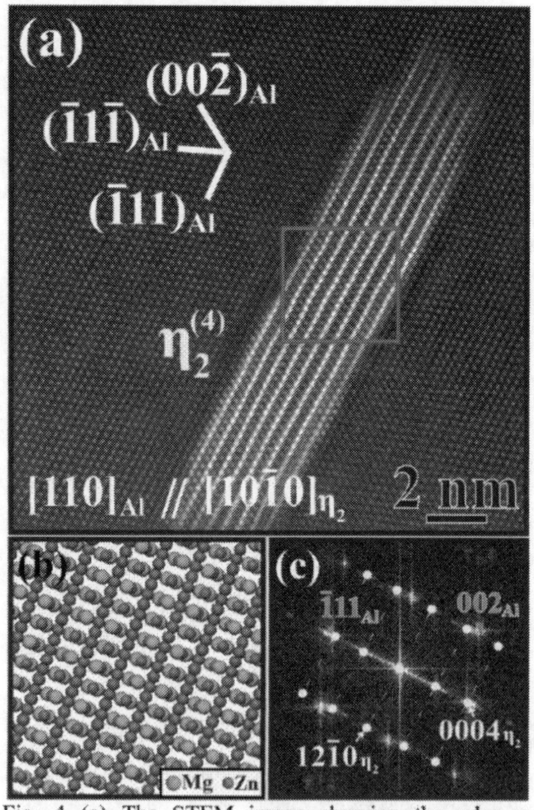

Fig. 4 (a) The STEM image showing the edge-on configuration of $\eta_2^{(4)}$ along the $[110]_{Al}$ zone axis [13]. (b) Stimulated atomic-stacking. (c) FFT diffractogram.

Fig. 5a shows that two edge-on configurations of two variants of η_2 (i.e., $\eta_2^{(2)}$ and $\eta_2^{(4)}$) exhibiting the sandwiched stacking structure on the $(0001)\eta_2$ // $(\bar{1}1\bar{1})_{Al}$ and $(0001)\eta_4$ // $(\bar{1}11)_{Al}$ habit planes, respectively. Their different size can be ascribed to the difference in the growth rate. Particularly, the joint area, i.e., the interweaved interface, between $\eta_2^{(2)}$ and $\eta_2^{(4)}$ possesses a twinning relationship with an angle of 109.5° between the $(00\bar{2})_{Al}$ plane, as marked by the red lines in Fig. 5a. Here, it can be ascribed to two phenomena. First, the collision of two variants of η_2 occurred as they grow, the so-called hard impingement [12, 14]. Secondly, these two η_2 would develop at the same interface of the nucleation stage. we temporarily call the intrinsic boundary.

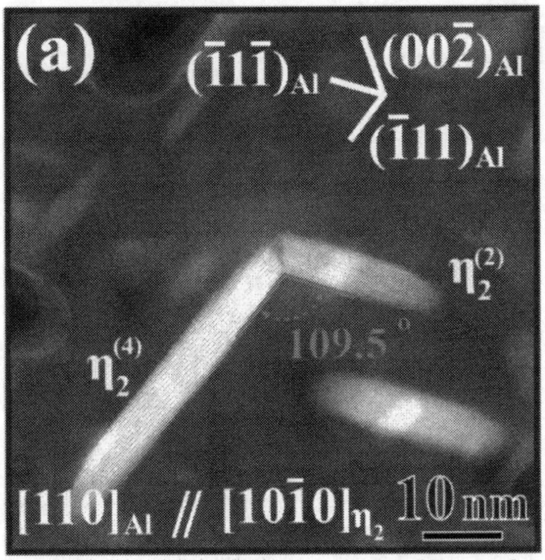

Fig. 5 (a) HAADF STEM image showing the two edge-on configurations of η_2 precipitate (i.e., $\eta_2^{(2)}$ and $\eta_2^{(4)}$) grown on the $(0001)\eta_2$ // $(\bar{1}1\bar{1})_{Al}$ and $(0001)\eta_4$ // $(\bar{1}11)_{Al}$ habit planes, respectively under the $[110]_{Al}$ // $[10\bar{1}0]\eta_2$ zone axis [13].

For the detailed investigation on the atomic arrangement and microstructural evolution in between two variants of η_2 (i.e., $\eta_2^{(2)}$ and $\eta_2^{(4)}$), the area with the nearly darkness Z-contrast (as marked by the red arrow in Fig 5b) can be recognized as segregation of vacancies or Mg solute atoms. This given area was supposedly ascribed to the early stage of nucleation. As the growth of these two variants of η_2, vacancies and Mg atoms are gradually annihilated by the substitutions of Zn and Cu solute atoms. Finally, the mismatched interface would turn into the well-matched interface (as marked by the solid dot of the red arrow in Fig. 5b). Whether the intrinsic boundary would be observed in 15 typed η under the observation of $[110]_{Al}$, $[001]_{Al}$ and $[11\bar{2}]_{Al}$ zone axes has yet to be reported.

Fig. 5 (b) A enlarged image of (a) revealing the twin relationship of an angle of 109.5° between the $(00\bar{2})_{Al}$ plane with respect to the $(0001)\eta_2^{(2)}$ // $(\bar{1}1\bar{1})_{Al}$ and $(0001)\eta_2^{(4)}$ // $(\bar{1}11)_{Al}$ habit planes [13].

Another case crystallographic relationship of different variants of η_2 is shown in Fig. 6a. Specially, the joint area (Fig. 6) is totally different from the end of two variants, as shown in Fig. 5. Along the zone axis of $[110]_{Al}$, the edge-on configurations of $\eta_2^{(2)}$ and $\eta_2^{(4)}$, grown on the $(0001)\eta_2$ // $(\bar{1}1\bar{1})_{Al}$ and $(0001)\eta_4$ // $(\bar{1}11)_{Al}$ habit planes, respectively, reveal the twinning relationship with an angle of 109.5° or 70.5° with respect to two different twinning planes in between these two variants.

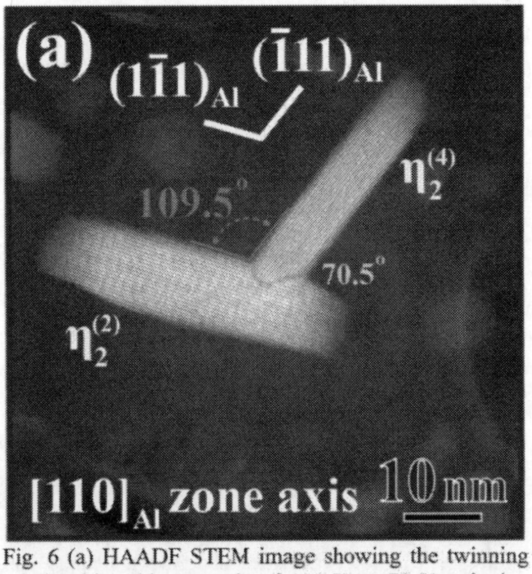

Fig. 6 (a) HAADF STEM image showing the twinning relationship with an angle of 109.5° or 70.5° twinning relationship with an angle of 109.5° respect to two edge-on configurations of $\eta_2^{(2)}$ and $\eta_2^{(4)}$ along the zone axis of $[110]_{Al}$.

Similarly, the detailed study on the atomic configurations in between $\eta_2^{(2)}$ and $\eta_2^{(4)}$ are illustrated in Fig. 6b. It has been found that the twin relationships of two angles of 109.5° and 70.5° between the $(1\bar{1}3)_{Al}$ and $(\bar{3}31)_{Al}$ planes of the Al matrix, respectively. The coherent twin boundaries would be caused by the nucleation and growth, instead of the effect of the hard impingement. It could be suggested that the nucleation site would be the partial segregation of Zn/Cu nearby the interfaces between the η_2 precipitate and the Al matrix with the abnormally strong bright Z-contrast, reported in the previous work [2]. In the present case, the nuclei would be located on the interfaces of $\eta_2^{(2)}$ and would follow the twinning relationship as the growth of precipitates.

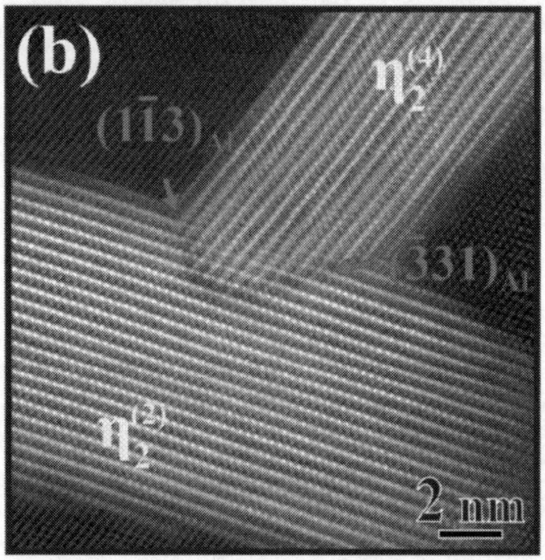

Fig. 6 (b) A enlarged image of (a) revealing the twin relationships of two angles of 109.5° and 70.5° between the $(1\bar{1}3)_{Al}$ and $(\bar{3}31)_{Al}$ planes with respect to $\eta_2^{(2)}$ and $\eta_2^{(4)}$ precipitates.

Furthermore, in Fig. 7, the edge-on configurations of $\eta_2^{(2)}$ and $\eta_2^{(4)}$ are specifically observed along the zone axis of $[\bar{1}\bar{1}0]_{Al}$ with respect to the resulted HAADF STEM image of Fig. 6. At the joint or intersected position, the crystallographic relationship in twinning would exist. However, one coherent twinning boundary, as indicated by the red arrow in Fig. 7, can be found. It can be suggested that the segregations of Zn/Cu to be the nucleation site of $\eta_2^{(4)}$ would locate on the part of the $\eta_2^{(2)}$/Al interfaces.

Conclusively, according to the orientation relationships of 15 typed η precipitates (η_1-η_{14}, and η_4') and the corresponding variants, their crystallographic relationships are related to be twinning. In the present work, along the zone axes of $[110]_{Al}$ and $[\bar{1}\bar{1}0]_{Al}$, it can be fount the twinning boundaries in between the edge-on configurations of $\eta_2^{(2)}$ and $\eta_2^{(4)}$.

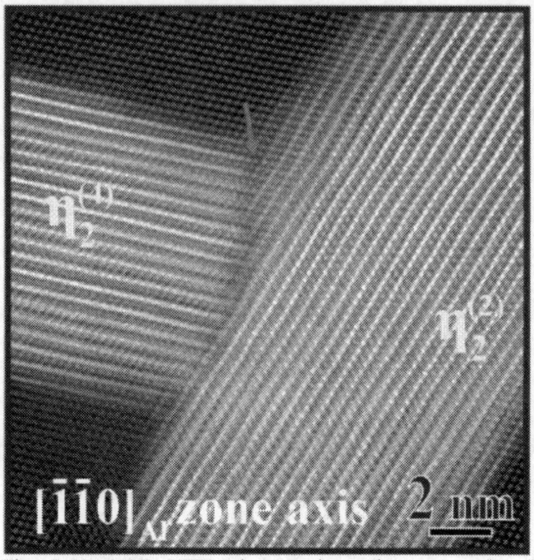

Fig. 7 HAADF STEM image along the zone axis of $[\bar{1}\bar{1}0]_{Al}$ with respect to that of Fig. 6 revealing the atomic configuration of the joint area in between $\eta_2^{(2)}$ and $\eta_2^{(4)}$.

4. Acknowledgements

This work was carried out with the financial support from Foundation for the Advancement of Outstanding Scholarship and Ministry of Science and Technology (Taiwan) under Contract MOST108-2622-E-002-016-CC2. The authors are grateful to Taiwan Instrument Research Institute and Materials Analysis Technology Inc. (MA-tek) for technical advice.

5. Reference

[1] T.-F. Chung, Y.-L. Yang, B.-M. Huang, Z. Shi, J. Lin, T. Ohmura, J.-R. Yang, Acta Mater. **149** (2018).

[2] T.-F. Chung, Y.-L. Yang, M. Shiojiri, C.-N. Hsiao, W.-C. Li, C.-S. Tsao, Z. Shi, J. Lin, J.-R. Yang, Acta Mater. **174** (2019).

[3] A. Deschamps, A. Bigot, F. Livet, P. Auger, Y. Brechet, D. Blavette, Philos. Mag. A **81** (2001).

[4] A. Deschamps, F. Bley, F. Livet, D. Fabregue, L. David, Philosophical Magazine **83** (2003).

[5] L. Couturier, A. Deschamps, F. De Geuser, F. Fazeli, W.J. Poole, Scripta Mater. **136** (2017).

[6] C.R. Hutchinson, F. de Geuser, Y. Chen, A. Deschamps, Acta Mater. **74** (2014).

[7] A. Bendo, K. Matsuda, A. Lervik, T. Tsuru, K. Nishimura, N. Nunomura, R. Holmestad, C.D. Marioara, K. Shimizu, H. Toda, Mater. Charact. (2019).

[8] H. Degischer, W. Lacom, A. Zahra, C. Zahra, Z. Metallk. **71** (1980).

[9] J. Gjønnes, C.J. Simensen, Acta. Metall. Mater. **18** (1970).

[10] S. Ringer, K. Hono, Mater. Charact. **44** (2000).

[11] J. da Costa Teixeira, D.G. Cram, L. Bourgeois, T.J. Bastow, A.J. Hill, C.R. Hutchinson, Acta Mater. **56** (2008).

[12] T.-F. Chung, Y.-L. Yang, C.-N. Hsiao, W.-C. Li, B.-M. Huang, C.-S. Tsao, Z. Shi, J. Lin, P.E. Fischione, T.

Ohmura, J.-R. Yang, International Journal of Lightweight Materials and Manufacture **1** (2018).
[13] T.-F. Chung, Y.-L. Yang, M. Shiojiri, C.-N. Hsiao, W.-C. Li, J.-R. Yang, Procedia Manufacturing **37** (2019).
[14] B. Decreus, A. Deschamps, F. De Geuser, P. Donnadieu, C. Sigli, M. Weyland, Acta Mater. **61** (2013).

In situ alloying of aluminium-based alloys by (multi-)wire-arc additive manufacturing

Thomas Klein*, *Alois* Birgmann and *Martin* Schnall

LKR Light Metals Technologies Ranshofen, Austrian Institute of Technology, 5282 Ranshofen, Austria

Abstract. Wire-arc additive manufacturing (WAAM) has received considerable attention in the past years due to advantages in terms of deposition rate, design freedom, buy-to-fly ratio and economic factors. This process can generally be conducted using conventional or near-conventional welding equipment to fabricate intricate but relatively large-scale structures. The present contribution explores options to utilize this novel process not only for manufacturing of particular aluminium structures, but to create the actual alloy composition during processing. Thereby, the possibilities of dual-wire techniques based on cold metal transfer (CMT) to create alloys in the welding process *in situ* is investigated. For this purpose, a modified CMT twin welding system is used with standard wires differing significantly in their alloying content. The characterization of the chemical compositions at different specimen positions suggests good chemical homogeneity after initial process optimization steps. The microstructural homogeneity is analysed by means of optical light microscopy and scanning electron microscopy. Quantified phase fractions underpin non-equilibrium solidification conditions, when compared to theoretical equilibrium predictions. The assessment of the performed analyses suggests that dual-wire processes are powerful in terms of enhancing achievable depositions rates as well as enabling *in situ* alloying. This approach might be expandable to multi-wire-based techniques.

1 Introduction

Wire-arc additive manufacturing (WAAM) is gaining increasing interest due to its manifold prospects for manufacturing medium complex structures with a high deposition rate [1-3]. In this process a metallic wire is fed into an energy source (here the electric arc) resulting in melting of the wire feedstock. The resulting droplet is deposited by a robotic or portal system according to a predefined track cumulatively resulting in the build-up of whole components or the modification of semi-finished products [4]. The iterative droplet deposition sequence results in an accumulation of heat – an intrinsic process characteristic – determining the alloy dependent microstructure formation sequence [5-8].
To increase the deposition rate and, more importantly, enable the creation of new, or locally differing alloy composition, feeding two or more wires into the electric arc is desirable as has been suggested for welding with consumable [9-11] and non-consumable electrodes [12,13]. However, achieving high quality material properties requires sufficient intermixing of the different alloys via adaption of processing parameters. Here we report on the chemical and microstructural characteristics obtained by *in situ* alloying Al-Si alloys after the initial process optimisation steps.
In this study we use the well-known Al-Si alloy system as a model system to qualify a dual-wire process based on the cold metal transfer (CMT) welding technology. Emphasis is placed on the effects triggered using differing chemical compositions of the feedstock wires to form a novel alloy during WAAM by *in situ* alloying and the resultant effects on microstructure.

2 Materials and experimental methods

Commercial aluminium welding wires, ML-AlSi7Mg and ML-AlSi12, were used with 1.2 mm diameter supplied by MIGAL.CO GmbH with the chemical composition given in Table 1.
The welded material was characterized using chemical analysis by optical emission spectroscopy (OES) (SPECTROMAXx 6 from SPECTRO Analytical Instruments). All compositional values are given in mass percent in this work. The resulting microstructures were analysed by optical light microscopy (OLM) (Olympus BX60) on metallographic specimens. Hardness measurements according to Vickers using a 100 g load were conducted to assess the effects of microstructural homogeneity on mechanical properties. Moreover, microstructural details were visualised by scanning electron microscopy (SEM) (Tescan Mira 3) at an acceleration voltage of 20 kV using back-scattered electron (BSE) mode. Local chemical compositions were analysed by energy dispersive X-ray spectroscopy (EDX) (EDAX Octane Elect) using an acceleration voltage of 20 kV. Additionally, phase fractions were

* Corresponding author: thomas.klein@ait.ac.at

https://doi.org/10.1051/matecconf/202032601003

calculated using Thermo-Calc software in combination with the TTAl8 database using the global minimization of the Gibbs free energy.

3 Processing

The CMT twin system (Fronius GmbH) visualised in Fig. 1 was used for the investigations. This setup combines two CMT-systems. Recent works provided evidence for the suitability of CMT for WAAM of various aluminium alloys [8,14-16]. The two torches (ML-AlSi7Mg and ML-AlSi12) were used with mean welding currents of 51 A and 65 A and mean voltages of 11.7 V and 15.9 V, respectively. The travel speed amounted to 5 mm/s and the wire feed rates were 3.5 m/min each. The CMT twin system was combined with an ABB IRB 4600 welding robot and an ABB IRB 500A workpiece positioner. For the present investigations Argon 5.0 (Linde Gas GmbH) was used.

Fig. 1. Dual-wire welding system with parallel wire feeding used in the present study inclusive welded cylindrical specimen on circular flat substrate.

Cylindrical specimens with a diameter of ≈ 70 mm, a height of ≈ 60 mm and a wall thickness of ≈ 10 mm were welded on flat substrates as visible in Fig. 1. The target composition was chosen as AlSi10Mg with a composition as given in Table 1. This composition is an exemplary alloy composition, which is not available as commercial welding wire. The detailed analyses of the subsequent subchapters were conducted using the process variant with the apparently best chemical and microstructural homogeneity as well as least porosity, which was obtained by parallel wire feeding (see Fig. 1) and utilization of an oscillating motion.

4 Material characterisation

4.1. Analysis of chemical composition

Cylindrical structures were welded and cut vertically providing a cross-section suitable for further analyses (schematically shown as inset in Fig. 2(a)). Across this section the chemical compositions were determined at three different heights: Bottom, middle, and top. In Fig. 2 the chemical profiles for the elements Al (a), Si and Mg (b) are shown. Alloying elements and impurities with comparably low concentrations are omitted for clarity. Observed variations in all three elements can be identified corresponding to one another. The chemical analyses suggest good chemical homogeneity throughout the specimen in both thickness and height directions. However, the Si target value was not reached and is slightly beneath an acceptable value [17], which could be adjusted by increasing the wire feed rate of the ML-AlSi12 welding wire but the actual concentration is not of relevance for the subsequent microstructural analyses.

4.2. Microstructure and hardness analysis

A representative OLM micrograph is depicted in Fig. 3 as hardly any microstructural variations were discernible at this scale. Clearly, the primary dendritic solidification is discernible with fairly homogeneously distributed eutectic constituents in the interdendritic regions.

Table 1. Chemical compositions of the welding wires used in the present work as well as the targeted final material composition. All values are given in mass percent.

	Al	Si	Fe	Cu	Mn	Mg	Zn	Ti
ML-AlSi7Mg	91.7	7.0	0.2	0.05	0.1	0.65	0.1	0.2
ML-AlSi12	86.5	12.0	0.6	0.3	0.15	0.1	0.2	0.15
Target	88.5	10.0	0.44	0.20	0.13	0.31	0.16	0.17

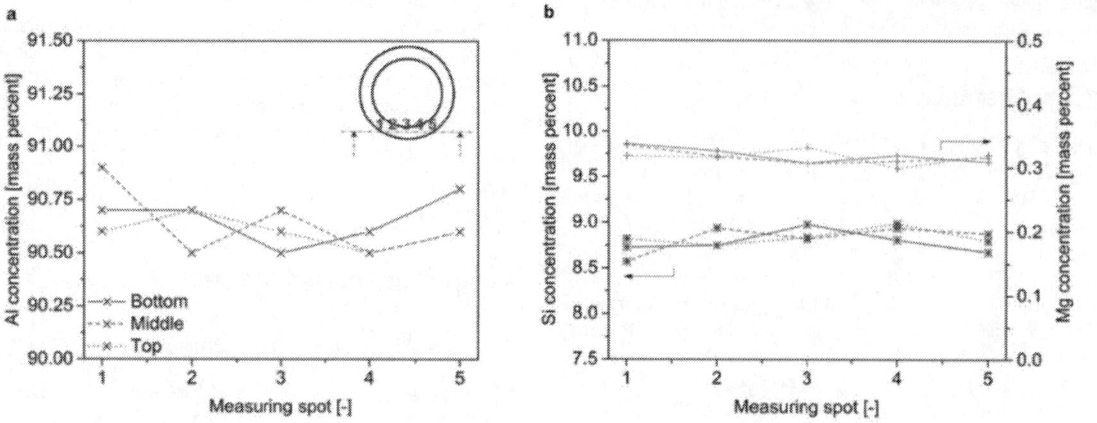

Fig. 2. Variation of chemical composition of the major alloying elements Al (a), Si and Mg (b). The cross-section prepared for chemical analysis is schematically visualised as inset in Fig. 2(a).

Additionally, some porosity is visible. Porosities in WAAM of aluminium are mainly caused by (i) insufficient shielding gas protection of the melt, (ii) differing hydrogen solubility of liquid and solid aluminium and (iii) insufficient liquid feeding upon solidification. Addressing (i) can be conducted by modification of the shielding gas flow, (ii) by increasing the melt and, thus, feedstock quality and (iii) by variation of WAAM parameters affecting solidification conditions or alloy design targeting grain refinement. In the present work insufficient inert gas shielding of the melt pool was identified as the major cause of porosities. Best results were obtained using a modified shielding gas nozzle as visible in Fig. 1, which may, however, be further improved.

The secondary dendrite arm spacing is in the range of ≈ 10 µm irrespective of the location within the specimen. According to Refs. [18,19] this magnitude suggests a cooling rate of ≈ 50 K/s during solidification in qualitative agreement with Ref. [5].

It is noted that the amount of eutectic constituent appears to be affected by the non-equilibrium solidification, i.e. by the high process-intrinsic cooling rates, dominating the solidification conditions during WAAM. Thermo-Calc calculations were performed for the mean actual chemical composition as determined by OES, which is assumed to correspond the microstructure as exemplarily visible in Fig. 3 due to its high homogeneity. It is further assumed that the phase fractions within the eutectic constituent are separated into α-Al and Si according to equilibrium condition, which appears reasonable, since eutectic reactions often prevail similarly as in thermodynamic equilibrium due the physical nature of this reaction. These calculations suggest an amount of primary α-Al grains of ≈ 32.2 mole percent in thermodynamic equilibrium at room temperature. Experimentally, this low amount of primary dendrites and corresponding high value of eutectic constituent is not observed in the WAAM processed specimen visualized in Fig. 3. This is unlikely to be due to chemical discrepancies (see Fig. 2). It is argued that the high solidification rate results in a non-equilibrium distribution of Si and the concomitant variation in the amounts of primary phases. These discrepancies might be equilibrated upon ensuing heat treatments.

Hardness values were measured at the bottom, middle and top section of the specimen. In these three regions values of 80 ± 3, 76 ± 3 and 79 ± 1 HV0.1, respectively, were determined. The slightly higher hardness in the close vicinity of the substrate is presumably due to an increased cooling rate in this specimen region.

4.3. Microstructure and phase formation

The microstructure visualized by SEM in BSE mode is depicted in Fig. 4 separated in three different specimen regions (bottom (a, d), middle (b, e), top (c, f)). Its general appearance is independent of the building direction. The microstructure with regard to primary dendrites and eutectic constituent reflects observations by OLM. Visually the microstructure is comparable to the microstructures reported in Refs. [20,21], which were fabricated by single wire WAAM but differs significantly from microstructures observed after selective laser melting (SLM) [22].The microstructures obtained after SLM are dominated by the severe non-equilibrium condition of the feedstock alloy powders and the subsequent intrinsic heat treatment being repetitive in nature, but due to the particularly small melt pool, cooling rates are much higher than during WAAM.

EDX analysis was used for the identification of additional constituents visible in Fig. 4. Clearly, at the specimen's top fewer bright interdendritic phases are present. In all specimen regions, these phases were identified to be rich in Al, Fe and Si and formed abundantly upon solidification [23]. However, Mg enrichments are discernible at the bottom and the middle sections in these phases or their close vicinity but were not present at the top. The presence of Mg containing intermetallic phases in conditions close to equilibrium fits well to theoretic predictions based on Thermo-Calc. Thus, it is concluded that the intrinsic heat treatment prevailing during WAAM provides sufficient thermal energy for diffusion resulting in the local accumulation of Mg. Hence, the specimen's top corresponding to a specimen condition far off equilibrium is free of these accumulations. Similar enrichments were detected in Ref. [19] during single wire CMT of an ML-AlSi7Mg0.6 alloy but were not discussed in this work.

5 Summary and outlook

In this work the feasibility of *in situ* alloying of aluminium-based alloys is proven. To this end, two aluminium wires with significantly varying Si content were used to obtain AlSi10Mg material during dual-wire-arc additive manufacturing. From the performed analyses the following major conclusions can be drawn:

- Process optimization enables good chemical homogeneity throughout the cross-section of the welded structures suggesting sufficient intermixing of the two feedstock materials.

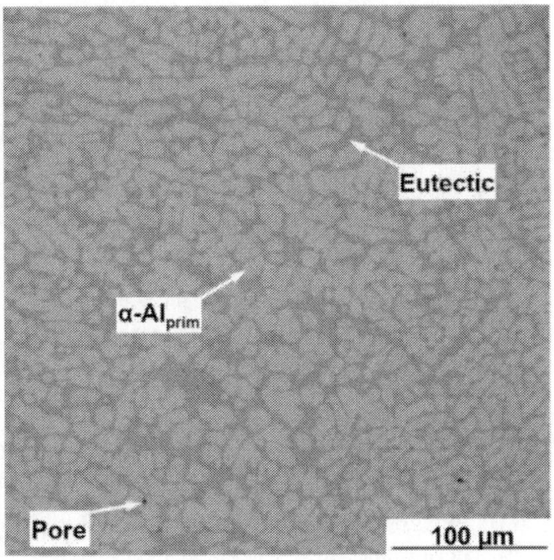

Fig. 3. Representative optical micrograph visualizing primary dendrites with the eutectic constituent in the interdendritic regions as well as small pores.

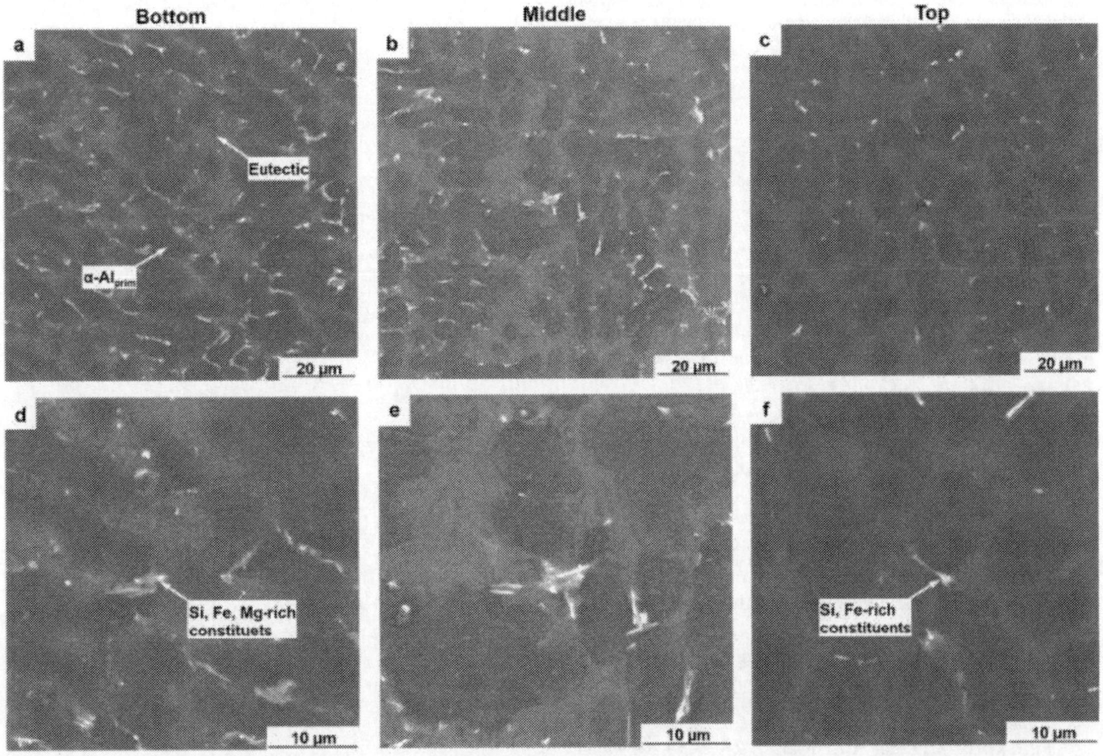

Fig. 4. Micrographs (SEM) taken from the specimen regions bottom (a, d), middle (b, e) and top (c, f). Primary grain and eutectic constituent morphology are similar in all regions. Fewer interdendritic intermetallic phases are visible at the specimen's top.

- The resultant microstructure is characterized by a fine-grained dendritic appearance resulting from the high cooling rates during solidification.
- Eutectic microstructural constituents appear uniformly distributed within the specimen material.
- Phase fractions observed experimentally deviate from equilibrium predictions resulting from pronounced non-equilibrium solidification conditions.
- SEM analysis provided evidence that the repeated heating results in varying microstructures from bottom to top. Particularly, it is apparent that fewer intermetallic phases are visible at the specimen's top, which is subject to the least heat input.
- EDX analysis shows that in all regions of the specimens, Al + Fe + Si-rich phases are discernible. However, the intrinsic heat treatment prevailing during processing results in the accumulation of Mg in their close vicinity. This phenomenon is proven by the absence of these Mg enrichments at the specimen's top.

The good homogeneity of the obtained specimen in terms of chemical composition and microstructures suggests that dual-wire approaches in WAAM are feasible and that the approach might be expandable to multi-wire techniques.

This research was funded by the Austrian Federal Ministry for Digital and Economic Affairs (BMDW) within the framework of the program "COIN (Cooperation & Innovation)" administered by the Austrian Research Promotion Agency (FFG) within the project "PAmelA" (grant no. 872667).

References

1. D. Ding, Z. Pan, D. Cuiuri, H. Li, Int. J. Adv. Manuf. Technol. **81**, 465 (2015)
2. B. Wu, Z. Pan, D. Ding, D. Cuiuri, H. Li, J. Xu, J. Norrish, J. Manuf. Proc. **35**, 127 (2018)
3. C.R. Cunningham, J.M. Flynn, A. Shokrani, V. Dhokia, S.T. Newman, Add. Manuf. 22, 672 (2018)
4. H. Lockett, J. Ding, S. Williams, F. Martina, J. Eng. Des. **28** (7-9), 568 (2017)
5. W. Ou, T. Mukherjee, G.L. Knapp, Y. Wie, T. DebRoy, Int. J. Heat Mass Trans. **127** 1084 (2018)
6. F. Hejripour, D.T. Valentine, D.K. Aidun, Int. J. Heat Mass Trans. **125**, 471 (2018)
7. K. Oyama, S. Diplas, M. M'hamdi, A.E. Gunnaes, A.S. Azar, Add. Manuf. **26**, 180 (2019)
8. T. Klein, M. Schnall, Int. J. Adv. Manuf. Techn. **108**, 235 (2020)
9. J. Gu, J. Bai, J. Ding, S. Williams, L. Wang, K. Liu, J. Mater. Proc. Techn. **262**, 210 (2018)
10. F. Martina, J. Ding, S. Williams, A. Caballero, G. Pardal, L. Quintino, Add. Manuf. **25**, 545 (2019)
11. Y. Wang, X. Chen, S. Konovalov, C. Su, A. Noor Siddiquee, N. Gangil, Appl. Surf. Sci. **487**, 1366 (2019)
12. Z. Qi, B. Cong, B. Qi, H. Sun, G. Zhao, J. Ding, J. Mater. Proc. Techn. **255**, 347 (2018)
13. A.E. Davis, C.I. Breheny, J. Fellowes, U. Nwankpa, F. Martina, J. Ding, T. Machry, P.B. Prangnell, Mater. Sci. Eng. A 765, 138289 (2019)
14. X. Fang, L. Zhang, G. Chen, X. Dang, K. Huang, L. Wang, B. Lu, Materials **11**, 2075 (2018)
15. J.S. Panchagnula, S. Simhambhatla, Robot Comput. Integr. Manuf. **46**, 193 (2018)
16. A. Horgar, F. Fostervoll, B. Nyhus, X. Ren, M. Eriksson, O.M. Akselsen, J. Mater. Proc. Techn. **259**, 68 (2018)
17. Schüle Metallgiesserei, *Technisches Merkblatt: Legierungen nach DIN EN 1706*
18. W. Kurz, D.J. Fisher, *Fundamentals of solidification*, 4th ed., Trans Tech Publications (1998)
19. A. Furtado Ferreira, J.A. de Castro, L. de Olivé Ferreira, Mater. Res. **20**, 68 (2017)
20. Q. Yang, C. Xia, Y. Deng, X. Li, H. Wang, Materials **12**, 2525 (2019)
21. C. Li, H. Gu, W. Wang, S. Wang, L. Ren, Z. Wang, Z. Ming, Y. Zhai, Appl. Sci. **10**, 70 (2020)
22. J. Wu, X.Q. Wang, W. Wang, M.M. Attallah, M.H. Loretto, Acta Mater. **117**, 311 (2016)
23. A. Dinsdale, C. Fang, Z. Que, Z. Fan, JOM **71**, 1731 (2019)

Grain boundary segregation and precipitation in an Al-Zn-Mg-Cu alloy

Huan Zhao [1*], *Baptiste Gault* [1,2], *Frédéric De Geuser*[3], *Dirk Ponge* [1], *Dierk Raabe* [1]

1. Max-Planck-Institut für Eisenforschung, Max-Planck-Str. 1, 40237, Düsseldorf, Germany
2. Department of Materials, Royal School of Mines, Imperial College, Exhibition Road, London SW7 2AZ, United Kingdom
3. Univ. Grenoble Alpes, CNRS, Grenoble INP, SIMaP, F-38000 Grenoble, France

Abstract. High strength Al-alloys are highly susceptible to intergranular embrittlement, which severely limits their lifetime. This article summarizes our recent work on the effect of solute segregation in the precipitation behavior at grain boundaries (GBs) compared to the grain interiors. Solute segregation could accelerate the precipitation behavior at GBs, which causes the formation of coarse precipitates and precipitate free zones along GBs. Furthermore, the interplay of solute segregation and the local structure at GBs has been considered. We show that the distinct segregation and precipitation behavior occurs within the same GB, which makes the GB excess of solutes at one facet significantly higher than the other facet. This paper enriches the current understanding on the role of chemistry and structure at GBs related to intergranular fracture and corrosion resistance in high strength Al-alloys.

1 Introduction

7xxx Al-alloys are widely employed in aerospace and automobile applications due to their low mass density and high strength ability since Sander and Meissner's findings in 1923 [1-3]. 7xxx series are alloyed with small amounts of Zn and Mg (Al-Zn-Mg system) or Zn, Mg and Cu (Al-Zn-Mg-Cu system). These alloys are mainly strengthened through precipitation hardening by forming nano-sized precipitates during aging, which further act as obstacles impeding dislocation motion [4]. 7xxx Al-alloys have a high nucleation rate of precipitates among the precipitation-hardening Al-alloys, which provide high specific strength. The precipitation behavior of Al-Zn-Mg-(Cu) alloys has been well studied and established as the sequence of the supersaturated solid solution to Guinier-Prestion (GP) zones, followed by metastable η' phases and stable η phases ($MgZn_2$) [5-7].

For many age-hardened Al-alloys, the precipitation reaction starts with the formation of coherent, nano-sized GP zones as found in the early work of Guinier and Preston [8, 9]. GP zones transform to metastable, semi-coherent η' precipitates during aging. η' phase has the hexagonal Laves structure with a lattice parameter of a=b=0.496 nm, c=1.402 nm [10, 11]. η' precipitates occur as plates on {111} planes of the Al-matrix, elongated along the <110> direction. The composition of η' precipitates is controversial in literature due to their metastability. The previous results show that they contain 30-40 at.% Zn, ~33 at.% Mg, 5-10 at.% Cu and balance Al, with variations on the alloy composition and aging treatment [7, 12, 13]. The η'-plates eventually transform to coarse, incoherent η phases during over-aging, accompanied by a decrease in strength. The equilibrium η phase can also heterogeneously nucleate at GBs, phase boundaries, and dislocations during quenching [14]. The η phase has a similar hexagonal Laves structure with the lattice parameter of a=0.5221 nm, c=0.8567 nm [5, 15, 16]. It is originally proposed that the chemical composition of the η phase is $MgZn_2$. While, it has been widely shown that Cu and/or Al atoms substitute some of the Zn sites and cause the chemical composition to $Mg(Zn,Al)_2$ for Al-Zn-Mg system and $Mg(Zn,Al,Cu)_2$ for Al-Zn-Mg-Cu system [7].

Despite the wide commercial applications of 7xxx Al-alloys, their usage is restricted by the resistance to stress corrosion cracking and intergranular embrittlement [17-20]. For this purpose, they are often used in a slightly overaged temper to provide better corrosion resistance while at a loss in strength. The embrittlement has been attributed to the formation of coarse, incoherent precipitates along GBs [21-23], as well as the precipitate-free zones (PFZs) adjacent to GBs [24, 25]. GBs differ in structure and properties from the grain interiors, which could cause solute segregation according to the Gibbs adsorption isotherm. Such equilibrium segregation phenomenon modifies the GB composition compared to the bulk, which can play an important role in the early stage precipitation and PFZ formation at GBs [26-30]. An experimental investigation on the details regarding

* Corresponding author: h.zhao@mpie.de

segregation and its influence on the precipitation behavior at GBs is required. Besides, there is a strong motivation to understand more about the basic aspects of the interplay between chemistry and local structure at GBs, which is essential in understanding the local segregation and precipitation behavior. In this article, we have reviewed our recent work in the application of atom probe, in conjunction with electron microscopy, electron backscatter diffraction to study the segregation and precipitation behavior in an Al-Zn-Mg-Cu alloy.

2 Experimental

The studied Al-Zn-Mg-Cu alloy was laboratory-cast and homogenized in a vacuum induction furnace. Table 1 summarizes the material's chemical composition measured by the wet chemical analysis. The homogenized ingot was then subsequently hot-rolled at 450 °C and solution heat-treated for 1 hour at 475 °C by water quenching. The as-quenched samples were immediately aged at 120°C for different time to follow the aging response according to the Vickers hardness measurements. Electron backscatter diffraction (EBSD) was used to locate high-angle GBs. Atom probe tomography (APT) specimens were then prepared from the GBs using a FEI Helios Plasma Focused Ion Beam (Plasma FIB) instrument with a Xe source. APT probing was conducted on a local electrode atom probe (LEAP 5000XS), under ultra-high vacuum conditions of 10^{-11} Torr with a 20% pulse fraction and a 250 kHz pulse rate, at a cryogenic temperature of 40 K. APT datasets were analyzed using the commercial software package IVAS 3.8.4. The reconstruction parameters were calibrated according to the crystallographic poles appearing on the detector hit map. Scanning transmission electron microscopy (STEM) characterization was performed on a spherical probe corrected FEI Titan Themis 60-300 STEM, operated at 300 keV. High angle annular dark field (HAADF) images, bright field (BF)-STEM were recorded.

Table 1. Chemical composition of an Al-Zn-Mg-Cu alloy

Alloy	Zn	Mg	Cu	Zr	Fe	Si	Al
(wt.%)	6.22	2.46	2.13	0.15	0.02	<0.01	rest

3. Results and discussions

3.1 Precipitation in the bulk

Fig. 1 shows the hardness curve from the as-quenched state (120±3 HV) to the peak aged state (185±5 HV) and further to the overaged state (171±3 HV). Representative APT analysis over the aging response are also shown insets, with iso-surfaces of 10 at.% Zn highlighting the Zn-Mg-rich precipitates [31]. Over the first 2 hours aging at 120 °C, nearly spherical Zn-Mg-rich zones with a high number density of 10^{24}–10^{25} m^{-3} form and develop into GP zones. The composition of the precipitates is approximately 75 at. % Al, 13 at. % Zn, 10 at. % Mg and 2 at. % Cu. 120°C for 24 hours aging produced the peak hardness. The overaged state was conducted by further aging the peak aged sample at a higher temperature of 180°C for 6 hours. Significant precipitate coarsening is investigated in the overaged state, which decreases the hardness.

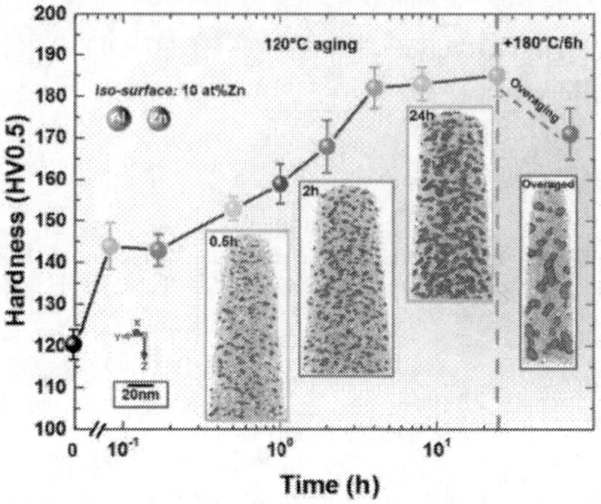

Fig. 1. Evolution of the hardness over the course of aging. Insets are APT analysis of the precipitates highlighted by 10 at. % Zn. (Reprinted with permission from Ref. 31)

3.2 Comparison of APT samples of Al-alloys prepared by Ga-FIB and Xe-FIB

We have then turned our attention to investigate the segregation effect on the precipitation behavior at GBs by APT. The site-specific APT specimen was commonly prepared by the lift-out method using a FEI focused ion beam with Ga+ source. This can become problematic for Al-alloys in the GB segregation study due to the fast diffusion of Ga in Al. Fig. 2 shows one representative example of the APT specimen of the peak aged Al-Zn-Mg-Cu alloy prepared on a Ga-FIB. The GB is visualized by the depletion of all solutes but the strong enrichment of Ga, while the GP zones are clearly shown in the adjacent bulk. This evidences that the presence of Ga during FIB preparation affects the precise quantification of solute segregation at GBs.

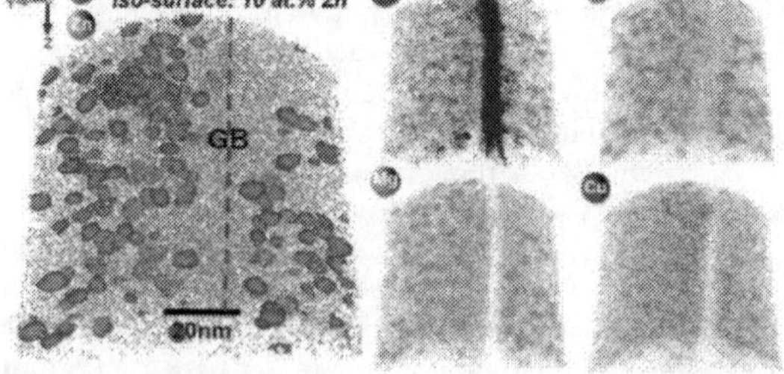

Fig. 2. APT analysis of a high-angle GB from a peak aged Al-Zn-Mg-Cu alloy prepared on a Ga-FIB, showing atom maps of the solutes and Ga.

The recent development of a FEI Helios Plasma focused ion beam (PFIB) using of alternative Xe source can alleviate this limitation. Fig. 3 is one example of the same peak aged material prepared by Xe-FIB and analyzed by APT. The GB is clearly visible in the atom maps with a region enriched with coarse precipitates as indicated by the red rectangle. The ions of Xe in the range of the corresponding mass spectrum are also displayed, with no enrichment of Xe observed at the GB [32].

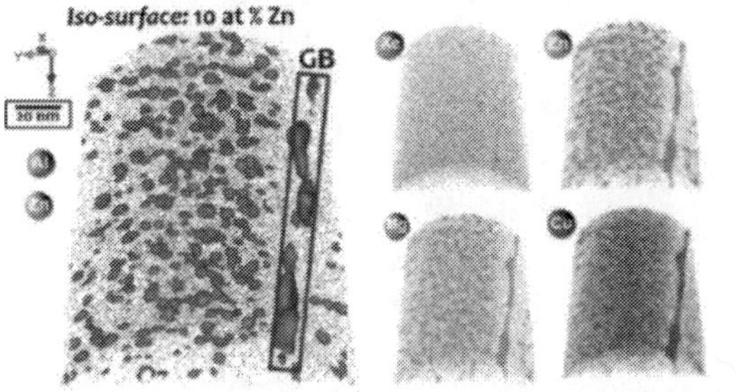

Fig. 3. APT reconstruction containing a high angle GB from a peak aged Al-Zn-Mg-Cu alloy prepared by PFIB.

3 Segregation and the early stage of precipitation at grain boundaries

Fig. 4 is an APT analysis of a high angle GB in the as-quenched Al-Zn-Mg-Cu alloy after solutionized at 475 °C. EBSD map (Fig. 4(a)) represents the GB where several APT samples were prepared. EBSD map (Fig. 4(a)) represents the GB where several APT samples were prepared. The density map in Fig. 4(b) shows the projection of the APT dataset. The {111} poles of two adjacent grains are indexed evidencing the existence of a GB in the analyzed APT dataset. Fig. 4(c) shows the atom

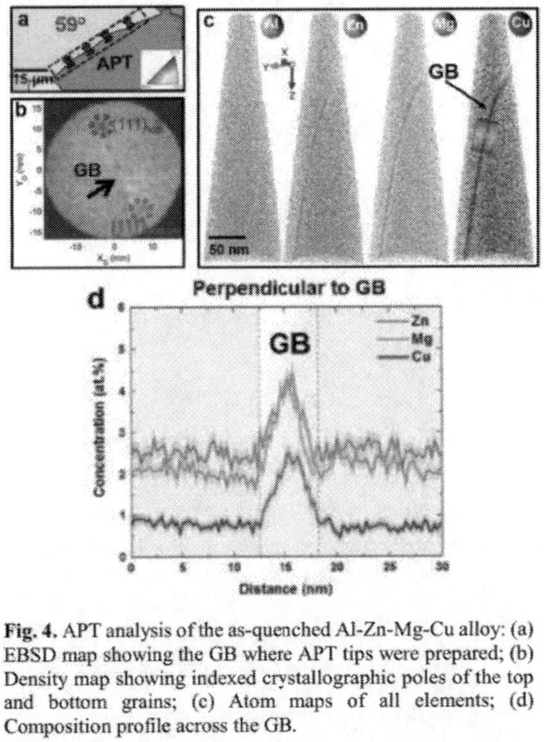

Fig. 4. APT analysis of the as-quenched Al-Zn-Mg-Cu alloy: (a) EBSD map showing the GB where APT tips were prepared; (b) Density map showing indexed crystallographic poles of the top and bottom grains; (c) Atom maps of all elements; (d) Composition profile across the GB.

maps of Al, Zn, Mg, and Cu, with the GB highlighted in terms of the solute enriched. We conducted GB composition analysis by adding a cylinder across the GB and the composition profile is shown in Fig. 4(d). The result shows that Zn, Mg, and Cu segregate to the GB with an enrichment factor of approximately 2. This evidences that all solute elements segregate to the GB already after quenching and provides quantitative evidence that solutes exhibit a strong tendency to be attracted to GBs.

Fig. 5 presents the APT analysis of the Al-Zn-Mg-Cu alloy after aging for 0.5 hours at 120°C. Nano-sized clusters are visualized by iso-surfaces of 6 at. % Zn. The GB is observed edge-on in Fig. 5(a), shown by the larger precipitate enriched region as indicated by the red arrow. Spherical Zn-Mg enriched precipitates are clearly revealed within the GB plane in Fig. 5(b). The precipitates formed within the GB are around 5 nm in size, i.e. larger than the zones with 3 nm in size distributed in the bulk. The composition profile across one GB precipitate (Fig. 5(c)) shows that it contains 60 at.% Al, 20 at.% Zn, 18 at.% Mg and 2 at.% Cu, which is close to the composition expected for GP zones ($Al_2ZnMgCu$) [6, 12, 33-36]. Typical precipitate-free zones with 10 nm in size concurrently form adjacent to the GB when the GP zones become visible, as indicated by the purple lines in Fig. 5(a). A corresponding composition profile taken across the GB in Fig. 5(d) in the region between the precipitates reveals that the solute content of Zn and Mg is about 1.5 at.% within the PFZs, which is lower than grain interior far away from the GB. This observation indicates that we do not only observe coarse precipitate formation on the GB but also solute depletion in the PFZs in the early stage aging. We observed the aging process at 120 °C for different time and hence propose that the solute enrichments at GBs not only accelerate the formation, but also the growth and coarsening of the GB precipitates [37]. Similar accelerated precipitation behavior of Mg-Zn rich phase was also observed at the Zr-dispersoids in the same alloy [38]. The difference in the precipitate composition between the bulk and GBs was further considered and their relationship with the corrosion properties of this alloy was discussed in our recent study [39].

3.4 Interplay of segregation and local structure at GBs

We recently reported on the interplay of chemistry and faceting at GBs in the same Al-Zn-Mg-Cu alloy [40] . The GBs show significant different segregation and precipitation behavior along the GB plane. Analysis of one of the APT measurements in the as-quenched alloy is shown here in Fig. 6. The location of the GB is highlighted, with a red box in Fig. 6(a), while iso-surfaces of 5 at.% Zn+Mg are shown in Fig. 6(b) for the cross-sectional slice from this region. Fig. 6(a) demonstrates that all solute species, i.e. Zn, Mg, and Cu, have already segregated to the GB after quenching. We see in Fig. 6(b) that this segregation follows periodic patterns throughout the entire GB. These segregation patterns are formed due to the preferential segregation of Zn and Mg to particular sites. The corresponding composition profile along the

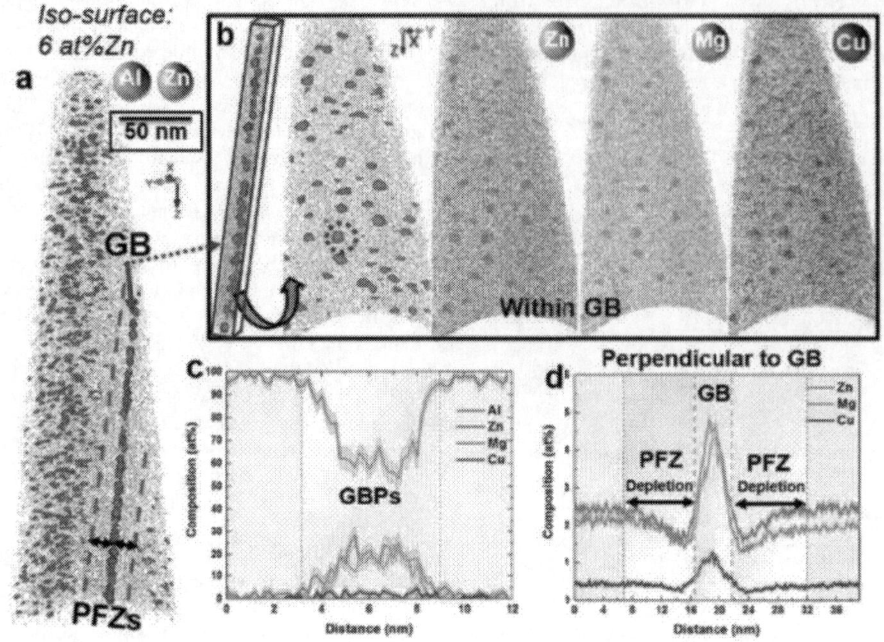

Fig. 5. APT analysis after aging for 0.5 hours at 120°C. (a) Precipitates visualized by iso-composition surfaces of 6 at% Zn; (b) Distribution of solutes along the GB plane; (c) 1D composition profile of the selected GB precipitate as indicated by the red circle; (d) 1D composition profile across the GB, devoid of GB precipitates. (Reprinted with permission from Ref. 37).

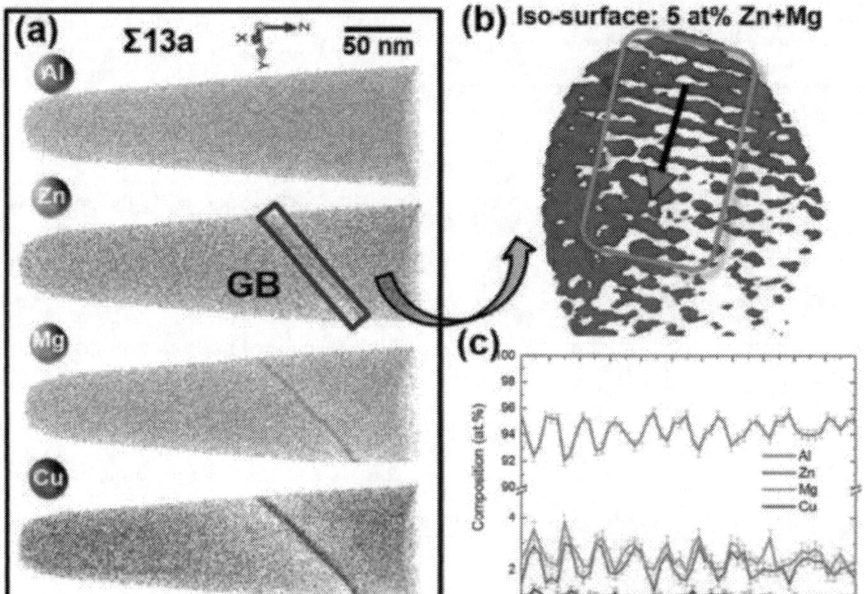

Fig. 6. (a) Atom maps of all elements of a Σ13a <110> GB in the as-quenched Al-Zn-Mg-Cu alloy; (b) Distribution of solutes within the GB plane (region indicated by the red rectangle in (a)); (c) Corresponding one-dimensional composition profile along the arrow in (b). (The Figure was reproduced from Ref. 40, for which is under the terms of the CC-Attribution 4.0 international license)

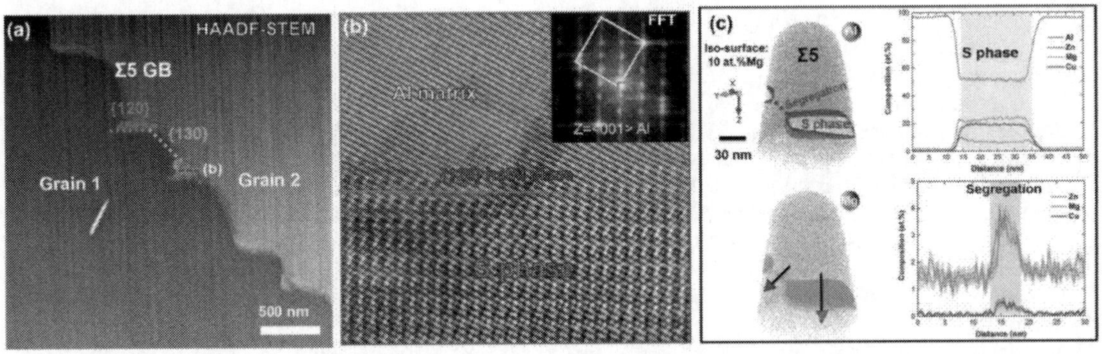

Fig. 7. (a) HAADF-STEM image of a Σ5 GB in the as-quenched Al-Zn-Mg-Cu alloy; (b) HAADF-STEM image of the precipitate highlighted by blue rectangle in (a), with inset the corresponding Fast Fourier transform (FFT) pattern; (c) APT analysis of the same GB on another location. The profiles showing the composition of the Mg-Cu enriched precipitates at the {120} facet, and the segregation at the {130} facet. (The Figure was reproduced from Ref. 40, for which is under the terms of the CC-Attribution 4.0 international license)

pattern axis is shown in Fig. 6(c). A periodic distribution is revealed, with 5 nm distance between the peaks. We revealed the interplay between the faceting and their chemical decoration with solutes at the Σ11 GB in the same alloy after solution heat treatment. The atom probe analysis shows that Mg and Zn follow similar periodic segregation patterns across the Σ11 GB plane, with a period of 6–10 nm. {200} and {111} facets alternating along the Σ11 GB were viewed by scanning transmission electron microscopy. The observed periodic patterns in the Σ13 GB might form due to the preferential segregation of Zn and Mg to particular facets, as indicated by our similar observations on the Σ11 GB. The linear segregation pattern along the GB plane is expected to affect the early stage of precipitation behavior during the subsequent aging.

Fig. 7(a) illustrates another investigation on a Σ5 GB in the as-quenched Al-Zn-Mg-Cu alloy. The GB was characterized by Z-contrast high-angle annular dark field images (HAADF)-STEM performed along the common tilt $\langle 001\rangle_{Al}$ zone axis. The GB configuration contains two sets of alternating facets along the entire length, with each facet highlighted by a colored dashed line in Fig. 7(a). The horizontal segments are symmetric {120} facets, and the other segments (in green) are asymmetric with a GB plane of {130}. The {120} facets bear roughly 300 nm lath-like precipitates. HAADF-STEM in Fig. 7(b) reveals that the precipitate has an orthorhombic crystal structure belonging to the cmcm group. It matches the structure of the S phase (Al_2CuMg) proposed by Perlitz and Westgren (PW model). It has an orientation relationship of $\langle 100\rangle_S // \langle 100\rangle_{Al}$, and an interface of $\{001\}_S // \{120\}_{Al}$ [41]. We performed APT analysis on the same Σ5 GB on another location, with the result shown in Fig.7(c). Two lathlike S phases enriched in Mg and Cu are visible on the {120} facets. APT crystallography analysis confirms that the habit plane of the S phase is the $\{120\}_{Al}$, which is consistent with the STEM observations. The composition profile across the S phase shows that approx. 23 at. % Mg, 18 at. % Cu, and 7 at. % Zn is enriched within the precipitate. We also observe Zn, Mg and Cu enrichment along the {130} grain boundary facets with the composition of 4 at.%, 4 at.% and, 0.5 at.% respectively. Previous studies on Al-Zn-Mg-Cu alloys showed that the S-phase can form on GBs during quenching, and the volume of the precipitates increases with decreasing cooling rates [42-45]. Investigations on Al-Cu-Mg alloys show that the S-phase forms laths on the $\{120\}_{Al}$ planes of the Σ5 GB , as its atomic arrangement in the {001} planes is identical to that on the $\{120\}_{Al}$ planes [41, 46]. Hence, it is reasonable to consider that S-phase preferentially heterogeneously nucleates at {120} facets rather than the {130} facets, which ensures good coherency and low free energy of interfaces.

4 Conclusions

In summary, we have reviewed our recent work in the application of atom probe, in conjunction with Xe-Focused Ion Beam, and electron microscopy to study the segregation and precipitation behavior in an Al-Zn-Mg-Cu alloy. We investigated the material from the solution heat treated state (475 °C), through the very early stages of aging to the peak aged state at 120 °C and further into the overaged regime at 180 °C. The process starts with solute enrichment on grain boundaries, the formation of GP zones in the solute-enriched grain boundary regions, and GP zones growth and transformation. The segregation of solutes to grain boundaries might accelerate the precipitation sequence during aging compared to the bulk. Segregation and precipitation are also strongly correlated with the local structures of GBs, which can cause the anisotropic segregation and preferential precipitation. The inhomogeneous distribution of the GB solutes can eventually reduce the GB cohesion, cause intergranular fracture, and affect the corrosion resistance of this alloy.

References

[1] E. Starke Jr, J. Staley, Progress in Aerospace Sciences, **32**, 131-172, (1996)

[2] L.F. Mondolfo, Aluminum alloys: structure and properties, Elsevier,2013.
[3] I. Polmear, Materials forum, 1-14, (2004)
[4] J.H. Chen, E. Costan, M.A. van Huis, Q. Xu, H.W. Zandbergen, Science, **312**, 416-419,(2006).
[5] L. Berg, J. Gjønnes, V. Hansen, X. Li, M. Knutson-Wedel, G. Waterloo, D. Schryvers, L. Wallenberg, Acta materialia, **49** (2001) 3443-3451.
[6] G. Sha, A. Cerezo, Acta Materialia, **52,** 4503-4516, (2004).
[7] T. Marlaud, A. Deschamps, F. Bley, W. Lefebvre, B. Baroux, Acta Materialia, **58** (2010) 4814-4826.
[8] A. Guinier, Nature, **142,** 569, (1938)
[9] G.D. Preston, The London, Edinburgh, and Dublin Philosophical Magazine and Journal of Science, **26,** 855-871, (2009)
[10] F. Cao, J. Zheng, Y. Jiang, B. Chen, Y. Wang, T. Hu, Acta Materialia, **164**, 207-219, (2019)
[11] X. Li, V. Hansen, J. Gjønnes, L. Wallenberg, Acta materialia, **47**, 2651-2659, (1999)
[12] K. Hono, N. Sano, T. Sakurai, Surface science, **266**,350-357, (1992)
[13] T. Marlaud, A. Deschamps, F. Bley, W. Lefebvre, B. Baroux, Acta Materialia, **58**, 248-260, (2010)
[14] A. Deschamps, Y. Bréchet, Materials Science and Engineering: A, **251**, 200-207, (1998)
[15] Y. Komura, K. Tokunaga, Acta Crystallographica Section B: Structural Crystallography and Crystal Chemistry, **36**,1548-1554, (1980)
[16] J. Gjønnes, C.J. Simensen, Acta Metallurgica, **18**,881-890, (1970)
[17] S.P. Knight, N. Birbilis, B.C. Muddle, A.R. Trueman, S.P. Lynch, Corrosion Science, **52** 4073-4080, (2010)
[18] R.M. Su, Y.D. Qu, R.D. Li, J.H. You, Materials Science, **51**, 372-380, (2015)
[19] A.C.U. Rao, V. Vasu, M. Govindaraju, K.V.S. Srinadh, Transactions of Nonferrous Metals Society of China, **26**, 1447-1471, (2016)
[20] P. Pao, M. Gao, R. Wei, Scripta metallurgica, **19** 265-270, (1985)
[21] A. Joshi, C. Shastry, M. Levy, Metallurgical Transactions A, **12**, 1081-1088, (1981)
[22] R. Viswanadham, T. Sun, J. Green, Metallurgical and Materials Transactions A, **11**, 85-89, (1980)
[23] A.K. Vasudevan, R. Doherty, Acta metallurgica, **35**, 1193-1219, (1987)
[24] T. Ogura, S. Hirosawa, A. Cerezo, T. Sato, Acta Materialia, **58**, 5714-5723, (2010).
[25] E.A. Starke, JOM, **22**, 54-63, (1970)
[26] D. Raabe, M. Herbig, S. Sandlöbes, Y. Li, D. Tytko, M. Kuzmina, D. Ponge, P.P. Choi, Current Opinion in Solid State and Materials Science, **18**, 253-261, (2014)
[27] T. Frolov, S.V. Divinski, M. Asta, Y. Mishin, Physical review letters, **110**, 255502, (2013)
[28] M. Kuzmina, D. Ponge, D. Raabe, Acta Materialia, **86**, 182-192, (2015)
[29] M.P. Harmer, Science, **332**, 182-183, (2011)
[30] P. Lejcek, Springer series in materials science, **136**, (2010)
[31] H. Zhao, B. Gault, D. Ponge, D. Raabe, F. De Geuser, Scripta Materialia, **154**, 106-110, (2018).
[32] Y. Xiao, J. Wehrs, H. Ma, T. Al-Samman, S. Korte-Kerzel, M. Göken, J. Michler, R. Spolenak, J.M. Wheeler, Scripta Materialia, **127**, 191-194, (2017).
[33] S. Ringer, K. Hono, Materials characterization, **44**, 101-131, (2000)
[34] P.V. Liddicoat, T. Honma, L.T. Stephenson, S.P. Ringer, Materials Science Forum, **519-521,** 555-560, (2006)
[35] T. Engdahl, V. Hansen, P. Warren, K. Stiller, Materials Science and Engineering: A, **327** 59-64, (2002)
[36] C. Wolverton, Acta Materialia, **49**, 3129-3142, (2001)
[37] H. Zhao, F. De Geuser, A. Kwiatkowski da Silva, A. Szczepaniak, B. Gault, D. Ponge, D. Raabe, Acta Materialia, **156**, 318-329, (2018)
[38] H. Zhao, Y. Chen, B. Gault, S.K. Makineni, D. Ponge, D. Raabe, Materialia, **10**, (2020)
[39] H. Zhao, B. Gault, D. Ponge, D. Raabe, Scripta Materialia, **188**, 269-273, (2020)
[40] H. Zhao, L. Huber, W. Lu, N.J. Peter, D. An, F. De Geuser, G. Dehm, D. Ponge, J. Neugebauer, B. Gault, D. Raabe, Phys Rev Lett, **124**, 106102, (2020)
[41] V. Radmilovic, R. Kilaas, U. Dahmen, G. Shiflet, Acta Materialia, 47, 3987-3997, (1999)
[42] Y. Zhang, B. Milkereit, O. Kessler, C. Schick, P.A. Rometsch, Journal of Alloys and Compounds, **584**, 581-589, (2014)
[43] B. Yang, B. Milkereit, Y. Zhang, P.A. Rometsch, O. Kessler, C. Schick, Materials Characterization, **120**, 30-37,(2016)
[44] D. Godard, P. Archambault, E. Aeby-Gautier, G. Lapasset, Acta Materialia, 50 (2002) 2319-2329.
[45] C. Mondal, A.K. Mukhopadhyay, Materials Science and Engineering: A, **391**, 367-376. (2005)
[46] G.B. Winkelman, K. Raviprasad, B.C. Muddle, Acta Materialia, **55**, 3213-3228, (2007)

 https://doi.org/10.1051/matecconf/202032601005

Detailed investigation of the shearing mechanism of β'' precipitates in Al-Mg-Si alloys

Emil Christiansen[1,2,*], *Calin Daniel* Marioara[3], *Inga Gudem* Ringdalen[3], *Ruben* Bjørge[3], *Bjørn* Holmedal[1,4], *Odd Sture* Hopperstad[1,2], and *Randi* Holmestad[1,5]

[1]Centre for Advanced Structural Analysis, NTNU – Norwegian University of Science and Technology — Norway.
[2]Department of Structural Engineering, Faculty of Engineering, NTNU – Norwegian University of Science and Technology — Norway.
[3]Materials and Nanotechnology, SINTEF Industry — Norway.
[4]Department of Materials Sceince and Engineering, Faculty of Natural Sciences, NTNU – Norwegian University of Science and Technology — Norway.
[5]Department of Physics, Faculty of Natural Sciences, NTNU – Norwegian University of Science and Technology — Norway.

Abstract. The mechanism behind shearing of β'' precipitates in Al-Mg-Si alloys during deformation is investigated by applying advanced transmission electron microscopy (TEM) techniques and frozen phonon multislice TEM image simulations on a selection of shearing configurations. In particular, the results indicate that the needle-like precipitates are sheared several times in single matrix Burgers vector steps. The multislice image simulations suggest that shearing events are most likely achieved in single Burgers vector steps, and there are some experimental evidence that the shearing planes are the matrix glide planes.

1 Introduction

During plastic deformation of precipitate strengthened alloys, dislocations either bypass or shear precipitates, depending on the precipitate phase, size, and shape [1]. In order to predict the strength of such alloys, it is important to understand the physical mechanisms of each process for various alloy systems and their precipitate phases.

The β'' precipitates are the main strengthening phase in the Al-Mg-Si alloy system (AA6xxx) and have composition $Mg_5Si_4Al_2$ [2]. They appear as long needles oriented along the matrix $\langle 100\rangle$ and usually with a monoclinic unit cell with $a_{\beta''} = 15.16$ Å, $b_{\beta''} = 4.05$ Å, $c_{\beta''} = 6.74$ Å, and $\beta = 105.3°$ with $(001) \parallel (010)_{\beta''}$, $[310] \parallel [001]_{\beta''}$, and $\left[\bar{2}30\right] \parallel [100]_{\beta''}$ [2–4]. Several studies indicate that these precipitates become sheared by gliding dislocations during deformation [5–9]. Because the precipitate crystal structure is different from the matrix structure, any matrix dislocation that glides through the precipitate phase will leave behind structural defects within and/or around the precipitate. The different matrix slip systems are not equivalent for a given precipitate and produce different results, depending on the slip system orientation to the precipitate. For instance, the geometrical shape of precipitate $(010)_{\beta''}$ cross-sections means that Burgers vectors with e.g. $[110]$ and $\left[1\bar{1}0\right]$ will produce interface steps and internal planar defects of different surface areas. This illustrates the complexity of how these precipitates are sheared, as a glide of one particular Burgers vector through a precipitate will be different from other Burgers vectors and produces different defects in the precipitate crystal structure. Hence, a variety in experimental results is to be expected, originating from the difference in crystal structures as discussed, but also due to a natural variation in precipitate shape, size, and experimental parameters such as thicknesses.

This study presents a combination of previously published experimental results (Christiansen *et al.* [9]) and recent frozen phonon multislice image simulation results. The results from the *post mortem* TEM experiments are explained and discussed in terms of the image simulation results. Together, the experimental and theoretical results indicate that β'' precipitates are sheared in single Burgers vector steps distributed along the needles. The results also form a basis for future works to better understand how this precipitate phase strengthens the industrially important AA6xxx alloys.

2 Materials and Methods

2.1 Materials

Cylindrical compression specimens, which measured 9 mm in diameter and 13 mm in height and were machined from an extruded AA6060 alloy with a composition given in table 1, were solution heat treated at 540 °C for 15 min in a salt bath, water quenched to room temperature, and left for 15 min to naturally age. They were then artificially aged to peak strength in an oil bath at 185 °C for 5 hours and air cooled to room temperature. The alloy has also been used in a number of studies investigating its

*e-mail: emil.christiansen@ntnu.no

Table 1. Nominal composition of AA6060 used in this study

	Fe	Si	Mg	Mn	Cr	Cu	Zn	Ti	Al
wt%	0.193	0.422	0.468	0.015	0.000	0.002	0.005	0.008	Bal
at%	0.093	0.406	0.520	0.007	0.000	0.001	0.002	0.005	Bal

crystallographic texture, grain size, and mechanical properties, and the reader is referred to these works for further details [10, 11]. The specimens were compressed to 20% engineering strain in an universal testing machine with a speed of 0.15 mm/min (strain rate of 2×10^{-4} s^{-1}). Thin foils for TEM investigations were prepared from sections of the extrusion plane close to the middle of the compression specimens, please see Christiansen *et al* [12] for more details.

2.2 Methods

2.2.1 Microscopy

The experimental results presented in this paper are taken from an earlier study and are openly accessible through online repositories [9, 13].

Conventional bright field TEM, high resolution TEM (HRTEM), and scanning precession electron diffraction (SPED) experiments were conducted on a *JEOL JEM2100F* operated at 200 kV. HRTEM images were acquired on a *Gatan 2k UltraScan CCD*, while SPED data with a frame rate of 25 Hz was acquired by an external *Allied StingRay* CCD and the *NanoMEGAS DigiSTAR* system. For SPED, a precession angle of 1° and a precession frequency of 100 Hz were used, together with a nominal probe size of 1 nm and a step size of 1.52 nm. Virtual bright-field and dark-field images were produced from the SPED data and selected components from non-negative matrix factorisation (NMF) analysis were overlaid to show precipitate needles lying in-plane in order to investigate the shearing planes and their configuration (i.e. distribution along needle-lengths). The reader is referred to Christiansen *et al* [9] for more details.

Annular dark field scanning TEM (ADF STEM) data were acquired on a double-corrected *JEOL ARM200F* with a cold-FEG operated at 200 kV. *SmartAlign* was used to limit scan distortions in the data by acquiring 10–12 fast frames and by applying non-rigid image registration algorithms afterwards [14]. A collection semi-angle of 48–206 mrad and a probe convergence semi-angle of 27.42 mrad were used to acquire atomically resolved scans with a pixel dwell-time of 2 μs and pixel sizes < 8 pm.

2.2.2 Image simulations

Frozen phonon multislice image simulations were performed using the *MULTEM* software package [15–17] with the parameters given in table 2. The simulation results presented in this paper are blurred with a Gaussian kernel of 1 Å full width at half maximum to simulate the spatial incoherency of the electron source. The atomistic models used in the simulations were prepared using a combination of density functional theory, rigid translations, and molecular dynamics. Virgin precipitate crystal structures were prepared by replicating a single β″ unit cell ($2 \times 100 \times 3$ unit cells) and inserting it into a matrix of aluminium ($20 \times 20 \times 100$ unit cells). The model thickness of 100 Al unit cells was chosen to match the average precipitate length in this alloy (~ 40 nm), although the specimen thicknesses in the experiments are ~ 60 nm. The β″ unit cell was taken from density functional theory calculations [18]. The virgin precipitates were then rigidly sheared on various slip systems and with different configurations (various distribution of slip along the needle length). Due to limitations in both density functional theory and molecular dynamics, it is not possible to relax the sheared structures themselves, and this is an inherent limitation in the following multislice image simulations. The matrix, however, was relaxed using molecular dynamics in LAMMPS and using the Al-Mg potential by Liu and Adams to get realistic strain fields in the nearby matrix [19, 20]. The two precipitate models used to produce the multislice simulation results presented in this paper are shown in in figure 1.

Table 2. Multislice simulation parameters

Parameter	Value	
Acceleration voltage	200 kV	
Detector response	homogeneous	
Potential sampling	2048×2048 pixels^2	
Model size	80.64×80.64 Å^2	
Model thickness	403.2 Å	
Slice thickness	2.025 Å	
Temperature	300 K	
Phonon configurations	20	
	STEM	HRTEM
Convergence semi-angle	27mrad	–
Defocus C_1	0 nm	−57 nm
Spherical aberration C_3	−300 nm	1.0 mm
Inner collection semi-angle	48 mrad	–
Outer collection semi-angle	206 mrad	–
Image pixel size	0.162 Å	0.039 Å

3 Results and Discussion

3.1 SPED investigations

Figure 2 presents SPED results of precipitates in the virgin material and in 20% compressed specimens. The SPED technique produces a wealth of data, and advanced machine learning routines are needed to extract relevant and important data. In this figure, selected NMF components have been coloured and overlaid their corresponding virtual bright field images (i.e. each pixel value is determined by the diffracted intensity to a region containing the direct beam of that pixel's precession electron diffraction pattern). The virtual bright field images produce a rich contrast and dislocations appear as dark lines, while the NMF results show variation in scattering to in-plane β″ precipitate reflections. Due to the similarity of scattering angles of in-plane and out-of-plane needles, out-of-plane

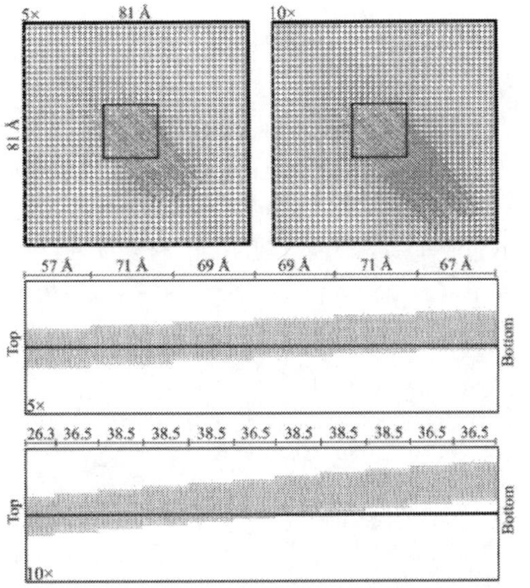

Figure 1. The two models used in the multislice simulations shown from the top and from the side with a [110] view (Al-atoms removed). The regions scanned in the STEM simulations are shown as boxes in the middle of the top-views. Al, Mg, and Si atoms are shown in grey, green, and brown, respectively.

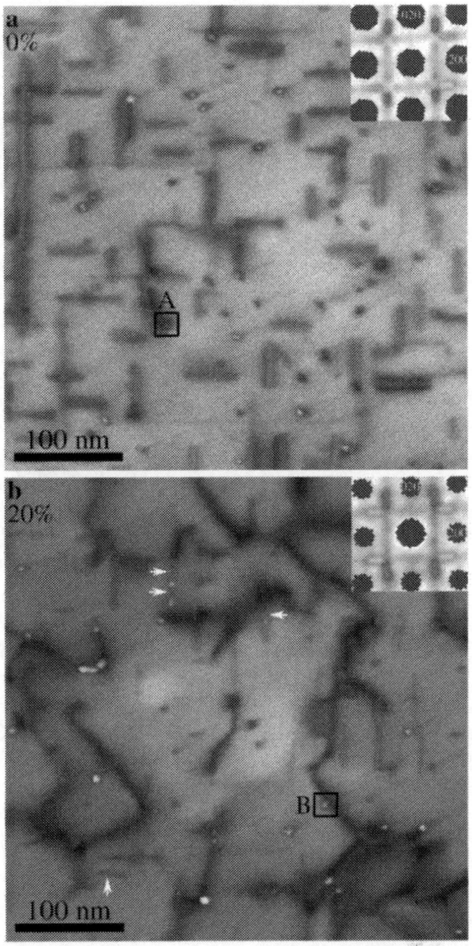

Figure 2. SPED results from virgin (**a**) and deformed (**b**) specimens. Red and green pixels scatter more significantly to red and green angles in the insets, respectively. Precipitates in deformed specimens contain dark regions (arrowheads) where scattering is reduced, indicating a disturbed crystal structure. The blue regions in the insets were not used in the NMF. Single-pixel precession electron diffraction patterns from out-of-plane precipitates marked A and B are shown in figure 3.

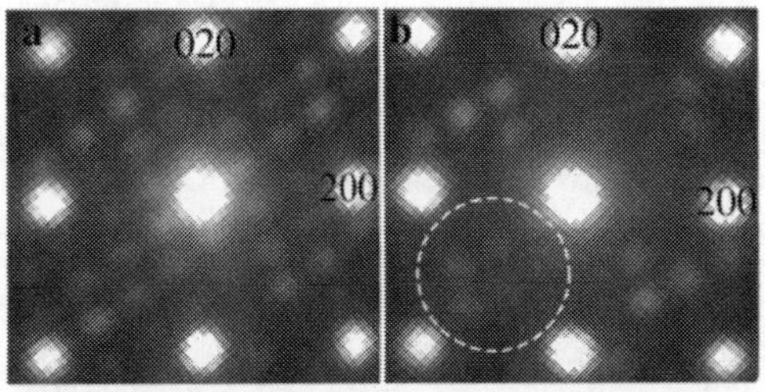

Figure 3. Single-pixel precession electron diffraction patterns from precipitates marked A (**a**) and B (**b**) in figure 2. Precipitates in deformed specimens tend to scatter less to certain reflections compared to precipitates in undeformed specimens.

precipitates are also present in the NMF results and become coloured.

It is clear that the deformed specimen contains dislocations that are pinned by precipitates in the material. It is also clear that the NMF intensities (i.e. the brightness of the colour in figure 2) are irregular for the in-plane needles. This means that the scattering from these regions is different compared to the regions where the intensity is higher. While the bright regions of the in-plane needles scatter as expected, the darker regions scatter differently. It is, however, important to point out that the NMF results do not imply how the scattering is different, they only indicate that there is a difference.

Considering the single-pixel precession electron diffraction patterns of out-of-plane precipitates presented in figure 3, however, it becomes clear that certain precipitate reflections are suppressed after deformation. In turn, this must mean that the precipitate crystal structure is disrupted along their lengths. Together, these results show that deformation disrupts the precipitate structure in wide regions and that these regions seem to be distributed randomly along the needle lengths. The width of the dark regions is likely due to closely spaced and inclined shearing planes (with respect to the electron beam travelling along [001]). This will be further discussed in section 3.3.

3.2 High resolution results and simulations

There is a large variation in HRTEM and ADF STEM results from both virgin and deformed specimens, owing to the complex image forming mechanisms of the techniques.

 https://doi.org/10.1051/matecconf/202032601005

In addition, the nature of the problem is also very complex, as each of the different twelve slip systems are different for the precipitate phase in addition to the possibility of glide on other planes, such as the (001) planes [21]. Hence, it is too ambitious to attempt to simulate and explain every experimental image. However, we have investigated shearing in a selection of different directions and planes with various distribution through the thickness of the model. There are three general trends. First, the influence of shearing plane (i.e. (001) versus (111)) on the appearance of simulated images is minimal. Second, different Burgers vectors produce different features in simulated images. In HRTEM, these features are most evident in the fast Fourier transforms (FFT) of the images, as different Burgers vectors alter the suppression of image Fourier components. For this paper, we focus on displacements of $\vec{b} = \vec{b}_{1\bar{1}0} = \frac{a}{2}\left[1\bar{1}0\right]$, as these capture the general trends of the features of both HRTEM and ADF STEM images. Third, distributed shearing events with an equal separation distance match better to experimental results than concentrated shearing events (i.e. either localised in a single plane producing one large step or localised in a narrow region producing many closely spaced single steps). Because there are many possible configurations, a full study of the influence of the through-thickness position of concentrated shearing events remains to be performed. In this paper, we consider results from models where the shearing events are distributed equally along the needle lengths, as this is a likely configuration with relatively few parameters.

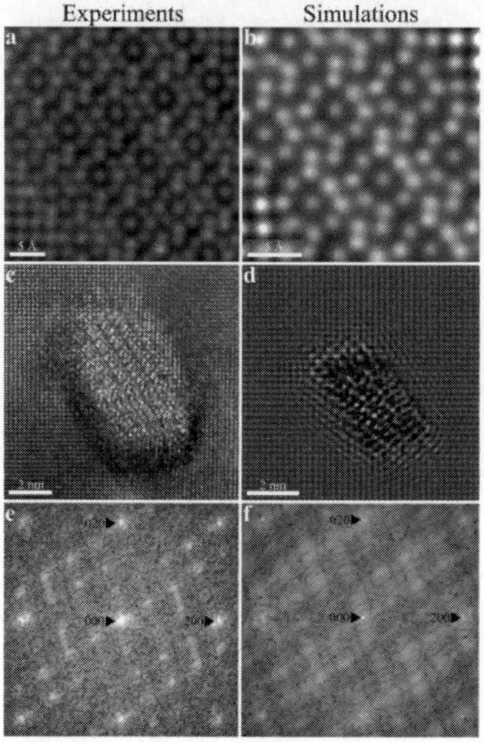

Figure 4. Comparison between experimental (**a**, **c**, **e**) results and multislice image simulation results from a model precipitate rigidly sheared 5 times along its length (**b**, **d**, **f**). ADF STEM results are shown in **a** and **b**, and HRTEM results are shown in **c** and **d**. FFT power spectra of the HRTEM images are given in **e** and **f**. Note that the experimental HRTEM and ADF STEM images are of different precipitates.

3.2.1 Five shearing events

Figure 4 presents a comparison between selected experimental HRTEM and ADF STEM results in addition to simulated results for a precipitate that is sheared by $\vec{b} = \frac{a}{2}\left[1\bar{1}0\right]$ rigid translations on five equally spaced (001) planes through the thickness. β'' unit cells can be clearly seen in both the experimental and simulated ADF STEM results, which are almost indistinguishable from images of virgin materials (not shown). This can be explained by the sequential image forming mechanism of ADF STEM, where the first few nanometres of the specimen decide most of the final image. The precipitate imaged in HRTEM is also similar to the HRTEM simulation, although direct comparison is not possible due to the HRTEM image forming processes. Because contrast reversals and the thickness-sensitivity of HRTEM make it challenging to compare the images directly, the FFT power spectra of the images are better suited for comparison instead. Both the experimental HRTEM and the simulated HRTEM images produce FFT power spectra where certain Fourier components are suppressed, similar to the suppression of reflections in the precession electron diffraction patterns shown in figure 3.

While the experimental HRTEM and ADF STEM images are of different precipitates, these results reveal two important points. First, a precipitate that is sheared several times along its length may appear very similar to virgin precipitates in ADF STEM, but the HRTEM images are greatly affected. This fits well with the large number of HRTEM images from deformed specimens that shows blurry precipitate cross-sections and FFT power spectra similar to the ones in figure 4e [13]. Second, precipitates that produce FFT power spectra such as the one given in figure 4e are likely sheared more than five times. Indeed, HRTEM image simulations of models with fewer number of shearing events reveal no noticeable suppression of Fourier components. Other Burgers vectors and shearing planes have also been tested, all indicating that 5 shearing events are needed to reproduce experimental HRTEM FFT power spectra.

3.2.2 Ten shearing events

Figure 4 presents experimental ADF STEM and HRTEM results and corresponding results from a multislice image simulation of a model with 10 shearing events distributed along the needle length. The shearing planes and directions are the same as for the simulations shown in figure 5. While the effect of more shearing events on the HRTEM simulation results is similar to the 5-shearing case (although more severe), the most clear changes appear in the ADF results. With 10 shearing events, the ADF STEM results show a disrupted β″ atomic configuration instead of the more or less ideal β″ structure obtained from the 5-shearing case. The disruption resembles two β″ unit cells shifted by a single Burgers vector relative to each other, and follows the β″ unit cell periodicity. Inspecting the simulated ADF STEM images through the thickness (not shown), reveals that the disruption appears before the second shearing event and remains relatively unchanged as the beam progresses through the model. This clearly shows that it is not the number of shearing events that is of importance for the ADF STEM images, but their spacing and their distance to the specimen surface. On the other hand, the full through-thickness configuration is important for the HRTEM images, as the suppression of Fourier components becomes more severe when there are more shearing events.

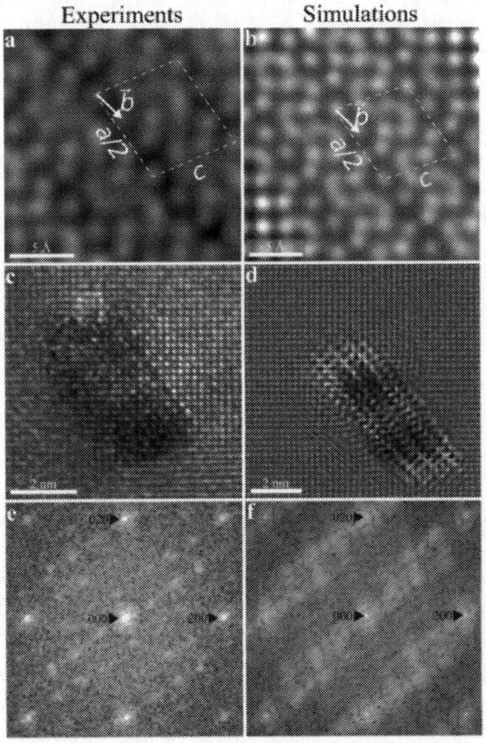

Figure 5. Comparison between experimental (**a**, **c**, **e**) results and multislice image simulation results from a model precipitate rigidly sheared 10 times along its length (**b**, **d**, **f**). ADF STEM results are shown in **a** and **b**, and HRTEM results are shown in **c** and **d**. FFT power spectra of the HRTEM images are given in **e** and **f**. Note that the experimental HRTEM and ADF STEM images are of different precipitates.

3.3 Deformation mechanisms

The β″ precipitates in the AA6060 alloy studied here are concluded to be sheared by dislocations during plastic deformation. In the following, the results are considered together to form a picture of likely deformation mechanisms.

When a β″ precipitate is sheared by a dislocation, the precipitate unit cells no longer match with each other across the shearing plane. This is evident from the different crystal structures, and is supported by SPED, HRTEM, and ADF STEM characterization. It is therefore reasonable to assume that atomic configurations close to the shear planes will shift and relax into new energy minima positions. Therefore, if the shearing plane itself is a planar defect, the effect of this defect will likely spread across nearby precipitate unit cells and disrupt the crystal structure locally. In turn, electron scattering from such regions will be altered and most likely reduced. We therefore suggest that the wide dark segments observed in SPED experiments are due to planar defects formed in the precipitates during deformation and their surrounding unit cell distortions. Given the relatively continuous contrast of the dark segments, we also suspect that the planar defects lie on planes inclined to the transverse direction of the needles (i.e. are not $(010)_{\beta''}$ planes).

For HRTEM images to develop the experimentally observed suppression of Fourier components, a number of shearing events must have taken place. At the same time, each of these shearing events must be spaced far enough apart for the electron beam to develop a marked phase shift. Considering the ADF STEM image in figure 5a, in which the precipitate has likely been sheared by a single dislocation close to the specimen surface (and possibly more times through its full thickness), it seems reasonable to conclude that shearing of β″ precipitates is achieved in single Burgers vector steps, and that these shearing events are spaced at least a few unit cells apart. If the shearing planes were spaced relatively far apart, the probability of having a shearing plane close to the specimen surface would be very low. Even though ADF STEM is not a statistical tool we have at least one image where this seems to be the case. In principle, it is also possible that the steps are some integer multiple of the matrix Burgers vector, rather than a single Burgers vector. However, this will require relatively large local plastic strains in order to achieve the number of planar defects that fits the SPED and HRTEM observations. Hence, it is likely that β″ precipitates are sheared in single Burgers vector steps on planes spaced a few nano meters apart. In some cases, dislocations may localise into narrow slip bands for a number of reasons, and this will then make the shearing planes more concentrated.

So far, we have not investigated the effect of combining different slip systems and other realistic shearing configurations, such as several bands of narrowly spaced shearing events using multislice image simulations. Such investigations would be useful to further understand the deformation mechanisms of β″ precipitates. In addition, the sheared precipitates were modelled as rigid translations of a perfect β″ needle, while we suspect that the planar defects distort the crystal lattice in their vicinity as well. Once more advanced atomic interaction potentials for molecular dynamics simulations are available, such effects could also be studied. For future work, we would suggest imaging the same precipitate in ADF STEM and HRTEM and to compare the images with detailed multislice image simulations.

4 Conclusions

Three advanced and complementary TEM techniques show that β″ precipitates in the AA6060 alloy are sheared by dislocations during deformation. Multislice image simulations aid the interpretation of the experimental results and show that several closely spaced shearing planes produce images similar to the experimental data. In terms of deformation mechanisms, we propose that β″ precipitates are sheared in several and equally spaced single Burgers vector steps on {111} planes.

Acknowledgements

This work was funded by the Research Council of Norway through SFI CASA (grant number 237885), NORTEM (grant number 197405), and Mind The Gap (231762). The simulations were carried out on the EPIC [22] infrastructure at the IDUN cluster at the NTNU HPC Group, while the atomistic models were created using the Sigma2 HPC clusters (grant numbers 269905/NN9158K).

References

[1] A. J. Ardell, Metallurgical Transactions A **16**, 2131–2165 (1985)

[2] T. Saito, E. A. Mørtsell, S. Wenner, C. D. Marioara, S. J. Andersen, J. Friis, K. Matsuda & R. Holmestad, Advanced Engineering Materials **20**, 1800125 (2018)

[3] G.A. Edwards, K. Stiller, G. L. Dunlop & M. J. Couper, Acta Materialia **46**, 3893–3904 (1998)

[4] S. J. Andersen, H. W. Zandbergen, J. Jansen, C. Træholt, U. Tundal & O. Reiso, Acta Materialia **46**, 3283–3298 (1998)

[5] F. Delmas, M. Vivas, P. Lours, M. J. Casanove, A. Couret & A. Coujou, Materials Science and Engineering A **340**, 286–291 (2003)

[6] W. J. Poole, X. Wang, D. J. Lloyd & J. D. Embury, Philosophical Magazine **85**, 3113–3135 (2005)

[7] K. Misumi, K. Kaneko, T. Nishiyama, T. Maeda, K. Yamada, K. Ikeda, M. Kikuchi, K. Takata, M. Saga & K. Ushioda, Journal of Alloys and Compounds **600**, 29–33 (2014)

[8] Ø. Ryen, B. Holmedal, K. Marthinsen & T. Furu, Materials Science and Engineering **89**, 012013 (2015)

[9] E. Christiansen, C. D. Marioara, B. Holmedal, O. S. Hopperstad, R. Holmestad, Scientific Reports **9**, 17446 (2019)

[10] M. Khadyko, S. Dumoulin, T. Børvik & O. S. Hopperstad, International Journal of Mechanical Sciences **88**, 25–36 (2014)

[11] B. H. Frodal, L. E. Dæhli, T. Børvik, O. S. Hopperstad, International Journal of Plasticity **188**, 36–69 (2019)

[12] E. Christiansen, C. D. Marioara, K. Marthinsen, O. S. Hopperstad, R. Holmestad, Materials Characterization **144**, 522–531 (2018)

[13] E. Christiansen, C. D. Marioara, B. Holmedal, O. S. Hopperstad & R. Holmestad, *Data for "Nano-scale characterisation of sheared β″ precipitates in a deformed Al-Mg-Si alloy"*, Dataset Version 1 archived in Zenodo, DOI: 10.5281/zenodo.2652906 (2019)

[14] L. Jones, H. Yang, T. J. Pennycook, M. S. J. Marshall, S. Van Aert, N. D. Browning, M. R. Castell & P. D. Nellist, Advanced Structural and Chemical Imaging **1**, 8 (2015)

[15] I. Lobato & D. Van Dyck, Ultramicroscopy **156**, 9–17 (2015)

[16] I. Lobato, S. Van Aert & J. Verbeeck, Ultramicroscopy **168**, 17–27 (2016)

[17] MULTEM, https://github.com/Ivanlh20/MULTEM

[18] P. H. Ninive, A. Strandlie, S. Guldbrandsen-Dahl, W. Lefebvre, C. D. Marioara, S. J. Andersen, J. Friis, R. Holmestad & O. M. Løvvik, Acta Materialia **69**, 126–134 (2014)

[19] S. Plimpton, Journal of Computational Physics **117**, 1–19 (1995)

[20] X. -Y. Liu & J. Adams, Acta Materialia **46**, 3467–3476 (1998)

[21] F. Delmas, J. Majimel, M. Vivas, G. Molenat, A. Couret & A. Coujou, Philosophical Magazine Letters **83**, 289–296 (2003).

[22] M. Själander, M. Jahre, G. Tufte, N. Reissmann, *EPIC: An Energy-Efficient, High-Performance GPGPU Computing Research Infrastructure*, arXiv: 1912.05848.

Using Artificial Intelligence to Aid Vehicle Lightweighting in Crashworthiness with Aluminum

Christopher P. Kohar[1,*], *Daniel S.* Connolly[1], *Timofei* Liusko[2], and *Kaan* Inal[1,2]

[1]Department of Mechanical and Mechatronics Engineering, University of Waterloo, Waterloo, Ontario, Canada, N2L 3G1
[2]Department of Systems Design Engineering, University of Waterloo, Ontario, Canada, N2L 3G1

Abstract. Significant efforts have been made in the automotive industry to reduce vehicle weight in order to improve vehicle fuel economy and reduce greenhouse gas emissions. New innovations in structural lightweight alloys and manufacturing techniques have allowed automakers to replace conventional steel with lighter aluminum structures. However, automakers have an enormous number of material and gauge thickness combinations to consider in the development process of the next generation production vehicle. Furthermore, the design combination of these materials and structures must not compromise the integrity of the vehicle during a vehicle collision. With the proliferation of inexpensive computational resources, automakers can now explore the effect of material selection on the crashworthiness of next-generation vehicles using computer simulations. While information from these simulations can be manually extracted, the vast amount of data lends itself to artificial intelligence (AI) techniques that can extract knowledge faster and provide more useful interpretations that can be convenient for designers and engineers. This work presents a framework for using artificial intelligence to aid the vehicle design cycle in crashworthiness using aluminum. Virtual experiments of a frontal crash condition of a pick-up truck are performed using finite element analysis to generate the data for this method. Different commercially available aluminum alloys and gauge thicknesses are varied in the virtual experiments. An advanced type of recurrent neural network is used to predict the time-series response of the occupant crash-pulse response, which is a key crashworthiness metric that is used for evaluating safety. This work highlights how automotive designs and engineers can leverage this framework to accelerate the development cycle of the next-generation lightweight vehicle.

1 Introduction

Over the last decade, significant attention has been placed on reducing greenhouse gas (GHG) emissions globally to combat the effects of climate change. The automotive and transportation industry is one of the major economic sectors that contributes substantially to GHG emissions. Many government mandates have been aimed at reducing GHG emissions. Structural lightweighting is a strategy that automakers employ to reduce energy consumption, which directly improves fuel economy, lowers operation costs, and overall reduces GHG emissions [1]. Automakers have pushed the manufacturing sector to develop new technologies in advanced manufacturing and lightweight materials engineering to be able to realize innovative designs for mass savings in production vehicles [2]. As a result, automakers now have several different material combinations and processes at their disposal to design and manufacture the lightweight vehicles of tomorrow.

However, automotive engineers must consider passenger safety in conjunction with vehicle lightweighting [3]. In the event of a vehicle collision, a crash pulse wave is generated and propagated throughout the vehicle. The shape and intensity of the crash pulse wave transmitted to the passenger directly correlates to the likeliness of a severe injury [4]. The transmission of the crash pulse wave can be mitigated through the systematic interaction of all vehicle components to either dissipate the energy through deformation of the structure or by strategically diverting the energy away from the passenger. This means that the development of new lightweight components cannot be generated in isolation. Although a designer can develop a new individual component from a new manufacturing process or material [5], this component may harm how the crash pulse wave propagates throughout the vehicle when integrated into the structure.

Therefore, automakers need to investigate how these innovative lightweight designs will affect the complete vehicle response during a collision. Although

* Corresponding author: cpkohar@uwaterloo.ca

prototypes of these new designs can be fabricated and tested, the time and cost associated with the trial-and-error process of development are no longer feasible for automakers in a rapidly evolving and competitive global economy. Automotive engineers now rely on numerical tools, such as finite element analysis, to perform virtual experiments of vehicles that are subjected to various crash scenarios before the prototyping and production stage.

With the proliferation of inexpensive computational resources, computer-aided engineering (CAE) software can now explore the effect of the various manufacturing processes and lightweight materials on the crash pulse wave of next-generation vehicles. While CAE engineers can extract knowledge and intuition from these studies, the vast amount of data that is generated from this process lends itself to artificial intelligence (AI) techniques that can extract knowledge and information orders of magnitude faster than the conventional manual methods. This means that AI techniques could relate the desired material properties and manufacturing process of a component to the crash pulse wave transmitted to a passenger. Once an AI tool has been trained on this data, an engineer can exploit the AI tool to predict and explore new combinations of materials and processing conditions that simultaneously reduces weight while improving passenger safety. Recent advancements within the field of AI can enable scientists and engineers to capture the time-dependent behaviour of a highly nonlinear phenomenon, such as vehicle crash, to accelerate the development of next-generation vehicles.

This work presents an AI framework that can be used to enhance the development cycle in vehicle lightweighting with an application to aluminum. This framework relies on virtual designs of experiments for a frontal crash test of a pick-up truck using finite element analysis to generate the training data for the AI system. Different commercially available aluminum alloys and gauge thicknesses are varied in the virtual experiments. Artificial neural networks are used to predict the relationship between vehicle mass and thickness. An advanced type of recurrent neural network, known as a long-short term memory neural network (LSTM-NN), is used to predict the crash-pulse response of an occupant during the impact. Comparisons of the AI framework to the results obtained from the finite element analysis are presented.

2 Proposed Framework and Architecture

The proposed framework follows the method of virtual design of experiments through the use of finite element (FE) simulations to improve vehicle safety while reducing vehicle mass. The procedure of the framework is summarized as follows:

1. *Problem definition*: Generate a finite element model of the problem of interest. Define the input parameters/variables and the output response.
2. *Sampling of problem domain*: Select combinations of input parameters that are statistically representative of the entire domain of interest and simulate their response.
3. *Define AI system*: Create an AI architecture that maps the input parameters to the desired output response.
4. *Training of AI system*: Select data from the sample set to calibrate the AI system. This step is often conducted in conjunction with Step 3 to select an architecture that accurately captures the system response iteratively.
5. *Prediction of AI system*: Use remaining data for prediction with the AI system. This serves as a blind test and evaluates the ability of the AI system to generalize the problem.

The problem of interest is the lightweighting of a pick-up truck using aluminum. The material thickness and mechanical properties, through the stress-strain curve, are varied throughout the vehicle simulations. This study is limited to a single frontal impact crash scenario; however, the proposed framework can be extended to capture an arbitrary number of load cases. The vehicle mass and the crash pulse are the desired output response that is to be determined by the AI system.

2.1 Representation of Mass

The mass of a vehicle can be computed directly from a computer-aided design (CAD) model and related to changes in geometric parameters (i.e., material thickness, connectors, assemblies, etc.). However, some of these geometric relationships can be related in a nonlinear manner. If the CAD model is not available, the mass of a vehicle can still be computed through the constructed finite element model. However, a mathematical function that relates the mass to the geometric properties needs to be identified. This work employs a feed-forward artificial neural network to map the geometric parameters to the mass of the vehicle.

2.1.1 Feed-forward artificial neural networks

Feed-forward artificial neural networks (ANN) are a class of AI that is commonly used for determining input-output relationships, which is designed to mimic the structuring of the human brain [6]. The values of each input parameter, x_i, are scaled by a corresponding weight, w_i, applied with an offset, b, and summed into a node called a neuron. Fig. 1 presents a schematic of a multi-weighted-input neuron.

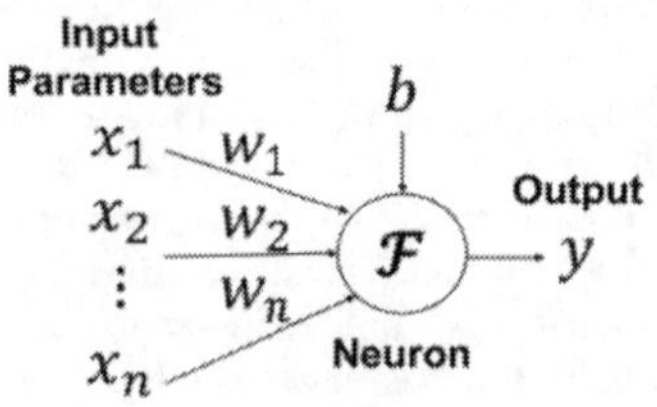

Fig. 1. Schematic of a multi-weighted-input neuron.

Each neuron is composed of a simple processing operation that has the natural property of storing

https://doi.org/10.1051/matecconf/202032601006

information through the weights [7]. The output of the neuron, y, is defined by an activation function, $\mathcal{F}$, such that

$$y = \mathcal{F}(w_i x_i + b) \tag{1}$$

It should be noted that the index notation implies a summation operation. Typically, these activation functions are selected such that the output is bounded between 0 and 1 to simulate the "firing" response of a neuron within the brain. These include the hyperbolic tangent, σ_c, or sigmoidal, σ_g, functions.

A series of neurons can be connected to the same inputs with unique weights to form a layer with multiple outputs; these outputs can be connected to another layer of neurons with accompanying weights to form a multi-layer ANN. Fig. 2 presents a schematic of a multi-layered ANN where y_i, is the output vector of interest. This architecture is known as "feed-forward" because the scaled values of the inputs propagate forward to the output. This scheme provides a massively parallel architecture that lends itself to efficient vectorized computational operations.

Each weight, $w_{ij}^{(l)}$, and bias, $b_j^{(l)}$, in the network is unique and is determined through a learning procedure called a learning algorithm. The objective of the learning algorithm is to determine the values of weights that minimize an error function, $f(\varepsilon_{rr})$, between the actual measurement of the desired output, $\tilde{y}_i$, and the predicted response from the ANN, such that

$$\min\big(f(\varepsilon_{rr})\big) = \min\big(f(\tilde{y}_i - y_i)\big) \tag{2}$$

Common learning algorithms for parameter identification include the back propagation [8], Levenberg-Marquardt [9, 10], and the Adam [11] methods.

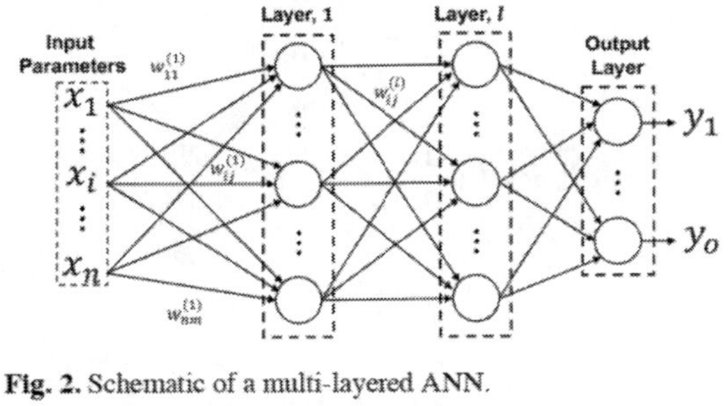

Fig. 2. Schematic of a multi-layered ANN.

2.2 Representation of Crash Pulse

The crash pulse response that is experienced by a vehicle occupant is the result of highly nonlinear interactions of the various components and structures during deformation. This deformation is governed by the mechanics and physics of solids, which can be determined through a series of partial differential equations. Fundamentally, this nonlinear response can then be represented by a transfer function, $\mathbf{\Sigma}$, that relates the time-series output response, $\boldsymbol{y}_t$, (i.e., the crash pulse response) to a time-series of inputs, $\boldsymbol{x}_t$, and their corresponding history. This is represented schematically in Fig. 3. In the application of this work, the transfer function is a single mathematical model that describes the deformation behaviour of a full vehicle in a crashworthiness scenario to predict the crash pulse response. Although the time-series response can be represented using a series of ANNs for each time increment [12], this would require a substantial number of ANNs and the necessary parameters to be identified. Furthermore, this approach would not capture the time-dependence of the problem that may produce a solution with less generalizability. Fortunately, there exist AI techniques that can capture time dependence. Recurrent neural networks (RNN) are an architecture that can predict sequences of data, such as time series. Unlike feed-forward ANNs, RNNs use information from the previous evaluation in the sequence to predict the next value. As such, this work employs a particular architecture of RNN called long-short term memory neural networks.

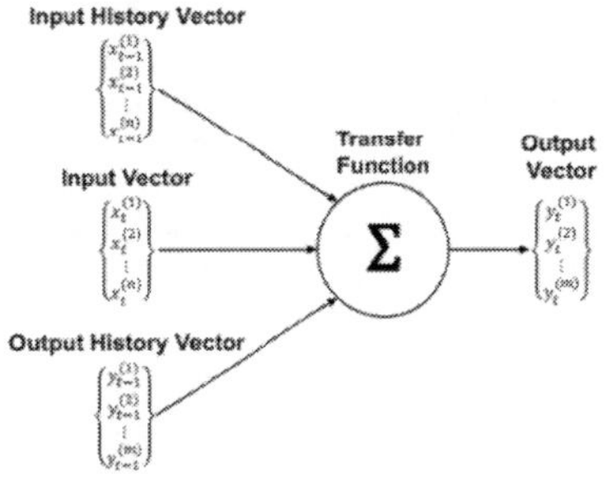

Fig. 3. Schematic of a transfer function.

2.2.1 Long-short term memory neural networks

Long-short term memory (LSTM) is a specific architecture of an RNN that has been effective in capturing long term temporal dependencies [13]. Fig. 4 presents the schematic of the LSTM cell. The cell receives the input parameter vector at the current time, $x_t \in \mathbb{R}^n$, and outputs the hidden vector, $h_t \in \mathbb{R}^h$, and the cell vector, $c_t \in \mathbb{R}^h$, where n is the number of inputs at each timestep and h is the number of hidden cells. It should be noted that the architecture of the LSTM cell requires the size of hidden and cell vectors to be the same. Furthermore, the cell receives the previous values of the hidden and cell vector, h_{t-1} and c_{t-1}, respectively. The fundamental concept in an LSTM architecture is the series of memory cells that regulate the flow of historical information (i.e., hidden and cell values) that enters and exits the cell.

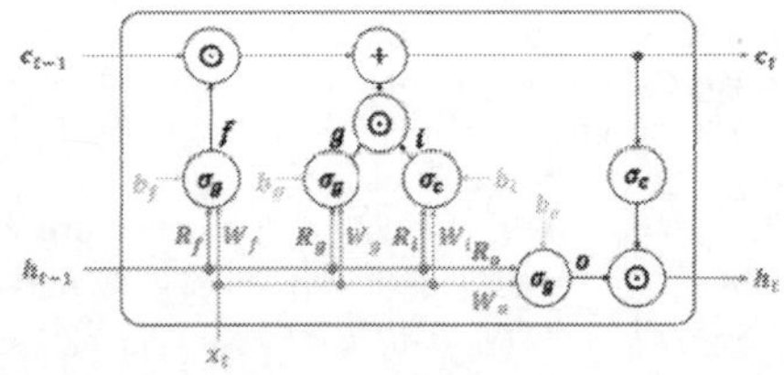

Fig. 4. Schematic of an LSTM cell.

These include an input gate, $\boldsymbol{i}$, forget gate, $\boldsymbol{f}$, cell candidate gate, $\boldsymbol{g}$, and an output gate, $\boldsymbol{o}$, that are governed by

$$\begin{aligned} \boldsymbol{f} &= \sigma_g(\boldsymbol{R}_f\boldsymbol{h}_{t-1} + \boldsymbol{W}_f\boldsymbol{x}_t + \boldsymbol{b}_f) \\ \boldsymbol{g} &= \sigma_g(\boldsymbol{R}_g\boldsymbol{h}_{t-1} + \boldsymbol{W}_g\boldsymbol{x}_t + \boldsymbol{b}_g) \\ \boldsymbol{i} &= \sigma_c(\boldsymbol{R}_i\boldsymbol{h}_{t-1} + \boldsymbol{W}_i\boldsymbol{x}_t + \boldsymbol{b}_i) \\ \boldsymbol{o} &= \sigma_g(\boldsymbol{R}_o\boldsymbol{h}_{t-1} + \boldsymbol{W}_o\boldsymbol{x}_t + \boldsymbol{b}_o) \end{aligned} \quad (3)$$

where $\boldsymbol{R} = [\boldsymbol{R}_f, \boldsymbol{R}_g, \boldsymbol{R}_i, \boldsymbol{R}_o], \boldsymbol{W} = [\boldsymbol{W}_f, \boldsymbol{W}_g, \boldsymbol{W}_i, \boldsymbol{W}_o]$ and $\boldsymbol{b} = [\boldsymbol{b}_f, \boldsymbol{b}_g, \boldsymbol{b}_i, \boldsymbol{b}_o]$ are weights and biases to be identified. The output vector size of all gate operators must be of size, h, such that $\boldsymbol{f}, \boldsymbol{g}, \boldsymbol{i}, \boldsymbol{o} \in \mathbb{R}^h$ and the size of the weights and biases vectors are $\boldsymbol{W}_j \in \mathbb{R}^{h \times n}$, $\boldsymbol{R}_j \in \mathbb{R}^{h \times h}$, $\boldsymbol{b}_j \in \mathbb{R}^h$. It should also be mentioned that the weights and biases do not evolve with time. The cell and hidden vector are updated according to

$$\begin{aligned} \boldsymbol{c}_t &= \boldsymbol{f} \odot \boldsymbol{c}_{t-1} + \boldsymbol{g} \odot \boldsymbol{i} \\ \boldsymbol{h}_t &= \boldsymbol{o} \odot \sigma_c(\boldsymbol{c}_t) \end{aligned} \quad (4)$$

where $\odot$ is the Hadamard product (element wise) operator. Finally, the desired output response of the time-series is obtained by decoding the hidden vector through the use of a feed-forward ANN. Fig. 5 presents a schematic of the LSTM-NN. Similarly, the weights in the LSTM cell and the weights in the feed-forward ANN do not evolve with time.

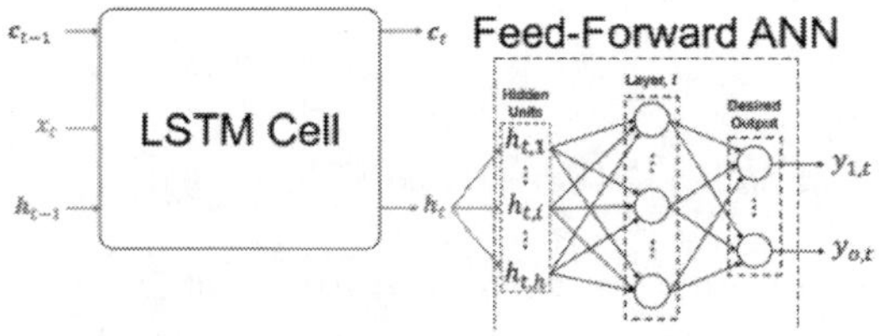

Fig. 5. Schematic of an LSTM-NN.

3 Problem Formulation

3.1 Finite Element Model

The full vehicle finite element (FE) model used in this study is a pick-up truck that was developed by George Washington University National Crash Analysis Center as part of several United States Department of Transportation (US-DOT) – National Highway Traffic Safety Administration (NHTSA) programs. Fig. 6 presents a schematic of the FE model and the setup of the pick-up truck. This FE model is constructed to replicate the testing conditions outlined in the United States New Car Assessment Program (US-NCAP) [14] for frontal collision into a concrete barrier. Each crash simulation was analyzed using a non-linear explicit dynamic formulation of the commercial FE software LS-DYNA – R9.3.0 [15] with double precision utilizing 16 × 2.7GHz Intel Xeon E5-2680 processors and parallel processing techniques. On average, each FE simulation required a wall clock of 2 hours 39 minutes 20 seconds (9560 seconds).

Fig. 6 also presents a typical deformed response of the FE model. The initial velocity of the vehicle in each simulation was 15.6 m/s [56.3 km/h]. The total simulation time was 150 milliseconds to allow for sufficient settling time after impact. The crash pulse was evaluated at the nodes of two solid elements (highlighted in Fig. 6b) that represent the location of an accelerometer for the occupant in the driver and passenger position in the vehicle. Each crash pulse measurement was evaluated in three-axes (x, y, z) that was sampled at 200 μs (5000Hz) and filtered using the SAE CFC 60 standard [16].

In total, the FE model consists of 875 groups of parts. In reality, an automaker would need to specify the material of all the components in the vehicle. For the purpose of this study, the components of the vehicle were grouped into four categories based on their structural functionality within a frontal crash scenario.

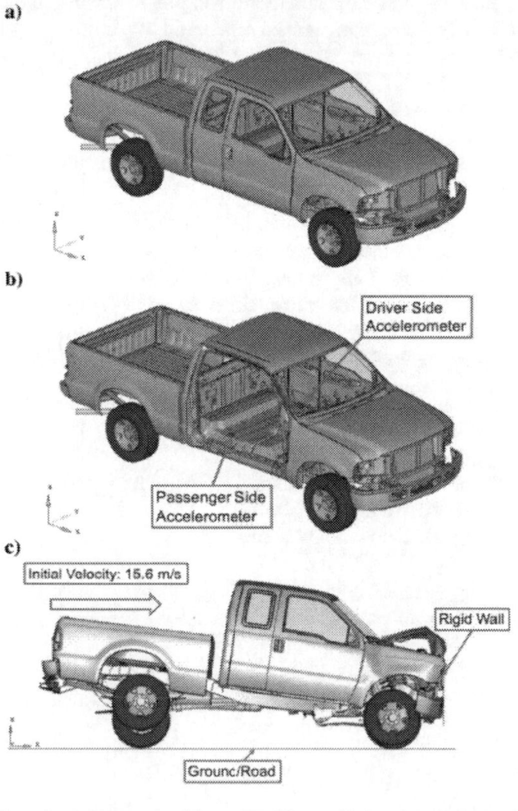

Fig. 6. a) Isometric View, b) View of Accelerometers, c) Representative Deformed Response for Pick-up Truck Model.

Fig. 7 presents a schematic of the component groupings for the pick-up truck model. Table 1 presents the list of component groupings used in this study. These groupings include the front rail (blue), front fenders (red), front end (green) and hood assembly (yellow). All remaining body-in-white components (i.e., doors, occupant cell, bed) were grouped into a single category. The materials in the components that are presented in light grey (Fig. 7) remained constant as the original FE model during this study. Therefore, a total of five component classifications are used in this study.

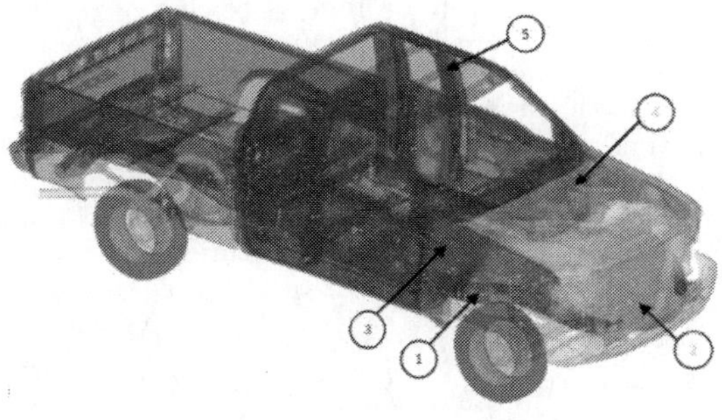

Fig. 7. Component Groupings for Pick-up Truck Model.

Table 1. List of Component Groupings for Pick-up Truck Model.

Component Groupings	**ID**
Front Rail Assembly	1
Front End Assembly	2
Front Fenders Assembly	3
Hood Assembly	4
Remaining Body-in-White	5

3.2 Definition of Material Properties

3.2.1 Constitutive Model

The behaviour of the materials that can vary in this study is assumed to follow an elasto-viscoplastic formulation that is governed by an isotropic von Mises hardening law. The total deformation rate, $\dot{\boldsymbol{\varepsilon}}$, can be decomposed into an elastic strain rate, $\dot{\boldsymbol{\varepsilon}}^*$, and plastic strain rate, $\dot{\boldsymbol{\varepsilon}}^P$, tensor, such that

$$\dot{\boldsymbol{\varepsilon}} = \dot{\boldsymbol{\varepsilon}}^* + \dot{\boldsymbol{\varepsilon}}^P \quad (5)$$

The deformation is considered to be elastic if the Cauchy stress tensor, $\boldsymbol{\sigma}$, satisfies the von Mises yield criterion such that

$$\Phi = \frac{1}{2}s_{ij}s_{ij} - \frac{{\sigma_y}^2}{3} \leq 0 \;\Rightarrow\; \dot{\boldsymbol{\varepsilon}}^P = 0 \quad (6)$$

where $\boldsymbol{s}$ is the deviatoric Cauchy stress tensor defined as $s_{ij} = \sigma_{ij} - \frac{1}{3}\sigma_{kk}\delta_{ij}$, δ_{ij} is the 2nd order identity tensor, and σ_y is the plastic flow stress required to generate yielding. The rate of Cauchy stress, $\dot{\boldsymbol{\sigma}}$, is related to the elastic strain rate through the 4th order isotropic elasticity tensor, $\boldsymbol{L}^{el}$, defined by

$$\dot{\boldsymbol{\sigma}} = \boldsymbol{L}^{el}\colon \dot{\boldsymbol{\varepsilon}}^* \quad (7)$$

which can be described by the elastic modulus, E, and Poisson ratio, ν. If yielding occurs, the plastic strain rate is governed by an associative flow rule defined as

$$\dot{\boldsymbol{\varepsilon}}^P = \dot{\bar{\varepsilon}}^P \frac{\partial \Phi}{\partial \boldsymbol{\sigma}}, \qquad \dot{\bar{\varepsilon}}^P \geq 0 \quad (8)$$

where $\dot{\bar{\varepsilon}}^P$ is the effective plastic strain rate and the accumulation of the total plastic strain follows $\bar{\varepsilon}^P = \int_0^t \dot{\bar{\varepsilon}}^P\,dt$. The consistency criterion during plastic flow governs that

$$\dot{\Phi} = 0 \quad (9)$$

The effective plastic strain rate is assumed to follow a Cowper-Symonds [17] strain rate sensitivity formulation governed by

$$\dot{\bar{\varepsilon}}^P = C\left(\frac{\Phi}{\sigma_y} - 1\right)^S \quad (10)$$

The plastic flow stress model is assumed to follow a Voce-type plasticity law [18] with linear hardening contribution defined by

$$\sigma_y = \sigma_0\{U_0 - (U_0 - 1)[\exp(-D\bar{\varepsilon}^P)]\} \quad (11)$$

where σ_0 is the material quasi-static yield stress, $U_0 \geq 1$ is the ratio of the saturation stress to quasi-static yield stress, and D is the Voce hardening saturation coefficient. Eqns. (5) - (11) form a series of simultaneous equations that can be incrementally solved using Newton-Raphson to identify $\Delta\bar{\varepsilon}^P$.

3.2.2 Application to Commercially Available Aluminum Sheet

As mentioned earlier, automakers can choose from a variety of sheet metals, such as aluminum, magnesium, and steel, to construct a vehicle. Various grades exist within each family of materials based on their alloying elements and material processing conditions. In particular, Fig. 8 presents a representation of the various automotive grades of aluminum alloys that are commercially available to vehicle designers. Each of these materials has fundamental differences in alloying elements and manufacturing processing that generates unique deformation behaviours that ultimately affect the mechanical response in the form of the stress-strain curve. These differences can be beneficial or detrimental to the response of a vehicle in a crashworthiness application. In addition, the manufacturing complexity and cost associated with producing these materials and integrating them into a vehicle is also a factor that automakers must carefully consider in their development.

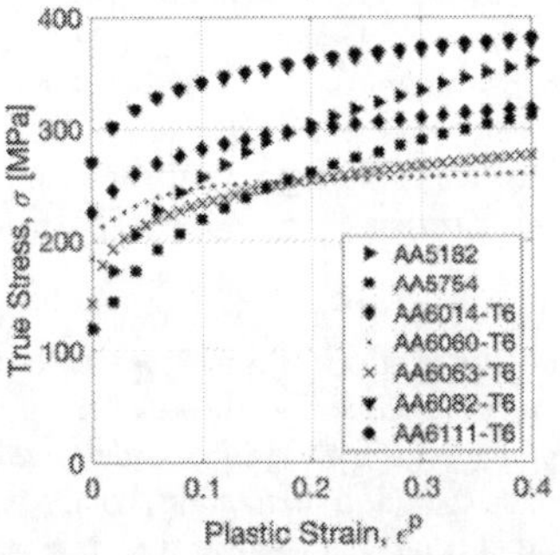

Fig. 8. Representative stress-strain response of various commercial automotive aluminum alloys.

3.3 Sampling of Component Grouping Thickness and Material Properties

Various combinations of part thicknesses and material parameters can be used to create virtual experiments of the frontal crash scenario, which can be used for data generation and machine learning. The original FE model assumed the majority of these components were fabricated using various grades of automotive sheet steel. Now, the direct substitution of the steel components with automotive grade aluminum will accomplish the goal of mass reduction because of the difference in density ($\rho_{steel} \approx 7800\ kg/m^3$ vs. $\rho_{al} \approx 2700\ kg/m^3$). However, the general strength differential between automotive grade aluminum and steel can have an adverse effect on the overall crashworthiness performance of the vehicle. Furthermore, the difference in elastic properties will also affect other vehicle performance metrics, such as vehicle stiffness and noise, vibration, and harshness (NVH). Nevertheless, the gauge thickness of the aluminum components can be increased to compensate for this reduction in strength. This enhancement in the geometric cross section will have a non-linear effect that can offset the added mass. As such, the sheet thickness of a part, i, within a component category, p, are scaled by the ratio of the density

$$t_i^{(p)} = (\rho_{steel}/\rho_{al})t_{0,i}^{(p)} \tag{12}$$

where $t_{0,i}$ is the initial thickness of the part defined in the FE model. Similar to material selection, an automaker would also need to specify the material thickness of each component. In the data generation, the thickness of each component category is sampled according to

$$t_i^{(n,p)} = \eta^{(n,p)} t_i^{(p)} = \eta^{(n)} (\rho_{steel}/\rho_{al}) t_{0,i}^{(p)} \tag{13}$$

where n is a sample in the data generation and $\eta^{(n,p)}$ scales the thickness of a component category, p. This scaling parameter is constrained, such that

$$0.80 \leq \eta^{(n,p)} \leq 1.25 \tag{14}$$

Next, the presented flow stress model has five unique parameters to describe the evolution of the stress-strain response of the material. Smerd et al. [19] have reported that automotive aluminum alloys generally exhibit low strain-rate sensitivity at room temperatures. Yet, aluminum alloys start to exhibit positive rate sensitivity at moderate and high strain-rates due to a change in temperature through adiabatic heating [3, 20]. As such, the Cowper-Symonds coefficients of $C = 9.39 \times 10^{10} s^{-1}$ and $P = 10.55$ are used for all materials in this study [21]. This reduces the number of free material parameters to three. Now, some of these parameter combinations will produce materials that are non-physical (i.e., a material with a low yield and no hardening or a high yield with too high hardening). To constrain the domain for this study, a sample combination of the material parameters must produce a stress-strain curve that is monotonically increasing by imposing

$$U_0^{(n,p)} \geq 1, \qquad D^{(n,p)} > 0 \tag{15}$$

The combination of material parameters must generate a stress-strain curve that lies within the upper and lower limits of the stress-strain responses presented in Fig. 8. The upper limits are formed by the AA6111-T6 response, while the lower limits are formed by AA5754 and AA6060-T6. Therefore, any sample must satisfy

$$\begin{aligned} \min\begin{pmatrix} \sigma_y^{AA6060-T6} \\ \sigma_y^{AA5754} \end{pmatrix} &\leq \sigma_y\left({\sigma_0}^{(n,p)}, U_0^{(n,p)}, D^{(n,p)}\right) \\ &\leq \max\left(\sigma_y^{AA6111-T6}\right) \forall\, \bar{\varepsilon}^P \\ &\in [0,0.4] \end{aligned} \tag{16}$$

This creates a convex hull of material parameters that satisfy these constraints. A multi-dimensional Latin hypercube algorithm is used to sample the convex hull of material parameters to generate 400 unique combinations of material coefficients for the 20 distinct parameters ((3 material parameters + 1 thickness coefficient) × 5 component categories).

4 Results and Discussion

4.1 Network Architecture and Training

Table 2 and Table 3 presents the network architectures. All models use a training/validation split of 80/20, where 320 samples are randomly selected for determining the weights in each model. All models used the Adam optimizer with an initial learning rate of 0.001 that converged using a mean squared error (MSE) function. The tanh activation function was used in the ANNs of each model. All models were trained using the Keras/Tensorflow libraries for Python using 2 x 1.4GHz nVidia Titan Xp (12GB GDDR5) GPUs with an nVidia GTX SLI HB Bridge.

Table 2. Architecture of ANN Model for Mass.

Dropout	0.00
Batch Size	32
# Layers in ANN	2
# Neurons in ANN	[40, 1]
# of Weights	281
# of Epochs	5000
Training Time [sec]	44.8

Table 3. Architecture of LSTM-NN Model for Accelerometers.

# Hidden Units	1536
Dropout	0.25
Batch Size	16
# Layers in ANN	2
# Neurons in ANN	[192, 6]
# of Weights	9.8×10^6
# of Epochs	1000
Training Time [sec]	3.8×10^3

https://doi.org/10.1051/matecconf/202032601006

Fig. 9 presents the convergence plots for the ANN and LSTM-NN models that were used for representing the vehicle mass and accelerometers, respectively. The training of the ANN model for mass converges quickly to low MSE magnitude ($> 10^{-5}$) with a relatively low number of neurons and depth. The validation error is also low with having a similar order of magnitude as the training error. This means that the model for mass should have excellent predictive capabilities. This is a result of the degree of nonlinearity in the mass being minor while varying only the section thickness [22]. However, this response can change significantly if the topology of the structure is varied. The training error of the LSTM-NN model for the accelerometers converged to an MSE magnitude of ~0.01. Although the MSE of the LSTM-NN is higher than the ANN model, the error is still rather low. It should be reiterated that the degree of nonlinearity is much higher in predicting the acceleration response. Nevertheless, the LSTM-NN model still shows good predictive capabilities in capturing the time-series response in acceleration for all axes and locations of interest.

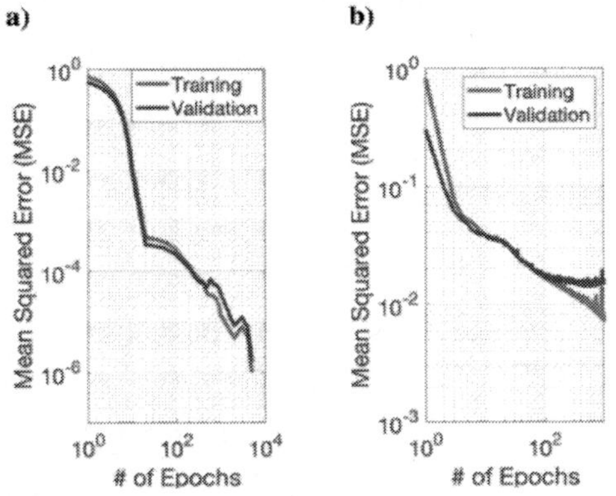

Fig. 9. Convergence plots for a) ANN model for Mass and b) LSTM-NN for Accelerometers.

To demonstrate this capability, a representative sample was taken from the validation set for prediction of the driver and passenger accelerations via the AI framework. Fig. 10 presents the acceleration response predicted by the AI framework for both the driver and passenger in all three axes. This sample has an MSE value of 0.0139. This figure shows the capability of the AI framework to simultaneously predict the response of the two nodes with good fidelity. The total time to predict the acceleration response of the driver and passenger using the AI framework was 6.77 milliseconds; this is significantly faster (1.41×10^{6}) than the time required to generate the result via the FE method. Now, the total number of nodes in the FE model is 7.34×10^{5}, which accounts for significant overhead for the FE solution. This methodology can be scaled to predict the acceleration of all nodes within the FE model. The ability of this AI framework to well predict the nodal accelerations offers an opportunity to predict the resulting displacement of all nodes through a simple integration of time. In turn, this suggests the prospect of predicting deformation, such as strain (i.e., the spatial gradient of nodal displacements), of a complete structure. This approach could allow for the real-time prediction of the deformation of structures that can be useful for intrusion measurements and occupant safety. However, it should be noted that this requires the use of the GPU to achieve these computational gains. Nevertheless, this significant reduction in computational time can aid scientists and engineers in developing lightweight solutions when coupled with an intelligent optimization framework [3].

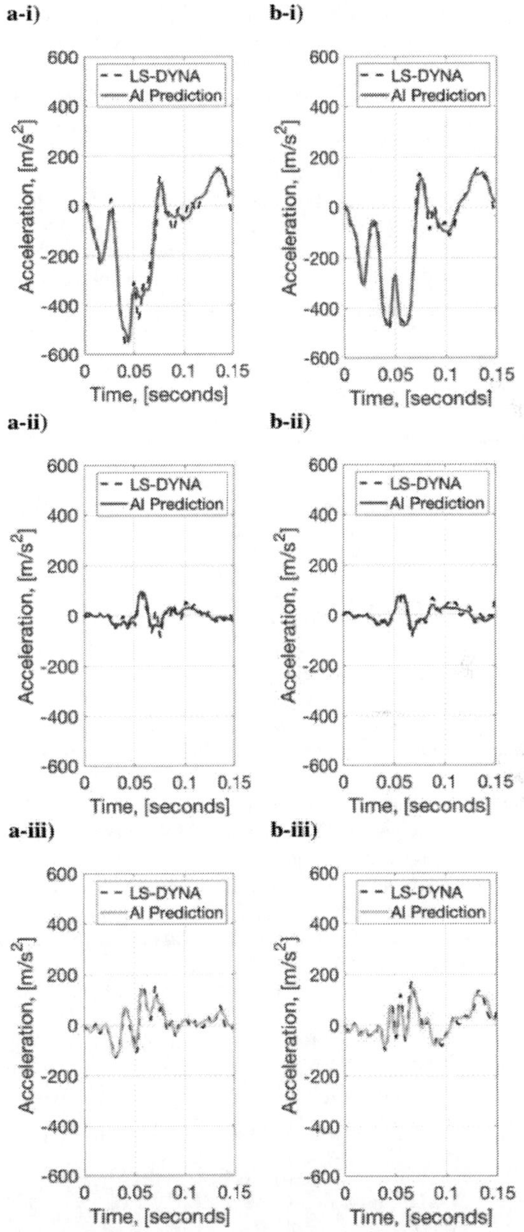

Fig. 10. Representative Sample of a) Driver and b) Passenger Acceleration Response in the i) X-direction, ii) Y-direction, iii) Z-direction from LS-DYNA and AI predictions.

5 Conclusion

This work presented an AI framework that can accelerate the development cycle in vehicle lightweighting. This framework was applied to the application of lightweighting a pick-up truck with the use of aluminum. This approach relied on a virtual design of experiments for a frontal crash test of the pick-up truck. The data for training the AI system was collected from finite element simulations of the vehicle under this loading condition. Different material properties that lie within the range of commercially available aluminum alloys along with various gauges of thickness were used for generating this data. An ANN was used to predict the mass of the vehicle, while an LSTM-NN was used to predict the crash pulse response that would be subjected to a driver and occupant. The AI approach was able to predict the occupant acceleration response up to 1.41×10^6 times faster than the conventional FE solution with minimal error (MSE < 0.01). Coupling this approach into an optimization framework would aid in the accelerated development of lightweight solutions. This framework also shows promise in the capability of predicting the entire deformation response of a vehicle.

This research was enabled in part by the support provided by Calcul Québec (www.calculquebec.ca) and Compute Canada (www.computecanada.ca). The authors also gratefully acknowledge Dr. HuiZhong Lu at Compute Canada - University of Sherbrooke for his technical support.

References

1. H. Kim, G. Keoleian, S. Skerlos, J Ind Ecol **15**(1), 64080 (2011).
2. A. Sachdev, R. Mishra, A. Mahato, A. Alpas, ICAA13 Pittsburgh, Springer, Cham, 609-622, (2012).
3. C. Kohar, A. Zhumagulov, A. Brahme, M. Worswick, R. Mishra, K. Inal, Int J Impact Eng **95**, 17-34 (2016).
4. A. Kullgren, M. Krafft, Å. Nygren, C. Tingvall, Accident Anal Prev **32**(2), 197-205 (2000).
5. C. Kohar, M. Mohammadi, R. Mishra, K. Inal, Thin Wall Struct **93**, 64-87 (2015).
6. K. Hornik, M. Stinchombe, W. Halbert, Neural Networks **3**, 551-560 (1990).
7. I. Aleksander, M. Helen, *An introduction to neural computing: 2nd edition*, (International Thomson Computer Press, London, 1995).
8. D. Rumelhart, G. Hinton, R. Williams, Nature **323**, 533-536 (1986).
9. K. Levenberg, Q Appl Math **2**, 164-168 (1944).
10. D. Marquardt, J Soc Ind Appl Math **11**(2), 431-441 (1963).
11. D. Kingma, J. Ba, arXiv preprint arXiv:1412.6980 (2014).
12. L. Lanzi, C. Bisagni, S. Ricci, Comput Struct **82**(1), 93-108 (2004).
13. K. Greff, R. Srivastava, J. Koutník, B. Steunebrink, J. Schmidhuber, IEEE T Neur Net Lear **28**(10), 2222-2232 (2016).
14. L. Hershman, SAE Technical Paper, 2001-06-0245 (2001).
15. J. Hallquist, Livermore Software Technology Corporation, Livermore, California (2006).
16. Society of Automotive Engineers, SAE Technical Report, J211-1 (2007).
17. G. Cowper, P. Symonds, Providence, Rhode Island (1957).
18. E. Voce, Metallurgia **51**(307), 219-226 (1955).
19. R. Smerd, S. Winkler, C. Salisbury, M. Worswick, D. Lloyd, M. Finn, Int J Impact Eng **32**, 541-560 (2005).
20. C. Kohar, A. Brahme, J. Imbert, R. Mishra, K. Inal, Int J Solid Struct **128**, 174-198 (2017).
21. S. Shu, N. Jones, Int J Crashworthiness **9**(2), 195-217 (2004).
22. G. Sun, G. Li, S. Zhou, H. Li, S. Hou, Q. Li, Struct Multidiscip O **44**(1), 99-110 (2011).

Unravelling the characteristics of Al-alloy corrosion at the atomic to nanometre scale by transmission electron microscopy

Shravan K. Kairy[1,*], and *Nick* Birbilis[2]

[1]Department of Materials Science and Engineering, Monash University, Clayton, Victoria, 3800, Australia.
[2]College of Engineering and Computer Science, The Australian National University, Acton, ACT, 2601, Australia.

Abstract. The localised corrosion associated with Mg_2Si in the Al-matrix of an Al-Mg-Si alloy was studied in 0.1 M NaCl at pH 6 by quasi in-situ transmission electron microscopy. Herein, physical imaging of corrosion at the atomic to nanometre scale was performed. Phase transformation and subsequent chemical composition variations associated with the localised corrosion of Mg_2Si were studied. It was observed that corrosion initiated upon Mg_2Si, often preferentially at the interface with the Al-matrix, and propagated until Mg_2Si was completely dealloyed by Mg-dissolution, resulting in an amorphous SiO-rich phase remnant. The SiO-rich remnant became electrochemically inert and did not initiate corrosion in the Al-matrix. This study provides a clear understanding on the localised corrosion of Al-alloys associated with Mg_2Si. In addition, the methodology followed in this study can also be applied to understand the role of precipitates and second phase particles in the localised corrosion of Al-alloy systems.

1 Introduction

Precipitation hardenable aluminium (Al) alloys such as 6xxx (Al-Mg-Si-(Cu)), 2xxx (Al-Cu-(Mg)) and 7xxx (Al-Zn-Mg-(Cu)) are strengthened by alloying additions and heat treatment processes that cause evolution of nanoscale precipitates in the Al-matrix [1-2]. However, such microstructures, which are intentionally generated for increasing strength, notably deteriorate the localised corrosion resistance of the alloys [1-5]. Nanoscale precipitates exhibit electrochemical characteristics that are different from the Al-matrix, resulting in the selective dissolution of precipitates or the surrounding Al-matrix, leading to pitting and/or intergranular corrosion (IGC) [3-7]. The extent of localised corrosion experienced by different Al-alloys is dependent on the type (and composition) of precipitates within the alloy, and therefore, a definitive understanding of the role of specific nanoscale precipitates in the localised corrosion of Al-alloys is essential. Transmission electron microscopy (TEM) has the ability to image corrosion up to nanometre to atomic scale, and simultaneously record any chemical variations and phase transformations that occur during the corrosion process via energy dispersive X-ray spectroscopy (EDXS) and electron diffraction, respectively. Herein, as an example, the role of Mg_2Si in the localised corrosion of Al-alloys was studied using a quasi in-situ TEM technique [8].

Nanoscale Mg_2Si precipitates play a dominant role in increasing the strength of Cu-free 6xxx Al-alloys [1-2, 9-12]. However, these alloys suffer IGC with the evolution of Mg_2Si precipitates along grain boundaries during ageing [13-17]. Electrochemical tests were performed upon synthesised Mg_2Si bulk specimens, and also through the electrochemical microcell technique on microscale Mg_2Si particles in the Al-matrix [18-19]. Based on the tests, it was identified that Mg_2Si was less noble than the Al-matrix in neutral and acidic NaCl [18-19]. A consensus on the dealloying of Mg_2Si by Mg-dissolution in acidic and neutral NaCl exist [8, 18-39]. However, a consensus on the role of dealloyed Mg_2Si in the localised corrosion do not exist. Of the limited studies on Mg_2Si exist to date, a few studies reported that following complete dealloying of Mg_2Si by Mg dissolution, the Si-rich remnants were enriched in O and did not take part in the corrosion process [21-22, 30-33]. In contrast, ambivalent electrochemical characteristics of Mg_2Si in Al-alloys was also reported [34-38]. It was observed that following initial dealloying, MgSi remnants became cathodic, more noble than the Al-matrix, resulting in the dissolution of the Al-matrix [34-38].

Herein, the localised corrosion associated with an Mg_2Si particle in the Al-matrix of a 6xxx Al-alloy in 0.1 M NaCl at pH 6 was studied at nanometre to atomic scale by a quasi in-situ TEM technique.

* Corresponding author: shravan.kairy@monash.edu

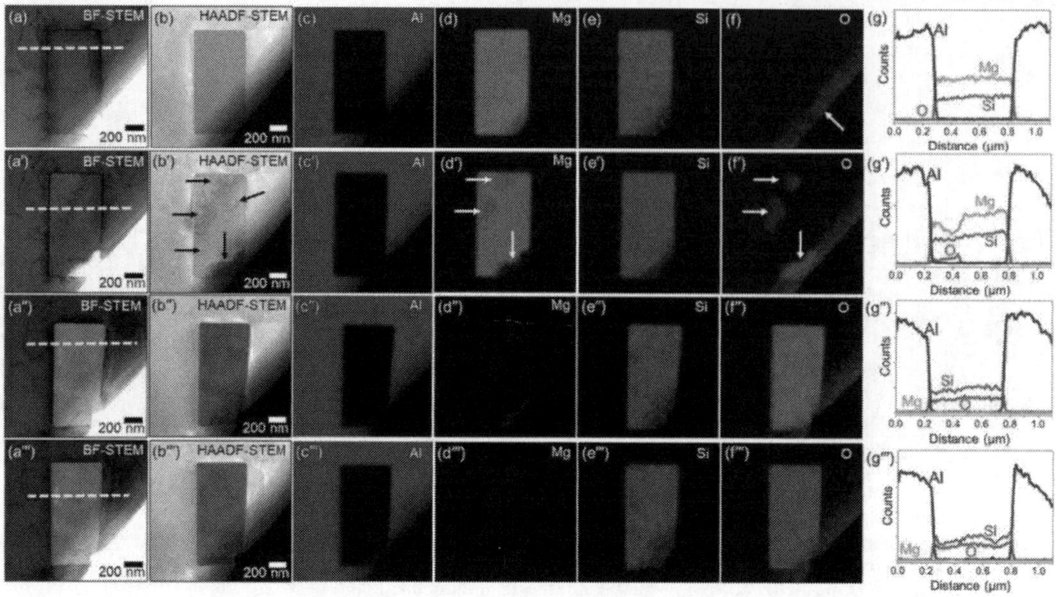

Fig. 1. BF- and HAADF-STEM images of an Mg_2Si particle in the Al-matrix of an Al-0.8Mg-0.6Si alloy, (a, b) prior to immersion, and following immersion in 0.1 M NaCl for (a′, b′) 3 s, (a″, b″) 30 s and (a‴, b‴) 30 min. Quantified EDXS maps in (c-f), (c′-f′), (c″-f″) and (c‴-f‴) correspond to (a, b), (a′, b′), (a″, b″) and (a‴, b‴), respectively. EDXS composition analysis along dashed lines in (a-a‴) are provided in (g-g‴), respectively. The arrow in (f) shows O enrichment in the Al-matrix at the edge of the TEM foil. Arrows in (b′) point to corroded sites within the Mg_2Si particle, and the Mg depletion and O enrichment at the corresponding sites are indicated by arrows in (d′) and (f′), respectively.

2 Experimental

2.1 Material

A 6xxx Al-alloy, Al-0.8Mg-0.6Si (% in mass fraction) was chosen for the present study. Alloy specimens were thermally treated in a PerkinElmer Pyris Diamond differential scanning calorimeter. The heat treatment involved solutionising at 540°C for 20 min and then cooling at 0.1 K/s to a room temperature. The detailed procedure was described elsewhere [40]. Such a thermal treatment resulted in the growth of monocrystalline Mg_2Si particles in the Al-matrix.

2.2 Transmission electron microscopy

A quasi in-situ TEM technique was applied to investigate the role of Mg_2Si in the localised corrosion of Al-alloys. Microstructural and microchemistry variations of an Mg_2Si particle in the Al-matrix were studied prior to immersion of a TEM specimen in 0.1 M NaCl at pH 6, and following immersion for 3 s, 30 s and 30 min. Specimens were subsequently cleaned in ethanol following immersion in NaCl, and vacuum dried prior to TEM observation.

A Gatan 691 precision ion polishing system was used for the TEM specimen preparation, which involved thinning of 3 mm diameter disc at - 80°C. In addition, a Gatan Solarus 950 advanced plasma system was used for plasma cleaning of TEM specimens for 2-10 min using an Ar and O_2 gas mixture, immediately prior to the specimen loading into a TEM. An FEI Tecnai G[2] F20 S-TWIN FEG TEM at 200 kV was operated in Scanning TEM (STEM) and TEM modes. In the STEM mode, bright field (BF) as well as high angle angular dark field (HAADF) images were collected. For microchemistry analysis, EDXS was performed with an X-flash X-ray detector using a Bruker Quantax 400. The collected EDXS hypermap data was then quantified by Cliff-Lorimer method using a Bruker Esprit1.9 X-ray software. In the TEM mode, electron diffraction to obtain selected area diffraction patterns (SADPs) and atomic resolution imaging were performed to study phase transformation associated with localised corrosion.

3 Results and Discussion

Scanning TEM imaging of an Mg_2Si particle in the Al-matrix and the corresponding chemical composition variations prior to and following a period of corrosion are provided in Fig. 1. In the pristine condition, i.e. prior to immersion in 0.1 M NaCl, a submicrometre sized Mg_2Si particle in the Al-matrix can be seen in Fig. 1 (a, b). The quantified chemical composition of the Mg_2Si particle and the surrounding Al-matrix as shown in Fig. 1 (c-f), revealed that the particle contained Mg and Si, with almost no O. The EDXS composition analysis over the dashed line in Fig. 1 (a) revealed that the particle contained higher Mg than Si, with negligible O content. The arrow in Fig. 1 (f) point to high O content in the Al-matrix, which is because of the sample preparation technique and is quite common observation, at the edge of a TEM foil.

To understand the origin of corrosion, the TEM specimen was immersed in 0.1 M NaCl for a short period of 3 s and the same location as in Fig. 1 (a, b) was observed under TEM, and the results are shown in Fig. 1 (a′-g′). Arrows in Fig. 1 (b′) point to the sites at which corrosion occurred. It can be observed that corrosion initiated on the Mg_2Si particle mostly at the interface of the Mg_2Si particle and the Al-matrix, revealing that the Mg_2Si particle is anodic with respect to the Al-matrix (arrows in Fig. 1. (b′)). In addition, corrosion initiation sites were also observed upon the Mg_2Si particle away from the Mg_2Si particle and Al-matrix interface (arrows in Fig. 1. (b′)). The Mg_2Si particle dealloyed by Mg dissolution and subsequently enriched in O (arrows in Fig. 1 (d′ and f′)). The chemical composition analysis along the dashed line in Fig. 1 (a′) clearly revealed the dissolution of Mg and enrichment of O, with negligible change in Si, at the sites where the Mg_2Si particle corroded. Further immersion in 0.1 M NaCl for 30 s led to complete dissolution of Mg from the Mg_2Si particle and the remnant was enriched in O, with negligible change in Si (Fig. 1 (a″-g″)). To further investigate the role of the SiO-rich remnant on the corrosion, the TEM foil was further immersed for an extended period of 30 min in 0.1 M NaCl. It was revealed that following complete dissolution of Mg from the Mg_2Si particle, the SiO-rich remnant became electrochemically inert and did not take part in the corrosion process, as evident from the negligible change in the chemical composition within the dealloyed Mg_2Si particle and the surrounding Al-matrix, following 30 s and 30 min (Fig. 1 (a″-g″) and (a‴-g‴)).

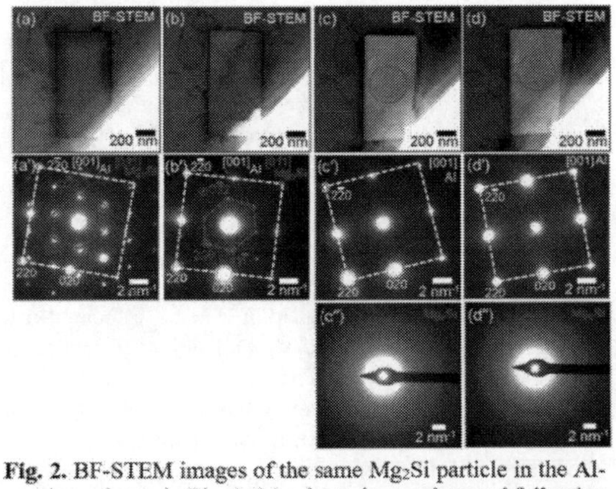

Fig. 2. BF-STEM images of the same Mg_2Si particle in the Al-matrix as shown in Fig. 1, (a) prior to immersion, and following immersion in 0.1 M NaCl for (b) 3 s, (c) 30 s and (d) 30 min. (a′-d′) are SADPs corresponding to (a-d), respectively. SADPs in (c″) and (d″) are collected from the region within the circles in (c) and (d), respectively.

To understand the phase transformation involved with the dealloying of the Mg_2Si particle following immersion in 0.1 M NaCl, electron diffraction studies were performed on the same Mg_2Si particle as shown in Fig. 1., prior to and following immersion, and the results are shown in Fig. 2. In the pristine condition, it was revealed that the orientation relationship of the Mg_2Si particle with the Al-matrix was $[001]_{Al}$ // $[011]_{Mg2Si}$ and $(1\bar{1}0)_{Al}$ // $(0\bar{1}1)_{Mg2Si}$ (Fig. 2 (a′)). The SADP obtained from Fig. 2 (b), i.e. following 3 s immersion, did not reveal any significant change in the corrosion-induced phase transformation, i.e. change in the chemistry and/or crystallographic structure, as the extent of corrosion is very small. A significant change in the electron diffraction was observed following complete dealloying of the Mg_2Si particle, i.e. following 30 s and 30 min (Fig. 2). Based on the results, it is quite evident that the crystalline Mg_2Si particle transformed to amorphous SiO-rich remnant, following dealloying by Mg dissolution.

Atomic resolution TEM imaging was performed along the $[001]_{Al}$ // $[011]_{Mg2Si}$ direction to understand the onset of corrosion at atomic scale and any influence of the SiO-rich remnant on the surrounding Al-matrix, and the results are provided in Fig. 3. Such a resolution is essential because the electron diffraction and chemical composition analysis will not be able to reveal significant details when the extent of corrosion is at the atomic scale. In the pristine condition, clear atomic columns or lattice fringes of the Mg_2Si particle and the Al-matrix were observed (Fig. 3 (a)). Following immersion for 3 s (Fig. 3 (b)), the lattice structure within the Mg_2Si particle transformed to amorphous, which was revealed to be of SiO-rich phase from the chemical composition analysis (Fig. 1 (a′-g′)), and the lattice structure of the Al-matrix did not change. A key finding from the present study is that the corrosion did not initiate uniformly upon the Mg_2Si particle at its periphery (Fig. 2 and 3 (b)). In fact, corrosion initiated randomly at few of the interfaces of the Mg_2Si particle and the Al-matrix, and propagated to completely dealloy the Mg_2Si particle. The direction of corrosion propagation was marked by a single arrowed line in Fig. 3 (b, b′). It is evident that with the propagation of corrosion, the crystalline Mg_2Si structure transforms to amorphous SiO-rich phase (Fig. 3 (b′)). Following complete dealloying of the Mg_2Si particle, i.e., after 30 s and 30 min immersion, the lattice structure of the Al-matrix at the interface of the amorphous SiO-rich phase remained unaffected (Fig. 3 (c, d and d′)), revealing that the SiO-rich remnant became electrochemically inert and did not initiate corrosion in the Al-matrix even at the atomic scale. The electrochemical behaviour of Mg_2Si in Al-alloys following extended period of exposure to 0.1 M NaCl at different pH was provided elsewhere [41]. In that study, it was revealed that the physical dimensions of Mg_2Si phase decreased following complete dealloying at pH 2, resulting in the formation of a trench at the Al-matrix and SiO-rich phase [41]. However, such significant change in physical dimensions was not observed following complete dealloying of Mg_2Si phase at pH 6 [41].

The results obtained from the present study aid in understanding the localised corrosion, i.e. pitting and IGC, of Al-alloys associated with Mg_2Si particles and precipitates. Upon exposure to NaCl, corrosion initiates on Mg_2Si, randomly at the interface of Mg_2Si and the Al-matrix, and also away from the interface (dotted arrows in Fig. 4 (a)). Corrosion propagates to completely dealloy Mg_2Si, leaving amorphous SiO-rich remnant that will become electrochemically inactive. As a result, pitting

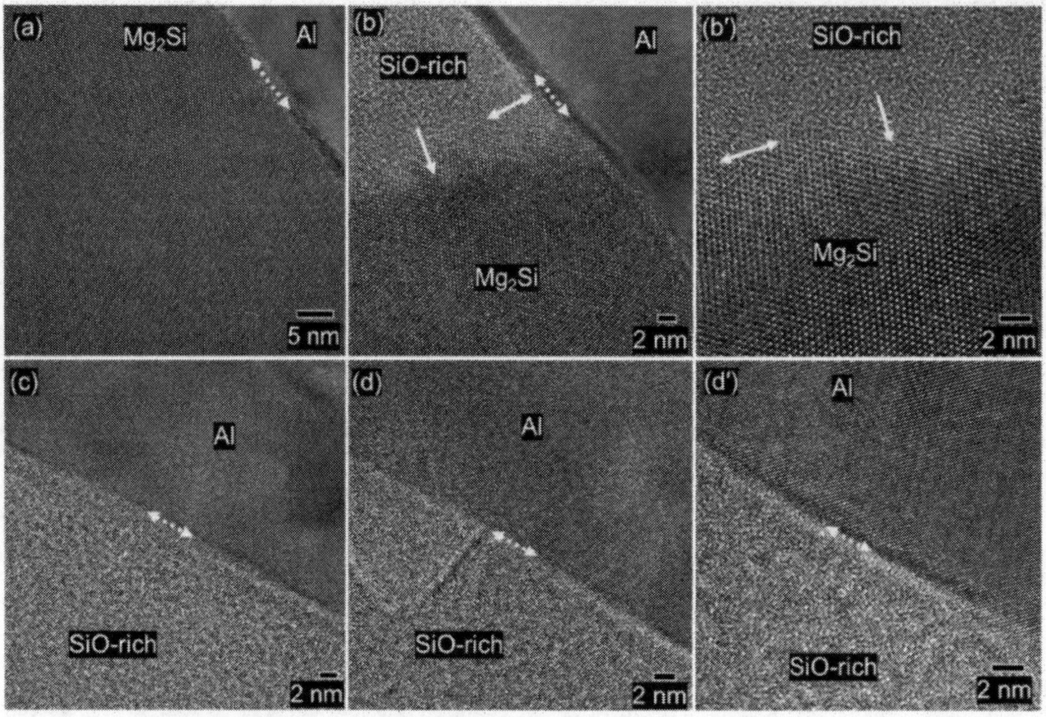

Fig. 3. Atomic resolution TEM images of an identical location showing Al-matrix, Mg_2Si and the interface between them, (a) prior to immersion, and following immersion in 0.1 M NaCl for (b) 3 s, (c) 30 s and (d) 30 min. (b′) shows atomic resolution Mg_2Si following immersion in 0.1 M NaCl for 3 s. (d′) shows higher magnification image of d. Double arrowed dotted lines in (a-d, d′) and solid lines in (b and b′) indicate interfaces between the phases. Whilst single arrowed solid line in (b) and (b′) indicate the direction of corrosion propagation.

will not be initiated within the Al-matrix following complete dealloying of Mg_2Si. Of the limited studies to date on IGC, a discrepancy exists in Cu-free alloys [14-17, 21-23, 42]. A few studies reported that upon artificial ageing, Cu-free 6xxx Al-alloys suffered IGC as result of the formation of Mg_2Si precipitates along grain boundaries [14-17, 42]. In contrast, these alloys were also reported to be immune to IGC though Mg_2Si precipitates form along grain boundaries [21-22]. From the present study, it can be understood that Mg_2Si precipitates along grain boundaries do not propagate IGC. However, the precipitates rapidly dealloy upon exposure during the IGC process, weakening grain boundaries. As a result, the propensity for IGC of Cu-free 6xxx Al-alloys depends on how closely the Mg_2Si precipitates are arranged along grain boundaries.

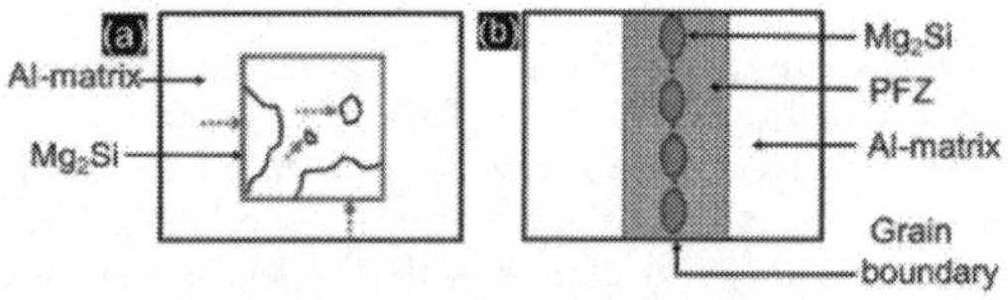

Fig. 4. Schematic microstructural model showing Mg_2Si (a) within the Al-matrix in a grain and (b) precipitates along a grain boundary. Dotted arrows in (a) point to corrosion initiation sites upon Mg_2Si. PFZ in (b) is a precipitate free zone, which is electrochemically more noble than Mg_2Si in 0.1 M NaCl at pH 6.

Conclusions

A quasi in-situ TEM method was employed to physically image corrosion initiation and propagation at the atomic to nanometre scale. The corresponding phase transformations (probed via electron diffraction) and chemical composition variations (probed via energy dispersive X-ray spectroscopy) that were associated with the localised corrosion of an Mg_2Si particle in the Al-matrix upon exposure to 0.1 M NaCl at pH 6 were studied. The following conclusions are drawn:

- Corrosion was initiated upon Mg_2Si either at its interface with the Al-matrix, but also was observed to initiate away from the interface with the Al-matrix.
- The corrosion process continued until the whole of Mg_2Si was completely dealloyed by Mg-dissolution, leaving an SiO-rich remnant.
- The SiO-rich remnant was determined herein to be amorphous and electrochemically inactive, and thus it did not take further part in, or initiate further, corrosion in the surrounding Al-matrix - even at the atomic scale.
- The research approach followed herein can be extended to other Al-alloy systems in understanding the corrosion phenomenon at nanometre to atomic scale.

Acknowledgements

The authors gratefully acknowledge P. Schumacher and B. Milkereit from University of Rostock, Germany, for the provision of the alloy. Microscopy was performed at the Monash Centre for Electron Microscopy (MCEM).

References

[1] I.J. Polmear, *Light alloys from traditional alloys to nanocrystals*, 4th ed. (Oxford, United Kingdom: Elsevier, 2006).
[2] J.F. Nie, "Physical metallurgy in light alloys", in *Physical metallurgy*, eds. D.E. Laughlin, K. Hono (Elsevier, Amsterdam, 2014).
[3] J.R. Davis, *Corrosion of aluminium and aluminium alloys*, 1st ed. (Materials Park: ASM International, 1999).
[4] S.K. Kairy, P.A. Rometsch, C.H.J. Davies, N. Birbilis, CORROSION **71**, 11 (2015) 1304-1307.
[5] S.K. Kairy, P.A. Rometsch, K. Diao, J.F. Nie, C.H.J. Davies, N. Birbilis, Electrochim. Acta **190** (2016) 92-103.
[6] S.K. Kairy, T. Alam, P.A. Rometsch, C.H.J. Davies, R. Banerjee, N. Birbilis, Metall. Mater. Trans. A **47**, 3 (2016) 985-989.
[7] S.K. Kairy, S. Turk, N. Birbilis, A. Shekhter, Corros. Sci. **143** (2018) 414-427.
[8] S.K. Kairy, P.A. Rometsch, C.H.J. Davies, N. Birbilis, CORROSION **73**, 1 (2017) 87-99.
[9] J.H. Chen, E. Costan, M.A. Van Huis, Q. Xu, H.W. Zandbergen, Science **312** (2006) 416-419.
[10] D.J. Chakrabarti, D.E. Laughlin, Prog. Mater. Sci. **49** (2004) 389-410.
[11] T. Saito, S. Muraishi, C.D. Marioara, S.J. Andersen, J. Royset, R. Holmestad, Metall. Mater. Trans. A **44** (2013) 4124-4135.
[12] A.K. Gupta, D.J. Lloyd, S.A. Court, Mater. Sci. Eng. A **316** (2001) 11-17.
[13] S.K. Kairy, P.A. Rometsch, C.H.J. Davies, N. Birbilis, CORROSION **73** (2017) 1280-1295.
[14] A. Laurino, E. Andrieu, J.P. Harouard, J. Lacaze, M.C. Lafont, G. Odemer, C. Blanc, J. Electrochem. Soc. 160 (2013) C569-C575.
[15] A. Laurino, E. Andrieu, J.P. Harouard, J. Lacaze, M.C. Lafont, G. Odemer, C. Blanc, ECS Trans. **41** (2012) 93-105.
[16] A. Shi, B.A. Shaw, E. Sikora, CORROSION **61** (2005) 534-547.
[17] A.K. Bhattamishra, K. Lal, Mater. Des. **18** (1997) 25-28.
[18] N. Birbilis, R.G. Buchheit, J. Electrochem. Soc. **152**, 4 (2005) B140-B151.
[19] R.K. Gupta, N.L. Sukiman, K.M. Fleming, M.A. Gibson, N. Birbilis, ECS Electrochem. Lett. **1**, 1 (2012) C1-C3.
[20] K.M. Fleming, A. Zhu, J.R. Scully, CORROSION **68** (2012) 1126-1145.
[21] M.H. Larsen, J.C. Walmsley, O. Lunder, K. Nisancioglu, J. Electrochem. Soc. **157** (2010) C61.
[22] G. Svenningsen, M.H. Larsen, J.H. Nordlien, K. Nisancioglu, Corros. Sci. **48** (2006) 3969-3987.
[23] W.J. Liang, P.A. Rometsch, L.F. Cao, N. Birbilis, Corros. Sci. **76** (2013) 119-128.
[24] J. Wloka, S. Virtanen, J. Electrochem. Soc. **154** (2007) C411-423.
[25] S. Jain, PhD Thesis, The Ohio State University, USA, 2006.
[26] C. Blanc, Y. Roques, G. Mankowski, Corros. Sci. **40** (1998) 1019-1035.
[27] O. Gharbi, N. Birbilis, J. Electrochem. Soc. **165** (2018) C497-C501.
[28] Y.T. Zhou, Y.N. Zan, X.X. Wei, B. Yang, B. Zhang, S.J. Zheng, X.H. Shao, J.H. Dong, X.K. Ma, B.L. Xiao, Q.Z. Wang, Z.Y. Ma, Corros. Sci. **153** (2019) 74-84.
[29] D.M. Sun, Y.M. Jiang, Q.W. Xiang, C. Zhong, J. Gong, L.H. Zhang, J. Li, Mater. Corr. **61** (2010) 105-110.
[30] K. Mizuno, A. Nylund, I. Olefjord, Corros. Sci. **43** (2001) 381-396.
[31] O. Knudsen, B.S. Tanem, A. Bjorgum, J. Mardalen, M. Hallenstvet, Corros. Sci. **46** (2004) 2081-2095.
[32] Y. Zhu, K. Sun, G.S. Frankel, J. Electrochem. Soc. **165** (2018) C807-820.
[33] K.A. Yasakau, M.L. Zheludkevich, S.V. Lamaka, M.G.S. Ferreira, Electrochim. Acta **52** (2007) 7651-7659.
[34] F. Eckermann, T. Suter, P.J. Uggowitzer, A. Afseth, P. Schmutz, Electrochim. Acta **54** (2008) 844-855.
[35] F. Zeng, Z. Wei, J. Li, X. Tan, Z. Zhang, Z. Zheng, Trans. Nonferrous Met. Soc. China **21** (2011) 2559-2567.
[36] Y. Wu, H. Liao, J. Mater. Sci. Technol. **29** (2013) 380-386.
[37] C. Li, J. Sun, Z. Li, Z. Gao, Y. Liu, L. Yu, H. Li, Mater. Charact. **122** (2016) 142-147.
[38] Z. Li, C. Li, Z. Gao, Y. Liu, X. Liu, Q. Guo, L. Yu, H. Li, Mater. Charact. **110** (2015) 170-174.
[39] L.L. Li, B. Zhang, B. Tian, Y. Zhou, J.Q. Wang, E.H. Han, W. Ke, J. Electrochem. Soc. **164** (2017) C240-C249.
[40] P. Schumacher, Ph.D. Thesis, University of Rostock, Germany, 2018.
[41] S.K. Kairy, N. Birbilis, CORROSION **76** (2020) 464-475.
[42] S.K. Kairy, P.A. Rometsch, C.H.J. Davies, N. Birbilis, CORROSION **73** (2017) 1280-1295.

Quench Sensitivity and Phase Transformation Kinetics of AlSi7MnMg High Pressure Vacuum Die Casting

Mengyun Liu[1], *Zhan Zhang*[1,*], *Francis Breton*[2] *and X.-Grant Chen*[1]

[1] Department of applied science, University of Quebec at Chicoutimi, Saguenay, QC, Canada, G7H 2B1
[2] Arvida Research and Development Centre, Rio Tinto, Saguenay, QC, Canada, G7S 4K8

Abstract. The quench sensitivity of an AlSi7MnMg alloy in high-pressure vacuum die (HPVD) casting was investigated by time-temperature-transformation and time-temperature-property diagrams with an interrupted quench technique. The quench sensitive temperature range of the alloy is from 260 to 430 °C and its nose temperature is 350 °C. The mechanical strength versus cooling rates of the HPVD casting was predicted using quench factor analysis method and verified by experimental results. The critical cooling rate is 6 °C/s to remain 95% of the maximal mechanical strength. The coefficients k_2 - k_5, related to the nucleation and precipitation kinetics of TTP curves, and phase transformation diagrams were determined. The precipitation of Mg_2Si phase in the castings was observed during isothermal treatment using transmission electron microscope. Moreover, the quench sensitivity and kinetics of the phase transformation of AlSi7MnMg alloy and AlSi10MnMg alloys were compared. It reveals that the quench sensitivity and phase transformation rate of the former are lower than that of the latter.

1. Introduction

AlSi7MnMg (Aural™-5) alloy has been widely applied in high pressure vacuum die (HPVD) castings due to its very good fluidity and castability, particularly suitable for thin-wall and large complex cast structural components [1]. The AlSi7MnMg die casting alloy is heat-treatable and weldable, and thus its mechanical properties can be greatly enhanced through heat treatment, because the vacuum system used during the die casting process removes the entrapped air in die cavity during filling and consequently limits the formation of blisters during solution heat treatment [2, 3].

Quenching is one of important steps in precipitation-hardening heat treatment processes. Excessive rapid quenching can introduce severe residual stress and may lead to distortion at thin sections. Insufficient quenching can result in a low level of supersaturated solutes in the matrix, and eventually reduce the strengthening effect during artificial aging. Therefore, it is important to select an optimal quenching rate in the heat treatment. Quench sensitivity is related to the formation of a non-hardening phase during quenching. Time-temperature-transformation (TTT) and time-temperature-precipitate (TTP) diagrams are broadly used to study the quench sensitivity because these diagrams deliver important information, such as the critical time and nose temperature for precipitation during quenching [4]. An interrupted quenching technique with series of isothermal heating tests has been applied to construct these diagrams [4]. Quench factor is defined as the severity of a quench and can be quantified to a single number using integrated cooling rate and TTP curves, and it has been used to predict properties under certain cooling rates [5]. By quenching samples in water at various temperatures, Zhang et al. [6] studied the quench sensitivity of Al-7%Si-0.4%Mg (A356) permanent mold (PM) castings. As the quench rate decreases from 250 °C/s to 0.5 °C/s, the ultimate tensile strength and yield strength of peak-aged A356 castings decreased by approximately 27 and 33%, respectively. Ma et al. [7] used the quench factor analysis method to predict the effect of quench rates on mechanical properties of A356 PM castings, and the predicted results were in good agreement with the experimental data. Recently, Liu et al. [8] investigated quench sensitivity of AlSi10MnMg PM and HPVD castings. They demonstrated that the casting process had an impact on quench sensitivity of castings. HPVD castings were more quench sensitive than PM castings mainly due to the fine grain size and large amount of grain boundaries in HPVD castings.

The aim of the study is to investigate the quench sensitivity, phase transformation kinetics, and the effect of the quench rate on mechanical properties of the AlSi7MnMg HPVD castings. Moreover, the characteristics of TTP diagrams of both AlSi7MnMg and AlSi10MnMg alloys are analyzed to understand the influence of alloy chemical compositions on the quench sensitivity.

* Corresponding author: zhan_zhang@uqac.ca

2. Experimental

Both AlSi7MnMg (Aural™-5) and AlSi10MnMg (Aural™-3) HPVD casting samples were produced by a cold chamber vacuum die casting machine (Buhler 26D, Markham, ON, Canada) equipped with a Castool vacuum system (Uxbridge, ON, Canada). The dimension of the cast plates is 220 mm x 65 mm with a thickness of 2.5 mm. Table 1 lists the chemical compositions of the two alloy samples analyzed with an optical emission spectrometer.

Table 1. Chemical composition (wt.%) of experimental alloys

Alloys	Si	Fe	Mn	Mg	Ti	Sr	Al
AlSi7MnMg	7.40	0.17	0.49	0.20	0.08	0.018	bal.
AlSi10MnMg	10.10	0.18	0.50	0.55	0.06	0.012	bal.

The size of the samples used in isothermal holding tests is 20 mm x 20 mm cut from HPVD cast plates. The samples were solution-heat-treated at 500 °C for 3 h, and then immediately transferred into a salt bath at 250-450 °C for different holding times. The salt bath temperature was continuously monitored and maintained at ± 1 °C of the targeted temperature. After isothermal holding, the samples were quenched in water at room temperature, and then artificially aged at 170 °C for 2.5h. An electrical conductivity tester (Sigmascope SMP10) was used to evaluate the electrical conductivity of as-quenched samples. Four measurements were conducted for each sample. Conventional metallographic polishing process was applied for the microstructure observation. Vickers hardness were measured using a Vickers hardness tester (Nextgen NG-1000CCD) with a load of 100 g and a dwell time of 15 s, according to the ASTM E92. The average values were reported from 10 measurements.

The microstructures of the samples were examined by an optical microscope coupled with an image analyzer (CLEMEX), and a scanning electron microscope (JSM-6480LV). The phase precipitation was observed using a transmission electron microscope (JEM-2100). TEM samples were prepared by a twin-jet electro-chemical polisher in a solution of 25% nitric acid in methanol at -25 °C. To verify predicted mechanical properties at various quench rates, the solution-treated samples were quenched in different media (still or forced air, fiberglass insulator, oil at different temperatures), followed by an aging treatment at 170 °C for 2.5 h. Then, the microhardness of the samples was measured to determine the mechanical properties at different quench rates.

3. Results and discussion

3.1 Time-temperature-transformation diagram

During isothermal treatment, supersaturated aluminum matrix would decompose to form precipitates, which is corresponding to the changes in electrical conductivity (EC) of samples. The EC evolution of the quenched AlSi7MnMg samples at different isothermal temperatures are shown in Fig. 1a. At the first stage, the EC increases rapidly then more slowly to finally reach a stable value during isothermal holding. At 350 °C, the EC increasing rate is faster than that at lower or higher temperatures. The EC of the quenched samples after solution treatment is 38.5% IACS, which represents a supersaturated solid solution state, whereas the EC of isothermal holding at 350 °C for 27 h increases to 43.5% IACS, which represents a complete decomposition of supersaturated solid solution. The time-temperature transformation (TTT) diagram was constructed using the values of 10%, 20%, 30% and 40% of the EC difference between the supersaturated and the complete decomposition samples. The measured data based on Fig. 1a and the mathematical fitted curves are shown in Fig. 1b. The TTT contours show a "C" shape. The nose temperature of the curves is 350 °C, and the critical time, corresponding to the minimum time required for 10% precipitate transformation, is 16 seconds (Fig. 1b). Moreover, the quench sensitive temperature range is 260-430 °C. In this temperature range, the EC varies rapidly.

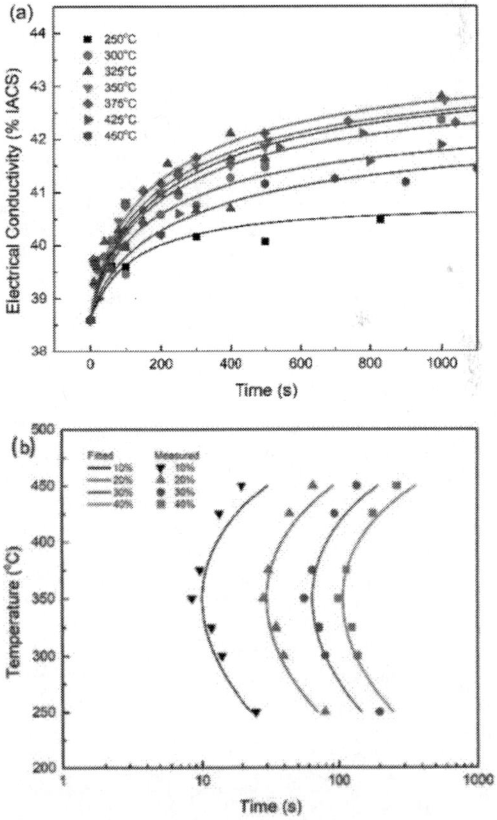

Fig.1. Electrical conductivity evolution at different isothermal temperatures as a function of holding time (a); TTT diagram of AlSi7MnMg HPVD castings (b). The lines in (b) are the fits to the experimental data by using Equation (1).

3.2. Time-temperature-property diagram and mechanical property prediction

The hardness varies with the evolution of precipitation in aluminum alloys during heat treatments. The hardness of the AlSi7MnMg HPVD castings under T6 condition (500 °C for 3 h followed by 170 °C for 2.5 h) is 73.3 HV, which is considered to be the highest hardness of cast samples, but it decreases with an increase of the holding time at isothermal temperatures. Fig. 2a illustrates the hardness evolution of cast samples during isothermal holding. Generally, the hardness decreases rapidly at the first stage, then the decrease rate is significantly slowdown. At 350 °C, the hardness decrease rate is faster than that at lower or higher temperatures. TTP curves can be mathematically described in Equation 1 [9]:

$$C_t(T) = -k_1 k_2 exp\left[\frac{k_3 k_4^2}{RT(k_4-T)^2}\right] exp\left(\frac{k_5}{RT}\right) \quad (1)$$

Where $C_t(T)$ is the critical time for a certain fraction of solutes to precipitate; note that the fraction of solutes precipitated corresponds to the fraction of hardness loss in TTP diagram; k_1 is the natural logarithm of the unprecipitated fraction during isothermal holding; k_2 is the constant associated with the reciprocal of the number of nucleation sites; k_3 is the constant corresponding to the nucleus energy; k_4 is the constant related to the solvus temperature; k_5 is the constant related to the activation energy for diffusion; R is the mole gas constant; T is the temperature (K). The coefficients in Equation 1 were determined by non-linear least squares regression based on the data in Fig. 2a and are listed in Table 2.

Table 2. Coefficients of TTP curves of experimental alloys.

Castings	k_2*(s)*	k_3 *(J/mol)*	k_4*(K)*	k_5*(J/mol)*
Aural-5	2.38×10^{-6}	2.87×10^{3}	9.45×10^{2}	7.42×10^{3}
Aural-3 [8]	1.92×10^{-6}	2.48×10^{3}	9.55×10^{2}	7.30×10^{3}

The TTP diagram with the contours of 99.5%, 95%, and 90% of the maximal hardness of the samples, is shown in Fig. 2b. The measured data in Fig. 2b are from the curves in Fig. 2a, and the mathematically fitted curves are based on the calculation with Equation 1. The contours in TTP diagram also show "C" shape. The nose temperature and quench sensitive temperature range in the TTP diagram are similar to those in TTT diagram. The critical time at the nose temperature is 2.4 s to obtain 99.5% maximum hardness.

The quench factor analysis is widely applied to predict the effect of cooling rate (quench rate) on mechanical property, and the quench factor τ, can be calculated using Equation 2 [9]:

$$\tau = \sum_{t_0}^{t_f} \frac{\Delta t}{C_t(T)} \quad (2)$$

Where $C_t(T)$ is the critical time for a specific amount of solutes to precipitate; t_0 is the time at the start of the quench; t_f is the time at the end of the quench; and Δt is time step. Then, the predicted properties at various cooling rates can be expressed as Equation 3 [9]:

$$\sigma = \sigma_{max} \exp(k_1 \tau) \quad (3)$$

Where σ is the predicted property; σ_{max} is the property attainable with an infinite quench rate; k_1 is ln (0.995). Equation 4 describes the property drop Δ%, as a function of the quench factor [10]:

$$\Delta\% = 1- \exp(k_1 \tau) \quad (4)$$

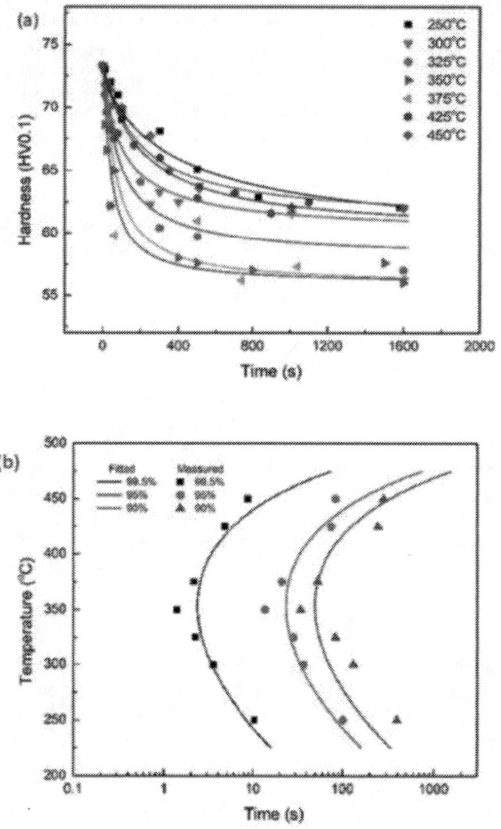

Fig. 2. Hardness evolution at different isothermal temperatures as function of holding time (a); TTP diagram of AlSi7MnMg HPVD samples (b).

Δt = 0.01s has been selected in Equation 2 because the temperature decrease should be less 25 °C during each time step to ensure the accuracy of the quench factor calculation [8, 11]. When the quench sensitive temperature range is considered in quench factor calculation, the effect of the temperature range on quench factor calculation is ignored [12]. In this study, the range of 260-430 °C is applied to the quench factor calculation. The effect of cooling rate (quench rate) on the quench factor and predicted hardness is illustrated in Fig. 3a. As the cooling rate increases, at the beginning, the quench factor decreases rapidly, but the decrease of the quench factor becomes slow when the cooling rate reaches a certain value. Conversely, the predicted hardness initially increases rapidly, then the increase becomes slow. Fig. 3b demonstrates the effect of the cooling rate on hardness drop. When the cooling rate increases, the hardness drop decreases. It can be seen that the hardness drop is less than 5% when the cooling rate is over 6 °C/s. This cooling rate is referred as the critical cooling rate, *Cr*. In the other words, when the cooling rate is higher than 6 °C/s, higher than 95% of maximal mechanical strength of the samples can be obtained. In Fig. 3b, the dark line represents the predicted data based on Equation 4, and the red dots are the measured hardness of the samples subjected to quench experiments at various cooling rates. The predicted results are basically consistent with experimental data.

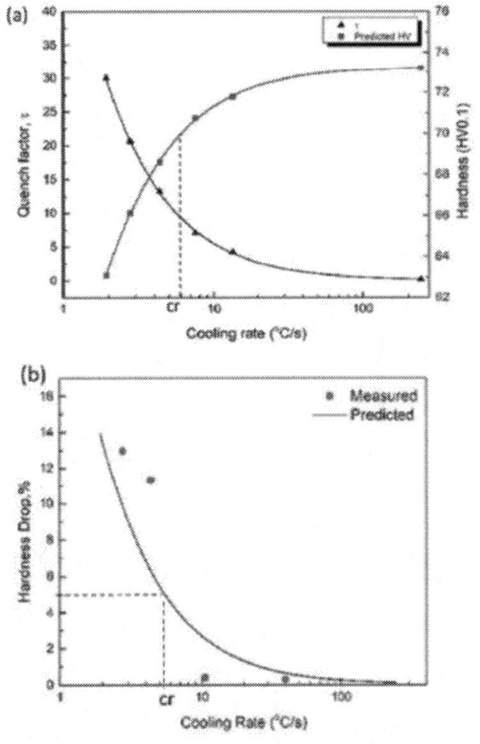

Fig. 3. Effect of cooling rate on quench factor and predicted hardness (a); Drop in the hardness vs. cooling rate of AlSi7MnMg samples (b).

3.3. Precipitation during isothermal treatment

Quench sensitivity of Al-Si-Mg die casting alloys is mainly determined by the precipitation of a non-hardening phase during quenching, which would negligibly affect the alloy strengthening during the subsequent aging treatment. The precipitates at the nose temperature (350 °C) for 300 s in the isothermal holding were examined, as shown in Fig. 4. Few of needle-like precipitates with length of 0.5-1 μm and a thickness of 20-50 nm were observed in the aluminum matrix (Fig. 4a), while a large number of particles with an equivalent diameter of 0.2-0.5 μm appears along the grain boundaries (Fig. 4b). These needle-like precipitates in the aluminum matrix and particles in the grain boundaries are identified as equilibrium β-Mg_2Si phase. It is evident that such large and coarse precipitates have only negligible effect on the alloy strength, and it causes a low level of solute concentrations (Mg and Si) in the matrix, resulting in a reduced hardening effect during subsequent aging.

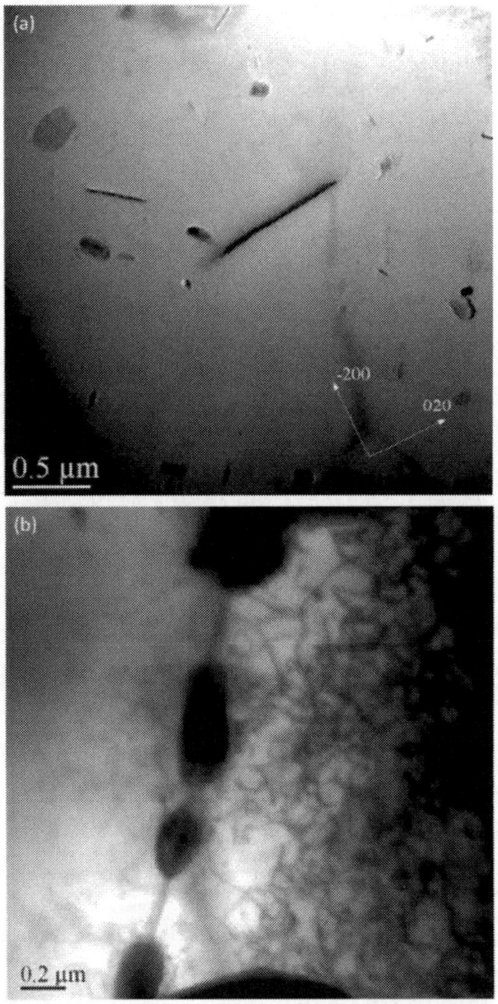

Fig. 4. TEM bright field images of the AlSi7MnMg HPVD sample isothermally treated at 350 °C for 300 s. Equilibrium β-Mg_2Si precipitated in the aluminum matrix (a) and along the grain boundary (b).

3.4. Phase transformation kinetics

Avrami kinetics equation, $\psi = 1 - \exp(-kt^n)$, is generally applied to describe the kinetics of a phase transformation [13, 14], and here, the phase transformation refers to large Mg_2Si particles precipitated from the supersaturated aluminum matrix, which would not contribute the alloy strengthening during aging treatment. Where ψ is the volume fraction of the precipitation at the time t, k is the Avrami constant depending on the nucleation rate and the growth rate, and n is the Avrami exponent which is depending on the nucleation mechanism.

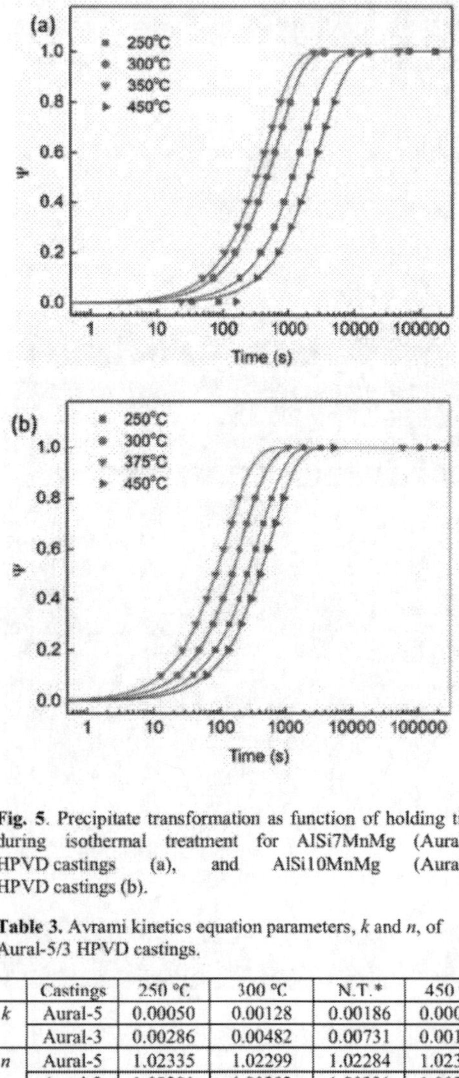

Fig. 5. Precipitate transformation as function of holding time during isothermal treatment for AlSi7MnMg (Aural-5) HPVD castings (a), and AlSi10MnMg (Aural-3) HPVD castings (b).

Table 3. Avrami kinetics equation parameters, *k* and *n*, of Aural-5/3 HPVD castings.

	Castings	250 °C	300 °C	N.T.*	450 °C
k	Aural-5	0.00050	0.00128	0.00186	0.00027
	Aural-3	0.00286	0.00482	0.00731	0.00176
n	Aural-5	1.02335	1.02299	1.02284	1.02362
	Aural-3	1.00201	1.00202	1.00331	1.00365

*N.T. - the nose temperature: 350 °C for Aural-5; 375 °C for Aural-3.

The precipitated volume fraction ψ at the time t can be obtained from the TTP curves in Fig. 2, which (the marks) are shown in Fig. 5a, and the relative experimental points can be also found in Fig. 2. These marks would be fitted by an iterative non-linear fitting procedure with Avrami kinetics equation. The phase transformation diagram, fitted time-dependent volume fraction of the precipitates, of AlSi7MnMg HPVD castings at 250 °C, 300 °C, 350 °C and 450 °C is illustrated in Fig. 5a, and Avrami constant and exponent, k and n, are listed in Table 3. As shown in Fig. 5a, the phase transformation rate ($d\psi/dt$) is low at the beginning of isothermal treatment, but the rate then increases greatly and reaches a high value after an approximately 0.2 volume fraction of precipitates. However, the phase transformation rate decreases gradually when the volume fraction of the transformed phases is more than 0.8. The values of k of AlSi7MnMg, listed in Table 3, vary with the isothermal temperatures. The sequence of the k value is that k at 350 °C (N.T.) > at 300 °C > at 250 °C > at 450 °C. Thus, the phase transformation rate at the nose temperature is larger than that at lower or higher temperatures because a large value of k is relative to a rapid phase transformation rate [14]. Moreover, the shortest time to perform a certain phase transformation in the castings is at 350 °C. For example, the time to perform 60% precipitate transformation is approximately 350 s at 350 °C (nose temperature), 500 s at 300 °C, 1500 s at 250 °C, and 2000 s at 450 °C, respectively. It further demonstrates that the maximum phase transformation rate in the isothermal treatment is at the nose temperature. The average phase transformation rate from 0.2 and 0.8 volume fraction is around $1\text{x}10^{-3}$ s^{-1} (volume fraction per second) at the nose temperature, 350 °C, for AlSi7MnMg.

The phase transformation diagram of AlSi10MnMg HPVD castings is determined based on the TTP diagram in the reference [8], as shown in Fig. 5b. And the values of k and n are listed in Table 3, respectively. The phase transformation tendency in AlSi10MnMg is similar to that in AlSi7MnMg. The k value at 375°C is larger than that at 300 °C, 250 °C and 450 °C, respectively. Compared with AlSi7MnMg, AlSi10MnMg has much larger k values, and at their nose temperatures the k value of AlSi10MnMg is about 4 times that of AlSi7MnMg. Moreover, the time to perform a certain amount of precipitate transformation for AlSi10MnMg is much shorter than that for AlSi7MnMg. For example, to perform the transformation of 60% precipitates, AlSi10MnMg spends approximately 100 s at 375 °C, 180 s at 300 °C, 250 s at 250 °C, and 500 s at 450 °C, respectively. The average phase transformation rate in the range from 0.2 to 0.8 volume fraction is $3.8\text{x}10^{-3}$ s^{-1}, at the nose temperature, 375 °C. The average phase transformation rate of AlSi10MnMg is much higher than that of AlSi7MnMg.

There are two kinds of precipitation mechanisms in phase transformation, one is diffusion-controlled process, and the other is nucleation-controlled process [14]. The former occurs at the temperature lower than the nose temperature, while the latter conducts at the temperature higher than the nose temperature. In diffusion-controlled precipitation, when the value of Avrami exponent n is approximately 1, precipitates have the morphologies of thickening of long cylinder or needle-like shapes whereas in nucleation-controlled precipitation, the precipitates are mainly formed at grain boundaries [14]. Therefore, according to the values of n in Tables 3 and 4, the needle-like precipitates would be present in AlSiMnMg alloys when the phase transformation at a temperature lower than the nose temperature, but if the phase transformation occurs at a temperature higher than the nose temperature, the precipitates would be formed along grain boundaries. Actually, Fig. 4 and the reference [8] show the precipitate morphology and distribution as described above.

3.5. Comparison of quench sensitivity and phase transformation kinetics between two alloys

AlSi7MnMg and AlSi10MnMg are two most widely used Al-Si-Mg alloys for high pressure die castings. Fig. 6 illustrates TTP contours of 99.5% and 90% of maximal hardness for two alloys. And the characteristics of the TTP contours are listed in Table 4. The critical time at the nose temperature in 99.5% of maximal hardness for AlSi7MnMg is longer than that for AlSi10MnMg. And the nose temperature of the former is lower than that of the latter. Moreover, the quench sensitive temperature range of the former is 10 °C narrower than that of the later. Similar phenomena were observed in the contours of 90% maximal hardness. In addition, the critical cooling rate of AlSi7MnMg is 6 °C/s, but the rate of AlSi10MnMg is 20 °C/s. Hence, the quench sensitivity of AlSi7MnMg is lower than that of AlSi10MnMg.

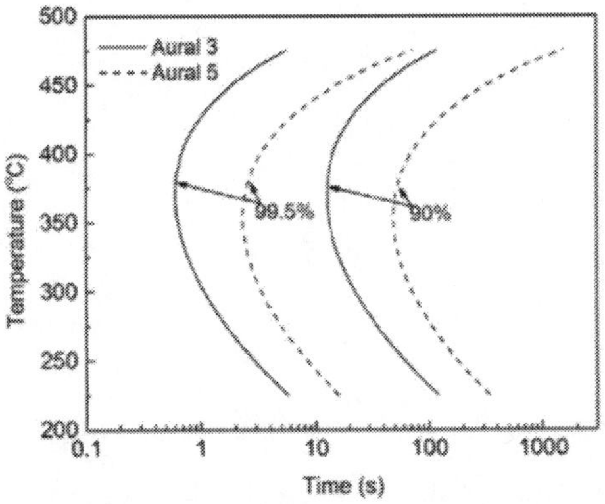

Fig. 6. TTP contours of 99.5% and 90% of maximum hardness of AlSi7MnMg (Aural™-5, dash line) and AlSi10MnMg (Aural™-3, solid line [8]).

Table 4. Parameters of 99.5% maximal hardness contours of Aural-3/5 HPVD castings.

Castings	C.T.*	N.T.**	T.R***
Aural-5	2.4 s	350 °C	260-430 °C
Aural-3 [8]	0.6 s	375 °C	270-450 °C

*C.T.: the critical time at the nose temperature; **N.T.: the nose temperature; ***T.R.: the quench sensitive temperature range.

According to the precipitate transformation kinetics of the two castings (Fig. 5), the average phase transformation rate of AlSi7MnMg is much slower than that of AlSi10MnMg. Moreover, the values of Avrami constant, k, corresponding to the nucleation rate and the growth rate, of AlSi7MnMg are much less than that of AlSi10MnMg at different isothermal temperatures.

Generally, the phase transformation and quench sensitivity are mainly determined by nucleation and growth of precipitates during heat treatment [14]. The concentration of magnesium (0.20 wt.%) and silicon (7.4 wt.%) in AlSi7MnMg is lower than that (0.55 wt.% Mg, 10.1 wt.% Si) in AlSi10MnMg, respectively (see Table 1). The silicon maximal solubility in aluminum at 500 °C is 0.8 wt.%. It is evident that magnesium solute atoms in AlSi7MnMg matrix is lower than that in AlSi10MnMg matrix due to the difference in magnesium concentration between the two alloys. Consequently, the driving force for precipitate nucleation in AlSi7MnMg is lower than that in AlSi10MnMg because the driving force for precipitate nucleation is proportional to the concentration of the solutes supersaturated in the matrix [14]. Moreover, based on results of EBSD grain orientation maps listed in Table 5, the grain size of AlSi7MnMg is larger than that of AlSi10MnMg, whereas the length of grain boundaries in the former is much shorter than that in the latter. Therefore, nucleation sites for precipitates in AlSi7MnMg would be less than that in AlSi10MnMg because the grain boundaries provided potential sites for the precipitate nucleation (see Fig. 4b). In addition, grain boundaries supply fast solute diffusion channels, hence, they facilitate the growth of phases during heat treatment.

Table 5. Grain size and grain boundary length of HPVD castings.

Castings	A.E.G.S.*	A.G.B.L**
Aural-5	10.3 μm	6.2 μm/μm²
Aural-3 [8]	4.4 μm	11.3 μm/μm²

*A.E.G.S: Average equivalent grain diameter; ** A.G.B.L: Average grain boundary length

Furthermore, according to the calculated coefficients of TTP curves of the two alloys (see Table 2), AlSi7MnMg exhibits higher values of k_2 (related to the reciprocal of the number of nucleation sites), k_3 (corresponding to the nucleus energy), and k_5 (related to the activation energy for diffusion), than AlSi10MnMg. It implies that the potential nucleation sites in AlSi7MnMg are less than that in AlSi10MnMg, and the nucleation and solute diffusion in the latter are easier and faster than that in the former. This further demonstrates that the nucleation and phase growth in AlSi7MnMg are more difficult and slower than that in AlSi10MnMg. Therefore, compared with AlSi10MnMg, the lower concentration of Mg and Si in AlSi7MnMg results in a lower nucleation driving force and larger grain size, consequently the quench sensitivity and phase transformation rate of AlSi7MnMg are lower than that of AlSi10MnMg.

4. Conclusions

1. The TTT and TTP diagrams of AlSi7MnMg (Aural™-5) HPVD castings were constructed. The nose temperature is 350 °C and the quench sensitive temperature range is 260-430 °C. Moreover, the coefficients, k_2-k_5, related to the nucleation and precipitation kinetics of TTP curves, were determined.

2. The cooling rates versus hardness of AlSi7MnMg HPVD castings were predicted and verified by the experimental results. The critical cooling rate of the HPVD castings is 6 °C/s. When the cooling rate is higher than 6 °C/s during quenching, more than 95% of the maximal mechanical strength can be obtained.

3. The phase transformation diagrams of AlSi7MnMg and AlSi10MnMg (Aural™-3) alloys were determined, and the Avrami constant *k*, and exponent *n*, of the two alloys were determined. The *k* values of AlSi7MnMg are much lower than that of AlSi10MnMg. The average phase transformation rate of the former is $1x10^{-3}$ s^{-1}, while the rate of the latter is $3.8x10^{-3}$ s^{-1} in the range from 0.2 to 0.8 phase volume fraction at the nose temperatures.

4. AlSi7MnMg exhibits lower quench sensitivity and slower phase transformation rate than AlSi10MnMg. It is mainly due to lower Mg and Si concentration in AlSi7MnMg.

Acknowledgments

The authors would like to acknowledge the financial support of the Natural Sciences and Engineering Research Council of Canada (NSERC) and Rio Tinto Aluminum through the Research Chair in the Metallurgy of Aluminum Transformation at University of Quebec at Chicoutimi.

References

1. Rio Tinto. Aluminium - Your Guide to Automotive Innovation. Available online: https://www.riotinto.com/documents/Aluminium_Automotive_innovation_brochure.pdf

2. M. Hartlieb, Die Cast. Eng., **57**, 40–43, (2013)

3. X.P. Niu, B.H. Hu, I. Pinwill, H. Li, (2000). J. Mater. Process. Technol. **105**, 119-127, (2000).

4. H. Wang, Y. Yi, S. Huang, J. Alloy Compd. **690**, 446–452, (2017).

5. J.T. Staley, Mater. Sci. Technol. **3**, 923-935, (1987).

6. D.L. Zhang, L. Zheng, (1996). Metall. Mater. Trans. A, **27**, 3983-3991, (1996).

7. S. Ma, M. Maniruzzaman, D. MacKenzie, R. Sisson, Metall. Mater. Trans. B, **38**, 583–589, (2007).

8. M.Y. Liu, Z. Zhang, F. Breton, X.G. Chen, Materials, **12**, 1-13, (2019).

9. J.W. Evancho, J.T. Staley, Metall. Trans. **5**(1), 43, (1974).

10. S. Liu, Q. Zhong, Y. Zhang, W. Liu, X. Zhang, Y. Deng, Mater. Des. **31**, 3116-3120, (2010).

11. G.E. Totten, G.M. Webster, E.C. Bates, In: Bains T, MacKenzie DS, editors. *Proceedings of the first international non-ferrous processing and technology conference*, 305–313, (1997).

12. S.D. Liu, X.M. Zhang, J.H. You, Z.B. Huang, C. Zhang, X.Y. Zhang, J. Nonferrous Met., **16**, 2034, (2006).

13. M. Avrami, J. Chem. Phys, **7**, 1103-1112, (1939).

14. D.A. Porter, K.E. Easterling, M.Y. Sherif, *Phase transformations in metals and alloys* (CRC Press, Boca Raton, FL, USA, 2009).

Formation of intermetallic phases during solidification in Al-Mg-Mn 5xxx alloys with various Mg levels

Ahmed Y. Algendy[1], *Kun Liu*[1]*, *X.-Grant Chen*[1]
[1]Department of Applied Science, University of Quebec at Chicoutimi, Saguenay (QC), Canada

Abstract. In the present study, four Al-Mg-Mn 5xxx alloys with different Mg levels (2-5 wt.%) were investigated for better understanding the evolution of intermetallic phases formed during solidification. Optical and scanning electron microscopes, electron backscattered diffraction and differential scanning calorimetry analyses in combination with thermodynamic calculation were used to identify various intermetallic phases. Results showed that the most dominant intermetallic phases are $Al_6(Mn,Fe)$, α-Al(Fe,Mn)Si, Al_3Fe, $Al_m(Mn,Fe)$ and Mg_2Si in experimental Al-Mg-Mn alloys, which is greatly dependant on the Mg levels. It is found that Chinese script α-Al(Fe,Mn)Si is the dominant iron-rich intermetallic phase for the alloys containing 2-3 wt.% Mg, while blocky $Al_6(Mn,Fe)$ and needle-like $Al_3(Mn,Fe)$ become the major phases for the alloy containing 4 wt.% Mg. Further increasing Mg content to 5 wt. %, the dominant phase transfers to blocky $Al_6(Mn,Fe)$ intermetallic. Meanwhile, the morphology of primary Mg_2Si is changed from well-branched to plate-like with increasing Mg contents. In addition, β-Al_3Mg_2 and τ-Al_6CuMg_4 eutectic phases have been observed in the alloys with 3-5 wt. % Mg. A comparison on various intermetallic phases from the Scheil simulation and the actual as-cast microstructure is provided.

Keywords: Al-Mg 5xxx alloys; Intermetallic phases; Mg_2Si; Solidification.

1 Introduction

Commercial Al-Mg-Mn 5xxx wrought alloys offer an attractive combination of high strength, good formability, weldability and excellent corrosion resistance, making them suitable for a wide spectrum of applications, such as structures, automotive industries and marine applications. [1-2]. During the design and manufacture processes of 5xxx alloys, the microstructure plays a significant role on the alloy properties, especially the type and distribution of intermetallic phases formed during solidification, which can be greatly affected by chemical compositions. Various researches have been performed to investigate the microstructure and phase identification during the solidification in 5xxx aluminum alloys. The main intermetallic phases formed during solidification of this series were reported to be $Al_6(Mn,Fe)$, α-Al(Fe,Mn)Si, Al_3Fe and Al_m(Mn, Fe) [3–11]. Alloys with 3% Mg content or higher display the formation of β-Al_5Mg_3 or τ-Al_6CuMg_4 [3]. Simensen *et al.* [4] proposed that the main intermetallic phases formed during solidification of Al-Mg-Mn AA5182 alloy are $Al_6(Mn,Fe)$, $Al_3(Mn,Fe,Cu)$, $Al_8(Mg,Cu)_5$, and Mg_2Si, while it is reported that the dominant iron-bearing particles are $Al_m(Fe,Mn)$ and $Al_3(Fe,Mn)$ instead of Al_6(Fe, Mn) and $Al_3(Fe,Mn)$ [5]. Engler *et al.* reported [8] that the as-cast microstructure of AA5454 alloy consists of two major different types of constituents $Al_6(Fe,Mn)$ and Mg_2Si with occasionally observed eutectic β-Al_5Mg_3. The solidified microstructure of AA5083 alloy is found to be composed of two main iron-rich phases, which are $Al_6(Mn,Fe)$, α-Al(Fe,Mn)Si with minor ε-$Al_{18}(Cr,Mn)_2Mg_3$ and eutectic β-Al_5Mg_3 or τ-Al_6CuMg_4 [6]. Studies showed that the dominant Fe-rich phases of the as-cast AA5754 alloy are changed to $Al_3(Fe,Mn)$ and α-Al(Fe,Mn)Si, and only $Al_3(Fe,Mn)$ phase with Fe:Mn ratio of about 5 in AA5182 alloy [7, 10]. However, most of literatures focused on the alloys with a high Mg range (4-5%), such as 5083, 5182 and 5457 alloys, and there are also controversies on the formation of different intermetallic phases.

Therefore, it is important to better understand the formation of various intermetallic phases formed during solidification in Al-Mg-Mn 5xxx alloys, especially at a wider range of Mg levels. In this study, the formation of intermetallic phases in Al-Mg-Mn 5xxx with various Mg contents (2-5 wt.%) were systematically studied, focusing on the identification and characterization using the combined analytic methods of optical microscope (OM), scanning electron microscope (SEM) equipped with an energy dispersive X-ray spectrometer (EDS) and electron backscattered diffraction (EBSD).

2 Methodology

Four experimental Al-Mg-Mn alloys with different Mg levels were prepared with commercial pure Al (99.8 wt.%), pure Mg (99.9 wt.%), Al-50 wt.% Si, Al-50 wt.% Cu, Al-25 wt.% Mn, Al-25 wt.% Fe, Al-20 wt.% Cr, and Al-5wt.%Ti-1wt.%B master alloys. Their chemical compositions

* Corresponding author: kun.liu@uqac.ca

analyzed using an optical emission spectrometer are shown in Table 1. For each batch, approx. 3 kg of material was melted in clay-graphite crucible in an electric resistance furnace. The temperature was maintained at 780 °C for 30 min after melting followed by degassing for 15 min. Then the melt was poured into a permanent steel mold preheated at 250 °C. The dimensions of the cast ingots were 30 mm × 40 mm × 80 mm.

Table 1: Chemical Compositions of Experimental Alloys

Alloys	Elements, wt.%						
	Mg	Mn	Si	Fe	Cu	Cr	Ti
M2	2	0.81	0.28	0.3	0.1	0.15	0.09
M3	3.3	0.81	0.28	0.25	0.1	0.16	0.09
M4	4	0.85	0.26	0.29	0.1	0.17	0.06
M5	5.2	0.78	0.26	0.31	0.1	0.18	0.1

After casting, the samples were taken from the center of the ingot, grinded and polished. OM, SEM with EDS as well as EBSD and the Channel 5 software were used to identify various intermetallic phases formed during solidification. Meanwhile, Thermo-Calc software was used to predict the intermetallic phases formation as a function of Mg contents in both the equilibrium and non equilibrium states, while a Differential Scanning Calorimeter (DSC, Perkin Elmer DSC 8000) was used to confirm the intermetallic formation. DSC samples were taken from the as-cast ingot and heated with a heating rate of 10 °C/min up to 750 °C.

3 Results and discussion

3.1 Thermo-Calc simulation

The phase diagram calculated from Thermo-Calc as a function of Mg contents with the actual chemical compositions of experimental alloys is shown in Fig. 1a. It can be observed that the most important intermetallic constituents involve Al_7Cr, Al_6Mn, α-Al(Fe,Mn)Si and Mg_2Si at the equilibrium condition. The dominant Fe-rich intermetallic phase at the low Mg level is α-Al(Fe,Mn)Si ($Al_{15}Si_2M_4$), while it is transferred to Al_6Mn with increasing Mg level. In addition, τ-phase (Al_6CuMg_4) and β-phase (Al_5Mg_3) can be formed at relatively low temperature. In order to simulate the non-equilibrium solidification, which is closer to our experiment condition, the solidification paths of four alloys are simulated with Scheil model, and the results of two alloys with 2% Mg (M2) and 5% Mg (M5) are shown in Fig. 1b and c. The Scheil simulations indicate the formation sequence of different types of intermetallic phases during solidification, including Al_7Cr, Al_6Mn, Al_3Fe ($Al_{13}Fe_4$), Mg_2Si and finally, τ-phase and β-phase. As shown in Fig. 1b and c, increasing the Mg content not only affect the formation temperature of intermetallic phases but also change the type and sequence of phases.

In M2 alloy (Fig. 1b), the solidification starts at 648 °C by the formation of α-Al dendrites (Reaction 1). Afterwards, Al_6Mn phase precipitates at 637 °C (Reaction 2) as α-Al dendrites continue to grow. Similarly, Al_3Fe ($Al_{13}Fe_4$) and α-Al(Fe,Mn)Si ($Al_{15}Si_2M_4$) phases start to precipitate at 608 °C and 604 °C (Reaction 3 and 4), respectively, followed by the precipitation of Mg_2Si at 580 °C (Reaction 5). At the near end of the solidification τ-phase (Reaction 7) may occur. On the other hand, in M5 alloy (Fig. 1c), the solidification starts by the precipitation of primary metastable Al_7Cr phase at temperature of 670 °C (Reaction 1) followed by the formation of α-Al dendrites (Reaction 2). Al_6Mn phase precipitates with decreasing temperature at 628 °C (Reaction 3) as the growth of α-Al dendrites continues, while the primary metastable Al_7Cr formed at higher temperature disappears with decreasing the solidification temperature. Afterwards, the precipitations of Al_3Fe ($Al_{13}Fe_4$) (Reaction 4) and Mg_2Si (Reaction 5) occur at 590 °C and 565 °C, respectively. Low-melting point τ-phase (Reaction 6) and β-phase (Reaction 7) appear at relatively low temperatures of 452 °C and 446 °C, respectively. Therefore, it can be found that Mg level has influence on the formation of intermetallic. It is observed that Al_6Mn, Al_3Fe ($Al_{13}Fe_4$) and α-Al(Fe,Mn)Si ($Al_{15}Si_2M_4$) are the main Fe-rich phases for M2 alloy as shown in Fig. 1b, while Al_6Mn and Al_3Fe ($Al_{13}Fe_4$) are important Fe-rich phases with increasing Mg content (M5 alloy Fig. 1c).

Fig. 2 shows the results of the thermodynamic simulations (Scheil model) in the form of the mass fraction of the stable phases as a function of temperature of alloys M2 and M5. It indicated that Al_6(Mn,Fe) formed as a major phase in both alloys. Further decrease the temperature a minor fraction of $Al_{13}Fe_4$ (Al_3Fe) and Mg_2Si start to form. Alloy M5, (Fig. 2b) shows the formation of β-phase at temperature below 446 °C where the solubility limit of Mg is reached. Notably, at relatively low temperature, the software accuracy lost, explaining the sharp increase of β-phase in alloy M5 at 446 °C. The majority and minority phases of experimental alloys compared with simulated results will be discussed later.

3.2 Intermetallic phases in the as-cast structure

As shown in Figs 1-2, a variety of intermetallic phases existed in Al-Mg-Mn alloys, therefore all intermetallic phases formed during solidification of the experimental alloys are examined and identified. Fig. 3 shows the as-cast microstructure of M2 alloy containing 2% Mg, which is consisted of α-Al dendrites as the matrix surrounded by two intermetallic phases with different morphology and contrast (Fig. 3a). Intermetallic particles that appear grey in the SEM (Fig. 3b) seem to contain heavy elements such as Fe, Mn and Cr, while the phase containing light elements such as Mg and Si displays dark. SEM-EDS analysis has been employed to verify the chemical compositions of different phases and the results are shown in Table 2. The majority of intermetallic phases with Chinese script morphology is α-Al(Fe,Cr,Mn)Si (see the marks of a1 and a2), while the dark branched phase is identified as Mg_2Si (marked as a3).

Fig. 4 displays the distribution of intermetallics in M3 alloy containing 3.3% Mg. Three different intermetallic constituents with different contrast are observed, as shown in Fig. 4a and 4b. The irregular bright phase is identified as α-Al(Fe,Mn)Si (marked as b1), while the grayish particles contain higher amount of Cu and Mg is classified as τ-Al_6CuMg_4 (see b2) [6–8], which is often found to be close to the black intermetallic Mg_2Si (see b3).

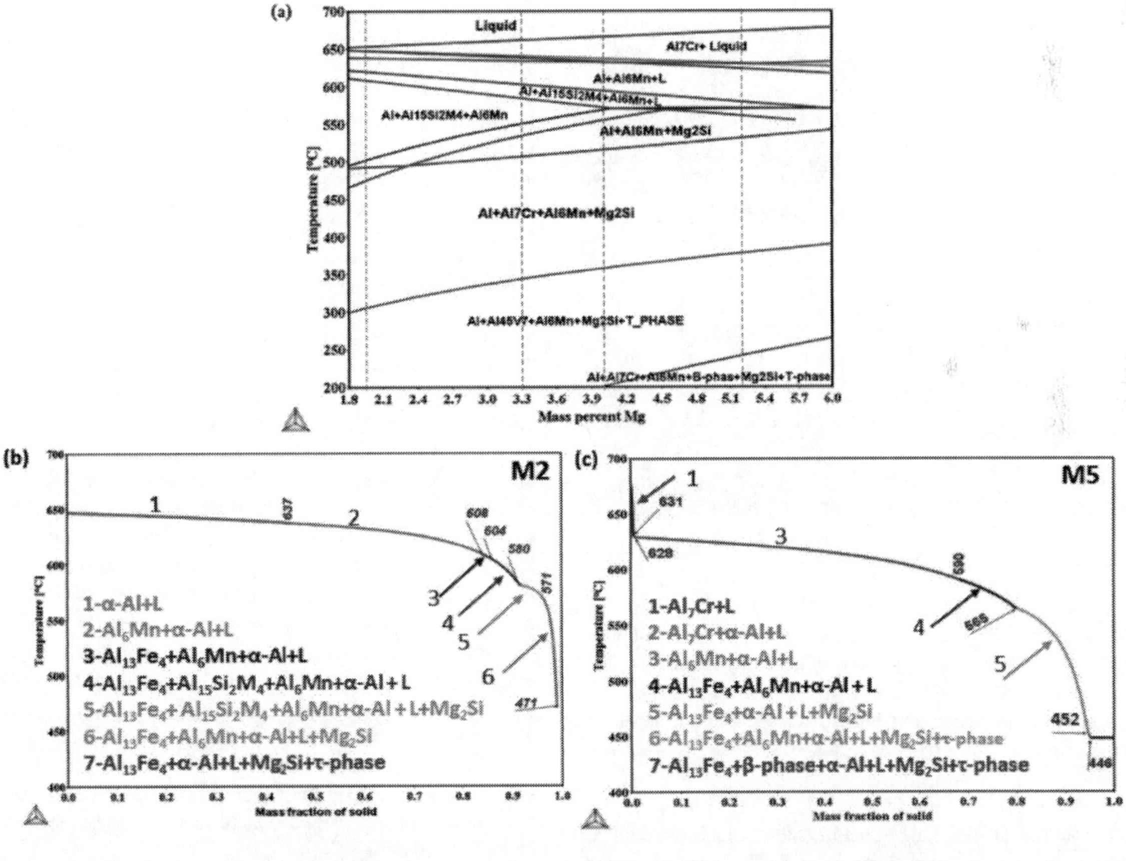

Fig. 1. Thermodynamic simulation for Al-Mg 5xxx alloys (a) The equilibrium phase diagram of Al-Mg-Mn alloy and (b and c) Scheil simulations showing the solidification sequence of different phases for M2 (b) and M5 (c) alloys.

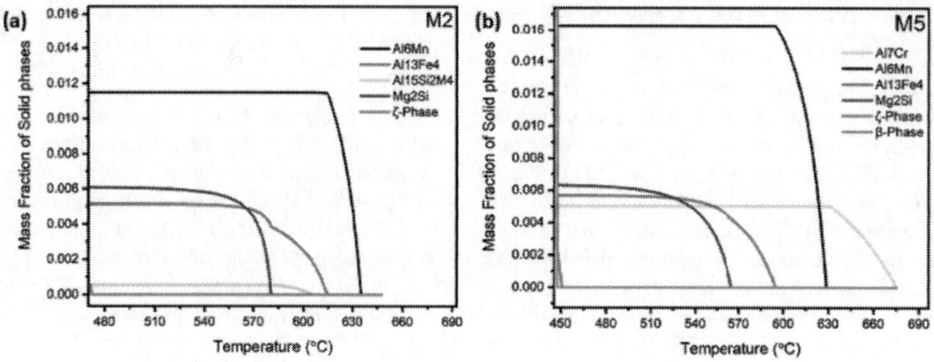

Fig. 2. Thermodynamic simulation from scheil model showing the mass fraction of different phases as a function of temperature of alloys M2 and M5.

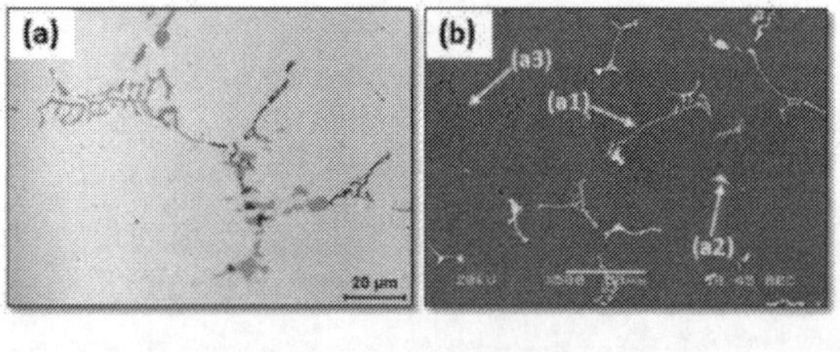

Fig. 3. As-cast microstructure of M2 alloy

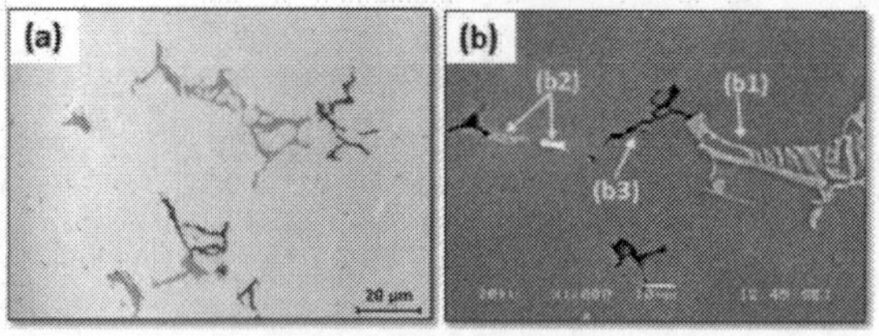

Fig. 4. As-cast microstructure of M3 alloy

The as-cast microstructure of M4 alloy containing 4% Mg is shown in Fig. 5, which is quite different with M2 and M3 alloys. Two needle-like phases with two different color (gray and dark) are observed in Fig. 5a together with a Chinese script phase and blocky phase. SEM images of Fig. 5b-e show the details of typical intermetallics observed in M4 alloy. The bulky bright particles in Fig. 5b are found to be comprised of Mn and Fe without Si, which can be classified as $Al_6(Mn,Fe)$ phase (marked as c1). Intermetallic particles that exhibit branched and Chinese script morphology with higher Si are identified as α-Al(Fe,Mn)Si (marked as c2). These two phases are further confirmed by the EBSD analysis (shown later) since their morphology are similar but with different crystal structure [7, 13]. Furthermore, the lighter phase with elongated long rod-shape was observed to be comprised of Mn and Fe but free of Si, and it is identified as $Al_3(Mn,Fe)$ (see c3). Occasionally, a minor Cr-rich phase with blocky shape were detected with high Mn, Cr and Mg contents but without Fe and Si, which is identified as ε-$Al_{18}(Cr,Mn)_2Mg_3$ (see c4). The other grayish intermetallic contains high amount of Cu and Mg may be classified as τ-Al_6CuMg_4 (see c5), while the dark particles observed with needle, flakes-like morphology are characterized as Mg_2Si (marked as c6).

The as-cast microstructure of M5 alloy is shown in Fig. 6. The major grey particles containing Mn and Fe but without Si with compacted morphology are proposed to be $Al_6(Mn,Fe)$ and $Al_m(Mn,Fe)$ (marked as d1), more details on these two phases will be discussed later using EBSD analysis. A minor irregular Fe-rich phase with high Si is found to be α-Al(Fe,Mn)Si (marked as d2). Occasionally, a Cr-rich phase with blocky shape is detected with relatively high Mn, Cr and Mg content but without Fe and Si, which is identified as ε-$Al_{18}(Cr,Mn)_2Mg_3$ (see. d3). In addition, bright/grayish phase is found to be close to Mg_2Si with high Cu and Mg, which can be classified as Cu-containing eutectic phase τ-Al_6CuMg_4 (see d4). Occasionally, a minor grey phase with rod shapes is observed to have high Cr and Ti and it is believed to be $Al_7(Cr,Ti)$ (see d6), which formed as a primary phase at the beginning of solidification at relatively high temperature (Reaction 1 in Fig. 1c). Finally, the dark phases with plate-like morphology are characterised as Mg_2Si (see d5).

As shown in Figs. 3-6 the formation of low eutectic phases β-Al_5Mg_3 or τ-Al_6CuMg_4 can be observed in the alloys with 3% Mg content or higher, such as M5 in Fig. 6. Besides, various Fe-rich intermetallics are observed in

https://doi.org/10.1051/matecconf/202032602002

experimental alloys. In the present study, DSC is further applied to confirm the formation of various intermetallic phases. As shown in Fig. 7 (Peak 1), a small exothermic peak is observed in the temperature range 447-449 °C, which is in the same range of β-phase formation [3] for alloy containing 5% Mg content. On the other hand, there is no peak detected in alloy with lower Mg (i.e. 2% Mg). Notably in Fig 1c, the simulation results for the formation temperatures of τ-and β-phases are 452 °C and 446 °C, respectively for alloy M5, which fall in the similar range observed by DSC in Fig. 6, confirming the formation of these phases in alloys with relatively high Mg content, as shown in Figs. 3b, 4e and 5d. Besides, it is found that the peak formation of Mg_2Si (Peak 2) shifts from 587 °C in M2 to 558 °C in M5. In addition, the formation temperature of Fe-intermetallic (Peak 3) is also different in two alloy, confirming the different dominant Fe-intermetallic in M2 and M5 (Fig. 3 and Fig. 6). In addition, the formation temperature of α-Al also changes from 647 °C in M2 to 631 °C in M5, which is corresponding to the Scheil simulation (Fig. 1b and Fig. 1c)

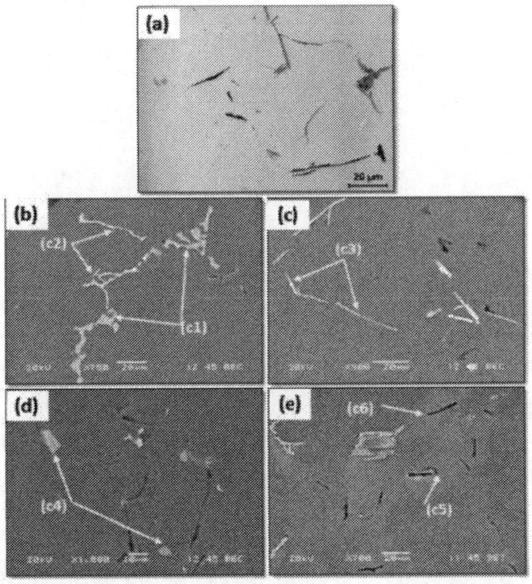

Fig. 5. Optical micrograph (a) and SEM images (b-e) showing different intermetallic phases in M4 alloy.

As shown in Figs. 2-6, the various intermetallic phases in alloys with different Mg levels have been studied and identified based on the color contrast, morphology and chemical composition (Table 2). Due to some similarities on the chemical composition and the morphology of intermetallic phases, especially on the Fe-rich phases, hence, the Fe-rich phases have been further confirmed by EBSD, which is based on the crystal structure of phase. As shown in Table 3, the various Fe-rich phases exhibit different crystal structure and lattice parameter [13], and therefore, collecting crystallography information from EBSD can improve the phase identification. As examples, the identification of Fe-rich phases in M4 and M5 alloys, in which the formation of Fe-rich is more complicated, are shown in Fig. 8 and 9. For each sample, more than 20 particles are randomly selected and measured.

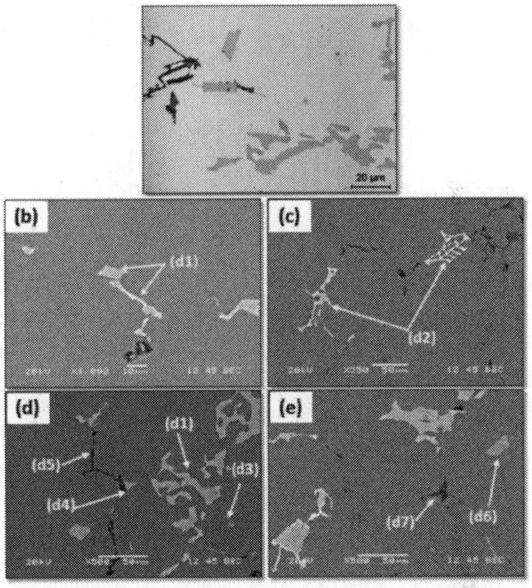

Fig. 6. Optical micrograph (a) and SEM Images (b-e) of M5 alloy showing various intermetallic phases.

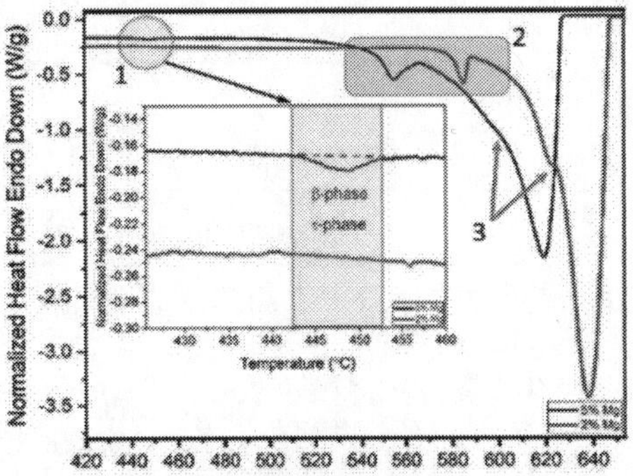

Fig. 7. DSC heating curves for two experimental alloy with different Mg levels, showing the formation of β-phase in the temperature range 443-450°C

Table 2. SEM-EDS results for various intermetallic phases found in as-cast microstructure of experimental alloys

Alloy ID		Intermetallic phases	Element, wt.%							
			Mn	Fe	Si	Mg	Cu	Cr	Ti	Al
M2	a1	α-Al(Fe,Mn)Si	10.3	13.8	5.9	0	0	1.6	0	78.9
	a2	α-Al(Fe,Mn)Si	7.8	11.4	5.8	0	0	0	0	83.6
	a3	Mg_2Si	0	0	0	20.2	16.3	0	0	63.5
M3	b1	α-Al(Fe,Mn)Si	8.3	8.9	5.2	1.2	0	0	0	76.3
	b3	τ-Al_6CuMg_4	0	0	0	23	16.7	0	0	60.3
	b2	Mg_2Si	0	0	3.7	10.3	0	0	0	86.1
M4	c1	Al_6(Mn,Fe)	11.4	11.3	0	0	0	0	0	77.3
	c2	α-Al(Fe,Mn)Si	9.7	11.4	4.01	1.3	0	0	0	73.5
	c3	Al_3(Fe,Mn)	9.2	8.9	0	3.2	0	0	0	78.7
	c4	ε-$Al_{18}(Cr,Mn)_2Mg_3$	5.2	0	0	10.1	0	6.7	0	78.1
	c5	τ-Al_6CuMg_4	0	0	0	17.8	9.5	0	0	72.7
	c6	Mg_2si	0	0	4	11.4	0	0	0	84.6
M5	d1	Al_m(Fe,Mn)	8.9	12.1	0	0	0	0	0	79.0
	d1	Al_6(Mn,Fe)	11.6	11.4	0	0	0	0	0	77.1
	d2	α-Al(Fe,Mn)Si	11.6	11.2	4.4	0	0	0	0	72.8
	d3	ε-$Al_{18}(Cr,Mn)_2Mg_3$	11.3	0	0	11.1	0	4.6	0	73
	d4	τ-Al_6CuMg_4	0	0	0	20.5	12.9	0	0	66.6
	d5	Mg_2si	0	0	15.8	18.4	0	0	0	65.8
	d6	Al_7(Cr,Ti)	0	0	0	11.0	0	7.9	5.8	75.2
	d7	β-Al_5Mg_3	0	0	0	13.3	0	0	0	86.8

Table 3 Crystallographic information of iron-rich intermetallics [13]

Phases	Crystal structure	Crystal parameters
α-Fe	body-centered cubic	a = b = c = 1.265 nm
β-Fe	tetragonal	a = 0.643 nm, b = 0.746 nm, c = 0.878 nm
Al_6Fe	Orthorhombic	a = 0.643 nm, b = 0.746 nm, c = 0.878 nm
Al_mFe	body-centered tetragonal	a = b = 0.884 nm, c = 2.16 nm
Al_3Fe	Monoclinic	a = 1.549 nm, b = 0.808 nm, c = 1.248 nm

https://doi.org/10.1051/matecconf/202032602002

Figs. 8 and 9 display the typical Fe-rich intermetallic phases found in M4 and M5 alloys with their corresponding patterns and simulated results. The mean angular deviation (MAD) between the experimental and calculated patterns indicates the accuracy of the solution given by the software (Channel 5), in which a smaller value of MAD means a higher accuracy between the experimental and simulated patterns. For a desirable accurate solution, the MAD value should be at least lowly than 0.7.

As shown in Fig. 8 (a1, a2 and a3), the MAD values for α-Al(Fe,Mn)Si, $Al_6(Mn,Fe)$ and $Al_3(Mn,Fe)$ are 0.351, 0.446 and 0.115, respectively, which is much lower than the critical value 0.7. Consequently, those three Fe-rich intermetallic phases in M4 alloy have been confirmed with the combined use of EBSD and SEM-EDS. Similarly, the formation of three Fe-rich intermetallics in M5 alloy, e.g. $Al_6(Mn,Fe)$, $Al_m(Mn,Fe)$ and α-Al(Fe,Mn), have also been confirmed by EBSD analysis with low MAD values, as shown in Fig. 9 (b1, b2 and b3).

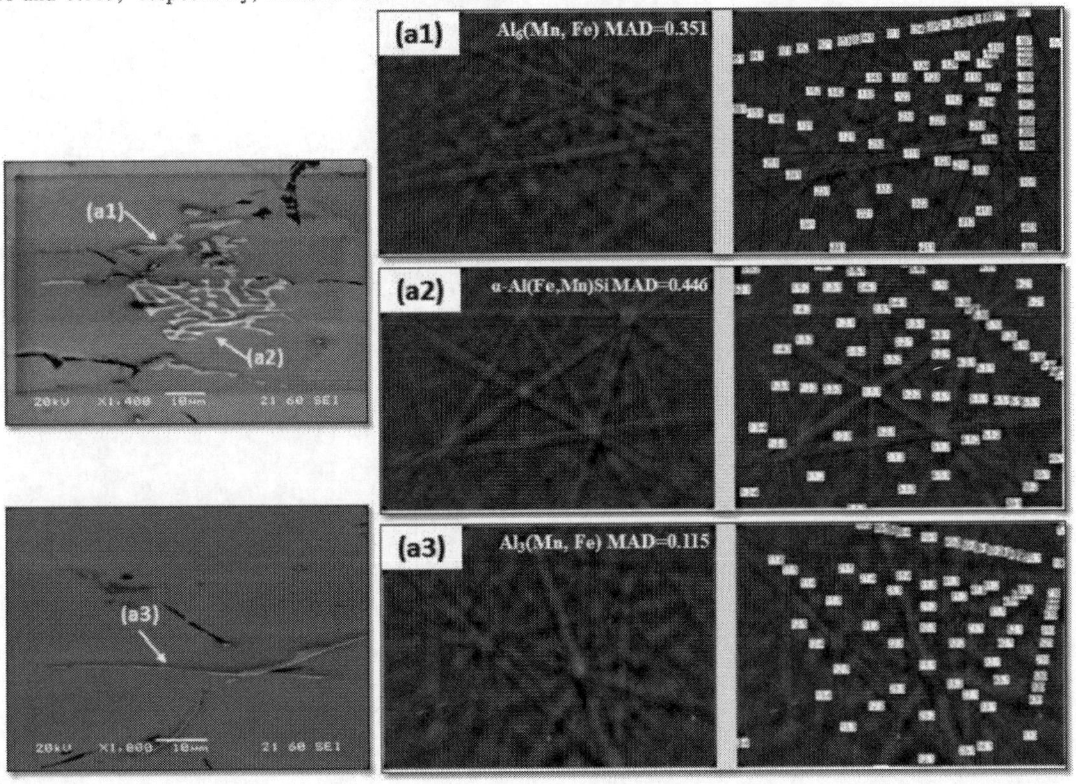

Fig. 8. SEM Micrographs and EBSD patterns of M4 alloy showing different intermetallic phases with corresponding patterns and simulated results

3.4 Discussion

As shown in Figs. 2-7 from the SEM and EBSD analysis (Figs. 8-9), it is confirmed that Al_6Mn, Al_mFe, α-Al(Fe,Mn)Si and Al_3Fe are the common Fe-rich phases present in the experimental Al–Mg–Mn alloys with various Mg levels. Generally, the solidification rate and chemical composition will have important influence on the as-cast microstructure. For instance, the higher solidification velocity and the presence of grain refinement help the formation of the metastable Al_mFe and Al_6Fe phases over Al_3Fe [11, 12]. In the present study, all the samples were taken from the center of the ingot where the solidification rate is almost the same.

Therefore, the formation of various intermetallic in experimental alloys is likely to be attributed to the chemical composition, especially the Mg levels. In the as-cast microstructure, the α-Al(Fe,Mn)Si becomes a dominant phase in the low Mg alloys (M2 and M3), while Al_6Mn, Al_3Fe, α-Al(Fe,Mn)Si are the common Fe-rich phases with increasing the Mg level in M4 alloy.

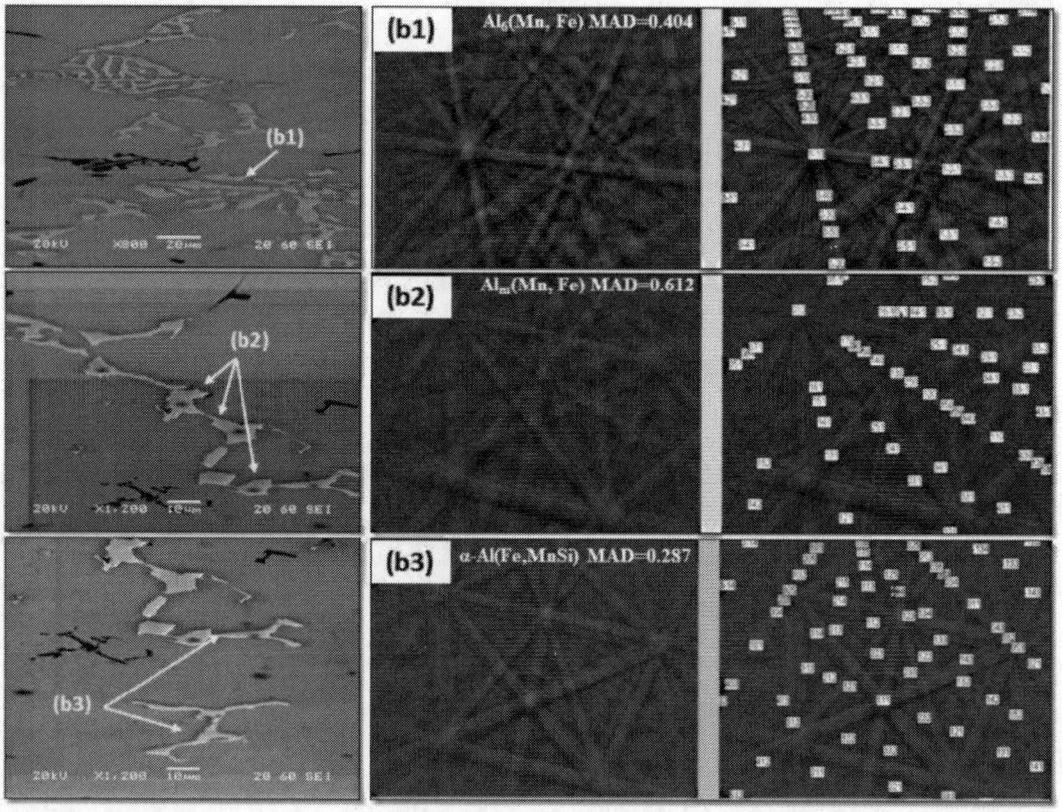

Fig. 9. SEM Micrographs and EBSD patterns of M5 alloy showing different intermetallic phases with corresponding patterns and simulated results.

Further increase in Mg content in M5 alloy, Al_6Mn become the major iron-rich phases in addition to a minor fraction of Al_mFe and α-Al(Fe,Mn)Si, which could be related to Mg content and the Si/Mg ratio.

In addition, it is also observed the morphology of primary Mg_2Si changed from well-branched to flake-like shape by increasing the Mg content, which may be due to the solidification behavior of the experimental alloys. As shown in Fig. 7 and Figs. 1b&c, increasing Mg content shifts the formation temperature of eutectic phases toward lower value, which can affect the growth rate, and the concentration of primary Mg_2Si.

A comparison on the formation of various intermetallic phases from Scheil simulation and real as-cast microstructure is provided in Table 4. In general, the results of Scheil simulation give an appropriate sequence of different intermetallic formed during solidification, in which Fe-rich intermetallic phases with different types precipitate first followed by Mg_2Si and finally with the low meting point eutectic phases.

As shown in Table 4, the Scheil simulations predicted that Al_6Mn is the major intermetallic phase for M2 and M3 alloys in addition to two minor Fe-rich phases of α-Al(Fe,Mn)Si and Al_3Fe, while α-Al(Fe,Mn)Si was found to be the dominant Fe-rich intermetallic phase in experimental as-cast microstructure. It is reported that the higher Si level in Al-Mg alloys encourage the transformation of the Fe-rich intermetallic from $Al_6(Fe,Mn)$ to α-Al(Fe,Mn)Si [9], suggesting that a relatively high Si/Mg ratio in low Mg alloys (M2 and M3) promotes the formation of α-Al(Fe,Mn)Si over Al_6Mn. On the other hand, the Scheil model predicted that $Al_6(Fe,Mn)$ is the major intermetallic phase with a minor Al_3Fe intermetallic in the high Mg alloys (M4 and M5), which is in a good agreement with the microscopic observation of as-cast microstructure in the corresponding two experimental alloys. The low Si/Mg ratio in combination of high Mn might stabilize the formation of $Al_6(Fe,Mn)$ over Al_3Fe. The presence of a minor Al_mFe in the as-cast microstructure of M5 alloy may be related to the higher Mg content or the presence of TiB_2 in the grain refiner, which act as a heterogeneous nucleation site for Al_mFe particles [5, 14-15].

Table 4 Summary and comparison on various intermetallic phases from Scheil simulation and real as-cast microstructure

Alloy	Scheil simulation		Experimental	
	Majority	Minority	Majority	Minority
M2	Al_6Mn	α-Al(Fe,Mn)Si ($Al_{15}Si_2M_4$) Al_3Fe ($Al_{13}Fe_4$) Mg_2Si τ-phase	α-Al(Fe,Mn)Si	Mg_2Si
M3	Al_6Mn	Al_3Fe ($Al_{13}Fe_4$) Mg_2Si β-phase τ-phase	α-Al(Fe,Mn)Si	Mg_2Si τ-phase
M4	Al_6Mn	Al_3Fe ($Al_{13}Fe_4$) Mg_2Si β-phase τ-phase	$Al_6(Fe,Mn)$ $Al_3(Fe,Mn)$	α-Al(Fe,Mn)Si Mg_2Si ε-phase τ-phase
M5	Al_6Mn	Al_3Fe ($Al_{13}Fe_4$) Mg_2Si β-phase τ-phase	Al_6Mn	$Al_7(Cr,Ti)$ α-Al(Fe,Mn)Si Al_mFe Mg_2Si ε-phase τ-phase β-phase

4. Conclusions

1) Intermetallic phases formed during solidification were identified using SEM-EDS and EBSD techniques in the present work. $Al_6(Mn,Fe)$, α-Al(Fe,Mn)Si, Al_3Fe, $Al_m(Mn,Fe)$ and Mg_2Si are the most common intermetallic phases in the as-cast microstructure of Al-Mg-Mn alloys depending on the Mg levels.
2) α-Al(Fe,Mn)Si with Chinese script morphology is the dominant Fe-rich intermetallic phase for the alloys containing 2-3 wt.% Mg, while the major intermetallic phases are $Al_6(Mn,Fe)$ with blocky morphology and $Al_3(Mn,Fe)$ with needle-like morphology in the alloy containing 4 wt.% Mg. Further increasing Mg content to 5 wt. %, $Al_6(Mn,Fe)$ becomes the dominant intermetallic phase with a minor $Al_m(Mn,Fe)$ phase.
3) Two low melting point eutectic phases of β-Al_3Mg_2 and τ-Al_6CuMg_4 are observed in the alloys with 3-5 wt. % Mg.
4) The morphology of Mg_2Si change from well-branched to plate-like morphology by increasing the Mg contents.
5) The comparison on the formation of various intermetallic phases from Scheil simulation and real as-cast microstructure is provided.

The authors would like to acknowledge the financial support of the Natural Sciences and Engineering Research Council of Canada (NSERC) and Rio Tinto Aluminum for their financial support through the Research Chair in the Metallurgy of Aluminum Transformation at University of Quebec in Chicoutimi.

References

1. W. S. Miller, L. Zhuang, J. Bottema, A. J. Wittebrood, P. De Smet, A. Haszler, A. Vieregge, Mater. Sci. Eng. A, **vol. 280**, no. 1, pp. 37–49, (2000).
2. J. Kauman, ASM Int., Ch. 6, (2000).
3. L. Bäckerud, E. Król, J. Tamminen, **vol. 2,** pp. 113–123, (1986).
4. C. J. Simensen and U. Södervall, Surf. Interface Anal., **vol. 30**, no. 1, pp. 309–314, (2000).
5. Y. J. Li and L. Arnberg, *Acta Mater.*, **vol. 52**, no. 9, pp. 2673–2681, 2004.
6. O. Engler and S. Miller-Jupp, J. Alloys Compd., **vol. 689**, pp. 998–1010, (2016).
7. O. Engler, K. Kuhnke, and J. Hasenclever, J. Alloys Compd., **vol. 728**, pp. 669–681, (2017).
8. O. Engler, Z. Liu, and K. Kuhnke, J. Alloys Compd., **vol. 560**, pp. 111–122, (2013).
9. Y. Liu, L. Luo, C. Han, L. Ou, J. Wang, and C. Liu, J. Mater. Sci. Technol., **vol. 32**, no. 4, pp. 305–312, (2016).
10. Z. F. Sundaram Kumar, H. B. Nadendla, G. M. Scamans, D. G. Eskin, stm media köthen, **vol. 28**, pp. 5–11, (2012).
11. Z. Zhang, G. Li X.-G. Chen, 13th International Conference on Aluminum Alloys. pp. 1857-1863, (2012).
12. C. M. Allen, K. A. Q. O'Reilly, B. Cantor, and P. V. Evans, Prog. Mater. Sci., **vol. 43**, no. 2, pp. 89–170, (1998).
13. K. Liu, X. Cao, X. G. Chen, Metall. and Mat. Trans. A, Phys. Metall. Mater. Sci, **vol. 44,** no 2, pp. 682-695, (2013).
14. W. T. Denholm, J. D. Esdaile, N. G. Siviour, and B. W. Wilson, Metall. Trans. A, Phys. Metall. Mater. Sci., **vol. 18 A**, no. 3, pp. 393–397, (1987).
15. M. W. Meredith, J. Worth, and R. G. Hamerton, Mater. Sci. Forum, **vol. 396–402**, no. 1, pp. 107–112, (2002).

Influence of transition elements (V, Zr and Mo) and cooling rate on the precipitation of dispersoids in Al-7Si-0.6Cu-0.35Mg foundry alloy

Dong Li[1], *Kun* Liu[1,*], and *X-Grant* Chen[1]
[1]Department of applied sciences, University of Quebec at Chicoutimi, Saguenay, (Quebec), Canada, G7H 2B1

Abstract. In the present work, individual/combined additions of transition elements (V, Zr and Mo) were introduced into Al-7Si-0.6Cu-0.35Mg foundry alloy at different cooling rates to study their influence on the precipitation behaviour of dispersoids. Results showed that both individual and combined additions of V, Zr, Mo lead to the formation of dispersoids but with different composition, morphology and number density during solution treatment. The addition of V produces the precipitation of both $(Al,Si)_3M$ dispersoids and α-dispersoids, while the Zr addition promotes $(Al,Si)_3M$ type dispersoids but inhibits the formation of α-Al(Mn,Fe)Si dispersoids. The addition of Mo effectively promotes α-Al(Mn,Mo,Fe)Si dispersoids and significantly reduces the dispersoid size and increase the number density of dispersoids. The combined addition of V, Zr and Mo produces the largest number of finer dispersoids among all five alloys studied, but the most dispersoids are $(Al,Si)_3M$. The $(Al,Si)_3M$ dispersoids and α-dispersoids have the rod-like and block-like morphologies, respectively. High cooling rate can generally refine the dispersoids and increase their number density, while it also increases the proportion of $(Al,Si)_3M$ dispersoids.

1 Introduction

Since commercially replacing cast iron with Al-Si foundry alloys in automotive engine applications, the weight of automobiles has been significantly reduced and consequently with better fuel efficiency. However, the mechanical properties of Al-Si alloys will significantly decrease with increasing temperature (above 200 °C) owing to the coarsening of conventional strengthening precipitates (mainly β'-Mg_2Si and θ'-Al_2Cu) [1, 2]. Therefore, how to improve the elevated-temperature properties of Al-Si foundry alloys is of primary importance for their sustainable development. On the other hand, the dispersion strengthening of thermally stable dispersoids has been proved to be one of the best strengthening ways for Al-Si alloys at elevated temperature [1, 3, 4].

According to the Lifshitz and Slyozov and Wagner theory, the ideal dispersed phase should have low interfacial energy (low lattice mismatch with matrix), low diffusivity and solubility limit, to prevent coarsening from bulk diffusion at the high temperature [5]. Since the transition elements like V, Zr and Mo have much lower diffusion rate (4.85×10^{-24} m^2 s^{-1}, 1.20×10^{-20} m^2 s^{-1} and 5.52×10^{-23} m^2 s^{-1} at 400°C respectively) in the α-Al matrix than the common used dispersoid initiator of Mn (6.24×10^{-19} m^2 s^{-1} at 400 °C) [6], the dispersoids containing these elements are expected to have better coarsening resistance.

The strengthening by nanoscale $L1_2$ metastable Al_3M type precipitates that can maintain their coherency with the matrix at elevated temperatures is often applied in aluminum wrought alloys [5-7]. However, the existence of Al_3Zr ($L1_2$) in Al-Si foundry alloys with high Si level (>3 wt.%) has rarely been observed [8-12]. It is widely accepted that the addition of Zr in Al-Si alloys leads to the formation of equilibrium rod-shaped $(Al,Si)_3Zr$ ($D0_{23}$) dispersoids [13, 14].

V is an extraordinarily sluggish diffuser in the α-Al matrix. Although Al_3V is not considered to be in equilibrium with α-Al [6], the $Al_3V_{1-x}M_x$ trialuminide has low lattice mismatch with the α-Al matrix and thereby potentially being thermally stable [15]. It is reported that when Zr, V, and Ti were simultaneously added into Al7Si1Cu0.5Mg alloy, the rod-shaped Zr,V,Ti-containing dispersoids precipitated[12]. V is also reported to be able to promote the precipitation of α-Al(Mn,V,Fe)Si dispersoids by substituting Mn of α-Al(Mn, Fe)Si [16, 17]. Therefore, it is expected that $(Al,Si)_3M$ and α-Al(Mn,Fe)Si dispersoids with modification of V are thermally stable.

It is recently discovered that Mo is a supreme dispersoid former in both aluminum foundry and wrought alloys [1, 3, 4, 18]. The combination of Mn and Mo can lead to a uniform distribution of partially coherent α-Al(Mn,Mo,Fe)Si dispersoids and reduce the dispersoid free zone, because of their opposite solid-liquid partition coefficient during solidification [18].

* Corresponding author: kun.liu@uqac.ca

High cooling rate can refine the as-cast microstructure but increase the area fraction of eutectics in Al-Mg-Si alloys due to the greater segregation effect of the solute atoms [19]. It can also reduce the size of intermetallic particles in Al-Si-Cu-Mg casting alloys [20]. High cooling rate with fine dendrite arm spacing generally results in high strengths and elongation as well as improved fatigue properties [20, 21].

As reviewed above, although several works have been done to study the effect of individual transition element on the microstructures and properties, the influence of combined V, Zr and Mo as well as the cooling rate on the dispersoids precipitation in Al-Si foundry alloys has not been systematically investigated. Therefore, the aim of the present work is to investigate the effects of individual and combined additions of V, Zr and Mo and cooling rate on cast microstructure and the precipitation behavior of dispersoids in Al-7Si-0.6Cu-0.35Mg alloys.

2 Experimental procedure

Five Al-Si-Cu-Mg foundry alloys with different V, Zr and Mo contents were prepared with commercial pure Al (99.6%), pure Mg (99.9%), Al-50%Si, Al-25%Mn, Al-5%Ti-1%B, Al-50%Cu, Al-10%Sr, Al-5%V, Al-15%Zr and Al-10%Mo master alloys. Their chemical compositions analyzed by an optical emission spectrometer are shown in Table 1. For each alloy, approximately 3.5 kg of materials were melted in a clay-graphite crucible using an electric resistance furnace. The liquid metal was maintained at approximately 750 °C for 30 min and then degassed for 15 min with high purity Ar. All five batches were grain-refined by adding the Al-5%Ti-1%B master alloy and modified by the addition of Al-10%Sr master alloy. The liquid metal was poured into the permanent mold preheated at 250 °C. In the present work, two different permanent molds were used to test the effect of the cooling rate on the evolution of microstructure; one is permanent steel mold with a low cooling rate (~ 2°C/s) and the other is permanent copper mold with a high cooling rate (~10 °C/s). The sample from the steel mold was "Y" block shape with a dimension of 80mm×40mm×30mm, which was designated as "Y" while the sample from the copper mold was the thin plate with a dimension of 100mm×80mm×4.5mm and designated as "S" in the present work. After casting, all samples were subjected a two-step solution treatment (500°C/2h + 520°C/2~12h) to evaluate the formation of dispersoids according to the literatures [22, 23].

Table 1. Chemical composition of experimental alloys (wt.%)

No.	Si	Fe	Mg	Mn	Ti	Cu	Sr	V	Zr	Mo	Al
0	7.3	0.13	0.36	0.18	0.19	0.61	0.01	-	-	-	Bal.
1	7.4	0.15	0.35	0.20	0.20	0.60	0.01	**0.28**	-	-	Bal.
2	7.2	0.13	0.35	0.18	0.19	0.59	0.01	-	**0.23**	-	Bal.
3	7.4	0.14	0.35	0.20	0.19	0.60	0.01	-	-	**0.31**	Bal.
4	6.8	0.12	0.33	0.20	0.19	0.60	0.01	**0.20**	**0.20**	**0.28**	Bal.

To investigate the evolution of microstructures, such as the formation of intermetallics and dispersoids, an optical microscope (OM), a scanning electron microscope (SEM) equipped with energy dispersive spectroscopy (EDS) and electron backscatter diffraction (EBSD) and a transmission electron microscope (TEM) operated at 200 kV were employed.

In addition, Vickers microhardness were measured to evaluate the room-temperature properties. The tests were conducted at room temperature by NG-1000 CCD micro-hardness test machine with a 10g load and a 20s dwelling time. 10 measurements were performed in the α-Al matrix to calculate the average hardness value for each sample. Electrical conductivity was measured by Sigmascope SMP10 electrical conductivity test device at room temperature, with 10 measurements for each sample.

3 Results and Discussion

3.1 As-cast and solutionized microstructures

In the present work, the as-cast microstructures of experimental alloys are observed to be composed of dendritic α-Al, eutectic silicon and various intermetallic phases, which exhibit a variety of types and morphologies and highly depend on the addition of transition elements. The high cooling rate (S-samples) produced finer microstructures including the second arm spacing (SDAS), Si particles and intermetallics, compared to the low cooling rate (Y-samples). The as-cast S samples have the average SDAS of 20 μm, while the average SDAS of Y samples is 35 μm. As an example, the typical as-cast microstructures of #4 alloy under two cooling rates are exemplarily given in Fig. 1, which displays the various intermetallics formed due to the combined addition of transition elements as well as the finer microstructures in S-sample (Fig. 1b).

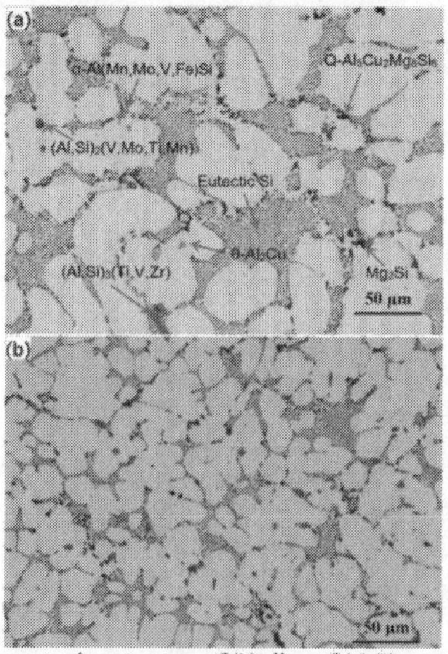

Fig. 1. As-cast microstructure of #4 alloy of (a) Y-sample and (b) S-sample.

The #0 base alloy has a major α-Al(Mn,Fe)Si intermetallic phase and a trace $(Al,Si)_3Ti$ intermetallic due to presence of Ti by the grain refinement addition. In V-added #1 alloy, in addition to the major α-Al(Mn,V,Fe)Si intermetallic, a $(Al,Si)_2(V,Ti,Mn)$ phase were found as suggested by [17, 24] with some Si substituted by Al and some V substituted by Ti and Mn on the basis of Si_2V phases because of their similar atomic radius [25]. Similarly, in Mo-added #3 alloy, some (Mo,Si)-rich $(Al,Si)_2(Mo,Ti,Mn)$ intermetallic particles were also found derived from $Mo(Al,Si)_2$ phase [26, 27]. In #4 alloy with the combined addition of V, Zr and Mo, three main intermetallic phases with complicate chemical composition were found, which are α-Al(Mn,Mo,V,Fe)Si phase, plate-like $(Al,Si)_3(Ti,V,Zr)$ and hexagonal $(Al,Si)_2(V,Mo,Ti,Mn)$. It should be mentioned that all five alloys contain Mg_2Si, and two Cu-containing θ-Al_2Cu and Q-$Al_5Cu_2Mg_8Si_6$ phases due to the presence of Mg, Si and Cu in experimental alloys. The details of various intermetallic phases found and identified in as-cast microstructures of five alloys are summarized in Table 2.

Table 2. Intermetallic phases in the as-cast microstructure of different alloys

Alloys	Intermetallic phases	
#0	α-Al(Mn,Fe)Si, $(Al,Si)_3Ti$	Mg_2Si, θ-Al_2Cu, Q-$Al_5Cu_2Mg_8Si_6$
#1	α-Al(Mn,V,Fe)Si, $(Al,Si)_3(Ti,V)$, $(Al,Si)_2(V,Ti,Mn)$	
#2	α-Al(Mn,Fe)Si, $(Al,Si)_3(Ti,Zr)$	
#3	α-Al(Mn,Mo,Fe)Si, $(Al,Si)_3Ti$, $(Al,Si)_2(Mo,Ti,Mn)$	
#4	α-Al(Mn,Mo,V,Fe)Si, $(Al,Si)_3(Ti,V,Zr)$, $(Al,Si)_2(V,Mo,Ti,Mn)$	

During two-step solution treatment, the eutectic Si particles were spheroidized, while all three low meting point Mg/Cu-containing intermetallic phases, including Mg_2Si, θ-Al_2Cu and Q-$Al_5Cu_2Mg_8Si_6$, were fully dissolved into the α-Al matrix. In general, the Fe/Ti/V/Zr/Mo-containing intermetallic phases (in the left column of Table 2) were the high melting point phases. They may be fragmented and partially dissolved during solution treatment at 520 °C but most of them remain in the microstructure after solution treatment. Fig. 2 gives an example of as-solutionized microstructure of the #4-Y sample, in which the α-Al(Mn,Mo,V,Fe)Si, $(Al,Si)_3(Ti,V,Zr)$ and $(Al,Si)_2(V,Mo,Ti,Mn)$ phases formed during solidification were still observed after solution treatment.

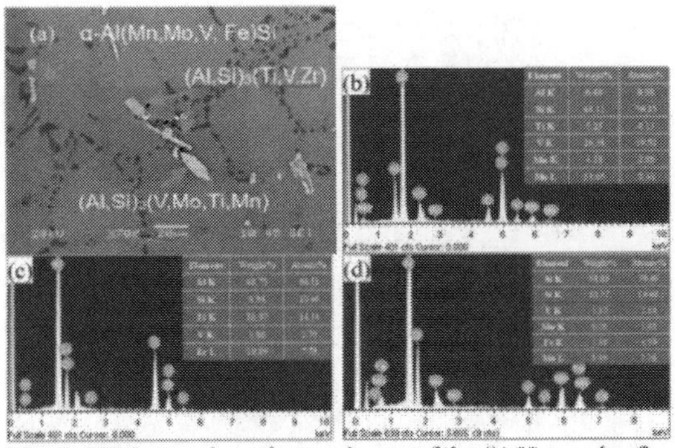

Fig. 2. SEM secondary electron image of the #4-Y sample after 520 °C solution treatment for 4 h (a) with EDS analysis results of intermetallics: (b) $(Al,Si)_2(V,Mo,Ti,Mn)$; (c) $(Al,Si)_3(Ti,V,Zr)$; (d) α-Al(Mn,Mo,V,Fe)Si.

The total volume fractions of remaining intermetallic phases after solution treatment in different alloys were measured to evaluate the effects of different solute elements and cooling rate on their formation, and results are shown in Fig. 3. It can be found that the #0 base alloy has the lowest volume fractions of intermetallics at both cooling rates. With the addition of transition elements, the volume fractions of intermetallics increase, while the high cooling rate reduces the formation of intermetallics and more transition elements can be retained in the α-Al matrix. It is evident that the #4 alloy with combined addition possesses the highest volume fraction of intermetallics, which has more than doubled intermetallic fraction relative to the base alloy. The Zr-added #2 alloy has the second highest volume fraction of intermetallics among all alloys but the largest intermetallics fraction among three individual transition element added alloys (#1 to #3) due to the low solute ability of Zr in Al.

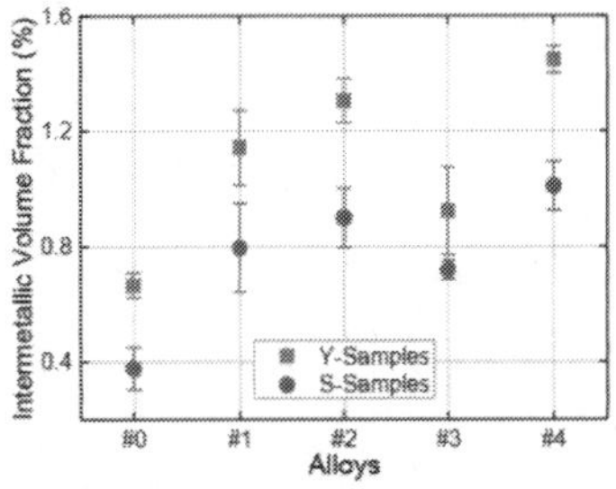

Fig. 3. Volume fraction of remained intermetallic phases after 520°C/4h solution treatment in various alloys

3.2 Formation of dispersoids during solution treatment

In Al-Si foundry alloys, the dispersoids usually form during solution treatment [1, 4, 18, 23]. To monitor the precipitation of dispersoids during solution treatment and to compare the hardening effect of the transition elements, Vickers microhardness (HV) and electrical conductivity (EC) as a function of holding time were measured and the results are shown in Fig. 4.

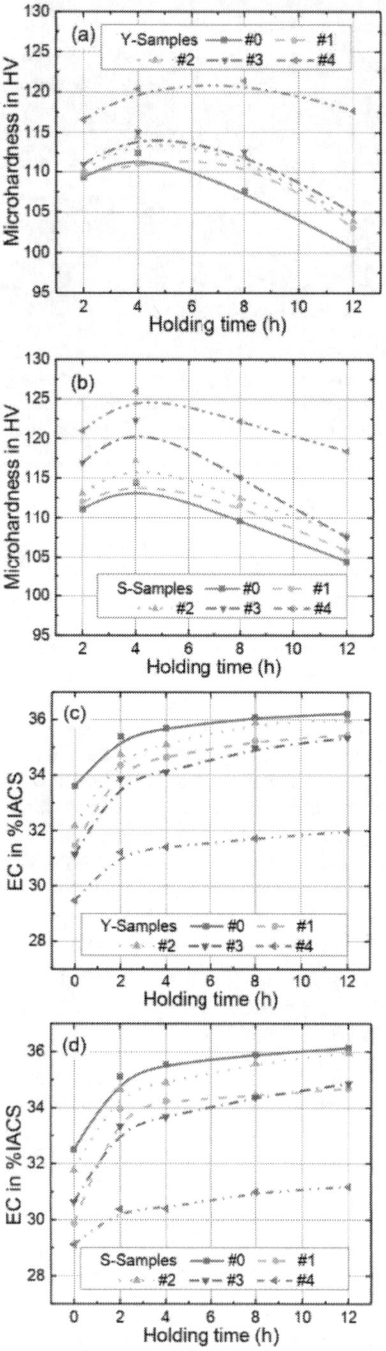

Fig. 4. Vickers microhardness (a, b) and electrical conductivity (c, d) as a function of holding time during solution treatment at 520 °C in different alloys

Except of the consumption in the intermetallic phases, all dispersoid-forming alloy elements (Mn, Fe, Si, Zr, V and Mo) are supersaturated in the as-cast microstructure. During solution treatment and with increasing holding time, both microhardness and EC gradually increase, indicating the dispersoids precipitate from the supersaturated aluminum matrix. At both cooling rates, the peak microhardness is reached after 4 h holding time (Fig. 4a-b), while the EC also approaches the maximum value (Fig. 4c-d), suggesting that the dispersoids are fully precipitated after 520°C/4h; thereby the 520°C/4h solution treatment is chosen for further metallographic investigation of the dispersoids. With prolonged holing time after 4 h, the values of microhardness of all five alloys exhibit a decreasing tendency, while the EC values reach a plateau, which is most likely attributed to that the dispersoids start to coarsen with prolonged holding time at 520 °C.

The microhardness differences between different alloys are mainly ascribed to the amount of precipitated dispersoids and their strengthening effect. As shown in Fig. 4a-b, the #4 alloy with combined addition of V, Zr and Mo has the highest microhardness among all of the alloys at any given holding time. The Mo-added #3 alloy has an obvious improvement, showing the second highest microhardness. The #1 and #2 alloys show only a slight improvement compared to the #0 alloy. In addition, the high cooling rate (S-samples) always has the higher microhardness compared to the low cooling rate (Y-samples) for each alloy, suggesting the higher amount of precipitated dispersoids due to the higher supersaturated level at the high cooling rate condition.

The dispersoids usually precipitate within dendrite cells (the dispersoid zone (DZ) in the intradendritic region). In the outside of the dispersoid zone, it forms the dispersoid free zone (DFZ), which locates closely to the dendrite boundaries and intermetallic particles due to the depleted alloying solutes at the area. Fig. 5 shows optical images giving an overview of the distribution of the dispersoid zones and dispersoid free zones after solution treatment in the #0, #1 and #3 alloys as examples. All of the samples are etched with 0.5% HF for 90 s in order to reveal the DZs and DFZs clearly. It can be found the volume of DZ is lower but with higher DFZ in #0 compared with #1 and #3 alloys while #3 has the maximum volume of DZ.

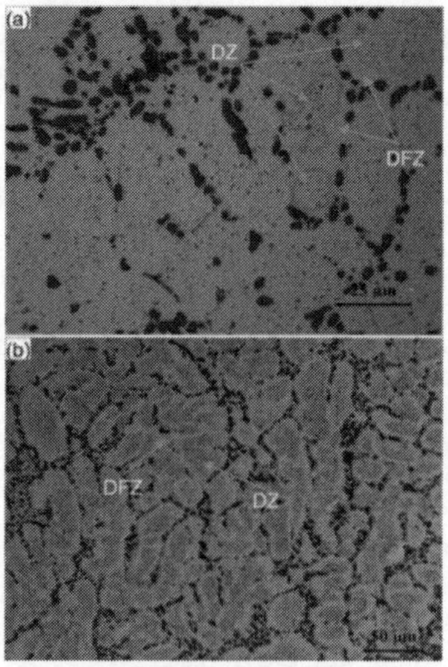

Fig. 5. Dispersoid zone (DZ) and dispersoid free zone (DFZ) formed during solution treatment (520°C/4h) (etched with 0.5%HF for 90s): (a) #0-S; (b) #1-S; (c) #3-S.

The measured area fractions of DZs in different alloys are shown in Table 3. Based on the measurements, it is clear to see that the cooling rates have limited influence on the fraction of DZ, while the added transition elements have strong effects on it, which can be related to the inverse of their partition coefficients (k_0 (Mn, Fe and Si) < 1 while k_0 (Ti, V, Zr and Mo) > 1 [6, 17, 18]). Therefore, the formation of DZ was increased with transition elements. The maximum DZ fraction was found in #3 alloy, around 73% with both molds, indicating that Mo is a strong DFZ eliminating element for Al-Si-Cu-Mg foundry alloys. The #1, #2 and #4 alloys have similar DZ fractions which are smaller than that of #3 alloy, because of their relatively larger DFZs around the dendrite boundaries, as we can see in Fig. 4b-c. However, they absolutely have larger DZs compared to #0 alloy whose DZs are even hard to be identified, as shown in Fig. 5a.

Table 3. Area fractions of DZs in solutionized alloys

DZ Fraction (%)	#0	#1	#2	#3	#4
Y-Samples	41±5	64±6	64±8	74±5	66±7
S-Samples	44±4	62±7	65±6	73±3	68±7

To study the details of the dispersoids (type, morphology, size and number density), TEM investigation is carried out for each alloy condition, and the distribution of dispersoids in DZ of experimental alloys is shown in Fig. 6.

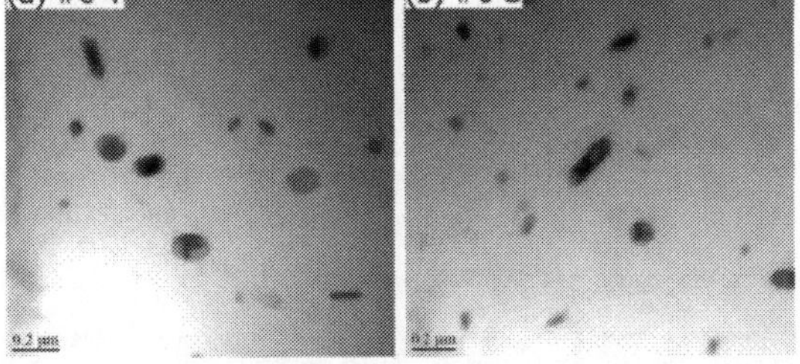

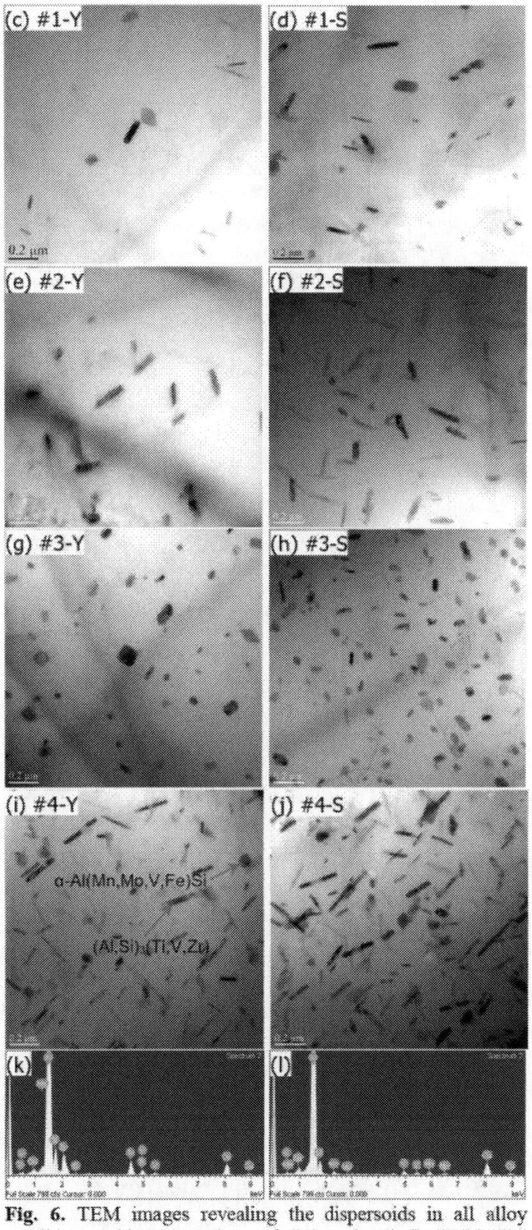

Fig. 6. TEM images revealing the dispersoids in all alloy conditions with Y-Samples and S-Samples (a-j); TEM-EDS results of typical $(Al,Si)_3(Ti,V,Zr)$ and α-Al(Mn,Mo,V,Fe)Si dispersoids (k-l).

To help the identification of the dispersoids, TEM-EDS was used to analyze the chemical compositions of different dispersoid types. In general, the dispersoids found under TEM can be divided into two groups: α-Al(Mn,Fe)Si type dispersoids with block-like morphology and $(Al,Si)_3M$ type dispersoids with rod/plate-like morphology and their distribution in experimental alloys are summarized in Table 4. The morphology difference between these two kinds of dispersoids may be explained by their respective cubic and $D0_{22}/D0_{23}$ tetragonal crystal structures [13, 18, 28, 29]. In addition, the overall equivalent diameter and number densities of the dispersoids were quantitatively analyzed based on the image analysis on TEM images, with ratios of the 2 types of dispersoids empirically estimated based on the TEM-EDS results in each alloy condition, as showed in Fig. 7.

Table 4. Dispersoids observed in TEM

Alloys	Dispersoids
#0	α-Al(Mn,Fe)Si
#1	α-Al(Mn,V,Fe)Si, $(Al,Si)_3(Ti,V)$
#2	$(Al,Si)_3(Ti,Zr)$*, α-Al(Mn,Fe)Si
#3	α-Al(Mn,Mo,Fe)Si
#4	$(Al,Si)_3(Ti,V,Zr)$*, α-Al(Mn,Mo,V, Fe)Si,

* Dominant dispersoids

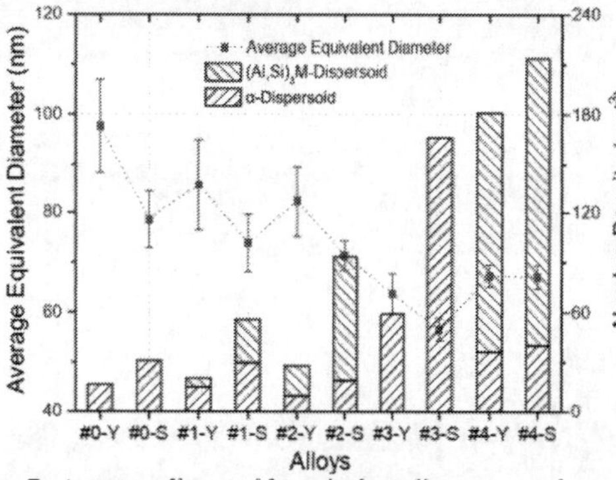

Fig. 7. Average dispersoid equivalent diameters and number densities in different alloys and molds

In the #0 base alloys, a number of block-like dispersoids occurred (Fig. 6a-b), which are identified as α-Al(Mn,Fe)Si, due to the presence of Mn in the base alloy [1, 19, 24]. However, the base alloy has the largest average dispersoid diameter (98 nm in Y-samples and 79 nm in S-samples) and the lowest number density compared with other alloys.

In the V-added #1 alloy, the number density of block-like α-dispersoids seems to be similar with base alloy but with smaller size. In addition, the rod-like $(Al,Si)_3(Ti,V)$ dispersoids also formed (Fig. 6c-d). The average dispersoid diameters are 86 nm in Y-samples and 74 nm in S-samples, which is moderately decreased relative to the #0 alloy. At the high cooling rate (S-samples), the proportion of $(Al,Si)_3(Ti,V)$ dispersoids is higher than Y sample, which is even similar to α-dispersoids.

In the Zr-added #2 alloy, a number of $(Al,Si)_3(Ti,Zr)$ with rod-like morphology formed instead of α-dispersoids (Fig. 6e-f). Since Zr has a very limited solubility in Al, 0.083 at.% of maximum equilibrium solid solubility and 0.0005 at.% at 400 °C [6], the number density of $(Al,Si)_3(Ti,Zr)$ dispersoids formed in the #2 alloy is relatively low at the lower cooling rate. However, the formation of $(Al,Si)_3(Ti,Zr)$ dispersoids is strongly promoted (Fig. 6f) at higher cooling rate (S-samples) and become the majority of dispersoids with a relatively high number density (Fig. 7) but the formation of α-Al(Mn,Fe)Si dispersoid seems to be inhibited. The average dispersoids diameter are 82 nm in Y-samples and 71 nm in S-samples, which is slightly smaller than #1 alloy.

The addition of Mo in the #3 alloy effectively promoted the formation of α-Al(Mn,Mo,Fe)Si dispersoids with block-like morphology, which are the only dispersoids found in the microstructure (Fig. 6g-h). The alloy has the lowest dispersoid diameter (65 nm in Y-samples and 58 nm in S-samples) among all five alloys. In addition, the number densities of dispersoids in the #3 alloy increase significantly compared to the #0, #1 and #2 alloys. Besides, the ideal interaction of Mo and Mn also lead to the largest DZ fraction as mentioned above.

The synergistic effect of the combined addition of V, Zr and Mo in the #4 alloy is quite complicated. Obviously, the alloy produces the largest number of fine dispersoids among all five alloys (Fig. 6i-j). The average dispersoid size is around 67 nm in both Y- and S-samples, which is remarkably finer than the #0, #1 and #2 alloys but slightly coarser than the #3 alloy. However, it has the highest number densities among all five alloys (Fig. 7). Probable owing to the combined effect of V and Zr, most of the dispersoids in the #4 alloys are rod-shape $(Al,Si)_3(Ti,V,Zr)$, which are preferred to precipitate around the center of dendrite cells and the promotion from Mo appears to be affected. As a result, the DZ fraction in #4 alloy is not as large as that in #3 alloy.

Regarding the number density of the dispersoids, the individual addition of V or Zr in the #1 and #2 alloys has a limited increase compared to the base alloy. The Mo addition and the combined addition of V, Zr and Mo have a large impact on the number density and these two alloys possess much higher number densities at both cooling rates than the #0, #1 and #2 alloys (Fig. 7). Owing to its highest number density of fine dispersoids, the combined addition of V, Zr and Mo induces the largest hardening effect among all five alloys. For individual additions of V, Zr or Mo, the sequence of hardening effect due to dispersoid strengthening is Mo > Zr > V (Fig. 4a-b).

The high cooling rate increases the number density and decreases the size of dispersoids in all 5 alloys. On the other hand, with the additions of V and Zr in #1, #2 and #4 alloys, it is also worthy to notice that the high cooling rate usually leads to a higher portion of rod-like $(Al,Si)_3M$ type dispersoids owing to the increase of V, Zr and Ti content in solid solution.

The authors would like to acknowledge the financial support of the Natural Sciences and Engineering Research Council of Canada (NSERC) and Rio Tinto Aluminum through the Research Chair in the Metallurgy of Aluminum Transformation at University of Quebec in Chicoutimi.

4 Conclusion

1) The addition of V in Al-7Si-0.6Cu-0.35Mg foundry alloy mainly produces the precipitation of

$(Al,Si)_3(Ti,V)$ dispersoids but has limited influence on α-Al(Mn,V,Fe)Si dispersoids during the solution treatment; while the addition of Zr greatly increase the volume of $(Al,Si)_3(Ti,Zr)$ dispersoids but inhibits the formation of α-Al(Mn,Fe)Si dispersoids. The individual addition of V or Zr fairly reduces the dispersoid size and moderately increases the number density of the dispersoids relative to the base alloy.

2) The addition of Mo effectively promotes the precipitation of α-Al(Mn,Mo,Fe)Si dispersoids and significantly reduces the dispersoid size, exhibiting the lowest dispersoid diameter among all five alloys studied. In addition, the number density of dispersoids in the alloy increases significantly compared to the V-added or Zr-added alloys.
3) The combined addition of V, Zr and Mo produces the largest number of fine dispersoids among all five alloys studied. Owing to the combined effect of V and Zr, most of the dispersoids in the alloys are rod-like $(Al,Si)_3(Ti,V,Zr)$. The alloy shows considerably higher hardening effect than all other alloys owing to its highest dispersoid strengthening contribution. For individual additions of V, Zr or Mo, the sequence of hardening effect due to dispersoid strengthening is Mo > Zr > V.
4) The high cooling rate not only decrease the SDAS and fraction of intermetallics in as-cast microstructure, but also refines the dispersoid size and increases the number density of dispersoids in all of the alloys studied during the solution treatment. In addition, it increases the proportion of $(Al,Si)_3M$ type dispersoids in the microstructure of V or Zr added alloys.

References

1. A. R. Farkoosh, X. Grant Chen, M. Pekguleryuz. Mater. Sci. Eng., A. **620,** 181-189 (2015).
2. M. Javidani, D. Larouche. Int. Mater. Rev. **59,** 132-158 (2014).
3. K. Liu, H. Ma, X. G. Chen. J. Alloys Compd. **694,** 354-365 (2017).
4. L. Jin, K. Liu, X. G. Chen. J. Mater. Eng. Perform. (2020).
5. M. Zedalis, M. Fine. Metall. Trans. A. **17,** 2187-2198 (1986).
6. K. E. Knipling. Development of a nanoscale precipitation-strengthened creep-resistant aluminum alloy containing trialuminide precipitates. DOCTOR OF PHILOSOPHY, Northwestern University, 2006.
7. D. Tsivoulas, J. Robson. Acta Mater. **93,** 73-86 (2015).
8. P. Sepehrband, R. Mahmudi, F. Khomamizadeh. Scripta Mater. **52,** 253-257 (2005).
9. R. Mahmudi, P. Sepehrband, H. Ghasemi. Mater. Lett. **60,** 2606-2610 (2006).
10. H. A. Elhadari, H. A. Patel, D. L. Chen, W. Kasprzak. Mater. Sci. Eng., A. **528,** 8128-8138 (2011).
11. B. Baradarani, R. Raiszadeh. Mater. Des. **32,** 935-940 (2011).
12. W. Kasprzak, B. S. Amirkhiz, M. Niewczas. J. Alloys Compd. **595,** 67-79 (2014).
13. T. Gao, A. Ceguerra, A. Breen, X. Liu, Y. Wu, S. Ringer. J. Alloys Compd. **674,** 125-130 (2016).
14. M. Rahimian, S. Amirkhanlou, P. Blake, S. Ji. Mater. Sci. Eng., A. **721,** 328-338 (2018).
15. K. E. Knipling. Microsc. Microanal. **21,** 359-360 (2015).
16. Y. Wu, H. Liao, K. Zhou. Mater. Sci. Eng., A. **602,** 41-48 (2014).
17. J. Rakhmonov, G. Timelli, A. Fabrizi, F. Bonollo. International Journal of Materials Research. **109,** 1099-1112 (2018).
18. A. R. Farkoosh, X. Grant Chen, M. Pekguleryuz. Mater. Sci. Eng., A. **627,** 127-138 (2015).
19. W. Fei, S. Kang. J. Mater. Sci. Lett. **14,** 1795-1797 (1995).
20. L. Ceschini, A. Morri, S. Toschi, S. Seifeddine, S. Messieri. Mater. Sci. Forum. **884,** 81-92 (2017).
21. B. Zhu, P. Leisner, S. Seifeddine, A. E. W. Jarfors. Surf. Interface Anal. **48,** 861-869 (2016).
22. A. Samuel, F. Samuel. Journal of Materials Science. **27,** 6533-6563 (1992).
23. S. Chen, K. Liu, X. G. Chen. J. Mater. Res. **34,** 3071-3081 (2019).
24. J. Rakhmonov, G. Timelli, F. Bonollo. Mater. Charact. **128,** 100-108 (2017).
25. B. Huber, H. S. Effenberger, K. W. Richter. Intermetallics. **18,** 606-615 (2010).
26. C. Guo, C. Li, P. J. Masset, Z. Du. Calphad. **36,** 100-109 (2012).
27. C. E. Ramberg, W. L. Worrell. J. Am. Ceram. Soc. **83,** 946-948 (2000).
28. S. Yamaguchi. Keikinzoku. **38,** 228-238 (1988).
29. Y. Li, A. Muggerud, A. Olsen, T. Furu. Acta Mater. **60,** 1004-1014 (2012).

Effect of ceramic particles on precipitation in an Al-Mg-Si alloy with silicon excess during ageing

Gwenaëlle Meyruey[1,*], *Véronique* Massardier[1], and *Michel* Perez[1]

[1]Univ. Lyon, INSA Lyon, MATEIS – UMR CNRS 5510, Bât. Saint-Exupéry, 25 avenue J. Capelle, F-69621 Villeurbanne Cedex, France

Abstract. Providing a good balance between lightweight and high mechanical properties, the aluminum-based metal matrix composites (MMC) became an interesting alternative for specific industrial applications. However, considering an Al-Mg-Si alloy with a high silicon excess, the ceramic particles added as reinforcement can act on : i) the precipitation kinetics of the coherent and semi-coherent phases, ii) the precipitation sequence of the alloy and iii) the loss of mechanical strength from a peak-aged microstructure obtained by a T6 condition. In order to understand the influence of reinforcement on these aspects, the composite was characterized during isothermal ageing between 100°C and 350°C, allowing us to propose an experimental Isothermal Transformation Curve. Compared to the unreinforced alloy, heterogeneous precipitation of disordered semi-coherent phases occurs on dislocations and the precipitation kinetics were found to be accelerated in the composite leading to an acceleration of the loss of strength from the T6 state, due to the precipitation of the Type-C phase. A study performed on a deformed alloy, demonstrated that the most of the differences observed between the unreinforced alloy and the composite can be explained by the high dislocation density generated in the matrix of the composite due to the presence of ceramic particles. Finally, the JMAK approach turned out to be a powerful tool to model the decrease in mechanical strength occurring during isothermal treatments from T6 state.

1 Introduction

The aluminum-based metal matrix composites (MMC) have been developed for specific applications such as aerospace, aeronautic and automotive fields [1-2]. The addition of ceramic particles to Al-Mg-Si alloys is a successful method to achieve a good combination between lightweight, high mechanical properties at elevated temperatures and good conductivity properties.
In some cases, silicon can be in excess in Al-Mg-Si alloys, since it enhances the forming properties and the alloy's yield strength. In particular, the silicon excess increases the maximum mechanical strength reached after a T6 treatment (8 hours at 170 °C) [3-4].
However, both the addition of silicon excess and ceramic particles affect the precipitation sequence and its kinetics during artificial ageing. The study of the impact of these two factors on microstructural modifications during ageing has a particular interest for industrial applications since it can lead to changes of mechanical properties during the use of the materials.

Al-Mg-Si alloys are characterized by their structural hardening property attributed to the precipitation of the coherent needle-shaped β'' phase during ageing. In the literature, several studies proved that the peak in mechanical strength, achieved after the T6 treatment, is higher and reached earlier, with an increase in the silicon excess [5-7]. However, it is generally accompanied by an acceleration of the over-ageing, resulting in a sudden loss of mechanical strength beyond the peak. The precipitation sequence is also affected by the silicon excess since it is marked by the formation of new semi-coherent precipitates, besides the conventional quasi-binary semi-coherent β' phase, named Type-A, Type-B and Type-C, distinguished by their crystallographic properties as detailed in the literature [8-14].

A recent study [15], carried out on the effect of silicon excess in an unreinforced Al-Mg-Si alloy on the precipitation kinetics and microstructural evolutions during isothermal ageing, between 100°C and 350°C, highlighted that precipitation can be correlated to ThermoElectric Power (TEP) and Vickers hardness. The treatment conditions leading to high variations of TEP and hardness were characterized by Conventional and High-Resolution Transmission Electron Microscopy (C-TEM and HR-TEM) and provided the following precipitation sequence for the investigated alloy [15]:

$$\boldsymbol{\beta'' \rightarrow \beta'' + \text{Type-B} \rightarrow \text{Type-B/Type-A/Type-C} / \beta' + \text{Si particles} \rightarrow \text{Type-A} + \text{Type-C} + \text{Si particles}}$$

In this previous work, the β'' phase was found to be responsible for the maximum of strength as expected. Its precipitation was associated with an increase in hardness to its maximum value while, at the same time, the TEP decreased to its minimum value. Further ageing leads to

* Corresponding author: gwenaelle.meyruey@insa-lyon.fr

the precipitation of the Type-B phase followed by the Type-A and the Type-C. The beginning of the semi-coherent precipitation was correlated with a decrease in hardness and a TEP increase.

Using the TEP and hardness variations during ageing, combined with the microstructural characterization, an isothermal transformation curve (Time Temperature Precipitation diagrams) was proposed in the considered temperature range.

Finally, the loss of alloy's strength was followed by hardness measurements from the T6 state during isothermal ageing. In this work, the Johnson-Mehl-Avrami-Kolmogorov (JMAK) formalism was an efficient approach to predict the hardness decrease.

In MMC, the ceramic particles affect the alloy microstructure during ageing. Namely, they lead to : i) a decrease in the grain size [16], ii) the presence of interfaces and residual stresses [17-18], and iii) a high dislocation density generated during cooling from solutionising temperature [19–21]. The effect of reinforcing particles on precipitation in Al-Mg-Si alloys has been widely studied in the literature. Based on the results obtained by TEP, hardness measurements, and Differential Scanning Calorimetry (DSC) [19,22–25], the presence of reinforcement seems to accelerate the β'' precipitation and it is explained by the abovementioned microstructural modifications [8,23,25–27]. By means of the same experimental techniques, an acceleration of the semi-coherent β' precipitation has been observed as well [34-36,38-40]. In this case, the high dislocation density is considered to be responsible for this acceleration since it provides preferential nucleation sites for the heterogeneous precipitation.

In spite of the numerous studies carried out on the effect of the ceramic particles on the Al-Mg-Si alloy behaviour during thermal ageing, the prediction of this behaviour remains difficult. Namely, several parameters (volume fraction, morphology, size and nature of reinforcement [16,28–30], matrix composition [31]...) have to be considered for the analysis of the results. In addition, in presence of silicon excess in the matrix, the effects were not extensively studied.

Considering this context, the aim of this work is to clarify how the ceramic particles act on the precipitation kinetics, the microstructural and the strength evolutions during the thermal treatment of an Al-Mg-Si alloy with silicon excess.

To this end, the same alloy as that studied in [15] reinforced with 10 % of carbide particles was investigated. In order to compare the isothermal transformation curve of the unreinforced and reinforced alloy, both TEP and hardness measurements were performed throughout ageing treatments. C-TEM/HR-TEM was also used for microstructural characterization and identification of the precipitates responsible for the main TEP and hardness variations observed.

In a second part, the Johnson-Mehl-Avrami-Kolmogorov (JMAK) formalism was used to predict the loss of strength during isothermal ageing between 100°C and 350°C from a T6 state and compare to the unreinforced alloy behaviour.

Lastly, the effect of a high dislocation density in the composite due to the presence of reinforcement was isolated considering a deformed alloy wherein dislocations were artificially introduced. Following the same procedure than for the reinforced alloy, the deformed alloy was characterized and its behaviour was compared with that of the composite.

2 Materials and experimental procedure

2.1 Material

The composite investigated in this work is an Al-Mg-Si alloy with silicon excess and reinforced with 10 % of carbide particles added to the molten metal. As illustrated in Fig. 1, the carbide particles, irregularly shaped, have a size between 15 and 20 µm. The thermal dilatation coefficient is of the order of $6x10^{-6}$ K^{-1}. The chemical composition of the alloy is Al-0.9 wt.% Mg_2Si-0.7 wt.% Si with a content in Fe and Mn of between 0.1-0.3 wt.% and 0.5-0.7 wt.% respectively.

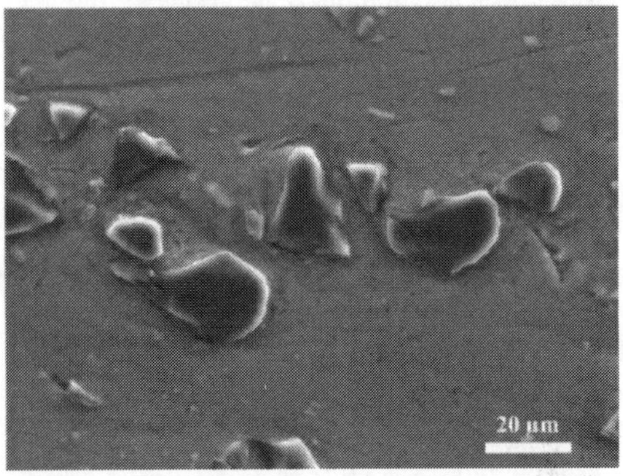

Fig. 1. SEM image of the reinforcement.

2.2 Thermal treatments

In order to follow the precipitation kinetics and the microstructural evolutions during ageing, the samples were, initially, solution treated for 1 hour at 540°C in a salt bath and water-quenched.

Artificial ageing was also carried out in salt bath at temperatures between 100°C to 350°C. The ageing was interrupted by water-quench after different treatment times to perform TEP or hardness measurements and microstructural characterization.

For the characterization of the loss of strength from a T6 state, a treatment of 8 hours at 170°C was made between homogenization and artificial ageing.

2.3 Experimental techniques

2.3.1 ThermoElectric Power (TEP)

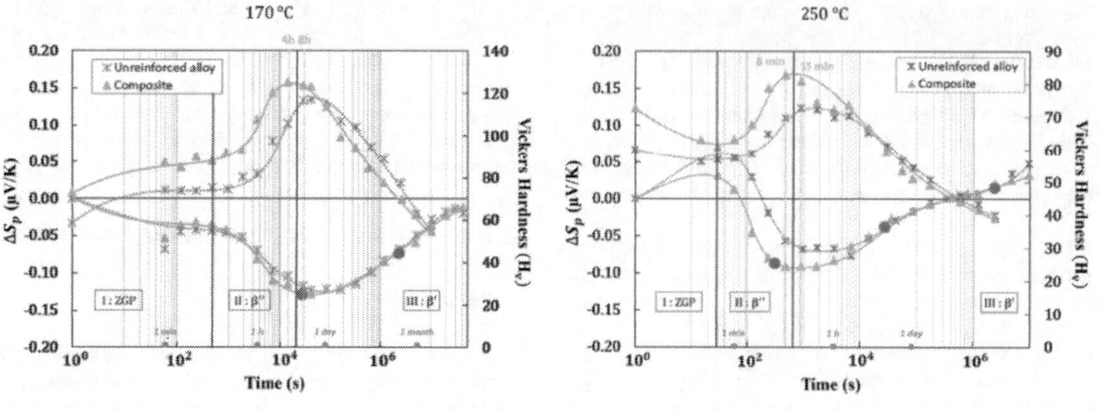

(a) (b)

Fig. 2. Comparison of the hardness and TEP evolution of the unreinforced alloy and the composite during isothermal ageing at (a) 170°C and (b) 250°C.

This technique was used to follow, indirectly, the microstructural transformations occurring during heat treatment. The measured TEP variations, noted ($\Delta S_p(T,t)$), during ageing for a time *t* at temperature *T* are the result of two main contributions: one due to the alloying elements in solid solution ($\Delta S_{ss}(T,t)$) and one due to the intrinsic effect of precipitates ($\Delta S_{pre}(T,t)$).

In this work, the TEP measurements were performed at room temperature on samples 70 mm long, 5 mm wide and 1 mm thick.

More details on the technique and on the influence of the parameters of the microstructure on the TEP can be found in paper [15].

Based on previous work [27], it was considered that the reinforcement does not affect the TEP variations measured in the matrix.

2.3.2 Hardness

The hardness measurements were performed on samples 5 mm thick with an applied load of 1 kg on a BUEHLER Micromet 500 apparatus. As in the case of the TEP measurements, the hardness measurements were performed after different treatments times. The values, presented in this work, are averaged values of 10 measurements.

2.3.3 Transmission Electron Microscopy (TEM)

The microstructural characterization by TEM was used to identify the phases responsible for the TEP and hardness variations observed during ageing treatments. Selecting the appropriate ageing states, the samples were first characterized by Conventional-TEM (C-TEM) in order to determine the morphology and the average size of precipitates. These observations were performed on a Jeol 2100 microscope equipped with a CCD camera Gatan Orius SC1000. Then, for more precise information, such as the crystallography and nature of precipitates the samples were characterized by High Resolution - TEM (HR-TEM). In this case, the observations were performed on a field-emission gun Jeol 2010 microscope.

The composite thin foils were mechanically thinned from the bulk sample after ageing treatment, and thinned by ion beam.

The precipitates were observed along a $<100>_{Al}$-zone axis. The identification of the precipitates was based on the numerical fast-Fourier transforms of the images obtained with the CCD camera.

3 Results

3.1 Characterization of the precipitation from the supersaturated solid solution state

3.1.1 Study of the TEP and hardness kinetics during isothermal ageing

TEP and hardness measurements were carried out during isothermal ageing and compared during a treatment at 170 °C and 250 °C after quench from the solid solution state for both the unreinforced alloy and the composite. Fig.2 illustrates the perfect correlation between the TEP and hardness kinetics; the microstructural modifications responsible for the decrease in TEP gives rise to a hardness increase and conversely, a hardness decrease corresponds to a TEP increase.

Due to their high sensitivity to microstructural modifications, these techniques reflects the complex transformations taking place during thermal treatment. In Fig.2, three domains can be defined:

- Domain I : TEP/hardness variations are related to the Guinier-Preston zones (GPZ). At low temperatures, the GPZ can precipitate and generate an hardness increase (TEP decrease) as observed in fig.2.(a) for an ageing at 170 °C. At higher temperatures, the GPZ already created during the quench after solution treatment are dissolved resulting in a hardness decrease (TEP increase) illustrated in Fig.2.(b).
- Domain II : marked by the hardness increase until the maximum value (respectively, TEP decrease until its

minimum). This significant variation, seen in Fig.2(a) and (b), can be attributed to the precipitation of the hardening phase, β''. The maximal value reached is higher at 170°C because this temperature corresponds to the conditions giving the finer microstructure and the higher density of β''.

- Domain III : defined by a sharp hardness decrease (TEP increase) corresponding to the precipitation of the semi-coherent phases.

Considering the domain I, no effect of ceramic particles can be deduced from the TEP/hardness measurements. The magnitude of the hardness/TEP variations and the kinetics are similar for the alloy and the composite, indicating that the density of GPZ formed could be more or less identical. In the domain II, the reinforcement leads to an acceleration of the kinetics and a higher hardness peak. In the domain III, the composite exhibits a faster loss of strength beyond the peak, meaning that the precipitation of semi-coherent phase would begin sooner.

In order to understand the microstructural transformation responsible for the difference of behaviour, C-TEM and HR-TEM characterization was used for specific ageing conditions: 8 hours and 1 month at 170°C, 6 minutes, 8 hours and 1 month at 250°C. These treatments can be located on the hardness/TEP kinetics of Fig. 2 thanks to the orange points

3.1.2 TEM characterization of the microstructure evolution

TEM characterization in peak-aged conditions highlighted that the homogeneous precipitation of β'' and the heterogeneous precipitation of a semi-coherent and disordered phase on dislocations occur simultaneously (Fig.3.) in the early stages of ageing of the reinforced alloy, while it has not been observed in the unreinforced alloy. As illustrated in Fig.4., after the peak hardness the loss of strength obtained after 1 month at 170°C and 8 hours at 250°C is mainly due to the precipitation of the Type-C phase, even if the microstructure contains also Type-B and β' phases. In over-aged conditions (1 month at 250°C), the microstructure consists in a mixture of Type-C and Type-A precipitates with the Type-A precipitate as main phase.

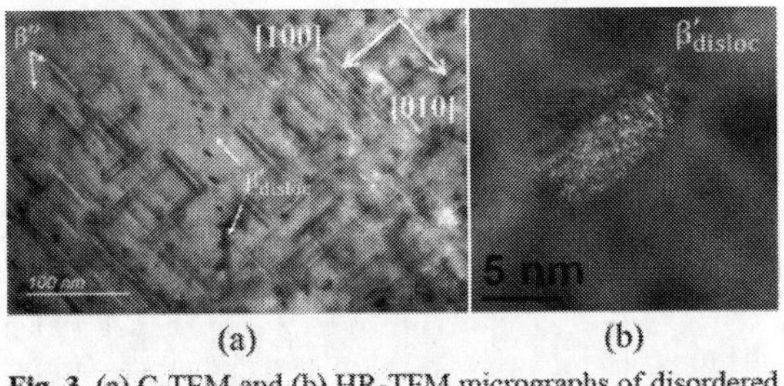

Fig. 3. (a) C-TEM and (b) HR-TEM micrographs of disordered β'_{disloc} phase observed after 6 minutes at 250°C in the

Based on the precipitates identified in the ageing states characterized by TEM, the following precipitation sequence could be proposed:

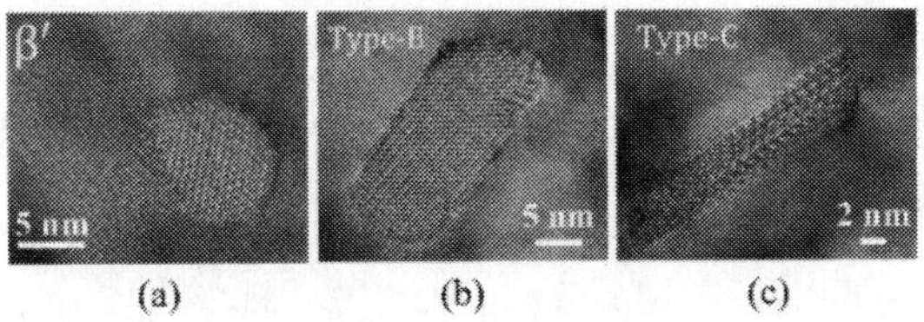

Fig.4. TEM micrographs of semi-coherent phases responsible for the loss of strength in the composite during isothermal ageing

Solid Solution → ZGP → $\boldsymbol{\beta''}$ + β'_{disloc} + β' → Type-B + β' + **Type-C** + Type-A → **Type-A** + Type-C

3.1.3 Isothermal Transformation Curve

The isothermal transformation curves for the unreinforced and reinforced material were built from the TEP kinetics obtained at different temperatures. They present two lines: the first line indicating the time corresponding to the beginning of the coherent precipitation (β'') and the second corresponding to the beginning of the semi-coherent precipitation. The method used to obtain the Isothermal Transformation Curve is described in paper [15]. Fig.5. reflects the acceleration of the coherent/semi-coherent precipitation in the composite and a decrease in the stability domains of the β'' phase and of the semi-coherent phases.

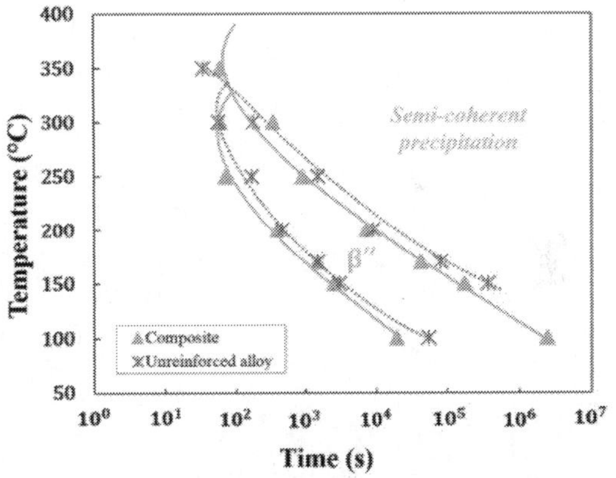

Fig.5. Comparison of the isothermal transformation curves of the unreinforced alloy and the composite.

3.2 Effect of reinforcement on the loss of strength during ageing from a T6 state

Following the same procedure as for the unreinforced alloy (described in paper [15]), the hardness $H(t)$ of the T6-treated composite was measured during isothermal ageing for $100°C \leq T \leq 350°C$. Similarly to the alloy behaviour detailed in [15], the hardness curves for all ageing treatments performed at T are stackable for an activation energy Q of 120 kJ/mol. The good superposition is due to the thermal activation of the phenomena leading to the hardness decrease and can be obtained thanks to a time-temperature equivalence based on an Arrhenius law. Afterwards, the hardness curves, due to their sigmoidal shape, were fitted by an isothermal Johnson-Mehl-Avrami-Kolomogorov (JMAK) approach defined by the following relation :

$$Y(T,t) = 1 - \exp(-(kt)^n) \quad (1)$$

$$\text{With } k = k_0.\exp\left(-\frac{Q}{RT}\right) \quad (2)$$

where $Y(T,t) = \frac{H(t)-H_{max}}{H_{min}-H_{max}}$ is the transformed fraction (0<Y<1) with H_{min} the minimum hardness value and H_{max} the maximum hardness value (T6 state), n the Avrami exponent, k_0 a constant and R the gas constant. After having determined the parameters (n, k_0 and Q) of this law for the composite material (Table 1), this approach allowed the proper description of the strength evolution and the mechanical property prediction for extremely long-term ageing as illustrated in Fig.6(a).

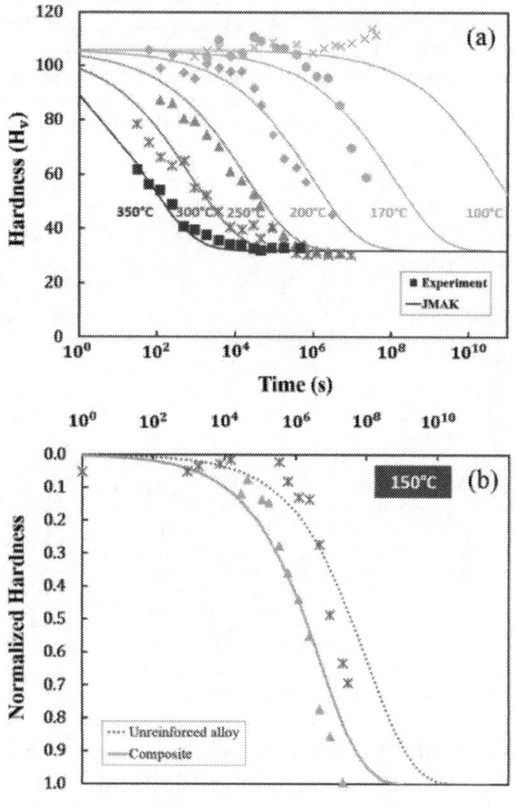

Fig.6. (a) Hardness decrease measured from a T6 state and fitted with JMAK law for different isothermal treatments of the composite material between 100°C and 350°C. (b) Comparison of the normalized hardness evolution of the alloy and of the composite for an ageing at 150°C.

Considering the parameters of Table 1, the normalized hardness kinetics of the unreinforced and reinforced alloys were plotted from the experimental data at 150 °C and modeled with the JMAK approach (Fig.6(b)). Fig.6(b) highlights the acceleration of the loss of strength due to the reinforcement added in the matrix. This result is correlated to the premature precipitation of semi-coherent phases in the composite observed by TEM characterization. The difference of behaviour is reflected in the lower activation energy value in the case of the composite. By comparison, the activation energies for the diffusion of magnesium and silicon in aluminum are, respectively, 130 kJ/mol and 124 kJ/mol. The activation energy of 160 kJ/mol estimated for the alloy suggests that the loss of strength could be governed by nucleation of semi-coherent phases and diffusion of the alloying elements. However, since the activation energy of the composite is very close to that for the diffusion of Mg and Si, it indicates that the strength loss is mainly controlled by the diffusion of these two elements. Furthermore, this phenomenon is enhanced in the matrix of the composite due its high dislocation density. The presence of dislocations results in: (i) preferential diffusion paths, accelerating both the coherent and semi-coherent precipitation and the coarsening phenomena, (ii) heterogeneous nucleation sites, promoting the nucleation of the semi-coherent phases. In order to validate the predominant role of dislocations in the differences observed between the alloy and the composite in this study, a deformed alloy (with dislocations artificially introduced by cold-rolling) was characterized.

Table 1. Summary of the JMAK parameters for the alloy (cold-rolled and undeformed) and the composite.

	Alloy	Deformed alloy	Composite
Q (kJ/mol)	160 (+/- 5)	135 (+/- 5)	120 (+/- 15)
n	0.35	0.34	0.38
k_0 (s^{-1})	5.25×10^{11}	2.5×10^{9}	9.91×10^{7}

3.3 Effect of the presence of dislocations in the alloy on microstructural evolutions during ageing

Due to the significant difference in the thermal expansion coefficients between the alloy and the ceramic particles, a high dislocation density is generated in the matrix of the composite material during cooling from the solutionizing temperature. With the aim of demonstrating that dislocations are mainly responsible for the differences observed between the alloy and the composite in this work, a deformed alloy was studied. To this end, an alloy sample was cold-rolled with a reduction ratio of 2.4 % after homogeneisation and water-quench.

The deformed alloy underwent an isothermal ageing at 170 °C interrupted to perform hardness measurements and the results were compared with the undeformed alloy and the composite. As shown in Fig.7, the deformed alloy exhibits the same kinetics as that of the composite and reaches the same maximum value.

https://doi.org/10.1051/matecconf/202032602004

From TEM observations of the deformed alloy aged 8 hours at 250 °C, it was established that the Type-C is the prevailing phase. As revealed for the composite, the Type-C is responsible for the loss of strength in the cold-rolled alloy wherein dislocations were introduced, while it was the Type-A for the undeformed alloy. This result is in good agreement with a previous study showing that dislocations promote the Type-C precipitates [32].

Starting from the T6 state, the hardness decrease during isothermal ageing can also be predicted with the JMAK law described above determining the JMAK parameters (given in Table 2). An increase in the dislocation density of the alloy leads to Q and k_0 values closer to those of the composite. These results confirm that the acceleration of the loss of strength in the composite mainly comes from the dislocations. Fig.8(b) compares the hardness decrease predicted by the JMAK for the undeformed alloy, the cold-rolled alloy and the composite during an ageing at 250 °C. The deformed alloy shows the same acceleration of the hardness decrease in the early stages of the treatment but, in overaged conditions, the curve reaches the same final value as for the undeformed alloy. This observation suggests that, recovery or recrystallization took place in the deformed alloy for long ageing times, and that other elements contributes to the hardening of composite in over-aged conditions.

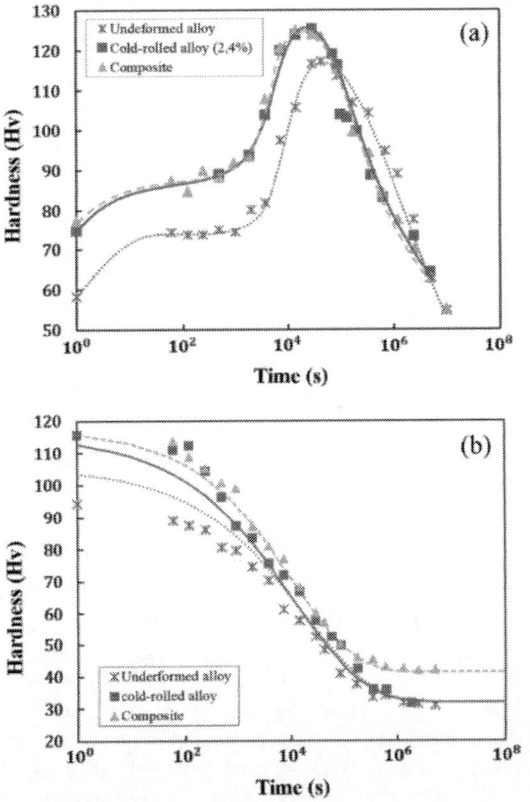

Fig.8. (a) Comparison of the hardness kinetics during ageing at 170 °C for the unreinforced alloy, deformed alloys and the composite. (b) Comparison of the decrease in hardness from the T6 state during ageing at 250°C predicted by the JMAK law for the undefomed alloy, the cold-rolled alloy and the composite.

4 Conclusion

The aim of this study was to understand how the ceramic particles act on the matrix microstructure during isothermal ageing and the consequences on mechanical properties considering an Al-Mg-Si alloy with silicon excess.

The reinforcement accelerates the precipitation of the coherent and semi-coherent phase leading to an acceleration of the loss of strength from the T6 state. The microstructure is more heterogeneous even in the early stages of ageing since dislocations induced by the particles lead to premature precipitation of semi-coherent phase and disordered precipitates. Studying a cold-rolled alloy, the major role of dislocations in the difference of behaviour observed between the alloy and the matrix has been confirmed. It was also shown that dislocations promote the precipitation of the Type-C phase, which is responsible for the loss of strength in the composite and deformed alloy while this phase was barely observed in the undeformed alloy.

The JMAK model is a powerful tool to predict the hardness decrease from the T6 state during isothermal ageing and considering the appropriate JMAK parameters defined in this study. This approach links the acceleration of the hardness decrease in the composite to a lower activation energy of the phenomenon explained by the higher dislocation density acting as diffusion path for alloying elements and promoting the nucleation of semi-coherent phase. The same approach applied to deformed alloy confirms the role of dislocation in the activation energy decrease and suggests that the dislocations are the main factor responsible for the acceleration of the loss in composite strength as long as the recovery phenomena do not come into play.

Acknowledgments

We would like to thank the Clym (Consortium Lyonnais de Microscopie) for access to the 2010F and 2100 microscope and to N. Blanchard for his kind assistance with the Conventional TEM observations.

References

1. S. Mazumdar, *State of composites industry*, Composite Manufacturing, 1–6 (2016)
2. L. Berreur, B. DeMaillard, S. Nosperger, *L'industrie française des matériaux composites*, Nodal Consultant (2002)
3. K. Fukui, M. Takeda, T. Endo, Journal Japan Inst Met, **46**, 880-884 (2006)
4. A.K. Gupta, D.J. Lloyd, S.A. Court, Mater Sci Eng, **316**, 11–17 (2001)
5. J.P. Lynch, L.M. Brown, M.H. Jacobs, Acta Metall, **30**, 1389-1395 (1982)
6. Y. Ohmori, L.C. Doan, K. Nakai, Mater Trans, **43**, 246-255 (2002)
7. L.C. Doan, K. Nakai, Y. Matsuura, S. Kobayashi, Y. Ohmori, Mater Trans, **43**, 1371–1380 (2002)
8. K. Matsuda, S. Tada, S. Ikeno, T. Sato, A. Kamio,

Scr Metall Mater, **32**, 1175–1180 (1995)

9. K. Matsuda, T. Naoi, K. Fujii, Y. Uetani, Mater Sci Eng A, **262**, 232–237 (1999)
10. K. Matsuda, Y. Sakaguchi, Y. Mitaya, Y. Uetani, T. Sato, A. Kamio, et al., Mater Sci, **35**, 179–189 (2000)
11. K. Matsuda, S. Ikeno, T. Sato, A. Kamio, Acta Metall, **34**, 1797–1802 (1996)
12. P. Derlet, S. Andersen, C. Marioara, A. Froseth, J Phys Condens Matter, **14**, 4011–4024 (2002)
13. A. Froseth, S. Andersen, C. Marioara, P. Derlet, R. Hoier, *Solving the Structure of the Phases in the Al-Mg-Si Alloy System with the Help of Ab Initio Modelling*, MRS Online Proceeding Libr Arch, (2002)
14. A. Froseth, R. Hoier, P. Derlet, S. Andersen, C. Marioara, Phys Rev B, **67**, 1-11 (2003)
15. G. Meyruey, V. Massardier, W. Lefebvre, M. Perez, Mater Sci Eng A, **27,** 92-105 (2018)
16. F. Humphreys, H. Lilholt, O. Pederson, *Deformation and annealing mechanisms in discontinuously reinforced metal-matrix*, Proc 9th Riso Int Symp Mech Phys Behav Met Ceram Compos, 51–74 (1988)
17. Z.M. Sun, J.B. Li, Z.G. Wang, W.J. Li, Acta Met Mater, 40, 2961–2966 (1992)
18. M. Suery, C. Teodosiu, L.F. Menezes, Mater Sci Eng, **167**, 97–105 (1993)
19. S. Suresh, T. Christman, Scr Metall, **23**, 1599–1602 (1989)
20. R.J. Arsenault, L. Wang, C.R. Feng, Acta Metall Mater, **39**, 47–57 (1991)
21. M. Vogelsang, R.J. Arsenault, R.M. Fisher, Metall Trans A, **17**, 379–389 (1986)
22. S. Abis, G. Donzelli, J Mater Sci Lett, **7**, 51–52 (1988)
23. T.G. Nieh, R.F. Karlak. Scr Metall, **8**, 25–28 (1984).
24. M. Hadianfard, Y. Mai, J. Healy, J Mater, **28**, 3665–3669 (1993)
25. M. Gupta, M. Surappa, Mater Res Bull, **30**, 1023–1030 (1995)
26. J.M. Papazian, Metall Trans A, **19**, 2945–2953 (1988).
27. D. Dafir, G. Guichon, R. Borrelly, S. Cardinal, P.F. Gobin, P. Merle. Mater Sci Eng A, **144**, 311–318 (1991)
28. N.K. Sharma, R.K. Misra, S. Sharma, Ceram Int, **43**, 513–522 (2017)
29. N. Ramakrishnan, Acta Mater, **44**, 69–77 (1996).
30. G. Meijer, F. Ellyin, Z. Xia, Compos Part B Eng, **31**, 29–37 (2000)
31. V. Massardier, P. Merle, Mater Sci Eng A, **249**, 109–120 (1998)
32. K. Matsuda, S. Shimizu, H. Gamada, Y. Uetani, F. Shinagawa, S. Ikeno, J Soc Mat Sci, 48, 10–15 (1999)

Influence of quench rate on multi-stage ageing of AA6014 alloy

Zi Yang[1,*], *Xiaohe* Jiang [1], *Xingpu* Zhang[1], *Meng* Liu[1], *Zeqin* Liang[2], *David* Leyvraz[2] and *John* Banhart[1]

[1]Helmholtz-Centre Berlin for Materials and Energy, Hahn-Meitner-Platz 1, 14109 Berlin, Germany
[2]Novelis R&T Centre, Sierre, Route des Laminoirs 15, 3960 Sierre, Switzerland

Abstract. The influence of quench rate after solution heat treatment on the microstructure in the as-quenched state and subsequent ageing kinetics of alloy AA6014 was investigated by means of transmission electron microscopy, positron annihilation lifetime spectroscopy and hardness measurements. Various ageing temperatures and stages were taken into consideration. Consistent with previous studies, we found that solute and vacancy supersaturation decrease during slow quenching due to precipitation and annihilation, respectively. Additionally, we observed cluster formation during cooling below 200 °C. As for the influence on ageing behaviour we observe different behaviour for high and low ageing temperature: Artificial ageing is more affected than pre-ageing and natural secondary ageing. The detrimental effect of natural ageing on paint-bake hardening also depends on the quench rate. Possible interpretations are associated with cluster formation during natural ageing and also during quenching. The influence of pre-ageing at different temperatures on subsequent ageing kinetics is similar for slower industrial-type quenching and for fast quenching, thus allowing to apply the findings from idealised quenching conditions to situations closer to real application.

1 Introduction

Precipitation hardening is the main method for strengthening the Al-Mg-Si alloys (6000 series). As the first step after solution heat treatment (SHT), quenching plays a crucial role as it ensures the high level of dissolved solute atoms and trapped excess vacancies necessary for precipitate formation during the ensuing natural or artificial ageing (NA/AA). In laboratory conditions, quenching is most often carried out in water, which is convenient, efficient, and reproducible. However, quenching in industrial production can be much slower due to much larger sample volumes and other factors. Investigating the quenching procedure is therefore a task of great practical relevance.

Studies of quench sensitivity have shown that slower quenching can limit the maximum hardenability of the Al-Mg-Si alloys during direct AA [1-3]. However, a peculiarity of this type of alloy appears when applying multi-stage ageing treatments, i.e. hardening during AA is dependent also on previous ageing treatments. A classic example is that for alloys containing more than 1 at.% solute atoms, NA at 'room temperature' introduces a detrimental effect on subsequent AA, while pre-ageing (PA) at intermediate temperatures, e.g. 100 °C, does not [4-6]. Moreover, PA also suppresses hardening during subsequent natural secondary ageing (NSA) and also enhances the final paint-bake (PB) response [7, 8], the effect of which is determined by the PA temperature [9]. However, whether the multi-stage ageing behaviour will be influenced by the quench conditions has been studied only recently, when the combined effect of quench rate and NA on AA was investigated on two different alloys [10, 11], but the mechanisms still remain elusive. Hence, in this study we aim to elucidate the influence of quench rate on the microstructure as well as on the following single-, two-, and multi-stage ageing treatments.

2 Experimental

Alloy AA6014 (0.65% Mg, 0.60% Si, 0.18% Fe, 0.08% Mn, 0.12% Cu, all in wt.%) was provided by Novelis Switzerland. Square samples with a geometry of ~10×10×1 mm^3 were prepared. SHT was conducted at 540 °C for 1 h in an air circulation furnace. Two principal types of quenches were performed. In the first type (T1), samples were quenched in one step either in ice-water (IWQ), in the air flow of a ventilator (ventilator cooling, or VC), or in resting air (air cooling, or AC) to the ambient temperature. In the second two-step type (T2), the process of VC or AC was interrupted at a given target temperature by applying a subsequent fast quench in ice water. 'VC_*x*' or 'AC_*x*' are used to label the samples after T2 quench, where *x* stands for a target temperature, which was controlled within ±15 °C for VC and ±5 °C for AC. Fig. 1 shows the temperature evolution during quenching, which is measured by a thermocouple inserted in the hole on the side of a dummy sample. Quench rates from 533 °C to 250 °C are in average >980 K/s for IWQ, ~27 K/s for VC, and ~5 K/s for AC, respectively. The VC quenching is close to industrial cooling conditions. Subsequent ageing was performed in an oil bath at (180.0 ± 0.2) °C for AA, and (100.0 ± 0.2) °C for PA or in a Peltier-cooled incubator at 20 ± 0.1 °C (NA and NSA). AA for 30 min at 180 °C

* Corresponding author: zi.yang@helmholtz-berlin.de

is used in this study to imitate the industrial paint-baking (PB). The time required for transferring samples from the quenching device to the oil bath for ageing is ~0.5 min. Single-stage (AA) and some multi-stage ageing treatments (NA+PB) were performed on all three T1-type cooled samples, while multi-stage ageing involving PA (PA+PB, PA+NSA, and PA+NSA+PB) were only performed on IWQ and VC samples. In the following, the abbreviation 'PA *x/y*' stands for PA at *x* °C for *y* min.

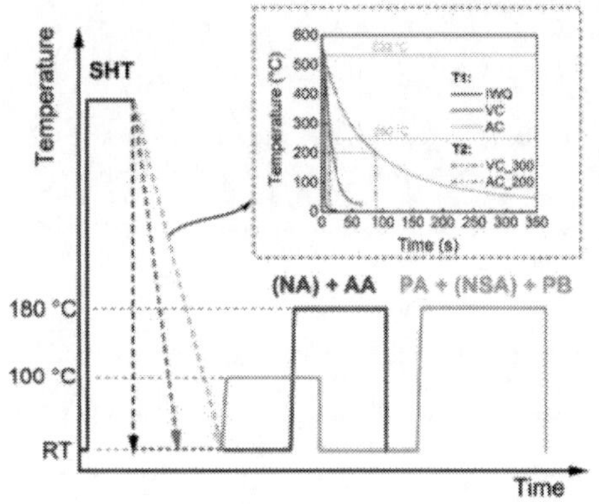

Fig. 1. Schematic illustration of the heat treatments applied to the alloy. Inset shows the measured temperature profile during various quenching conditions.

Brinell hardness was measured on a hardness tester (Qness, model 60M) using a load of 10 kgf for 10 s and a tungsten carbide indenter of 1 mm diameter. Eight indentations were performed on each sample for calculating the average, which ensures a standard error of <1 HBW for >98% of the measurements. TEM characterisation was performed on a Philips CM30 unit operated at 300 kV and equipped with an energy dispersive X-ray (EDX) spectrometer. The electron beam was aligned along a 〈100〉 direction of the Al matrix. TEM foils were prepared by mechanical thinning and then twin-jet electro-polishing in an electrolyte mixture of 75% methanol and 25% nitric acid. Positron annihilation lifetime spectroscopy (PALS) was performed after various quenches to probe the microstructure evolution during quenching. Such technique measures the lifetimes of positrons after production from the source and before annihilating with electrons in the material. Therefore, this technique possesses a unique sensitivity to the local electron density change, such as caused by solute clusters and, in particular, vacancies, because there very low electron density presents and thus markedly higher positron lifetimes are resulted. In this study, positrons were produced by a $^{22}NaCO_3$ source enveloped in kapton foil, which was sandwiched by two square samples during measurement. The resolution (FWHM) of the spectrometer was ~200 ps, and software LT9 was used to determine the one-component lifetimes, which represents an weighted average of positron lifetimes in all microstructural features.

3 Results

3.1 Transmission electron microscopy

Microstructure after all three T1-type quenches were observed under TEM. Two main types of secondary particles, namely P1 and P2, were identified according to their compositions analysed by EDX, as shown in Fig. 2. In particular, P1 particles, containing Fe, Si and Mn, were observed in all samples, while P2 particles which contain Mg and Si can be observed only in VC and AC conditions. P1 particles are mostly spherical or ellipsoidal, whereas P2 particles are mostly rod-shaped. P2 particles are found mainly growing on the P1 particles aligning along 〈100〉 directions of the Al matrix, and some of them are also found at grain boundaries.

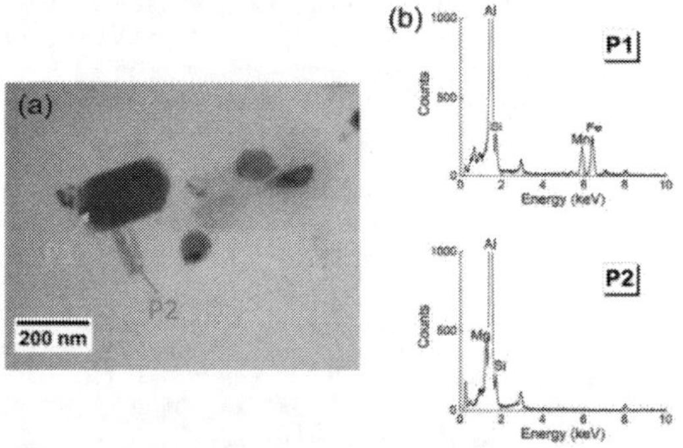

Fig. 2. (a) TEM micrograph of AC sample. (b) EDX analyses of the two particles in (a).

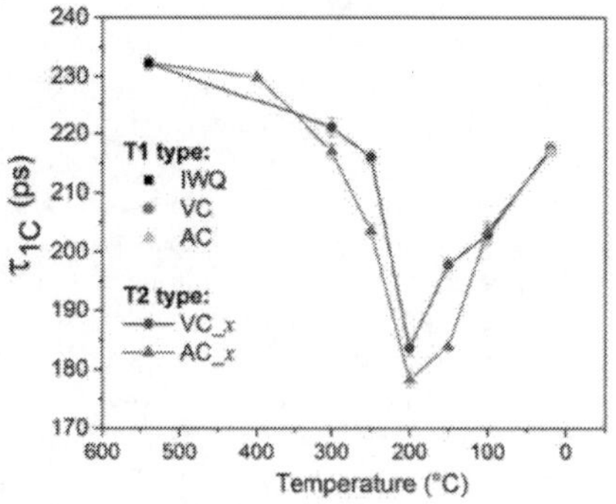

Fig. 3. One-component positron lifetimes after various T1 and T2 quenches.

3.2 Positron lifetime measurements

One component positron lifetimes τ_{1C} were measured after T1 and T2 type quenches (Fig. 3). After T1 type quenches, a positron lifetime of ~232 ps is obtained in IWQ sample, and ~217 ps is obtained in both VC and AC samples. Assuming the positron lifetime change during the fast quenching in ice water is negligible, the curves in Fig. 3 represent the lifetime evolution during VC and AC quenching. It is seen that during both quench routes, τ_{1C} first decrease and then increase. The transition between the two stages is at 200 °C. The lowest lifetime is ~185 ps for VC_200 and ~179 ps for AC_200.

3.3 Hardness measurements

3.3.1 After direct artificial ageing

Hardening during AA of the alloy after SHT and various quench conditions is given in Fig. 4. For samples after T1-type quenches (complete quench to ambient temperature), the initial hardnesses are the same within ±1 HBW. The times needed to reach peak age are also similar for all conditions (4 h), while the peak hardnesses are lower for slower cooling, with 112 HBW, 108 HBW and 99 HBW for IWQ, VC, and AC, respectively. In contrast to the later stages, early stage hardening (i.e. up to 5 min) appears to be slightly stronger for the VC and AC samples than for IWQ. Two T2-type quenched conditions are subjected to AA, namely VC_200 and AC_200. They show slightly lower hardnesses (max. 2 HBW) than VC and AC samples in the as-quenched condition, but identical hardnesses after PB, and at peak age condition.

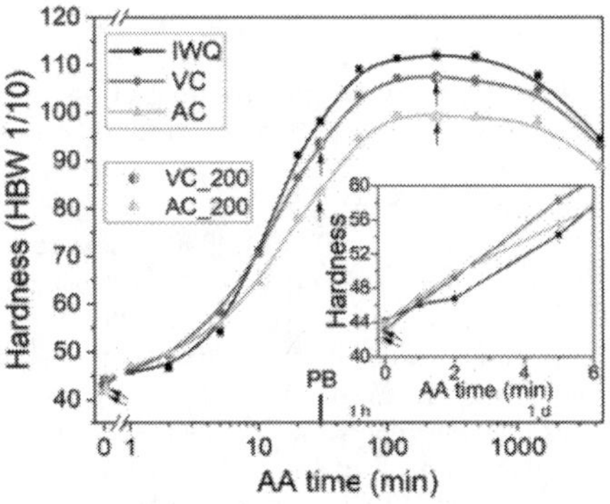

Fig. 4. Hardness of the alloy during AA after various T1 quenches and two T2 quenches. Inserted graph shows early stage AA on a linear time scale. Arrows denote the T2-type quenched samples.

3.3.2 After natural ageing and paint-baking

When nominally no NA is applied, PB leads to hardnesses of ~98 HBW, ~93 HBW, and ~83 HBW for IWQ, VC, and AC, respectively. As NA time increases, PB hardness initially declines rapidly and then levels off to a similar level (Fig. 5a). The absolute decrease of PB hardness after NA relative to the initial value is smaller for the slower cooled samples. Fig. 5b shows the relative PB decrease as a function of hardness increase during NA [12]. The influence of NA on PB hardening can be divided into two stages. During the first, the PB hardness decrease is significant and approximately proportional to the NA hardness increase. During the second, PB hardness decreases much slower and eventually stagnates although NA still hardens the material. The transition between the stages occurs at different NA hardnesses, whereas the corresponding NA times are quite similar for all quenched conditions, namely 90 min or 120 min.

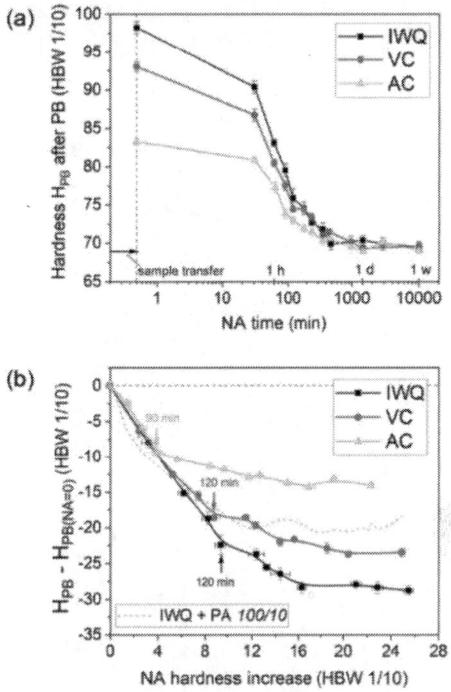

Fig. 5. (a) PB hardness after various NA times. First points are after NA for ~0.5 min, the interval for transferring samples from the cooling device to the oil bath for ageing. (b) Decrease of PB hardness plotted as a function of the hardness increase after NA (from Ref. [12]). NA time is an implicit parameter in both (b). The blue dashed curve in (b) is the corresponding curve of an ice-water quenched sample pre-aged at 100 °C for 10 min and then NSA taken from our previous study [9].

3.3.3 After pre-ageing, natural secondary ageing, and/or paint-baking

Applying PA effectively suppresses hardening during subsequent NSA, the extent of which depends on PA temperature and time (Fig. 6a) but hardly on whether IWQ or VC has been employed (the small offset ±1 HBW might reflect effects of different alloy batches). Longer PA at 100 °C (from 10 min to 4 h) causes a stronger suppression on NSA. When similar PA hardness is achieved by PA at 100 °C for 4 h and 140 °C for 30 min, NSA hardening kinetics for PA conducted at 100 °C is slower.

IWQ and VC samples behave differently when a final PB treatment is performed after PA (Fig. 6b, 1st, 3rd and 5th columns). For the same ageing condition, VC samples age to a lower PB hardness than IWQ samples. These hardness differences are reduced when 4 weeks of NSA are introduced after PA (2nd and 4th columns). When comparing PA treatments at different temperatures that lead to similar hardness, PA at 140 °C appears to be superior for achieving a higher PB hardness for both IWQ and VC samples (3rd vs. 5th column). Therefore, the finding previously found for IWQ samples [9] is still applicable to slower cooled samples, i.e. at lower PA temperature the alloy shows slower NSA hardening while at higher PA temperature a higher PB potential is encountered

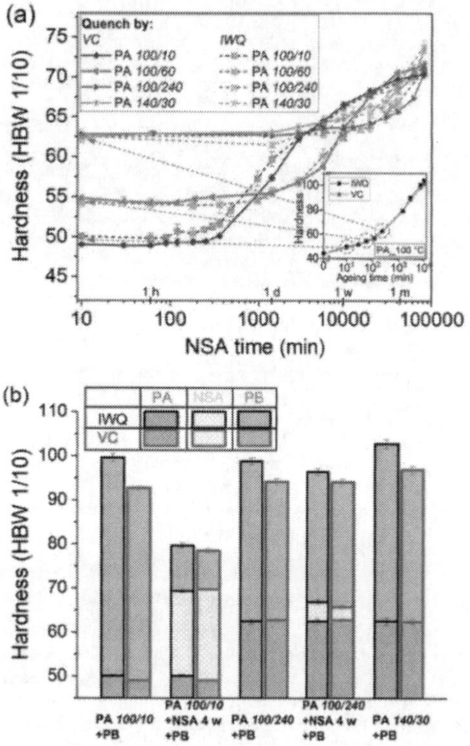

Fig. 6. Comparison of the hardness of IWQ and VC samples (a) during NSA after various PA treatments, and (b) after PB of various aged states. Inserted graph in (a) shows hardness during PA at 100 °C up to 1 week. Data for IWQ taken from Ref. [9].

4 Discussion

4.1 Influence of quenching conditions on as-quenched states

The microstructure after quenching is markedly influenced by the quench rate as primarily reflected by the TEM and PALS results. Fe-containing particles (P1 in Fig. 2a) found in all samples are dispersoids typically formed during homogenization [13, 14], whereas the Mg-Si containing particles (P2) are only found in VC and AC samples but not for IWQ, indicating that they are precipitates formed during quenching. Fast quenching in IWQ suppresses the formation of such precipitates. By measuring the in-situ heat evolution during cooling, Milkereit et al. characterised two precipitation reactions [1, 2]: A high-temperature reaction occurring typically from 500 °C down to ~380 °C represents the precipitation of equilibrium β phase, while a low-temperature reaction happens from 380 °C down to ~250 °C and stands for precipitation of β' and B$'$ phases. The high-temperature reaction is negligible for quench rates higher than 0.17 K/s [1]. Similar results were obtained by Strobel et al. [11]. This and the observation that the shape of the precipitates we find is closer to β' and B$'$ as described in Ref. [1] instead of to the typical plate shape for β imply that for our quenching conditions, the low-temperature reaction is dominant. As precipitates are mostly attached to dispersoids these seem to be their nucleation sites in accordance with the literature [15-18].

One component positron lifetime τ_{1C} represents a weighted average of positron lifetimes in various microstructural features, most importantly in mono-vacancies (~245 ps), solute clusters (~210 to 220 ps), and bulk Al (≤160 ps) [19]. After fast quenching (IWQ), the high lifetime (~232 ps) indicates a very high vacancy concentration so that positrons are annihilated mostly in vacancies. During slow quenching (VC and AC) down to 200 °C, lifetime decreases to a level close to 160 ps. This can only be caused by a substantial reduction of vacancies due to their high mobility at high temperature and consequently annihilating at vacancy sinks under the thermodynamic driving force. A rough estimation shows the vacancy concentration after VC and AC are at least 1 – 2 orders of magnitude lower than after IWQ [12]. Further cooling to room temperature increases τ_{1C} to 217 ps, which should be due to clustering or nucleation of precipitates at this temperature regime. Fig. 4 shows that the clustering contribution to hardness increase during cooling is at the most 2 HBW (AC_200 compared with AC) or even unnoticeable (VC_200 compared with VC). This shows that positron lifetime is much more sensitive to the formation of very small clusters than hardness, possibly because initial small clusters can trap positrons but contribute little to hardness.

In summary, slower quenching reduces the vacancy site fraction markedly, lowers the solute supersaturation owing to pre-precipitation, and prompts solute clustering primarily below 200 °C.

4.2 Influence of quenching conditions on subsequent ageing

4.2.1 Artificial ageing hardening

Slower quenching leads to a higher loss of (i) dissolved solutes and (ii) vacancies, both negatively affecting the formation of precipitates during AA. (iii) The properties of the clusters formed during quenching could also differ and affect AA, depending on the type of clusters. It has been assumed that in solute-rich Al-Mg-Si alloys, clusters formed at PA temperature will act as nuclei of strengthening precipitates (mainly β'') during AA, whereas the clusters formed at NA temperature will cause a detrimental effect on AA [20]. From the PALS measurements (Fig. 3), clustering phenomena during cooling are observed mainly below 200 °C. Thus, AA after VC_200 (and AC_200) is an interesting reference state as only little prior precipitation has taken place between 200 °C and 20 °C. The PB hardness as well as the peak-ageing hardness of VC_200 (and AC_200) treated samples are similar to those of the VC (and AC) samples in which clustering has taken place (Fig. 2). Cluster formation during quenching has therefore little negative impact on subsequent AA, which also implies that these clusters are primarily PA clusters that can develop into β'', most likely because most of them are formed notably above 'room temperature' during cooling.

Next, the influence of the three factors mentioned above on AA kinetics is discussed. The AA process is divided into various stages as schematically represented in Fig. 7 according to the evolution of hardness after two quenches, namely IWQ and VC. A possible scenario of the evolution of solute supersaturation (SS) and vacancy site fraction (VF) is deduced qualitatively. In the early stage AA (I in Fig. 7), a slightly higher hardening is observed for VC samples than for IWQ samples. As discussed above, SS and VF are both lower for VC sample than for the IWQ sample after quenching, and both factors should negatively influence the hardening rate. Therefore, it must be the clusters formed during VC that enhanced the precipitate formation during AA by acting as nuclei or precursors.

As ageing proceeds, the importance of these quenched-in clusters vanishes because more nuclei are gradually formed during AA. After that – in late stage I or early stage II – the hardening rate is dominated by the available solutes and vacancies. Thus, the hardness of IWQ samples increases faster and eventually exceeds the hardness of VC samples, until the largest difference is reached after ~30 min of AA (end of stage II). Subsequent ageing further hardens the material to the peak-ageing state, during which the difference between the two hardening curves remains approximately constant (stage III). This indicates that the SS and VF in two samples are very similar in this stage. In other words, the extra solute supersaturation in the IWQ sample has been fully transformed to the higher hardness at the latest by the end of stage II. However, the question remains which role the higher vacancy site fraction in IWQ than in VC during stage I and II plays. During AA at 180 °C, the vacancy concentration after quenching will move from highly excess towards the much lower thermal equilibrium. The positive influence on hardening kinetics of the high fraction of excess vacancies quenched-in by IWQ exists only until thermal equilibrium has been reached. If the equilibrium is reached rapidly in the early stage of AA (denoted as route '1' in Fig. 7) as predicted by simulations [21, 22], then the faster hardening in stage II for IWQ would be primarily due to the higher SS. However, if vacancy annihilation is not that fast (route '2' in Fig. 7), e.g. delayed by vacancy trapping in clusters, and a considerably higher vacancy concentration still continues to exist in IWQ samples than in VC samples for a few tens of minutes, then the higher hardening in stage II for IWQ would be caused by both higher SS and VF resulting from faster quenching. Which scenario is more likely cannot be deduced from the current measurements and remains to be further explored.

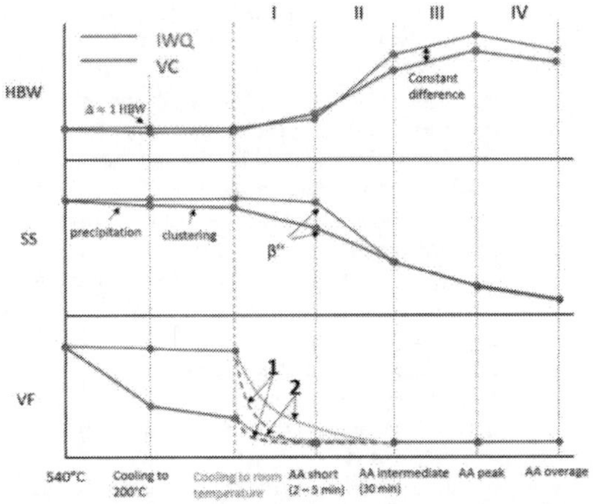

Fig. 7. Schematic graph of the evolution of hardness (HBW), solute supersaturation (SS), and vacancy site fraction (VF) during AA. '1' and '2' label the two possible vacancy evolution routes. This graph is meant to visualise the discussion and the lower two panels are not based on quantitative data.

4.2.2 Paint-baking hardening after natural ageing

The term 'negative effect' refers to the delayed AA hardening kinetics after NA and, more importantly in practice, the compromised PB hardness. The origin of the effect is widely attributed to the formation of NA clusters, although the exact mechanism is still in controversy. Serizawa et al. [20] and Aruga et al. [23] argue that the NA clusters are difficult to either grow into β'' or dissolve during AA, thus decreasing the supersaturation of solutes and consequently hardening during AA. Zandbergen et al. [24, 25] argue that the increase of the nucleation barrier for the β'' phase due to a decreased solute concentration after NA (even if only a few %) markedly suppresses nucleation of the β'' phase. Pogatscher et al. [26] attribute the retardation of AA hardening to the depletion of mobile excess vacancies which are "imprisoned" by the clusters formed during NA. To discuss which mechanism is more likely is out of the scope of the current study, but these quoted ideas allow us to discuss the distinct negative effects (Fig. 5 & Fig. 6b) introduced by different quench rates, i.e. why PB hardnesses are more similar after longer NA.

Fig. 5b now shows that the loss of PB hardness with progressing NA is not uniform but can be roughly divided into two stages and where the decrease mainly comes from the first stage. A similar slope for all the curves in this stage implies that the negative effect in differently quenched samples share some similarities. Each NA hardening by 1 HBW reduced PB hardness by ~2.3 HBW. The times for the first stage are all within the initial 120 min, pointing at initial NA cluster formation having the biggest effect on PB hardening. Such clusters are formed more in fast quenched sample than in slow quenched sample as the hardness at the end of first stage is increasing with higher quench rate and consequently the decrease of PB hardness is larger too. The effects of these clusters can be two-fold. First, they lock solutes which will be temporarily not available for further precipitation. Second, they influence the mobile vacancy concentrations by trapping as described above. When there are few mobile vacancies available, the precipitation rate will be limited by the number of these vacancies which transport the solutes to precipitates. Thus, after long NA clustering, PB does not result in any hardening regardless of quench condition.

The formation of PA clusters during slow quenching could be another reason. PA is well known to suppress NA clustering and the negative effect. The corresponding decrease of PB hardness against the NSA hardening of a typical PA treatment after IWQ (PA at 100 °C for 10 min) from our previous study [9] is shown in Fig. 4b. Similar to slow quenching, PA also shows a reduced negative effect despite of the same hardness increase due to NA/NSA clustering. It has been discussed (Sec. 4.2.1) that these clusters are more important for the early stage AA when hardening is low. As hardness increases, the advantage of having these nuclei for slower quenched samples diminishes as a high amount of precipitates are formed also in the IWQ sample. Therefore, as PB response decreases due to progressing NA clustering, the importance of quenched-in PA clusters increases, which results in a smaller PB hardness loss.

4.2.3 Multi-stage ageing

Comparison of multi-stage ageing hardness after IWQ and VC suggests the PA and NSA hardening are almost independent of quenching conditions. Since vacancies are important for the NA clustering, it needs to be explained why the large vacancy difference after IWQ and VC influences little to NSA hardening. This might be due to two reasons: 1) the difference of NA clustering kinetics after various quenching conditions is much smaller than expected, possibly associated to vacancy clustering or strong vacancy trapping at early stage of NA [12]; 2) the vacancy difference is narrowed after PA as distinct vacancy concentrations are moving towards the same equilibrium concentration during the process. Consequently, the NSA clustering depends less on quench rate but more on the PA condition, the effect of which depends primarily on PA temperature [9]. Final PB is lower for slower-quenched samples as PB is more dependent on solute supersaturation, which is lower in those samples.

5 Conclusions

We have investigated the influence of quenching conditions (quench rate: IWQ>VC>AC) after SHT on the microstructure and subsequent multi-stage age-hardening kinetics of an AA6014 alloy by means of TEM, hardness, and PALS measurements. We find:

- During slow quenching (VC or AC), precipitates are formed primarily on dispersoids and at grain boundaries unlike in the fast quenched IWQ sample. More vacancies annihilate during quenching before reaching 200 °C for slower quenching, as observed by positron lifetime measurements. Clusters are primarily formed during quenching below 200 °C.
- The AA hardness for slower quenched samples is initially slightly higher compared to faster quenched samples due to the formation of PA clusters or nuclei for precipitates during quenching, but later in technologically more important later stages of AA falls back due to a lower solute supersaturation.
- PB hardening is higher for faster quenched samples when no NA or no NSA is applied. However, the influence of quenching rate gradually diminishes as NA or NSA becomes longer. Less formation of clusters in the initial NA stage and the formation of PA clusters during quenching may contribute to this effect.
- Multi-stage age hardening for samples after IWQ and VC (simulated industrial quench) are similar, with almost identical PA+NSA hardening and only slightly lower PB hardening for VC samples. The trade-off effect previously found for samples after IWQ still applies for VC conditions, i.e., PA at lower temperature enhances the stability during subsequent NSA but decreases the maximum PB hardening potential and vice versa.

The industry can benefit from this study as most of the findings in laboratory regarding the study of multi-stage ageing can be now directly applied to the industrial conditions, only to consider a slightly higher PB hardening loss in industrial practice.

Acknowledgements

The authors would like to thank Mrs. Christiane Förster for TEM sample preparation and Dr. Anna Manzoni for her help with the TEM measurements.

References

1. B. Milkereit, N. Wanderka, C. Schick and O. Kessler, Materials Science and Engineering A **550**, 87-96 (2012)
2. B. Milkereit and M. J. Starink, Materials & Design **76**, 117-129 (2015)
3. J. L. Cavazos and R. Colas, Materials Science and Engineering A **363**, 171-178 (2003)
4. P. Brenner and H. Kostron, Zeitschrift für Metallkunde **31**, 89-97 (1939)
5. D. W. Pashley, J. W. Rhodes and A. Sendorek, Journal of the Institute of Metals London **94**, 41-49 (1966)
6. M. Saga, Y. Sasaki, M. Kikuchi, Z. Yan and M. Matsuo, Materials Science Forum **217**, 821-826 (1996)
7. L. Zhen and S. B. Kang, Scripta Materialia **36**, 1089-1094 (1997)
8. Y. Takaki, Y. Aruga, M. Kozuka and T. Sato, Materials Science Forum **794-796**, 1026-1031 (2014)

9. Z. Yang, Z. Liang, D. Leyvraz and J. Banhart, Materialia **7**, 100413 (2019)
10. K. Strobel, M. D. H. Lay, M. A. Easton, L. Sweet, S. Zhu, N. C. Parson and A. J. Hill, Materials Characterization **111**, 43-52 (2016)
11. K. Strobel, M. A. Easton, M. D. H. Lay, P. A. Rometsch, S. Zhu, L. Sweet, N. C. Parson and A. J. Hill, Metallurgical and Materials Transaction A **50A**, 1957-1969 (2019)
12. Z. Yang, X. Jiang, X. Zhang, M. Liu, Z. Liang, D. Leyvraz and J. Banhart, Scripta Materialia (accepted for publishing)
13. H. Hirasawa, Scripta Metallurgica **9**, 955-958 (1975)
14. L. Lodgaard and N. Ryum, Materials Science and Engineering A **283**, 144-152 (2000)
15. A. Dons and O. Lohne, MRS Proceedings **21**, 723-728 (1983)
16. S. Zajac, B. Bengtsson and C. Jönsson, Materials Science Forum **396-402**, 399-404 (2002)
17. K. Strobel, M. A. Easton, L. Sweet, M. J. Couper and J. F. Nie, Materials Transactions **52**, 914-919 (2011)
18. P. A. Rometsch, S. C. Wang, A. Harriss, P. J. Gregson and M. J. Starink, Materials Science Forum **396-4**, 655-660 (2002)
19. M. Liu, J. Cizek, C. S. T. Chang and J. Banhart, Acta Materialia **91**, 355–364 (2015)
20. A. Serizawa, S. Hirosawa and T. Sato, Metallurgical and Materials Transactions A **39A**, 245-251 (2008)
21. S. Pogatscher, E. Kozeschnik, H. Antrekowitsch, M. Werinos, S. S. A. Gerstl, J. F. Loffler and P. J. Uggowitzer, Scripta Materialia **89**, 53-56 (2014)
22. A. Falahati, P. Lang and E. Kozeschnik, Materials Science Forum **706-709**, 317-322 (2012)
23. Y. Aruga, M. Kozuka, Y. Takaki and T. Sato, Materials Science and Engineering: A **631**, 86-96 (2015)
24. M. W. Zandbergen, Q. Xu, A. Cerezo and G. D. W. Smith, Acta Materialia **101**, 136-148 (2015)
25. M. W. Zandbergen, Q. Xu, A. Cerezo and G. D. W. Smith, Data in Brief **5**, 626-641 (2015)
26. S. Pogatscher, H. Antrekowitsch, H. Leitner, T. Ebner and P. J. Uggowitzer, Acta Materialia **59**, 3352-3363 (2011)

Evolution of microstructure and elevated-temperature properties during hot rolling and post-rolling annealing process in Al-Mg-Si 6082 alloys

K. Liu*, *C.* Li, *and X.-G.* Chen

Department of Applied Science, University of Quebec at Chicoutimi, Saguenay, Quebec, Canada

Abstract. In the present work, the hot rolling at 400°C and post-rolling annealing at 500°C were applied on heat-treated Al-Mg-Si 6082 alloys with different Mn contents to study the evolution of microstructure and elevated-temperature properties. During the pre-heat treatment before rolling (400°C/2h), a number of fine rod-like dispersoids formed in Mn-containing alloy while only high volume of larger Mg_2Si particles was observed in the matrix of base alloy free of Mn. After hot rolling, the morphology of dispersoids was transferred from rod-like to spherical with finer size and increased number density while the dispersoids gradually coarsened during post-rolling annealing in Mn-containing alloy. The full recrystallization was completed after 1-2 h during annealing in the base alloy, while only partial recrystallization was observed in Mn-containing alloy. The micro-hardness at room temperature and the tensile yield strength at 300°C firstly increased from as-rolled condition to the initial stage of annealing (1 h) for both alloys, which was likely attributed to the dissolution of Mg_2Si during the beginning of annealing. With further increasing annealing time (2-8 h), both the microhardness at room temperature and the elevated-temperature strengths of the base alloy remained similar, while they were slightly decreased in Mn-containing alloy owing to the partially recrystallization and coarsening of dispersoids. However, the elevated-temperature strengths were always higher in Mn-containing alloy than the base alloy while their differences between two alloys was reducing with prolonging the post-rolling annealing time. The tensile fracture surface was observed to be ductile for all the conditions of both alloys but the dimples in Mn-containing alloy were finer and much more uniformly distributed.

Keywords: Al-Mg-Si 6082 alloy; hot rolling ; post -rolling annealing; dispersoids, elevated-temperature properties

1. Introduction

In recent years, the increasing demands in the legislation about the exhaust emissions place higher requirement on the elevated-temperature (250-350°C) performance of materials in automobile industrials, in which the Al-Mg-Si 6xxx aluminum alloys are widely used. As well-known, the mechanical properties of precipitation-strengthening 6xxx aluminum alloys will sharply drop at elevated temperature (e.g. 300°C) under traditional heat treatment (solution treatment followed by artificial aging treatment) due to the fast-coarsening process of precipitates [1]. Therefore, how to improve the elevated-temperature properties of Al-Mg-Si 6xxx aluminum alloys has been highlighted as one of the industrial key concerns. On the other hand, the introduce of thermal-stable dispersoids during the proper heat treatment is proved to greatly improve the elevated-temperature properties in 3xxx and 4xxx alloys [2-5]. Though the evolution of dispersoids are also investigated in Al-Mg-Si 6xxx alloys [6-8], most of them focused on the influence of dispersoids on the recrystallization behaviors at room temperature and the role of dispersoids on the elevated-temperature mechanical properties of 6xxx aluminum alloys is rarely discussed.

Recently, our study has shown that the formation of Mn-containing dispersoids in Al-Mg-Si 6082 alloys is of greatly benefit for the enhancement of Yield Strength (YS) at 300°C, which is increasing from 30 MPa in alloy free of dispersoids to 70 MPa in alloy containing dispersoids after treated at 400°C/2h [9]. Meanwhile, it is also proved that the formation the dispersoids can greatly inhibit the dynamic recrystallization during the hot rolling process [9, 10]. However, limited work has been performed on the elevated-temperature properties as well as the evolution of dispersoids during the hot-rolling and the post-rolling annealing process in Al-Mg-Si 6xxx alloys, which is of significance to their industrial applications.

In the present work, the emphasis was put on the evolution of microstructure and YS at 300°C of Al-Mg-Si 6082 alloy containing various Mn during the hot rolling and post-rolling annealing process with the aid of optical and electron microscopy. Meanwhile, the fracture surface after tensile test at 300°C was analyzed to

* Corresponding author: kun.liu@uqac.ca

discover the role of dispersoids on the failure mode at elevated temperature.

2. Experimental

In the present work, two Al-Mg-Si 6xxx alloys, which are free of Mn (Alloy A) and containing 0.5 wt.% Mn (Alloy B), were cast (Table 1) and then heat treated at 400°C/2h according to our previous study[9, 11]. The hot-rolling was performed at 400°C with the reduction of 90% (from 30 mm to 3 mm). More details about the hot-rolling process can be found in [9]. After rolling, the sample were further annealed at 500°C for 1, 2, 4 and 8h, which was defined as post-rolling annealing process.

Table 1 Chemical composition of alloys in present work

Alloy #	Element（wt. %）				
	Mg	Si	Mn	Fe	Al
A	0.99	0.66	**0**	0.26	Bal.
B	1.07	0.67	**0.52**	0.27	Bal.

The YS at 300°C under various conditions were measured from the tensile tests using Gleeble thermomechanical testing unit and the average value of 3 samples was calculated. The heating rate was controlled at 2°C/S with a holding time of 3 min at target temperature and the strain rate was set as 0.001 S^{-1}. The Vickers microhardness was measured at room temperature with a load of 10g and a dwelling time of 20s. The indentations were assured on the interdendrite cell to evaluate the effect of precipitation/dispersoids on the properties and the average value was calculated from 15 measurements for each condition.

The evolution of microstructure during the rolling and annealing processes, such as the precipitation of Mg_2Si and dispersoids, were characterized by Optical microscopy (OM), Scanning electron microscopy (SEM) and Transmission electron microscopy (TEM) while the fracture surface after tensile test was observed with SEM. Samples were etched with 0.5% HF for 30s to clearer display the large Mg_2Si precipitates and dispersoids. The grain structure during rolling and annealing process was mapped and analyzed by Electron Back-Scatter Diffraction (EBSD) with a step size of 2 μm.

3. Results and Discussion

3.1 Microstructure before and after rolling

According to our previous study [9, 11], it is proved that the dispersoids have been fully precipitated to the stable condition after 400°C/2h. Therefore, Fig.1 shows the microstructure of experimental alloys after 400°C/2h, which was defined as the pre-heat treatment before rolling. As shown in Fig. 1a, high volume of large Mg_2Si particles have precipitated in the matrix besides the Al-Fe-Si intermetallics in Alloy A free of Mn (base alloy), which can be due to the relative low heat treatment temperature (400°C).

However, the volume of Mg_2Si precipitates in the matrix was greatly reduced in Alloy B (Fig.1 b) and they were dominantly located at the dispersoids free zone (DFZ, more details about DFZ can be found in [9]) while some smaller Mg_2Si were occasionally observed in the interdendrite cell. On the other hand, high volume of fine dispersoids in dispersoids zone (DZ) was observed in Alloy B. As shown in Fig. 1c, high quantity dispersoids were precipitated with the morphology of cubic and rods, which were reported to be α-Al(MnFe)Si dispersoids [9]. The dispersoids was reported to nucleate on the pre-formed Mg_2Si and then decrease the volume of Mg_2Si during the heat treatment [6, 12], explaining the less Mg_2Si in the interdendrite cell in Alloy B. However, little dispersoids precipitated at the DFZ due to the lower super saturated solution level of Mn [2], therefore, some lager Mg_2Si can still be observed in DFZ (Fig. 1b).

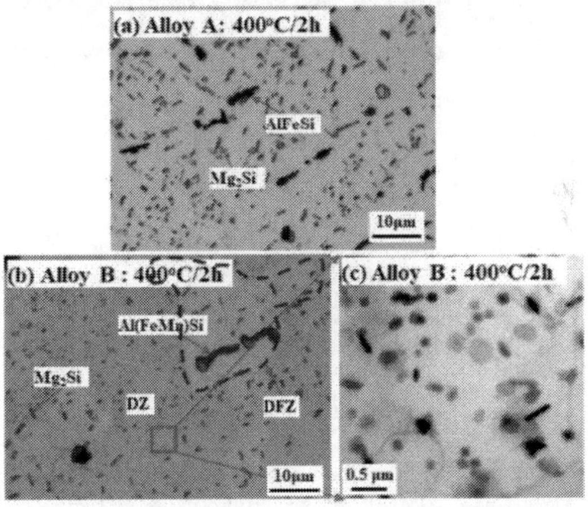

Fig. 1. Microstructure of experimental alloys after 400°C/2h (after etched): (a) Alloy A and (b, c) Alloy B

After pre-heat treatment (400°C/2h), the experimental alloys were hot-rolled at 400°C and the as-rolled microstructure was illustrated in Fig. 2. For Alloy A (Fig. 2a-2b), the Al-Fe-Si intermetallics were fragmented to smaller particles while high volume of Mg_2Si can still be observed (Fig. 2b). As shown in Fig. 1a, the size of Mg_2Si before rolling was so bigger that they were hardly to be dissolved into matrix during the rolling. However, it seems they were also fragmented to smaller size due to the high deformation during the rolling.

Similar to Alloy A, the Al-Fe-Si intermetallics were also fragmentated in Alloy B (Fig. 2c). However, much less Mg_2Si can be observed while the dispersoids were still present in the matrix (Fig. 2d). As shown in Fig. 1b, though lower volume of Mg_2Si were remaining in Alloy B after 400°C/2h, their size was much smaller than that in Alloy A (Fig. 1a), leading to it easier for them to dissolve into matrix during the hot-rolling process. Meanwhile, it is interesting to be noted that the dispersoids after hot-rolling (Fig. 2e) is greatly different with that before rolling (Fig. 1c) that the morphology of dispersoids were much more spherical and their size was finer with higher number density after rolling (Fig. 2e). The reason of this evolution of dispersoids is still under investigation but the favourable conditions promoting precipitation induced by the dislocations or sub-grain boundaries formed during the hot rolling was a plausible cause [13].

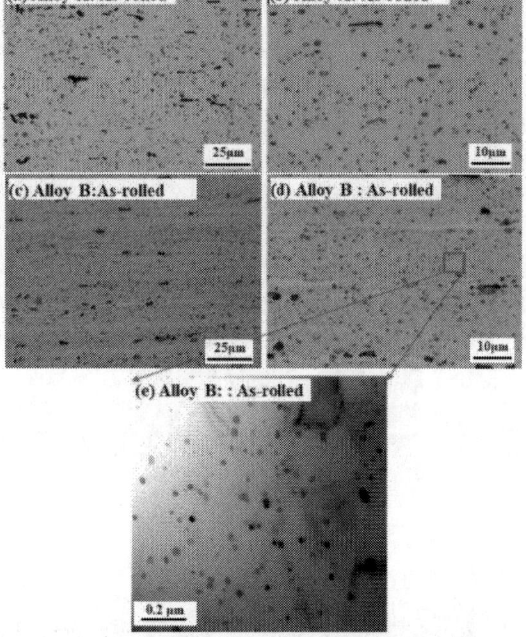

Fig. 2. As-rolled microstructure of experimental alloys (after etching): (a-b) Alloy A and (c-d) Alloy B

Fig. 3 displays the as-rolled grain structure of experimental alloys from the EBSD mapping analysis. It can be found that the as-rolled grain in Alloy A is almost equiaxed while some elongated grains in the rolling direction can still be observed (Fig. 3a), indicating that the partial or close-fully recrystallization during the rolling process in Alloy A occurred. This was also confirmed by the scattered and weaker texture from the {111} pole figure in Fig. 3a. As shown in Fig. 1a and 2b, only coarse Mg_2Si particles and fragmented Al-Fe-Si particles present in Alloy A, which have limited retarding effect on recrystallization during the hot-rolling process. Therefore, the recrystallized microstructure (equiaxed grains) was mostly observed in as-rolled condition of Alloy A.

However, the as-rolled microstructure was different in Alloy B (Fig. 3b) that the lamellar structure was dominate with deformed bands and highly elongated grains. Meanwhile, the {111} pole figure also showed the strong rolling texture, which was similar to the Brass and S texture [14]. This can be attributed to the precipitation of high volume of dispersoids in Alloy B, which have the strong pinning effect on the recrystallization during the hot-rolling [2, 9].

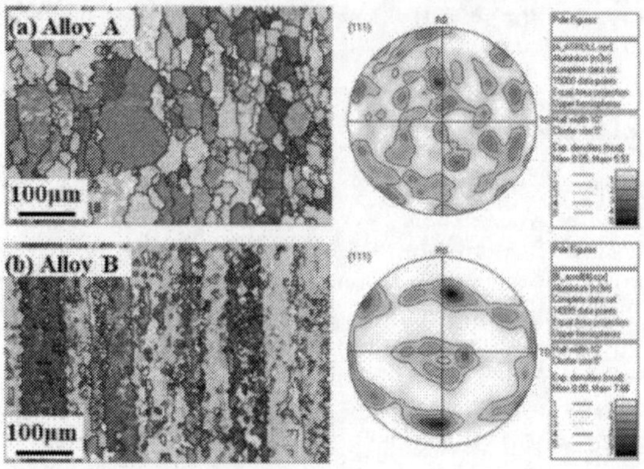

Fig. 3. As-rolled grain structure and the {111} pole figure from EBSD mapping of experimental alloys

3.2 Microstructure during the post-rolling annealing process

As shown in Figs. 2 and 3, different microstructure and grain structure was observed in Alloy A and Alloy B, which will also affect the microstructures during the post-rolling annealing process. Fig. 4 shows the annealing microstructure of experimental alloys after 1h and 8h at 500°C. For Alloy A, it can be seen that only the Al-Fe-Si intermetallics remains in the alloy while almost all the Mg_2Si particles present in Fig. 2b has dissolved into matrix after 1 h annealed at 500°C, as shown in Fig. 4a. The microstructure keeps similar even after 500°C/8h (Fig. 4b) that the matrix is clean with only fragmented Al-Fe-Si intermetallics.

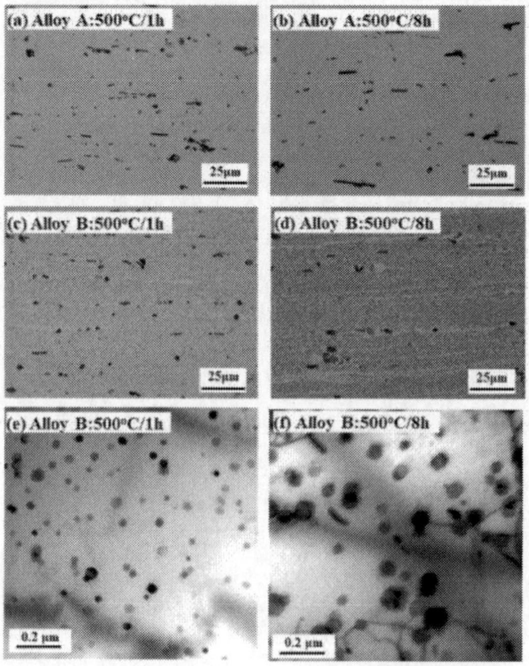

Fig. 4. Annealed microstructure of experimental alloys after 1h and 8h at 500°C

Compared with the as-rolled microstructure of Alloy B in Fig. 2d, all the Mg_2Si in Alloy B after 500°C/1h has also been fully dissolved into matrix (Fig. 4c), which is similar to Alloy A. This can be attributed to the higher annealing temperature (500°C). Meanwhile, it is observed that the dispersoids coarsened during the annealing process. As shown in Fig. 4d of annealed after 500°C/8h, the color of dispersoids zone was much darker, which was further confirmed by the TEM observation in Fig. 4e and 4f. Compared with the dispersoids in as-rolled condition (Fig. 2e), the dispersoids was gradually coarsened after 500°C/1h (Fig. 4e) and they are even much bigger after 500°C/8h (Fig. 4f). The coarsening process of dispersoids during the annealing process can be related to the formed and stored dislocations or sub-grains during the hot-rolling, which was favorite for the nucleation and coarsening of dispersoids [12, 13].

Fig.5 shows the evolution of grain structure of experimental alloys during the post-rolling annealing process, in which the differences can be observed between two alloys. As shown in Fig. 5a-5d, the equiaxed grains were always present in Alloy A free of Mn and the fully recrystallization was assumed to complete only after 500°C/2h (Fig. 5b), representing as the uniformed distributed equiaxed grains and they were gradually growing with the increase of annealing time, as shown in Fig. 5c-5d.

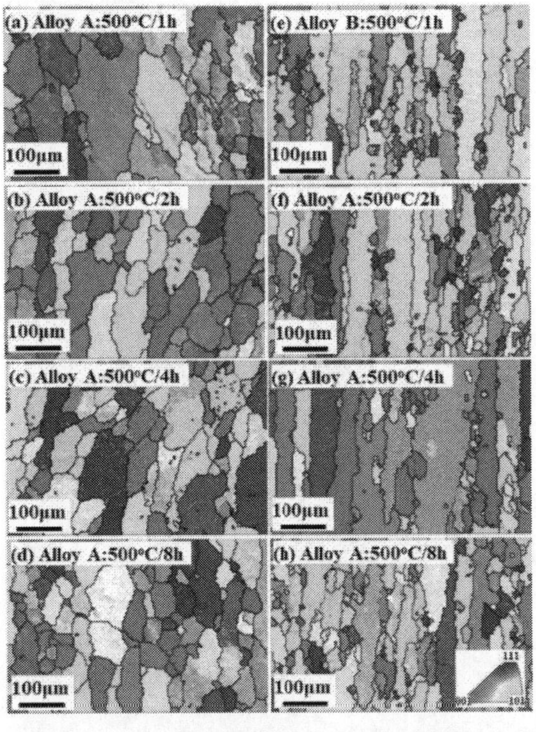

Fig. 5. Evolution of grain structure of experimental alloys during the post-rolling annealing process: (a-d) Alloy and (e-h) Alloy B

However, different grain structures from Alloy A were observed in Alloy B containing dispersoids due to the addition of Mn. As shown in Fig. 5e-5g, the similar lamellar structure to the as-rolled condition (Fig. 3b) with elongated grains and deformation bands was dominate until to 500°C/4h (Fig. 5g) thought some fine recrystallized grains can be observed after 500°C/2h (Fig. 5f) and 500°C/4h (Fig. 5g). With prolonged annealing time (500°C/8h in Fig. 5h), more recrystallized grains were observed between the elongated grains, indicating the partial discontinuous recrystallization occurred [15]. However, the volume fraction of recrystallization was still lower, even after 500°C/8h (< 20 vol.%), and the elongated recovery grains were still dominant structure in Alloy B during the post-rolling annealing process, which is greatly different with the fully recrystallized structure in Alloy A.

The different recrystallization behavior between two experimental alloys can be explained by the formation of dispersoids in Alloy B since Mg_2Si were already dissolved into matrix in both alloys (Fig. 4). In Alloy A, no precipitation or dispersoids were present (Fig. 4a-b) in the matrix and then the recrystallization was easier to occur and even it has fully completed after 500°C/2h. However, there were high density of fine dispersoids formed after pre-heat treatment (Fig. 1c) and they were even much finer with increased number density after rolling (Fig. 2e), which were reported to strongly inhibit the recrystallization through increasing the Zener drag force due to its finer size and higher number density [4, 16]. Therefore, the recrystallization in Alloy B was greatly retarded (Fig. 4e) while it was gradually happened after 500°C/8h due to the reduced Zener drag force from coarsened of dispersoids (Fig. 4f).

3.3 Evolution of elevated-temperature properties during the post-rolling annealing process

As shown in Figs. 4 and 5, it can be observed that the grain structure as well as the formation of precipitates and dispersoids were greatly different between Alloy A and Alloy B after rolling and post-rolling annulling process, which would also lead to the various mechanical properties. In the present work, the tensile tests at 300°C were performed. According to the evolution of microstructure shown in Figs. 4 and 5, the tensile train-stress curves under conditions of as-rolled, 500°C/1h and 500°C/8h were shown in Fig. 6. It can be found the total strain at failure under all these conditions of experimental alloys was in the range 0.3-0.4 (except Alloy A after 500°C/1h), indicating the excellent ductility of experimental 6xxx aluminum alloys at 300°C. Meanwhile, it was also observed that the tensile stress in Alloy B was always higher than that in Alloy A during the tests in all the conditions. As shown in Fig. 6a, the ultimate tensile strength (UTS) at 300°C of Alloy B was measured to be 78 MPa, which was the almost doubled of that in Alloy A (42 MPa). During the post-rolling annealing process, it seemed that the extent of differences on the UTS was decreasing with increasing annealing time. For instance, the differences of UTS between two alloys decreased to 13 MPa after 500°C/1h (Fig. 6b) and further to 7 MPa after 500°C/8h (Fig. 6c).

Furthermore, it was more interesting to notice that the total tensile strain of Alloy B was still higher than Alloy A, especially at as-rolled condition (Fig. 6a) and the beginning of annealing process (Fig. 6b) while the strength of Alloy B was also higher than Alloy A, which was similar to the results in Al-Si 356 alloys that Mo-containing dispersoids improved both the strength and ductility [17].

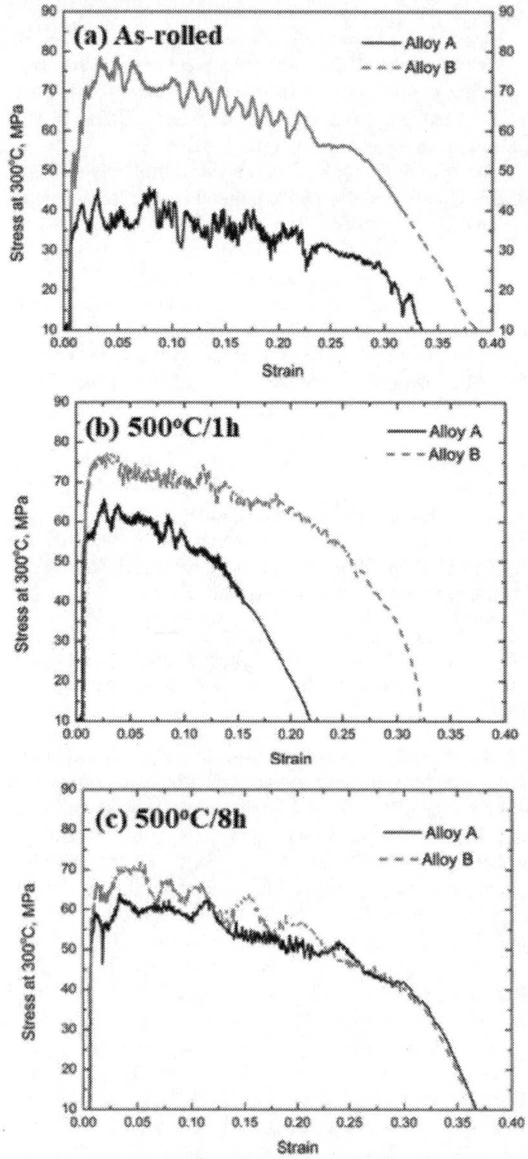

Fig. 6. Tensile strain-stress curves at 300°C of experimental alloys at: (a) as-rolled; (b) 500°C/1h and (c) 500°C/8h

In order to fully evaluate the evolution of mechanical properties of experimental alloys, Fig. 7 displayed the evolution of microhardness at room temperature (RT) as well as the tensile YS at 300°C experimental alloys during the rolling as well as the post-rolling annealing process. As shown in Fig. 7, the evolution of microhardness at room temperature and the tensile YS at 300°C was generally similar for both experimental alloys that it was firstly increased from as-rolled condition to 500°C/1h and then stabilized (Alloy A) or gradually decreased (Alloy B) with prolonging of the post-rolling annealing time, which was agreed with the tendency shown in Fig. 6. For instance, the microhardness of Alloy A increased from 56 HV at as-rolled condition to 77 HV after 500°C/1h and then reduced to 74-75 after 500°C/8h while the YS at 300°C was initially increased from 35 MPa at as-rolled condition to 56 MPa after 500°C/1h and then stabilized at 56-57 MPa until to 500°C/8h. However, it was observed that both the microhardness at RT and the YS at 300°C in Alloy B were always higher than that in Alloy A, especially at as-rolled condition. As shown in Fig. 7, the as-rolled microhardness at RT and YS at 300°C of Alloy B were 80 HV and 54 MPa, respectively, which were much higher than that of Alloy A (56 HV and 35 MPa). However, the differences of mechanical properties between two experimental alloys were reducing with prolonging the annealing time during the post-rolling annealing process. For instance, the differences of YS at 300°C between Alloy A and Alloy B was decreasing from 19 MPa at as-rolled condition to 11 MPa after 500°C/1h and 7 MPa after 500°C/2-4h and then 5 MPa after 500°C/8h.

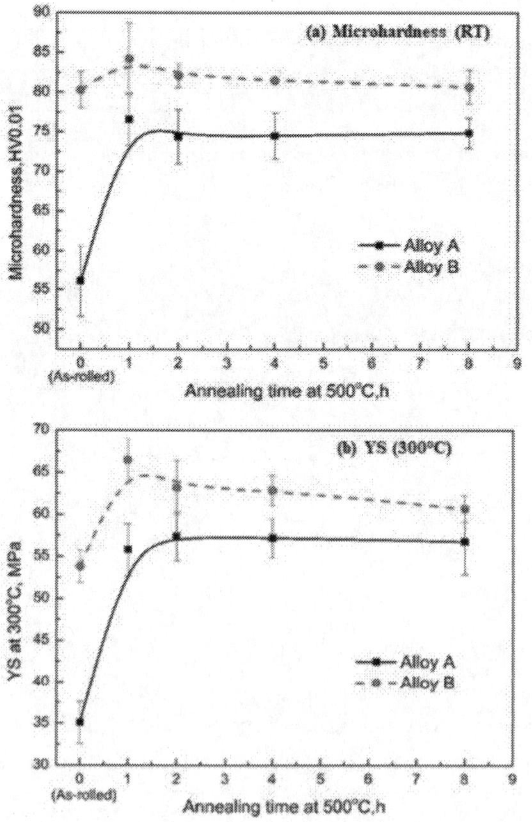

Fig. 7. Evolution of mechanical properties of experimentally alloys during the post-rolling annealing process: (a) Microhardness at RT and (b) tensile YS at 300°C

The evolution of mechanical properties was greatly related to the change of microstructures during the post-rolling annealing process. As shown in Fig. 2 and Fig. 4, high volume of Mg_2Si was observed to fully dissolve into the matrix after 500°C/1h, which can be the likely reason for the initial increase of microhardness and YS from as-rolled condition to 500°C/1h. However, it can be found that much higher volume of Mg_2Si in Alloy A were dissolved than Alloy B after 500°C/1h, explaining the higher increasing degree of mechanical properties of Alloy A at this stage. After 500°C/1-2h, the fully recrystallization was observed in Alloy A followed by a minor growth of grain, leading to the stabilization of microhardness at RT and YS at 300°C while the partially recrystallization and coarsening of dispersoids was occurred in Alloy B with increasing post-rolling annealing time, resulting in its minor reduced mechanical properties during the annealing process. However, the contribution of dispersoids on the mechanical properties can be clearly demonstrated by the difference between both the microhardness at RT and YS at 300°C. For instance, the difference of YS at 300°C at as-rolled condition between Alloy B and Alloy A was 19 MPa due to the high quantity of finer dispersoids as well as the higher Mg and Si solid solution strengthening induced by the formation of dispersoids (Figs. 2e and 4e) and it was still 5 MPa after 500°C/8h though the dispersoids has greatly coarsened (Fig. 4f). It should not be underestimated this "5 MPa" different on the YS at 300°C since every "1 MPa" counts for the mechanical properties at elevated temperature. For instance, it was reported that the creep resistance of materials could be improved one order with increasing of 3 MPa on the threshold stress at 300°C [2, 18].

The fracture surface after tensile test at 300°C of experimental alloys was analyzed using SEM and shown in Fig. 8.

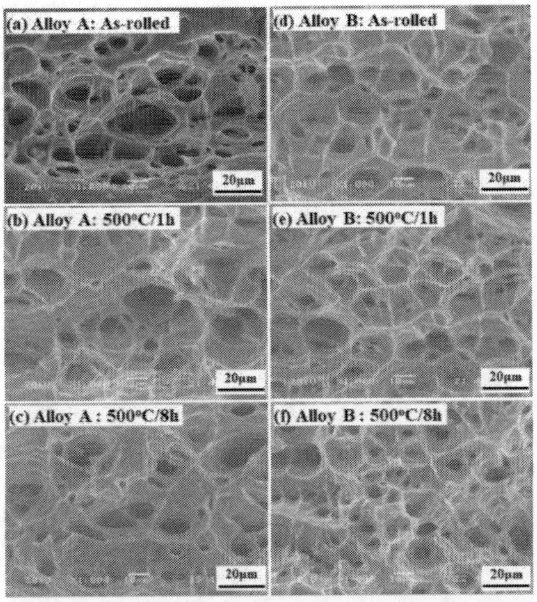

Fig. 8. Fracture surface of experimental alloys after various conditions

It can be found that the failure mode of both experimental alloys at all conditions was ductile due to the high tensile temperature (300°C), representing by the formation of dimples at the fracture surface. However, the dimples in Alloy B were smaller and more uniform than that in Alloy A, which can be attributed to the homogenous deformation due to the presence of dispersoids [17]. During the tensile test, the effect of dispersoids-dislocation interactions became more remarkable due to the increase of dislocation density [19]. It was reported that dispersoids can change the slip system by means of cross-slip at high temperature when the dislocations were trapped by dispersoids, making the homogenization of slip [20], explaining of much more uniformly distributed dimples in Alloy B. Besides, it was also proposed that the homogenization deformation can be highly promoted by converting the dislocation movement from the planar slip to a wavy slip mode through cross slip [21], which would also benefit for the higher ductility. Therefore, the formation of dispersoids was the likely reason for both higher strength and better ductility in Alloy B.

4. Conclusions

In the present work, the evolution of microstructure and elevated-temperature strength in 6xxx alloys containing various Mn was investigated during hot-rolling and post-rolling annealing processes and the following conclusions can be drawn:

(1). After the pre-heat treatment before rolling (400°C/2h), high volume of larger Mg_2Si was precipitated in the base alloy free of Mn while a number of fine rod-like dispersoids were dominate precipitates with lower volume and smaller Mg_2Si in Mn-containing alloy.

(2). During the hot rolling, the dispersoids was transfer from rod-like to the spherical and most of smaller Mg_2Si has dissolved into matrix in Mn-containing alloy. However, high volume of fragmented larger Mg_2Si was still remaining in base alloy free of Mn.

(3). During the post-rolling annealing process at 500°C, fully recrystallization was observed after 1-2h in base alloy free of Mn while only partial recrystallization was observed in Mn-containing alloys, even after 8h.

(4). The elevated-temperature strength and microhardness at room temperature was firstly increase from as-rolled to 500°C/1h followed by a plateau or a gradually reduction in both alloys. However, they were always higher in Mn-containing alloy than the base alloy and the biggest difference was observed at as-rolled condition while the difference was becoming smaller with increasing post-rolling annealing time.

(5). Tensile fracture surface at 300°C was observed to be ductile with the presence of dimples under all the conditions of both experimental alloys, which is confirmed by the high total strain at failure (~ 0.35). However, the dimples were smaller and more uniformly distributed in Mn-containing alloy, leading to its higher ductility.

The authors would like to acknowledge the financial support of the Natural Sciences and Engineering Research Council of Canada (NSERC) and Rio Tinto Aluminum through the Research Chair in the Metallurgy of Aluminum Transformation at University of Quebec in Chicoutimi.

Reference

[1] J. G. Kaufman: *Properties of aluminum alloys : tensile, creep, and fatigue data at high and low temperatures*, ASM International ; Aluminum Association, Materials Park, Ohio; Washington, D.C.,1999.
[2] K. Liu and X. G. Chen, Mater. Des. **84,** 340 (2015).
[3] K. Liu, H. Ma and X. G. Chen, J. Alloys Compd. **694,** 354 (2017).
[4] L. Jin, K. Liu and X.-G. Chen, J. Mater. Eng. Perform. **29,** 126 (2020).
[5] Z. Li, Z. Zhang and X. G. Chen, Mater. Sci. Eng., A **708,** 383 (2017).
[6] L. Lodgaard and N. Ryum, Mater. Sci. Eng., A **283,** 144 (2000).
[7] R. Hu, T. Ogura, H. Tezuka, T. Sato and Q. Liu, J. Mater. Sci. Technol. **26,** 237 (2010).
[8] F. Hichem and G. Rebai, Appl. Phys. A **119,** 285 (2015).
[9] C. Li, K. Liu and X. G. Chen, J. Mater. Sci. Technol. **39,** 135 (2020).
[10] X. Qian, N. Parson and X. G. Chen, Mater. Sci. Eng., A **764,** 138253 (2019).
[11] C. Li, K. Liu, N. Parson and X.-G. Chen. The effect of heat treatment on precipitation behavior of dispersoids in Al-Mg-Si Alloy In: M. W. a. M. Brochu, editor. Proceeding of 16th international Aluminum Alloys Conference (ICAA16). Mcgill University, Montreal, Canada: Candian Institute of Mining, Metallurgy & Petroleum; 2018.
[12] Z. Li, Z. Zhang and X. G. Chen, Metall. Mater. Trans. A **49,** 5799 (2018).
[13] J. D. Robson, T. Hill and N. Kamp, Mater. Sci. Forum **794-796,** 697 (2014).
[14] S. Roy, S. S. D, S. Suwas, S. Kumar and K. Chattopadhyay, Mater. Sci. Eng., A **528,** 8469 (2011).
[15] H. Jazaeri and F. J. Humphreys, Acta Mater. **52,** 3251 (2004).
[16] E. Nes, N. Ryum and O. Hunderi, Acta Metall. **33,** 11 (1985).
[17] S. Chen, K. Liu and X. G. Chen, J. Mater. Res. **34,** 3071 (2019).
[18] A. R. Farkoosh, X. G. Chen and M. Pekguleryuz, Mater. Sci. Eng., A **627,** 127 (2015).
[19] O. Sitdikov, E. Avtokratova, T. Sakai and K. Tsuzaki, Metall. Mater. Trans. A **44,** 1087 (2013).
[20] S. W. Nam and D. H. Lee, Met. Mater. Int. **6,** 13 (2000).
[21] K. C. Kim and S. W. Nam, Mater. Sci. Eng., A **244,** 257 (1998).

https://doi.org/10.1051/matecconf/202032603002

Relationship between mechanical properties and age-hardening behavior around grain boundaries of Al-Mg-Si alloy

Takuya Hashimoto[1,*], *Ken-ichi* Ikeda[2], and *Seiji* Miura[2]

[1]Graduate School of Engineering, Hokkaido University, Kita13, Nishi8, Kita-ku, Sapporo, Hokkaido, 060-8628, Japan
[2]Faculty of Engineering, Hokkaido University, Kita13, Nishi8, Kita-ku, Sapporo, Hokkaido, 060-8628, Japan

Abstract. Nanoindentation tests were conducted near the grain boundary (GB) of the Al-Mg-Si alloy, and the influence of GB character on the aging precipitation behavior and the mechanical properties was confirmed. After obtaining the GB characters by electron back scattered diffraction (EBSD) analysis, nanoindentation tests were carried out on under-aged, peak-aged, and over-aged samples. And then, the indentation areas were observed by back scattered electrons imaging (BSE) in order to identify indentation positions with respect to the GB. In this study, for the GB character, focusing on the rotation angle, the high-angle GB (HAGB) and the low-angle GB (LAGB) were selected. In addition, coincident site lattice GBs (CSL) were selected as the special GB. In the 180°C under-aged samples, the nano-hardness near GB is higher than that far from GB, while 180°C peak-aged samples, the nano-hardness is lower than that far from GB. Then the range near the GB where the hardness changes was larger at HAGB than at LAGB and CSL3. This suggests that the GB character affects the aging precipitation behavior and mechanical properties.

1 Introduction

Al-Mg-Si alloys are superior in workability and corrosion resistance as well as bake hardenability (age hardenability) compared to other aluminum alloys and are used as body panels for automobiles. Bake hardenability means that the hardness increases by heat treatment of paint bake after molding the body of an automobile, and it is considered that the expression of this hardenability is attributed to aging precipitates formed during heat treatment. Therefore, when developing products utilizing the age-hardening properties of Al-Mg-Si alloys, it is necessary to understand the effects of aging precipitates on strength.

When an aging heat treatment is applied to an Al-Mg-Si alloy, it is proposed that the aging precipitates transition in the following process [1-2]:

$$\text{SSSS} \rightarrow \text{Mg-Si cluster} \rightarrow \text{GP zone} \rightarrow \beta'' \rightarrow \beta' \rightarrow \beta.$$

Precipitates that are predominantly precipitated in the alloy differ depending on the aging temperature and time, and β″ is a precipitate involved in the automobiles body manufacturing process.

As the heat treatment is continued, β″ eventually transits to β', a rod-like semi-coherent precipitate, and to plate-like non-coherent precipitate, β (Mg_2Si), the final stable phase. They have a smaller contribution to strength as compared to β″ and are rarely used as the main reinforcing phase. However, β' and β themselves are considered to precipitate on the grain boundaries depending on the heat treatment temperature and time. In order to clarify the intergranular fracture during the tensile tests and the effect of GB precipitates on the improvement of alloy strength, it is necessary to clarify the characteristics of the GB precipitates. Therefore, it is necessary to clarify the nature of GB precipitates.

As the strength within each grain interior of the polycrystalline material increases, embrittlement due to GB fracture occurs, and the effect of the GB cannot be ignored when using structural materials. Matsuda *et al.* [3] reported the relationship between the Schmid factor for the tensile axis and the stress transfer coefficient in the age-hardened Al-Mg-Si-based alloy, which is expanding its application as a lightweight structural material. Cai *et al.* [4] also systematically classified the precipitate-free zone (PFZ), which is one of the factors preventing the increase in strength, in the age hardening type Al-Zn-Mg-Cu alloy, which has the highest strength among aluminum alloys, based on GB character. The width of the PFZ tends to be wider as the quenching speed decreases [5], which reduces the strength of the material. In order to reduce the PFZ width, it is necessary to narrow the PFZ width by increasing the quenching rate.

Here, the GB character is the property of the GB classified by the relative orientation (rotation angle / axis) of the adjacent crystal grains and the grain interface orientation, and this property is known to greatly affect various properties of a material. Random GBs, which are ordinary GBs, have a high energy structure and are likely to cause GB degradation phenomena. On the other hand, there is a grain boundary with a regular arrangement of atoms and few gaps, called the corresponding grain boundary. On the other hand, there is a type of grain

* Corresponding author: takuya_hashimoto@eis.hokudai.ac.jp

Table 1
Chemical compositions of the alloys utilized in this work (mass%).

Al	Si	Fe	Mn	Mg	Zn	Cr	Ti	Mg/Si
Bal.	0.344	0.042	0.001	0.621	0.003	0.001	0.011	2.1

boundary called the coincidence site lattice boundary (CSL GB), in which atoms are regularly arranged with a small number of gaps between them. In particular, the $\Sigma 3$ GB, which is classified as CSL GB, has low GB energy. It is also considered to be less prone to GB degradation phenomena in aluminium [6]. Therefore, GB engineering (GBE), which is a technique intended to improve the characteristics of polycrystalline materials by introducing such corresponding grain boundaries into many materials, has been developed [7]. As an application example Shimada *et al.* [8] reported that in SUS304L, random GBs that could be the starting point of corrosion and fracture were reduced, some CSL GBs were introduced, and the corrosion rate was suppressed to one-quarter of that of the base metal. As mentioned above, attention has been paid to studies relating the properties of materials and the factors affecting them to the GB character. However, at present, there are many points to be discussed about the influence of the GB character on the mechanical properties through the aging precipitation process.

In this study, Al-Mg-Si based alloys subjected to under-aging, and peak-aging heat treatments and pure aluminum from which Mg and Si were removed from the Al alloy were subjected to nanoindentation test and transmission electron microscopy (TEM) observation of near GBs. And the influence of GB character on aging precipitation behavior and mechanical properties was confirmed.

2 Experimental procedure

2.1. Under-aged heat treatment

The chemical compositions of the alloys used in this study are listed in Table 1. The forged material was homogenized at 530°C for 6 hours, and then hot-rolled to a thickness of 4 mm. After that, a sample that had been subjected to cold rolling to have a thickness of 4 mm to 1 mm was used as a sample and was cut so that RD × TD was 15 mm × 10 mm to obtain a sample.

The samples were solution treated at 550°C for 1 hour and subsequently quenched into iced water to produce a supersaturated solid solution (SSSS). Then, the average grain size was about 350 μm. Thereafter, The SSSS samples were artificially aged at 180°C in an oil bath. After being taken out of the oil bath at the under-aged stage and quenched, mechanical polishing and buff polishing were immediately conducted, and finishing treatment was carried out by electrolytic polishing. As the electrolyte, a solution mixed at a ratio of perchloric acid: ethylene glycol: ethanol = 1: 2: 7 was used.

2.2. Acquisition of GB character distribution

GB character distribution classified by Brandon's criterion [9] was obtained by electron back scattered diffraction (EBSD) analysis using TSL solution OIM D.C. attached to scanning electron microscopy (FE-SEM) , JSM-6500F. Focusing on the rotation angle, the high-angle (HAGB), low-angle (LAGB) were selected. In addition, CSL GBs were selected as the special GB. Table 2 summarizes the names of each grain boundary in this study. The step size was selected 10 or 3 μm. Thereafter, a back scattered electron (BSE) image was obtained by SEM observation, and the microstructure was confirmed.

2.3 Nanoindentation test

The relative distance from the indentation position was confirmed from the grain map and the BSE image, and a nanoindentation test (TI950 Tribo indenter) was conducted near the selected GBs. The maximum load was set to 500 μN, the interval between measurement points was set to 1, 2, or 3 μm, the loading rate was set to 50 μN/s, and the keeping time was set to 10 s. The indention areas were observed by the BSE image. The distance to the GB was calculated as the vertical distance to the GB.

2.4 Additional peak-aged heat treatment and other heat treatment

After the nanoindentation test for 180°C and 250°C under-aged samples, aging at the same temperature was added to produce a peak-aged samples, and the same GB as the under-aged sample was subjected to the test described in section 2.3. Here, the aging conditions and names of each sample are summarized in Table 3.

2.5 Microstructure observation

Φ3 mm disk-shaped samples for transmission electron microscopy (TEM) observation were prepared by twin-jet electrolytic polishing. As the electrolytic solution, a

Table 2
Selected grain boundaries and their misorientation angle.

GB	High-angle	Low-angle	$\Sigma 3$
Name	HAGB	LAGB	CSL3
angle[°]	15~180	2~15	60.00±8.66
axis	-	-	[111]

Table 3
Aging conditions for each sample.

Name	180U	180P
temperature [°C]	180	180
time [h]	1	8

Table 4 Grain boundary character distribution (GBCD) and grain size of 180°C aged and SSSS samples.

sample	GBCD [%]					Grain size[μm]
	HAGB	LAGB	CSL3	CSL5	otherCSL	
180°C	69.1	20.2	2.02	2.85	5.11	341.8±136.7
SSSS	73.4	14.6	2.42	2.19	7.75	384.9±184.0

solution mixed at a ratio of nitric acid: methanol = 1: 4 was used. TEM observation (JEM-2000FX) was carried out each sample to observe the microstructures within the grain interior and at the GBs.

3 Results

3.1 Mechanical test

Table 4 shows the grain boundary character distribution (GBCD) and the grain size of aged sample at 180°C, an aging heat treatment and a supersaturated solid solution sample prepared for comparison. After aging heat treatment, GBCD and grain size of both samples are almost same. From this comparison, it is considered that the microstructure is not significantly affected by the aging heat treatment.

Fig. 1 shows the nano-hardness of 180°C under and peak aged samples. Fig. 1(a) and (d) show the nano-hardness around HAGB of under-aged and peak-aged samples, respectively. The average hardness in the grain interior of the peak-aged sample is larger than that of the under-aged sample. The hardness near the GB of the under-aged sample is larger than that of grain interior. In the peak-aged sample, the hardness near the GB is smaller than that of its grain interior. Furthermore, it can be confirmed that the range near the LAGB and the $\Sigma 3$ CSL GB where the hardness increases, or decreases is smaller than that near the HAGB.

3.2 Microstructure

Fig. 2 shows TEM images of the sample aged at 180°C. In 180U (Fig. 2(a)), a linear contrast was observed in the matrix, suggesting the precipitation of β″. In 180P (Fig. 2(b)), many needle-shaped contrasts along the <100> direction were uniform in the matrix. And the precipitation of β″ is suggested. Its density is higher than that of the 180U grain interior.

Precipitates were distributed on the GBs, and large precipitates were sparsely distributed in 180P (Fig. 2(c)) compared to 180U. In addition, the PFZ is formed along the GB under any aging conditions, and the precipitates outside the PFZ are larger than the precipitates in the grain interior. At 180P, it can be confirmed that the precipitates are dispersed at a lower density than in the grain interior.

4 Discussion

4.1 Local deformation behavior around grain boundary

In Al-Mg-Si based alloys, it is known that the strengthening phase after aging at 180°C is β″. Fig. 2 (a) and (b) show that the under-aged sample also contains β" precipitates that can be confirmed by diffraction contrast. However, the number density is smaller than that of peak-aged sample, which means that Mg-Si clusters and GP zones, which are not visible in the diffraction contrast, are

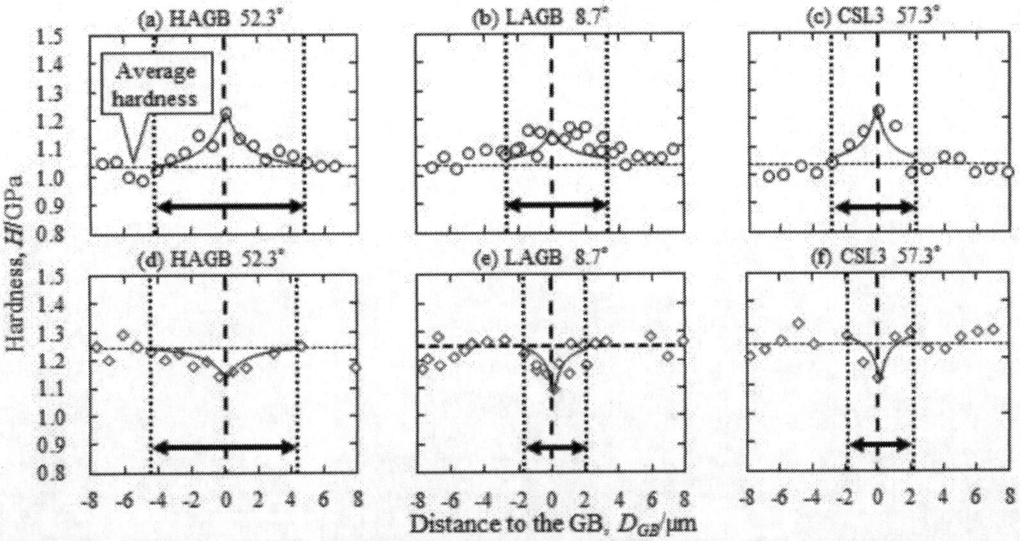

Fig. 1 Nano-hardness around grain boundaries at aging temperature of 180°C; (a) at under-aged around HAGB, (b) at under-aged around LAGB, (c) at under-aged around CSL3, (d) at peak-aged around HAGB, (e) at peak-aged around LAGB and (f) at peak-aged around CSL3.

considered to exist densely in the grains. With the increase in the aging time, many Mg-Si clusters and the GP zone become β" and the number density of β" increased. Therefore, it can be understood that the average hardness in the grain interior increased at the peak-aged sample than at the under-aged sample.

In Fig. 2 (a) and (c), some larger precipitates than the intragranular precipitates were observed in a region just outside the PFZ. This indicates "Precipitate Sparse Zone: PSZ" and "Transition-area" reported by Ma *et al* and Ogura *et al.* [10-11]. In the Al-Zn-Mg based alloy, Ogura *et al.* reported that during aging heat treatment, the concentrations of Zn and Mg dissolved in the matrix decreased as the distance to the GB decreases, and that increasing the aging time also reduced the average Zn and Mg concentrations in the PFZ. This indicates that the solute concentration of the matrix shifts to a lower concentration near the GB than in the grain interior, and the solute concentration further decreases by increasing the aging time. Miyazaki *et al.* [12-13] systematically investigated the change in composition with composition using the "composition gradient method". Miyazaki *et al.* reported that in Cu-Ti and Ni-Al alloys, the solute concentration in the matrix decreased as it approached the phase interface, and the precipitates on the lower concentration region were more sparsely distributed than the precipitates on the higher concentration region. This is because there is a dependence between the equilibrium composition at the matrix/precipitate interface and the size of the precipitate, and the precipitation on the low concentration side follows the Gibbs-Thomson equation [14] .

As a result, the critical radius in the precipitation increases, so that the precipitate size increases on the low concentration region. In addition, it is conceivable that precipitates are coarsely distributed by reducing the mass of solute atoms used for precipitation because of decreasing the solute concentration.

Based on the above, "PSZ" confirmed from the results of this experiment is considered to be formed in the following process.

(1) The solute atoms diffuse to the GBs and form GB precipitates.

(2) The solute concentration in the matrix near the GBs decreases due to the formation of GB precipitates.

(3) According to the Gibbs-Thomson relation, large precipitates nucleate at low concentration region.

(4) In the low-concentration region, precipitates are coarsely distributed because there are fewer solute atoms in the matrix used for precipitation than in the high-concentration region.

From these consideration, it is considered that larger precipitates were nucleated and coarsely distributed because the concentration near the GBs further decreased with increasing aging time at 180P. Therefore, it is considered that the amount of precipitation strengthening decreased in PSZ near the GB, and the nano-hardness decreased as compared with the nano-hardness in the intragranular region. On the other hand, in the grain interior of 180U, the precipitation was not sufficient due to the short aging time, and the parallelogram-type β" which is known as the strengthening phase had not been formed. It is considered that the precipitates dominated by the Orowan-type strengthening mechanism were more occupied than in the grain interior, so that in the grain interior, solid solution strengthening and slight precipitation strengthening acted. In PSZ, the amount of solid solution strengthening decreased, but the amount of precipitation strengthening acted greatly. From this assumption, it is considered that the hardness near the GB showed a relatively higher value than the hardness in the grain interior.

In addition, since there is no precipitate in the PFZ, it is considered that only the solid solution strengthening is affected, but no decrease in hardness on the PFZ in 180U was confirmed. This is because the indentation size is much larger than the PFZ width, and even if the indentation just above the PFZ is affected by the precipitates in the GBs, PSZ and grain interior.

4.2 Effect of grain boundary characters on precipitation behavior

From the results of 180U and 180P, the behavior of mechanical properties near the GB was different depending on the GB character.

Cai *et al.* [4] suggests that low-angle GBs with a misorientation of 11 ° or less have the maximum diffusion activation energy of solute atoms among all types of GBs and are random GBs. It is conceivable that the GB diffusion in the LAGB is much more difficult than in the HAGB. In addition, the atomic matching of $\Sigma 3$ is good, so that solute atoms are less likely to diffuse than random

Fig. 2 TEM images of the aged samples, (a): 180U, (b): grain interior in 180P, (c): around GB in 180P.

GBs. When the GB is regarded as a dislocation array, it is considered that the random GB contains more nucleation sites than the LAGB and low-Σ CSL GB. Therefore, the volume fraction of the precipitate on the random GB is larger than that of the LAGB or low-Σ CSL GB. As a result, the PFZ width around HAGB also increases.

From the results of nanoindentation on the 180°C aged sample, PSZ of the hardness near LAGB and CSL3 is smaller than that of HAGB because the volume fraction of precipitates on GB is small and the PFZ width is narrow. This is probably due to the small influence of the precipitate in PSZ.

From the above results, it is suggested that influence of the GB character on the aging precipitation behavior exists.

Furthermore, in this experiment, the influence of the GB plane normal were not considered, so further investigation is needed in the future.

5 Conclusion

Nano-indentation and TEM observation were performed on Al-Mg-Si alloys to investigate how GB characteristics affect aging precipitation behavior and local mechanical behavior. The summary of the results obtained is as follows.

(1) At 180°C aging, in the under-aged stage, the density of the precipitates having high resistance increased on the grain boundaries and PSZ, so that the nano-hardness increased as decreasing the distance to the GBs. In the peak-aged stage, large precipitates nucleated due to the decrease in the solute concentration near the GBs. The coarse distribution of the precipitate reduced the amount of precipitation strengthening, and decreased the hardness as approaching the GB.

(2) The volume fraction of precipitates is smaller on LAGB and CSL3 than on HAGBs due to the difficulty of diffusion of solute atoms. Therefore, the PFZ width and the PSZ become narrower, and the range affected by precipitates on the PSZ around LAGB and CSL3 becomes narrower than around HAGB.

References

1. G. A. Edwards, K. Stiller, G. L. Dunlop, M. J. Couper, Acta Materialia, **46**, 11 (1998).
2. M. Murayama, K. Hono, Acta Materialia, **47**, 5 (1999).
3. K. Matsuda, J. Tsukiyama, Y. Uetani, S. Ikeno, Journal of Japan Institute of Light Metals, **58**, 8 (2008)[in Japanese].
4. B. Cai, B. L. Adams, T. W. Nelson, Acta materialia, **55** (2007).
5. P. N. T. Unwin, G. W. Lorimer and R. B. Nicholson: Acta Met., **17** (1969).
6. K. Saitoh, H. Kitagawa, A. Nakatani and S. Ogata: Journal of the Society of Materials Science, Japan, **46** (1997)[in Japanese].
7. T. Watanabe, Res Mechanica, 11 (1984).
8. M. Shimada, H. Kokawa, Z. J. Wang, Y. S. Sato, I. Karibe, Acta Materialia, **50** (2002).
9. D. G. Brandon, Acta marerialia, 14 (1993).
10. P. P. Ma, C. H. Liu, Q. Y. Chen, Q. Wang, L. H. Zhan, J. J. Li: Journal of Materials Science & Technology **46** (2020).
11. T. Ogura, S. Hirosawa, T. Sato, Science and Technology of Adovanced Materials, **5** (2004).
12. T. Miyazaki, T. Koyama, and S. Kobayashi, Metallurgical and Materials Transactions, **27A** (1996).
13. T. Miyazaki, S. Kobayashi, and T. Koyama, Metallurgical and Materials Transactions, **30A** (1999).
14. J. W. Martins and R. D. Doherty : *Stability of Microstructure in Metallic System*, Cambridge University Press, Cambridge, United Kingdom, 1976

Study on Microstructure and Mechanical Properties of Al-Mg-Si-Cu Aluminium Alloy with High Ductility

Guanxia Xue [1], *Gu* Zhong [1,*], *Shipeng* Lin [1], *Hu-Tian* Li [2], *Xinghui* Gui [1], and *Lei* Zhang [3,4]

[1]Chinalco Materials Application Research Institute Co., Ltd., Suzhou Branch, Suzhou 215026, China
[2]Chinalco Materials Application Research Institute Co., Ltd., Beijing 102209, China
[3]Biomanufacturing Center, Department of Mechanical Engineering, Tsinghua University, Beijing 100084, China
[4]Biomanufacturing and Rapid Forming Technology Key Laboratory of Beijing, Beijing 100084, China

Abstract. A new type of Al-Mg-Si-Cu aluminium alloy with high ductility was studied in the present work. The microstructure features and mechanical properties of this alloy were systematically characterized by scanning electron microscopy (SEM), electron back-scatter Diffraction (EBSD), high resolution transmission electron microscopy (HRTEM) and tensile and fatigue test. The percentage of sub-grain boundary under forging and aging process reaches up to 72% which can be attributed to the suppression of recrystallization by the nano-sized AlMnCrSi dispersoids. The combination of mechanical properties of the new alloy product in aged state showed that the ductility keeps in the range of 15~18%, yield strength and tensile strength are 310MPa and 380MPa respectively, fatigue strength ranges from 130MPa to 135MPa. It presents more excellent properties than commercial 6061 alloy for the nano-sized AlMnCrSi dispersoids, initial-β" precipitates and high percentage of sub-grain boundary.

Keywords: Microstructure; Mechanical Property; Al-Mg-Si-Cu Aluminum alloy; High Ductility

1. Introduction

6xxx series aluminum alloys are extensively used as structural parts of vehicles due to their excellent combination of properties, such as medium strength, formability, good corrosion resistance, weldability, and low manufacture cost [1-3]. Traditional forged vehicle parts are made of commercial 6061 alloy, which possess yield strength, tensile strength and ductility up to 280MPa, 320MPa, 10%-12% respectively. In consideration of energy conservation and environmental protection, lightweight is the development trends in automotive industry. Nowadays, in order to reduce weight of the vehicle forging parts, studies focus on developing high strength aluminum alloy through alloying compositions optimization, heat treatment and processing technology [4-8]. The microstructure and mechanical properties of Al–Mg–Si–Cu alloy with high manganese content are studied. The dominant strengthening precipitates comprising the needle-shaped pre-β″or β″ ($(Al+Mg)_5Si_6$ or Mg_5Si_6) and lath-shaped Q′ ($Al_3Cu_2Mg_9Si_7$) phases are identified in the T6 temper. With the increase of magnesium content, S (Al_2CuMg) phase is promoted to precipitate to give an enhancement in strength. The yield strength of the examined alloys with high manganese content is up to 380MPa, which is 50% higher than that of commercial 6061 alloy. It is considered that, in addition to the strengthening precipitates, Mn dispersoids generating the dispersion hardening effect and the homogeneous deformation contribute a lot to the favorable mechanical properties [9.10].

Traditional forging process of vehicle parts is relatively complicated, usually more than two-step. For example, forged hubs of 22 inches commercial vehicle of commercial 6061 alloy are produced by two-step forge and one-step spinning process [11]. However, in order to further save cost, improve fatigue resistance, and lightweight auto parts of 6xxx series aluminum alloy by using one-step forging process, developing high toughness Al–Mg–Si–Cu alloy is necessary.

In this paper, a new type of Al-Mg-Si-Cu aluminum alloys with high ductility was studied. The microstructure features and mechanical properties of this aluminum alloy were systematically determined by SEM, HRTEM and tensile test respectively.

2. Experimental procedure

2.1 Material preparation

The chemical composition of the new Al–Mg–Si–Cu alloy studied is listed in Table 1. The content of excess Si was controlled from 0.1 to 0.5. The material was prepared as follows.

* Corresponding author: zghust@qq.com

Table 1. The chemical composition of the Al–Mg–Si–Cu alloy used in the present work (wt.%).

Element	Si	Fe	Mg	Mn	Cu	Ti	Cr	Al
Content	0.746	0.141	0.73	0.653	0.427	0.024	0.268	Bal

(i) The billets of this new alloys were prepared experimentally by direct-chill (DC) casting, which was of 154 mm in diameter and 2500 mm in length. All billets were machined to make samples with 130 mm in diameter and 300 mm in length.

(ii) In order to eliminate elemental micro-segregation and form Mn and Cr containing dispersoids homogeneously, the homogenization treatment at 560°C for 5h subsequently air-cooled was carried out for these samples.

(iii) After homogenization treatment, samples were hot forged at 450°C~500°C to make 120 mm thick plates with the reduction ratio of 60%.

(iv) The forged plates were solution treated at 560°C for 3 h and quenched into water at room temperature. The subsequent aging for a short-time was performed at 160°C~170°C for 2h~3h in an air furnace.

2.2 Characterization

Microstructural observation of the studied Al–Mg–Si–Cu alloy was performed by using a 15 kV JSM-6480 type scanning electron microscope (SEM) and the type of intermetallic phases was identified by energy dispersive system (EDS). The effect of homogenization treatment on the percentage of sub-grain boundary was determined by examining hot forged specimens using electron back-scatter diffraction (EBSD) technique. The surface of the examined specimens was mechanical polished and electrochemical polishing sequentially. The experimental data were collected using a JSM-6480 type scanning electron microscope operating at 25kV with a TSL orientation imaging system. OIM analysis 4.6 software was used to analyze the EBSD map. The scanned area was typically 100um×240um with a step size of 0.8um. At least three different areas were selected and the reported results were the average values. The morphology and distribution of nano-precipitates of AlMnCrSi dispersoids during the homogenization annealing and forged process and aging treatment were investigated using transmission electron microscopy (TEM), JEOL-2100F. The morphology and distribution of aged precipitates were investigated by using high resolution transmission electron microscopy (HRTEM) with the symmetric condition along Al (001) axes. Thin foils for TEM study were prepared by first mechanical polishing to a thin foil about 80 um and then electro-polishing in a methanol solution mixed with 30% nitric acid and 70% alcohol. To evaluate the strength and ductility of the examined alloys, tensile tests were carried out. The tensile tests were conducted on an Instron 5967 30KN Materials testing machine operated at a constant crosshead speed with an initial strain rate of 5×10^{-4} s^{-1}. The tensile specimens were machined perpendicular to the forging direction as GB/T228.1-2010 standard with the gauge length of 50 mm. The schematic of tensile specimen is depicted in Fig. 1(a). As for tensile test, three specimens were tested and the average value was taken. High cycle fatigue (HCF) tests were obtained for constant amplitude loading with sinusoidal wave forms. The fatigue life tests (S-N curves) were carried out on specimens (midsection diameter 5.0 mm). These tests were performed at R = - 1 using MTS Landmark 370.10 fatigue testing machine with a frequency of 60 Hz. The S-N curves were measured in laboratory air. The schematic of fatigue specimen is depicted in Fig. 1(b). The final surface finish of the fatigue specimen after machining requires 0.32 and no further operation is carried out.

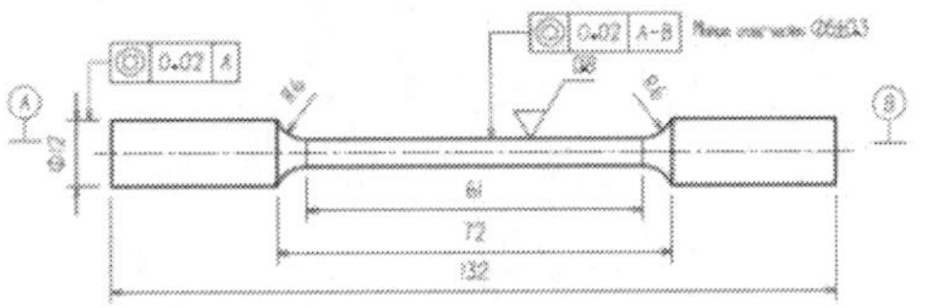

(a)

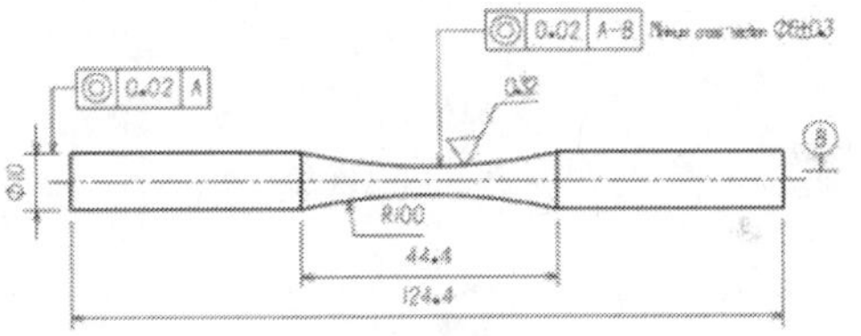

(b)

Fig.1. Schematics of the test specimens
(a) tensile specimen; (b) fatigue specimen.

3 Results and discussion

3.1 Microstructural characterization

The as cast microstructures of the new Al–Mg–Si–Cu alloy is shown in Fig.2. Substantial distinct intermetallic constituents are observed in the α-Al inter-dendritic regions, forming almost continuous network. Fig. 3 shows higher magnification microstructure of the different intermetallic constituents, which contain three intermetallic phases: Mg_2Si, Al(Fe,Mn,Cr)Si and AlCuMgSi. They were identified by a combination of morphology and EDS analysis (Table 2). The Al(Fe,Mn,Cr)Si intermetallic compound, which commonly known as α-phase, with a typical feathery structure or lamellar morphology, is the dominant phase in this new Al–Mg–Si–Cu alloy. Mg_2Si phase appears as needle-shaped or Chinese script-like structure. A small

* Corresponding author: zghust@qq.com

quantity of AlCuMgSi phases with round or oval-shaped morphology were observed to form as separate phase in the inter-dendritic regions.

In as cast Al–Mg–Si–Cu alloys, the remaining content of Si which does not form intermetallic compound Mg_2Si is defined as excess Si. Generally, the more excess Si the alloy contains, the higher tensile strength and yield strength will be [9]. In our opinion, a small amount of excess Si, ranging from 0.1 to 0.5 wt.%, is favorable for improving the ductility of the Al–Mg–Si–Cu alloy while ensuring the tensile strength and yield strength of the alloy. It can be found that, in Fig.3, in as cast microstructure of the Al–Mg–Si–Cu alloy, two intermetallic phases with excess Si are Al (Fe,Mn,Cr)Si and AlCuMgSi. In Table. 2, based on the EDS analysis, the average composition of the Al (Fe,Mn,Cr)Si phase are 72.33wt.% Al, 20.78 wt.% (Mn+Cr+Fe), and 6.89 wt.% Si. It is close to α-$Al_{15}(Fe,Mn)_3$-Si_2 as reported in literature[11]. The AlCuMgSi phase has a variable copper content from 5.64 to 10.01 mass% depending on Mg/Si ratio.

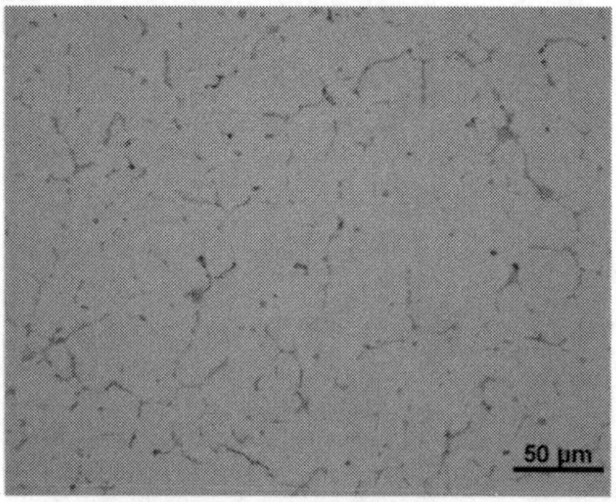

Fig.2. Microstructure of the new Al–Mg–Si–Cu alloy in as-cast state

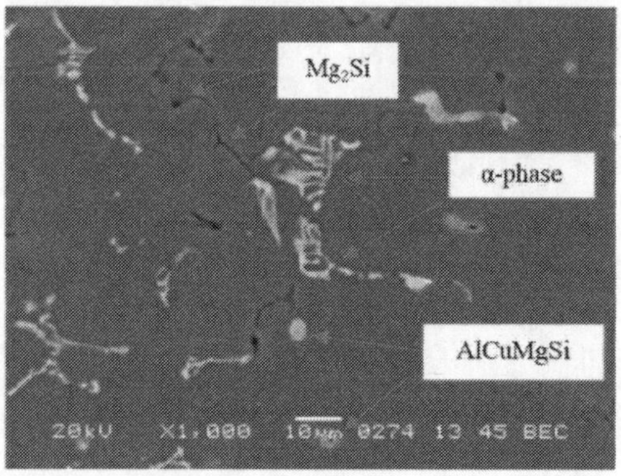

Fig.3. Higher magnification microstructure of the new Al–Mg–Si–Cu alloy showing the different morphology of intermetallic constituents.

Table 2. Composition of the experimentally observed intermetallic phases in Fig. 3 (wt.%).

Phase	Mg	Si	Cu	Cr	Mn	Fe	Al
Mg_2Si	8.44	11.77	-	-	-	-	79.79
	8.99	12.74	0.49	-	-	-	77.78
α-phase	1.00	7.01	1.21	2.22	8.00	9.20	71.37
	0.96	3.40	0.99	1.10	5.76	6.57	81.22
AlCuMgSi	4.17	10.57	8.14	-	-	-	77.12
	2.00	8.80	10.01	-	-	-	79.19

The microstructure of this new alloy after homogenization annealing is shown in Fig. 4. The majority of Mg_2Si and the all of AlCuMgSi phases were dissolved into the Al matrix. It can be seen from Fig. 5 that there was a fragmentation and spheroidization of the α-phase with distribution along boundaries after homogenization annealing.

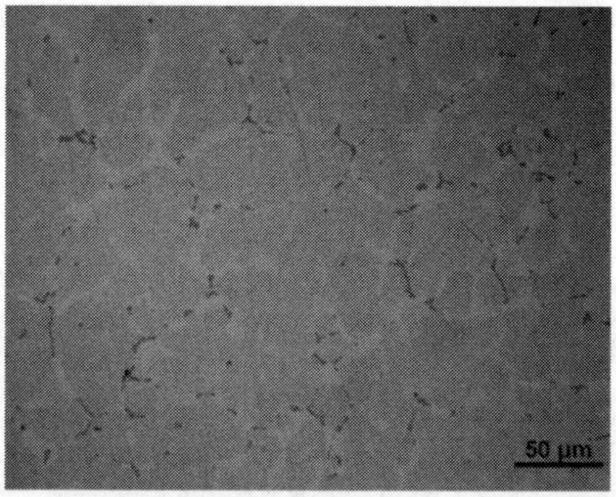

Fig.4. Microstructure of the new Al–Mg–Si–Cu alloys after homogenization annealing at 560°C for 5h.

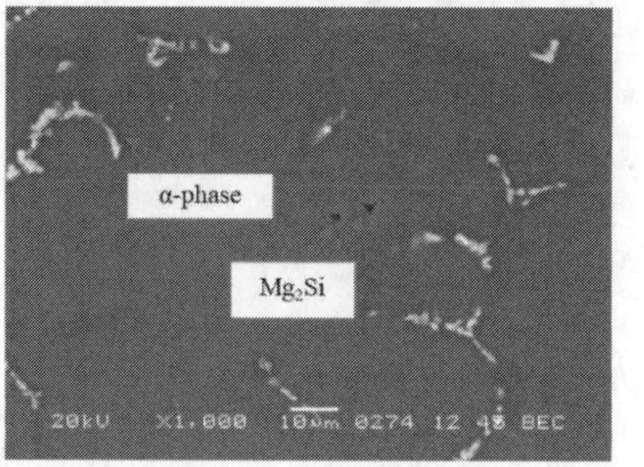

Fig.5. Higher magnification microstructure of the new Al–Mg–Si–Cu alloy after homogenization annealing at 560 °C for 5h showing an obvious fragments of α-Al(Fe,Mn,Cr)Si phase.

The EBSD micrograph of the new Al–Mg–Si–Cu alloys after hot forging with homogenization heating rate 7°C/min is shown in Fig.6. The different color contrast in the EBSD maps is corresponding to different grain orientation. The severely fragmented and elongated sub-grains like thin ribbon are observed. Through statistical analysis, the percentage of sub-grain boundary of the

https://doi.org/10.1051/matecconf/202032603003

new Al–Mg–Si–Cu is 72%, which is 20% higher than that of 6061 alloy. The higher percentage of sub-grain boundary, the higher ductility of the new Al–Mg–Si–Cu alloy will be. It is responsible for the highly uniform precipitation and dispersion characteristics of dispersoids during homogenization annealing in the new Al–Mg–Si–Cu alloy with appropriate amount of excess Si (in the range of 0.1-0.5wt.%), which play a critical role in preventing the nucleation of recrystallized grains during forging process and solution treatment [12]. The uniform precipitation dispersoids during homogenization annealing in the new Al–Mg–Si–Cu alloy with appropriate amount of excess Si will be illustrated with HRTEM observation in detail.

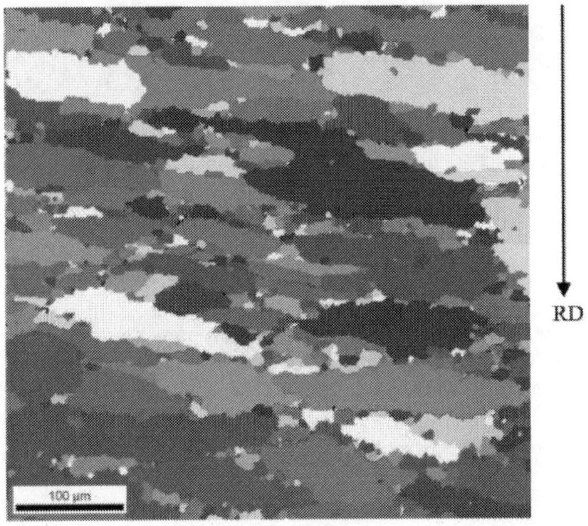

Fig.6. EBSD micrographs of the new Al–Mg–Si–Cu alloy after hot forging process.

3.2 HRTEM observations in material preparation

Fig. 7a shows a TEM bright-field images obtained from this new Al–Mg–Si–Cu alloy homogenized at 560°C for 5 h and subsequently air-cooled. It can be found that numerous rod-like or cube-like nano-precipitates of AlMnCrSi dispersoids in size of 100-500nm, distribute dispersedly in Al matrix. Fig. 7b shows a TEM bright-field images obtained from this new Al–Mg–Si–Cu alloy after forging process. During the forging process, the grains are broken into small ones relatively. Simultaneously, the size and distribution of the nano-sized AlMnCrSi dispersoids also change. It can be seen, that most of the AlMnCrSi dispersoids locate in grain boundary and its size reduce to 50-200nm. The effect of nano-sized AlMnCrSi dispersoids on pinning grain boundary is better for preventing the nucleation of recrystallized grains during forging process and make grains smaller. Therefore, the percentage of sub-grain of the new Al–Mg–Si–Cu alloys after forging process is higher than the percentage of sub-grain of 6061 alloy. Fig. 7c shows a TEM bright-field images obtained from this new Al–Mg–Si–Cu alloy after aging treatment. During the aging treatment, the size and shape of the remained nano-sized AlMnCrSi dispersoids also change simultaneously. It can be seen, that the shape of the AlMnCrSi dispersoids has become rounded and its size reduce to 20-100nm furtherly.

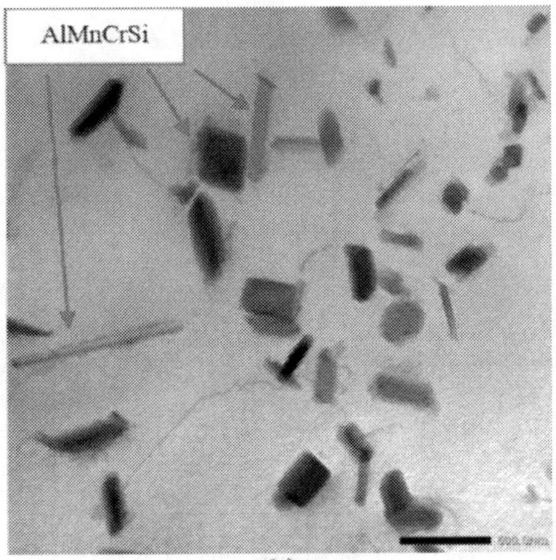

(a)

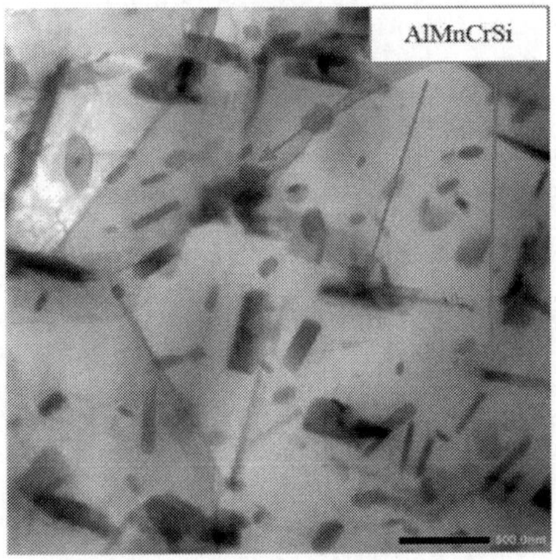

(b)

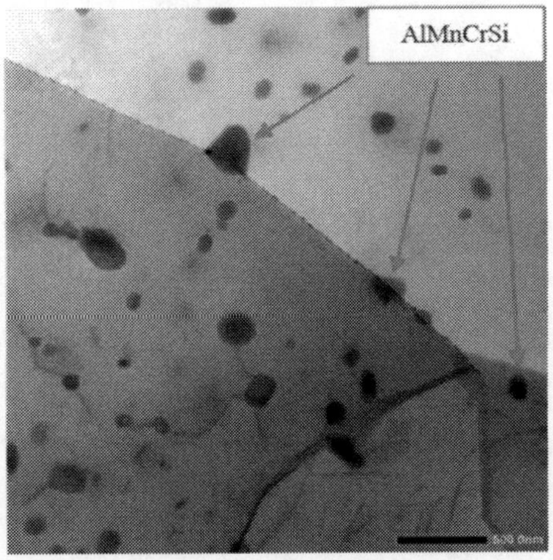

(c)

Fig. 7. TEM micrographs of AlMnCrSi dispersoids in the new Al–Mg–Si–Cu alloy: (a) dispersion distribution after homogenization annealing; (b) uniform distribution in the matrix after forge process; (c) uniform distribution in the matrix after aging treatment.

Commonly, the precipitation sequence in the matrix of 6xxx series alloys after aging treatment is as follows: Supersaturated solid solution – clusters – GP zones and initial-β" – pre-β", β" and Q" – Q' (partially coherent version of Q) and β (or β') – Q and β. Meanwhile, for Al–Mg–Si–Cu alloys, two types of strengthening phases, needle-shaped pre-β" and lath-shaped Q' precipitates, would precipitate in the T6 temper at 170-180°C [10,11,15,16]. However, the precipitated phase of this new Al–Mg–Si–Cu alloy taken an aging of low-temperature and shot-time treatment is different compared to T6 temper of 6061 alloy.

Fig. 8(a) shows a TEM bright-field images obtained from this new Al–Mg–Si–Cu alloy aged at 160°C~170°C for 2h~3h. A great number of aged nano-precipitates, AlMnCrSi dispersoids, are delineated by strain-field contrast. HRTEM micrographs further reveals only one type of strengthening phase, needle-shaped initial-β" precipitates, whose composition is $Mg_2Si_3Al_6$ (Fig. 8b). The initial-β" precipitate is generally considered as the precursor of pre-β" ($(Al+Mg)_5Si_6$) or β" (Mg_5Si_6) [10,14-16]. The needle-shaped initial-β" precipitate is oriented along the Al <001> direction. As shown in Fig. 8b, the cross section of aged nano-precipitates is observed making its diameter measurable accurately. The average diameter of initial-β" precipitate is about 2–3 nm. In addition, the needle-shaped initial-β" precipitate oriented along the Al <001> direction is also observed lying in Fig. 8b, which is nearly 8-10nm length.

Furthermore, the remained nano-sized AlMnCrSi dispersoids after the aging treatment distributed dispersedly and uniformly in the grain and on the grain boundary. When the fatigue crack propagates, these AlMnCrSi dispersoids can prevent crack growth further. So, the fatigue life of the new Al–Mg–Si–Cu alloy products may be improved.

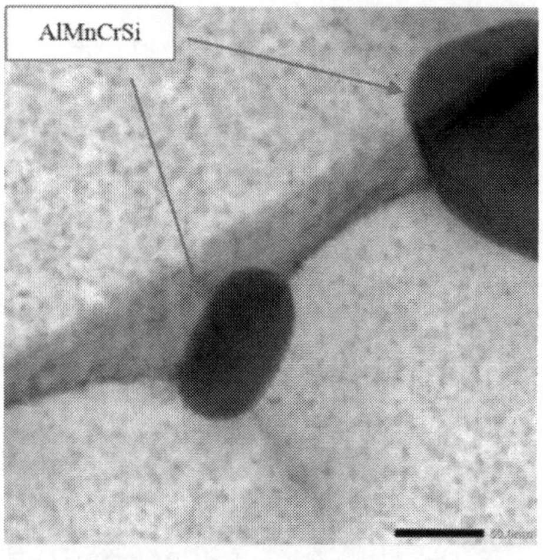

(a)

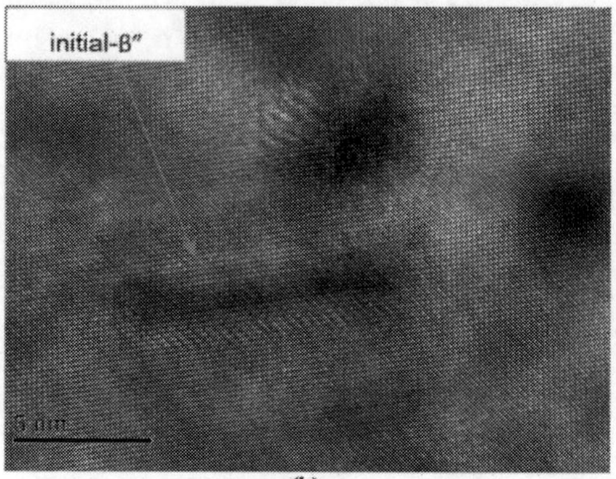

(b)

Fig. 8. TEM images of the new Al–Mg–Si–Cu alloy: (a) bright-field micrograph along with [001]Al; (b) HRTEM micrographs of needle-shaped initial-β" precipitates imaged along their axes.

3.3 Mechanical properties

The combination of mechanical properties of the new Al–Mg–Si–Cu alloy in aged state are given in Table 3. Compared with the commercial 6061 alloy, both strength and ductility of this new alloy increase significantly. Table 3 shows that the elongation is stable in the range from 15% to18%, yield strength and tensile strength are 310MPa and 380Mpa, respectively. In summary, the ductility and strength of this new Al-Mg-Si-Cu aluminum alloy are 35 percent and 11 percent higher than the commercial 6061 alloy.

The excellent ductility of this new alloy can be attributed to the uniform precipitation and dispersion characteristics of nano-sized AlMnCrSi dispersoids during homogenization annealing process and then the nano-sized AlMnCrSi dispersoids pinned the grain boundary to improve the ratio of sub-grain during the forging and heat treatment process, subsequently. As a result, the ratio of sub-grain boundary in the annealing microstructure, forging microstructure and age-state microstructure can be improved. So, the ductility of the Al–Mg–Si–Cu alloy is improved while ensuring its tensile strength and yield strength. Furthermore, the needle-shaped phase of the initial-β" precipitates during the aging treatment can further improve the ductility. Moreover, nano-sized AlMnCrSi dispersoids also can improve the fatigue life of the new Al–Mg–Si–Cu alloy products as mentioned above in section 3.2.

Table 3. Mechanical properties of the new Al-Mg-Si-Cu alloys and commercial 6061 alloy

Alloy		Ultimate tensile strength (MPa)	Yield strength (MPa)	Elongation (%)	Fatigue strength (MPa)
New alloy	Test 1	391	320	16.0	135
	Test 2	387	310	16.3	134

	Test 3	385	312	18.0	132
6061 alloy		290	255	13.0	118

4. Conclusion

1. The as cast microstructures of the new Al–Mg–Si–Cu alloy contain three intermetallic phases: Mg_2Si, Al(Fe,Mn,Cr)Si and AlCuMgSi, Which present in nano-sized rounded AlMnCrSi dispersoids about 20-100nm in size, and initial-β" precipitates about 2–3 nm in diameter and nearly 8-10nm in length respectively after solution and short time aging treatment. The percentage of sub-grain boundary of the new Al-Mg-Si-Cu aluminum alloys under the forging process reach up to 72%.
2. The combination of mechanical properties of the new Al-Mg-Si-Cu aluminum alloys product in aged state show that the ductility is in a range from 15% to 18%, yield strength and tensile strength are 310MPa and 380MPa respectively, fatigue strength is range from 130MPa to 135MPa.
3. In summary, the ductility and strength of this new Al-Mg-Si-Cu aluminum alloys are 35 percent and 11 percent higher than commercial 6061 alloy. The new type of Al-Mg-Si-Cu aluminum alloys with high ductility can be used in forged wheel of commercial vehicle with one forging process.

This work was funded by Project 2016YFB0101603 supported by National Key Research and Development Program.

References

1. D. Gao, W. Nonferrous Met. 11, 174-176 (2019).
2. D. Maisonnette, M. Suery, D. Nelias, P. Chaudet, Mater. Sci. Eng. A, 528, 2718 (2011).
3. SK. Panigrahi, R. Jayaganthan, J Alloys Compd, 470, 285(2009).
4. H. Liu, Reaearch on Liquid Forging and Heat Treatment of High Strength and Toughness Aluminum Alloy Wheel: [D]. Harbin: Harbin Institute of Technology, 2015.
5. S.K. Panigrahi, R. Jayaganthan, J ALLOY COMPD, 470, 285-288 (2009).
6. S. Zhou, Study on fatigue characteristics and fracture behavior of high strength and high toughness AlMgSiCu alloy sheet: [D]. Guangxi: Guangxi University of Science and Technology, 2019.
7. D. Maisonnette, M. Suery, D. Nelias, P. Chaudet, Mater. Sci. Eng., A, **528,** 2718-2724 (2011).
8. A. Loucif, R.B. Figueiredo, T. Baudin, Mater. Sci. Eng., A, **527,** 4864-4869 (2010).
9. Y. Han, K. Ma, L. Li, H. Nagaumi. Mater. Des. **39,** 418–424 (2012).
10. J. H. Chen, C.H. Liu, Chin. J. Nonferrous Met, **21**,10(2011).
11. C. He, W. Nonferrous Met. 3, 282-283 (2018)
12. M. Murayama, K. Hono, W.F. Miao, and D.E. Laughlin, Metal. Mater. Trans. A, **32A,** 239(2001).
13. Y. Han, K. Ma, H. Nagaumi, TMS, 418–424 (2012).
14. X. Wang, S. Esmaeili, D. J. Lloyd, Metal. Mater. Trans. A, **37A,** 2006—2691 (2006).
15. G.C. Weatherly, A. Perovic, N.K. Mukhopadhyay, Metal. Mater. Trans. A, **32A,** 213(2001).
16. J. H. Chen, E. Costan, M. A. van Huis, Q. Xu, H. W. Zandbergen, SCI. **312**, 416-419 (2006).

MATEC Web of Conferences **326**, 03004 (2020) https://doi.org/10.1051/matecconf/202032603004
ICAA17

A New Approach to Implement High Strength 7075 Aluminum for Automotive Application

Cedric Wu[1*], *Yudie* Yuan[1], and *Rajeev* Kamat[1]

[1]Novelis Global Research and Technology Centre, 1950 Vaughn Rd NW, Kennesaw, GA, USA

Abstract. Aluminum alloys offer high specific strength than advanced high strength steels, making them preferred candidates for automotive light weighting. Among them, AA7075 aluminum alloy offers significantly higher strength than 5xxx and 6xxx alloys and is considered an attractive candidate by automotive OEMs for structural applications such as door intrusion beams, B pillars etc. There are several challenges in implementing AA7075, such as long artificial aging time to reach peak strength, joining method and corrosion resistance. In this study, an artificial aging practice that significantly reduces aging time was explored and its influence on mechanical properties of AA7075 was investigated in comparison with conventional peak age practice. In addition, this practice offers a potential solution for joining through self-piercing riveting. Moreover, the effect of artificial aging on corrosion, specifically intergranular corrosion (IGC) and stress corrosion cracking (SCC) was evaluated. The results are discussed with in depth analysis and correlation with microstructure.

1. Introduction

Today, automotive OEMs are constantly pursuing ways to improve vehicle fuel economy to bring more value to customers and reduce carbon footprint to combat global warming and meet more stringent government regulations. Light weighting through implementing aluminum ranked among most preferred choices by automotive industry in a recent survey [1]. Aluminum alloys offer many benefits such as higher specific strength, better corrosion resistance in comparison with steels.

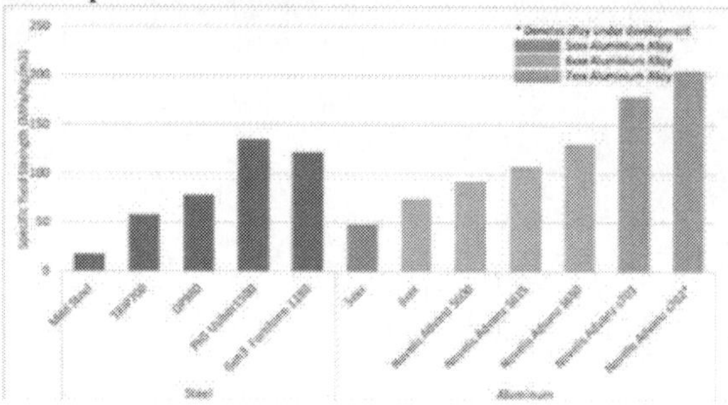

Figure 1: Comparison of specific strength between aluminum alloys and steels.

7xxx family of alloys is a high strength, heat treatable aluminum alloy primarily comprised of Al-Zn-Mg-Cu. It offers specific strength that are 10-30% higher than press hardened steels (PHS). Figure 1. Historically 7xxx alloys were widely used in the aerospace industry for applications like fuselage stringers and lower wing skins [2]. In recent years, new applications emerged in the automotive industry for components that requires high strength such as door intrusion beams, bumper reinforcements, B pillar outers and floor cross-members [3, 4, 5]. In order to fully utilize its high strength, 7xxx alloys need to be artificially aged to peak aged condition (T6 temper). As a result, the ductility of the alloy in T6 temper is reduced making it difficult to form in conventional room temperature stamping or joined mechanically. On the other hand the 7xxx alloy typically exhibits unstable T4 behavior leading to steady strength increase at room temperature. In order to improve AA7075's formability, extensive research has been done to look into warm forming 7xxx aluminum [6, 7, 8]. In a warm forming scenario, the material is initially in peak aged condition, i.e. T6. 7xxx alloy at elevated temperature between 200°C to 260°C has shown improvement in formability compared to room temperature. However, exposure to high temperature leading to over aging, the component's mechanical properties are typically reduced and lower than peak aged condition. Alternatively, others have investigated hot stamping route for 7xxx aluminum. [4, 9] The concept of hot stamping aluminum is similar to that of hot stamped boron steel. The difference being 7xxx series aluminum starts in F temper, is solutionized in a furnace, subsequencially formed and quenched rapidly in a water cooled stamping die. It requires a post forming artifical aging cycle to achieve high strength. Conventionally, in order to achieve full hardened T6 temper, AA7075 alloy requires long artificial aging time (i.e. 24 hours). Such a long practice makes implemention of AA7075 difficult and can result in higher cost especially for high volume production. These challenges become limiting factors especially in automotive industry. Several heat treatment methods have been proposed and investigated by researchers such as T616 temper, high temperature pre-precipitation (HTPP) Retrogression Re-aging (RRA) and 2 step aging aiming to improve the alloy's formability and

* Corresponding author: cedric.wu@novelis.com

reduce artificial aging time [10, 11, 12]. Second area of concern with 7xxx aluminum usage comes from corrosion resistance which is preceived to be lower than 5xxx and 6xxx alloys. It has been reported in the literature [14-19] that 7xxx series aluminum alloys are susceptible to localized corrosion such as pitting, intergranular corrosion and stress corrosion cracking in aqueous solution containing particularly chloride ions. The corrosion behaviour of the 7xxx alloys was found to be strongly dependent on alloy composition and resulting microstructure, which is the evolution of material fabrication processes including thermo-mechanical history. The focus of this paper is to provide a comprehensive solution to implement Novelis Advanz™ s701 (AA7075) alloy for complex automotive components through hot stamping process. An accelerated artifical aging practice tailored for automotive process is proposed. It will reduce total aging time and associated processing cost. In addition, the study investigates the microstrucual evolution of the material at various stages of artificial aging. As a result, s701 can achieve similar in service strength levels as conventional aging practice. Furthermore, the proposed artifical aging practice enables s701 to be mechanically joined with self-piercing rivet (SPR). Additionally corrosion performance of the alloy, specifically IGC and SCC in the proposed temper was investigated.

2. Experiment

The material used in this article was Novelis Advanz™ s701 alloy. Its chemical composition falls within AA7075 range, (Table 1) presented in weight percent.

Table 1 Chemical Composition of Novelis Advanz™ s701

	Si	Fe	Cu	Mn	Mg	Cr	Zn	Ti	Al
AA7075	0.4	0.5	1.2-2.0	0.3	2.1-2.9	0.18-0.28	5.1-6.1	0.02	Rem.

S701 alloy was DC cast, homogenized, scalped, hot rolled and cold rolled to 2.0mm final gauge in Novelis Korea Plant. Lab evaluations were carried out at Novelis Global Research and Technology Centre (NGRTC).

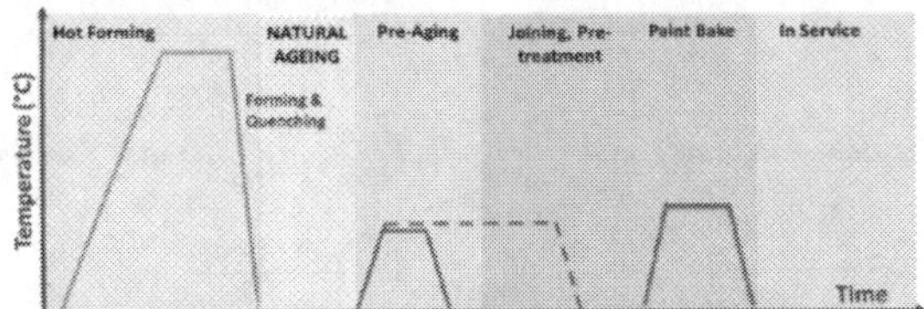

Figure 2: Schematic of s701 alloy hot stamping process flow path with 2 artificial aging routes: T4x to T6x and standard T6 to T6+PB.

Starting with an as cold rolled F temper, the material was solutionised at 480°C for 5 minutes followed by quenching in temperature controlled water bath to simulate quench rate > 350°C/sec similar during hot stamping process [13]. The material was allowed to naturally age in ambient temperature for 1 week before artificially aging to various tempers. Figure 2 shows potential hot stamping process flow path for s701 alloy with both conventional and accelerated artificial aging. Tempers and heat treatment conditions are defined below:

- F: As fabricated sheet metal after hot rolling and cold rolling.
- W: Solutionised and freshly quenched, unstable temper.
- T4x: Artificially aged between 100-120°C for 1 hour, under-aged and intermediate temper.
- T6: Artificially aged at 125°C for 24 hours, peak aged.
- T6x: T4x followed by aging at 180°C for 0.5 hour, accelerated aging, in service temper.
- T6+PB: T6 followed by aging at 180°C for 0.5 hour, conventional practice, in service temper.

Full size tensile specimens were tested as per ASTM B557 to determine various tempers' mechanical properties. Metallurgical samples were cut parallel to rolling direction and mounted in long transverse direction. Mounted samples were ground, polished using standard metallographic sample preparation procedure. Mounted samples were etched using Barker's agent to reveal the grain structure. Mounts were observed under a Zeiss Axio metallurgical microscope. A polarized filter was used to reveal the grain structure. Scanning Electron Microscopy (SEM) was done using a Philips XL30 S FEG-SEM at an operating voltage of 20kV. Transmission Electron Microscopy (TEM) samples were prepared using electro polishing at Natural Resources Canada. Samples were analysed using FEI's Tecnai Osiris TEM equipped with X-FEG gun at 200 keV. Samples were tested for intergranular corrosion (IGC) as per ASTM G110 specification with immersion time of 24 hours. Stress corrosion cracking (SCC) was tested as per ASTM G44 alternate immersion test in 3.5% NaCl solution at 0% and 80% yield stress for a duration of 40 days. Samples were pulled after completion of SCC test to determine the residual stress as per ASTM G139. Self-Piercing Rivet (SPR) tests were conducted at Stanley® joining technology laboratory in Troy, MI. The stack up in this study comprises of 2.0 mm Novelis Advanz™ s615 T4 material on top and 2.0 mm Advanz™ s701 in T6, T4x with 0 days and 32 days of natural aging respectively at the bottom of the stack. Through multiple preliminary trials, a C5.3x6.0 H4 semi-tubular rivet and M260 468 die combination was deemed to have given the best result for this particular stack up and selected.

3. Results and Discussion

3.1 Microstructure

Figure 3 shows s701 alloy's microstructure in F temper and T6 temper respectively.

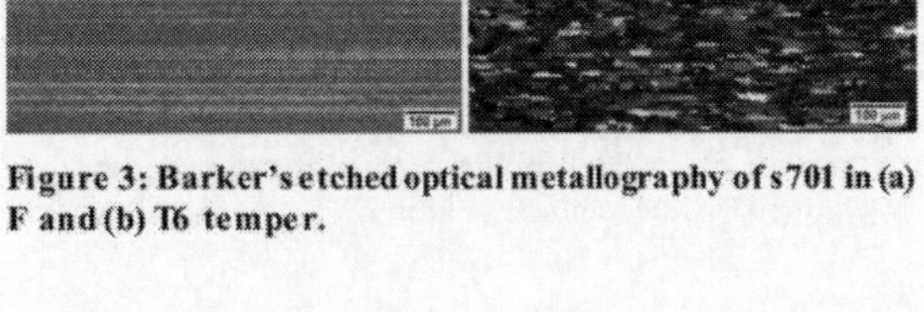

Figure 3: Barker's etched optical metallography of s701 in (a) F and (b) T6 temper.

The alloy in its initial state of a typical aluminum hot stamping process is F temper, which has deformed grains with fibrous structure shown in Figure 3a. During the solution heat treatment, the alloy was brought to above its solvus temperature where the coarse precipitates formed during processing were effectively dissolved into solid solution and the material recrystallized as shown in Figure 3b.For any quaternary Al-Zn-Mg-Cu alloy system, it typically comprises a complex system of precipitation phases and constituent particles. Primary strengthening phase in 7075 alloy is M phase (η=$MgZn_2$). Its precursor GP zone and metastable semi-coherent η' phase are the main governing precipitates responsible for high strength [14]. In Cu bearing 7xxx alloys such as s701, Cu may replace Zn atoms to form η-$Mg(Zn,Al,Cu)_2$ [15]. In addition, there are other undesired phases present. For example, T phase, nominally $Al_2Mg_3Zn_3$; S phase, nominally Al_2CuMg; θ phase, nominally Al_2Cu and Z phase, norminally Mg_2Zn_{11}. The amount of major equillibrium phases (M,T,S) are essentially determined by factors such as Zn:Mg ratio, casting practice homogenization temperature and time, cooling rate and artificial aging temperature [16]. Besides precipitates, there are also coarse constituent particles such as Mg_2Si and Al_7Cu_2Fe, which originated in casting process. These phases are insoluble and incoherent to the matrix. Such coarse constituents are in size range of 5-30 microns, depending on processing conditions. They are visible under microscopy as illustrated in Figure 4. These coarse constituents are insoluble during solutionization step of hot stamping process and are detrimental to the materials' corrosion behaviour [17].

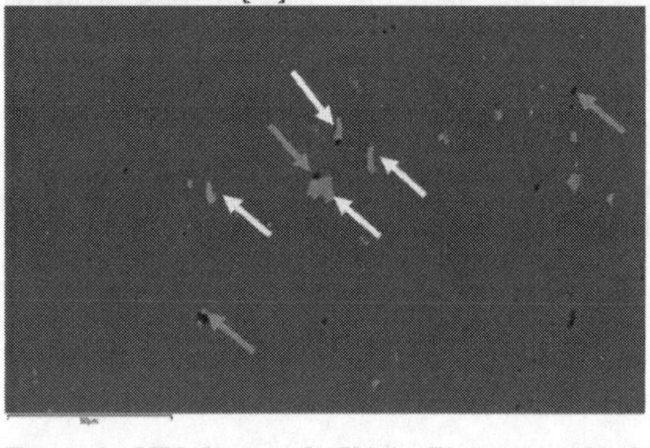

Figure 4: SEM image of s701 in T6 temper, showing constituent particles. Blue arrow indicates Mg_2Si and yellow arrow indicates Al_7Cu_2Fe.

The precipitation sequence in AA7075 alloy can be described as:

SSSS-> Clusters -> G.P. zone -> η' -> η.

In its maximum strength condition, G.P. zones are in the size of 20 to 35 Å in diameter along with metastable η' transition precipitate. Increase in the natural aging interval or preaging at a lower temperature can promote the development of G.P. zones, consequently provide a finer precipitate structure and facilitate the alloy reach peak strength when artificially aged. Chrome containing dispersoids, nominally Al_7Cr are often found along the grain boundaries, they are typically triangular, round or rod in shape, 80-120nm in size.(Figure 7) These dispersoids can pin the grain boundaries and inhibit recrystallization during high temperature processing. Chrome containing dispersoids would increase the alloy's fracture toughness better than Mn based dispersoids. [18]

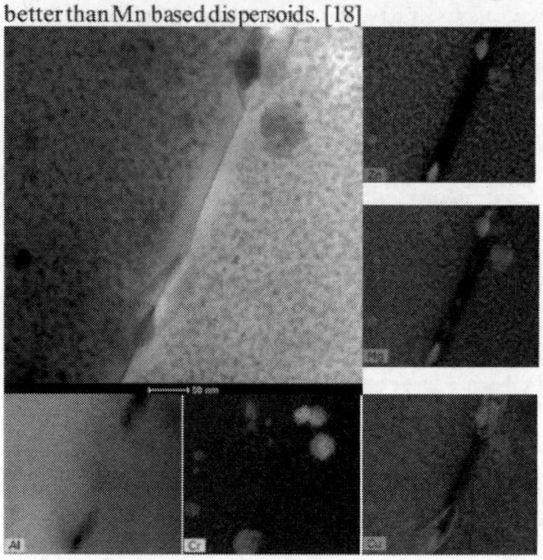

Figure 5: HAADF and elemental map showing typical s701 microstructure in T6+PB condition.

Figure 5 shows High Angle Annular Dark-Field Imaging (HAADF) of typical s701 alloy microstructure in T6+PB temper. Al_7Cr dispersoids can be seen at the grain boundaries. Coarse η-$Mg(Zn,Al,Cu)_2$ was found to be discontinuously nucleated on Al_7Cr dispersids. High volume density of η'precipitates can be obvserved in the matrix. Precipitation free zone (PFZ) are measured 30-40 nm in width.

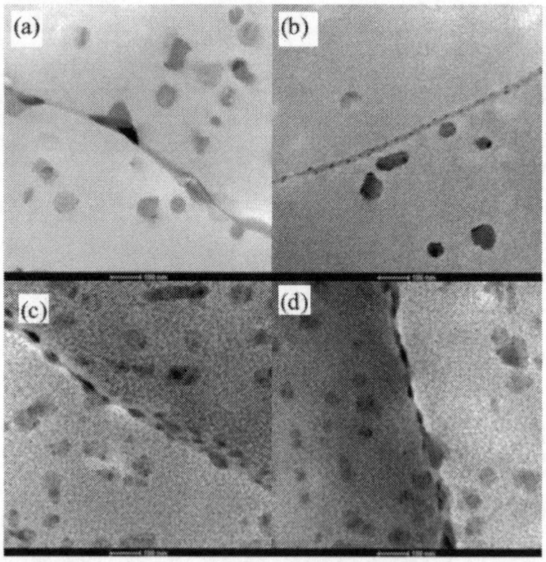

Figure 6: Bright Field (BF) image of typical s701 microstructure under various heat treatment conditions. (a) W temper; (b) T4x temper; (c) T6x temper and (d) Standard T6+PB temper.

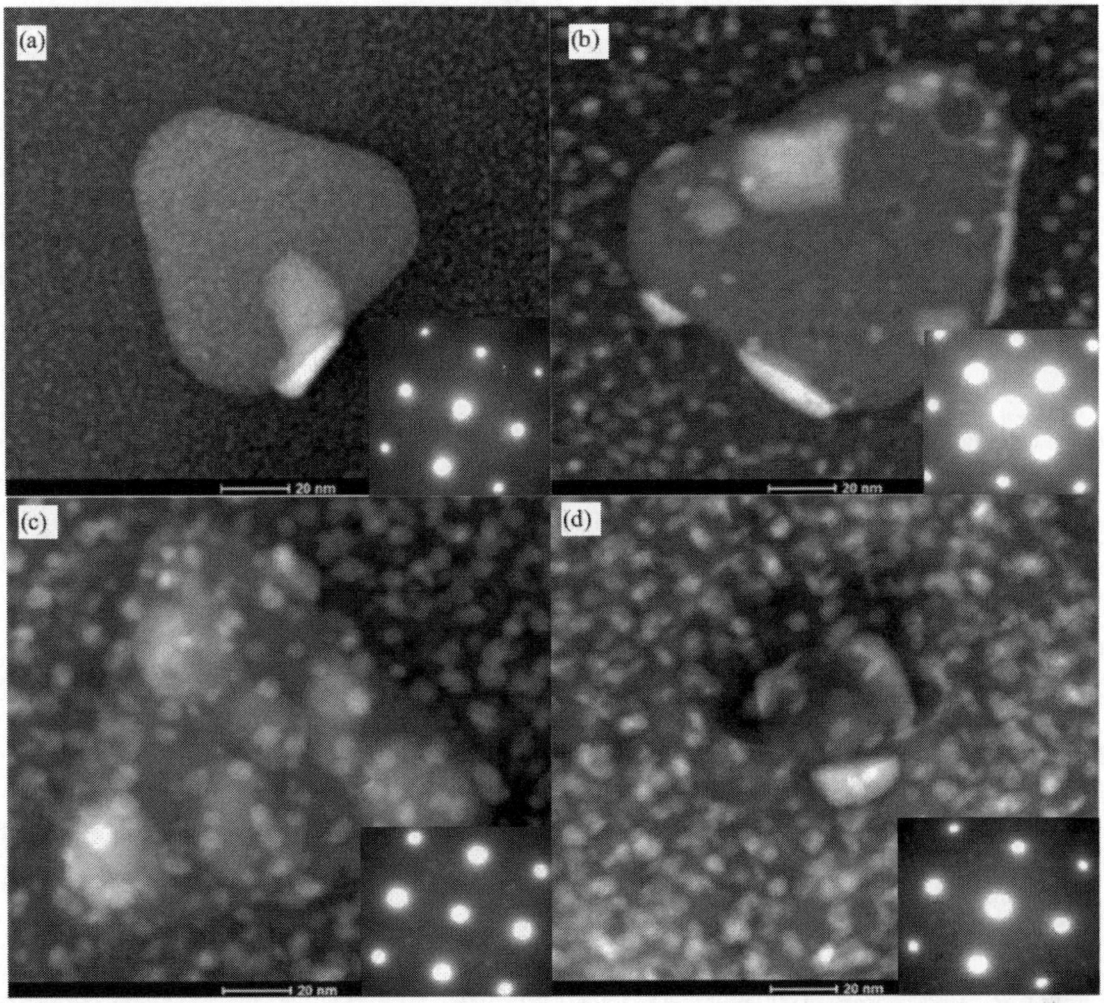

Figure 7: HAADF with a Chrome containing dispersoid in the centre of field of view and SAD obtained along [001] zone axis of (a) W temper + 1 month of NA s701, (b) T4x temper, (c) T6x temper and (d) Standard T6+PB temper.

Figure 6a shows s701 microstructure in W temper, The grains are free of any precipiates since all the solute were effectively dissolved into solid solution during solution heat treatment. However, high resolution images shows clusters and early stage nano size GP zone are present (Figure 7a) in the matrix. The clusters and GP zones formed after hot stamping process with natural aging, in this case 1 month of NA. In addition, the quench rate used during preperation of the sample resembles that of during hot stamping, there are evidence of Al_7Cr dispersoids at grain boundary along with coarse η-$Mg(Zn,Al,Cu)_2$ phase. Artificial aging of the material brings the alloy into T4x temper. The positions of super lattice reflection of GP zones and η' precipitates are presented in a four folded pattern at the vicinity of 2/3 {220} positions [19]. The super lattice reflections were first observed in T4x condition. The precipitates also became more evident and emerges from the matrix as shown in Figure 6b. High magnification image show η' precipitates in the range of 5 nm, as illustrated in Figure 7b. Figure 8 shows high resolution TEM with beam ||[001], showing η' precipitates. The transition state precipitate is coherent to the matrix making them the primary strengthening mechanism for s701 alloy.

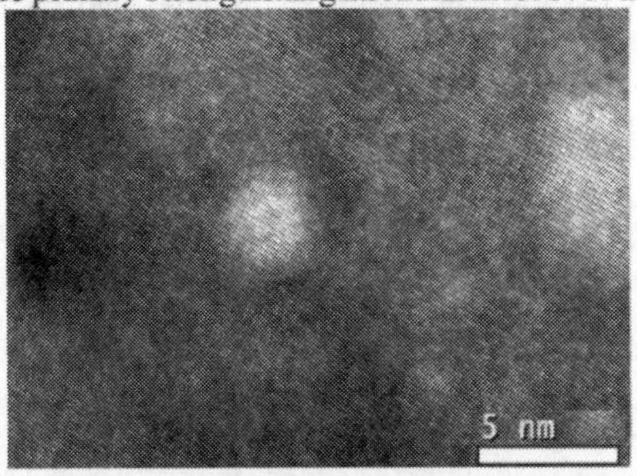

Figure 8: HRTEM with beam ||[001], showing T4x s701 sample's η' precipitates coherent with matrix.

The formation of GP zones and η' transition phases during first aging step to a certain size and distribution is critical to the alloy's next aging response. They would act as precursor, allowing precipitates to transform during subsequent artificial aging process. Figure 6c and d compares T6x temper with standard T6+PB temper, both presumably represent the material in service condition. One can no longer delineate between the two conditions from a microstructural level. In both conditions, grain boundaries are well decorated with discontinuous η' precipitation. PFZ are in the size of 40-50 nm. Precipitates are approximately 5-10 nm in size, larger and higher number density compared to earlier stage tempers. Coarse η-$Mg(Zn,Al,Cu)_2$ phases are found to have a tendency to nucleate on Al_7Cr in all four tempers examined. The coarse precipitates are larger size than they are in the matrix and grew during aging. These coarse precipitates are incoherent to the matrix and do not contribute to strengthening of the alloy. Furthermore, it would reduce the amount of solute available in the matrix and hinder the alloy's aging response. It should be pointed out that such phenomenon would vary depending on quench rate during hot stamping process. If the quench rate is reduced further, undesirable coarse precipitates could become more evident. Having more of these coarse precipitates at the grain boundary is detrimental to the alloy's strength at peak age. These coarse precipitates are heterogeneously distributed along grain boundaries makes the alloy prone to corrosion attack. Thus, attaining a faster quench rate during hot stamping of AA7075 aluminum alloy is desirable to avoid the nose of the CCT curve, resulting in higher strength and better corrosion resistance. Quench rate from a hot stamping process with water cooled die could achieve 350~400°C/sec which is sufficient for AA7075 [4].

3.2 Mechanical Properties

The beauty of the proposed accelerated aging practice is the ability to achieve strength comparable to conventional T6+PB process route while significantly reduce the amount of artificial aging time after hot stamping. Figure 9 compares mechanical property of s701 material in various tempers. F temper properties are governed by the upstream processing conditions such as homogenization, hot rolling, hot band coiling temperature, the amount of reduction during hot and cold rolling etc. F temper has the lowest mechanical property in terms of elongation due to the fibrous deformation microstructure and coarse uncontrolled precipitates. W temper is an unstable temper after hot stamping process. Its microstructure would continue to change as previously discussed. Therefore, the mechanical property of s701 alloy would continue to increase overtime as the material continues to naturally age. The data presented here is material's mechanical property immediately after solutionization, i.e. within 24 hours. The alloy in the W temper has a low yield strength of 240MPa with 20% total elongation. However, the unstable nature of the alloy making it hard to utilize it in standard stamping operations. It would be possible to take advantage of the good ductility if the material can be stabilized. After extensive research and experiment at Novelis, practice of 100-120°C for 1 hour was selected as T4x heat treatment for s701. S701 in its T4x condition has intermediate mechanical properties that fall between W and its final in service temper T6x. The initial yield strength in T4x is about 400 MPa. Total elongation of the alloy is 18%, which only exhibits slight decrease in comparison with W temper. Thus providing an opportunity for mechanical fastening such as SPR, discussed in section 3.4. The proposed accelerated aging practice (T6x) bring the alloy to T6x temper by utilizing paint bake cycle in automotive BIW production process. It will quickly grow the precipitate size and distribution to an optimum level and resulted in having similar microstructure with the conventional T6+PB condition. As a result, its mechanical properties are similar to that of a standard T6+PB practice, in terms of yield strength, tensile strength and elongation.

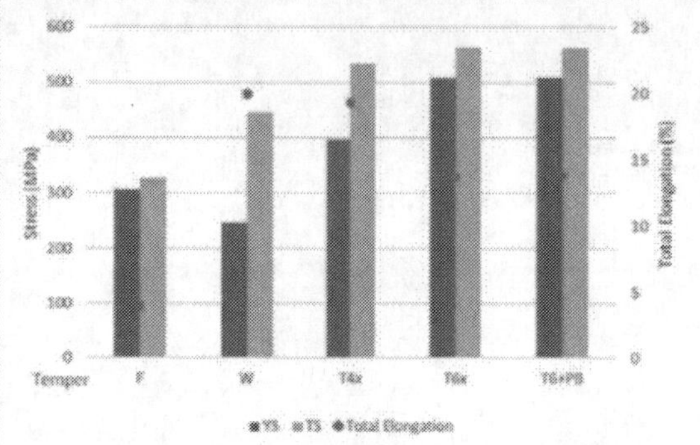

Figure 9: Mechanical properties of s701 in various tempers.

In previous microstructural analysis, it was obvious that the first heat treatment step transitioned clusters into GP zones and nano scale η' precipitates. The precipitates are still in transition period with smaller size and lower number density than T6x. It suggests that the material may have a propensity to continue to naturally age in T4x condition, similar to a 6xxx alloy. Figure 10 shows comparison of s701 natural aging behavior between W temper and T4x temper for up to 90 days. S701 alloy ages more rapidly in W temper than it does in T4x condition (Figure 10).

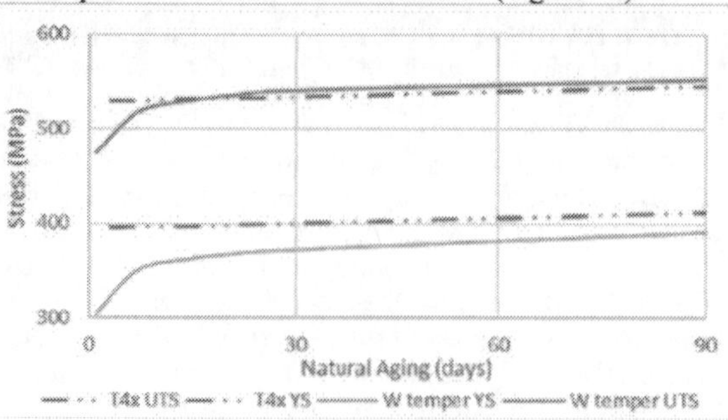

Figure 10: Natural aging trend comparison between T4x and W temper

Within a 90-day period, its yield strength and tensile strength increased by 87 and 77 MPa respectively. In contrast, within the same time frame s701 in T4x condition, yield strength and tensile strength both increased by 15 MPa. Thus it is evident s701 in T4x is a stables state. Its mechanical properties can be guaranteed with good ductility to allow certain mechanical joining/deformation process. In a typical automotive production scenario, Body In White (BIW) will go through multiple coating steps such as electro-coat (eCoat), top coat etc. The total paint bake time and temperature combination would vary between different OEMs. In addition, there could be some temperature variations depending on the location and adjacency of the parts. An artificial aging of 180°C for 30 minutes is a simulated single step paint bake cycle intended for laboratory use. Furthermore, during production, the paint line maybe stalled/slowed due to mechanical issues. The BIW may require rework to remove minor imperfections and need to be sent through the paint line again. Thus, the robustness of the structural components to meet target mechanical properties under extended exposure at paint bake temperature becomes critical. Figure 11 shows simulated T6x temper going through additional paint bake cycles up to 2 times. It maintained over 500 MPa in yield strength and did not exhibit any strength loss despite experiencing multiple paint bake cycles. The finding shows that proposed accelerated aging practice has similar level of robustness as conventional 7075 T6 [11].

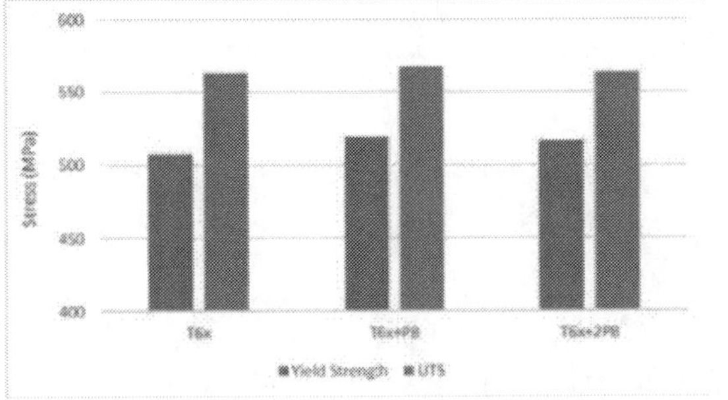

Figure 11: s701 multiple paint bake response after proposed T6x temper.

3.3 Corrosion Results

S701 alloy would only be exposed to corrosive environment in its final in-service temper, i.e. T6x or T6+PB, therefore corrosion testing was carried out in these tempers. Two specific types of corrosion tests were conducted: Intergranular corrosion resistance (IGC) and Stress corrosion cracking (SCC). IGC of s701 alloy in simulated service temper T6x and T6+PB was studied by immersion test per ASTM G110. The cross-section of specimens subjected to 24 hour full exposure is shown in Figure 12.

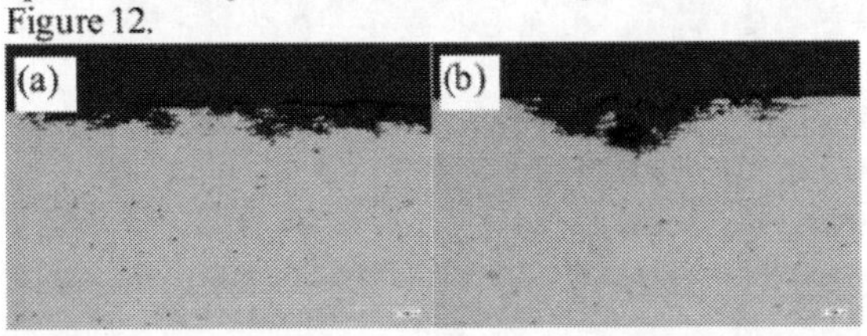

Figure 12: Metallographic cross section of s701 in (a) standard T6+PB and (b) T6x condition after 24 hours immersion test.

Corrosion depth was measured and summarized in Table 2. Overall, the two tempers behave similarly in IGC performance with T6x temper showing slightly higher in corrosion depth. The corrosion depth exhibited for s701 is comparable to that of a high Cu 6xxx alloy used in automotive applications, presumably can fulfill IGC acceptance criteria in terms of maximum and average corrosion depth.

Table 2: Summary of IGC results in G110 test (24 h exposure)

Alloy	Temper	Corrosion depth evaluation		
		Max. µm	Average, µm	STDEV, µm
s701	T6+PB	84	41	17
	T6x	103	57	24

The alternate immersion test (G44) for characterizing SCC ended with no specimens fracturing in-situ during 40 days exposure. Therefore, strength loss with respect to the bulk material property is used to compare SCC resistance for two tempers, as shown in Figure 13. The average residual strength in specimens stressed to 80% of their yield strength showed very little drop (~3%) in comparison with the unstressed ones for either temper (T6x or T6 + PB).

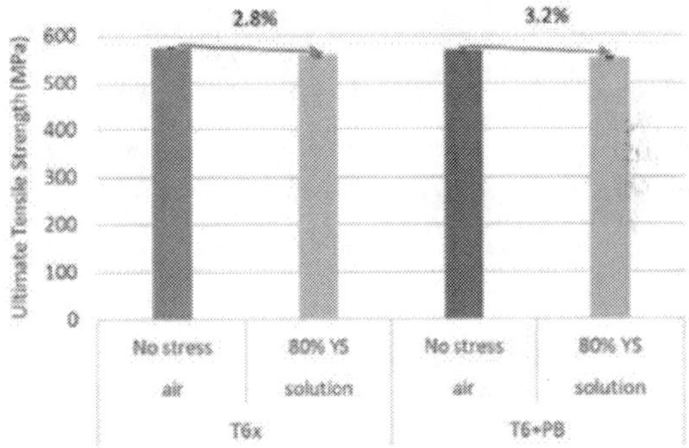

Figure 13: Comparison of ultimate tensile strength between T6x and T6+PB specimens before and after SCC test

The fracture surfaces of the specimens in air and in corrosive NaCl solution were examined by SEM. Figure 14 displays representative fractographies of T6x and T6+PB tempers tested in air and in solution respectively. In air, the specimens of T6+PB and T6 temper possess predominant dimple fracture surface, characterized by ductile fracture appearance. Representative SEM fractography of the T6+PB is shown in Figure 14a, where cup and cone dimples with varying sizes are observed. Figure 14b was taken from the exact same location under SEM but in backscattered electron (BSE) mode that can detect secondary particles present in dimples. It can be seen from that the main dimples (e.g. regions indicated by red arrows in Figure 14a,b are traces of voids formation around second-phase particles; EDS analysis determined that these bright particles are Cu and Fe containing ones (e.g. Al_7Cu_2Fe). Figure 14c-f shows fractographs of the specimens in T6+PB and T6x tempers tested in solution. In contrast to specimens tested in air, the T6+PB and T6x undergone alternate immersion in 3.5% NaCl solution for 40 days show a completely brittle fracture surface that is characterized by inter-granular cracks and a mixture of transgranular and intergranular cleavage fracture. Furthermore, some corrosion products are also visible as brighter particles due to charging on the fracture surface (orange arrows in Figure 14c and f).

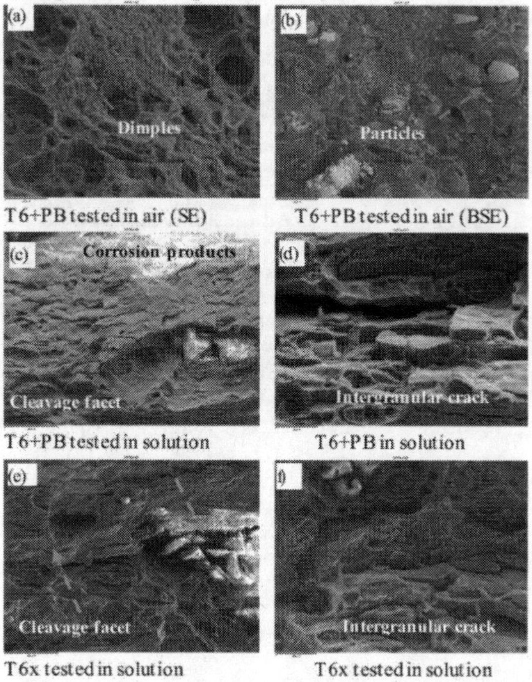

Figure 14: SEM fractographs of the s701 in T6x and T6+PB tempers tested in air and in solution (per ASTMG44 for 40 days)

The intergranular cracking for the temper T6+PB and T6x suggested that tne grain boundaries acted as the most preferential and active corrosion path for stress corrosion cracking due to the dissolution of densely distributed η' precipitates distributed along the grain boundaries (Figure 14c and d). It is worth noting here again that both the T6+PB and T6x specimens passed the defined SCC-G44 test and only showed less than 5% drop in ultimate strength. Therefore, it is to be inferred that the T6x temper exhibits similar SCC resistance to the T6+PB temper. Improved localized corrosion (such as pitting) and SCC resistance in overaged temper as compared to peak-edged T6 temper are widely reported [20, 21, 22]. In addition, the material were tested in bare condition, with relatively high stress level of 80% yield stress. In reality, the components in BIW would have an layer of eCoat which will provide an extra layer of protection. Proper stress level need to be determined based on component function and simulated drive cycle load cases. Potential exposure to corrosive environment would also be different depending on location of the component in a vehicle.

3.4 SPR Joining

Self-piercing rivet (SPR) is a joining method commonly used in aluminum intensive and mixed materials vehicles. However, conventional s701 alloy in standard T6 temper was challenging to join using SPR due to low ductility [23]. Since s701 material demonstrated a stable mechanical property in its T4x condition with good elongation, it opens an opportunity for it to be SPR joined. For the ease of production, in most SPR applications, the desired stack up is with a thicker and more formable alloy on the bottom of the stack. Previous experience suggests it was easier to form a joint with 7xxx alloys located on the top of stack. Hence, the current study emphasized on a solution having s701 alloy positioned at the bottom stack. When s701 is riveted in T6 condition, the rivet buckled after it pierced through the top layer into the high strength s701 on the bottom stack, as illustrated in Figure 15a,b. The material also developed button crack due do its poor ductility. Under this scenario the SPR joint was considered as unacceptable. However, when the temper of s701 on the bottom stack changed to T4x temper in Figure 15 c-f, due to the alloy's improved ductility, there was no sign of button crack. Cross-section reveals the rivet has good amount of interlock ≥0.2mm and no concern of bottom thinning. Gaps under the rivet head were completely filled, alleviated corrosion concerns with potential moisture trapped underneath. The tests were repeated at several time intervals from 0 day to up to 32 days of natural aging to address concern of changing properties in T4x and all resulted in formation of a good joint.

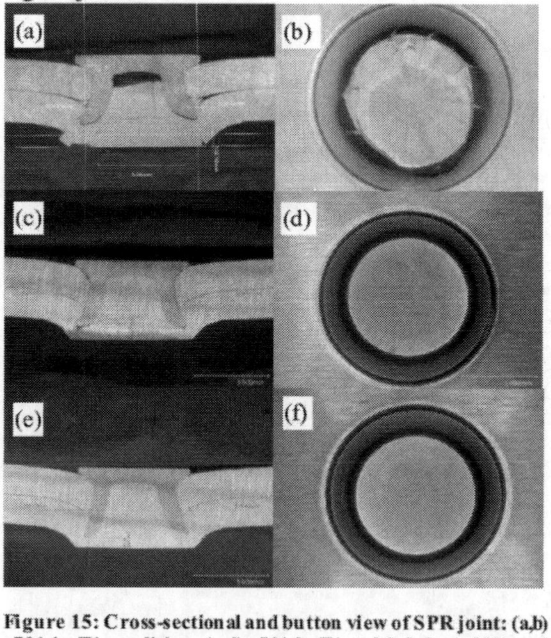

Figure 15: Cross-sectional and button view of SPR joint: (a,b) s701 in T6 condition; (c,d) s701 in T4x with 0 day of NA and (e,f) s701 in T4x with 32 days of NA.

Components made with s701 alloy could be SPR joined with other materials in its T4x state. The component could achieve full strength potential after paint baking process. Hence, s701 in T4x temper with good mechanical property stability and improved ductility opens the opportunity to mechanically fastened using SPR even when positioned in the bottom stack. In real applications, actual alloy grades, stack up and gauge combinations may vary depending on applications. For instance, in a potential B pillar application in an aluminum intensive vehicle, it was likely to have a 6xxx body side outer, 6xxx B pillar outer and 7xxx B pillar inner forming a 3T stack. The rivet and die selection would need to be adjusted to accommodate each scenario.

3.5 Trial results

One recent example of implementing high strength s701 alloy in an automotive application is Nio Inc. on their ES6 vehicle. There are a total of seven part including ski rails, B pillar reinforcements LH/RH and 3 floor cross members as illustrated in Figure 16.

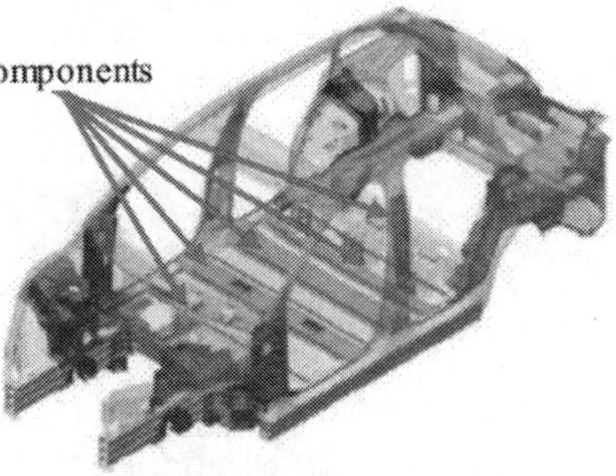

Figure 16: Overview of AA7075 alloy applications in Nio ES6.

These seven 7075 components replaced parts previously made with 6xxx alloys in ES8 with a total mass of 7.7 kg/vehicle. The high strength s701 alloy allowed all parts to be down gauged by 10-25%. As a result, an average of 20% weight saving was achieved [3]. Novelis and Nio jointly developed the hot stamping process, artificial aging practice and corrosion mitigation strategy. A B pillar reinforcement made with Novelis s701 alloy after hot stamping was sectioned and evaluated to determine its mechanical properties. Figure 17 shows the mechanical properties from various locations of hot stamped part using sub-size samples due to part geometry.

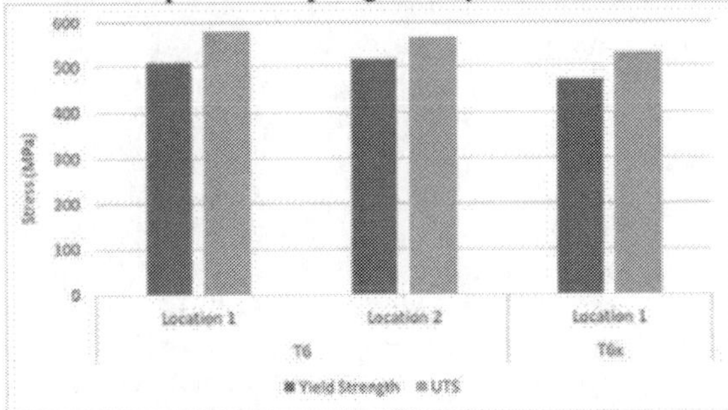

Figure 17: Mechanical property from parts made with Novelis s701 alloy.

The part was able to achieve 500MPa yield strength in T6 condition and 480MPa yield strength in T6x temper under a simulated paint bake condition of 185°C/20 minutes. A conservative approach employed by customers to simulate metal exposure during paint bake. The mechanical properties achieved were higher than targets and acceptable. The stamped part also shown good SPR performance with 0% crack rate when joined with adjacent aluminum components. In summary, through hot stamping process and accelerated aging practice, Nio is able to reduce their production cycle time, cost and improve SPR performance and successfully implemented AA7075.

4. Conclusions

S701 (AA7075) alloy can be a preferred solution for automotive light weighting especially to replace advance high strength steels (AHSS). It can be used for intrusion prevention and strength driven applications such as door intrusion beams, B pillars, floor cross-members etc.
S701 can be formed into complex geometry using hot stamping process.
An accelerated aging practice tailored to automotive process has been demonstrated. It will reduce artificial aging time from 24 hours to 1 hour.
The localized corrosion test results of the new accelerated ageing practice showed pitting corrosion morphology with the maximum pit depth less than 150 μm.
The new accelerated aging practice exhibited similar SCC resistance as compared to conventional in-service temper T6+PB.
The new proposed accelerated aging practice stabilizes the material after hot stamping and maintains good ductility. This allows s701 alloy to be mechanically joined using SPR and possibly other joining techniques with deformation involved.
With the new proposed accelerated aging practice, S701 alloy can achieve similar strength level as conventional aging and is robust when exposed to automotive paint bake cycle with extended time.

Acknowledgement

The authors gratefully acknowledge the contributions of the following:
Duane Bendzinski, Rahul Kulkarni, Raj Talla, Rashmi Mohanty, Yan Lin, Gan Ming, WonJun Kim, Kevin Ryan, Anna Janoff, Kyle Dasch, Jason Leszczewicz and all colleagues from Novelis Asia and NGRTC who supported the project.
Babak Shalchi from Canmet MATERIALS, Canada for his support on the TEM work.
David Head and Matthew Smith from Stanley Engineering for their SPR experiment.
Gary Denton, Leon Kaunitz Xuan Tang, Zhong Chen and Daniel Saltzmann from Nio for their support on hot stamping trials.

References

[1] WARDSAUTO, "2015 WARDSAUTO Survey," Dupont, 2015.

[2] J. Stanley and D. Lege, "Advances in aluminum alloy products for structural applications in transportation," *Journal De Physique IV, v3,* pp. 179-190, 1993.

[3] G. Denton, "The NIO ES8 & ES6," in *Europe Car Body*, Bad Nauheim, 2019.

[4] N.R.Harrison and S.G.Luckey, "Hot Stamping of a B-Pillar Outer from High Strength Aluminum Sheet AA7075," *SAE International*, pp. doi:10.4271/2014-01-0981, 2014.

[5] J. Parnell and L.Blayden, "Forming High Strength Bumpers from Aluminum Sheet," *SAE Technical Paper 780141*, p. doi:10.4271/780141, 1978.

[6] H. Wang, Y.B.Luo, P. Friedman, M.H.Chen and L.Gao, "Warm forming behavior of high strength aluminum alloy AA7075," *Trans. Nonferrous Met. Soc. China*, pp. 1-7, 2012.

[7] J.M.Papazian, "The effects of warm working on aluminum alloy 705-T651," *Materials Science and Engineering*, pp. 223-230, 1981.

[8] M.Kumar, N.Sotirov and C.M.Chimani, "Investigations on warm forming of AW-7020-T6 alloy sheet," *Materials Processing Technology 214*, pp. 1769-1776, 2014.

[9] A.Kraly, D. Vipava, F. Grabner, N. Ross and M. Kumar, "Hot Stamping Process of Aluminum 7xxx Sheets, Simulative Investigation," in *Neuere Entwicklungen in der Blechumformung*, Stuttgart, 2016.

[10] J. Li, Z. Peng, C. Li, Z. Jia, W. Chen and Z. Zheng, "Mechanical properties, corrosion behaviors and microstructures of 7075 aluminum alloy with various aging treatments," *Trans. Nonferrous Met. Soc. China 18*, pp. 755-762, 2008.

[11] N. Harrison, S. Luckey, B. Cappuccilli and G. Kridli, "Paint Bake Influence on AA705 and AA7085," *SAE Technical Paper*, pp. doi:10.4271/2017-01-1265, 2017-01-1265.

[12] J. Park, "Influence of retrogression and re-ageing on the strength and stress corrosion resistance of aluminium alloy 7075-T6," *Mater Sci Eng A*, pp. 223-231, 1988.

[13] A.Keci, N. Harrison and S. Luckey, "Experimental Evaluation of the Quench Rate of AA7075," *SAE Technical Paper*, pp. 10.4271/2014-01-0984, 2014.

[14] L.K.Berg, D. Schryvers and L. Wallenberg, "GP-Zones in Al-Zn-Mg Alloys and their role in Artificial Aging," *Acta Mater*, p. vol4, 2001.

[15] J. Hatch, "Aluminum: Properties and Physical Metallurgy 1st edition," 1984, p. 51.

[16] S. Lim, I. Eun and S. Nam, "Control of Equilibrium Phases (M,T,S) in the Modified Aluminum Alloy 7175 for Thick Forging Applications," *Materials Transactions, Vol. 44 No.1*, pp. 181-187, 2003.

[17] A. Chemin, D. Marques, L. Bisanha, A. Motheo and W. F. Ruchert, "Influence of Al7Cu2Fe intermetallic particles on the localized corrosion of high strength aluminum alloys," *Materials and Design*, pp. 118-123, 2014.

[18] G.Liu, G.Zhang, X.Ding, K.Chen and J. Sun, "Dependence of fracture toughness on multiscale second phase particles in high strength Al alloys," *Materials Science and Technology*, no. 19, pp. 887-896, 2003.

[19] R.Ghiaasiaan, B. Amirkhiz and S.Shankar, "Quantitative metallogrpahy of precipitating and secondary phases after strengthening treatment of net shaped casting of Al-Zn-Mg-Cu (7000) alloys," *Materials Science & Engineering A*, vol. 698, pp. 206-217, 2017.

[20] F.Andreatta, H. Herryn and J. Wit, "Corrosion behaviour of different tempers of AA705 aluminium alloy," *Electrochimica Acta*, pp. 2851-2862, 2004.

[21] P. Rout, M.M.Ghosh and K.S.Ghosh, "Effect of solution pH on electrochemical and stress corrosion cracking behaviour of a 7150 Al-Zn-Mg-Cu alloy," *Materials Science & Engineering A*, pp. 156-165, 2014.

[22] C.Altenbach, S. C., U.A.Mercado, J.P.Suuronen, D.Zander and G.Requena, "Synchrotron-based holotomography and X-ray fluorescence study on the stress corrosion cracking behavior of the peak-aged 7075 aluminum alloy," *Journal of Alloys and Compounds*, p. 152722, 2020.

[23] M. Jackel, T. Grimm, R. Niegsch and W. Drossel, "Overview of Current Challenges in Self-Pierce Riveting of Lightweight Materials," in *18th International Conference on Experimental Mechanics*, Brussels, 2018.

[24] J. Staley, "Effect of Natural Aging Interval, Preaging and Heating Rate on the Aging Characteristics of 7075," Alcoa Laboratories, Pitsburg, 1975.

[25] X. Varone, "New Aluminum Alloys in Electric Vehicles," in *Materials in Car Body Engineering*, Bad Nauheim, 2018.

[26] A.C.U.Rao, V.Vasu, M.Govindaraju and K. Saisrinadh, "Stress corrosion cracking behaviour of 7xxx aluminum alloys: A literature review," *Trans. Nonferrous Met. Soc. China*, pp. 1447-1471, 2016.

[27] F.E.Watkinson and J. Scully, "The Stress-Corrosion Cracking of a High Purity Al-6Zn-3Mg Alloy," *Corrosion Science*, pp. 905-924, 1972.

[28] J.Zuo, L. Hou, J.Shi, H. Cui, L. Zhuang and J. Zhang, "Enchanced plasticity and corrosion resistance of high strength Al-Zn-Mg-Cu alloy processed by an improved thermomechanical processing," *Journal of Alloys and Compounds*, pp. 220-230, 2017.

[29] N. Holroyd and G.M.Scamans, "Stress Corrosion Cracking in Al-Zn-Mg-Cu aluminum alloys in saline environments," *Metallurgical and Materials Transactions A*, pp. 1230-1253, 2013.

[30] S.P.Knight, N.Birbilis, B.C.Muddle, A. Trueman and S.P.Lynch, "Corrolations between intergranular stress corrosion cracking, grain-boundary microchemistr, and grain-boundary electrochemistry for Al-Zn-Mg-Cu alloys," *Corrosion Science*, pp. 4073-4080, 2010.

Effect of processing parameters on the formability of recycle friendly AA5754 alloy

Sazol Das[1*], *Matthew* Heyen[1], *John* Ho[1], and *ChangOok* Son[1]

[1]Novelis Global Research and Technology Center, Kennesaw, GA 30144, USA

Abstract. AA5xxx series Al-Mg alloys possess good combination of high specific strength-to-weight ratio, formability and corrosion resistance, which makes them attractive to the automakers for their light weighting needs. Increasingly the automakers are demanding sustainable materials. Developing aluminum alloys with increased recycled content is becoming imperative. However, increasing the recycled content can negatively impact the overall formability and joinability of the alloy. Formability is important in the shaping of complex parts and it is a key requirement in automotive manufacturing. Similarly, the other key requirement for automotive sheet is joinability. Self-piercing riveting (SPR) technology is increasingly being used for joining. In this study, the process optimization of high recycle content AA5754 alloy's for formability and rivetability will be discussed. Controlling the annealing heat treatment to produce optimum combination of grain size along with balanced recrystallized and rolling texture to improve the SPR joint configuration will be presented.

1 Introduction

Aluminum alloys have attracted considerable interest from the automotive industry in the last two decades as manufacturers seek to design lightweight vehicles with improved fuel efficiency and reduced vehicle emissions [1]. Materials for lightweight construction in automotive manufacturing must meet complex requirements. However, it is essential to combine good formability with in-service strength, excellent corrosion resistance, joinability, and crash worthiness [2]. The automotive structural inners and outers can be a sustainable material of choice due to its high value scrap from stamping trimmings as well as end-of-life vehicle scrap. However, a high level of Fe, Mn, Si and other minor alloying elements resulting from from scrap mixing can decrease the overall formability of the alloy. For example, Fe has a very limited solubility (0.052 wt. %) at the eutectic temperature, which leads to excess Fe to form constituent particles with Mn and Si that deteriorates the mechanical and fracture behavior of the alloys during forming [3 – 4]. Thus, the success of using aluminum as weight saving material depends on developing a very clear understanding of physical metallurgy behind it.

Automakers prefer AA5754 aluminum alloy for inner and structural applications in vehicles due to its good combination of moderate strength and formability. AA5754 aluminum alloy's strengthen form solid solution strengthening and can be easily formed in fully annealed condition (O Temper). However, the formability can be negatively affected by the increased scrap usage in the remelting process. A scrap metal containing high Fe, Mn and Si, can lead to poor forming performance [3]. To explore the full potential of recycled content (Fe, Mn and Si) on the formability, AA5754 alloy was designed and processed with two different levels of Fe, Mn, Si and other minor alloying element.

2 Procedures and experiments

Industrial scale full size direct chill (DC) ingots of AA5754 were cast with two different levels of minor alloying additions. The chemical compositions of these ingots were determined by inductively coupled plasma analysis (ICP) and the ranges are reported in Table 1. Variant B (A with increased Fe, Mn, Si etc.) contained more minor alloying elements compared to variant A.

Table 1. Chemical compositions of the ingots for both alloys.

	Si	Fe	Cu	Mn	Mg	Cr
Alloy A	0.05 – 0.10	0.05 – 0.10	0.01 – 0.05	0.20 – 0.30	3.10 – 3.20	0.01 – 0.03
Alloy B	0.08 – 0.18	0.08 – 0.18	0.06 – 0.10	0.25 – 0.40	3.15 – 3.30	0.04 – 0.08

Ingots were scalped, homogenized, hot rolled (HR) and cold rolled (CR) using standard practice. The cold rolled samples were then continuously annealed using standard peak metal temperature (PMT) as well as 20, 40, 50 80°C higher PMT over the standard practice to obtain O temper. Standard (ASTM B557 2" GL) tensile tests were

* Corresponding author: sazol.das@novelis.adityabirla.com

performed for both variants in the longitudinal (L), transverse (T) and diagonal (D) orientations. Self-piercing riveting (SPR) for both alloys was performed using Henrob rivet gun to evaluate SPR performance and the impact of recycled content on formability.

Optical microscopy (OM) was used to conduct a detailed microstructural study of the size and distribution of constituent particles in the final gauge samples. To observe the grain structure under OM samples were electrolytically etched using Barker's reagent. Grain sizes were measured using ASTM standard for all conditions. Various microstructural features, including the different types and spatial distributions of constituent phases, and their relative sizes and area fractions, were characterized in the O - temper samples using a Philips XL-40 field emission gun scanning electron microscope (FEG-SEM) in back-scattered electron (BE) imaging mode together with an Oxford Instruments energy dispersive X-ray spectroscopy (EDXS) system. Texture component of these O temper samples were analyzed to determine ratio between recrystallization (α – fiber) and deformation texture (β – fiber) using Panalitycal texture goniometer.

3 Results and discussions

Detailed mechanical property, riveting, microstructural features and texture component data obtained from the study are summarized below with critical observations.

3.1 Mechanical properties

Tensile properties in O temper after standard annealing practice shows some noticeable difference in strength and elongation for addition of recycled content in Alloy B. Alloy B measured 10 – 15 MPa higher strength and 1 – 2% lower elongation compared to alloy A (Fig. 1).

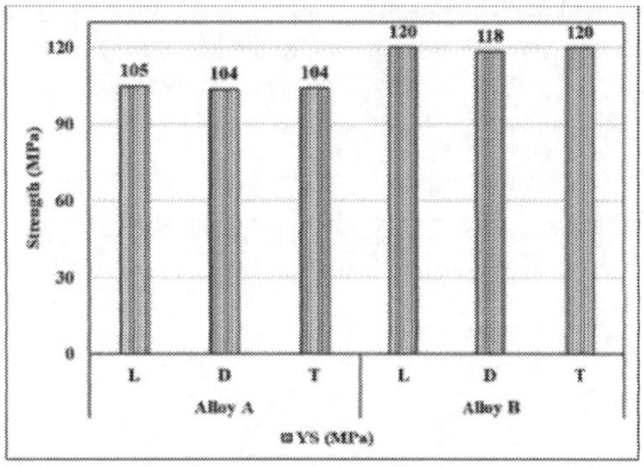

a)

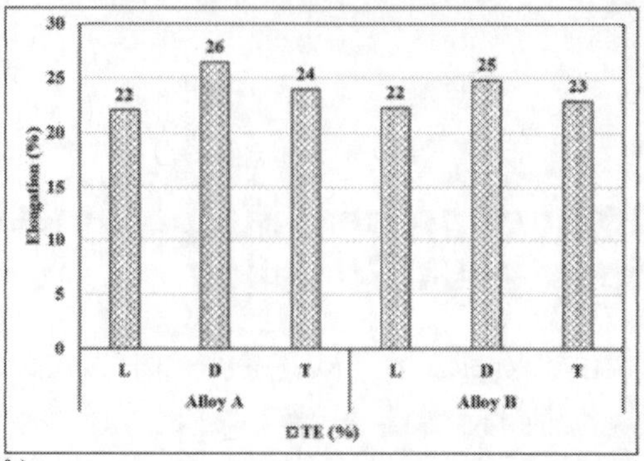

b)

Fig. 1. Tensile properties of both alloys after standard annealing practice in L, T and D orientation: a) yield strength and b) elongation.

The increase in strength with lower elongation is caused by the change in recrystallization behavior due to addition of minor alloying additions (Mn and Cr). Mn and Cr are both known to form dispersoids during homogenization and the dispersoids inhibit recrystallization by pinning the grain boundaries [5]. Strength and elongation of alloy B after various PMT annealing are shown for L orientation in Fig. 2.

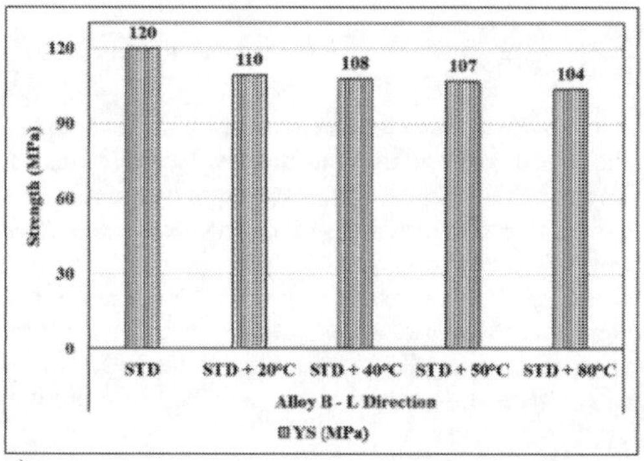

a)

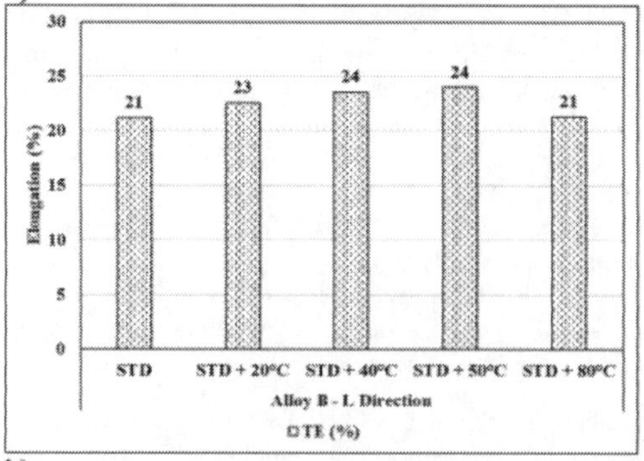

b)

Fig. 2. Tensile properties of alloy B after various PMT annealing in L orientation: a) yield strength and b) elongation.

High temperature annealing appears to lower the strength gradually. This might be due to the change in recrystallization and grain coarsening. However, elongation changes with increasing annealing PMT was not consistent. Drop in the elongation was observed after highest annealing PMT. This change in elongation is the resultant effect of fast heating rate, which might alter the grain size or texture within the bulk material.

3.2 Joining assessment

Figure 3 shows cross-sections of both alloys after SPR assembly in 3 –stack configuration where AA6451 occupies top two layers and AA5754 sits at the bottom to reflect structural joints in a vehicle. Standard annealing practice shows alloy B suffers from bottom thinning whereas alloy A retains more than 100 um bottom layer thickness after riveting (Fig. 3a and 3b). SPR for alloy B was performed with high annealing PMT's and their cross-sections showed better material retention at the bottom layer compared to standard annealing for 40°C higher PMT (Fig. 3c). However, the bottom thinning get more severe for 80°C higher PMT (Fig. 3d).

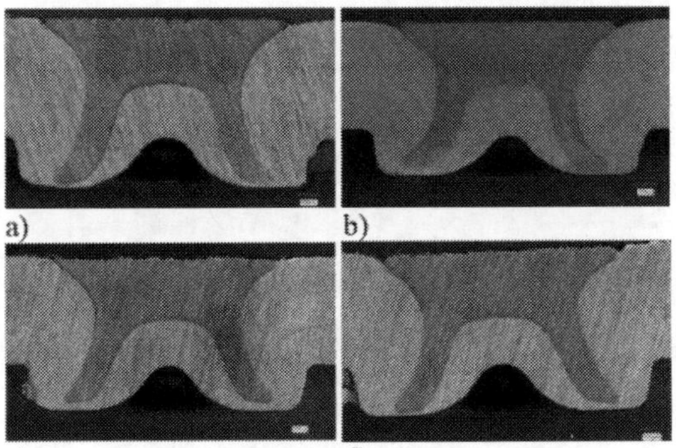
a) b) c) d)

Fig. 3. SPR cross-section after various PMT annealing: a) alloy A – standard, b) alloy – B standard, c) alloy B – standard + 40°C and d) alloy B – standard + 80°C.

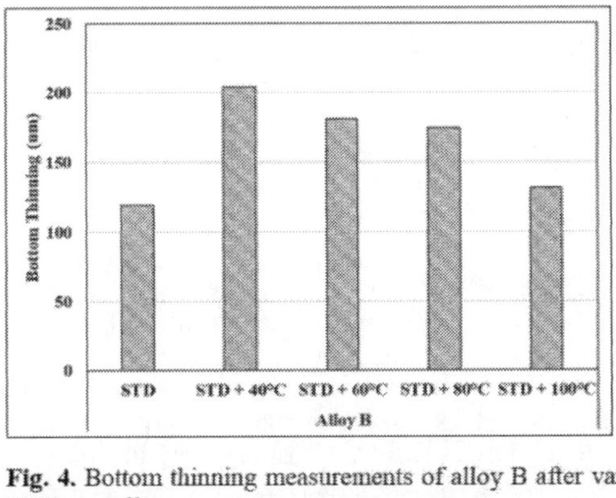

Fig. 4. Bottom thinning measurements of alloy B after various PMT annealing.

Fig. 4 is showing bottom thinning measurement for alloy B with various annealing temperatures. Bottom thinning gets better with increased annealing PMT over the standard practice. However, drop in thinning observed with highest annealing PMT (100°C higher than standard). Similar observations were made for elongation for alloy B with various annealing temperatures. Detailed microstructural evaluation including grain size and texture analyses were done to understand the mechanism for this thinning behaviour.

3.3 Microstructural observations

Particle distributions and grain structures from OM observation are shown for both alloys in Fig. 5 with standard annealing practice.

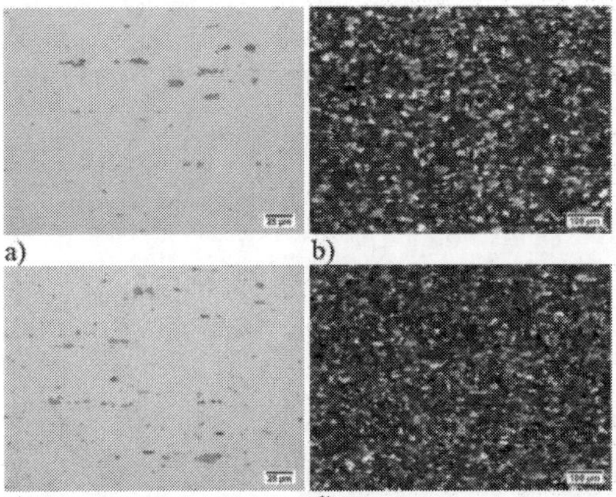
a) b) c) d)

Fig. 5. Microstructure in longitudinal cross-section after standard annealing practice for alloy A: a) particle distribution, b) grain structure and alloy B: c) particle distribution, d) grain structure

Alloy B shows coarse Fe constituent features compared with alloy A. The increase in Fe – constituents are due to high recycled content (Fe, Mn, Si etc.) in alloy B compared with alloy A. Grain sizes of alloy B were about 2 – 3 µm smaller compared to alloy A for the same reason. More Fe and Mn results in the formation of more constituent particles, which in turn provides heterogeneous nucleation sites for particle-stimulated nucleation (*PSN*) of recrystallized grains during annealing. This results in an overall finer grain size distribution in alloy B. Particles and grain structure of alloy B were analyzed after high temperature annealing (Fig. 6).

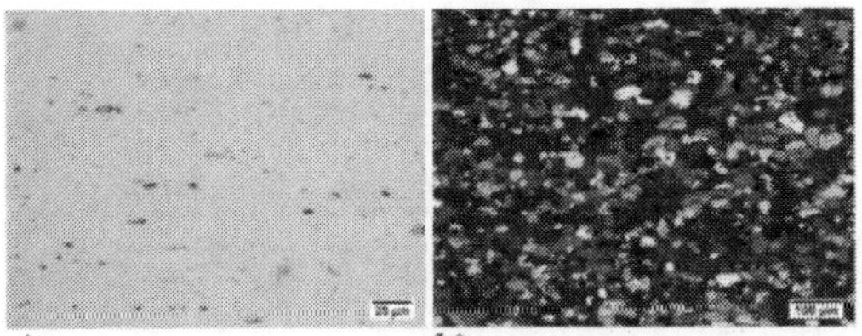
a) b)

 https://doi.org/10.1051/matecconf/202032603005

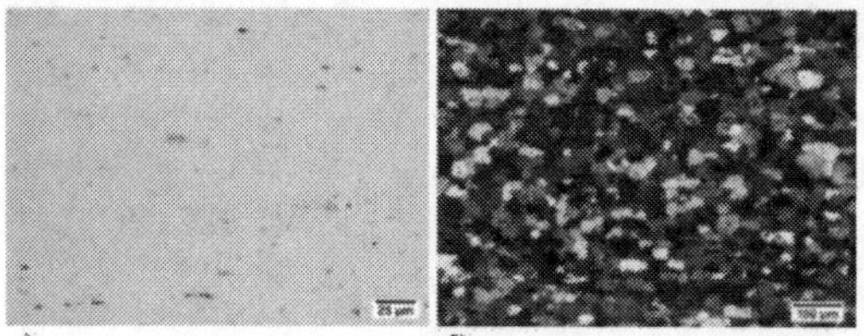

c) d)

Fig. 6. Microstructure in longitudinal cross-section after various PMT annealing of alloy B for std + 40°C: a) particle distribution b) grain structure and for std + 80°C B: c) particle distribution, d) grain structure.

There was negligible effect of high annealing temperature on particle distribution, which was expected. However, grain coarsening (around 14 µm compared to 10 µm for standard practice) observed for very high temperature annealing.

High-resolution SEM images in back-scattered electron (BE) contrast were taken to observe the morphology of Fe constituents between alloy A and B. Higher Fe and Mn results in the formation of coarse constituents (grey contrast) in alloy B with some Mg_2Si (dark contrast) particles due higher Si content (Fig. 7).

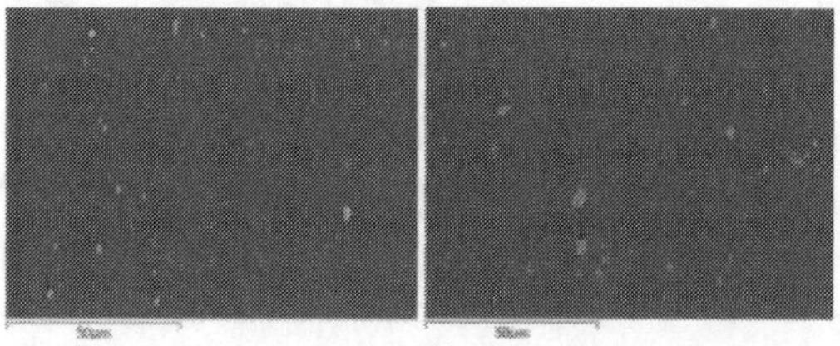

a) b)

Fig. 7. SEM BE images in longitudinal cross-section after standard annealing for both alloys: a) alloy A and b) alloy B

The constituent phases in alloys A and B were quantified using Feature Detection and Characterization (FDC) analysis from 50 random fields for each sample and processed to obtain size and number distribution. Fig. 8 shows constituents phase area fraction and number density and size distribution against equivalent circular diameter (ECD) for both alloys in final gauge after standard annealing practice. Alloy B shows large numbers of coarse constituent particles compared to alloy A as seen in micrographs (Fig. 5, 7).

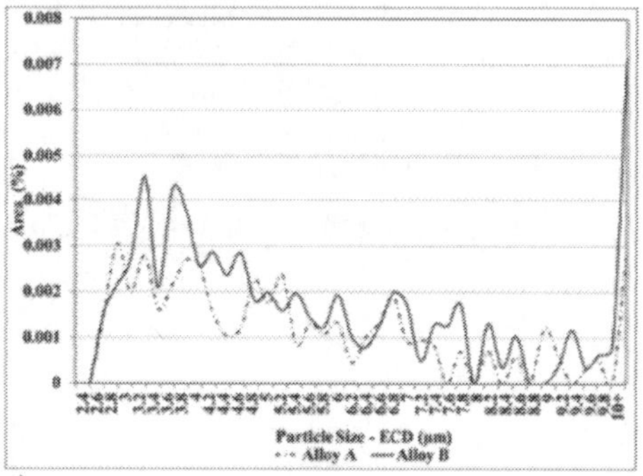

a)

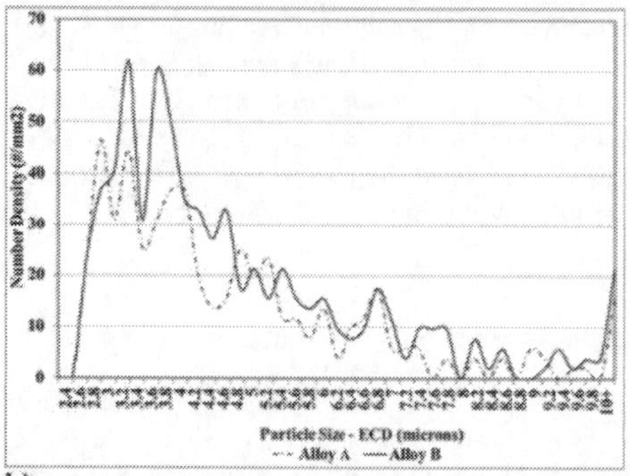

b)

Fig. 8. Constituent particle distribution after standard annealing practice for both alloys: a) area fraction and b) number density.

Recrystallization, α – fiber (cube and goss) and deformation, β – fiber (brass, copper, S) texture component for bulk microstructure were measured for alloy B after different annealing temperature. Fig. 9 shows that recrystallization texture drops after high temperature annealing compared to standard annealing practice.

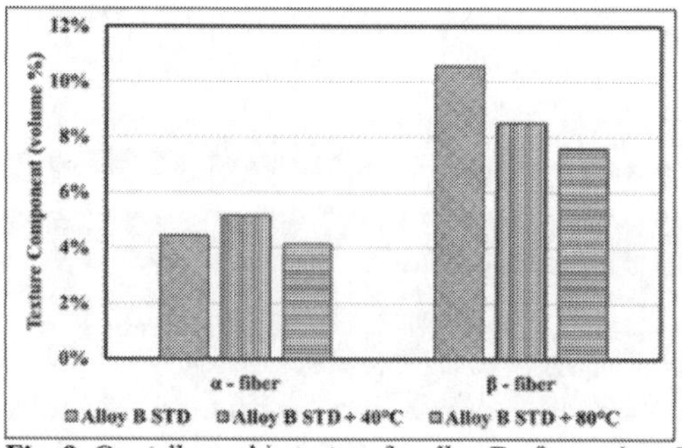

Fig. 9. Crystallographic texture for alloy B after various PMT annealing.

However, deformation texture continue to drop at high annealing temperatures. A good balance of recrystallization and deformation texture is necessary for better formability, which explains why elongation and bottom thinning get worse after high annealing temperature.

4 Conclusion

Increase in recycled content can lead to formation of coarse Fe constituent particles along with finer grains due to PSN and grain boundary pinning by dispersoids formed during homogenization. These dispersoids can also inhibit recrystallization, which results in high strength and low elongation. Forming and joinability of recycled materials are compromised if processed through standard processing routes. Raising annealing temperature can help to produce a balanced recrystallization and deformation texture, which results in better SPR performance. Too high annealing temperature on the contrary will lead to grain growth and

https://doi.org/10.1051/matecconf/202032603005

an imbalance in the bulk texture resulting in lower formability.

The authors would like to acknowledge the expert support from staff in the metallography, electron optics, application technology, mechanical testing laboratories at the Novelis Global Research and Technology Center (NGRTC) and the support of operations personnel at the Novelis Oswego works.

References

1. L. Mattei et al., Mater. Sci. Eng. A **559,** 812 (2013)
2. L. E. Elend et al., Automotive Circle International, Bad Nauheim, Germany, May (2011)
3. J. Sarkar et al., Mater. Sci. Eng., A **316,** 52 (2001)
4. W. B. Lievers et al., Mater. Sci. Eng. A **361,** 312 (2003)
5. G. W. Zhang et al., Mater. Sci. For. **877**, 172 (2016)

Effect of Cu and Mg on the corrosion behavior of 4004/Al-Mn-Cu-Mg-Si/4004 aluminum alloy brazing sheet

Xiaojie Jin[1], Pizhi Zhao[2,*], Shijie Guo[1,*], Wenjing Zhang[1], Nianmei Han[1], Yunlong Zhang[3], Guojun Wang[3]

[1]Suzhou Branch, Chinalco Materials Application Research Institute Co., Ltd., Suzhou 215026, China
[2]Chinalco Materials Application Research Institute Co., Ltd., Beijing 102209, China
[3]Northeast Light Alloy Co., Ltd., Harbin 150060, China

Abstract. Higher strength and better corrosion resistance are necessary to enable the thickness reduction of the aluminum alloy brazing sheets. In this work, an age-hardenable Al-Mn-Cu-Mg-Si core alloy was developed by the addition of Mg, Si and Cu. The corrosion behavior of 4004/Al-Mn-Cu-Mg-Si/4004 aluminum alloy brazing sheet in sea water acidified accelerated test (SWAAT) was investigated and compared with the traditional 4004/3003/4004 aluminum alloy brazing sheet. The microstructure was characterized by optical microscopy (OM), scanning electron microscopy (SEM) and electron probe microanalysis (EPMA). After brazing, it was found that the microstructures of two aluminum alloy brazing sheets from outer to the center were precipitate-free zone (PFZ), the band of dense precipitates (BDP) and core material, respectively. In particular, both Cu and Mg concentration gradients were formed in the BDP region of 4004/Al-Mn-Cu-Mg-Si/4004. Moreover, the SWAAT results showed exfoliation corrosion in 4004/Al-Mn-Cu-Mg-Si/4004, while in 4004/3003/4004, severe inter-granular corrosion in core material was observed after corrosion for 584 h. The theoretical electric potential distribution showed that the presence of Cu and Mg concentration gradient promoted the formation of the electric potential trough in the BDP region, which may be the reason for the occurrence of exfoliation corrosion in 4004/Al-Mn-Cu-Mg-Si/4004.
Keywords: 4004/Al-Mn-Cu-Mg-Si/4004, Aluminum alloy brazing sheet, Corrosion, Copper concentration gradient

1 Introduction

Aluminum alloy brazing sheets are commonly used for heat exchangers in the automotive industry, due to their high thermal conductivity, excellent mechanical properties, good corrosion resistance combined with low specific weight [1-3]. Aluminum alloy brazing sheets usually have a multilayered structure – two (or one) AA4xxx clad alloy sheets are assembled to a sandwich with an AA3xxx core alloy in between. The core alloy provides sufficient strength against deformation or collapsed while the clad alloy acts as brazing alloy to achieve a metallic bond between the components. Aiming at the reduction of the thickness, such sheets essentially require improved properties in terms of strength and corrosion resistance [3].

To achieve a higher strength of the sheet on the one hand, some methods have been developed, such as the addition of solid solution strengthening elements [4] and the formation of finely dispersed precipitates [5, 6]. Due to the fact that the brazing process is usually operated at elevated temperatures (~600 ℃) for a short time (which is equivalent to the solid solution heat treatment), thus, in principle, it is possible to develop age-hardening AA3xxx core alloy to enhance the strength of aluminum alloy brazing sheets. It has been reported that the addition of Mg and Si in AA3xxx core alloy followed by the control of cooling rate after brazing, makes it an ideal candidate for this application [7-9].

On the other hand, in order to improve corrosion performance of aluminum alloy brazing sheets, long life alloys with a Cu-containing AA3xxx core alloy, have been developed and used for many years [10-12]. In such sheets, Si atoms diffuse from the cladding into the core during the brazing process, promoting the precipitation of Mn out of solid solution in the core adjacent to the cladding and subsequently the formation of the band of dense precipitates (BDP). Simultaneously, Cu atoms in the core adjacent to the cladding also diffuse into the cladding. Due to the low solution level of Mn and Cu, the electric potential of the BDP region is lower than that of the normal core material, which produces a sacrificial layer on the rest of the core material, and finally the corrosion resistance is remarkably improved.

The aim of this work was to design an age-hardenable aluminum alloy brazing sheet with excellent corrosion resistance. The Al-Mn-Cu-Mg-Si core alloy was prepared by adding Mg, Si and Cu to a traditional 3003 aluminum alloy. The corrosion behavior of 4004/Al-Mn-Cu-Mg-Si/4004 aluminum alloy brazing sheet in sea water acetic acid test (SWAAT) was investigated and compared with the traditional

* Corresponding author: zhaopizhi@cmari.com; guoshijie@cmari.com

4004/3003/4004. Based on the results, we intend to contribute to the understanding of the corrosion mechanism involved.

2 Experimental

2.1 Material preparation

The investigated sheets with the size of approximately 100 mm×60 mm were cut from a rolled sheet consisting of an Al-Mn-Cu-Mg-Si core alloy (~2 mm thick) and two AA4004 clad alloy (~0.25 mm thick). The comparative material includes slightly thinner sheets, i.e. ~1.7 mm for AA3003 core alloy and ~0.15 mm for AA4004 clad alloy. The chemical compositions were shown in Table 1. The brazing experiment was conducted at 610 °C for 15 min in a vacuum brazing furnace, furnace cooled to 400 °C and then air cooled to room temperature.

2.2 Corrosion test

The corrosion properties of two aluminum alloy brazing sheets were evaluated by the sea water acidified accelerated test (SWAAT). The SWAAT was conducted in a cyclic corrosion test chamber (Ascott, CC450ip) according to the ASTM G85 A3 standard. The electrolyte was synthetic seawater (ASTM D1141), acidified at 49 °C to a pH of 2.8 using acetic acid. The corrosion tests took 64~584 h, depending on the experiments.

2.3 Characterization analysis

After brazing, the cross-sectional microstructures of two aluminum alloy brazing sheets were analysed using optical microscopy (OM) and scanning electron microscopy (SEM) with backscatter electron (BSE) imaging. Element line analysis on the cross-sections has also been performed by means of electron probe microanalysis (EPMA) on a Shimadzu EPMA-1720 system. After SWAAT, the surface and the cross-sectional corrosion morphologies of the sheets were investigated by OM.

3 Results and Discussion

3.1 Microstructure

The cross-sectional microstructures after brazing are shown in Fig. 1. It is found that large precipitation-free α-Al grains, i.e. precipitate-free zone (PFZ), are formed on the surface of two aluminum alloy brazing sheets due to the liquid film migration [11, 12]. The average thicknesses of PFZs in 4004/Al-Mn-Cu-Mg-Si/4004 and 4004/3003/4004 aluminum alloy brazing sheets are ~139 and ~51 μm, respectively. These α-Al grain boundaries, however, provide fast-diffusion paths for the solutes during the brazing process, leading to the formation of needle-like Si phases along the boundaries, as shown in the inset of Fig. 1. Besides, no BDP region can be observed beneath the PFZ region, as shown in Fig. 1.

Fig. 2 shows the microstructure of core material adjacent to and away from the PFZ region, respectively. It is obvious that large numbers of finely dispersed precipitates are present in the core material of two aluminum alloy brazing sheets. The statistical data shows that, for 4004/Al-Mn-Cu-Mg-Si/4004, the number densities of finely dispersed precipitates in core material adjacent to and away from the PFZ region are $\sim 1.92\times 10^6$ and $\sim 1.67\times 10^6$ mm^{-2}, respectively; while for 4004/3003/4004, they are $\sim 1.20\times 10^6$ and $\sim 0.55\times 10^6$ mm^{-2}, respectively. Hence, the number density of finely dispersed precipitates is higher in core material adjacent to the PFZ region than that away from the PFZ region in both cases. This phenomenon is due to the in-diffusion of Si from the cladding to the core material and the subsequent precipitation of densely-distributed dispersoids. BDP region can hardly be observed in Fig. 1, but BSE images provide more evidences, see the BDP region beneath the PFZ region as shown in Fig. 2. In addition, due to the higher Si content, more precipitates are observed in the core material of 4004/Al-Mn-Cu-Mg-Si/4004 than 4004/3003/4004 – in accordance with H. Ahmed et.al [10].

Based on the above, it can be found that the microstructures of two aluminum alloy brazing sheets from outer to the center after brazing are both precipitate-free zone (PFZ), the band of dense precipitates (BDP) and core material, respectively.

3.2 Corrosion behavior

Fig. 3 shows the macroscopic surface morphologies of two aluminum alloy brazing sheets after SWAAT for 64, 288 and 584 h, respectively. As shown in Figs. 3a, b, at the initial stage of corrosion (64 h), more corrosion pits are found on the 4004/Al-Mn-Cu-Mg-Si/4004, compared with 4004/3003/4004. When corrosion proceeds (288 h),

Table 1. Chemical compositions (wt.%) of the alloys investigated

Alloy	Mn	Si	Fe	Mg	Cu	Ti	Al
Al-Mn-Cu-Mg-Si	1.1-1.5	0.3-0.8	0.1-0.6	0.3-0.8	0.3-0.8	0.01-0.05	balance
AA4004	-	8-10	0.1-03	1.0-1.6	-	0.010	balance
AA3003	1.116	0.060	0.503	0.002	0.070	0.024	balance

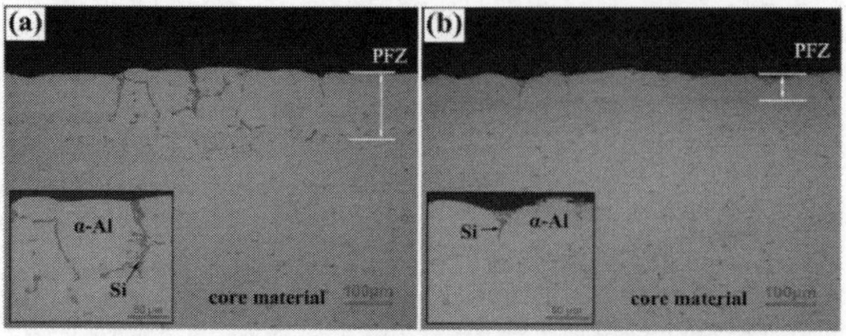

Fig. 1. Cross-sectional morphologies of (a) 4004/Al-Mn-Cu-Mg-Si/4004 and (b) 4004/3003/4004 aluminum alloy brazing sheets after brazing

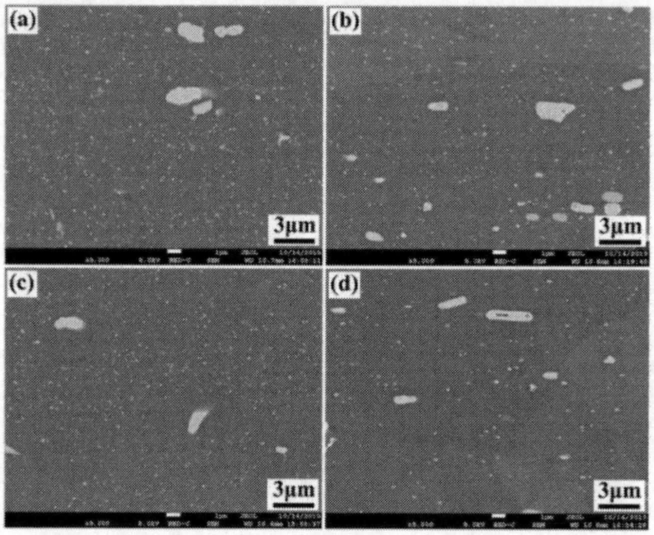

Fig. 2. BSE images of core material (a, b) adjacent to and (c, d) away from PFZ region: (a, c) 4004/Al-Mn-Cu-Mg-Si/4004 and (b, d) 4004/3003/4004 aluminum alloy brazing sheets

local exfoliation is observed on the surface of 4004/Al-Mn-Cu-Mg-Si/4004, while crater-like corrosion pits are present on the surface of 4004/3003/4004, see in Figs. 3c, d. As the corrosion time further increases to 584 h, the area of exfoliation increases significantly on the surface of 4004/Al-Mn-Cu-Mg-Si/4004, Fig. 3e. Meanwhile, the area of crater-like corrosion pits also increases on the surface of 4004/3003/4004, Fig. 3f.

Fig. 4 displays the cross-sectional morphologies of 4004/Al-Mn-Cu-Mg-Si/4004 and 4004/3003/4004 aluminum alloy brazing sheets after corrosion. At the initial stage of corrosion (64 h), the corrosion quickly goes along primary α-Al grain boundaries and reaches the core material of two aluminum alloy brazing sheets, as shown in Figs. 4a, d. After 288 h of corrosion, the PFZ region is gradually peeled away from the core material of 4004/Al-Mn-Cu-Mg-Si/4004, see in Fig. 4b, while for 4004/3003/4004, the corrosion directly spreads into the core material and the corrosion depth reaches ~800 μm, as shown in Fig. 4e. After 584 h of corrosion, the PFZ region continues to be peeled away, and the corrosion in depth direction (the sheet thickness direction) is still not found in 4004/Al-Mn-Cu-Mg-Si/4004, as shown in Fig. 4c. In contrast, the corrosion depth further increases in 4004/3003/4004 and the maximum depth reaches ~1150 μm, as shown in Fig. 4f. High magnification OM image in Fig. 4f also reveals that the corrosion of core material in 4004/3003/4004 is in the form of inter-granular corrosion (IGC). Therefore, 4004/Al-Mn-Cu-Mg-Si/4004 aluminum alloy brazing sheet developed in this work presents a different form of corrosion in comparison with 4004/3003/4004.

3.3 The corrosion mechanism

The above SWAAT results show that the corrosion behaviors of 4004/Al-Mn-Cu-Mg-Si/4004 and 4004/3003/4004 aluminum alloy brazing sheets are entirely different – the former demonstrates typical exfoliation corrosion while the latter rather shows IGC. The corrosion behavior usually depends on various factors, such as the distribution of alloying elements, grain size, the nature and morphology of precipitates. In this work, we focus on the effect of alloying element distribution on corrosion behavior.

During the brazing process, the diffusion of solutes

https://doi.org/10.1051/matecconf/202032604001

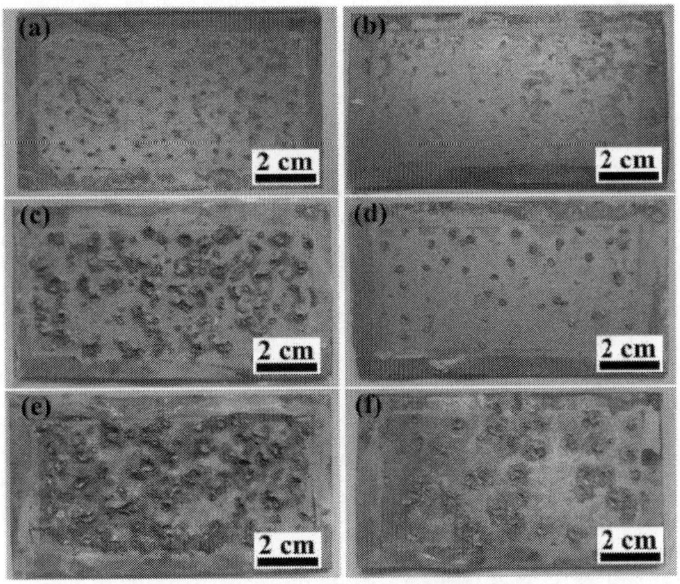

Fig. 3. Macroscopic surface morphologies of (a, c, e) 4004/Al-Mn-Cu-Mg-Si/4004 and (b, d, f) 4004/3003/4004 aluminum alloy brazing sheets after corrosion for (a, b) 64, (c, d) 288 and (e, f) 584 h

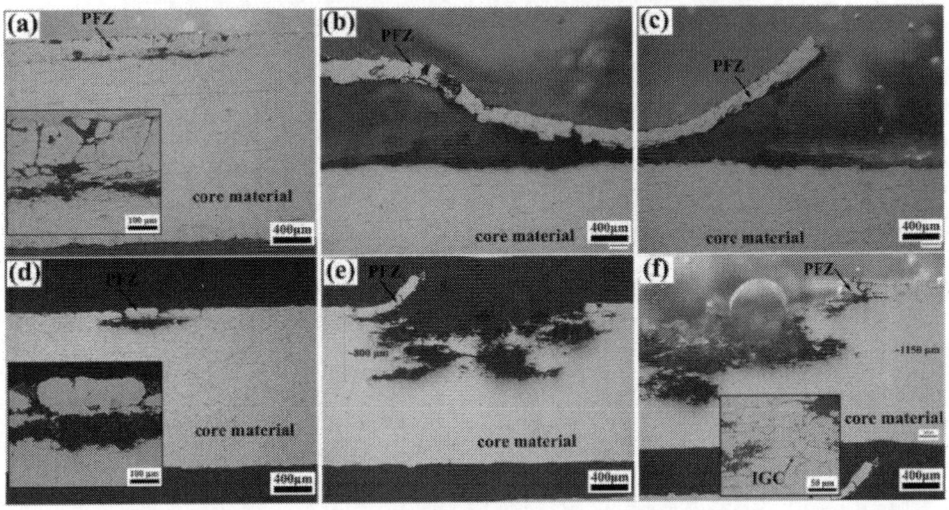

Fig. 4. Cross-sectional morphologies of (a-c) 4004/Al-Mn-Cu-Mg-Si/4004 and (d-f) 4004/3003/4004 aluminum alloy brazing sheets after corrosion for (a, d) 68, (b, e) 288 and (c, f) 584 h

determines the distribution of the alloying elements in the aluminum alloy brazing sheet. The electric potential of such an alloy, however, mainly depends on the type and the concentration of the alloying element in the solid solution. For AA3xxx, the electric potential can be theoretically estimated as follows [12-14]:

$$E(mV)=-755+54\times[Mn]+48\times[Cu]+20\times[Si]-12\times[Mg] \quad (1)$$

where *[M]* is the alloying element concentration in solid solution. According to Eq. (1), and except for Mg, a higher Mn, Cu and Si content leads to a higher *E*. Hence, the diffusion of alloying elements during brazing may cause the formation of electric potential difference between different regions. It has been reported that when the electric potential of the BDP region is lower than that of the core material, the exfoliation corrosion will set in [4].

To demonstrate the effect of alloying element concentration on the corrosion properties, the EPMA line analysis of Mn, Si, Cu and Mg in the cross-sections of 4004/Al-Mn-Cu-Mg-Si/4004 and 4004/3003/4004 aluminum alloy brazing sheets after brazing is given in Fig. 5. As shown in Figs. 5a, b, the Mn concentration profiles of two aluminum alloy brazing sheets show that the Mn content in the core material and BDP region is almost the same. There is no obvious difference in the Mn concentration profile between two sheets. In Figs.

* Corresponding author: zhaopizhi@cmari.com; guoshijie@cmari.com

5c,.d, the Si concentration profile is very similar between

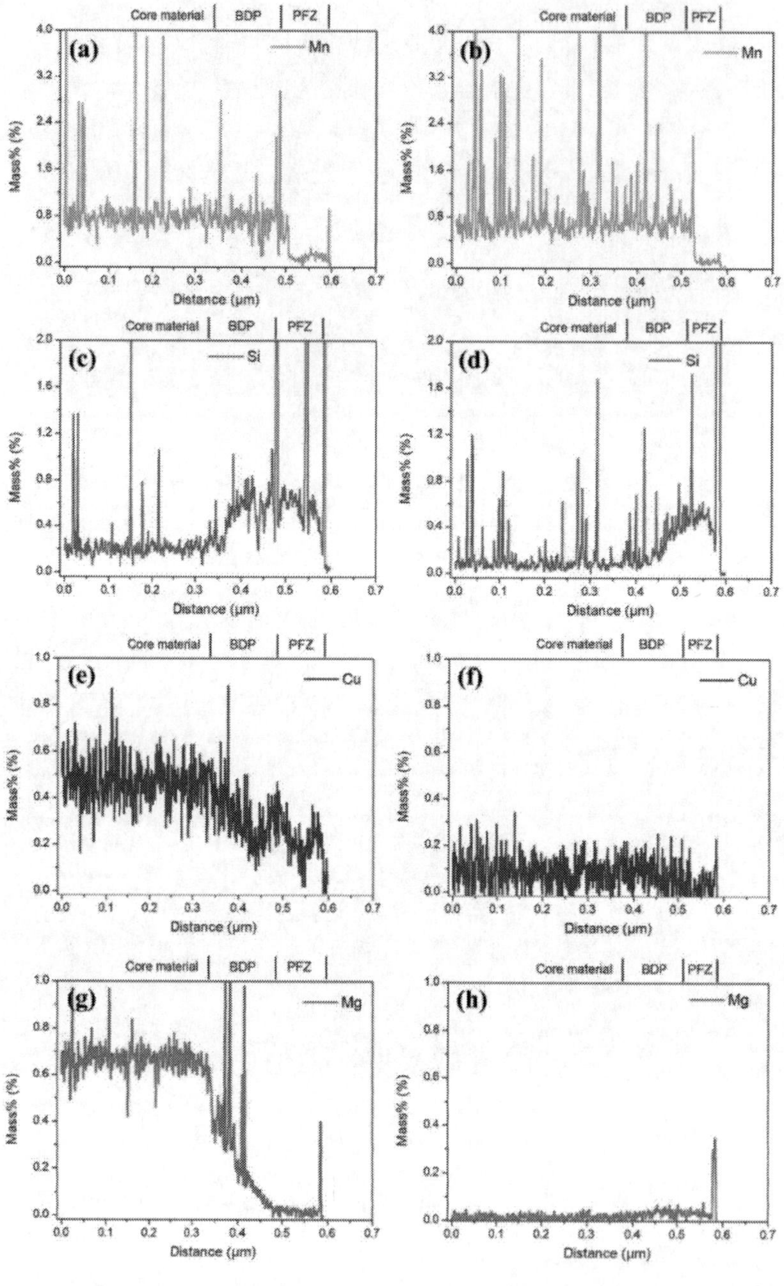

Fig. 5. EPMA line analysis of Mn, Si, Cu and Mg in (a, c, e, g) 4004/Al-Mn-Cu-Mg-Si/4004 and (b, d, f, h) 4004/3003/4004 aluminum alloy brazing sheets after brazing

two sheets, and the enrichment of Si is both observed in the BDP and PFZ regions. Due to the out-diffusion, both Cu and Mg concentration gradients are formed in the BDP region of 4004/Al-Mn-Cu-Mg-Si/4004 (Figs. 5e, g). In contrast, neither Cu nor Mg concentration gradient can be observed in 4004/3003/4004, due to the low Cu and Mg in the core material (Fig. 5f, h).

In Fig. 6, the theoretical electric potential distribution in the cross section of two aluminum alloy brazing sheets is calculated based on the EPMA measurement of the alloying element concentration in solid solution and Eq. (1). It is shown that a distinct potential trough is formed in the BDP region of 4004/Al-Mn-Cu-Mg-Si/4004, as indicated by the arrow in Fig. 6. Usually, the corrosion preferentially happens in the adjacent zone with lower electric potential. Therefore, when the corrosion reaches this potential trough, the corrosion in depth direction is likely to be retarded, and finally the exfoliation corrosion occurs, as shown in Figs. 4a-c. Such a trough, however, cannot be found in the case of 4004/3003/4004, and Fig. 4f shows only IGC regions in the core material. Moreover, it appears that the Mn and Si concentration profiles are almost same in two aluminum brazing sheets, as shown in Figs. 5a-d. Hence, the electric potential distribution difference probably correlates with the distribution of Cu and Mg.

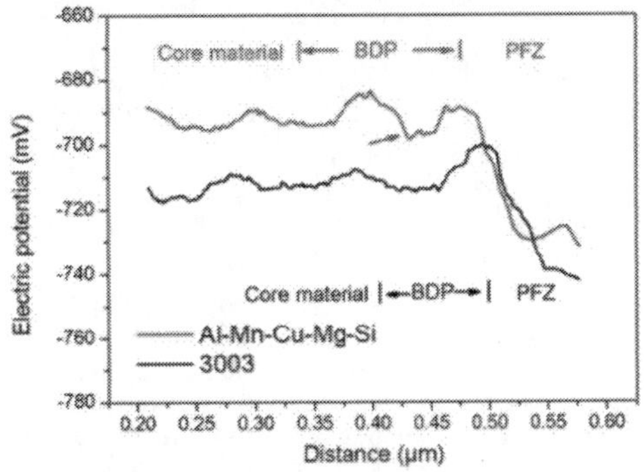

Fig. 6. Theoretical electric potential distribution in the cross section of 4004/Al-Mn-Cu-Mg-Si/4004 and 4004/3003/4004 aluminum alloy brazing sheets

4 Conclusions

(1) After brazing, the microstructures of 4004/Al-Mn-Cu-Mg-Si/4004 and 4004/3003/4004 aluminum alloy brazing sheets from outer to the center were both primary α-Al (PFZ), the band of dense precipitates (BDP) and core material, respectively. Meanwhile, both Cu and Mg concentration gradients were formed in the BDP regions of 4004/Al-Mn-Cu-Mg-Si/4004.

(2) The SWAAT results showed that 4004/Al-Mn-Cu-Mg-Si/4004 presented a different form of corrosion in comparison with 4004/3003/4004. The former showed typical exfoliation corrosion, while the latter showed rather IGC in the core material.

(3) The theoretical electric potential distribution showed that the presence of Cu and Mg concentration gradient promoted the formation of the electric potential trough in the BDP region, which may be the reason for the occurrence of exfoliation corrosion in 4004/Al-Mn-Cu-Mg-Si/4004.

The authors are grateful to the Technology Foundation of Chinalco for supporting this work.

References

[1] P.K. Velu, *Study of the effect of brazing on mechanical properties of aluminum alloys for automotive heat exchangers*, *Diss.*, 2017.

[2] F. Norouzi Afshar, J.H.W. de Wit, H. Terryn, J.M.C. Mol, Corros. Sci., **58**, 242-250 (2012).

[3] C. Liu, X. Xue, X. Chen, L. Li, C. Xia, Z. Zhong, D. Zhou, J. Mater. Eng. Perform., **26(11)**, 5563-5570 (2017).

[4] R. Westergård, S. Norgren, S. Wass, SAE Tech. Pap., **No.2005-01-2012** (2005).

[5] Y.J. Li, A.M.F. Muggerud, A. Olsen, T. Furu, Acta Mater., **60(3)**, 1004-1014 (2012).

[6] K. Liu, X.G. Chen, Mater. Des., **84**, 340-350 (2015).

[7] A. Magnusson, K. Schölin, B. Mannerskog, T. Stenqvist, A. Örtnäs, SAE Tech. Pap., **No. 971786** (1997).

[8] H.S. Goodrich, Z.M. Connor, G. Murty, SAE Tech. Pap., **No. 2001-01-1725** (2001).

[9] T. Stenqvist, K. Bång, S. Kahl, A. Contet, O. Karlsson, Mater. Sci. Forum, **794-796**, 1239-1244 (2014).

[10] H. Ahmed, J. Kadali, K. Gatenby, A. Howells, Mater. Sci. Forum, **794-796**, 1103-1108 (2014).

[11] A.J. Wittebrood, *Microstructural changes in brazing sheet due to solid-liquid interaction*, *Diss.*, 2009.

[12] L. Guo, J. Wang, W. Hu, D. Zhou, Surf. Rev Lett., **26(07)**, 1850224 (2019).

[13] C. Vargel, *Corrosion of aluminium*, Elsevier, 2004.

[14] J.R. Davis, *Corrosion of aluminum and aluminum alloys*, ASM International, 1999.

Ultrasound induced fragmentation of primary Al_3Zr crystals

Abhinav Priyadarshi[1,*], *Tungky* Subroto[2], *Marcello* Conte[3], *Paul* Prentice[4], *Koulis* Pericleous[5], *Dmitry* Eskin[2,6], *John* Durodola[1], and *Iakovos* Tzanakis[1,7]

[1] Faculty of Technology, Design and Environment, Oxford Brookes University, Oxford OX33 1HX, United Kingdom
[2] Brunel Centre for Advance Solidification Technology (BCAST), Brunel University London, Uxbridge UB8 3PH, United Kingdom
[3] Anton Paar TriTec SA, Vernets 6, 2035 Corcelles, Switzerland
[4] Cavitation Laboratory, School of Engineering, University of Glasgow, Glasgow G12 8QQ, United Kingdom
[5] Computational Science and Engineering Group (CSEG), Department of Mathematics, University of Greenwich, London SE10 9LS, United Kingdom
[6] Tomsk State University, Tomsk 634050, Russia
[7] Department of Materials, University of Oxford, Oxford OX1 3PH, United Kingdom

Abstract. Ultrasonic cavitation melt treatment (UST) of aluminium alloys has received considerable attention in the metal industry due to its simple and effective processing response. The refined primary intermetallic phases formed in the treated alloys during controlled solidification, govern alloy structural and mechanical properties for applications in the automotive and aerospace industries. Since the UST is performed close to the liquidus temperatures of the alloys, understanding the refinement mechanism of the primary intermetallic phases has been beset by difficulties in imaging and handling of liquid metals. In this paper, the sonofragmentation behaviour of primary intermetallic Al_3Zr crystals extracted from the matrix of an Al-3 wt% Zr alloy and fixed on a solid substrate was investigated. The intermetallics were exposed to cavitation action in deionized water at 24 kHz of ultrasound frequency. The fragmentation mechanism from the nearby collapsing cavitation bubbles was studied with in-situ high speed imaging. Results revealed that the main fragmentation mechanism is associated with the propagation of shock wave emissions from the collapsing bubble clouds in the vicinity of the crystal. The mechanical properties of the Al_3Zr phase determined previously were used for the fracture analysis. It was found that an Al_3Zr intermetallic undergoes low cycle fatigue fracture due to the continuous interaction with the shock wave pressure. The magnitude of the resulting shear stress that leads to intermetallic fragmentation was found to be in the range of 0.6 – 1 MPa.

Keywords: Ultrasonic melt treatment, intermetallic crystal, high speed imaging, cavitation, fragmentation, deflection

1 Introduction

Ultrasound induced cavitation and its possible benefits in liquid metal processing has received considerable attention from both the academic and industrial communities since the 1950's. Ultrasonic melt treatment (UST) being an eco-friendly, sustainable and economical processing route offers several advantages in terms of degassing, enhanced heterogeneous nucleation and structural refinement of the as-cast product resulting in improved quality of the material [1,2].

Primary intermetallics of finer size and shape formed in the Al alloys are highly desirable to augment heterogeneous nucleation in the alloy melt during solidification and so obtain microstructural refinement leading to enhancements in mechanical properties. Although induced ultrasonic cavitation in metallic melts has been proven to be the cause of structural refinement of various Al alloys [3], the fundamental understanding of the mechanism by which UST promotes fragmentation and the corresponding nucleating effects to obtain finer grain structures is still deficient. Previously the effects of UST in light alloy melts have only been studied using ex-situ (i.e. after the treatment) characterization techniques to analyse the ultrasonically treated materials. In recent years, in-situ characterization methods have received much attention for UST performance visualization in real time conditions [4–8]. The two most common in-situ experimental techniques for characterizing materials processed using UST are; (i) high-speed optical imaging of transparent organic liquids/melts, and (ii) X-ray synchrotron radiography of liquid metals. The former technique is much more widely established due to low temperature processing

* Corresponding author: abhinav.priyadarshi-2018@brookes.ac.uk

and transparent nature of the treated samples. The latter technique allow the real time observation of the cavitation bubbles and corresponding phenomena specifically growth rate, average radius and their distribution. However, due to handling and processing difficulties of analysing real metallic melts and limited field of view for capturing the dynamic effects of multi-phase interactions, a common technique is to use optically transparent liquids such as water to replicate the cavitation conditions and monitor the interaction with the solid phases during treatment. In-situ optical imaging studies of solidifying organic transparent alloys under the presence of ultrasound have proven to be effective for analysing grain nucleation by fragmentation of evolving dendritic structures accelerated by the oscillation of stable and transient cavitation bubbles [4,6]. Specifically, Shu et al. [5] found that the rate of fragmentation of growing dendrites in transparent organic alloy systems can be either slow or violent and depends primarily on the type of cavitation bubbles. Lately, in-situ synchrotron X-ray imaging has also been applied to studying real liquid metals under the influence of different external fields [9–11]. Growth rate, average radius and distribution of cavitation bubble sizes in Al-10 wt% Cu alloy was studied by Xu et al. [9] and Mi et al. [12]. Tzanakis et al. [13] provided the first direct evidence of instantaneous re-filling of a micro-capillary channel using Al-10 wt% Cu alloy melt confirming the previously postulated ultrasonic capillary effect (UCE). Dynamic collapse of a cavitation bubble in multiphase liquid flow in a Bi-8 wt% alloy melt has been observed by Tan et al. [14]. Although, observation of cavitation bubbles and their dynamic behaviour under the influence of ultrasound in real and transparent organic melts have been conclusive to a certain extent, understanding of the direct interaction of ultrasound with the dispersed or agglomerated solid phase is still lacking. Wagterveld et al. [15] imaged the influence of acoustic cavitation on suspended calcite crystals in saturated $CaCO_3$ solution and demonstrated that fracture of single calcite crystals is induced by the inception and collapse of cavitation cluster and acoustic streaming. Wang et al. [16] noticed that the fracture of intermetallics by the action of nearby cavitation bubbles is not an instantaneous process and requires substantial time to occur. In addition to handling and processing difficulties in analysing real metallic melts through in-situ X-ray synchrotron technique, the radiography method offers a very limited field of view for capturing the dynamic effects of multi-phase interactions.

In this paper, following an approach used by Wang et al. [16] the cavitation-induced fragmentation of extracted primary Al_3Zr crystals under the influence of a 24 kHz ultrasonic excitation signal has been investigated in deionized water by high-speed imaging. The fracture mechanism of a single intermetallic crystal has been elucidated using the recorded images and the stress-deflection theory. The induced stress has also been compared with the crack propagation studies conducted earlier.

2 Methodology

2.1 Sample preparation

Pure Al and a master alloy (Al-5wt% Zr) were smelted to produce about 350 grams of an Al-3wt% Zr alloy. The cast alloy was then re-melted using an electric arc furnace and slowly cooled in a cylindrical graphite crucible of 50 mm diameter following a thermal cycle as discussed in [17]. Al-3wt.% Zr alloy cubes of dimension 5 x 5 x 5 mm were cut from the solidified ingot using a rotating silicon carbide blade.

Extraction of the primary Al_3Zr crystals was done by immersing the alloy in a 15% NaOH water solution for 24 hrs. Subsequently, the Al matrix was completely dissolved leaving only the primary crystals based on the following chemical reaction:

$$2Al + 2NaOH + 2H_2O \rightarrow 2NaAlO_4 + 3H_2$$

The intermetallic crystals were then filtered out from the solution and were carefully rinsed with ethanol and left to dry out prior to the sonofragmentation studies. The optical micrographs of an Al-3wt% Zr alloy and extracted Al_3Zr intermetallic particles are displayed in Fig. 1.

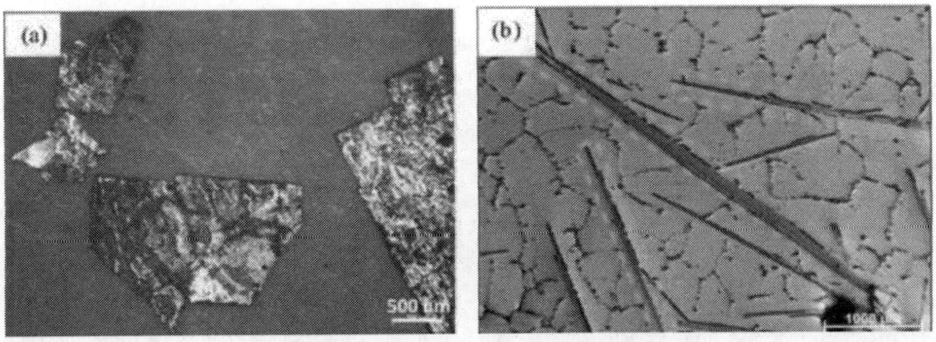

Fig. 1. Morphological images of Al_3Zr (a) extracted crystals, (b) intermetallics embedded (highlighted in green) in Al-3wt%Zr alloy matrix.

2.2 Experimental setup

The chemically extracted intermetallic particles were fixed on a steel base with a superglue adhesive and placed in a glass container of dimension 7.5 cm x 7.5 cm x 10 cm. The intermetallic was strategically positioned 2-3 mm below the piezoelectric transducer with a titanium sonotrode tip of 3 mm diameter with power density of 460 W/cm^2 (Hielscher UP200S processor) operating at a frequency of 24 kHz. The UP200S ultrasonic processor handbook enlists the detailed configuration of the system [18]. The ultrasound excitation was employed at a selected amplitude of 210 µm and the experiments were conducted in de-ionized water at room temperature.

Before capturing the fragmentation phenomenon, monitoring of the cavitation field just below the sonotrode tip and within 2 mm distance where the intermetallic crystals were later mounted was performed using a Hyper Vision HPV X2 (Shimadzu, Japan) high speed video camera and images were recorded at 1M fps in order to capture the fast shock wave propagation from the collapsing bubbles under synchronous 10 ns laser pulse illumination as in [19]. A In-situ interaction of intermetallic particles and ultrasound induced cavitation was filmed using a high-speed camera (Photron SA-Z) operating at 100,000 fps adequate to capture the fragmentation sequence of intermetallics. The camera lens was placed at a distance of 165 mm from the ultrasound source to have a full focussed observation of the interaction plane. For imaging with maximum illumination of the interaction plane, a multi LED flash light (GS Vitec) was used illuminating both the front and rear of the tank. The schematic of the experimental setup is illustrated in Fig. 2.

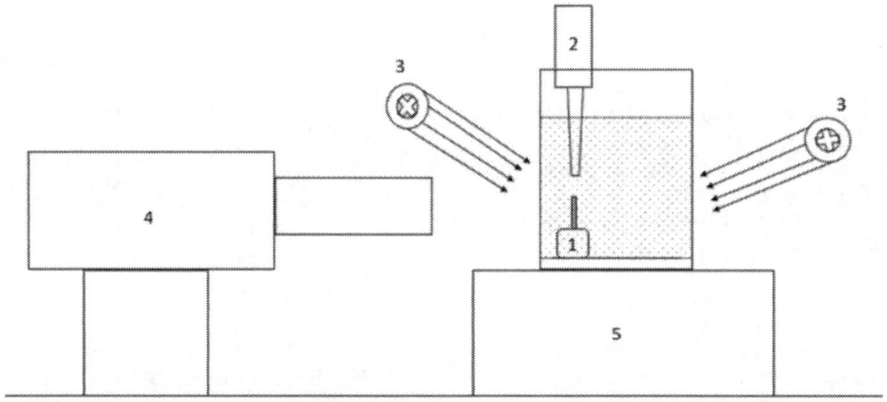

Fig. 2. A schematic of the in-situ high-speed imaging setup involving, (1) extracted Al_3Zr intermetallic, (2) ultrasonic processor with horn, (3) LED flash lamp, (4) high speed camera, and (5) movable platform.

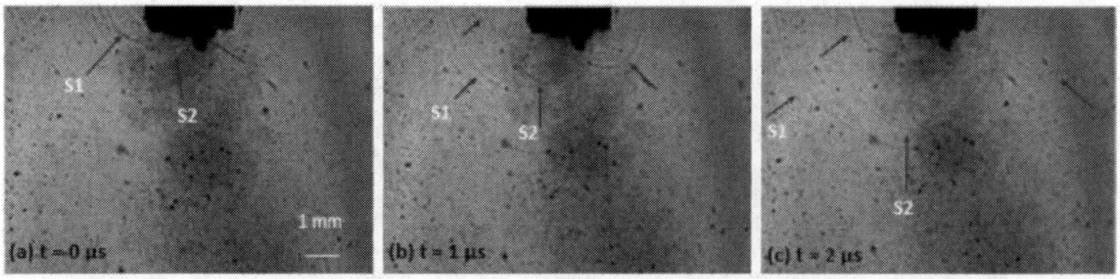

Fig. 3. Representative high-speed images of ultrasound induced cavitation and emitted shock waves recorded at 1M fps.

https://doi.org/10.1051/matecconf/202032604002

3 Results and discussion

3.1 Ultrasonic cavitation and shock wave emission

Figure 3 shows recorded images of ultrasound induced cavitation cloud and propagating shock wave fronts from the imploding bubbles. From the sequence of images, it is evident that introduction of an ultrasonic wave in a liquid medium leads to the development of acoustic cavitation cloud and emission of periodic high energy shock waves reaching pressures of several GPa and shock velocities up to 4000 m/s [20,21]. However, most of the shock wave energy is released within the first few hundred micrometres from the bubble rim [21]. Fig. 3(a-c) represents the movement of shock waves marked as S1 and S2 (indicated with blue curves and arrows) at definite intervals. It can be seen from the images that as the shock wave S1 propagates further away, another shock wave emerges from a thick cavitation cloud near the ultrasonic horn and the progression continues. Using frame by frame images, the velocity of S1 was calculated at a radial distance of approximately 3 mm and was found out to be almost 1650 m/s. It has been frequently observed that these emitted shock wave are responsible for micro-damage on any solid surface present in their vicinity [22]. Other effects such as micro-streaming and turbulences have also been found to attack the solid interfaces aggressively causing fragmentation and erosion of the material [23].

3.2 Fragmentation of primary Al_3Zr particles

To better understand and observe the effect of shock wave fronts on the fragmentation of the intermetallic crystals, the imaging was carried out at a comparatively lower frame rate i.e. 100000 fps Figure 4(a-l) represents the fragmentation sequence of the two Al_3Zr crystals, one placed in a perpendicular plane to the other. The displayed sequences of fragmentation images are representative of at least 10 of similar and reproducible observations. Only carefully chosen images haven been included for brevity of the manuscript. The first frame at $t = 0\ \mu s$, shows the two crystals positioned at right angles to each other. From here onwards, the crystal on the left will be referred to as side facing (SF) crystals and the crystal on the right will be mentioned as front facing (FF) crystal to avoid any ambiguity for the reader. Figure 4a shows the two well developed and illuminated tabular-plated crystals with similar dimensions; roughly 3 mm x 2.5 mm and a thickness of 60 - 100 µm. The detailed morphology of these primary Al_3Zr crystals can be found elsewhere [24]. The first frame also shows the SF crystal having a small notch on the edge (marked in red). It should be noted that the visibility and details of the notch/crack are limited by the camera resolution based on the present high-speed images. The ultrasonic device was subsequently switched on generating a cluster of cavitating bubbles across the sonotrode tip as depicted in Fig. 4b. As soon as the bubble cloud starts to propagate towards the crystal, slight deflection of the tip of SF is observed (marked in yellow). Since this sequence of images was recorded using white light illumination, it was difficult to capture the propagation of shock waves due to wavelength restrictions. After about 4 ms, the SF crystal above the small notch (indicated with the arrow mark) starts to deflect vigorously due to continuous emission of shock waves from the oscillating and imploding bubble cloud. At the same time, a fine crack was also formed on the FF crystal (marked in red). With the continuous oscillation of the ultrasonic horn tip, the cavitation cloud became bigger, simultaneously moving towards the crystal causing both the notch and crack to enlarge and grow in size [Fig. 4(e-f)] until it completely separated off the parent crystal (Fig. 4g). Note that up till this period the cavitation cloud had not even reached those locations of crystal imperfections from where the notch/crack began to grow indicating that shock waves (reaching ahead of the cloud) have the potential to fragment the intermetallic crystals rapidly and violently. It is also important to understand that the occurrence of the notches/cracks in the crystal results from geometrical irregularities structural defects and micro-cracks arising from the residual stresses in the intermetallic. It is also interesting to note that once the crack initiates and reaches its critical length, the crystal fails immediately in just 270 µs confirming the extremely brittle nature of the intermetallic as observed by authors in [25]. It was also revealed that the shock waves emitted from the collapse of a single cavitation bubble are primarily responsible for the fragmentation of the solid interface present nearby. At around $t = 10\ ms$, the cavitation field further grows and encapsulates half of the crystal (SF and FF). Figure 4i shows the real time snapshot of the maximum deflection induced in the SF crystal (marked in yellow). With a continued oscillating cavitation field, the SF crystal experienced cyclic fatigue owing to developed shear stresses before completely fragmenting at $t = 12.42\ ms$ as shown in Fig. 4j. At the same instant, a crack can be seen on FF crystal (marked with red arrow) which also eventually propagates and breaks off (Fig. 4k). Overall, this whole process of intermetallic failure can be attributed to the combined effect of cavitation bubble collapses and emitted shock waves. From the sequence of high speed images, it can also be deduced that the first fracture happens in just 86 cycles of ultrasonic vibrations representative of low cycle fatigue failure. The crystal disintegrates into micron sized particles in few acoustic cycles upon sonofragmentation

* Corresponding author: abhinav.priyadarshi-2018@brookes.ac.uk

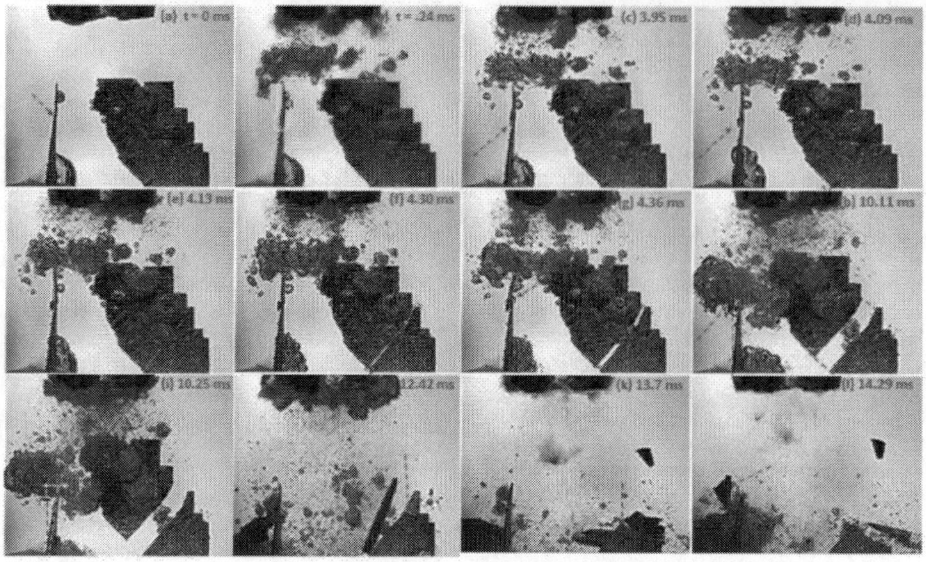

Fig. 4. Real-time high-speed photographic sequence capturing ultrasound induced fragmentation of Al_3Zr intermetallic crystal at 100,000 fps.

3.3 Application of stress-deflection theory

It has been previously observed that for an intermetallic crystal with a pre-existing notch/crack to fail completely, the required tensile stress should be in the range of 20-30 MPa and the shock pressure amplitude generated from a single (laser-induced) bubble is around 30-40 MPa at a distance of 2-3 mm [25]. However, in the case of ultrasound, the pressure amplitude of the induced cavitation field is expected to be strongly reduced owing to decrease in acoustic radiation resistance (real part of acoustic radiation impedance) [26]. The acoustic pressure was measured at a distance of 2 mm (the position of the intermetallic crystal from the sonotrode surface) using a calibrated fibre optic hydrophone system and was found to be around 1 ± 0.2 MPa.

The reason for this relatively low value of pressure compared to the pressure obtained from the single bubble collapse [25] can be attributed to the cavitation shielding [27] and the decrease in the speed of sound and the density of the surrounding medium due to the presence of bubble clouds under the sonotrode as observed by Yasui et al. [26].

In order to determine the magnitude of the shear stress acting on the tip of SF crystal (Fig. 4i), the crystal was considered to be a rectangular plate cantilever of length L, width, b and thickness d, as illustrated in Fig. 5 for the sake of geometrical simplicity. The maximum shear stress was mathematically evaluated based on the deflection observed from the frame by frame high speed images and the corresponding pixel size using the following equation [28]:

$$\tau_{max} = \frac{FQ}{Ib} \quad (1)$$

where, F is the transverse shear force obtained from the maximum deflection (δ_{max}) of the cantilever, Q is the first moment of area, I is the moment of inertia and b is the width of the crystal. For the stress calculation, the elastic modulus (E) of Al_3Zr crystal was taken from previously conducted nanoindentation measurements as 200 GPa [25].

* Corresponding author: abhinav.priyadarshi-2018@brookes.ac.uk

https://doi.org/10.1051/matecconf/202032604002

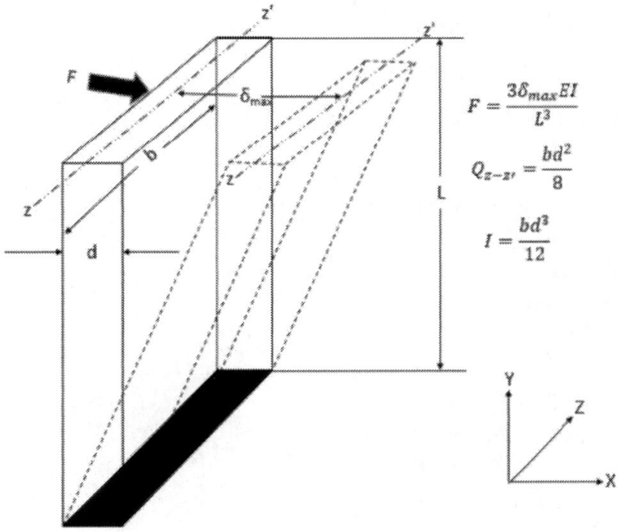

Fig. 5. Schematic of one end fixed crystal acted upon by transverse shear force F.

Using Eq. 1, and assuming that the maximum shear force is acting exactly on the tip of the crystal, the corresponding maximum shear stress produced on the intermetallic is estimated to be about 0.77 MPa while the measured cyclic acoustic pressure at that location was found to be slightly higher inducing low cycle fatigue within the intermetallic leading to fragmentation. The shear stress developed at the tip of the crystal was confirmed after 10 such observations and the average stress was found to be 0.8 ± 0.2 MPa. Nevertheless, this approximation was established for a constant load, which is not illustrative of the influence of continuous pressure pulses generated by the shock front on the crystal. Moreover, the acoustic pressure generated by the ultrasound induced cavitation bubbles cannot elucidate the fragmentation alone. Additional effects of pulsating cavitation bubbles on the crystal surface, effect of shock waves released and the liquid jet upon collapse, also need to be considered, accelerating the fragmentation.

4 Conclusions

Sonofragmentation experiments of primary Al_3Zr intermetallic crystals extracted from an Al-3wt.% Zr alloy were conducted in water. The fracture mechanism was elucidated using in-situ high speed imaging. The shock wave induced stress on the crystal was quantified using the defection fracture mechanics approach. It was confirmed that the high energy shock wave resulting from cavitation bubble collapse is mainly responsible for the fragmentation of the crystal. The Al_3Zr intermetallic undergoes cyclic deflection and eventually fails in a typical brittle manner upon interaction with the propagating shock front and cavitation bubble clouds. Crystal failure occurs in 80-100 acoustic cycles implying a low cycle fatigue fracture mechanism. The acoustic pressure amplitude at a distance of approximately 3 mm was found to be approximately 1 MPa, which is sufficient for fragmenting an intermetallic present nearby.

The authors would like to deeply acknowledge the financial support received from UK Engineering and Physical Sciences Research Council (EPSRC) grant UltraMelt2 (EP/R011095/1, EP/R011001/1 and EP/R011044/1).

References

1. O. V. Abramov, *High-Intensity Ultrasonics : Theory and Industrial Applications* (Gordon and Breach Science Publishers, Amsterdam, The Netherlands, 1998).

2. G. I. Ėskin and D. G. Eskin, *Ultrasonic Treatment of Light Alloy Melts* (CRC Press, Boca Raton, Florida, 2015).

3. G. I. Eskin, Ultrason. Sonochem. **8**, 319 (2001).

4. G. M. Swallowe, J. E. Field, C. S. Rees, and A. Duckworth, Acta Metall. 37, 961 (1989).

5. D. Shu, B. Sun, J. Mi, and P. S. Grant, Metall. Mater. Trans. A **43**, 3755 (2012).

6. R. Chow, R. Blindt, A. Kamp, P. Grocutt, and R. Chivers, Ultrason. Sonochem. **11**, 245 (2004).

7. W. W. Xu, I. Tzanakis, P. Srirangam, S. Terzi, W. U. Mirihanage, D. G. Eskin, R. H. Mathiesen, A. P. Horsfield, and P. D. Lee, (Springer International Publishing, Cham, 2015), pp. 61–66.

* Corresponding author: abhinav.priyadarshi-2018@brookes.ac.uk

8. I. Tzanakis, W. W. Xu, G. S. B. Lebon, D. G. Eskin, K. Pericleous, and P. D. Lee, Phys. Procedia **70**, 841 (2015).

9. W. W. Xu, I. Tzanakis, P. Srirangam, W. U. Mirihanage, D. G. Eskin, A. J. Bodey, and P. D. Lee, Ultrason. Sonochem. **31**, 355 (2016).

10. E. Liotti, A. Lui, R. Vincent, S. Kumar, Z. Guo, T. Connolley, I. P. Dolbnya, M. Hart, L. Arnberg, R. H. Mathiesen, and P. S. Grant, Acta Mater. **70**, 228 (2014).

11. D. Ruvalcaba, R. H. Mathiesen, D. G. Eskin, L. Arnberg, and L. Katgerman, Acta Mater. **55**, 4287 (2007).

12. J. Mi, D. Tan, and T. L. Lee, Metall. Mater. Trans. B Process Metall. Mater. Process. Sci. **46**, 1615 (2015).

13. I. Tzanakis, W. W. Xu, D. G. Eskin, P. D. Lee, and N. Kotsovinos, Ultrason. Sonochem. **27**, 72 (2015).

14. D. Tan, T. L. Lee, J. C. Khong, T. Connolley, K. Fezzaa, and J. Mi, Metall. Mater. Trans. A Phys. Metall. Mater. Sci. **46**, 2851 (2015).

15. R. M. Wagterveld, L. Boels, M. J. Mayer, and G. J. Witkamp, Ultrason. Sonochem. **18**, 216 (2011).

16. F. Wang, I. Tzanakis, D. Eskin, J. Mi, and T. Connolley, Ultrason. Sonochem. **39**, 66 (2017).

17. A. Priyadarshi, T. Subroto, M. Conte, K. Pericelous, D. Eskin, P. Prentice, and I. Tzanakis, in *Miner. Met. Mater. Ser.* (Springer, 2020), pp. 168–173.

18. "Instruction manual UP200S/ UP400s," *Hielscher*, 2007.http://www.bendarygroup.com/images/instruction_manual_up200_400s_2007_ultrasonics.pdf.

19. K. Johansen, J. H. Song, K. Johnston, and P. Prentice, Ultrasonics **73**, 144 (2017).

20. R. Pecha and B. Gompf, Phys. Rev. Lett. **84**, 1328 (2000).

21. E. A. Brujan, T. Ikeda, and Y. Matsumoto, Exp. Therm. Fluid Sci. **32**, 1188 (2008).

22. T. Prozorov, R. Prozorov, and K. S. Suslick, J. Am. Chem. Soc. **126**, 13890 (2004).

23. K. S. Suslick and G. J. Price, Annu. Rev. Mater. Sci. **29**, 295 (1999).

24. S. Zhen and G. J. Davies, J. Cryst. Growth **64**, 407 (1983).

25. K. Yasui, Y. Iida, T. Tuziuti, T. Kozuka, and A. Towata, Phys. Rev. E - Stat. Nonlinear, Soft Matter Phys. **77**, (2008).

26. G. S. B. Lebon, I. Tzanakis, K. Pericleous, and D. Eskin, Ultrason. Sonochem. **42**, 411 (2018).

27. J. M. Gere and B. J. Goodno, *Mechanics of Materials* (Cengage Learning, 2013).

Formation Mechanism of Strip Defects in Anodizing AA5252 Aluminium Alloy Sheets

Chong Gao[1,2], *Zhongyu Jiang*[3], *Zhenshan Liu*[1,2,*], *Pizhi Zhao*[1,2,*], *Ruiyin Huang*[3], *Yongfu Wu*[1,2] and *Kangcai Yu*[1,2]

[1]Chinalco Materials Application Research Institute Co. Ltd., Institute of light alloy materials, Beijing 102209, China
[2]Chinalco Research Institute of Science and Technology Co. Ltd., Beijing 102209, China
[3]Chinalco Ruimin Co. Ltd., Fuzhou 350015, China

Abstract. Processed by sulfuric acid anodizing, several unexpected strip defects exhibited on the anodizing film of AA5252 aluminium alloy sheets. Their formation mechanism was studied in detail. The normal zone and strip zone were compared, with respect to surface brightness, porosity of anodizing film, and microstructures of the corresponding aluminium substrate tested by spectroscopic colorimeter and SEM, and SEM-EBSD. Results showed that the brightness of strip zone was lower than that of normal zone. Additionally, compared with normal zone, the porosity of anodizing film was higher, and cubic grains fewer, which located in the corresponding surface layer of aluminium substrate in strip zone. In order to further clarify the effect of grain orientation on the brightness of anodizing film, single crystal aluminium sheet with (100), (110) or (111) orientation was anodized. Results showed that the anodized (100) oriented specimen, i.e., cubic grain was brighter, whose anodizing film had lower porosity, compared with (110) and (111) oriented ones. Therefore, the formation mechanism of strip defects was proposed as uneven distribution of cubic grains. The zones with fewer cubic orientated grains were darker, and were visible as strip defects after anodizing.

1 Introduction

Aluminium and its alloys have attractive attention because of their low density, high strength, good corrosion resistance and electrical conductivity etc. Meanwhile, aluminium alloy materials processed by colour anodizing could present a variety of colours and excellent corrosion protection, they are widely utilized as exterior structural components in 3C devices, including computer, cell phone and consumer electronic. The surface appearance is a critical characteristic for these components. Unfortunately, strip defects often become visible after anodizing, rather than in the rolled condition. Thus, it difficult to identify the original causes of the defects. To date, to achieve reproducible and well-controlled surface appearance of anodized aluminium alloy is still a challenging issue.

Strip defect is typically characterized by narrow bands with a different brightness from the surrounding material, usually darker than the normal zone. The anodizing film formed in sulfuric acid solution consists of a thin compact barrier layer and a thick layer consisting of hexagonally shaped cells. The barrier layer is adjacent to the aluminium substrate, whose thickness is about 1.1 nm/V with respect to the applied voltage. And in the other layer the hexagonal cell is a central pore perpendicular to the substrate surface [1, 2].

Previous researches [3-5] indicated that the strip defects not only correlate strongly with the anodizing conditions (voltage, temperature, composition and acid concentration of the bath), but also depend on the microstructure of the aluminium substrate, i.e. impurity distribution, grain size, texture and surface morphology. Y. Ma et al. [6] suggested that Mg_2Si phase in AA6063 aluminium alloy was preferentially oxidized at low anodizing voltages, and resulted in the development of a rough alloy/film interface and the incorporation of silicon into the porous anodic film, which led to the strip defects. H.L. Zhu et al. [7] found that the formation of a heterogeneous distribution of grain boundary grooves is a critical factor for strip defects. The grooves are mainly influenced by the distribution of Mg_2Si precipitates and Fe-rich intermetallic particles on the surface of 6060 aluminium alloy sheets.

In this study, a high purity AA5252 aluminium alloy sheet was selected as the research target, because the effect of secondary phases, like Mg_2Si precipitates and Fe-rich intermetallic particles, could be excluded in this investigation. Therefore, the study on the effect of strip defects focused on the grain structure in the anodized AA5252 aluminium alloy sheets. The correlation among the strip defect, the porosity of anodizing film, and the different grain structures was systematically studied, and the formation mechanism of strip defects was discussed.

2 Experimental procedure

2.1 Materials preparation

Materials for the present study were AA 5252 aluminium alloy sheets with gauge of 0.6 mm and H22 temper. The composition of the alloy is given in Table 1. They were trial-produced on industrial scale, AA5252 alloy ingots were manufactured by direct chill (DC) casting and subjected to homogenization, and subsequently rolled to necessary gauge by hot rolling and cold rolling. These sheets were blasted and subsequently anodized in 20 wt.% sulfuric acid at 25 °C for 25 minutes using direct current (DC) with 1.4 A·dm^{-2}, then immersed in blue dye solution, used to manufacture backplane parts of cell phone in batch commercially. Specimens as large as 80 mm× 50 mm was cut by electrical discharge machining

* Corresponding author: zhaopizhi@cmari.com, liuzhenshan@cmari.com

from the area without plastic deformation in the backplane.

Table 1. Chemical composition of AA5252 aluminium alloy sheets (wt./%).

Si	Fe	Cu	Mn	Mg	Zn	Bal.
≤0.02	≤0.04	≤0.10	≤0.10	2.2~2.8	≤0.05	Al

Furthermore, (100), (110) and (111) orientated single crystals of pure aluminium (Al ≥ 99.99 wt.%) were prepared through Bridgeman method, to understand the effect of grain orientations on the anodic films properties. Square specimens with the size of 10 mm × 10 mm were prepared for the investigation. The orientations of the samples were verified by X-ray diffraction (XRD), as shown in Fig.1. Before the anodization, specimens were mirror polished. They were anodized in 20 wt.% sulfuric acid at 25 °C for 25 minutes using direct current (DC) with 1.4 A·dm^{-2}, subsequently sealed in boiled water and dried in air.

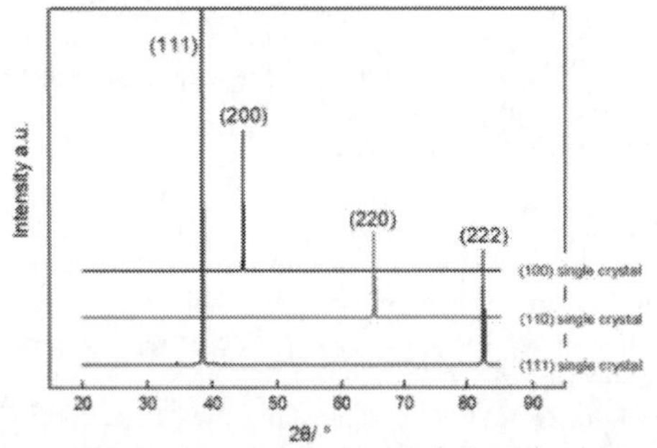

Fig. 1. XRD spectrum of single crystal aluminium alloys.

2.2 Microstructure analysis

The appearance of anodized specimens was recorded in D65 light source with a viewing angle of about 45°. The surface brightness L* of the anodized aluminium was measured by a 3NH YS3060 high-precision spectrophotometer. The value L* ranges from 0 to 100 represent for black-and-white. The porosity of anodizing film refers to the percentage of pore area in the observed total area. It was investigated quantitatively by a Hitachi SU8010 ultra high-resolution field emission scanning electron microscope (SEM) from the top view. The thickness of anodizing film was measured quantitatively by a Fischer FMP 40 equipped with a FTA3.3H probe. To reveal the microstructure on the surface of aluminium substrate, the anodized AA5252 aluminium alloy specimen cut from cell phone backplane was ground 1 min with sandpaper (1200#), and then mirror polished for 3 min to remove the anodizing film. The specimen thickness reduced around 10~15 μm, which is slightly over the thickness of anodizing film. For comparison, secondary phases and grain structures of AA5252 aluminium alloy sheets both located under the strip defects and the normal zone were evaluated respectively by a JEOL JSM-7800F SEM equipped with Electron Backscatter Diffraction (EBSD).

3 Results and discussion

3.1 Strip defects

Fig.2 presented a typical anodized AA5252 aluminium alloy part of cell phone. The thickness of anodizing film was 8.9 ± 0.4 μm. Several strip defects appeared on the surface. Strip defects showed strip patterns with random distribution along the rolling direction. Strip zones were dark, their width less than 1.0 mm, length up to tens of millimetres. The brightness L* of the normal zone and strip zone was 87.93±0.11 and 87.59±0.10 respectively, the deviation △L*=0.34. This indicated that the strip zone was darker than normal zones, and was visual slightly on the appearance of anodized AA5252 aluminium alloy sheets.

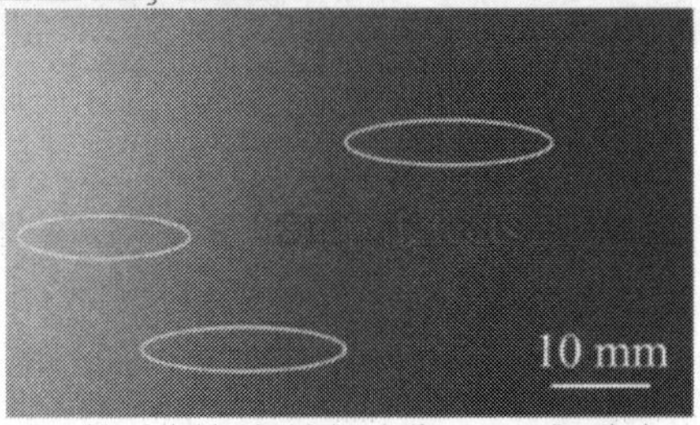

Fig. 2. Anodized 5252 aluminium alloy part of cell phone with strip defects.

Fig.3 showed the morphologies of anodizing film on the anodized AA5252 aluminium alloy part of cell phone. Many round or oval pores were located in the anodizing film. And the porosities of the film in the normal zone and strip zone was 14.4% and 17.1%, respectively.

Fig.4 showed the relationship between the surface brightness and film porosity, both in normal zone and strip zone of anodized AA5252 aluminium alloy sheets. Compared with the normal zone, the surface brightness of strip zone was lower.

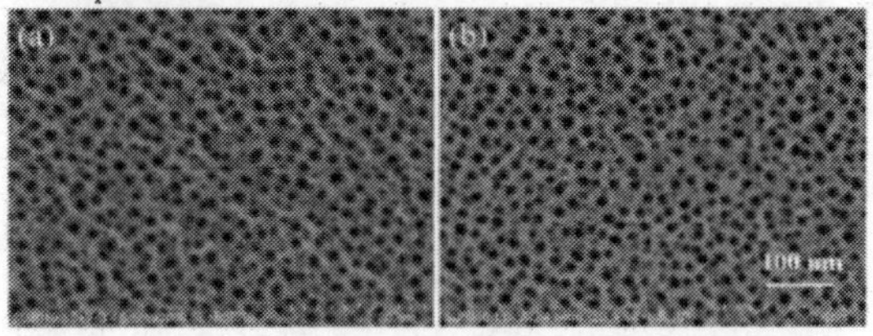

Fig. 3. Anodizing film morphologies of anodized AA5252 aluminium alloy part of cell phone: (a) normal zone with porosity of 14.4%; (b) strip zone with porosity of 17.1%.

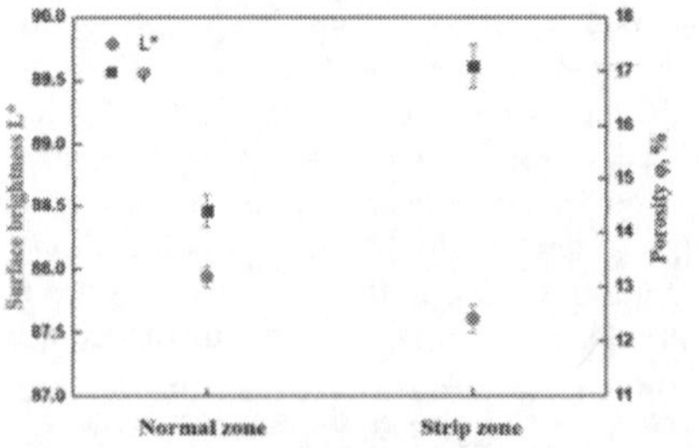

Fig. 4. Surface brightness and film porosity of normal zone and strip zone.

3.2 Microstructures

Fig.5 showed the distribution of secondary phases in AA5252 aluminium alloys. Several white secondary phases with an equivalent circle diameter less than 1.5 μm located randomly in both the normal zone and strip zone. And they were identified as Fe-rich intermetallic particles by EDS detection. Statistical results showed that the area proportions of the secondary phases in normal zone and strip zone were similar to each other, 0.028% and 0.022% respectively. Hence, the effect of secondary phases on the strip defects can be ignored.

Fig.5. Distribution of secondary phases in AA5252 aluminium alloy sheet: (a) normal zone with area proportion of 0.028%; (b) strip zone with area proportion of 0.022%.

Fig.6 showed grain structures in AA5252 aluminium alloy. The equiaxed and elongated grains with large amount of substructures (2°< misorientation < 15°) distributed in both the normal zone and strip zone. Statistical results showed that the average grain size in normal zone and strip zone was 16 μm and 17 μm, respectively, and the difference between them was not significant.

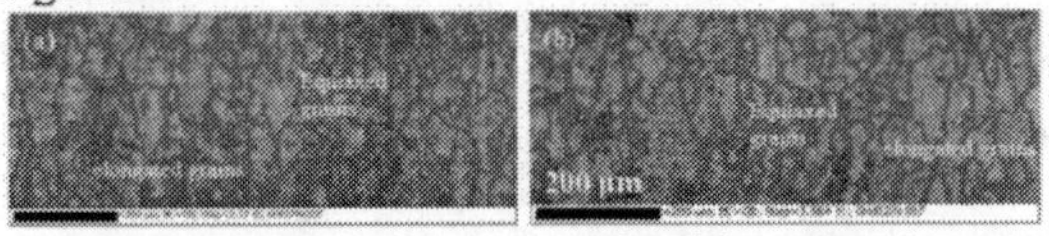

Fig.6. Grain structures in AA5252 aluminium alloy sheet: (a) normal zone with grain size of 16 μm; (b) strip zone with grain size of 17 μm.

Fig.7 demonstrated the typical grain orientations in AA5252 aluminium alloys. Table 2 listed the area proportion of typical texture components. Fig.7a and Fig.7b gave the distribution of typical oriented grains, Copper, S, Brass and Goss oriented grain distributed uniformly along the rolling direction both in the normal zone and strip zone. However, the area proportions of cubic oriented grain in normal zone and strip zone were11.3% and 5.6%, respectively (Fig.7c and Fig.7d), indicating an uneven distributed in AA5252 aluminium alloy sheets.

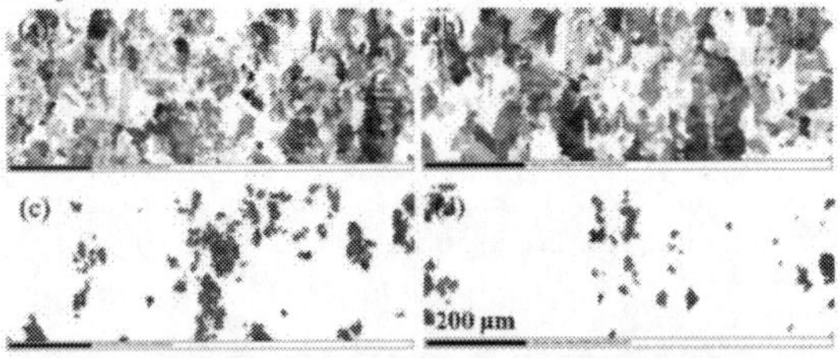

Fig.7. Typical oriented grains in (a) normal zone and (b) strip zone (Blue for Copper, purple for S, yellow for Brass, green for Goss, red for Cube, white for remainder); Cubic oriented grains in (c) normal zone and (d) strip zone.

Table 2. Area proportion of typical grain orientations in AA5252 aluminium alloy sheets.

ZONE	Copper {112} <111>	S {123} <634>	Brass {011} <211>	Goss {011} <100>	Cube {001} <100>	Remainder
Normal	8.6	10.7	19.6	5.4	11.3	Bal.
Strip	4.3	18.6	17.7	8.0	5.6	Bal.

3.3 Characterization of anodizing film in single crystals with different orientations

Furthermore, in order to confirm the effect of grain orientation on the anodizing film, the single crystal specimens with the (100), (110) to (111) grain orientation were anodized with the same anodizing conditions and studied.

Fig. 8 showed the surface and cross-section morphologies of the anodizing film with different grain orientations. From the top view, the pore size of the anodizing film of (100), (110) and (111) oriented single crystal aluminium was measured as 9.5 nm, 8.6 nm and 8.3 nm, respectively. The porosity was 13.6%, 14.7% and 14.1%, respectively. It can be seen from the cross-section that the angle between the alignment direction of (100), (110) and (111) oriented single crystal anodizing pore and the horizontal direction of aluminium substrate was 90°, 76° and 84°, respectively. And the anodizing film thickness was 10.7 μm, 12.1 μm and 12.0 μm, respectively. Compared with (110) and (111) oriented crystals, the anodizing film on (100) oriented crystal had a lower porosity, and the nanopores seemed more regular. These phenomena were consistent with previous publications [8, 9].

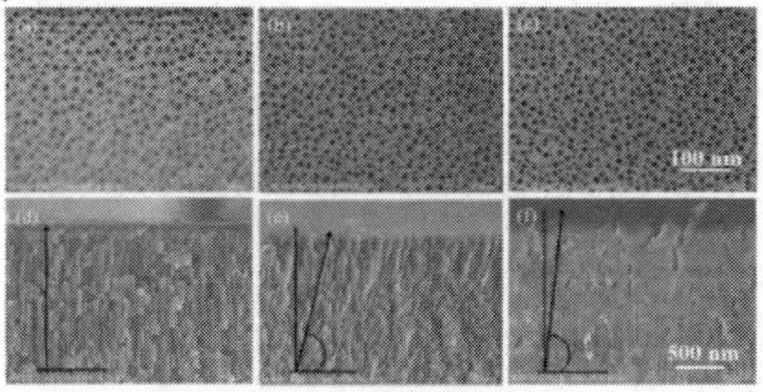

Fig.8. Surficial and cross-section morphologies of anodizing film of single crystals with different grain orientations: (a), (d) 100; (b), (e) 110; (c), (f) 111.

Fig.9 showed the relationship between the surface brightness and porosity of anodized single crystals with different orientations. It can be seen that the surface brightness of anodizing film with (100), (110) and (111) orientation were 93.40, 92.71 and 92.49, respectively; And the porosity of anodizing film with (100), (110) and (111) orientation were 13.6%, 14.7% and 14.1%, respectively. Compared with (110) and (111) oriented crystals, the higher brightness of anodizing film on (100) oriented crystal had a lower porosity.

https://doi.org/10.1051/matecconf/202032604003

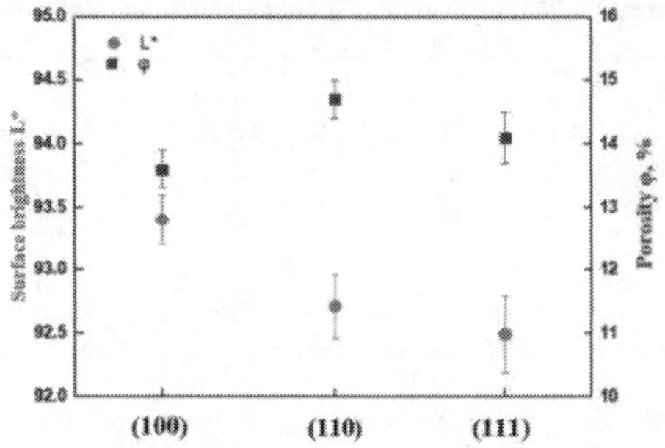

Fig.9. Surface brightness and porosity of anodized single crystal aluminium specimens.

3.4 Discussion

Surface brightness of anodized aluminium comes from two mechanisms: the reflected light from anodizing film surface and the reflected light from anodizing film interior. Therefore, the surface brightness was higher, because of the lower porosity and thinner anodizing film in (100) oriented grain, compared with other oriented grains.

Fig.10 gave the sketch maps for the formation mechanism of strip defects in AA5252 sheets. The amount of cubic orientated grains, i.e., (100) oriented grains, in the normal zone was relatively higher than that in strip zone, leading to a lower porosity and a thinner anodizing film. One hand, the lower porosity led to more reflected lights from anodizing film surface, and less proportion of incident lights entered into the pores; On the other hand, the thinner anodizing film reduced the absorption of reflected light in the interior of anodizing pores.

In contrast , the cubic orientated grains appear in lower fraction in the strip zone, leading to a higher porosity and a thicker anodizing film. For the same beam of incident lights, intensity of reflected light from surfaces decreased. Meanwhile, reflected light decreased further due to absorption by the interior of anodizing pores.

To sum up, the formation of strip defects on the surface of anodized AA5252 aluminium alloy sheets was caused by the uneven distribution of cubic orientated grains. The lower proportion of cubic orientated grains in strip zone, leading to dark after anodizing, compared with normal zone with higher proportion of cubic orientated grains.

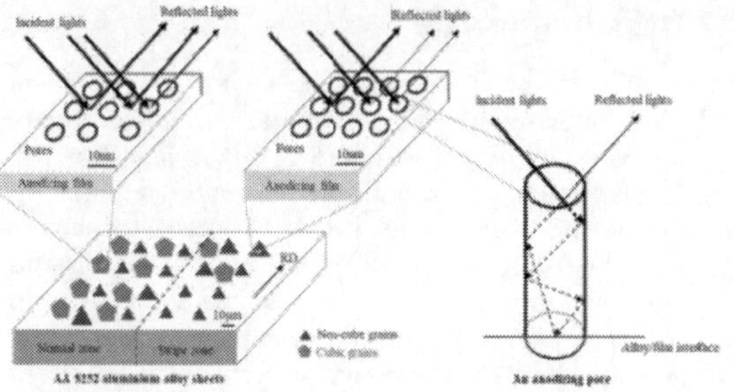

Fig.10. Sketch maps for the formation mechanism of strip defects.

4 Conclusions

(1) For the anodized AA5252 aluminium alloy sheet, strip defects zone was darker than normal zone. The strip defects had a higher porosity level in the anodizing film with fewer cubic orientated grains.

(2) The (100) oriented single crystal aluminium after anodizing was brighter than (110) and (111) oriented single crystal aluminium, because its anodizing film is thinner and has lower porosity.

(3) The forming mechanism of strip defects was supposed as uneven distribution of cubic orientated grains. The zones with fewer cubic orientated grains were darker, and were visible as strip defects after anodizing.

References

1. C.K. Chung, C.H. Tsai, C.R. Hsu, E.H. Kuo, Y. Chen, I.C. Chung. Corros. Sci. **125** (2017) 40-47.
2. A. Nemcova, I. Kubena, M. Smid, H. Habazaki, P. Skeldon, G.E. Thompson. J. Solid State Electr. **20** (4) (2016) 1155-1165.
3. Y. Zhu, S. Wang, Q. Yang, F. Zhou. J. Mater. Eng. Perform. **23** (9) (2014) 3389-3399.
4. L.F. Hu, D.M. Chen, F.R. Shi, S.P. Chen. Surf. Interface Anal. **48** (12) (2016) 1299-1304.
5. J. Evertsson, F. Bertram, L. Rullik, G. Harlow, E. Lundgren. J. Electroanal Chem. **799** (2017) 556-562.
6. Y. Ma, X. Zhou, J. Wang, G.E. Thompson, W. Huang, J.O. Nilsson, M. Gustavsson, A. Crispin. J. Electrochem Soc. **161** (6) (2014) C312-C320.
7. H.L. Zhu, X.Q. Zhang, M.J. Couper, A.K. Dahle. Metall. Mater. Trans. A. **40** (13) (2009) 3264-3275.
8. G. Beck, R. Bretzler. Mater. Chem. Phys. **128** (3) (2011) 383-387.
9. X. Zhou, G.E. Thompson, P. Skeldon. Electrochimi. Acta. **53** (18) (2008) 5684-5691.

Effect of pre-aging deformation on the intergranular corrosion resistance of 6000 series aluminium alloys

Otto Lunder[1,2*], *Julie* Grønvold[2], *Calin Daniel* Marioara[1], *Sigurd* Wenner[1], and *Trond* Furu[2,3]

[1]SINTEF Industry, NO-7465 Trondheim, Norway
[2]Norwegian University of Science and Technology, NO-7491 Trondheim, Norway
[3]Norsk Hydro ASA, NO-0283 Oslo, Norway

Abstract. Aluminium alloys in the 6000 series may become susceptible to intergranular corrosion (IGC) by unfavourable thermomechanical processing in combination with presence of Cu and a low Mg/Si ratio. The IGC susceptibility has been related to the segregation of Cu during artificial ageing to form a nearly continuous Cu film along the grain boundaries with an anodic solute depleted zone adjacent to it. In this work, extruded alloys 6005 and C28 as well as two variants of 6016 automotive sheet were subjected to various degrees of deformation by rolling or stretching before artificial aging. Accelerated IGC testing showed that pre-deformation significantly improved the IGC resistance of the Cu containing extruded alloys, while no improvement was observed for the 6016 sheet materials containing < 0.01 wt% Cu. High resolution scanning transmission electron microscopy (STEM) of pre-deformed 6005 alloy showed precipitation of Cu-containing Q' phase on dislocations. As a result, the amount of Cu available for diffusion to the grain boundaries is reduced, which might explain why a beneficial effect of pre-deformation on the IGC resistance was observed only for the Cu-containing materials.

1 Introduction

Heat treatable 6000 series aluminium alloys have several attractive properties but may become susceptible to intergranular corrosion as a result of unfavourable thermomechanical treatment or alloying [1-3]. In a series of papers, Svenningsen et al. [4-6] and Larsen et al. [7, 8] showed the importance of heat treatment on the IGC resistance of 6000 series alloys with low Cu content. The presence of Cu increased the IGC susceptibility as a result of aging by forming a continuous Cu film and noble Q-phase ($Al_5Cu_2Mg_8Si_6$) precipitates along the grain boundaries and a precipitate free zone (PFZ) depleted in Cu and Si adjacent to it. The depletion of Cu and Si in the PFZ is expected to form a continuous anodic path along the grain boundaries, thus providing micro-galvanic coupling as the local driving force for IGC propagation. These results were recently confirmed by Kairy et al. [9, 10] who combined STEM and atom probe tomography (APT) to verify the existence of a Cu "wetting layer", estimated to be about 2 nm in thickness, along the grain boundaries. The role of Si/Mg ratio, Cu content and aging condition on IGC resistance and hardness was further reviewed through a systematic investigation of nine custom made 6000 series alloy. Based on this work, a generalized microstructure-based model for IGC susceptibility was proposed [10].

Additional progress in understanding initiation and propagation of IGC was made by Kumari et al. [11, 12]. In particular, it was shown that Cu enriches continuously by dealloying at corroding sites and becomes the dominating cathode both at the external surface and in the interior IGC filaments.

In general, results from previous studies indicate that even slight variations in alloy composition (particularly Cu content and Si/Mg ratio) and thermo-mechanical processing can significantly affect the susceptibility to IGC. Extruded profiles are often straightened or subjected to different forms of deformation before age hardening, which changes the precipitation kinetics. Earlier work [13] showed that pre-deformation reduced the time to reach maximum yield strength during artificial aging. Moreover, the precipitates formed in the pre-deformed material were coarser, with a lower number density but higher volume fraction than in the undeformed material.

Results from a recent study [14] indicated a significant improvement of the IGC resistance of an AA6005A alloy containing 0.14 wt% Cu by including a pre-deformation step before aging. High resolution STEM imaging showed that the pre-deformation caused Cu containing precipitates to nucleate at dislocation lines. Furthermore, pre-deformation resulted in a higher amount of Cu locked into bulk precipitates than in the undeformed material. Hence, the results indicated a correlation between increased Cu absorption into precipitates, reduced Cu concentration at the grain boundaries and an improved IGC resistance [14]. The purpose of the present work is to further verify the effect of pre-deformation on the IGC resistance by examining a

* Corresponding author: Otto.Lunder@sintef.no

few 6000 series aluminium alloys with different Cu content and Si/Mg ratio.

2 Experimental

2.1 Materials

Table 1 shows the chemical composition of the materials investigated. Extruded 6005 alloy profiles (150 mm x 3.9 mm) supplied by Hydro were solution heat treated at 540°C for 30 min, water quenched and pre-deformed 1, 5 and 10% by rolling or stretching, respectively, before ageing at 185°C for 5 h. The C28 alloy was cast into 95 mm billets at Hydro, homogenized at 575°C for 2 h and extruded to 25 mm x 3.9 mm profiles in an 8 MN laboratory press at SINTEF, then solution heat treated at 540°C for 30 min, water quenched, pre-deformed and aged under similar conditions as the 6005 alloy, cf. Figure 1a. In addition, a separate series of 6005 and C28 specimens were deformed 1, 5 and 10% by rolling or stretching *after* ageing (instead of before ageing); these are referred to as post-deformed. Figure 1b shows the thermomechanical processing of the two variants (denoted A and B) of 1 mm thick 6016 automotive sheet. After solution heat treatment at 560°C for 5 min, the materials were cooled by forced air, pre-baked at 90°C for 24 h, stored for at least one week at room temperature, pre-deformed and finally aged at 185°C for 20 min to simulate a paint bake cycle.

Table 1. Composition (wt%) of the test materials.

Alloy	Si	Fe	Cu	Mn	Mg	Cr
6005	0.62	0.19	0.15	0.16	0.56	< 0.001
C28	0.72	0.20	0.20	0.51	0.83	0.16
6016 A	1.49	0.19	0.002	0.078	0.38	0.001
6016 B	1.33	0.20	0.010	0.064	0.30	0.004

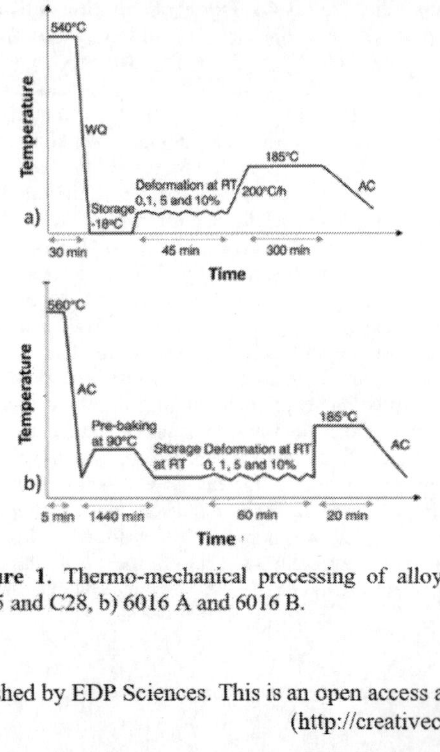

Figure 1. Thermo-mechanical processing of alloys a) 6005 and C28, b) 6016 A and 6016 B.

2.2 Intergranular corrosion testing

IGC testing was performed according to the ISO 11846 (method B) standard. The procedure included cleaning of 25 mm x 20 mm size samples by immersion for 3 min in 7.5% NaOH solution at 50-60°C, followed by desmutting in concentrated nitric acid for 2 min, rinsing in water and drying. The IGC test involved immersion in an acidified salt solution (3 wt% NaCl + 1 vol% HCl) for 24 h. The ratio of the solution volume to the total sample surface area was kept constant at 20 ml/cm^2 for all tests. After IGC testing, the samples were rinsed under running water, immersed in concentrated nitric acid for 2 min to remove corrosion products, rinsed in distilled water and air dried. All samples were weighed before and after IGC testing to assess the mass loss caused by corrosion.

2.3 Microstructure characterization

Corrosion tested samples were cut along the fabrication direction, mounted in epoxy resin, ground and polished, and the cross-section area examined by use of a Leica MEF4M optical microscope to assess the extent, type and maximum depth of localised attacks. Vickers hardness measurements were performed on a Zwick Roell Indentec hardness testing machine, using a load of 1 kg and loading time of 15 s on the as-manufactured surface. High resolution scanning transmission electron microscope (STEM) images of the precipitate microstructure and grain boundaries were acquired by use of a JEOL ARM200F operating at 200 kV. Thin foil samples from a region close to the original surface were electropolished using a Struers TenuPol-5 unit, with a 1/3 nitric acid + 2/3 methanol electrolyte.

3 Results

3.1 Grain structure and hardness

Figure 2 shows optical micrographs of the undeformed grain structure of the alloys. All materials were fully recrystallized as a result of solution heat treatment. However, the materials exhibited significant differences in grain size. The hardness of aged alloys 6005 and C28 was about 100 HV and 112 HV, respectively, practically independent of the degree of pre-deformation (results not shown). For both the 6016 sheets, the hardness after aging increased from about 82 HV for the undeformed material to 96 HV for the 10% deformed material.

3.2 Accelerated IGC testing

Metallographic cross section images from representative locations on IGC tested samples are shown in Figure 3. The undeformed 6005 alloy sample (Figure 3a) exhibited uniform IGC, characterized by narrow and knife-edge sharp attacks along all the grain boundaries at the surface. The 1% pre-deformed sample showed a similar behaviour. However, pre-deformation to 5% or 10% (Figure 3b) considerably reduced the extent of IGC.

Attacks became more localised and a there was a significant decrease in penetration depth compared to the undeformed material. A similar effect of pre-deformation was also observed for the C28 alloy. While 1% pre-deformation had little effect, a 5% or 10% pre-deformation by rolling changed the corrosion morphology from IGC to pitting (Figure 3c and 3d).

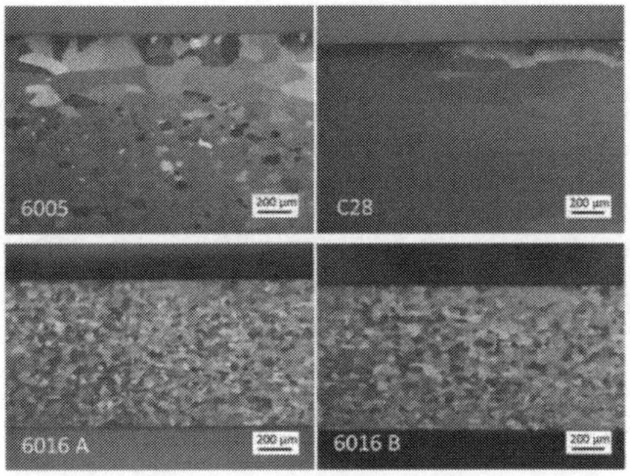

Figure 2. Grain structure of the alloys in longitudinal cross section.

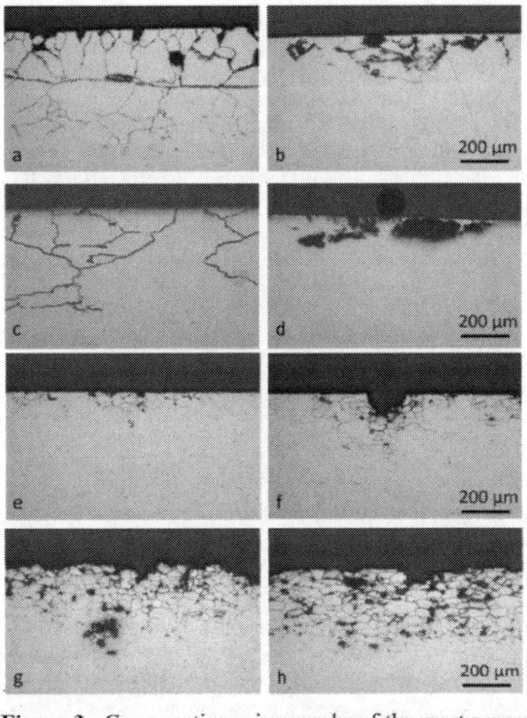

Figure 3. Cross section micrographs of the most severe attacks on undeformed (a, c, e, g) and 10% pre-rolled (b, d, f, h) samples after IGC testing: a, b) 6005, c, d) C28, e, f) 6016 A, g, h) 6016 B.

Contrary to the above results, pre-deformation of the 6016 A (Figure 3e and 3f) and 6016 B (Figure 3g and 3h) alloys had no beneficial effect on the corrosion behaviour. In fact, pre-deformation of the 6016 A alloy resulted in an increased susceptibility to IGC. The weight loss also doubled relative to the undeformed material. For the 6016 B alloy, very high corrosion rate and severe IGC occurred over the entire sample surface, independent of the degree of pre-deformation. The effect of pre-deformation on the maximum penetration depth after IGC testing is shown for all samples in Figure 4.

Samples deformed up to 10% by rolling or stretching *after* aging showed no significant change in IGC susceptibility relative to the undeformed 6005 and C28 samples. Figure 5 shows cross section micrographs of 10% post-deformed and IGC tested 6005 and C28 alloy samples. All post-deformed samples showed similar corrosion morphology as the undeformed material. Maximum depth of IGC attack and corrosion rate (determined from weight loss measurements) were also practically unaffected by post-deformation processing.

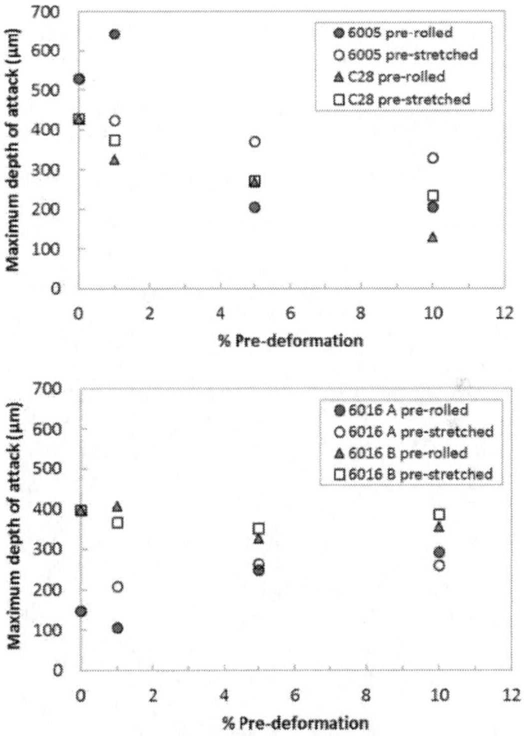

Figure 4. Influence of the degree of pre-aging deformation on the maximum depth of localised corrosion attack of IGC tested samples.

3.3 TEM characterization

TEM analysis of alloy 6005 in the undeformed and 10% pre-rolled condition showed a homogeneous bulk precipitate distribution in the undeformed material (Figure 6a), while in the pre-deformed material precipitation was more inhomogeneous and occurred preferentially on dislocation lines (Figure 6b and 6c), as expected. High-angle annular dark-field imaging

(HAADF) STEM images are shown in Figure 6d-e, with Cu atomic columns appearing bright, and Al/Mg/Si columns appearing darker.

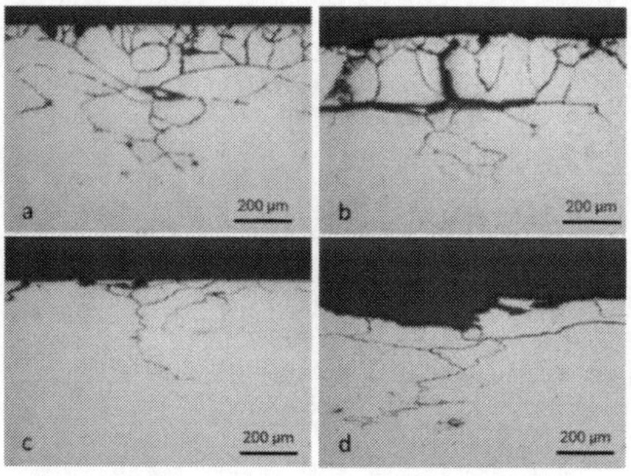

Figure 5. Cross section micrographs of IGC tested alloys 6005 (a, b) and C28 (c, d), deformed 10% by rolling (a, c) or stretching (b, d) after artificial aging.

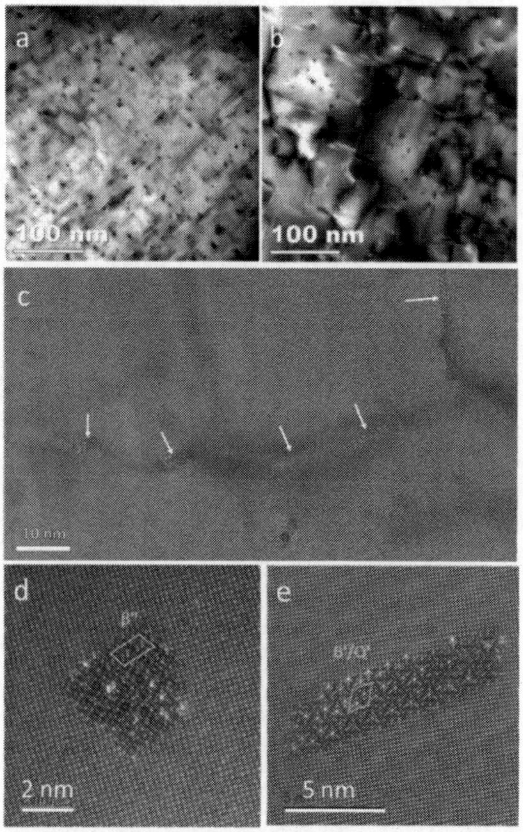

Figure 6. Bright field TEM overview images of a) undeformed and b) 10% pre-rolled 6005 alloy aged for 5 h at 185°C; c) HAADF-STEM image of Cu-containing precipitates nucleated along dislocation line (arrows) in the pre-deformed material; and individual precipitates in the d) undeformed and e) pre-deformed material.

Most of the precipitates in the undeformed material were disordered β" particles with low Cu content (Figure 6d). In the pre-deformed condition dislocations were present, which nucleate coarser, more ordered and more Cu-rich Q' precipitates (Figure 6e). The density of dislocations increases with degree of deformation, leading to a higher fraction of the heterogeneously nucleated kind of precipitates [14].

Although precipitate statistics was not presently made for the pre-deformed material, earlier work [13] on alloy 6060 has shown that pre-deformation results in a lower number density but higher precipitate volume fraction than in undeformed material. Hence, based on the present and earlier findings it seems clear that under similar aging conditions pre-deformation causes more Cu to become incorporated in precipitates than in the undeformed material. However, STEM/EDS element mapping (results not shown here) indicated that a continuous Cu film still formed along some of the grain boundaries in the pre-deformed material. Cu also segregated to interfaces between the Al matrix and Fe-rich primary particles.

4 Discussion

The present work confirms and supplements earlier results [14] showing a beneficial effect of pre-deformation on the IGC resistance of 6000 series alloys with low Cu content. Results revealed that deformation of the Cu containing 6005 and C28 alloys by rolling or stretching before aging significantly reduced the IGC susceptibility of these materials. It was further shown that the dislocations introduced by the pre-deformation acted as heterogeneous nucleation sites for precipitation of Cu-containing Q' phase particles, in accordance with previous findings [13, 14]. The increased amount of Cu locked in precipitates in the pre-deformed material implies that the amount of Cu available for diffusion to form a Cu film and cathodic precipitates along the grain boundaries is reduced, which should contribute to reduced IGC susceptibility. The results obtained for alloys 6005 and C28 (Figure 4) indicate that a 5% pre-deformation is enough to significantly reduce the maximum depth of IGC attacks after accelerated testing. Pre-rolling appeared to be more efficient than pre-stretching in reducing the penetration depth, possibly related to differences in the deformation distribution across the material thickness.

Extensive evidence from previous studies [4-10] suggests that the IGC susceptibility of the 6005 and C28 alloys can be ascribed to the formation of a continuous Cu film and cathodic Q phase precipitates along the grain boundaries, accompanied by an adjacent solute depleted PFZ which constitute the anodic path for IGC. Evidently, the introduction of dislocations before aging modified the precipitate microstructure and local distribution positively with respect to IGC. However, deformation applied *after* aging caused no change in corrosion behaviour relative to the undeformed materials, demonstrating that the presence of dislocations *per se* had negligible influence on the corrosion mechanism.

Since pre-deformation results in coarser precipitates with a higher volume fraction than in undeformed material [13], the precipitate structure might be regarded more like that of an overaged material. It has been observed [5] that AlMgSi(Cu) alloys may become susceptible to pitting in overaged temper due to coarsening of matrix precipitates and solute depletion of the bulk matrix. The Al matrix then becomes more similar to the PFZ in composition and the preferred pathway for anodic dissolution, i.e. the PFZ, becomes less distinct. The transition from IGC to pitting with increasing degree of pre-deformation was particularly observed for the C28 alloy (Figure 3d). Moreover, the improved corrosion resistance of alloys 6005 and C28 by pre-deformation was obtained without any significant change in hardness. Apparently, the work hardening introduced by rolling or stretching compensated for the coarser precipitate structure formed by aging of the pre-deformed material.

Pre-deformation of the 6016 alloys did not influence the IGC susceptibility much, which is consistent with the above hypothesis since these materials were practically Cu free but have a very high Si/Mg ratio to enable adequate hardening during a paint baking cycle. For these alloys, the IGC susceptibility is probably controlled mainly by the excess Si rather than the presence of a small amount of impurity Cu. It is well known [8, 10] that the severity of IGC increases with an increase in the Si/Mg ratio of 6000 series alloys, related to the formation of a Si-depleted zone adjacent to the grain boundaries after aging [10]. However, as seen from Figures 3 and 4, the 6016 B material was clearly inferior to 6016 A in the IGC test despite a relatively small difference in Si/Mg ratio between the two. The effective Si/Mg ratio in 6016 A and 6016 B was estimated to be 3.7 and 4.1, respectively, when subtracting the amount of Si incorporated into Fe-containing primary particles and dispersoids. In addition to a slightly higher Si/Mg ratio, 6016 B contained more Cu (0.010 %) than 6016 A (0.002 %). Hence, it cannot be ruled out that the presence of very low Cu concentrations contribute to the IGC susceptibility in alloys with high Si/Mg ratio. Work is in progress to further understand the role of Cu and Si/Mg ratio in 6016 alloys.

5 Conclusion

The IGC resistance of extruded AlMgSi alloys with low Cu content can be considerably improved by pre-deforming the material 5-10% before aging, with minimal change in hardness. The favourable effect of pre-deformation on corrosion properties is attributed to an increased amount of Cu locked in precipitates nucleated on dislocation lines in the pre-deformed microstructure, causing less Cu to be available for diffusion to the grain boundaries. Pre-deformation did not improve the IGC resistance of Cu-free (< 0.010%) sheet materials with very high Si/Mg ratio, where IGC susceptibility is associated with Si depletion adjacent to the grain boundary.

6 Acknowledgements

This work was performed as part of the Norwegian national project entitled, *Fundamentals of Intergranular Corrosion in Aluminium Alloys (FICAL)*, supported by The Research Council of Norway (contract no. 247598), Hydro, Gränges, Benteler Automotive Raufoss and Steertec Raufoss. The (S)TEM work was conducted on the NORTEM (NFR: 197405) infrastructure at the TEM Gemini Centre, Trondheim, Norway.

7 References

1. H. P. Godard, W.B. Jepson, M.R. Bothwell, R.L. Kane, *The Corrosion of Light Metals* (John Wiley, New York, 1967).
2. L. F. Mondolfo, *Aluminium Alloys: Structure and Properties* (Butterworths, London, 1976).
3. J. E. Hatch, *Aluminium – Properties and Physical Metallurgy* (American Society for Metals, Ohio, 1984).
4. G. Svenningsen, J. E. Lein, A. Bjørgum, J. H. Nordlien, Y. Yu, K. Nisancioglu, Corros. Sci., **48**, 226 (2006).
5. G. Svenningsen, M. H. Larsen, J. C. Walmsley, J. H. Nordlien, K. Nisancioglu, Corros. Sci., **48**, 1528 (2006).
6. G. Svenningsen, M. H. Larsen, J. H. Nordlien, K. Nisancioglu, Corros. Sci., **48**, 3969 (2006).
7. M. H. Larsen, J. C. Walmsley, O. Lunder, R. H. Mathiesen, K. Nisancioglu, J. Electrochem. Soc., **155**, C550 (2008).
8. M. H. Larsen, J. C. Walmsley, O. Lunder, K. Nisancioglu, J. Electrochem. Soc., **157**, C61 (2010).
9. S. K. Kairy, T. Alam, P. A. Rometsch, C. H. J. Davies, R. Banerjee, N. Birbilis, Metall. Mat. Trans. A, 47A, 985 (2016).
10. S. K. Kairy, P. A. Rometsch, C. H. J. Davies, N. Birbilis, Corrosion, **73**, 1280 (2017).
11. S. Kumari, S. Wenner, J. C. Walmsley, O. Lunder, K. Nisancioglu, J. Electrochem. Soc., **166**, C3114 (2019).
12. S. Kumari,S. Wenner, J. C. Walmsley, O. Lunder, K. Nisancioglu, Corros. Sci., 158, 108090 (2019).
13. K. Teichmann, C. D. Marioara, S. Andersen, K. Marthinsen, Metall. Mat. Trans. A, 43A, 4006 (2012).
14. C. D. Marioara, A. Lervik, J. Grønvold, O. Lunder, S. Wenner, T. Furu, R. Holmestad, Metall. Mat. Trans. A, **49**, 5146 (2018).

Multifunctional aluminium surfaces – Laser-structured micropatterns with ice-repellent, superhydrophobic and easy-to-clean properties

Stephan Milles[1,*], *Marcos* Soldera[1, 2], *Bogdan* Voisiat[1] and *Andrés Fabián* Lasagni[1, 3]

[1] Institute of Manufacturing Science and Engineering, Technische Universität Dresden , Dresden, Germany
[2] PROBIEN-CONICET, Universidad Nacional del Comahue, Department of Electrical Engineering, Neuquén, Argentina
[3] Fraunhofer Institute for Material and Beam Technology IWS, Dresden, Germany

Abstract. In this work, the fabrication of multifunctional periodic microstructures on pure aluminium is presented. Three different geometries were fabricated with feature sizes ranging between 7 µm and 50 µm by using laser-based microstructuring methods. In detail, nanosecond pulsed direct laser writing and picosecond pulsed direct laser interference patterning were used with infrared laser radiation. The wetting characteristics of these structures were investigated performing static water contact angle measurements as well as by measuring the contact angle hysteresis and the sliding angle. The final wetting results show constant static contact angles above 150°, permitting the water droplets to roll off the substrate as well as collecting contamination at the same time. This self-cleaning effect led to a reduction of up to 94% of the spread of 1 µm sized manganese oxide particles. In addition, the freezing time required for droplets laying on the patterned surfaces was increased nearly by 300% at a temperature of 20 °C below zero. Finally, the results are compared to finite element simulations of heat transfer.

1 Introduction

A water droplet that is rolling off a slightly tilted surface is an outcome of the surface chemistry and topography as it was already observed on the lotus flower (nelumbo nucifera) many years ago [1]. This so-called superhydrophobic property results in many beneficial characteristics like high corrosion resistance, ice-repellence and self-cleaning, which gained an increasing demand for many industrial markets like the aviation, automotive, food or cooling industries [2]. A frequently used material in these fields is aluminium, as well as its alloys, due to its high strength and low weight [3].

Today, there is a wide variety of methods to modify the chemistry and/or topography in order to fabricate superhydrophobic surfaces on aluminum. For instance, by chemical etching (where the substrates are processed by a multi-step process in different acid baths), applying a silane coating onto the substrate or by micro-structuring the material using cutting tools like disk cutters [4–6]. In particular, water-repelling and self-cleaning surface are interesting for the food industry, where chemical cleaning should be avoided for hygienic reasons. On technical surfaces, pollution and sand particles can lead to increased wear of the components [7]. Therefore, a self-cleaning surface capable to remove particles from its surface by rinsing droplets and without applying any additional cleaning agents is beneficial. Besides that, not all superhydrophobic surfaces present an ice-repellent function [8]. The characteristic of preventing the formation of ice on the surface can be for instance achieved by thermo-mechanical treatments [9,10].

Besides the above mentioned surface treatments, laser technology presents a promising alternative to create a functional surface. Using Direct Laser Writing (DLW) well-defined surface topographies can be produced by ablating the material locally using a single laser beam which is focused on the material surface [11]. Due to the focusing limitations of the beam, the resolution of the fabricated patterns is usually restricted down to 10-20 µm, depending on the laser wavelength, focusing objectives and beam diameter.

An alternative method to fabricate surface patterns on a metallic surface with a resolution below 10 µm is Direct Laser Interference Patterning (DLIP). In this case, a single laser beam is divided into at least two beams which are overlapped on the surface of the material. If two laser beams are used, this leads to the formation of a line-like periodic intensity distribution, with an spatial period Λ (repetitive distance between the interference maxima or minima) that can be calculated using Equation 1:

$$\Lambda = \lambda / [2 \sin (\beta /2)], \qquad (1)$$

where λ is a wavelength of the used laser radiation and β is the angle between two interfering beams.

* Corresponding author: stephan.milles@tu-dresden.de

The geometry of the produced pattern depends on the number of overlapping beams [12,13]. The DLIP technology has been already used in a broad variety of applications due to the flexibility of materials that can be treated as well as the different patterns that can be created. For example, DLIP was used for manufacturing holographic structures as well as for producing security elements or textured surfaces for antibacterial applications [14–16]. However, to the best of our knowledge, there is a lack of research in the field of multifunctional microstructures on Al surfaces fabricated with laser-based methods, which includes three functionalities, namely superhydrophobicity, self-cleaning, and anti-icing.

In the presented study, we utilized DLW and DLIP technologies to create three different types of periodic structures. Nanosecond pulsed DLW is used for producing a mesh-like geometry with a feature size of 50 µm. In addition, picosecond two-beam DLIP is implemented for the fabrication of pillar-like structures with a spatial period of 7 µm. For producing a hierarchical structure, both technologies are combined. The time-dependent wetting characteristic of the microstructures is investigated by water contact angle measurements for 36 days together with the contact angle hysteresis and the sliding angle. The self-cleaning efficiency is investigated by an self-developed optical evaluation method. The influence of the microstructure on the icing-process is also examined under temperature conditions of -20°C. Finally, the experiments are additionally supported by finite element method simulations [17,18].

2 Materials and methods

2.1 Material

Samples consisting of pure aluminum 99.5 % (EN AW-1050) were used in this study. The sheets were electrolytically polished before the laser treatment obtaining a surface roughness (RMS) of 101 nm ± 10 nm.

2.2 Laser microstructuring procedure

For DLW, a workstation (GF machining solutions P 600) equipped with an Ytterbium fiber laser emitting at 1064 nm wavelength radiation with a maximal output power of 30 W was utilized. The pulse duration was set to 14 ns and the used repetition rate was 30 kHz. The beam was guided by an optical scanner with a speed of 250 mm/s and focused by a 254 mm focal length f-theta objective resulting in a focal diameter of 70 µm on the aluminium surface. The applied laser fluence was 1.1 J/cm². The fabricated mesh-like structure consists of three consecutively formed line-like structures, whose orientation is rotated by 60° respectively to each other. The distance between the lines was set to 50 µm [17].

The DLIP technology was applied to fabricate periodic 7 µm pillar-like structures, utilizing a solid-state Nd:YVO_4 laser (Edgewave PX 200, Germany) emitting 1064 nm wavelength with 10 ps pulse duration. The repetition rate was 30 kHz with a maximum output power of 10 W. The system is equipped with a commercial available DLIP optical head (manufactured by Fraunhofer IWS, Germany). Using this head, the line-like intensity profile, which extends over a circular shaped area with a diameter of ~ 140 µm, is created [17]. The laser fluence was set to ~ 1.9 J/cm². In order to pattern larger areas, the sample was moved in x- and y-directions by using high-precision stages (Aerotech, USA) with an accuracy of ± 2.5 µm. In order to obtain pillar-like structures, the samples had to be structured twice by rotating the sample at 90° between the processing steps. Further information about the procedure has been published elsewhere [18,19].

The hierarchical DLW+DLIP structures were manufactured in a two-step process where the previously mentioned technologies were combined and conducted subsequently.

2.4 Surface characterization

The topography of the processed samples was analyzed using confocal microscopy with an 150x objective (Sensofar S Neox, Spain), resulting in 2 nm and 140 nm vertical and lateral resolutions, respectively. Scanning Electron Microscopy were also used (ZEISS Supra 40VP, Germany) at an operating voltage of 5.0 kV.

The static water contact angle (WCA), contact angle hysteresis (Hys) and sliding angle (SA) measurements were performed using a contact angle system (Krüss DSA 100 S, Germany), which is equipped with a special temperature chamber (Krüss TC40, Germany) based on a Peltier-element. For the measurements, 8 µl droplets of deionized water were used. The static water contact angles were calculated using the Young - Laplace and the tangent drop profile fitting methods and each point is the average of five measurements. In all experiments, the droplets were positioned automatically on the surface. For the ice-repellence tests at -20 °C, the freezing process was monitored and recorded by a CCD camera with a frame rate of 50 fps [18].

The heat transfer between the cold surface and the textured Al surface was simulated using finite element method with a commercial software (COMSOL Multiphysics 5.3.). More details regarding the simulation can be found in [18]. This model approach has been compared with the results of the experiments. The simulations are well suited for a deeper understanding of the heat flux between the different layers, namely the cold aluminium surface, the air inclusions due to the microstructure and the water droplet. However, they are limited with respect to the influence of thermal conductivity before and after the laser processing.

The self-cleaning functionality was evaluated using a self-developed setup including a CCD camera. 10 µl deionized water droplets were released at a tilting angle of 15° on the laser-treated substrates which were contaminated with 1 µm manganese oxide particles (MnO_2). Then, a picture was taken after every single droplet and the number of remaining particles on the

 https://doi.org/10.1051/matecconf/202032604005

area of interest on the specimen was calculated using a image analysis software (ImageJ, Version 1.52a) [20].

3 Results and discussion

For investigating various topographies with specific feature sizes regarding their wettability, self-cleaning and the ice-repellent properties, three different textures were produced. In the following section, the fabrication of these patterns is described in detail, namely, (i) mesh-like structure fabricated using DLW, (ii) pillar-like structure fabricated by DLIP and (iii) a hierarchical multiscaled structure fabricated combining of DLW and DLIP.

3.1. Fabrication of periodic structures using DLW and DLIP technology

Mesh-like structures with a feature size of 50 µm were fabricated on Al substrates using DLW technology, based on a comprehensive previous study [17]. The scanning electron microscope (SEM) image (Fig. 1a) shows a homogeneous distributed pattern in which the third scanned line-like structure is dominant. The confocal microscope (CF) image, presented as an insert in Fig. 1a, reveals the topography of the structure, which shows on average a structure depth of 36.2 µm. This texture emerges due to the nanosecond-pulsed laser process, where the thermal diffusion length for aluminum is in the µm-range, leading to an increase of molten material and in consequence to the deep structures in this order of magnitude [21,22].

Later, the DLIP technology was used for producing significantly smaller structures compared to the previous described mesh-like pattern[23]. In order to produce 7 µm pillar-like structures with a two-beam interference setup, the sample was irradiated twice, with a 90° rotation between the irradiation steps [19]. Due to the pulsed manufacturing process, large areas can be processed as shown in Fig. 1b. The resulting structure depth was in this case 4.5 µm on average (Fig. 1b). In contrast to the ns-pulsed DLW process, the DLIP structures were fabricated utilizing a picosecond-pulsed (ps) laser which provides a thermal diffusion length in the nm-range. As a result the amount of molten and recasted material is significantly reduced [24].

Finally, both above-described technologies were combined for fabricating a multiscale hierarchical pattern. Applying the same process parameters, the substrate was firstly irradiated with DLW (50 µm mesh-like) and afterward using DLIP (7.0 µm pillar-like) resulting in the structure depicted in Fig. 1c. In the image it is also shown, that the facets of the DLW structure were not significantly ablated by the DLIP process due to their steep slope (see arrow in Fig. 1c). The resulting topographies were evaluated by CF microscopy and revealed structure depths of 33.8 µm for the DLW structures and ~4.2 µm for the DLIP structures situated on top of DLW structure (insert Fig. 1c).

3.2. Wetting characteristics

For examining the wetting characteristics of the three fabricated structures measurements of the static water contact angles (CA), the contact angle hysteresis (Hys) and the sliding angle were performed. In general, wetting is roughly divided into three regimes: hydrophilic (is defined by CAs below 90°), hydrophobic (describes the range between 90° and 150°) and superhydrophobic (represents CAs above 150°) [25].

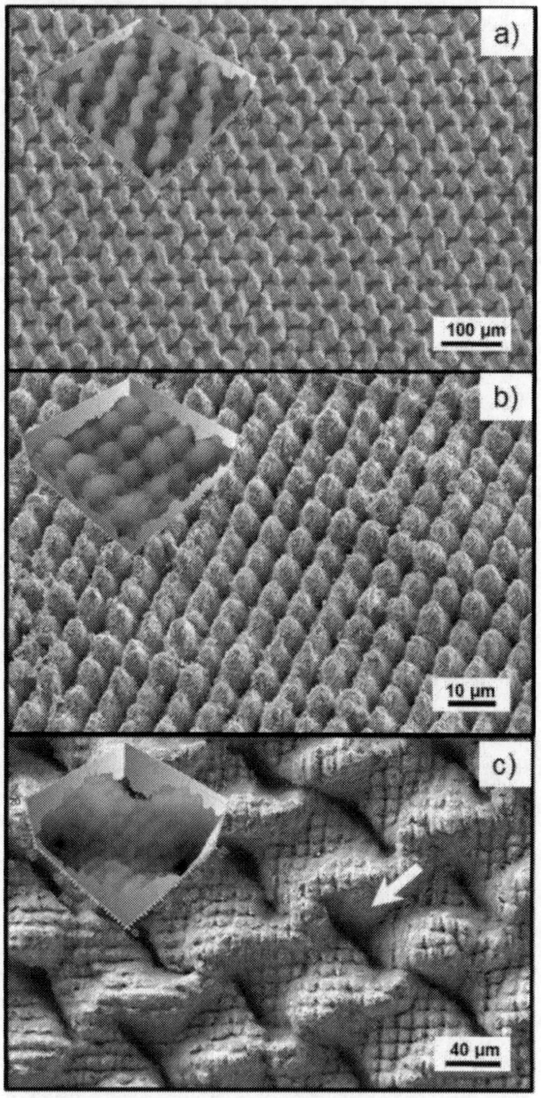

Fig. 1. SEM images of (a) mesh-like DLW structures with a feature size of 50 µm The utilized laser fluence was 1.1 J/cm², 14 ns pulse duration, 30 kHz repetition rate and the scan speed was 250 mm/s. Ten scan repetitions were used. (b) Pillar-like DLIP structures with a spatial of 7.0 µm fabricated with a fluence of 1.9 J/cm², a pulse overlap of 99% and a repetition rate of 1 kHz. (c) Hierarchical structure fabricated by combining DLW+DLIP. The laser wavelength for all experiments was 1064 nm. Confocal images of each structure are given in the top left corner, respectively.

The treated Al surfaces using DLW, DLIP and DLW+DLIP sand an untreated reference were constantly characterized during 36 days period. The obtained results are shown in Fig. 2a. The untreated reference presented a constant wetting characteristic with a CA of ~93° (see lower insert in Fig. 2a), which is common in aluminum surfaces [17,26]. In contrast, it was observed that all laser-treated substrates showed a constant increase in the CAs with time.

The initial low CA can be explained due to a significant change of the surface chemistry after the laser treatment. Then, the increase of carbon content at the Al-surface led to an increase of non-polar sites and thus to the raising of CA [27,28]. Later, the substrates became all hydrophobic and turned to a steady superhydrophobic state with a CA higher than 151° after 13 days (see upper insert Fig. 2a).

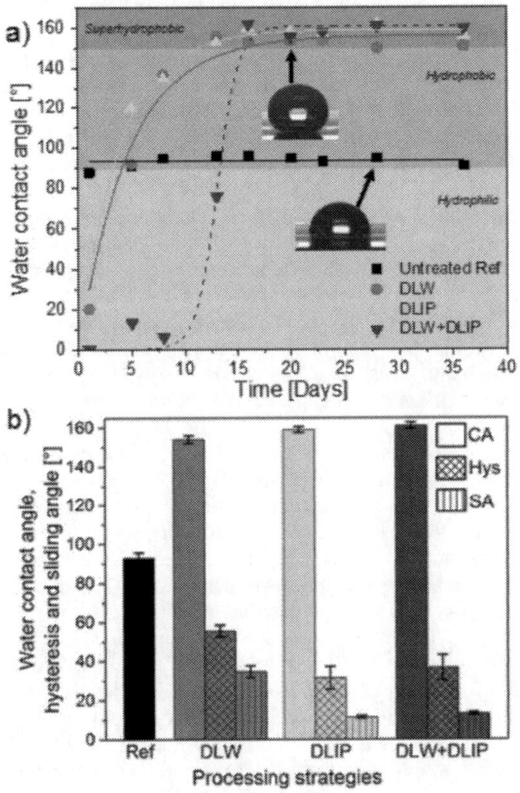

Fig. 2. Static water contact angle as a function of time of an untreated reference sample and of the laser processed samples, the fits are to guide the eye (a), the final contact angle after 95 days, the contact angle hysteresis and the sliding angle, respectively (b) [18].

Mostly responsible for this wetting state is, besides the surface chemistry, the fabricated topography which creates air cushions between the structure and the water droplet leading to the Cassie-Baxter wetting condition [18,29].

Besides the static contact angle, the wetting characteristic of a superhydrophobic surface was further analyzed by measuring the sliding angle (SA), which is the tilting angle of the substrate where the droplet starts to slide. In addition, also the contact angle hysteresis (Hys) was evaluated, defined as the difference between the advancing and the receding contact angle during the sliding of a droplet [30,31]. To evaluate the time-dependency of the reached superhydrophobic conditions, the static contact angle was measured again after 95 days. As shown in Fig. 2b, the CA of all laser treated samples remained above 150°. Additionally, the measured sliding angles were 34° for the DLW, 12° for DLIP and 13° for the hierarchical DLW+DLIP treated substrates. The contact angle hysteresis were 55° (DLW), 31° (DLIP) and 36° (DLW+DLIP), respectively (Fig. 2b) [18]. The significantly higher values for the DLW processed samples could indicate a Wenzel-controlled wetting regime instead of Cassie-Baxter [29]. For SA and Hys was not measureable for the untreated reference due to the static CA of 91°.

3.3 Self-cleaning efficiency

The method described in section 2.4 was applied for evaluating the self-cleaning efficiency of the different superhydrophobic surfaces.

In this case, the samples were contaminated with 1 µm manganese oxide particles (MnO_2) and held at a constant tilting angle of 15°. Then, 10 µl water droplets were deposited over the contaminated area. After each droplet rolled down the sample, the remaining relative contamination of manganese oxide particles was optically measured. The results are shown in Fig. 3. In the Fig. 3 insert, an SEM image of a single MnO_2 particle is shown.

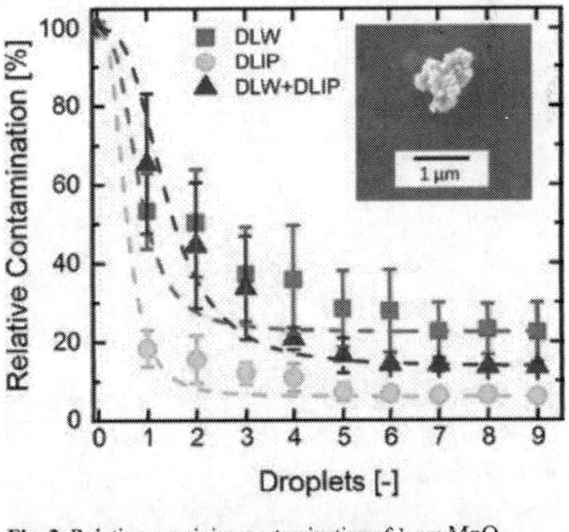

Fig. 3. Relative remaining contamination of 1 µm MnO_2 particles as a function of the released 10µl water droplets on the DLW, DLIP and DLW+DLIP structured surfaces for a tilting of the sample of 15°. The insert shows an SEM image of a single MnO_2 particle [20].

As it can be seen from Fig. 3, all surfaces showed a self-cleaning property due to the decrease of contamination as the droplets were sequentially released. However, significant differences can be found between the laser treated surface. For instance, the number of remaining particles after the first droplet was 63%, 52% and 20% for the DLW+DLIP, DLW and DLIP treated samples, respectively. It should be emphasized that this is the contamination after the first droplet has passed the surface compared to the original contamination. The amount of particles was further reduced with an increasing number of droplets. The final contamination amount after 9 droplets, was 12% for DLW+DLIP, 21% for DLW and 6% for DLIP, showing again a better performance for the DLIP structures.

For the DLW treated Al samples, this behaviour could be partially explained due to the relative high Hys (55°) and an SA (34°) values (Fig. 2b).

For a more in-depth understanding of the interaction between the single particles and the laser-structured surfaces, SEM images of polluted substrates are shown in Fig. 4. As it can be seen, single particles as well as agglomerated particles are located at the large trenches of the DLW structures (Fig. 4a). In consequence, a single rolling droplet cannot reach into the deep DLW structures and pick up the trapped particles. This also explains why the amount of contamination after 9 droplets remains higher on the DLW structures compared to the DLIP structures, which have a shallower structure depth of 4.2 µm.

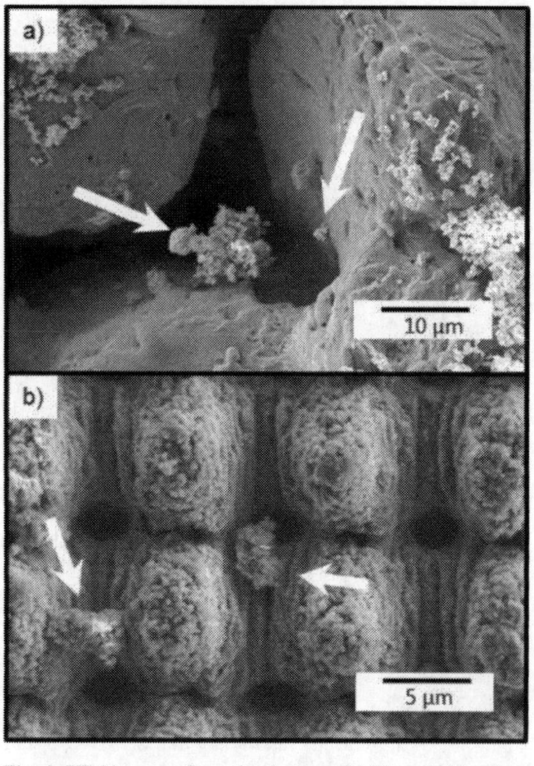

Fig. 4. SEM images of trenches between DLW-mesh (a) and DLIP pillar-like structures (b) contaminated with 1 µm MnO_2 particles. The arrows indicate single and agglomerated particles.

In contrast, the MnO_2 particles are not significantly trapped by DLIP structures (Fig. 4b) and they cannot agglomerate into big clusters due to the small spatial period of 7.0 µm. In addition, the superhydrophobic conditions are more pronounced in this sample, characterized by a SA of 13° and a Hys of 31°.

The self-cleaning efficiency of the hierarchical DLW+DLIP structure lies between DLW and DLIP showing a 13% of contamination after the 9 water droplets.. Further test are require in the future in order to evaluate the influence of (i) particle's material and (ii) size, as well as (iii) the tilting angle and (iv) water droplet volume.

3.4 Ice-formation characterization

Since not all superhydrophobic surfaces show ice-repellent or anti-icing characteristics, the manufactured structures were tested regarding the formation of ice at a temperature of -20°C [8,18].

By plotting the freezing time as a function of the static CA (Fig. 5), a linear correlation regardless of the used laser-technology, and resulting topography, was observed, as it was already described elsewhere [32,33].

Two main factors explain this effect. First, an increase of the CA implies a reduction of the contact area between the droplet and the Al-surface. Therefore, leading to a lower heat transfer between both materials. In consequence, a delayed freezing process is observed (Fig. 5).

In addition to that, the air that is trapped between the structures and bootom droplet surface could also have an insulating effect. This hypothesis was evaluated by heat transfer simulations using FEM.

These results are also shown in Fig. 5, by the data ploted with a star-shape. The solid stars correspond to the simulation results assuming a structured surface with a spatial period of 50 µm and a structure height of 35 µm (comparable to the DLW mesh-like structure geometries), while the open stars shows the simulated freezing time considering that the surface is perfectly flat, i.e. the whole droplet bottom surface is in contact with the cold aluminium surface [18]. It is observed that the experimental data is satisfactorily followed by the numerical model, despite some fluctuations, probably due to deviations of the initial droplet volume from the nominal volume of 8 µl. The simulation results also suggest that the insulation effect by the trapped air has a minor influence on the freezing time and therefore can be neglected [18].

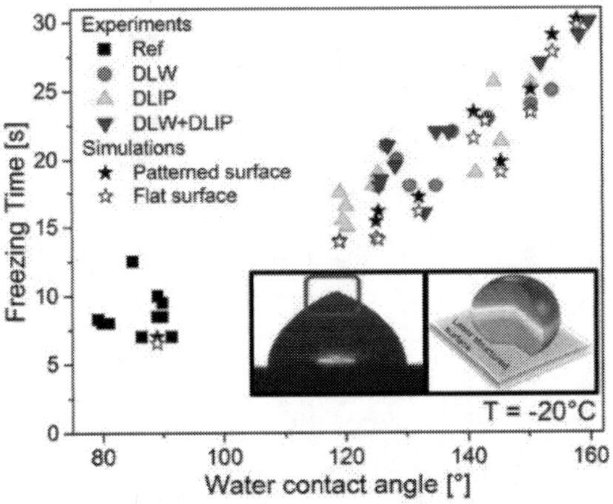

Fig. 5. Experimental and simulated freezing time of 8 µl droplets as a function of the static water contact angle at -20°C of patterned and flat surfaces. The left insert shows a frozen droplet on an untreated reference, the right insert depicts an impression of the heat transfer simulation inside a droplet.

Finally, an overview of the average freezing time of ten droplets placed onto each fabricated surface is presented in Fig. 6 (the error bars represent the respective standard deviation).

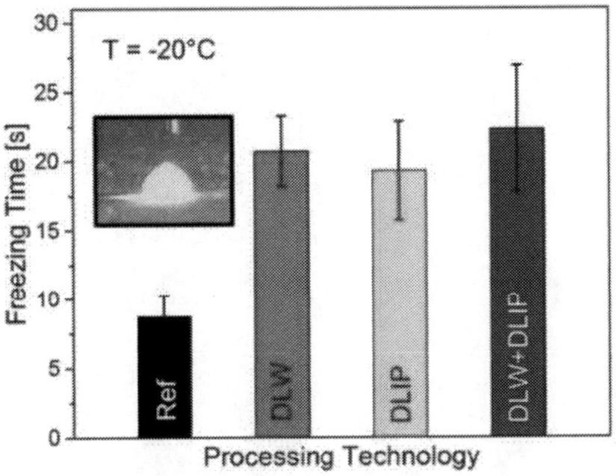

Fig. 6. Average freezing time of the droplets on an untreated reference, DLW, DLIP and DLW+DLIP structures at -20°C, the insert depicts the frozen droplet illuminated by the characterization machine.

The droplets froze the fastest on the untreated reference, after 8.7 s. Remarkably, droplets on the DLW+DLIP structure froze after 22.3 s, representing a nearly three-fold increase compared to the reference. The insert in Fig. 6 depicts an impression of a frozen 8 µl droplet on a reference sample illuminated by the drop shape analyzer.

4 Summary and conclusions

In this work, mesh-like, pillar-like, and multiscaled hierarchical structures were fabricated on pure aluminum using direct laser writing, direct laser interference patterning and a combination of both methods. The wetting characteristics were measured during 36 days period and revealed a superhydrophobic behavior for all laser structured samples with static water contact angles higher than 151° after at least 13 days. Additionally, the DLIP and DLW+DLIP treated substrates presented low sliding angles of 12° and 13°, respectively. This water repellent characteristic is closely linked to the effective self-cleaning property observed when the samples were covered with 1 µm manganese oxide particles and water droplets were let to roll down over the surface. Due to the significantly larger 50 µm mesh-like DLW structures, the contamination particles got trapped in between the trenches and agglomerated. Finally, the ice-repellent functionality at -20°C was demonstrated and the freezing time was increased from 8.7 up to 22.3 s for the hierarchical pattern.

This work was carried out in the framework of the Reinhart Koselleck project (323477257), which has received funding from the German Research Foundation (German: Deutsche Forschungsgemeinschaft DFG). The work of A.L. is also supported by the German Research Foundation (DFG) under Excellence Initiative program by the German federal and state government to promote top-level research at German universities. M. S. acknowledges the support of the Alexander von Humboldt Foundation.

References

[1] H.J. Ensikat, P. Ditsche-Kuru, C. Neinhuis, and W. Barthlott, *Beilstein J. Nanotechnol.*, **2**, 152–161 (2011).
[2] P. Varshney, S.S. Mohapatra, and A. Kumar, *Biomimetics*, **2**, 2 (2017).
[3] J.G. Buijnsters, R. Zhong, N. Tsyntsaru, and J.-P. Celis, *ACS Appl. Mater. Interfaces*, **5**, 3224–3233 (2013).
[4] R. Liao, Z. Zuo, C. Guo, Y. Yuan, and A. Zhuang, *Applied Surface Science*, **317**, 701–709 (2014).
[5] J. Lomga, P. Varshney, D. Nanda, M. Satapathy, S.S. Mohapatra, and A. Kumar, *Journal of Alloys and Compounds*, **702**, 161–170 (2017).
[6] J. Zhu and X. Hu, *J Coat Technol Res*, **16**, 249–255 (2019).
[7] M.A. Ramadan, *Friction*, **6**, 457–463 (2018).
[8] M. Nosonovsky and V. Hejazi, *ACS Nano*, **6**, 8488–8491 (2012).
[9] T.Y. Zhao, P.R. Jones, and N.A. Patankar, *Sci Rep*, **9**, 1–12 (2019).
[10] M.J. Kreder, J. Alvarenga, P. Kim, and J. Aizenberg, *Nat Rev Mater*, **1**, 1–15 (2016).
[11] L.R. de Lara, R. Jagdheesh, and J.L. Ocaña, *Materials Letters*, **184**, 100–103 (2016).
[12] L. Gao, W. Zhou, Y. Wang, S. Wang, C. Bai, S. Li, B. Liu, J. Wang, and Y.L. Li, *Optik*, **127**, 5211–5214 (2016).
[13] C. Zwahr, B. Voisiat, A. Welle, D. Günther, and A.F. Lasagni, *Advanced Engineering Materials*, **20**, 1800160 (2018).
[14] F. Rößler, T. Kunze, and A.F. Lasagni, *Opt. Express, OE*, **25**, 22959–22970 (2017).
[15] R. Helbig, D. Günther, J. Friedrichs, F. Rößler, A. Lasagni, and C. Werner, *Biomaterials Science*, **4**, 1074–1078 (2016).

[16] B.-H. Ha, S. Ahn, H.-S. Bae, H.-S. Kang, J. Kim, and J. Noh, *Optics and Lasers in Engineering*, **107**, 214–220 (2018).

[17] S. Milles, B. Voisiat, M. Nitschke, and A.F. Lasagni, *Journal of Materials Processing Technology*, **270**, 142–151 (2019).

[18] S. Milles, M. Soldera, B. Voisiat, and A.F. Lasagni, *Sci Rep*, **9**, 1–13 (2019).

[19] A.I. Aguilar-Morales, S. Alamri, and A.F. Lasagni, *Journal of Materials Processing Technology*, **252**, 313–321 (2018).

[20] S. Milles, M. Soldera, T. Kuntze, and A.F. Lasagni, *In submission*, (2020).

[21] J.I. Ahuir-Torres, M.A. Arenas, W. Perrie, and J. de Damborenea, *Optics and Lasers in Engineering*, **103**, 100–109 (2018).

[22] A. Semerok, B. Sallé, J.-F. Wagner, and G. Petite, *Laser and Particle Beams*, **20**, 67–72 (2002).

[23] M. Bieda, M. Siebold, and A.F. Lasagni, *Applied Surface Science*, **C**, 175–182 (2016).

[24] B.N. Chichkov, C. Momma, S. Nolte, F. von Alvensleben, and A. Tünnermann, *Appl. Phys. A*, **63**, 109–115 (1996).

[25] K.-Y. Law, *J. Phys. Chem. Lett.*, **5**, 686–688 (2014).

[26] M. Thieme, R. Frenzel, S. Schmidt, F. Simon, A. Hennig, H. Worch, K. Lunkwitz, and D. Scharnweber, *Advanced Engineering Materials*, **3**, 691–695 (2001).

[27] E. Louvis, P. Fox, and C.J. Sutcliffe, *Journal of Materials Processing Technology*, **211**, 275–284 (2011).

[28] R. Jagdheesh, M. Diaz, and J.L. Ocaña, *RSC Adv.*, **6**, 72933–72941 (2016).

[29] A.B.D. Cassie and S. Baxter, *Trans. Faraday Soc.*, **40**, 546–551 (1944).

[30] E. Chibowski and F. Gonzalez-Caballero, *Journal of Adhesion Science and Technology*, **7**, 1195–1209 (1993).

[31] B. He, J. Lee, and N.A. Patankar, *Colloids and Surfaces A: Physicochemical and Engineering Aspects*, **248**, 101–104 (2004).

[32] S. Tabakova and F. Feuillebois, *Journal of Colloid and Interface Science*, **272**, 225–234 (2004).

[33] H. Zhang, Y. Zhao, R. Lv, and C. Yang, *International Journal of Thermal Sciences*, **101**, 59–67 (2016).

Modelling Hydrogen Embrittlement using Density Functional Theory:

A theoretical approach to understanding environmentally assisted cracking in 7xxx series aluminium alloys

Benjamin T. Wilson[1,*], *Joseph D.* Robson[1,**], and *Christopher P.* Race[1,***]

[1]University of Manchester, Department of Materials

Abstract. The effects of H segregation to a Σ11 symmetric tilt Al grain boundary are investigated using atomistic simulations, as part of a wider study on cracking in 7xxx series alloys. Density functional theory based simulations of uniaxial straining of grain boundaries containing 11 different concentrations of H were performed under the cohesive zone fracture mechanics framework. The theoretical strength of grain boundaries is shown to be supressed by H segregation, and the cause of this is attributed to the prevention of the formation of Al ligaments across grain boundaries. Segregated concentrations of relevant alloying elements (Zn, Mg, and Cu) show minimal impact on the H embrittlement process investigated, namely H enhanced decohesion (HEDE). Further modelling, of H transport and grain boundary precipitates, is required to confirm the validity of the HEDE mechanism in the case of 7xxx alloys.

1 Introduction

7xxx series aluminium alloys, Al-Zn-Mg-(Cu), are used as high performance structural materials - AA7075 for instance carries an ultimate tensile strength 572MPa in the peak aged T6 temper [1]. However, material degradation arising from a collection of phenomena generally labelled as environmentally assisted cracking (EAC) has been a key challenge in the development of these alloy systems [2]; for alloys used by the aircraft industry, EAC is an intergranular phenomenon [3]. Considerable study within metallurgy has delivered two fundamental causes of EAC, namely anodic dissolution (AD) and hydrogen embrittlement (HE), and the problem has been recorded in a variety of environments where the relevant atomospheric conditions are reportedly humidity and chloride content [4].

Investigationg HE processes in Al alloys is a challenging endeavour because of the low equilibrium lattice solubility of H, and Al's inability to form stable hydrides [2]. Whilst evidence for H inducing cracks is limited, partitioning of H has been shown to degrade the material's resistance to crack growth. Detection of H's influence on localised effects has recently been achieved through advanced imaging techniques such as X-ray synchotron tomography [5] which confirms significant H enrichment at grain boundaries (GBs) and precipitate interfaces during deformation. A theoretical approach to understanding HE can circumvent experimental challenges by predicting material behaviour using fundamental physical and chemical principles [6] - instead of probing a material through successive experimental techniques, this work aims to model from first principles the relationship between intergranular fracture and H in 7xxx alloys by using electronic structure calculations of the cohesive strength of grain boundaries.

*e-mail: benjamin.wilson@manchester.ac.uk
**e-mail: joseph.robson@manchester.ac.uk
***e-mail: christopher.race@manchester.ac.uk

2 Theory

2.1 Hydrogen embrittlement

Proposed HE mechanisms are often speculative due to the difficulty in quantifying hydrogen levels accurately, although four main mechanisms are often discussed in the context of Al alloys. Hydriding has been detected in some Mg containing phases of aluminium and could potentially lead to brittle GBs [7]. Hydrogen-enhanced local plasticity (HELP) is a complex phenomenon involving high localised H concentrations increasing dislocation mobility by shielding dislocation interactions. Another related model of HE involves adsorbed H (i.e. external H) inducing the emission of dislocations [8]. The fourth mechanism is hydrogen-enhanced decohesion (HEDE) which posits that segregated H at the GB compromises cohesive forces between grains by disrupting bonds across the interface [9].

The HEDE mechanism itself requires several distinct steps [2]. H must be produced in large enough quantities to impact crack growth in any measurable way - a bare Al crack tip in the presence of water vapour will oxidise to provide a source of atomic H, but passivation will quickly inhibit this process. H transport along the GB has a low flux, meaning that under the HEDE model H is absorbed into the metal matrix and diffuses to the highly stressed region just ahead of the crack tip, Figure 1 shows a schematic of this process. If H in this region substantially embrittles the GB then the crack will advance, gen-

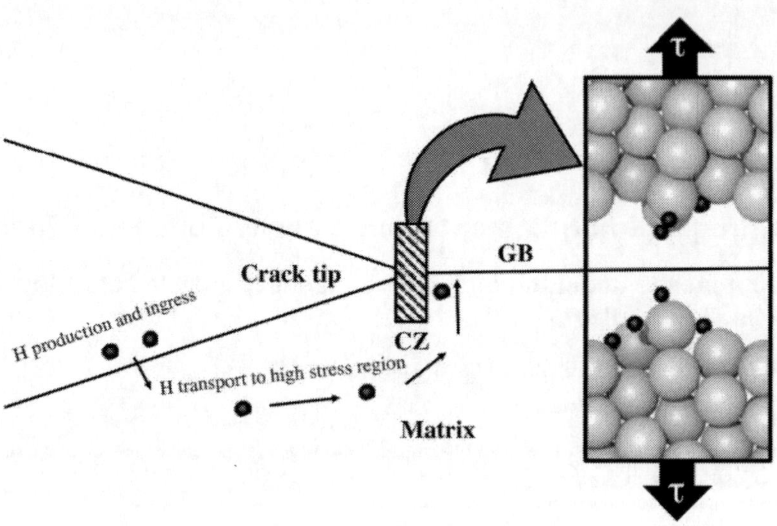

Figure 1. Schematic diagram of H enhanced decohesion ahead of a crack tip. H is generated on fresh crack surfaces and is transported to the high stress region ahead of the crack tip, which leads to H embrittlement induced crack growth. The cohesive zone (CZ) is a small region immediately ahead of the crack tip and is simulated in this work using density functional theory.

erating fresh surfaces for H production and the process is repeated.

2.2 Cohesive zone fracture mechanics

For fracture events displaying limited plasticity, such as those found in EAC, cohesive zone (CZ) modelling is a useful tool which introduces a virtual region at the crack tip called a CZ [10]; Figure 1 shows a CZ in a HEDE model. The CZ captures separating atomic planes immediately ahead of the crack tip and crack growth is seen as the successive failures of CZs along the crack path. First-principles simulations can be used to model the separation of atomic planes, or decohesion, as the stretching and breaking of molecular bonds which exist across the crack path [11]. In the context of HEDE, a CZ will be weaker when decorated with H. The CZ model assumes linearly elastic behaviour in the regions surrounding a CZ. A macroscopic load on the material will translate into a microscopic traction $\tau(u)$ which, in the CZ approximation, is a unixial stress applied perpendicular to the GB plane; the separation of two grains can be quantifed by a displacement u. The energy of decohesion as a function of displacement $\chi(u)$ is evaluated in (1).

$$\chi(u) = \int_0^u \tau(u')\, \mathrm{d}u'. \qquad (1)$$

A CZ fails when the bonding across its interface is completely broken i.e. the two surfaces are displaced past their interaction range u_f, with the work of separation (the energy required to complete the decohesion process) defined as $W = \chi(u_f)$.

Modern electronic structure calculations, or ab-initio simulations, avoid the requirement for difficult to measure experimental parameters as an input, and instead rely on fundamental quantum mechanics based principles to predict the strength performance of materials, and have been effective in calculating the traction separation laws of CZ models [12]. These calculations are routinely performed using Density Functional Theory (DFT) which approximates electron interactions through the use of appropriate exchange-correlation potentials. Application of DFT reflects a trade-off between fundamental complexity of atomistic behaviour and computational feasibility - finite temperature effects are often ignored in these simulations[13]. DFT based ab initio tensile test simulations are used to obtain W, u_f, the maximum traction τ^* required for fracture, and the displacement u^* at τ^*, properties which are collectively referred to as a traction-separation law for a CZ [14]. Once traction-separation laws have been evaluated they can be implemented in larger scale continuum models to constitute a full CZ based fracture mechanics model [15].

3 Method

3.1 Electronic structure calculations

We use the Vienna Ab-initio simulation package (VASP) [16–19] with the projector augmented wave method [20, 21] for DFT calculations. A Monkhorst-Pack k-point mesh [22] of 5 × 13 × 2 k-points and a plane wave cut off energy of 400 eV was found to be sufficient for an accuracy of 1 meV/atom. The PBE GGA exchange-correlation functional [23, 24] is used throughout, along with Methfessel-Paxton smearing of 200 meV [25]. Electronic structure is optimised using the Block-Davidson method [26] with a convergence criterion of 0.1 meV.

Our grain boundary was constructed for DFT calculations using the coincident site lattice model [27] to generate a supercell containing two $\Sigma 11(\bar{1}13)[110]$ symmetric

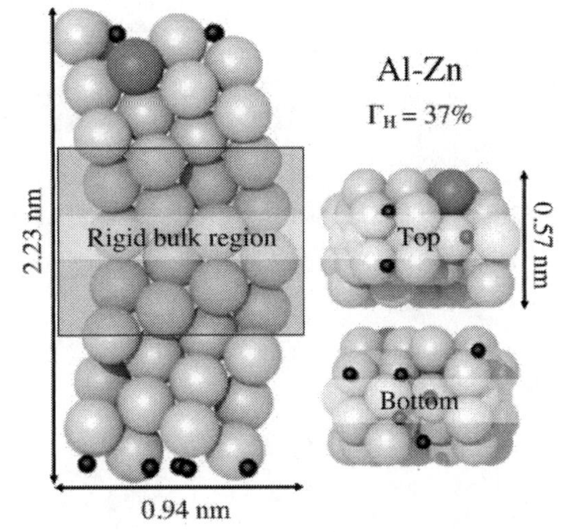

Figure 2. DFT simulation cell of two Al Σ11 GBs containing a Zn segregant atom (dark-grey) and a high concentration of H atoms (black) at the GB to be broken during tensile test simulations.

tilt boundaries with a misorientation angle of 129.5°. The pure Al cell contains 70 atoms (35 per grain) and is volume optimised by fitting ten energy calculations of the cell for a range of lattice parameters to the Birch-Murnaghan equation of state [28]. Three more structures, each containing a cosegregant of Cu, Mg, or Zn in its most energetically favourable substitutional site on the GB, are further optimised along the GB normal axis through fitting the energies of 6 different cell lengths to a quadratic function. Figure 2 shows an example structure containing a Zn atom as a cosegregant. Ionic positions are optimised through a conjugate gradient algorithm [29, 30] where ions are moved with respect to calculations of forces at each step, this has a convergence condition of 1meV per atom.

$N_H = 0$ to 10 hydrogen atoms are included in multiple configurations interstitial voids on the GB plane, leading to a total of 44 closed GB structures for DFT simulations. The arrangement of H atoms was determined through a systematic process of finding the most energetically favourable site for each additional H atom, excluding the option of H occupying a void already containing H. Given that there are 12 atoms on the GB in H-free structures, the GB H concentration can be defined as $\Gamma_H = N_H/(12 + N_H)$.

3.2 Ab-initio molecular dynamics

Most ab-initio tensile tests involve a predetermined failure mode, i.e. simulations have an artificially defined fracture path of bonds to break as grains are displaced [31]. For clean Al GBs the correct fracture path is easily identified as the GB plane, but with a high concentration of impurities the failure mode is unclear. To overcome this issue, we developed an ab-initio molecular dynamics tensile test. First, a small amount of vacuum (0.025 nm) is inserted on the GB plane. Second, an electronic structure calculation is peformed and ionic forces are calculated. Ions are moved by 0.025 nm in the directions of force, and ion positions are then optimised using the conjugate gradient algorithm. Finally, another slab of vacuum (0.025nm) is added and the process continues for 25 displacements u_i corresponding to each vacuum slab added. To speed up calculations, the central region of the cell (containing the untested GB) is not included in the ionic position optimisation, see Figure 2. The decohesion energy is calculated using $\chi(u_i) = E(u_i) - E(0)$, where $E(u_i)$ is the energy of a cell at displacement u_i. Traction is obtained using central differencing of the decohesion energies with respect to displacement. It can be hard to decide when interactions across the displaced GB are negligible so a convention of $u_f = 2W/\tau^*$ is adopted to define the decohesion separation when the surfaces are deemed to be no longer interacting [32].

A total of 44 dynamic ab-initio tensile tests are performed, each with fixed Γ_H. However small amounts of traction $\tau < \tau^*$ may occur at appreciable time scales to allow for changes in Γ_H [33]. To understand the potential for small displacements to drive H transport to a GB, differences in solution enthalpy for H atoms at a clean unbroken GB and a GB separated by u containing Γ_H are also calculated as $\Delta\mu_H(u, \Gamma_H)$.

4 Results and Discussion

4.1 Decohesion energy

Figure 3 shows the work of separation for all 44 tensile tests. For $\Gamma_H = 0\%$ the Σ11 Al grain boundary has $W = 15.0$ eV/nm^2, which is significantly higher than the result of 12.53 eV/nm^2 from [31]. This discrepancy is explained by [31] not accounting for ligamentation across the boundary through a dynamic tensile test procedure. Results in the literature often employ the rigid grain shift (RGS) method which takes a GB cell and peforms a DFT energy calculation at a range of separations. The RGS method models the interaction of two surfaces from their initial grain boundary equilibrium state to their non-interacting free surface state where $\chi(u)$ fits a universal binding energy curve. This method simulates an extremely high strain rate, and most research includes a surface relaxation at each of increment separation - however, these surface relaxations do not substitute our dynamic modelling used in this study. The decohesion energy $\chi(u)$ from the dynamic model can be seen in Figure 4 with Mg as a cosegregant. Energy curves do not follow the smooth path found in RGS simulations, which can be explained by discrete structural changes such as the formation (and dissolution) of ligament structures, and by sudden changes in the arrangement of surface H atoms.

The results of Figure 3 closely match those found in [34] which looked at a range of Γ_H concentrations for Σ3,5, and 9 GBs and included higher Γ_H than those explored in this work. In [34] it was reported that in some GBs, a high Γ_H can lead to a negative W, although the tensile testing simulation was not performed and instead a simple initial

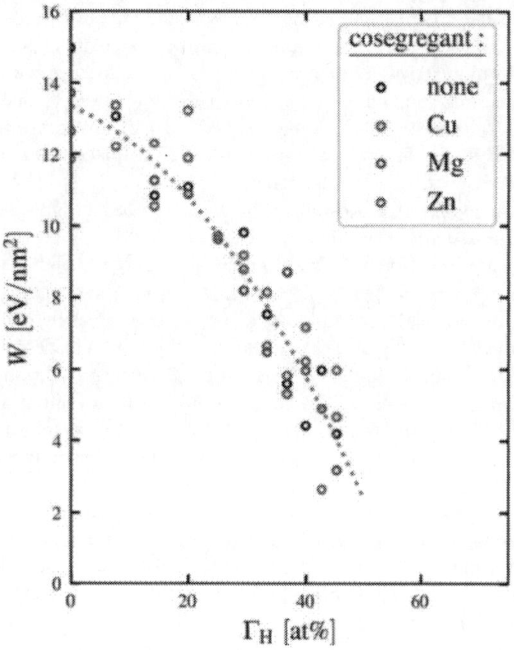

Figure 3. Work of separation W for different GB H concentrations Γ_H extracted from ab initio tensile test data. Each colour represents the presence of a cosegregant species. The grey dotted line is a least-squares 2nd order polynomial fit of the data.

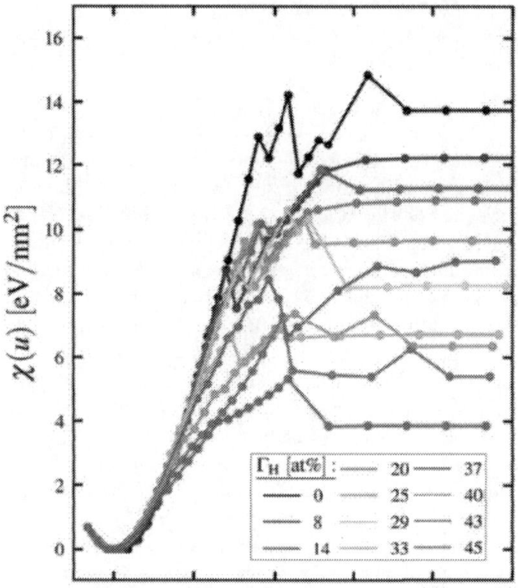

Figure 4. Decohesion energy χ as a function of displacement u for an Al $\Sigma 11$ GB with Mg segregation for different concentrations of GB H (Γ_H).

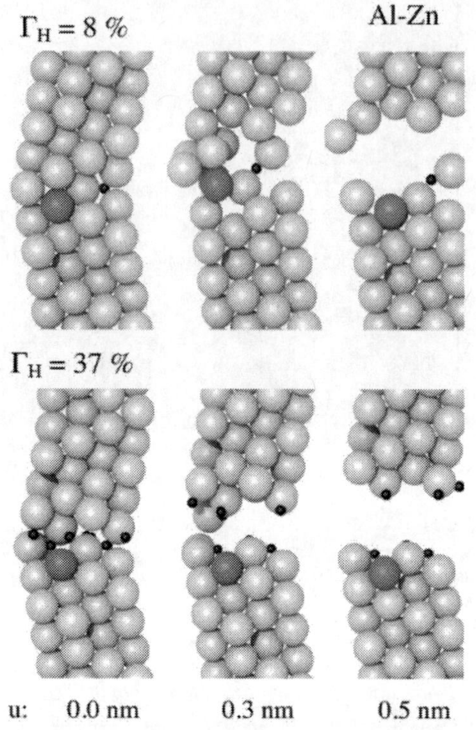

Figure 5. Snapshots of an Al (grey) $\Sigma 11$ GB with Zn segregation (dark grey) at three different magnitudes of displacement u, for a high and low GB H (black) concentration Γ_H.

closed GB and completely separated GB calculation was performed to derive W. This simpler approach hides the possible effects of ligamentation on the final free surface structure and doesn't obtain traction-separation laws for a CZ model of fracture.

Each cosegregant was found to have a small embrittling effect. A 2011 study on $\Sigma 5$ GBs [35] calculated the best possible substitutional site to place a Zn ion on the boundary in order to model the effects of Zn segregation on boundary cohesion - this study reported that with the Zn atom at different sites, the embrittlement could change from positive to negative; this result was attributed to changes in the lengths of bonds that naturally arise from the substitution of Zn atoms. Charge density difference plots from [35] demonstrate charge depletion which corresponds to preferred fracture paths where weaker bonds will break quickly. Also calculated in [35] is the electron density of states which show that Zn-Al bonds are metallic in character. However, results in [35] show that Zn has a lower bonding charge density than Al, therefore they are weaker than the Al-Al bond which is replaced after substitution. This supports the finding that Zn is an embrittler. On the other hand, in [35] they found that the reduction in W caused by Zn was very small, only 3%. With our dynamic tensile test of $\Sigma 11$, Zn embrittlement is found to be significantly more potent with a reduction in W of 20%.

Another more recent study [36] found strong segregation tendencies for Mg and Cu to the Σ5 GB, however Cu in fact preferentially occupies interstitial sites rather than substitutional. Mg acts as an embrittler and shows similiar charge depletion characteristics to Zn calculated in [35]. In contrast, the interstitial Cu segregants strengthen cohesion across the interface by making new Al-Cu bonds. Whilst the Cu atoms disrupt the mechanical integrity of the GB structure, their chemical bonding effect dominates leading to overall resistance from interfacial separation; in our work, Cu was limited to substitutional sites only, which negates this bonding effect.

The key observation from our atomistic tensile testing is the effect H has on the crystal structure around the GB under strain. For systems containing no H, separation of grains results in significant structural changes around the GB. Effectively, GB atoms transform into ligament structures to form bridges across the vacuum, see Figure 5. The presence of H quite clearly disrupts the formation of such structures, and crystals maintain their face-centered cubic structure throughout a test. This property is common to all cosegregants and provides a possible explanation for changes in W for different Γ_H. There is a stark drop in W for increasing H concentrations. We know from RGS tests in the literature that W is reduced when ligaments are artificially prevented from forming, so it is presumed that the strength of low Γ_H GBs is derived from these ligaments. There are no significant differences with respect to cosegregants, other than the alloying embrittlement already discussed at $\Gamma_H = 0$.

4.2 Traction-separation laws

The corresponding $\tau(u)$ curves were obtained by taking the derivative of $\chi(u)$ using central differences with respect to displacement. Due to the discontinuous decohesion energy curves, the form of traction-separation calculated from finite differences is not a smooth curve such as those found using RGS methods. However, the properties of interest for traction-separation laws, maximum (or critical) traction τ^* and the critical displacement u^* and final displacement u_f, can be extracted.

Figure 6 shows the maximum traction τ^* from each tensile test. A small embrittling effect from the cosegregants is observed at $\Gamma_H = 0$ where there is a small decrease in τ^* for Zn and Cu, and a slightly larger decrease for Mg. There is again a marked reduction in τ^* with increasing H content. From $\Gamma_H = 0$ to 45% there is a reduction in τ^* of over 50%. A simple 2nd order polynomial fit of the data roughly follows the data's trend in the region of Γ_H studied, where Cu cosegration shows improved strength over pure Al and Zn cosegregation, but Mg cosegregation tends to lead to weaker boundaries. At $\Gamma_H = 0$, this result is in agreement with [36], and shows that this property is also true for $\Gamma_H > 0$.

Combining the results for W and τ^* allows us to estimate the decohesion separations u_f, Figure 7. Whilst there is a lot of noise in the data, combining the quadratic fit from Figures 3 & 6 shows how the decohesion separation will likely change with Γ_H. Furthermore Figure 8

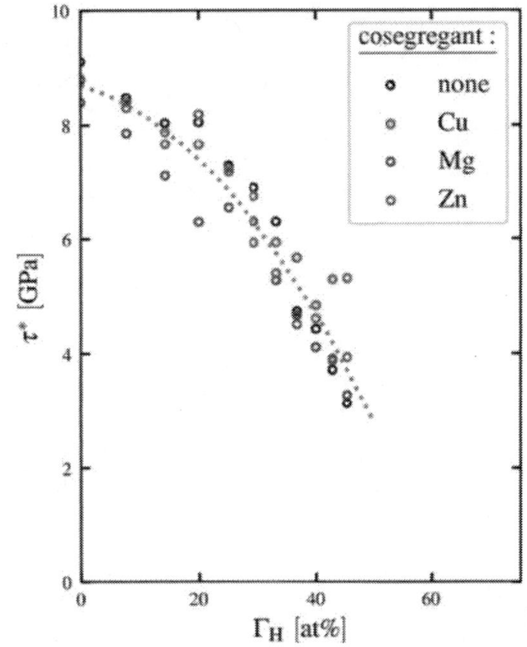

Figure 6. Maximum traction from all tensile tests of an Al Σ11 GB. The grey dotted line is a least-squares 2nd order polynomial fit of the data.

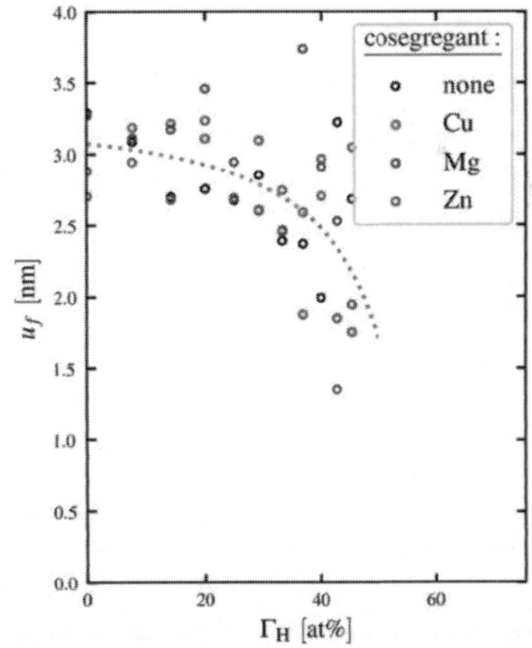

Figure 7. Final separations u_f extracted from calculations of τ^* and W for all tensile tests. The grey dotted line is derived from combining the trendlines of Figure 3 and 6 according to $u_f = 2W/\tau^*$.

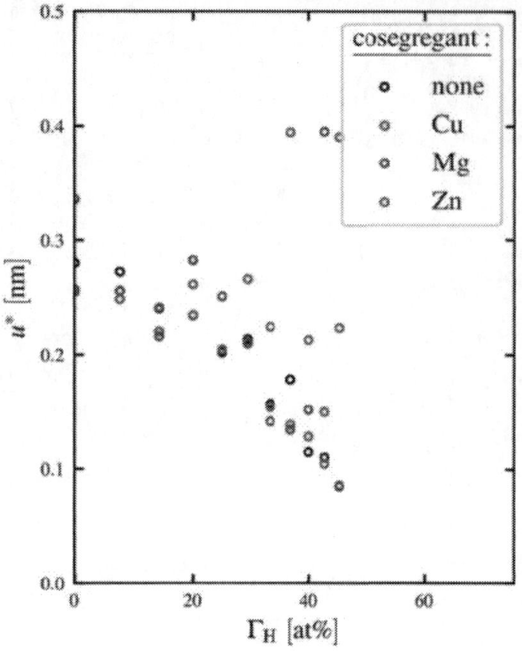

Figure 8. Critical separations u^* for all tensile tests, calculated as the separation at $\tau = \tau^*$, or the critical traction.

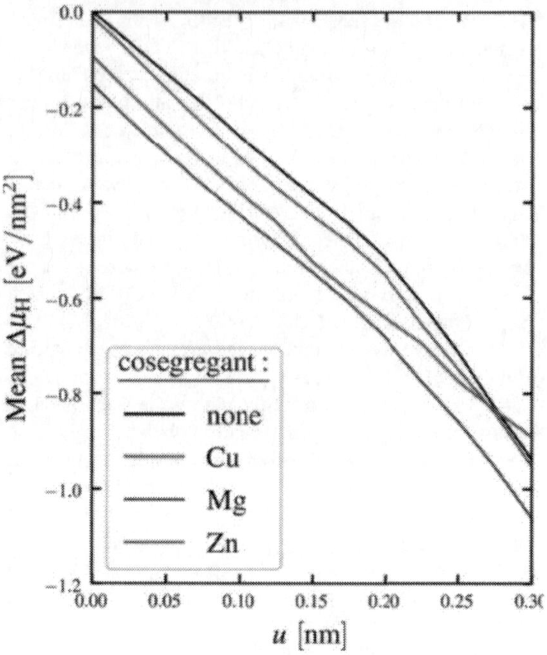

Figure 9. Changes in enthalpy of solution for H at a Σ11 Al GB averaged across all H concentrations studied in this work.

shows the grain separation at $\tau^*(= \tau(u^*))$, also known as the critical displacement u^*. Ignoring the three anamolous results showing values of approximately 0.4 nm, another decrease is observed with increasing Γ_H. These distances play in important part in continuum mechanical modelling of fracture that use traction separation laws as input data. Additionally, in scenarios with variable Γ_H, a strained GB could reach critical displacement without increasing traction by decohesion induced H segregation.

4.3 H segregation

During the tensile tests, traction gradually increases to τ^* at u^* and then decreases to u^f. Before u^* if Γ_H were to increase, less traction will be required to displace the grains further. The enthalpy of solution for H was found to not be influenced by Γ_H, i.e. the energy to dissolve a H atom into the GB is not affected by how much H is already present in the boundary for the range of concentrations used in this study. [34] found a similiar effect, although they also report that for high Γ_H the segregation energy, and enthalpy of solution at the GB, will decrease due to an expansion of the GB caused by Γ_H. In this study, these expansions were initially neglected and subsequently accounted for in the dynamic tensile test by shifting $\chi(u)$ so that its minimum is centered on $u = 0$.

Figure 9 shows the change in enthalpy of solution for H from a closed GB with no segregants as a fuction of displacement, shown are mean values for each possible cosegregant averaged over all considered Γ_H. Mg and Cu cosegregants act to increase hydrogen segregation by a small amount, but separation has a significant impact. For small displacements, due to tractions less than τ^*, there is a strong trend in reducing the enthalpy of solution for H, which would drive further segregation, meaning less force is required for continued decohesion. To put these results in perspective, H has been shown using DFT calculations [37] to preferentially occupy vacancies over interstitial voids in bulk face-centered cubic Al with a difference of $\Delta\mu_H$ = -0.4eV per vacancy.

By combining these results with models of H transport and crack growth rates, an assessment can be made as to the validity of the HEDE process in EAC. Experimental data obtained from desorption techniques in [38] suggests that there is no tendency for H segregation to GBs (with $u = 0$) that aren't associated with cracking. For displacements $u < u^*$ a question arises as to the effects of strain rate on both Γ_H and the formation of ligament structures. In the fast fracture limit, the CZ will break according to RGS simulations with a constant Γ_H and conversely, in the slow fracture limit Γ_H will increase in accordance with Figure 9. Since DFT simulations suggest lattice stiffenening of elastic moduli from H in solid solution in bulk Al FCC structures, H undergoing transport to the GB but not quite on the GB plane could also enhance decohesion.

5 Conclusion

We have presented results which simulate decohesion of a CZ model of fracture in Al alloys, using a broad range of segregated H concentrations on a Σ11 symmetric tilt GB. The methods we used capture the complex phenomenon of ligament structures forming between grains during microscopic loading of the CZ by using a dynamic ab-initio tensile test. We found that the strength of the GB is derived from the formation of ligaments, and that dissolved H at the GB can disrupt their formation leading to H embrittlement. The work of separation W was found to decrease quickly when the H concentration at the GB, Γ_H, increases - for a clean Al GB, W was calculated to be 15 eV/nm^2 but the same boudary containing Γ_H=45% has a much lower W of only 4.2 eV/nm^2. The effects of H on W was consistent with cosegregation of alloying elements Zn, Mg or Cu.

Traction-separation laws were derived for 10 concentrations of H and we found the same trend in the maximum traction τ^* as for W. Much weaker correlations occur for critical and final displacements, although a definite trend exists with u^* and u_f both showing some decrease with respect to increasing Γ_H. We also found a strong correlation between displacements of grains and H segregation, where the average enthalpy of solution for a H atom was found to drop by as much as 1eV with a displacement of 0.3nm; an extant Γ_H did not seem to affectt the enthalpy of solution of additional H atoms.

There is one small noteable difference in segregation behaviour for a pure GB and each of the cosegregated GBs, namely that the presence of Mg lowers the enthalpy of solution of H - other than this, cosegregation does not influence HEDE. Our results encourage further work to demonstrate the practicality of these theoretical discoveries by use of a multiscale model to include H transport rates and crack growth rates. Our results show how small displacements of grains could potentially drive further H segregation and induce embrittlement, so a model predicting the rate of H transport required to significantly change Γ_H during experimental crack growth timeframes is essential to confirm the validity of a HEDE mechanism. On top of this, H production and ingress into the material, along with the presence of GB precipitates, must also be considered as part of the overall HEDE process in EAC of 7xxx alloys.

Acknowledgements

BW is supported EPSRC Centre for Doctoral Training in Advanced Metallic Systems [EP/5022634/1] and Airbus. CR is supported by a University Researchn Fellowship of the Royal Society. High performance computing was carried out using the University of Manchester's CSF3. Images showing atomic positions are produced using OVITO [39].

References

[1] *ASM Handbook Volume 2: Properties and Selection: Nonferrous Alloys and Special-Purpose Materials* (ASM International, ISBN 9780871703781, 1990)

[2] J. R. Scully, G. A. Young, S. W. Smith, *Gaseous Hydrogen Embrittlement of Materials in Energy Technologies: The Problem, its Characterisation and Effects on Particular Alloy Classes* (Woodhead Publishing, ISBN 9781845696771, 2012) 707-768

[3] E. Schwarzenböck, E. Ollivier, A. Garner, A. Cassell, T. Hack, Z. Barrett, C. Engel, T.L. Burnett, N.J.H. Holroyd, J.D. Robson, P.B. Prangnell, S.A.S. Airbus, R. Point, M. Bellonte, F.-B. Cedex, Corrosion Science **171**, 108701 (2020)

[4] R. Hermann, Corrosion, **44**, 685 (1988)

[5] K. Shimizu, H. Toda, H. Fujihara, K. Hirayama, Engineering Fracture Mechanics, **216**, 106503 (2019)

[6] A. Tehranchi, W. A. Curtin, Engineering Fracture Mechanics, **216**, 106502 (2019)

[7] A. V. Bochkaryova et al, IOP Conf. Ser.: Mater. Sci. Eng., **71**, 012057 (2015)

[8] S. P. Lynch, Journal of Materials Science, **20**, 3329 (1985)

[9] W. Gerberich, *Gaseous Hydrogen Embrittlement of Materials in Energy Technologies: Mechanisms, Modelling and Future Developments* (Woodhead Publishing, ISBN 9780857095367, 2012) 209-246

[10] K. Park, G. H. Paulino, ASME: Applied Mechanics Reviews, **64**, 060802 (2013)

[11] R. A. Enrique, A. Van der Ven, Journal of the Mechanics and Physics of Solids, **107**, 494 (2017)

[12] M. Sob, M. Friak, D. Legut, J. Fiala, V. Vitek, Materials Science and Engineering A, **148**, 378-389 (2004)

[13] A. T. Paxton, Materials Science and Technology, **30**, 1063-1070 (2014)

[14] A. Van der Ven, G. Ceder, Acta Materialia, **52**, 1223-1235 (2004)

[15] X. Wei, C. Dong, Z. Chen, X. Li, RSC Advances, **6**, 27282 (2016)

[16] G. Kresse, J. Hafner, Physical Review B, **47**, 558 (1993)

[17] G. Kresse, J. Hafner, Physical Review B, **49**, 14251 (1994)

[18] G. Kresse, J. Furthmuller, Computational Material Science, **6**, 15 (1996)

[19] G. Kresse, J. Furthmuller, Physical Review B, **54**, 11169 (1996)

[20] P. E. Blochl, Physical Review B, **50**, 17953 (1994)

[21] G. Kresse, D. Joubert, Physical Review B, **59**, 1758 (1999)

[22] H. J. Monkhorst, J. D. Pack, Physical Review B, **13**, 5188 (1976)

[23] J. P. Perdew, K. Burke, M. Ernzerhof, Physical Review Letters, **77**, 3865 (1996)

[24] J. P. Perdew, K. Burke, M. Ernzerhof, Physical Review Letters, **78**, 1396 (1997)

[25] M. Methfessel, A. T. Paxton, Physical Review B, **40**, 3616 (1989)

[26] E. R. Davidson, *Methods in Computational Molecular Physics* (NATO Advanced Study Institute, Series C, New York, 1983) 95

[27] H. Lambert, A. Fekete, J.R. Kermode, A.D. Vita, Computer Physics Communications, **232**, 256 (2018)

[28] F. Birch, Physical Review, **71**, 809-824 (1947)

[29] M. P. Teter, M. C. Payne, D. C. Allan, Physical Review B, **40**, 12255 (1989)

[30] D. M. Bylander, L. Kleinman, S. Lee, Physical Review B, **42**, 1394 (1990)

[31] R. Janisch, N. Ahmed, A. Hartmaier, Physical Review B - Condensed Matter and Materials Physics, **81**, 4 (2010)

[32] A. M. Tahir, R. Janisch and A. Hartmaier, Modelling Simul. Mater. Sci. Eng., **21**, 075005 (2013)

[33] Y. Mishin, P. Sofronis, J. L. Bassani, Acta Materialia, **50**, 3609-3622 (2002)

[34] M. Yamaguchi, K. I. Ebihara, M. Itakura, T. Tsuru, K. Matsuda, H. Toda, Computational Materials Science, **156**, 368 (2019)

[35] S. Zhang, O. Y. Kontsevoi, A. J. Freeman, G. B. Olson, Acta Materialia, **59**, 6155 (2011)

[36] D. Zhao, O. M. Løvvik, K. Marthinsen, Y. Li, Acta Materialia, **145**, 235 (2018)

[37] L. Gang, E. Kaxiras, Physical Review Letters, **94**, 155501 (2005)

[38] G. A. Young, J. R. Scully, Acta Materialia, **46**, 6337-6349 (1998)

[39] A. Stukowski, Modelling Simul. Mater. Sci. Eng., **18**, 015012 (2010)

Comparison of long-term natural aging to artificial aging in Duralumin.

Magali Brunet[1,*], *Benoit* Malard[2], *Nicolas* Ratel-Ramond[1], *Christophe* Deshayes[1], *Bénédicte* Warot-Fonrose[1], *Philippe* Sciau[1] and *Joël* Douin[1]

[1] CEMES, CNRS, Université de Toulouse, 29 rue Jeanne Marvig, BP 94347 31055 Toulouse Cedex 4, France
[2] CIRIMAT, Université de Toulouse, CNRS, INP Toulouse, UPS, 4 allée Emile Monso, 31030 Toulouse, France

Abstract: The understanding of long-term aging of aeronautical materials, in particular aluminium alloys used in the fuselage and structure of aircraft is of extreme importance for airline fleets. In this work, a plate from an old aircraft (Breguet) was retrieved and studied in terms of microstructure and mechanical properties. A comparison was made between this naturally-aged alloy and a modern alloy on which different artificial aging conditions were applied. The old alloy exhibits a precipitation of θ-Al_2Cu at grain boundaries and of Ω-Al_2Cu on dispersoids. This non-expected nanostructure for an alloy in T4 state was attributed to the heat that the plate experienced during the aircraft cycles. However, it is shown that this aging is reversible (after a solution treatment). Moreover, the very long time of outdoors exposure seems to have caused intergranular corrosion causing the early failure during tensile tests on some of the specimens. The artificial aging (low temperature, 100°C for up to 10,000h) applied on the modern 2017A alloy did not allow to reproduce the nanostructure of the old plate, meaning that isothermal conditions for artificial aging might not be appropriate in this case.

1. Introduction

Long-term behaviour of aeronautical materials is of crucial importance for industries as airline fleets are aging, with many aircraft reaching their design service goal [1]. Alteration due to fatigue and corrosion are the main concerns. The main causes come from service environment: operating temperatures, loads, ambient environmental conditions, moisture and fluid exposures, radiation, maintenance and ground handling [2]. Properties of materials degrade with time when exposed to such environment. Properties generally monitored are: mechanical static properties (yield stress, tensile strength, hardness and fracture toughness), resistance to fatigue, fatigue-corrosion and corrosion resistance [3].

The approach to evaluate the response of materials to long-term exposure in a service environment, is to impose either accelerated testing to assess mechanisms involving accumulation of damage, for example creep and high-temperature fatigue, or accelerated aging to generate end-of-life microstructure by increasing temperature or loads through furnace treatments and/or cycling. For age hardening aluminium alloys, exposure to temperature can introduce microstructural changes, whether at microscale by the formation of subgrains or at lower scale with for instance the nucleation and growth of new nanostructure phases and the change in dislocation density. These changes will lead to decrease in strength, fracture toughness or fatigue resistance.

For 2xxx alloy family, several studies were reported where long-term aging was simulated with temperature exposures [4-7]. In 1999, a study of long-term stability was done by Braun et al. [4] on 2024-T351 alloys. The authors attempted to verify whether the change of microstructure in artificially aged alloys (aging temperatures between 85°C and 150°C) had an influence on mechanical properties and corrosion behaviour. The tested sheets were aluminum clad and uncladded 1.5 mm thick sheets for fuselage. An interesting result was the fact that tensile strength, ductility and fracture toughness were not affected by the thermal exposure at 85°C for 5,800 hours, at 100°C for 720 hours and at 120°C for 3 hours. Only corrosion behaviour was deteriorated as intergranular corrosion appeared. This was related to the grain boundary precipitation occurring at these temperatures. For Al-Cu-Li alloys of the first generation in T8 state, which are known to have a unstable microstructure during long term ageing in service [8], Deschamps et al. applied low temperatures for long times: 1,000 h at 100°C or 3,000 h at 85°C [7]. They observed an evolution of the yield stress during aging at 85°C and were able to link it to the microstructure evolution, i.e. precipitation of metastable phases in the volumes between T1 precipitates.

Although accelerated aging proved to provide information on the possible evolution of materials over long-term, data from the real case are required to confirm or infirm the predictions based on artificial aging for aeronautic materials. Indeed, cycles endured by aircraft during its life-time are not isothermal. To do so, an interesting approach is to observe the microstructure and measuring the macroscopic properties of alloys retrieved on end-of-life aircraft and compare them to an equivalent alloy on which artificial aging is being conducted. There have been

* Corresponding author : magali.brunet@cemes.fr

few attempts of such studies in the literature. Salimon at al. [9] have studied aged D16AT alloys (Russian nomenclature, equivalent of 2024-T4) retrieved from the fuselage skin of three Russian TU-154 aircraft which flew respectively 11, 15 and 20 years. They showed that for uncorroded parts, mechanical characteristics (yield stress, tensile strength and elongation to fracture) did not evolve. Low-stress, high-cycle fatigue did not cause dramatical structure changes in the bulk material. Only the near-surface layer was influenced by high-cycle fatigue, giving rise to micro-cracks and fracture of oxide film and thus corrosion. Schmidt et al. [1] studied 2024-T3 sheets taken from an Airbus 300 fuselage with 18 years of service usage (75% of its design service goal). Here too, although the old sheets exhibited a high susceptibility to intergranular corrosion, all properties tested (strength, elongation, fracture toughness, crack growth rate) were equal or even slightly better than the required properties. Our team demonstrated recently the interest in studying materials from old aircrafts (kept in museum or associations) [10]. Two plates recovered from a Breguet Sahara Deux-Ponts which first flew in 1958, were studied and compared. It appeared that the mechanical properties (yield stress, tensile strength and elongation) of these 60 years-old plates still conform to the standards of that time. Prudhomme et al. [6] compared mechanical properties of AA 2024 from a decommissioned A320 aircraft (wing panels) to an artificially aged alloy. They found similar results for an end-of-life material and an alloy artificially aged for 55h at 150°C. This aging treatment was chosen from an equivalence rule based on the Arrhenius law. The assumption is that homogeneous precipitation of second phases resulting from decomposition of a solid solution is controlled by the diffusion of copper atoms. The activation energy for copper diffusion is defined thanks to the model developed by Khan and Starink [11] for Al-Cu-Mg alloys. With this equivalence rule, 55h at 150°C should thus correspond to 100,000h exposure at 80°C, a service life of an aircraft. As a result, Prudhomme et al. showed that fatigue lives were indeed comparable as well as fracture mechanics.

In order to verify whether long-term artificial aging is representative of natural aging in the case of Duralumin, this article presents a comparison between an old alloy retrieved from the Breguet Sahara Deux-Ponts and a modern equivalent alloy 2017A. Different heat treatments were applied to assess the aging on the old alloy, i.e. solution treatment and artificial aging at 180°C. Long-term artificial aging at 100°C was applied to the modern alloy 2017A to simulate long-term natural aging (in-situ). Mechanical tests (hardness, tensile tests) followed by scanning electron microscope (SEM) fractography were conducted. Transmission electron microscope (TEM) was used to observe the nanostructure at different stages and to interpret the macroscopic behaviour of the artificially and naturally-aged alloys.

2. Experimental

2.1. Materials

A plate labelled L8 (1.6 mm thick) recovered from the engine nacelle of a Breguet Sahara Deux-Ponts already observed in [10] is the naturally-aged plate. A commercial 1 mm thick plate made of 2017A alloy in state T4 was bought: it is the modern alloy on which artificial aging will be conducted. On this alloy, a comprehensive study of the precipitation was previously conducted as it constitutes the first step to properly understand further aging [13].

Table 1 reports the elementary composition for each alloy measured by ICP-OES (Inductively Coupled Plasma - Optical Emission Spectrometry). The old plate is made of A-U4G (French designation used in 1950 for Duralumin). Compared with the equivalent modern alloy, it has a lower content of copper and a higher content of magnesium.

To assess aging, different aging treatments were applied, as described in Table 2. For the plate collected on the aircraft, labelled "AR" for as-received, no pre-treatment was applied. We know however from Breguet standards [14] that these pieces underwent upon fabrication a T4 treatment: solution treatment followed by water quench and room temperature aging. For this plate, the AR state corresponds thus to a T4 + 10 years of service + 50 years in outdoor conditions. The following treatments described in the table were all applied on the as-received alloys in the laboratory.

Table 1. Elemental composition of alloys obtained by ICP-OES

	Al	Cu	Mg	Mn	Si	Fe	Ti	Zn	Cr
2017A-T4	Base	4.32 ±0.08	0.68 ±0.01	0.611 ±0.002	0.618 ±0.006	0.34 ±0.01	0.043 ±0.001	0.20 ±0.04	0.029 ±0.003
L8 (A-U4G-T4)	Base	3.88 ±0.05	0.83 ±0.01	0.522 ±0.007	0.60 ±0.01	0.53 ±0.04	0.063 ±0.001	0.045 ±0.004	0.023 ±0.002

Table 2. Condition states studied in this work

Label	**State**
AR	As-received
180°C/8h	Thermally aged at 180°C for 8h
ST	Solution treatment: 500°C for 1h, water quenched and room temperature aged
ST-180°C/8h	Solution treatment: 500°C for 1h, water quenched and thermally aged at 180°C for 8h
100°C/10,000h	Thermally aged at 100°C for 10,000h

https://doi.org/10.1051/matecconf/202032604007

2.2. Characterization techniques

Tensile specimens were cut in each plate: 4 in rolling direction and 4 in the transverse direction. The gauge length/reduced section and width of the tensile specimen are 20 mm and 3 mm respectively. The thickness of the 2017A plate and the L8 plate are 1 mm and 1.6 mm respectively. Tensile tests were performed at a strain rate of 0.02 s^{-1}, on a Zwick Z030 with a MTS extensometer. SEM pictures of fracture surfaces were taken on a JEOL 6490.

Micro hardness measurements were done with a Buehler Omnimet series 2100, following the ASTM E384 standard. Average results on 5 indents with a load of 300g during 10s were plotted. For these measurements, the alloys were cut along three perpendicular orientations: rolling direction (RD), normal direction (ND) and transverse direction (TD) and each specimen was embedded in an epoxy resin and mechanically polished on water-lubricated abrasive papers (silicon carbide) then on polishing cloths with diamond paste.

For TEM observations, the preparation was the following: to bring the sample thickness down to 25 µm, a mechanical polishing on SiC paper from 600 down to 2400 grade was performed. The specimens were then electrochemically thinned using a Tenupol-5 Struers apparatus operating at 60 V in a solution of methanol and nitric acid (3:1) at -15°C. Observations of the nanostructure were carried out on a JEOL 2010 operating at 200 kV. Bright field (BF) images were taken with an orientation slightly off the $[100]_{Al}$ or the $[110]_{Al}$ zone axis. The length of nanoprecipitates (platelets) were estimated on BF images by averaging 30 measurements.

SAXS measurements were performed on a Xeuss® equipment (Xenocs) equipped with a microsource with copper anode. Samples were thinned down to 100 µm. 3 samples-to-detector distances were considered (40 cm, 1.2 m and 2.5m).

3. Results

3.1. As-received state

3.1.1. Mechanical properties

The engineering stress-strain curves for the modern plate (2017A) and the old plate (L8), both in the as-received state are shown in Figure 1 for two directions: rolling direction (RD) and transverse direction (TD). For L8, a circular plate, it was not possible to identify straightforwardly the directions [10]: a labelling D1 and D2 was thus used. They are supposedly the rolling and transverse direction respectively. Figure 1.a reveals that tensile strength curves exhibit good reproducibility for 2017A. On the old alloy however, a heterogeneous behaviour is observed. Early failure occurs quite often in direction D2, the estimated transverse direction. Table 1 summarizes the main mechanical properties. If early-failed specimens are not taken into account, the yield strength (YS), the ultimate tensile strength (UTS) and the elongation are conformed to the expectation for such alloy [15].

Table 1. Mechanical properties extracted from the stress-strain curves for L8 and 2017A plates in as-received state.

		E (GPa)	YS (MPa)	UTS (MPa)	True UTS (MPa)	Elongation (%)
L8-AR	D1	64±12	322±1	457±1	524±7	15±2
	D2	54±2	277±15	398±32	434±54	9±4
2017 A-AR	RD	57±3	314±1	439±1	512±8	17±2
	TD	65±5	290±2	433±1	506±2	17±1

3.1.1. Fracture surface analysis

Fracture surface analysis was carried out by SEM on post-mortem specimens. Figure 2 presents images representative of each alloy. On both alloys, equiaxed dimples were observed, typically present for ductile fracture caused by uniaxial tensile load. This type of fracture processes by microvoid coalescence. Microvoids nucleate at particles such as precipitates. After nucleation, the voids grow in the direction of the applied tensile stress. The dimples caused by coarse intermetallics are clearly visible in the microstructure of both alloys. Coarse intermetallics in 2017A-type alloys are well-identified: AlMnFeSi mostly (large, grey particles with faceted morphology) together with Al_2Cu (round and bright) and Mg_2Si (dark). They are quite numerous in both samples. Extremely small dimples were also present which could be caused by intragranular precipitation.

In the old alloy (L8), on all the early-failed specimens occurring mainly in the transverse direction, decohesive fracture along grain boundaries were observed (see framed area in Figure 2c and d.). Intergranular corrosion, unspotted before testing on the specimens, seems to be the main cause of this rupture mode. It is known that alloys of the 2xxx series are susceptible to stress-corrosion cracking and that the sensitivity depends strongly on the grain orientation. The transverse direction is thus the most susceptible to cracking, which is observed here.

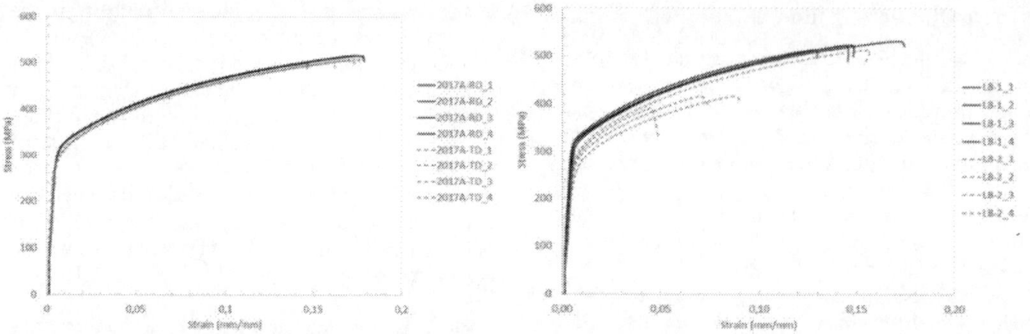

Fig. 1. True Stress/Strain curve in two orientations a) RD and TD for 2017A and b) D1 and D2 for L8 in as-received state.

Fig. 2. SEM photographs of fracture surface in back-electron scattering mode for a) 2017A-AR, b) L8-AR. Images c) and d) in secondary electron mode, show fracture due to intergranular corrosion on L8-AR in the transverse direction.

3.2. Assessment of aging on old plate (L8)

3.2.1. Microhardness

Microhardness was measured versus the aging time at 180°C for both plates, in as-received state (AR) and after a solution treatment (ST). Because the materials exhibit similar hardness in three orientation (RD, TD and ND), only hardness in RD is plotted in Figure 3.
Artificial aging (180°C) was first applied on the as-received samples. For 2017A alloy, in the early stages of the treatment at 180°C, the hardness is decreasing due to the dissolution of Guinier-Preston Zones (GPZ), already present in T4 stage. Then it increases up to a maximum (peak hardening) reaching 136 HV0.3 after 8h before decreasing (over-aging stage). L8-AR exhibits a peculiar hardness behaviour upon artificial aging at 180°C. The hardness indeed does not reach a maximum after 8h at 180°C, but stays at its original value around 120 HV0.3. The attribution of such a poor behaviour in hardness to an alteration due to the history of the plate during the plane service is a possible explanation.
To verify if this behaviour is due to a long-term natural aging and to assess the reversibility of the

phenomenon, the hardness of the old plate was re-measured after being solution treated at 500°C for 1 hour, quenched in water and maturated again. After this treatment, hardness of L8-ST is 123 HV0.3, close to the hardness before treatment.
Aging treatment at 180°C was then re-applied. In this case, hardness of the old alloy follows the behaviour of the alloy 2017A, reaching a maximum after 8h (see curve L8-ST in Figure 3). It can be concluded that the history of the Breguet's plate was erased. Its aging is thus reversible.

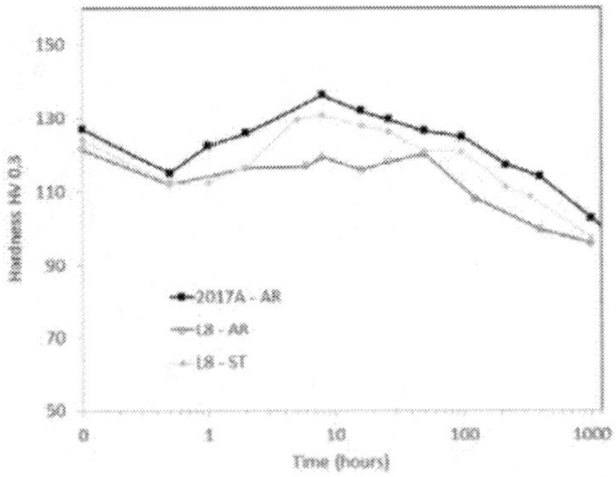

Fig. 3. Hardness HV0.3 measurement in the rolling direction of modern alloy 2017A and L8 (A-U4G) when aging at 180°C from an as-received state (AR) and after solution treatment (ST).

The reasons for this behaviour in the AR state are investigated with the observation of the micro and nanostructure.

3.2.2. Microstructure

On L8-AR, TEM observations (see Figure 4.a and 4.b) reveal a precipitation at grain boundaries and the presence of a precipitation on dispersoids. The precipitation is marked by arrows in BF-TEM image (Figure 4a) and circles in the corresponding selected area diffraction (SAD) pattern. This phase corresponds to **Ω**-Al_2Cu phase, as identified in [13] on similar Duralumin. No other precipitation is observed in this sample by TEM but it was verified by small angle X-ray scattering (SAXS) that GPZ/clusters are present in this state (see Figure 5). The results show indeed two main contributions: one in the low q-range ($0.002 < q < 0.1$ Å^{-1}) with a q^{-4} contribution, attributed to large precipitates (dispersoids) and one at high q-range ($0.1 < q < 1$ Å^{-1}), a plateau attributed to GPZ or small clusters in the alloy matrix.
At grain boundaries, the precipitation nature is Al_2Cu. Precipitation on dispersoids and at grain boundaries could be the sign of the heat experienced by the plate during its life-time. After solution treatment (ST), precipitation on dispersoids and at grain boundaries has disappeared: see Figure 4.c and 4.d.

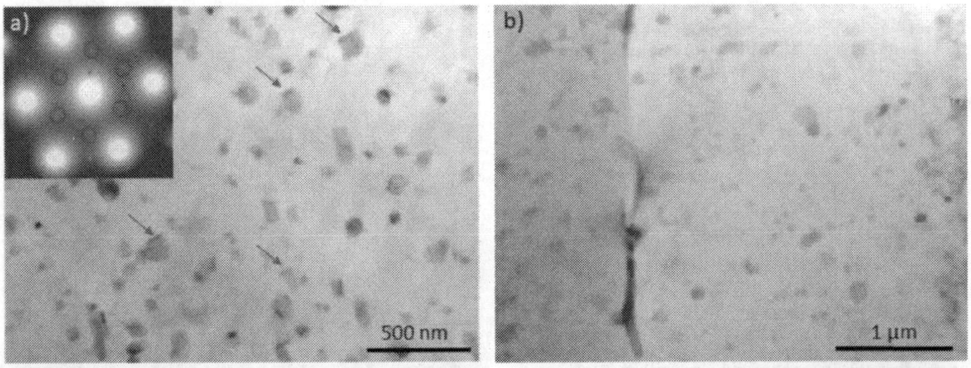

Fig. 4. BF-TEM images in [110]$_{Al}$ zone axis with corresponding SAD patterns of a) and b) L8-AR; c) and d) L8-ST.

 https://doi.org/10.1051/matecconf/202032604007

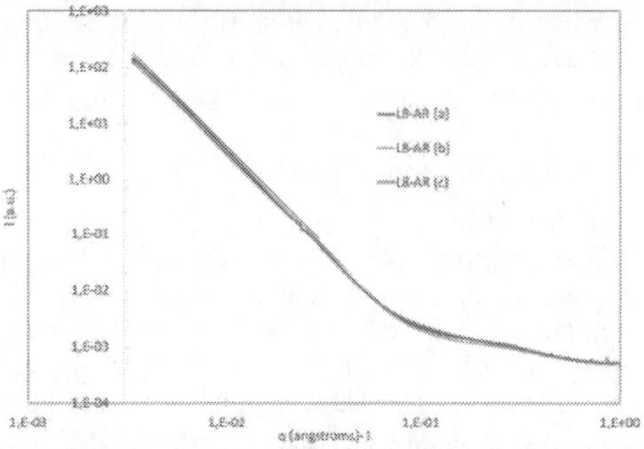

Fig. 5. SAXS curves for L8 (3 different zones) in the as-received state.

The nanostructure is then observed for 2017A and L8 after treatment at 180°C for 8h corresponding to peak hardening. It is shown in Figure 6. The observations of the old plate (Figure 6.a) reveals the presence of a variety of phases:

- the hardening θ"-Al_3Cu phase identified in the SAD patterns thanks to the discontinuous streaks along $[200]_{Al}$ [16],
- the θ'-Al_2Cu phase (platelets) identified thanks to faint spots $(110)_{\theta'}$ at ½ of $(220)_{Al}$,
- the Q'(Q)-AlCuMgSi phase which are rods observed end-on in the $[100]_{Al}$ direction in the BF images,
- the Ω-Al_2Cu phase with diffraction spots at 1/3 and 2/3 of the $[022]_{Al}$ orientation.

This precipitates zoology is equivalent to the one found in the modern 2017A alloy in the same aging conditions (Figure 6.b and [13]). Measurements of the hardening precipitates (θ"-Al_3Cu and θ'-Al_2Cu platelets) are reported in Table 3. On average, platelets are 86 ± 29 nm long for L8_180°C/8h and 60 ± 15 nm long for 2017A_180°C/8h. The difference is not significant enough to conclude. However, a much higher number of dislocations are found in the old plate indicating some work-hardening of the L8 plate during its shaping or during its lifespan.
After ST followed by aging at 180°C for 8h, the nanostructure is observed again: see Figure 6.c. The platelets are measured: in the old plate, they are 65 ± 18 nm, and dislocations have disappeared.

Table 3. Measurements of hardening precipitates (Platelets from θ sequence)

Alloy	Platelets length (nm)
2017A_180°C/8h	60 ± 14
2017A_ST-180°C/8h	50 ± 15
L8_180°C/8h	86 ± 29
L8_ST-180°C/8h	65 ± 18

These results show that hardening precipitation occurs after artificial aging (180°C-8h) in L8, in a similar manner as the 2017A alloy whether in the as-received state or after solution treatment. This means that the peculiar behaviour of the L8 in hardness during thermal aging (observed in Figure 2) is not due to a lack of hardening precipitation. The reason for the hardness stagnation on the old plate is thus suspected to be found at microscale.

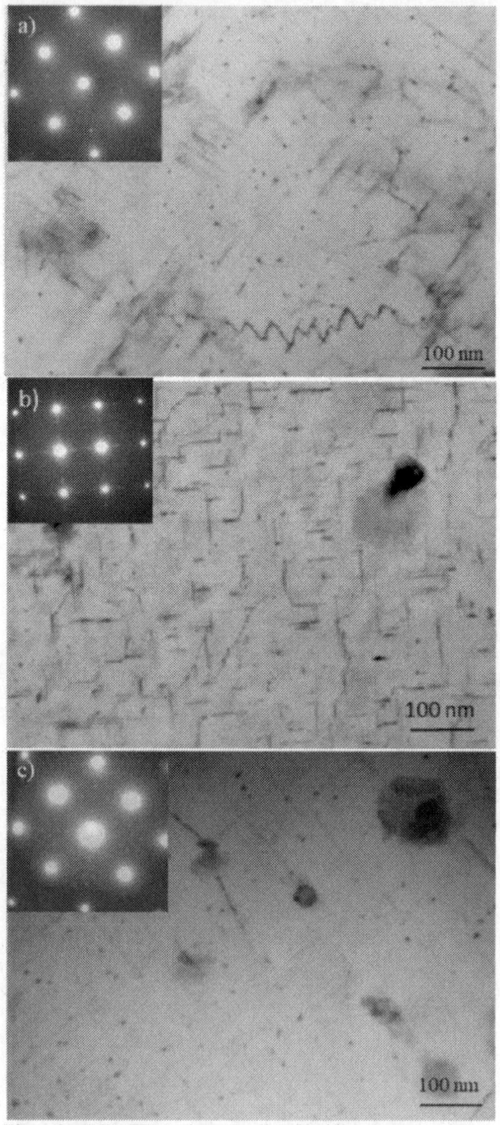

Fig. 6. BF TEM images in $[110]_{Al}$ zone axis a) L8_180°C/8h; b) 2017A_180°C/8h; c) L8_ST-180°C/8h

3.3. Long-term aging on modern alloy 2017A

As mentioned before, hardening precipitation in these alloys of the 2xxx family is considered as being controlled by the diffusion of copper atoms. The time-temperature equivalences can be used based on the copper diffusion and Arrhenius law [11]. In a recent PhD thesis work [17], several alloys among which

https://doi.org/10.1051/matecconf/202032604007

2024, were artificially aged at different temperatures in the range from 85°C to 250°C for different durations varying from 1 to 10,000h. Time-temperature equivalences based on microstructure evolution were found to be accurate for pre and post peak-hardening. Experimental data (mechanical properties) followed the predicted behaviour based on the evolution of the S-phase fraction.
Here, low-temperature artificial ageing (100°C) was applied on the modern 2017A alloy until 10,000h to simulate long-term aging. As hardening is based on copper diffusion, the assumption is made that the precipitation along the θ sequence should follow the same law. The time-temperature equivalence was then used with an activation energy for copper diffusion equal to 133 kJ/mol. In this case, it was calculated that the low temperature aging (100°C, 10,000h) should be equivalent to 5h at 180°C.

3.3.1. Microhardness measurement

For 100°C, hardness reaches 145 HV0.3 after 10,000h which is significantly higher than peak aging at 180°C (136 HV0.3). The higher hardness obtained at low temperature aging is classically observed in age-hardening alloys [18].

3.3.2. Microstructure

The microstructure corresponding to an aging time of 10 000h at 100°C is shown in Figure 7 and the one for 5h at 180°C in Figure 8.

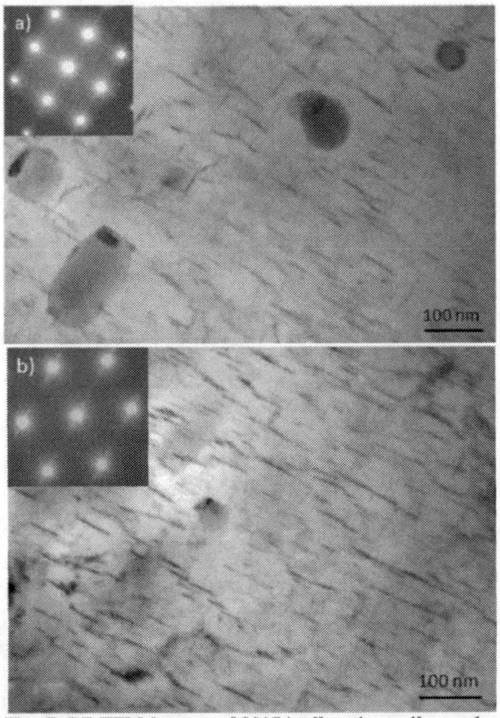

Fig. 7. BF-TEM images of 2017A alloy thermally aged at 100°C for 10,000h in a) $[100]_{Al}$ zone axis and b) in $[110]_{Al}$ zone axis.

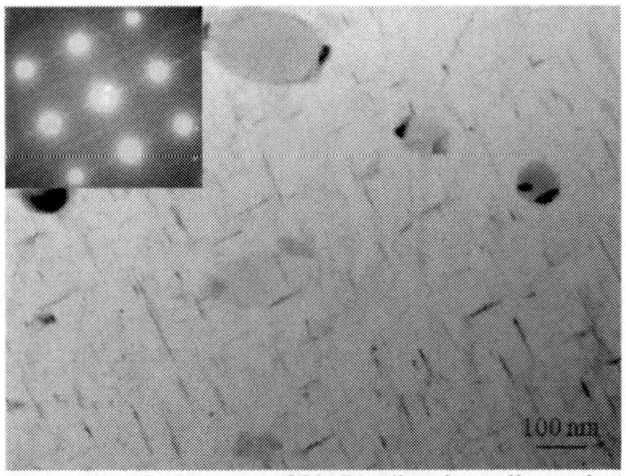

Fig. 8. BF-TEM images of 2017A alloy thermally aged at 180°C for 5h in $[100]_{Al}$ zone axis.

Table 4. Measurements of hardening precipitates (Platelets from θ sequence)

Alloy	Platelets length (nm)
2017A_180°C/5h	55 ± 17
2017A_100°C/10,000h	49 ± 21

In the SAD patterns in two zone axes ($[100]_{Al}$ and $[110]_{Al}$), for the 2017A alloy heat treated at 100°C for 10,000h, we observe discontinuous streaks along $[200]_{Al}$ with maximum intensities at $(100)_{Al}$ corresponding to the θ''-Al_3Cu phase. The θ'-Al_2Cu precipitates, if present, have a low volume density. The Q'(Q)-AlCuMgSi phase was observed in the BF images (rods seen end-on oriented along $[001]_{Al}$ direction) but the precipitates are also very few. The diffraction patterns, similar as the ones seen in [16], indicate that the alloy is just before peak hardening. Figure 8 provides the confirmation of a nanostructure very close to the one previously presented (for 8h at 180°C). The length of the hardening precipitates in the two aging conditions, is of the same order of magnitude (see Table 4). A major difference can be found for the Ω-Al_2Cu phase: this phase was hardly seen in the 2017A alloy aged at 100°C for 10,000h: diffraction spots are not visible in the SAD patterns (inserts of Figure 7.a and 7.b). However, the phase can be seen on dispersoids in BF-TEM images. It is concluded that the phase is present but in very little quantities and with no significant influence.

4. Discussion

4.1. Long-term natural aging

L8 plate exhibits a peculiar mechanical behaviour and microstructure in the AR conditions, i.e. after 10 years of service and 50 years in outdoor environment. Limited elongation (or early failure) was explained by intergranular rupture mode due most probably to intergranular corrosion present in the plate. Precipitation at grain boundaries (Al_2Cu) was indeed observed on the AR plate by TEM. This is clearly the sign of a heat exposure during the aircraft life-time.

This precipitation induces a fragilization of the material upon exposure to environment (50 years outdoors). Another precipitation was observed in L8-AR: the Ω-Al_2Cu phase, which grows on dispersoids. But this precipitation has no hardening effect on the alloy in the AR state: overall L8-AR exhibits a similar hardness as 2017A. Upon artificial treatment (180°C/8h), hardness of the old plate did not reach a maximum, as opposed to the modern alloy 2017A, although hardening precipitation was observed inside the grains by TEM. We believe that microhardness is thus affected by precipitation at grain boundaries. In the AR-state, the high dislocation density observed in L8 is the sign of work hardening that compensate for the loss of hardness due to intergranular precipitation. However, this intergranular precipitation decreases the potential for further hardening in the old plate upon aging.

When a solution treatment is applied: precipitation on grain boundaries and on dispersoids (Ω-Al_2Cu) disappear. The solution treatment erases thus the long-term aging of the plate. Consequently, the L8 behaves according to the modern alloy. The aging is reversible.

4.2. Artificial aging at low temperature

Artificial aging at low temperature (100°C) was applied on 2017A alloy to simulate natural aging. If time-temperature equivalence based on Arrhenius law is used, 10,000h at 100°C should correspond to 5h at 180°C. However, differences in hardness and in microstructure can be noticed between these two aging conditions.

The different age hardening responses at different aging temperature can be explained with thermodynamic considerations [18]. The free energy change per unit volume, i.e. the driving force for precipitate nucleation, is strongly influenced by the alloy composition and the aging temperature. It is higher for low temperature which makes the activity energy barrier to be the lowest, considering that the change of elastic strain energy and the change of interfacial energy are negligible versus aging temperature in this first phase of GPZ and θ" nucleation. The maximum nucleation rate and therefore the maximum density of precipitates is achieved for lower temperatures, hence the maximum hardness for aging temperature of 100°C. From the TEM observations on the 2017A thermally aged at 100°C for 10,000h, the presence of coherent precipitates (θ"-Al_3Cu) indicates that homogeneous nucleation is dominant. Although the observed platelets are of the same length as for the 180°C/5h conditions, the density of these precipitates was not precisely estimated but it appears higher after aging at 100°C than after aging at 180°C.

Another difference is noted between the two aging conditions: the presence of Ω-Al_2Cu phase was very rarely spotted after aging at low temperature (100°C). In AA2519 alloy (Al-5.6Cu-0.3Mn-0.3Mg-0.15Zr wt.%), Zuiko and Kaibyshev [19] showed that the precipitation of Ω-Al_2Cu phase happens right after the θ"-Al_3Cu phase precipitation and before the precipitation of the θ'-Al_2Cu phase. This is due to the free energy of the Ω-Al_2Cu phase, which is higher than that of the θ"-Al_3Cu phase. As it was shown that at 100°C/10,000h, the alloy is just before peak-hardening and the θ"-Al_3Cu phase is in the majority. It might require longer time at this temperature to generate the Ω-Al_2Cu phase.

Comparing 2017A artificially aged at 100°C with L8-AR, it has to be said that it was not possible to reproduce a similar nanostructure with a low-temperature aging. Long-term natural aging implies non-isothermal exposure (due to the aircraft take-off and landing cycles) which can indeed produce unpredicted nanostructure. The presence of the Ω-Al_2Cu phase on L8 might thus be explained by cycles of high-low temperatures experienced by the plate near the engine. Another hypothesis could also be found in the alloy composition itself: the slightly higher content of magnesium in L8 (0.83 wt.%) compared with that of the 2017A (0.68 wt.%) could help trigger this phase in L8 (always upon a heat treatment). It was shown indeed by Zuiko and Kaibyshev [19] that magnesium enhances the precipitation of transient phases.

5. Conclusions

From different measurements and observations at various scales on two aluminium alloys, an old one (L8) and a modern one (2017A), the following conclusions can be drawn:

- The static mechanical behaviour remains globally unchanged: upon deformation, L8 behaves similarly as 2017A alloy except for some specimens affected by intergranular precipitation and probably intergranular corrosion which caused early fracture.
- The hardness of L8 in the AR state, although close to 2017A's one, is not following the same hardening curve upon aging at 180°C. However, TEM studies reveal that hardening precipitation in L8 occurs in a similar manner as in 2017A.
  A solution treatment (500°C/1h followed by a quench) allowed to erase the particular microstructure on L8-AR, that is precipitation at grain boundaries and on dispersoids. After this, the hardness follows a typical curve with a peak-hardening after 8h.
- Time-temperature equivalence is not entirely confirmed between 180°C and 100°C: hardening precipitation (platelets) is of the same order of magnitude in length but hardness is different. This is explained by a higher volume fraction of precipitates at 100°C. Furthermore, the Ω-Al_2Cu phase is hardly present at 100°C: this aging temperature is thus below the Ω solvus.

- Long-term artificial aging on modern alloy was not able to reproduce the aging of the naturally-aged alloy (in-situ). This is due to the non-isothermal nature of the aging in the real case.

Acknowledgement

The authors wish to thank Pierre Roblin from FR FERMAT, Université de Toulouse, CNRS, Toulouse, France, for SAXS measurements and the volunteers of *Les Ailes Anciennes Toulouse* for providing us with the plate collected on the Breguet Sahara 765 aircraft. This study has been partially supported through the grant NanoX n° ANR-17-EURE-0009 in the framework of the "Programme des Investissements d'Avenir".

References

[1] H.-J. Schmidt, B. Schmidt-Brandecker, and H. Trey, in *The third joint FAA/DoD/NASA Conference on aging aircraft*Albuquerque, New Mexico, September 20–23 (1999)).
[2] National Research Council, The National Academies Press, Washington, DC., 1996, p. 65.
[3] G. Hénaff, F. Menan, and G. Odemer, Engineering Fracture Mechanics **77**, 1975 (2010).
[4] R. Braun, C. Juricic, and G. Tempus, Progress in Mechanical Behaviour of Materials (Icm8), Vol 2: Material Properties, 597 (1999).
[5] N. D. Alexopoulos, Z. Velonaki, C. I. Stergiou, and S. K. Kourkoulis, Corrosion Science **102**, 413 (2016).
[6] M. Prudhomme, F. Billy, J. Alexis, G. Benoit, F. Hamon, C. Larignon, G. Odemer, C. Blanc, and G. Hénaff, International Journal of Fatigue **107**, 60 (2018).
[7] A. Deschamps, M. Garcia, J. Chevy, B. Davo, and F. De Geuser, Acta Materialia **122**, 32 (2017).
[8] D. Ortiz *et al.*, Engineering Failure Analysis **13**, 170 (2006).
[9] S. R. Salimon, A. I. Salimon, and K. A.M., in *Proceedings of IMechE*2008), pp. 339.
[10] A. Cochard, K. Zhu, S. Joulie, J. Douin, J. Huez, L. Robbiola, P. Sciau, and M. Brunet, Materials Science and Engineering: A **690**, 259 (2017).
[11] I. N. Khan and M. J. Starink, Materials Science and Technology **24**, 1403 (2008).
[13] M. Brunet, B. Malard, N. Ratel-Ramond, C. Deshayes, S. Joulié, B. Warot-Fonrose, Ph. Sciau, J. Douin, F. De Geuser, A. Deschamps, Materialia **8**, 100429 (2019).
[14] Service Technique de l'Aéronautique section Avions, (Centre de Documentation du Musée de l'Air et de l'Espace du Bourget, 1956).
[15] J. G. Kaufman, *Properties of aluminum alloys, fatigue data and the effects of temperature, product form and processing* Materials Park, Ohio, USA, 2008).
[16] Z. Shen, Q. Ding, C. Liu, J. Wang, H. Tian, J. Li, and Z. Zhang, Journal of Materials Science & Technology **33**, 1159 (2017).
[17] L. Briez, PhD, Mines Paris-Tech University, France, 2018.
[18] J.-F. Nie, in *Physical Metallurgy (Fifth Edition)*, edited by D. E. Laughlin, and K. Hono (Elsevier, Oxford, 2014), pp. 2009.
[19] I. Zuiko and R. Kaibyshev, Journal of Alloys and Compounds **759**, 108 (2018).

Post corrosion tensile strength and failure of dissimilar friction stir welded aluminium alloys

Madhav Raturi [1,*] and *Anirban* Bhattacharya[1]

[1]Department of Mechanical Engineering, Indian Institute of Technology Patna, Bihar, India

Abstract. The present study efforts towards appraising the effects of corrosion on the tensile and fracture behaviour of dissimilar friction stir welding (FSW) of aluminium alloys. Three different dissimilar FSW joints obtained between AA6061-T6 and AA7075-T651, AA6061-T6 and AA2014-T6, AA7075-T651 and AA2014-T6, using threaded pin profile with three flat faces (TIF) tool at rotational speed of 1200 rpm and welding speed of 98 mm/min. The maximum joint tensile strength was achieved for AA7075-AA2014 joints followed by AA6061-AA2014 and least recorded for AA6061-AA7075 for as obtained FSW joints (non-corroded). The joints are further immersed into a corrosive solution for 1, 2, 7 and 14 days duration. The corrosion occurred all over the joint but much accelerated rate of exfoliation corrosion exists away from stir zone near the confluence of heat affected zone and base material irrespective of the advancing or retreating side. With increase in corrosion time the location of tensile failure shifted towards corroded region (AA6061-T6) instead of stir zone in dissimilar weld joint AA6061-AA2014, whereas it remained unchanged for other two joints. The fractured surfaces of AA6061-AA2014 FSW joints reveals the articulated view of pits and fracture morphology advocating the loss in YS, UTS and % elongation with increases in immersion duration.

1 Introduction

Aluminium alloys due to their low weight to strength ratio make them excellent for aerospace, marine and other applications. Along with the optimized mechanical properties, corrosion susceptibility of these alloys is also required to be reduced and worth investigation. Friction stir welding (FSW) evolved as a promising solid-state joining process for aluminium and other alloys. A non-consumable, rotating tool plastically deforms the work piece material generating sufficient heat for desired material flow, mixing and joining predominantly through solid state diffusion. Srinivasan et al. [1] studied corrosion susceptibility to stress corrosion cracking (SSC) of friction stir welded AA7075-AA6056 by performing tensile test in air and NaCl solution (3.5%). They reported that TMAZ/HAZ of AA7075 is susceptible to SCC, although fracture location is governed by both mechanical stresses as well as corrosive environments. Birbilis et al. [2] investigated the electrochemical and metallurgical behaviour of Al_7Cu_2Fe particles in AA7075-T651 alloys. They found that 65% of the constituent particle population is of Al_7Cu_2Fe and are of comparatively large size. Mentioned particles served as a local cathode during localized corrosion of AA7075-T651, developing large pits at the particle-matrix interface. Aluminium alloys of 7XXX series are more susceptible to corrosion in comparison to 2XXX and 6XXX series appealing several researchers to find the cause and remedy to its corrosion behaviour. Marlaud et al. [3] quantified corrosion susceptibility of 7XXX series with different heat treatment conditions using EXCO test and electrochemical test. They suggested over aging as the remedy to reduce localized inter-granular corrosion however found no considerable change in the results between two different test methods. Lahovary et al. [4] studied the galvanic behaviour between copper rich intermetallic and its adjacent matrix within AA2024-T351 alloy. For corrosion test, the test coupons were immersed in NaCl solution of 1 M for 24 hours. The probable cause for intergranular corrosion was found to be the dissolution of copper rich precipitates like Al_2Cu and Al_2CuMg. Difference in galvanic behaviour of constituent particles can't be apprehended as lone reason for corrosion as the studies proven grain or constituent particle size can also affect corrosion behaviour. Yan et al. [5] put forward their idea of grain boundary optimization to control corrosion resistance in Al-Mg alloys. Their study witnessed that better corrosion resistance and precipitation in Al-Mg alloy is controlled by grain boundary (GB) misorientation angle. They concluded that low angle GBs are immune and high angle GBs are prone to grain boundary acid attacks. In the similar context, Zuo et al. [6] recognized changes in corrosion susceptibilities of Al-Zn-Mg-Cu alloy by bringing changes in rolling process. They found that controlling grain structure can change corrosion behaviour. The corrosion potentials (E_{corr}) reported for AA2014 is -768 mV [7], for AA6061 is -807 mV [8] and for AA7075 the value is -831 mV [9]. The E_{corr} value signifies the potential at which

* Corresponding author: madhav_1821me04@iitp.ac.in

corrosion begins and more negative value notifies the poor resistance to corrosion.

In FSW process, by virtue of the process, regions in-and-around the weld nugget zone undergo different magnitude of deformation and/or thermal cycle thereby leading to grain size variations. Microstructural inhomogeneity at different regions in FSW joint is expected to exhibit variation in corrosion behaviour. Thus, the present work is conducted to analyse the behavioural change in corrosion for three dissimilar FSW joints by changing the adjoining alloy combinations and corrosion periods. The study is successively completed by first obtaining friction stir welds between AA6061-T6 and AA7075-T651, AA6061-T6 and AA2014-T6, AA7075-T651 and AA2014-T6. The dissimilar welds obtained then are investigated for susceptibility and extent of corrosion with increasing immersion time and its effect on the tensile strength and fracture behaviour.

2 Materials and methods

The three different friction stir welds AA6061-AA7075, AA6061-AA2014 and AA7075-AA2014 obtained by keeping earlier in advancing side (AS) and later in retreating side (RS). Each alloy plate of 6.1 mm thickness was joined on a vertical milling machine by clamping on a suitably designed fixture. The tool made from H13 material having pin of right-hand thread with intermittent flat surfaces (TIF) (3 flat faces cut at 120° apart) was used for FSW. The tool shoulder diameter is 20 mm whereas the pin major diameter and length are 6 mm and 5.9 mm respectively. Shoulder penetration of 0.1-0.2 mm was provided for all joints. The friction stir joining is done at 1200 rpm tool rotation and 98 mm/min welding speed for all joints. Total ten tensile samples were obtained from each dissimilar joined weld plate and two samples each were kept for different immersion timings in corrosive solution. Figure 1(a) portrays the schematic representation of the order in which samples are being obtained for tensile test and for optical micrographs. The tensile samples (transverse to weld line) cut as per (ASTM E8) dimension from the welded plates. The samples were then dipped in NaCl solution for 1, 2, 7 and 14 days for observing the site, extent of corrosion and later tested for its effect on joint tensile strength. The samples were cleaned using etch cleaner and concentrated HNO_3 to decontaminate the surface before dipping in the saline solution of NaCl (57 g) + H_2O (1 L) + H_2O_2 (10 mL) prepared for corrosion. For the tensile samples, only the gage length portion was exposed to the corrosive environment from all four sides and rest is masked using spray paint. The samples which were dipped for more than 2 days the corrosive solution is changed after every two days. The samples were then taken out of the solution after the decided time period and cleaned for corrosive residues. For removing the corrosion products samples were washed using distilled water and then dipped in cleaning reagent prepared from H_3PO_4 (50 mL) + CrO_3 (20 g) + H_2O (1 L) at 90 °C for 5 to 10 minutes and then were taken out, washed using distilled water and air dried. Afterwards the samples were dipped in heated concentrated HNO_3 for 1 minute as a counter measure to remove corrosion product (if any) and washed using distilled water and air dried. The samples were then ultrasonicated in ethanol. The fresh (without any corrosion) and corroded tensile samples were tested for yield strength (YS), ultimate tensile strength (UTS) and % elongation. Uniaxial tensile tests of transverse specimens were conducted at cross head speed of 0.5 mm/min on ultimate tensile testing machine (Zwick/Roell Z50, capacity 50 kN). Contact type extensometer was used to record YS at 0.2% offset strain. Macroscopic observations of failure of the joints were performed using an optical microscope and fractographic studies were carried out on field emission scanning electron microscope (Zeiss Gemini500).

Table 1. YS, UTS and % elongation of friction stir welds before and after immersion for different time periods

FSW	Immersion time	YS (MPa)	UTS (MPa)	Elongation (%)
AA6061-AA7075	Without immersion	133±11	193±1	4.1±0.3
	1 day	109±3	189±5	4.15±0.05
	2 days	108±3	183.5±0.5	4.1±0.4
	7 days	113.5±0.5	192±1	3.9±0
	14 days	118.5±0.5	193.5±1.5	4.3±2
AA6061-AA2014	Without immersion	108.5±0.5	182.5±3.5	4.15±0.15
	1 day	106.5±1.5	177.5±1.5	3.95±0.25
	2 days	105±2	172±1	3.65±0.15
	7 days	105±0	175.5±0.5	4.05±0.25
	14 days	103±1	164±4	3.5±3
AA7075-AA2014	Without immersion	189±6	291.5±25.5	2.3±0.4
	1 day	181.5±4.5	252±2	1.7±0
	2 days	162±9	238.5±6.5	1.55±0.15
	7 days	176.5±1.5	270.5±7.5	1.7±0.2
	14 days	190±2	273±4	1.9±1

3 Results and discussion

The tensile properties of the FSW joints before and after immersion for different duration are given in Table 1 and

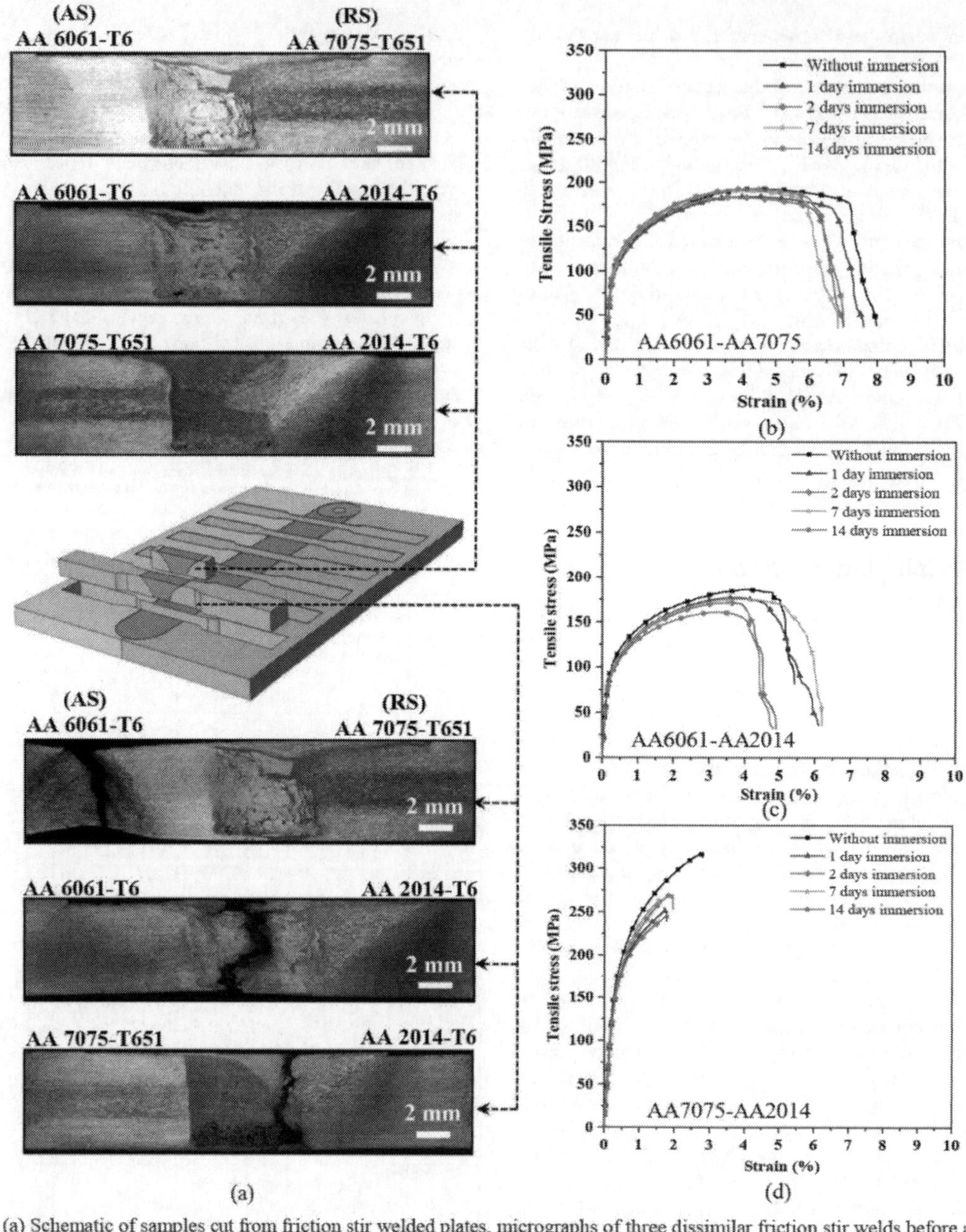

Fig. 1. (a) Schematic of samples cut from friction stir welded plates, micrographs of three dissimilar friction stir welds before and after tensile test of samples without immersion; tensile stress-strain plot for FSW joints between (b) AA6061-T6 and AA7075-T651 (c) AA6061-T6 and AA2014-T6 (d) AA7075-T651 and AA2014-T6 before and after immersion for different duration

the stress-strain plots are provided in Fig 1. Figure 1 also represents the optical microscopic images of the joints (nugget region) before and after tensile tests for the samples without immersion. As inferred from Table 1 and stress-strain plots (refer Fig 1. (b), (c), (d)) that the maximum UTS and YS was achieved for AA7075-AA2014 as compared to AA6061-AA2014 and AA6061-AA7075 dissimilar welds pre-corrosion. However, % elongation is noted minimum for AA7075-AA2014 joints compared to other two cases. For AA6061-AA7075 and AA7075- AA2014 there is no specific trend followed for different corrosion periods. Whereas for AA6061-AA2014 FSW joints loss in YS as well as UTS is noted with increase in corrosion period. The tensile failure of the joints (pre-corrosion) initiated and propagated from the stir zone for AA7075-AA2014 and AA6061-AA2014 FSW joints whereas failure in AA6061-AA7075 weld occurred from the advancing side (AA6061-T6). Figure 2 represents the corroded micrographs for all three dissimilar FSW joints after immersion for 1, 2, 7 and 14 days. The corrosion takes place all over the joint however minimum corrosion is observed at core of the nugget stir zone (SZ). There exists a region away from SZ, near the confluence of

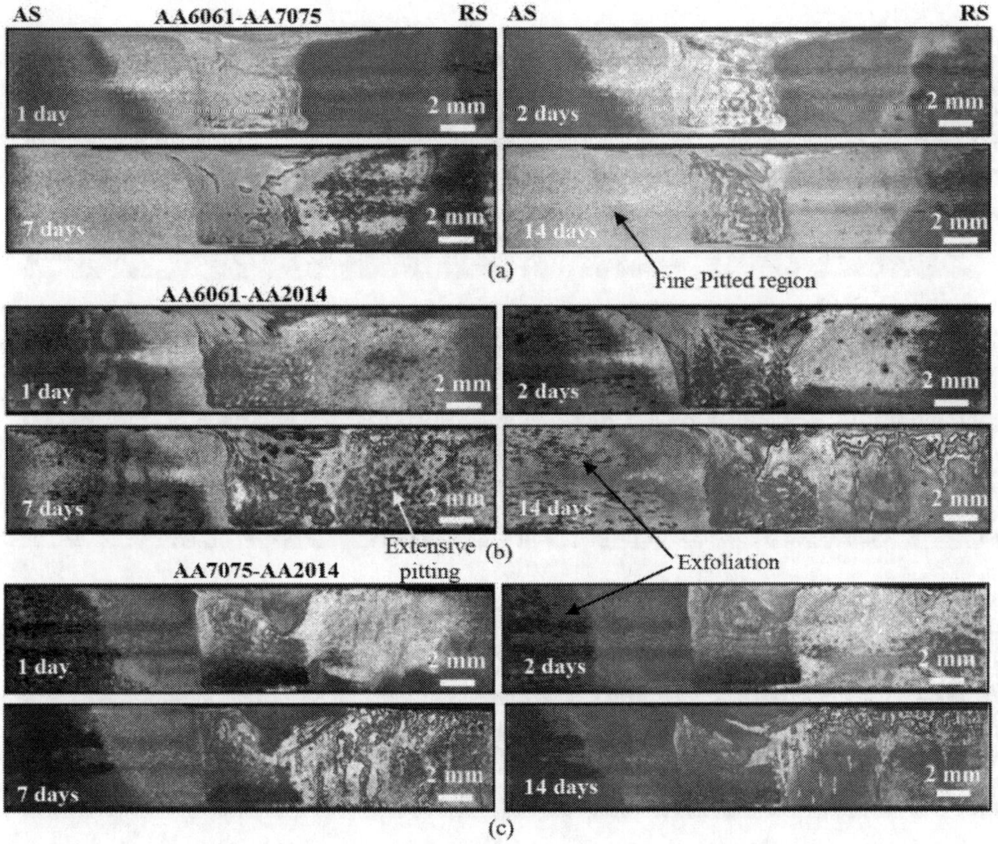

Fig. 2. Optical micrographs of corroded dissimilar friction stir welds after four different immersion periods for (a) AA6061-T6 and AA7075-T651 (b) AA6061-T6 and AA2014-T6 (c) AA7075-T651 and AA2014-T6 FSW joints

heat affected zone (HAZ) and base metal which experienced much accelerated rate of exfoliation corrosion (EFC). Once the weld joints are exposed to corrosive solution the extensive EFC is particularly observed on the AA7075-T651 side in two dissimilar welds (refer Fig.2 (a) and (c)). Whereas, for the dissimilar weld of AA6061-AA2014 it was observed in the AA6061-T6 side (refer Fig.2 (b)). Figure 2(a) shows the progressive corrosion after four different immersion durations for AA6061-AA7075. The extent of pitting corrosion can be observed increasing with the immersion time in the retreating side (AA7075-T651) of the weld. It may also be noted that the distance between SZ and extensive corrosion zone is more as compared to that of other two set of welds.

Thus, in AA6061-AA7075 FSW joints, the SZ and surrounding thermo-mechanically affected zone (TMAZ) is less prone to corrosion attack in NaCl solution. Whereas the extensive pitting is noted beyond HAZ till immersion for 2 days and HAZ regions also gets corroded after 7 days of immersion. In AA6061-AA2014 FSW joint the excessive pitting and EFC was observed on AS (AA6061-T6) with its extent increasing with immersion duration whereas on RS (AA2014-T6) metallic lusters are noted. Thus, it can be inferred from both cases that exfoliation corrosion sight is independent of the side in which alloy is placed. The corrosion took place as per the galvanic series in which 7XXX series is more anodic than 6XXX series and least is 2XXX series among all three. The AA7075 alloy with higher negative E_{corr} value justifies its extensive corrosion trending towards base metal. The lowest negative value of E_{corr} in AA2014 alloy conserves it from substantial material loss due to EFC as in other two alloys. Figure 2(c) shows the progressive corrosion of AA7075-AA2014 place in the AS (AA7075-T651) of the joint in regions beyond TMAZ and HAZ with increasing immersion time. However, the SZ of the nugget remains unattacked till 2 days of immersion and mild corrosion are observed in this zone after 7 or 14 days of immersion. The source of the Al_7Cu_2Fe particles in AA7075-T651 is due to the presence of iron impurities and their large size makes them prone to corrosion. The progressive corrosion in AA6061-AA2014 (refer Fig.2 (b)) dissimilar weld shows excess pitting and exfoliation corrosion in the advancing side (AA6061). The Mg_2Si is the common precipitates in AA6061-T6 and possess low corrosion potential compared to Al-matrix and acts as anode. During localized galvanic corrosion these particles get corroded fast leaving pits that further get exfoliated, although the rate of EFC is comparatively low in AA6061-T6 as compared to that of AA7075-T651. Figure 3 (a) shows the fracture location of tensile samples after immersion

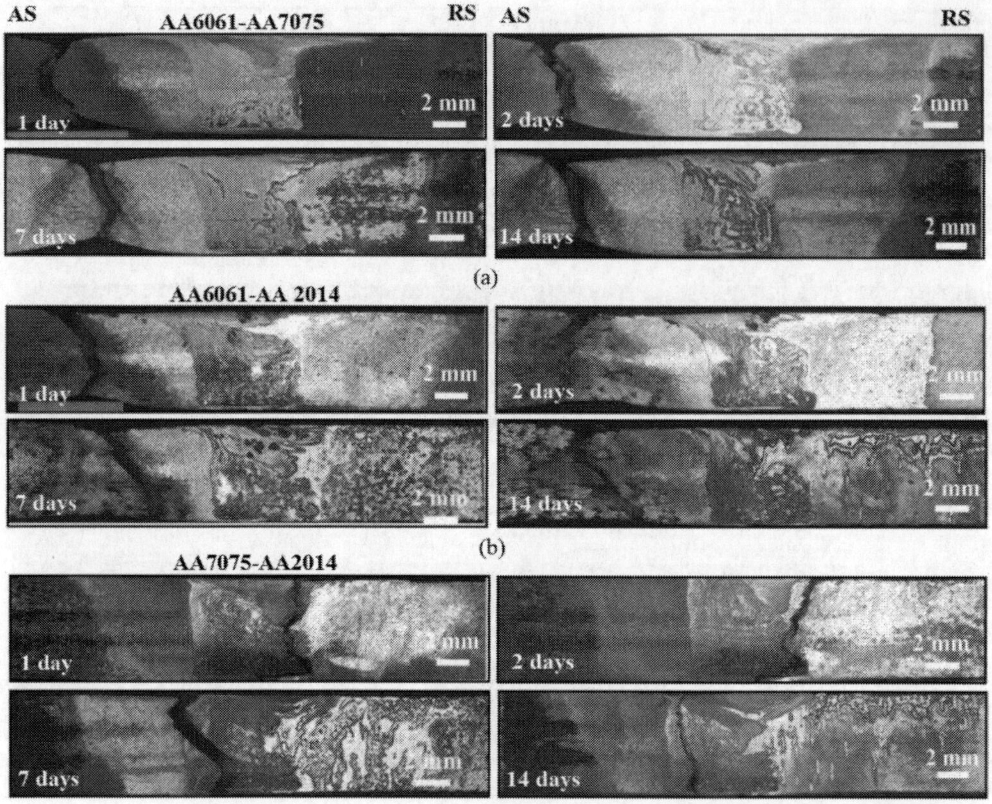

(c)

Fig. 3. Optical micrographs showing fractured locations of corroded dissimilar friction stir welds after different immersion periods of (a) AA6061-T6 and AA7075-T651 (b) AA6061-T6 and AA2014-T6 (c) AA7075-T651 and AA2014-T6 FSW joints

and progressive corrosion for different durations of AA6061-AA7075 FSW joints. With the increase in immersion duration the extent of corrosion gets increased but still the fracture location remains same. The UTS and % elongation is also same after corrosion and thus it can be inferred that advancing side (AA6061) participated maximum in the gage length elongation and remains the weakest among SZ and extensively corroded region. For the specific combination and the time duration for which the immersion was employed, the base metal (AA6061) is the weakest portion and thus it can sustain corrosive environment for longer periods without getting compromised on strength. However, the tensile testing of AA6061-AA2014 FSW joints after corrosion as shown in Fig. 3 (b) narrates a different nature with shifting in fracture location. The change is brought in even after getting exposed to the corrosive environment for 1 day i.e. least time period in this study. It can also be observed that with increase in immersion time the extent of corrosion is increasing and the EFC morphology is growing in AA6061-T6 (AS) side towards the HAZ. Prominent pitting and metallic lustres are noted on AA2014-T6 after 1 week immersion period. It can be also observed that from Fig 3 (b) that the AA6061 portion present in the SZ starts corroding after 2 days of immersion. UTS and % elongation tend to decrease with corrosion time as the populating pits accelerated the fracture propagation in AA6061 side of the joint. In AA7075-AA2014 dissimilar weld the extent of EFC is much higher than any of the other two combination as can be noted from Fig 3 (c). Observing the black region in AS (AA7075-T651) indicates EFC getting populated and lastly covering the whole width and narrowing the overall cross section from all four sides. Even after the narrowing of section due to corrosion, the fracture occurs from the SZ, although the side of fracture shifted from AA2014 towards AA7075 side of the SZ. This might be possibly due to the very narrow void at the bottom through which the corrosive solution sieved in, affecting the strength of joint at the SZ. Thus, for AA7075-AA2014 dissimilar weld, SZ remains the region of minimum strength possibly due to the presence of micro-voids in the SZ and the larger immersion duration will be required to weaken the joint from other location.

Figure 4 represents the SEM fractographs obtained for tensile fractured surface after the progressive corrosion of the dissimilar welds of AA6061-AA7075. The fracture occurred on the same location in AA6061-T6 for all the specimens irrespective of sample without

 https://doi.org/10.1051/matecconf/202032604008

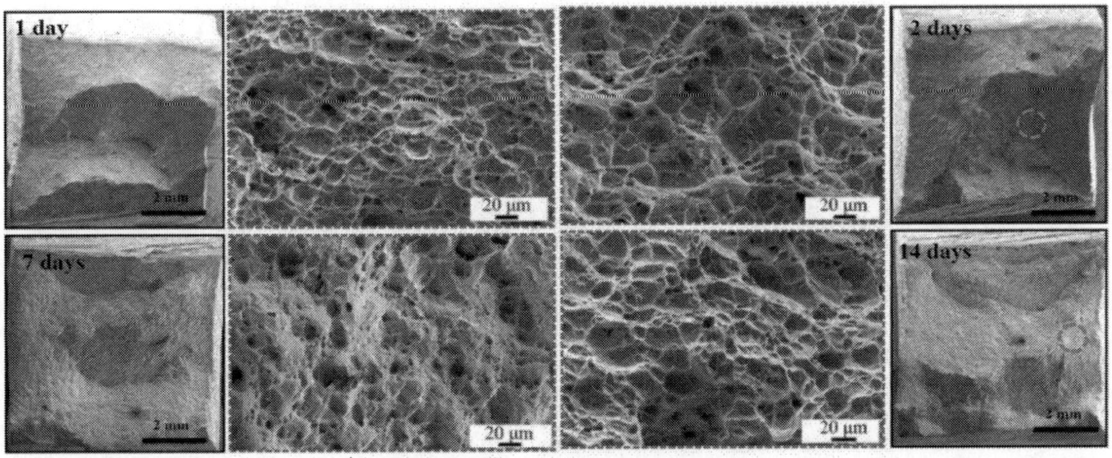

Fig. 4. FE-SEM micrographs of fractured surfaces of corroded AA6061-AA7075 dissimilar friction stir weld for four different immersion periods

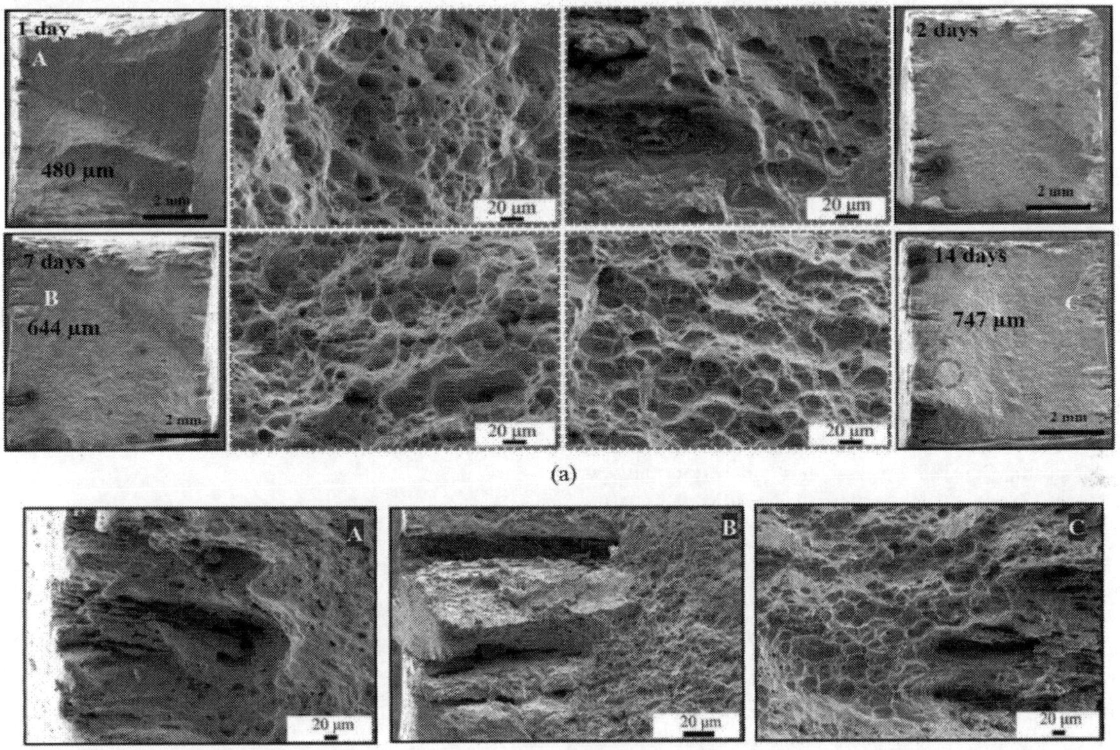

Fig. 5. FE-SEM fractographs of (a) corroded AA6061-AA2014 dissimilar friction stir weld for four different immersion periods (b) fracture morphology near the deep pitted region

or with corrosion for different durations. Thus, it can be inferred that corrosion time for which specimens are being exposed didn't affect the weld strength or fracture morphology. The fracture morphology and behaviour only get effected either due to the alloy constituents or due to the heating effects produced during welding and the metallurgical consequences. The fracture morphologies for all corrosive conditions show similar equiaxed dimple distributions and cup-cone failure can be observed for all the joints. The presence of secondary phase reinforcing particles of Mg_2Si is evident from different studies that work as the stress raisers in the weld joint [10-12]. The high magnification fracture images show broad population of microscopic voids of differing shape and size which resulted in local ductile mechanisms. These typical fracture surfaces are the evidence of existence of ductile failure with sufficient necking before failure.

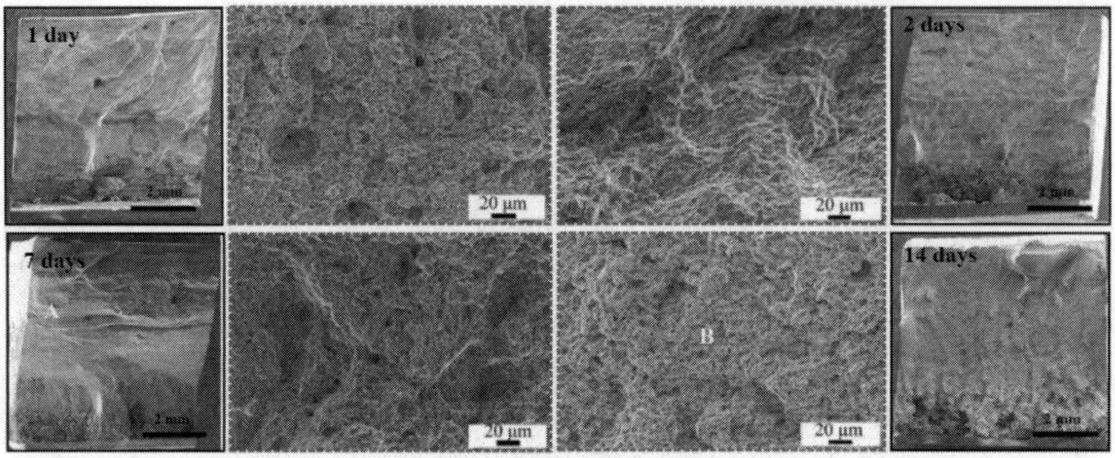

(a)

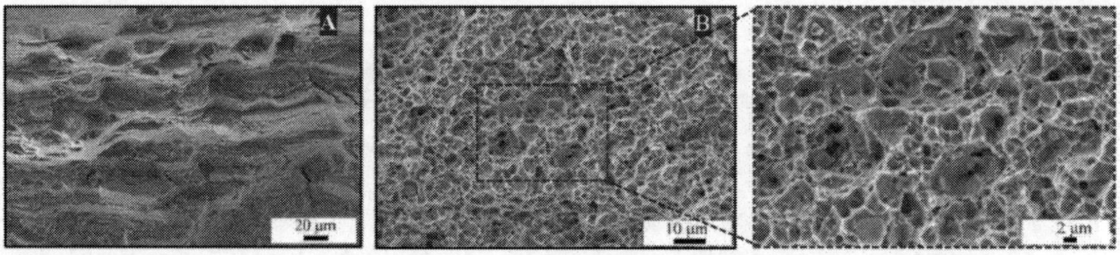

(b)

Fig. 6. FE-SEM micrographs of fractured surfaces of (a) corroded dissimilar AA7075-AA2014 friction stir weld for four different immersion periods (b) fracture morphology of highlighted regions at higher magnifications

Figure 5(a) represents the SEM fractographs obtained for tensile fractured surface after the progressive corrosion of the dissimilar welds of AA6061-AA2014. The location of fracture is on the advancing side (AA6061-T6) as for AA6061-AA7075 and thus similar fracture mechanisms have taken place. Although there is compositional similarity and thus failure morphology, mechanism might be similar but the difference can be noted in the existence and extent of pits and EFC (refer Fig. 5 (b)) that possibly is the crack initiation site. Thus, multiple fracture initiation took place from all the corners which is due the pits formed during corrosion. The representative pit depth is measured around 480 µm, 644 µm, 747 µm in three fractured surfaces obtained after the immersion time of 1, 7 and 14 days respectively. With immersion time, depth of pits increased at certain locations and subsequently, loss in strength recorded. These fractured sample also reveals that the pitting is more progressive from the sides and less from the top or bottom of the joint cross section.

Figure 6(a) shows extremely dense and uniformly distributed dimples with very narrow intermittent zone. All the fractures occurred near the SZ for all AA7075-AA2014 dissimilar welds. The fracture location is fine grain dominated region and thus very dense dimple region can be expected. The density of dimples can be compared as all are under same magnifications. Region 'A' spotted in the fracture surface shows stepwise failure and this region is common to all the fractured surfaces in Fig. 6 (a), signifying shear failure. Although a very different fracture morphology at the bottom of every joint can be observed due to the presence of small voids at the bottom due to insufficient joining. Figure 6(b) helps in observing the fracture morphology with dimples of low depth and separated by bright, narrow ridges at some intermittent locations.

4 Conclusions

- The maximum UTS and YS was achieved for AA7075-AA2014 as compared to AA6061-AA2014 and AA6061-AA7075 dissimilar welds without any corrosion treatment whereas minimum % elongation is noted for AA7075-AA2014 joints compared to other two cases.
- The corrosion takes place all over the joint but the much accelerated rate of corrosion exists away from stir zone near the confluence of HAZ and base metal. From the welds it is inferred that exfoliation corrosion sight is independent of the side in which alloy is placed.
- In AA6061-AA2014 weld combination the excessive pitting and exfoliation corrosion is in the advancing side (AA6061) and its extent increases with increase in immersion duration. Also the extent of pitting corrosion increased for AA7075 irrespective of the side in which it is placed.
- The corroded and tensile tested dissimilar weld of AA6061-AA2014 shows the shifting of fracture

https://doi.org/10.1051/matecconf/202032604008

location after getting exposed to the corrosive environment for the least time period whereas it remains unchanged for the other two welds.

Declaration

Authors declare no potential conflict of interest.

References

1. P.B. Srinivasan, W. Dietzel, R. Zettler, J.F. Dos Santos, V. Sivan, Mater. Sci. Eng. A **392**, 292 (2005)
2. N. Birbilis, M.K. Cavanaugh, R.G. Buchheit, Corros Sci. **48,** 4205 (2006)
3. T. Marlaud, B. Malki, C. Henon, A. Deschamps, B. Baroux, Corros. Sci. 53, 3139 (2011)
4. M.L. de Bonfils-Lahovary, L. Laffont, C. Blanc, Corros Sci. **119**, 60 (2017)
5. J. Yan, N.M. Heckman, L. Velasco, A.M. Hodge, Sci Rep. **6**, 1 (2016)
6. J. Zuo, L. Hou, L. Shi, H. Cui, L. Zhuang, J. Zhang, J Alloy Compd. **716**, 220 (2017)
7. S. Sinhmar, D.K. Dwivedi, Mater. Sci. Eng. A **684**, 413 (2017)
8. M.A. Dominguez-Crespo, A.M. Torres-Huerta, S.E. Rodil, E. Ramírez-Meneses, G.G. Suarez-Velazquez, M.A. Hernández-Pérez, Electrochim. Acta. **55**, 598 (2009)
9. F.S. da Silva, N. Cinca, S. Dosta, I.G. Cano, M. Couto, J.M. Guilemany, A.V. Benedetti, Corros. Sci. **136**, 231 (2018)
10. S. Lim, S. Kim, C.G. Lee, S. Kim, Metall. Mater. Trans. A **35**, 2829 (2004)
11. A. Heidarzadeh, H. Khodaverdizadeh, A. Mahmoudi, A.E. Nazari, Mater. Des. **37**, 166 (2012)
12. A.H. Feng, D.L. Chen, Z.Y. Ma, Metall. Mater. Trans. A **41,** 2626 (2010)

https://doi.org/10.1051/matecconf/202032604009

Assessment of hydrogen embrittlement susceptibility of an Al-Cu-Mg alloy in humid air

Toshiaki Manaka[1,*], *Takahiro* Suzuki[2], *Keisuke* Hiyama[3], *Junya* Kobayashi[4], *Shigeru* Kuromoto[4], and *Goroh* Itoh[4]

[1]Department of Environmental Materials Engineering, National Institute of Technology (KOSEN), Niihama College, 7-1 Yagumo-cho, Niihama, Ehime 792-8580 Japan
[2]Undergraduate Student, Department of Mechanical System Engineering, College of Engineering, Ibaraki University, 4-12-1, Nakanarusawa-Cho, Hitachi, Ibaraki 316-8511 Japan
[3]Graduate Student, Major in Mechanical Systems Engineering, Graduate School of Science and Engineering, Ibaraki University, 4-12-1, Nakanarusawa-Cho, Hitachi, Ibaraki 316-8511 Japan
[4]Department of Mechanical Systems Engineering, College of Engineering, Ibaraki University, 4-12-1, Nakanarusawa-Cho, Hitachi, Ibaraki 316-8511 Japan

Abstract. In the present study, we investigated the hydrogen embrittlement susceptibility of Al-4%Cu-1.5%Mg alloys subjected to several heat treatments by means of SSRT tensile test and humid gas stress corrosion cracking(HG-SCC) test. For SSRT tensile test, the tensile test pieces were cut from cold-rolled sheets of 1mm thickness. The test pieces were solution-treated at 500 °C for 1h, water-quenched and aged at 140oC for 72h or 360h. SSRT tensile test was performed in two environments, humid air (HA) and dry nitrogen gas (DNG) at a strain rate of $1.39\times10^{-6}s^{-1}$. Fracture surfaces were observed with a scanning electron microscopy(SEM). For HG-SCC test, compact tension(CT) test pieces were cut from hot-rolled plate of 6mm thickness. The CT test pieces were solution-treated, water-quenched and aged at 190 °C for 9h, 50°C for 96h or 140°C for 72h. HG-SCC test was carried out based on High Pressure Institute of Japan standards; HPIS E103:2018. The pre-cracked CT specimens with stress loading were kept for 90 days in two environments, HA and DNG. After 90 days, in order to observe whether cracks propagated due to HG-SCC, the specimens were loaded up to fracture rapidly, followed by SEM observation. Tensile properties obtained by SSRT tensile tests were almost the same in two environments. Also, fracture surfaces were not affected by test environments. Moreover, in HG-SCC tests, crack propagation was not observed at each test conditions. Therefore, Al-4%Cu-1.5%Mg alloy had high-resistance to hydrogen embrittlement.

1 Introduction

Al-Zn-Mg and Al-Cu-Mg alloys have been used as structural components due to their high strength. It is well known that these alloys are susceptible to stress corrosion cracking (SCC), which is based on intergranular cracking. The crack propagation mechanism of SCC depends on the alloy system. In Al-Zn-Mg alloys, hydrogen embrittlement dominates the crack propagation. In Al-Cu-Mg alloys, on the other hand, the crack propagates by anodic dissolution along grain boundaries [1]. It has been reported that the peak-aged Al-Zn-Mg ternary alloy showed degradation of ductility when a slow strain rate technique (SSRT) test was performed in humid air [2], which was attributed to hydrogen embrittlement [2-5]. On the other hand, the peak-aged Al-Cu-Mg ternary alloy did not show such a degradation in humid air [2].

The reason why the Al-Cu-Mg alloy shows resistance to hydrogen embrittlement is still unclear. We need to reveal this in order to suppress hydrogen embrittlement in high strength aluminium alloys. We have investigated the behaviour of hydrogen in Al-Cu-Mg alloys using hydrogen microprint technique and SSRT tensile test [6,7].

The hydrogen embrittlement susceptibility of Al-Zn-Mg alloys depends on heat treatment [8,9]. As far, effect of heat treatment on hydrogen embrittlement of Al-Cu-Mg alloys has not been reported. In the present study, we investigated the hydrogen embrittlement susceptibility of Al-4%Cu-1.5%Mg alloys subjected to several heat treatments by means of SSRT tensile test and humid gas stress corrosion cracking(HG-SCC) test.

2 Experimental procedure

2.1 Specimen

The material used in this study was an Al-Cu-Mg ternary alloy. The chemical composition is shown in Table 1. In order to investigate the origin of high-resistance to hydrogen embrittlement in this alloy system, the amount of impurity elements such as iron and silicon, and manganese that is recrystallization inhibitor element have been minimized. Optical microscopic image after

* Corresponding author: manaka@mat.niihama-nct.ac.jp

https://doi.org/10.1051/matecconf/202032604009

heat treatment is shown in **Fig.1**. Recrystallized grains were observed and some of second phase particles remained.

Table 1 Chemical composition of the specimen in mass%.

Si	Fe	Cu	Mn	Mg	Cr	Ti	Al
<0.01	0.01	4.0	<0.01	1.5	<0.01	0.02	Bal.

Fig.1 Optical microscopic image of the Al-4%Cu-1.5%Mg alloy after solution treatment.

2.2 SSRT tensile test

Sheet-type tensile test pieces with a gauge length of 12mm along rolling direction, width of 4mm and thickness of 1mm were cut from cold-rolled sheets. The test pieces were solution-treated at 500 °C for 1h and then water-quenched. The solution-treated specimens were aged at 140°C for 72h or 360h. The specimens were polished with abrasive papers to #2000 before SSRT tensile tests.

SSRT tensile tests were conducted at an initial strain rate of 1.39×10^{-6} s^{-1} at 25 °C in two environments: humid air (HA) with relative humidity (RH) of 90 % and dry nitrogen gas (DNG) atmosphere at ambient pressure with RH of 5% or less. When tensile testing is carried out in humid air, hydrogen uptake occurs resulting from the breakdown of the surface oxide and the reaction between the bare aluminum surface and water vapor. After tensile tests, fracture surfaces of the specimens were examined with a scanning electron microscopy (SEM).

2.3 HG-SCC test

HG-SCC test has been proposed to evaluate hydrogen embrittlement of aluminium alloys in humid air [10, 11]. In this method, hydrogen embrittlement susceptibility is judged by observing crack propagation of the pre-cracked specimens with stress loading in humid air.

From the hot-rolled plate, compact tension test pieces with a thickness of 6mm were cut by electric discharge machining. They were solution-treated at 500 °C for 1h, water-quenched and then aged at 190°C for 9h, 140°C for 72h or 50°C for 96h. After heat treatment, specimen surfaces were mirror-finished by buffing.

Pre-crack was introduced by fatigue machine and a constant stress intensity factor K_{IApp}, equivalent to 0.2% proof stress at a crack tip of 1mm length, was applied by a constant displacement method. Then, the test pieces were kept in two environments: HA and dry air with relative humidity of 10% or less. After holding the specimens for 30d, crack propagation on the surface was observed with an optical microscope, and then test pieces were subjected to fatigue loading in order to introduce post-crack with length longer than 1 mm, followed by rapid loading up to fracture. Fracture surfaces of the specimens were examined with SEM, and the region between the two fatigue crack surfaces regarded as HG-SCC surface.

3 Results and discussion

Figure 2 shows stress-strain curves of the specimens at a strain rate of 1.39×10^{-6} s^{-1}. Degradation of tensile properties was not observed in both aging conditions. Figure 3 shows SEM images of fracture surfaces of the specimens aged at 140 °C for 72h. In both test environments, fracture surfaces were covered with numerous dimples, which indicates the fracture resulted from ductile manner. In the specimen aged at 140 °C for 360h, intergranular fracture with fine dimples and ledges was observed as shown in Fig.4.

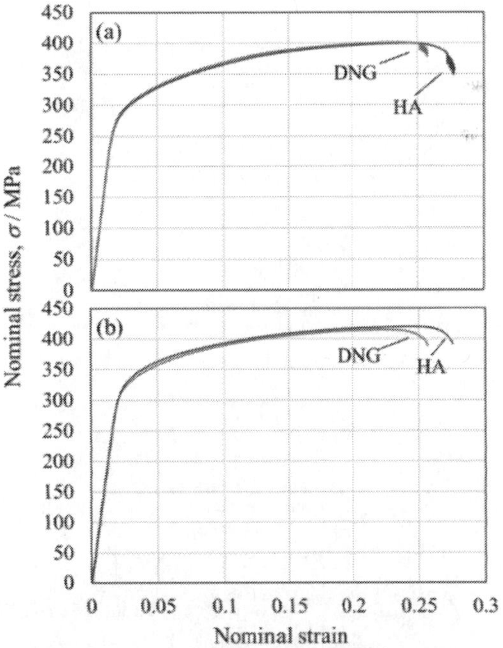

Fig.2 Stress-strain curves of the Al-4%Cu-1.5%Mg alloy tested in humid air (HA) and dry nitrogen gas (DNG) at a strain rate of 1.39×10^{-6} s^{-1}. The specimens were aged at 140°C for (a) 72h and (b) 360h, respectively.

https://doi.org/10.1051/matecconf/202032604009

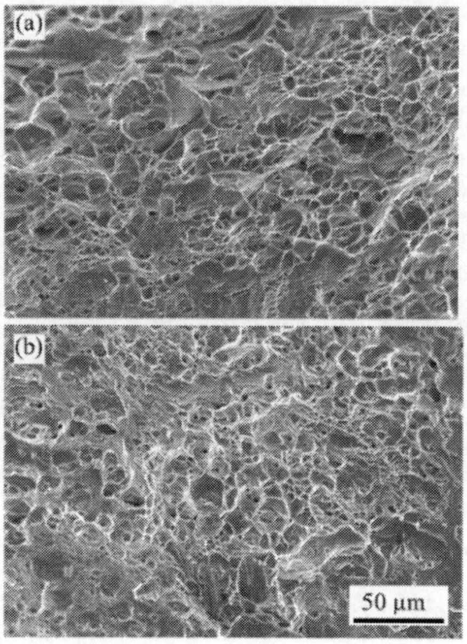

Fig.3 SEM images showing fracture surfaces of the Al-4%Cu-1.5% Mg alloy aged at 140 °C for 72h after SSRT tensile test in (a) HA and (b) DNG, respectively.

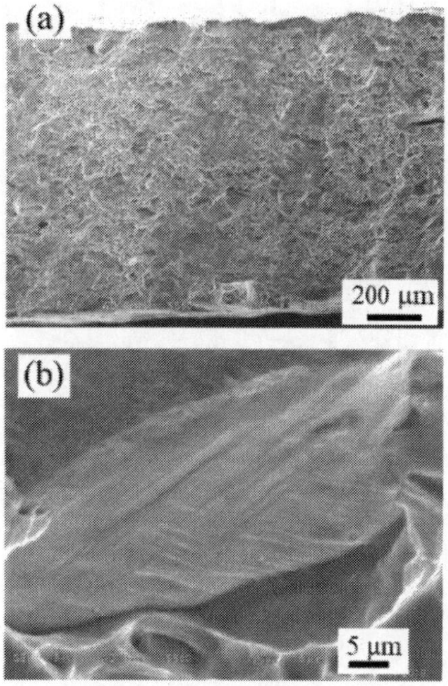

Fig.4 SEM images showing fracture surfaces of the Al-4%Cu-1.5% Mg alloy aged at 140 °C for 360h after SSRT tensile test in HA. (a): low magnification, (b)enlargement of intergranular fracture surface.

Intergranular fracture surfaces with ledges were observed in the specimens tested in DNG as well as in HA. When fine precipitates on grain boundaries and narrow precipitate free zones are present, large stress concentration at the matrix slip band termination at the grain boundaries causes intergranular fracture with ledges [12]. Therefore, the results of SSRT tensile test and fracture surface observation indicated that the Al-Cu-Mg alloy aged at 140 °C had high-resistance to hydrogen embrittlement.

Figure 5 shows SEM images of fracture surfaces of the specimen aged at 190°C for 9h after HG-SCC test in HA. Crack propagation was not observed on any thickness position. Although the SEM images are not shown here, the fracture surface observation confirmed the crack did not propagate in other test conditions. Thus, the Al-Cu-Mg alloy had high-resistance to hydrogen embrittlement examined by HG-SCC test as well as SSRT tensile test.

The origin of high-resistance to hydrogen embrittlement of Al-Cu-Mg alloy has not been realized in the present study. To reveal this, hydrogen behavior in the Al-Cu-Mg alloy is need to be investigated by means of thermal desorption analysis, which is a method to measure hydrogen content and trapping site.

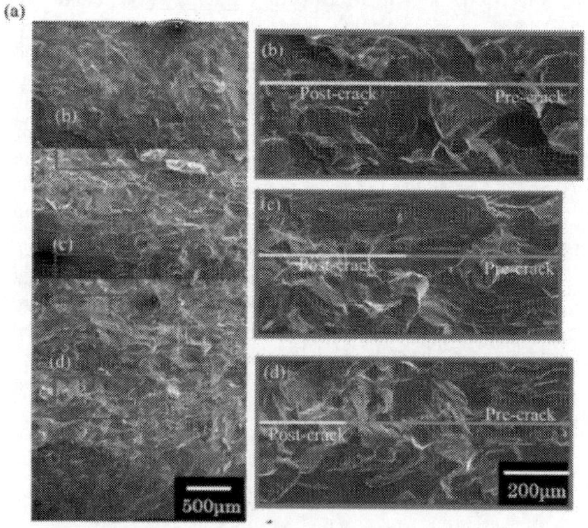

Fig.5 SEM images showing fracture surfaces of the Al-4%Cu-1.5% Mg alloy aged at 190 °C for 9h after HG-SCC test in HA. (a): low magnification image near boundary between pre-crack and post-crack regions, (b)~(d): enlargement of squares in (a).

4 Conclusions

In the present study, hydrogen embrittlement susceptibility of Al-4%Cu-1.5%Mg alloy was examined by means of SSRT tensile test and HG-SCC test. In SSRT tensile test, degradation of tensile properties was not observed in any aging conditions and fracture surfaces were not affected by test environments. Also, in

https://doi.org/10.1051/matecconf/202032604009

HG-SCC test, the crack propagation was not observed in any test conditions. Therefore, the Al-Cu-Mg alloy had high-resistance to hydrogen embrittlement.

This work was financially supported by JSPS KAKENHI Grant Number 18K14018 and the Light Metal Educational Foundation.

References

1. T. D. Burleigh, Corros. Sci. **47**, 89-98 (1991)
2. K. Ichitani, K. Koyama, Furukawa-Sky Review, 20-28 (2009)
3. S. Osaki, J. Ikeda, K. Kinoshita, Y. Sasaki, J.JILM **56**, 721-727 (2006)
4. T. Manaka, G. Itoh, J.JILM **67**, 67-71 (2017)
5. T. Manaka, M.Todai, M. Wada, J.JILM **68**, 615-620 (2018)
6. T. Manaka, G. Itoh, *Proceedings of International Hydrogen Conference (IHC 2016):Materials Performance in Hydrogen Environments* (ASME Press, 2017)
7. T. Manaka, M. Yoshida, *Proceedings of 14th INALCO2019*, 57-58 (2019)
8. G. A. Young, J. R. Scully, Metall. Mater. Trans. A **33,** 1167-1181 (2002)
9. S. Osaki, S. Haruyama, T. Koga, K. Kaminishi, J. JILM **63**, 57-64 (2013)
10. G. Itoh, Zairyo-to-Kankyo **65**, 432-437 (2016)
11. G. Itoh, T. Ogawa, A. Kurumada, S. Kuramoto, J. Kobayashi, S. Asami, *Proceedings of 14th INALCO2019*, 75-76 (2019)
12. S. Kuramoto, G. Itoh, M. Kanno, Metall. Mater. Trans. A **27,** 3081-3088 (1996)

Study of Recrystallization Kinetics of 1565ch Aluminum Alloy during hot deformation

Vasiliy Yashin[1,2], *Alexander* Drits[2], *Evgenii* Aryshenskii[1*], *Ilya* Latushkin[1,2] and *Ekaterina* Chitnaeva[1,2]

[1]Samara National Research University, 443086 Samara, Russia
[2]JSC "Arconic SMZ", 443051 Samara, Russia

Study is funded by a grant of the Russian Science Foundation, project 18-79-10099.

Abstract. The present study addresses recrystallization process in Al-Mg-Mn-Zn-Zr system alloy samples. The samples are collected from cast ingot, produced by casting to industrial DC mold, and homogenized based on standard industrial practice. After that the samples were rolled with different hot rolling schedules. Rolled samples were annealed at different temperatures and their resultant microstructure was examined using optical microscope. During the study new grains nuclei generation rate and their growth rate were determined, analytical records, describing recrystallization kinetics, were obtained, main differences, specific to this alloy recrystallization in 350 °C-450 °C temperature range, were identified.

1 Introduction

Multiple studies address recrystallization process [1-9], processes, occurring during cold worked metal annealing [3], as well as processes occurring during hot deformation of preliminary worked over structure [10-13] are well studied and described in detail, however, only few studies deal with recrystallization during early stages of hot deformation [1,14,15]. They use diverging data, producing invalid results, if applied in calculations of hot rolling initial passes. The causes stem from the following features specific to hot rolling initial passes: low strain level (ranging from 0.04 to 0.35, whereas researches mainly focus on the strain level, exceeding 0.4), low strain rate (ranging from 1 to 10 s^{-1}), high temperature (400-500 °C) and cast structure, characterized by large interdendritic distance (about 200 mcm).

However, structure evolution process study is invalid without understanding of structural modifications, occurring during the very early deformation phase. There is a trend to slab thickness increase, hence deformation non-uniformity will build up during multi-pass rolling, initial 5÷10 passes may involve full ingot surface recrystallization and completely intact as-cast structure in the ingot midsection. In this case the produced workpiece will be characterized not only by high mechanical properties anisotropy, it will demonstrate high scrap potential, eg., laminations and cracks, this is specifically true for hard alloys. It is difficult to predict the structure during hot rolling with high deformation centre, as the following multiple factors shall be accounted for: workpiece heating during plastic deformation and cooling down due to roll contact, strain level and rate heterogeneity, with max values in near-contact zone and min values in the slab axis, inter-deformation interval time varying from pass to pass and along the specimen length. The above necessitates recrystallization processes describing as mathematical model, which can be applied as a function for rolling process computation using FEM mode.

Recrystallization of magnesium rich aluminum alloy with transition elements addition presents special interest. Thus, zirconium, added to aluminum alloys for as-cast structure refinement [12 – 17], as well as for product strength improvement, is widely applied. Strengthening is achieved due to two factors: dispersion strengthening and structural strengthening. Heating temperature and hot deformation schedules control enable managing precipitation of fine particles (sized about 10 nm), coherent with aluminum matrix [18, 19], from the solid solution. Fine particles number control, combined with Mg2Al and Al6(MnFe) type coarse intermetallic particles, occurring in this alloy group and functioning as new grains nuclei [20, 21], can be the key to structure refinement during recrystallization. Such concept was efficiently deployed to design a new alloy 1565ch with average amount of transition elements [22]. 1565ch alloy belongs to weldable alloys of Al-Mg system and can be described as versatile material, demonstrating optimal combination of strength, plasticity and special features, dictated by its application field. Having plasticity characteristics, comparable with 1550 and 5083 alloys, this alloy is 20-30% superior in strength. 1565ch alloy is recommended for manufacturing of freight cars and tankers, carrying bulk or hazardous products, as well as articles and machinery, operating at cryogenic temperatures (down to minus 196 °C). Compared to conventional alloys, the innovative

* Corresponding author: ar-evgenii@ya.ru

1565ch alloy has the optimal combination of mechanical properties, demonstrated both in low and high temperature conditions. Due to its improved plasticity, as compared to 1560 and 1561 alloys, it is more appropriate for body structures fabrication.
This study addresses recrystallization kinetics in 1565ch samples, having initial as-cast structure, and processed with the above range of temperature-rate parameters.

2 Research methodology

10x40x300 samples, collected from 1565ch (for chemical composition refer to Table 1) DC cast, homogenized slab (8 h homogenization with 485-500 °C metal temperature) were used in the experiment, the samples were rolled in the lab mill, following the practices, specified in the Table 2. After rolling completion the samples were immediately transferred to the water tank for structure freezing. Max sample transfer time was 2 s.

Table 1. Test slab chemical composition.

Chemical element	Composition, % (weight)							
	Al	Cu	Mg	Mn	Fe	Si	Zn	Zr
1565ch	Base	0.08	5.75	0.77	0.17	0.08	0.67	0.11

Table 2. Experiment variables

Practice No.	ε, Ln(h0/hk)	$\dot{\varepsilon}$, s^{-1}	T_{def}, °C	Z
1	0.34	3	350	6.63E+14
2	0.35	8	350	1.77E+15
3	0.34	3	450	6.88E+12
4	0.34	8	450	1.84E+13
5	0.36	3	350	6.63E+14
6	0.56	8	350	1.77E+15
7	0.36	3	400	5.70E+13
8	0.54	8	400	1.52E+14
9	0.36	3	450	6.88E+12
10	0.54	8	450	1.84E+13
11	0.51	3	500	1.09E+12

Rolled strips were cut into 5x10x10 samples, which were annealed at 350, 400, 450 °C temperature settings with different soaking durations, for heating time reduction annealing was performed in saltpeter, heating rate was 20 °C/s.

The obtained samples structure was examined using optical metallography methods with 50X magnification. The samples were prepared by electrolytic polishing in fluoroboric electrolyte. Grain size was evaluated using secant method as per GOST 21073.2.

Scanning electron microscope FEI InspectS with back-scatter electron detector (BSED) having 10 nm image resolution and integrated energy dispersive X-ray micro-analyser EDAX, enabling identifying elements from B to U, was used for particles analysis in 10 fields with 10°000X magnification. ImageScope software complex was applied for images processing and phases counting.

General post-deformation recrystallization kinetics is described by Avrami equation

$$Xrex=1-exp[-0{,}693(t/t_{0,5})^n] \quad (1)$$

where $t_{0,5}$ is the time, required to achieve 50% recrystallization; is current post-deformation soaking time; n is the index with the values ranging from 1 to 2, depending on the material.

For single-phase materials time $t_{0,5}$, required to recrystallize 50% of the deformed material volume, is determined by the equation

$$t_{0,5}=Ad_0{}^{r}\,\varepsilon^{-p}\,Z^{q}\,exp(Q_{rex}/RT) \quad (2)$$

Here A, p, q, r are material constants; d_0 is initial grain size; ε is strain value; Q_{rex} is recrystallization activation energy; T is absolute isothermal time; R is gas constant; Z is Zener-Hollomon parameter, dependent on the strain rate ($\dot{\varepsilon}$), temperature (T_{def}) and activation energy of the process, managing strain rate (Q_{def}):

$$Z=\dot{\varepsilon}exp(Q_{def}/RT_d) \quad (3)$$

The situation is more complicated for the alloys with phases inclusions. Nano-size dispersoids efficiently retard grain boundaries mobility, hence recrystallization processes, while coarse inclusions (over 1 mcm) can serve as new recrystallized grains nucleation centres, thus accelerating recrystallization kinetics.

The most rigorous method of recrystallization kinetics study is to study microstructure of the quenched deformed samples with various post-deformation soaking periods. The example of such method application is the study [13], addressing hot rolled alloys AA5182 (Al–4,5%Mg–0,31%Mn–0,21%Fe) and AA5052 (Al–2,5%Mg–0,07%Mn–0,17%Cr–0,27%Fe) with fully recrystallized initial structure. It demonstrated, that alloy AA5182 static recrystallization was much faster, than in AA5052 alloy. Recrystallization kinetics difference is explained by higher magnesium content (higher level of strain hardening) and smaller initial grains in AA5182 alloy, as well as presence of fine chrome intermetallic particles in AA5052 alloy.

To account for dispersoids contribution in 1565ch alloy recrystallization kinetics, recrystallization process was divided into two sub-processes, i.e., new grains nucleation and grains growth:

- specific rate of new grains nucleation (per 1 mcm2 area)

$$\dot{N}=\Delta N/(\Delta\tau\cdot S) \quad (4)$$

- grain growth rate $\dot{D}=\Delta D/\Delta\tau$ (5)

New grains nucleation rate and grains growth rate were determined by metallographic examination of the structure at various soaking time settings, assuming these values are time independent. Grain increment or grain size change were determined for S=5.72 mm^2 area with growth rate computation, target values were obtained by achieved values division by soaking duration.

Equation for 1565ch alloy can be presented in the following form:

$X(\tau)=1-exp(-K(\tau)^n)$

where $K=\pi\,\dot{N}\dot{D}^3$ and $n=4$;

$\dot{N}=7,1\cdot10^{24}\cdot\varepsilon^{0.26}\,Z^{0.1}\,exp(-165059/RT_{rec})$

$$\dot{D}=\mu\cdot(P_D-P_Z) \quad (6)$$

where μ is boundary mobility, dependent on its atomic structure;

P_d- migration driving force;

P_z – retarding force.

Temperature dependence of the boundary mobility μ is described by the following equation:

$$\mu=1,9\cdot10^{-7}exp(-85068/RT_{rec}) \quad (7)$$

Primary recrystallization driving force P_D is the decrease of excess volumetric energy, accumulated during plastic deformation. Primarily, it is defined by the difference in dislocation density ΔN_d before and after deformation:

$$P_D\approx k\Delta N_d\,GB^2 \quad (8)$$

where k is the factor, considering the nature of dislocations distribution and ranging from 0.1 to 1; G is shear module, equal to $2.5\cdot10^{10}$ N/m^2; B is Burgers vector 0.286 nm, ΔN_d is a function of strain degree, determined by the following formula [23]:

$$dN_d/d\varepsilon=k_1(N_d+k_p(f/r)^2)^{0.5}-k_2N_d \quad (9)$$

where f and r are proeutectoid constituent particles volume fraction and size;
k_p is the coefficient, responsible for accumulation of dislocations on proeutectoid constituent particles.
Coefficients k_1, k_2 and k_p values were computed by numerical solution of differential equation (9):

$k_p=2\cdot10^5$;

$k_1=(0.95+0.44C^{0.5})\;10^8$, $м^{-1}$;

$k_2=7.2-0.75C^{0.5}$.

For integration of evolution equation (9) initial dislocation density value $N_{d0}=10^{10}$ m^{-2} was applied (such density value is typical for thermally treated castings and alloys, recrystallized after deformation).

Presence of dispersion phases affects boundaries migration retardation. Boundaries migration by dispersion phases is caused by boundaries and overall system energy decrease as a result of dispersion phases location at the boundary.

In case specific number of N_d dispersoids is present in the structure (2÷4 atomic percent [14]) the force, retarding boundaries mobility, will be computed as follows:

$$P_Z\approx3f\gamma_{ep}/d \quad (7)$$

where f - is particles volume fraction, equal to $f=4\pi\cdot r^3{}_0\cdot N_d/3$;
N_d - is particles count per 1m^3 and size r_0, calculated using the data from Table 3.
γ_b - is inter-grain boundary energy ≈0.5 J·m^{-2} [14].

Zener was the first to present this relationship. It means, that the larger particles volume fraction is, the smaller the particles are, and the higher γ_b is, the stronger particles dragging effect is. It is clear, that boundary migration is possible only, if $P_D>P_Z$.

3 Research results and discussion

Prior to testing, the samples were examined using scanning microscope to determine the intermetallic particles number, size and chemical composition; grain microstructure was examined using optical microscope. Figure 1a illustrates cellular-dendritic pattern of the alloy crystallization, microstructure has distinctive dendritic branches. Average grain size is ~ 200 mcm. Additions of 0.67 weight % Zn and 0.11 weight % Zr do not result in significant change of workpiece as-cast structure as compared to similar alloys of Al-Mg system. Primary intermetallic particles of crystallization origin with the sizes, ranging from 1 mcm to 31 mcm can be observed in the ingots. Energy dispersive X-ray micro-analysis identified Al3Mg2, Al6(Fe, Mn) intermetallic particles and Mg2Si magnesium silicide, located along grain boundaries. No zirconium or zinc containing additional phases were identified in 1565ch alloy during micro-analysis, thus it can be assumed, that these elements are either totally dissolved in aluminum matrix or precipitated as dispersoids, sized below 1 µm.

Thus, 1565ch alloy as-cast structure presents solid solution of alloying elements, having limited solubility in aluminum matrix. Below the specific temperature the solution becomes supersaturated and decomposes with the second phase dispersion particles precipitation. Precipitates, occurring in the alloy (see Figure. 1, Table 3) change density and distribution pattern of the dislocations, generated during deformation, thus producing significant effect on the future recrystallization, decreasing its initiation temperature, as a rule [14]. In general, particles function as barriers for the dislocations, migrating in the matrix during deformation. It leads to local increase of dislocations density in matrix micro-volumes, adjoining the particles, growing excess of the same sign dislocations, hence increasing crystalline lattice local bending in these areas. During heating these areas become preferential zones of recrystallization nucleation. The larger the particles and the less regular shape they have, the stronger local bending is and the lower strain values are, causing such bending effect [14].

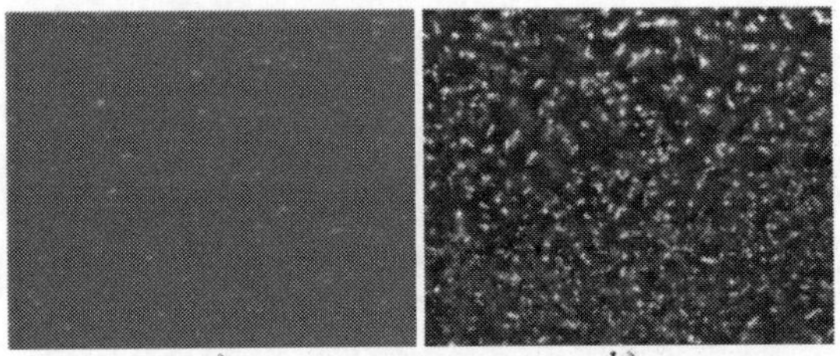

a) b)

Fig. 1. Cast 1565ch alloy sample examination using scanning microscope BSED. a) x250 magnification b) x10 000 magnification

Table 3. Intermetallic particles number in 1565ch sample

1565ch			
Specific fraction of intermetallic particles (f), %		Size (nominal diameter), mcm	
Intermetallic particles	Dispersoids	Intermetallic particles	Dispersoids
0.9341	13	6.24	0.148

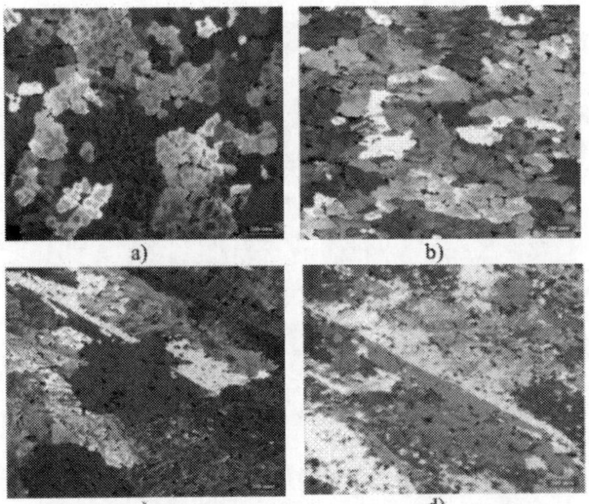

a) b) c) d)

Fig. 2. 1565ch samples examination using optical microscope: a) - cast; b) – deformed ε=0,34 Z=6.88 10^{12}; c) – annealed at 350 °C, 6 min; d) – annealed at 350 °C, 10 min

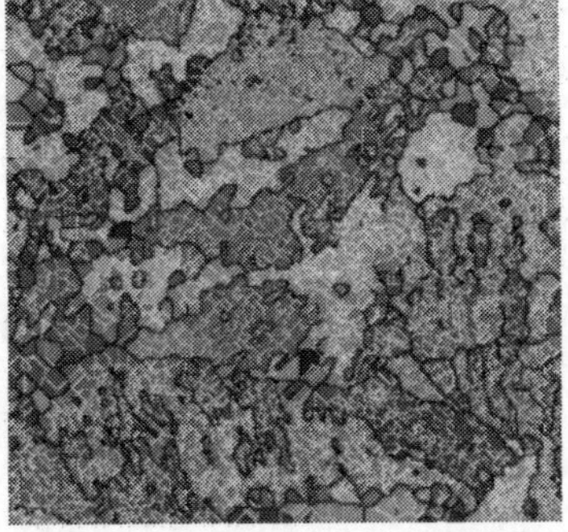

Fig. 3. Orientation analysis (EBSD). Alloy 1565, ε=0.52; Z=5.70E+13

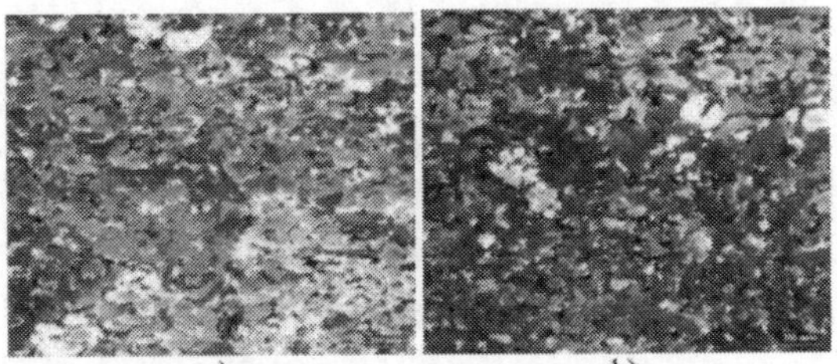

a) b)

Fig. 4. Optical microscope examination of alloy 1565ch, ε=0.52; Z=5.70E+13 samples: a) annealing at 400 °C, 2 min soaking, b) annealing at 400 °C, 10 min soaking.

Dendritic structures are preserved in an ingot at about 0.5 strain degree, crystals are elongated in the rolling direction (see Figure 2b). Examination using scanning microscope with integrated EBSD tool demonstrates grain division into defined subgrains (see Figure 3). Let us analyse grain evolution at annealing temperature of 350 °C, using the sample, treated at 450 °C, with strain degree of 0.34, and 3 s^{-1} strain rate. First recrystallization grains are observed after 6-minute soaking (see Figure 2c). Nuclei number increases after 10-minute soaking (2d).

High rate of new grains nucleation is typical during annealing at 400 °C (see Figure 4), at the same time further grain growth is significantly delayed to the extent of complete process halt, new grains shape is far from ideal hexagon, which is typical for homogenous recrystallization.

High rate of 1565ch alloy nucleation can be explained by PSN mechanism contribution [18]. If the heating continues after primary recrystallization nucleation, further structural modifications are associated with nuclei and grains high-angle boundaries migration. During this stage the second phase particles, represented by closely spaced fine aluminum matrix coherent particles, complicate recrystallization centres and grain boundaries migration. Particles enveloping is associated with local boundary curvature changes and boundary extension, i.e., surface energy buildup. While recrystallization centres still form at relatively low temperature, nuclei growth, as well as secondary recrystallization initiation, tend towards higher temperature. In 1565ch two recrystallization stages – nucleation and nuclei growth – are separated by a significant temperature-time interval. 1565ch alloy structure has multiple dispersion particles (see Figure 1 and Table 3), presenting a barrier for grain boundaries spreading. Upon reaching such barrier, the grain boundary stops, and the grain continues growing in the direction without barrier constraints, therefore new grains have filigree boundary shape, and the structure presents the combination of as-cast dendrites with visible deformed and recrystallized structure zones. The above statement is well illustrated by the plot in Figure 5.

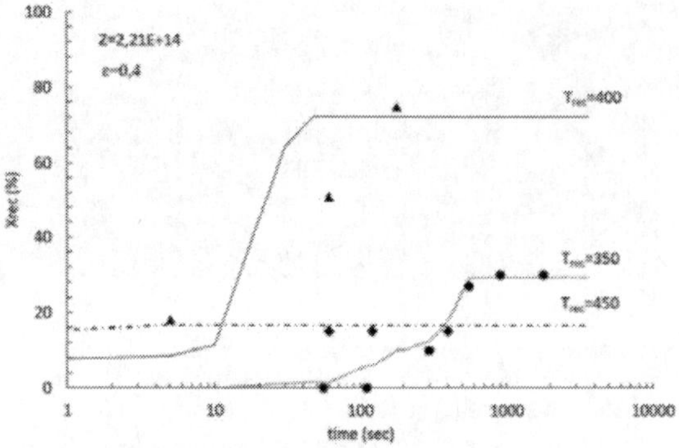

Fig. 5. 1565ch recrystallization kinetics plot
(▲-Trec=400 °C; ●- Trec=350 °C; ♦-Trec=450 °C)

Actual data, established by the structure assessment using optical microscope, are indicated with markers, while lines indicate theoretical calculation, based on the model, presented in the previous section.

4 Conclusions

In the process of research plots were constructed and coefficients were computed for Johnson – Mehl – Avrami – Kolmogorov type analytical equation, describing recrystallization kinetics of 1565ch alloy, rolled from as-cast condition using various practices with the following rolling parameters:

Temperature 300÷500 °C;

Strain level range is 0.14÷0.56;

Strain rates are 3 and 8 s^{-1};

Initial workpiece grain size is 200 mcm.

The obtained data allow the conclusion, that 1565ch alloy recrystallization process with low degree of as-cast structure deformation (0.14÷0.56) is characterized by a number of specific features, such as high rate of new grains nucleation and low rate of grains growth to the extent of such process total blocking. Optimal recrystallization temperature is established – 400 °C, it enables over 75% process progression, incubation period increases in case of temperature decrease (350 °C), and recrystallization halts, producing about 30% estimated volume of recrystallized structure, in case of temperature rise (450 °C) the structure is saturated with small new grains nuclei, probably, formed during heating, but due to high recovery and polygonization intensity, recrystallization driving force is reduced to zero, the process terminates, producing combined structure with max 20% recrystallized grains volume.

References

1. J. Hirsch, Fundamentals of Aluminium Metallurgy, **1** (2011)
2. J. Hirsch, Light Metals warrendale proceedings, **5** (2008)
3. C. Schäfer, V. Mohles, G. Gottstein, Applications of Texture Analysis, **1** (2008)
4. O. Engler, L. Löchte, K. F. Karhausen, Materials Science Forum, **495** (2005)
5. O. Engler, Materials science forum, **550** (2007)
6. O. Engler, H. E. Vatne, JOM. **50,** 6 (1998)
7. J. Hirsch, Materials Science Forum, **702** (2012)
8. M. A. Golovnin, R. F. Iskhakov, III International Scientific School for Young Scientists "Material Science and Physics of Light Metals", (2014)
9. V. K. Orlov, V. G. Drozd, M. A. Sarafanov, Journal Proizvodstvo Prokata. **4** (2016)
10. E. V. Aryshenskii, M. S. Tepterev, I. A. Latushkin Electrical Engineering. Power Industry. Machine Building, **1** (2014)
11. V.V. Yashin, V.Yu. Aryshensky, I.A. Latushkin, M.S. Tepterev, Non-ferrous Metals Journal, **7** (2018)
12. H. Zhang, D. S. Peng, L. B. Yang, L. P. Meng, Trans Nonferrous Met Soc China, **11,** 3 (2001)
13. M. A. Wells, I. V. Samarasekera, J. K. Brimacombe, E. B. Hawbolt, D. J. Lloyd, Metallurgical and Materials Transactions B, **29,** 3 (1998)
14. S.S. Gorelik, S.V. Dobatkin, L.M. Kaputkina Recrystallization of metals and alloys, (2005)
15. B. Koohbor, International journal of material forming, **7,** 1 (2014)
16. A. Nam, V. Yashin, E. Aryshenskii, A. Zinoviev, R. Kawalla, Materials Science Forum. – Trans Tech Publications, **918** (2018)
17. E.V. Aryshenskii, A.F. Grechnikova, V.V. Yashin, Electrical Engineering. Power Industry. Machine Building, **1** (2014)
18. F. J. Humphreys, M. Hatherly, Recrystallization and related annealing phenomena, (2012)
19. O. Dalland, E. Nes, Acta materialia, **44** 4 (1996)
20. H. E. Vatne, E. Nes, O. Daaland, Materials science forum, **157** (1994)
21. H. E. Vatne, E. Nes, Scripta Metallurgica et Materialia, **30,** 3 (1994)
22. M. Avrami, The Journal of chemical physics, **7** 12 (1939)
23. A.Yu. Churyumov Aluminum alloys ultimate yield strength and deformation hardening computation using structure parameters. Extended abstract of Ph.D (Engineering) Dissertation. Moscow. 2008.

Temper rolling to control texture and earing in aluminium alloy AA 5050A

Olaf Engler[1,*], and *Kristian* Knarbakk [2]

[1] Hydro Aluminium Rolled Products GmbH, Research and Development Bonn, P.O. Box 2468, D – 53014 Bonn, Germany
[2] Hydro Aluminium Rolled Products AS, Technical Customer Service Rolled Products Holmestrand, N – 3081 Holmestrand, Norway

Abstract. The development of crystallographic texture during the thermo-mechanical processing of aluminium sheet is known to result in the formation of pronounced plastic anisotropy, including the well-known earing phenomenon. In the present study we track the evolution of texture, microstructure and earing profiles in sheets of Al alloy AA 5050A during down-stream processing according to a process route resulting in temper H16 after cold rolling and temper H44 after lacquering. This process, which includes interannealing followed by a mild final pass of temper rolling, was designed for producing medium-strength sheet with low earing properties. Besides the experimental characterization of the evolution of microstructure, texture and resulting earing profiles along the process chain, means to optimize the processing by adapting the intermediate thickness are addressed.

1 Introduction

Aluminium alloy AA 5050A (ISO AlMg1.5(D)) is a medium strength Al alloy with about 1.5% Mg which is used in household applications, as tubes for automotive gas and oil lines, welded irrigation pipes and in several packaging applications [1]. Like all the AA 5xxx-series Al-Mg alloys, the strength of AA 5050A can be increased by cold working, typically by cold rolling, to the strain-hardened tempers H1x. However, work hardening of Al alloy sheet is commonly accompanied by the formation of significant plastic anisotropy. It is well documented that plastic anisotropy, including the well-known earing phenomenon, is attributed to the formation of preferred crystallographic orientation, or texture, during the thermo-mechanical production of the sheet [2–5]. The characteristic earing profiles of deep drawn cups form because the sheet texture gives rise to different radial elongations in different directions of the blank. In a rolled Al sheet with a pronounced deformation texture the ears are usually observed at the four positions ±45° to the rolling direction around the drawn cup. The height of the ears, and hence the percentage of earing, generally increase with the amount of cold rolling applied. In the soft annealed, recrystallized state ears are often found at angles of 0° and 90° to the former rolling direction.

The occurrence of earing during deep drawing of a textured sheet can cause major problems in the production of Al containers, closures or beverage cans. Hence, for such applications often low anisotropy is required, with the percentage of earing being controlled to be below 5%, and for certain critical applications even below 3%, in order to minimize material loss during trimming and prevent failure of heavily textured blanks.

In industrial practice, earing is reduced by producing mixed textures comprising both rolling and recrystallization texture components. That is to say, the respective 0°/90° and ±45° earing behaviour is superimposed, which results in low anisotropy with balanced earing profiles. This may be achieved by producing sheet in temper H14 or H16 with medium strength and very low earing properties through an intermediate annealing before a light final rolling pass, commonly referred to as temper rolling [6–9].

In the present paper it is demonstrated how texture control can be applied during industrial processing of Al alloy AA 5050A to provide low-earing grade sheet for a technical packaging application. Furthermore, means to improve earing properties through an adjustment of the processing parameters are discussed together with the resulting textural effects.

2 Experimental

The material investigated in this study was Al alloy AA 5050A, containing about 1.2% Mg and 0.1% Mn. Hot strip (or re-roll) of this alloy was produced by standard industrial routes in Hydro's rolling mill in Holmestrand, Norway, consisting of direct-chill (DC) casting of large ingots, pre-heating to 575 °C, followed by break-down and tandem hot rolling (Fig. 1). The hot strip with a thickness of 5 mm was coiled and cooled down to room temperature. Optionally, the coiled hot strip was subjected to a batch annealing treatment for 2 h at 350 °C in order to enforce recrystallization of the – non-recrystallized – hot strip (see below). Subsequently the

* Corresponding author: olaf.engler@hydro.com

material was cold rolled in several passes to a final gauge of 0.4 mm. The medium strength H16 temper with low earing at final gauge was achieved by subjecting the material to an intermediate annealing treatment at a thickness between 0.5 and 0.6 mm. Thus, the temper rolling pass ranged from 30 to 35% thickness reduction. Finally, the final gauge sheet was coated with a thin epoxy lacquer; curing of this lacquer for about 20 s at 285 °C led to a softening to temper condition H44.

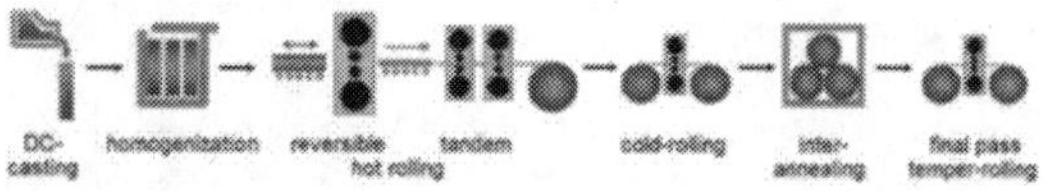

Fig. 1. Scheme of the processing of Al alloy AA 5050A sheet with low earing.

Samples for microstructural analysis and mechanical tests were taken from the hot strip, before and after intermediate annealing, and at final gauge before (H16) and after lacquering (H44). Mechanical properties of the sheets were determined by conventional tensile tests according to EN 10002-1. Cup drawing tests were performed on a hydraulic press following the guidelines of norm EN 1669. For the sheets with thicknesses in the range from 0.4 to 0.6 mm blanks with a diameter of 60 mm were deformed with a punch of 33 mm, resulting in a drawing ratio of ~1.8. From the 5 mm hot strip larger cups were produced, with a punch and blank diameter of 50 mm and 100 mm, respectively. For a quantitative assessment of the earing behaviour, the earing profiles $h(\alpha)$ were probed in steps of 1° with a mechanical set-up developed by Huxley Bertram Engineering Ltd., Cottenham, UK. Here, α indicates the angle with respect to the former sheet rolling direction. The results of two to three cups were averaged to minimize experimental scatter.

Because of the orthotropic sample symmetry of rolled strip, the earing profiles are principally symmetric with respect to the rolling direction (RD) and transverse direction (TD). Hence, the earing profile may be symmetrized ("mirrored") with respect to the RD and TD. For better comparability of different cups (e.g. with different sheet thicknesses or different drawing ratios), the earing profiles $h(\alpha)$ were normalized by the average cup height according to:

$$h^*(\alpha) = h(\alpha) \cdot \frac{\alpha_{max}}{\sum_\alpha h(\alpha)\,\Delta\alpha} \tag{1}$$

For a quantitative assessment of earing often the mean earing value, Z, is determined, which is defined as:

$$Z = \frac{\bar{h}_{ears} - \bar{h}_{troughs}}{\bar{h}_{troughs}} \tag{2}$$

where the values $\bar{h}_{ears}$ and $\bar{h}_{troughs}$ correspond respectively to the average values of all ears and all troughs of a given cup profile. For mixed textures with six or eight ears the delta earing value, ΔZ, may also be useful, which is defined as:

$$\Delta Z = \frac{2\bar{h}_{45^\circ} - (\bar{h}_{0^\circ} + \bar{h}_{90^\circ})}{\bar{h}_{0^\circ} + \bar{h}_{90^\circ}} \tag{3}$$

The microstructure of the different samples was studied by optical metallography. Longitudinal sections of the sheets were prepared and polished according to standard metallographical techniques. The grain structure was revealed by anodical oxidation and subsequent investigation under polarized light.

The crystallographic texture of the samples was analysed by means of electron back-scattered diffraction (EBSD) in a LEO 1530 field emission scanning electron microscope (SEM), equipped with a HKL NORDLYS II EBSD detector from Oxford Instruments plc. For this purpose, longitudinal sections of the sheet samples were prepared by mechanical polishing followed by electro-polishing. Areas of about 800 × 450 µm² were scanned with a step width of typically 1 µm (hot strip 2 or 3 µm). From the EBSD data of the various samples orientation distribution functions (ODF) $f(g)$ were computed by associating each orientation $g = \{\varphi_1, \Phi, \varphi_2\}$ of the orientation map with a Gauss-shaped peak with a (half) scatter width $\psi_0 = 5°$ in Euler angle space. ODFs are presented in form of iso-intensity lines in three characteristic sections through the Euler space, viz. $\varphi_2 = 45°$, 65° and 90°. With this representation we implicitly assume orthotropic sample symmetry, as given by the three orthogonal sample axes rolling direction, transverse direction and sheet normal direction. Further details of EBSD measurements, analysis and interpretation can be found in Ref. [10].

3. Experimental results

3.1 Standard processing route

The evolution of microstructure and texture during the various steps of the standard processing of low-earing grade AA 5050A sheet from hot strip to final gauge in tempers H16 before and H44 after lacquering is presented in Figs. 2 and 3. In parallel with the development of crystallographic texture during the various steps of materials processing the plastic anisotropy of the material changed as well, and Fig. 4 shows the (symmetrized and normalized) earing profiles during the various down-stream processing steps.

Fig. 2(a) shows the microstructure of the as-received AA 5050A hot strip. The material displayed a coarse, heavily layered microstructure where the grains were strongly elongated along the hot rolling direction. Evidently, the microstructure did not recrystallize during the final hot rolling passes. The texture of the hot strip is presented in Fig. 3(a), showing a pronounced texture fibre running from the Cu-orientation {112}<111> at $(\varphi_1,\Phi,\varphi_2)$ = (90°,30°,45°) through the S-orientation {123}<634> at about (57°,33°,65°) to the Bs-orientation {011}<211> at (35°,45°,90°) through orientation space. This is the typical rolling texture of Al sheet, see e.g. [3,8,9]. Furthermore, the ODF comprised weak intensities of the cube-orientation {001}<100> at (0°,0°,90°), pointing at beginning recrystallization during hot rolling or, more likely, during the subsequent cooling of the coiled hot strip to ambient temperature.

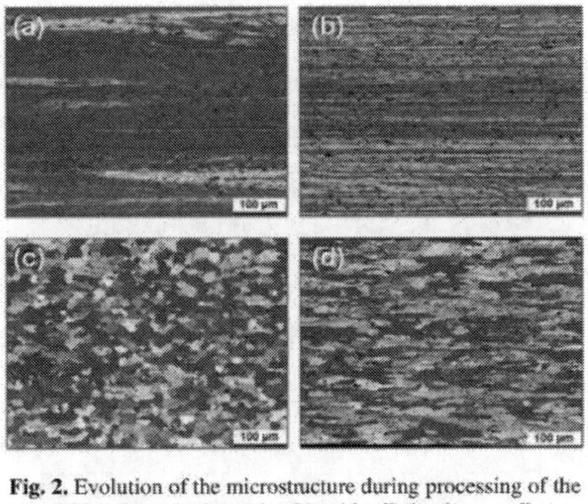

Fig. 2. Evolution of the microstructure during processing of the AA 5050A sheet; (a) hot strip, (b) cold rolled to intermediate gauge, (c) interannealed at intermediate gauge, (d) cold rolled to final gauge, i.e. temper H16 (optical metallography, longitudinal sections, magn. 200×).

From Fig. 4 it appears that the non-recrystallized hot strip had a weak eight-ear profile with ears under 0°, 45°, 90°, etc. (black curve). This earing behaviour is in accord with the observation of a mixed cube and rolling texture shown in Fig. 3(a).

Cold rolling to an intermediate gauge of 0.61 mm led to a further refinement of the elongated band structure (Fig. 2(b)), in that the thickness of the deformed bands in the through-thickness direction was significantly reduced. Simultaneously, the characteristic rolling texture orientations along the β-fibre sharpened, while the remaining orientations diminished (Fig. 3(b)). These texture changes are accompanied by an appreciable increase of the 45° ears at the expense of the original 0°/90° ears (Fig. 4, red curve), resulting in a Z value of as much as 8.5%.

Soft annealing at intermediate gauge gave rise to a recrystallization of the material. Fig. 2(c) shows a fully recrystallized microstructure consisting of grains with an average size of approximately 15 µm, which were slightly elongated along the former rolling direction. The resultant recrystallization texture was rather weak, consisting of the cube-orientation {001}<100> at (0°,0°,90°) and some intensities along the former S-orientation (Fig. 3(c)), which are commonly referred to as R-orientation. Such mixed cube + R textures are the typical recrystallization textures of many cold rolled, soft annealed Al alloys (e.g. [11–14]). Accordingly, the inter-annealed material showed a very smooth eight-ear profile (Fig. 4, green curve); the characteristic earing values were $Z = 1.6\%$ (see Eq. 2) and $\Delta Z = 0.6\%$ (Eq. 3).

The final temper rolling pass to final thickness led to a significant rise of the yield strength to temper H16. This temper rolling pass was again characterized by an elongation of the microstructure (Fig. 2(d)). Note that the aspect ratio of the individual grains was fairly small, about 5:1, which agrees well with the level of rolling reduction of about 35% during the temper rolling pass. Accordingly, the texture showed only minor changes, in

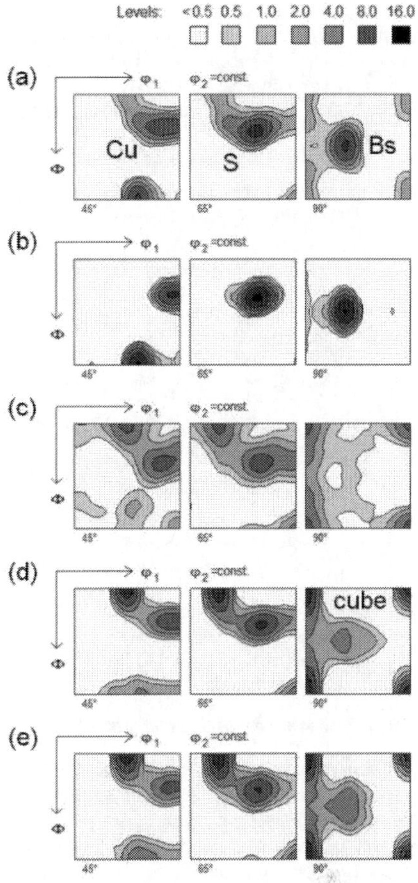

Fig. 3. Evolution of the texture during processing of the AA 5050A sheet; (a) hot strip, (b) cold rolled to intermediate gauge, (c) interannealed, (d) cold rolled to final gauge, i.e. temper H16, (e) lacquered, i.e. temper H44 (EBSD, ODF sections $\varphi_2 = 45°, 65°$ and 90°).

that the former rolling texture orientation Cu, S and Bs slightly re-increased while the cube-orientation of the recrystallization texture remained largely unchanged (Fig. 3(d)). This minor sharpening of the rolling texture orientations gave rise to a slight re-increase of the 45° ears (Fig. 4, blue curve), but the overall earing remained low with the characteristic earing values Z and ΔZ of 2.7% and 2.2%, respectively.

The final lacquering operation, more specifically, the curing of the lacquer at elevated temperature, gave rise to a softening of the material to temper H44. Since recovery at these moderately elevated temperatures did not involve any grain boundary reactions, no changes in the grain structure were detected, although detailed analysis at higher magnification in the SEM or TEM will undoubtedly reveal microstructural reactions, most notably reorganization of dislocation structures into cell or subgrain structures (e.g. [15]). The texture showed a minor sharpening of the cube texture (Fig. 3(e)) with respect to the H16 state (Fig. 3(d)). This minor texture sharpening gave rise to a slight re-increase of the 0° and 90° ears, while the 45° ears remained roughly constant (Fig. 4, cyan curve) which, in turn, led to a decline of the value ΔZ to 0.6%, while the earing value Z remained constant at 2.7%.

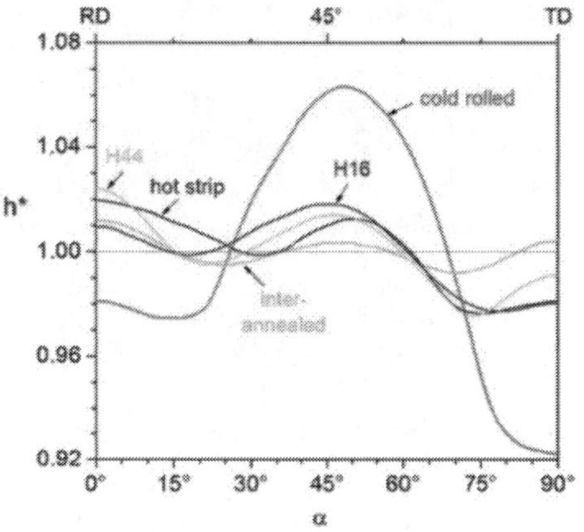

Fig. 4. Evolution of the earing profiles during processing of the AA 5050A sheet (symmetrized and normalized).

3.2 Modified processing to lower earing

The discussion of the results in the previous section has shown that the earing behaviour at final gauge – albeit being rather low – was dominated by ±45° ears. This may be seen from the positive value of ΔZ which, according to Eq. 3, implies that the height of the 45° ears exceeded that of the 0°/90° ears. Apparently, the evolution of the rolling texture orientations during the temper rolling pass could only partially be offset by the intensities of the cube+R recrystallization texture after interannealing. Hence, in order to further improve the final gauge earing behaviour two process modifications were probed, (i) lowered intermediate gauge and (ii) hot strip annealing.

In the *first trial* the intermediate thickness was lowered from 0.61 mm to 0.57 mm, i.e. the rolling reduction during the final temper rolling pass was lowered from 35% to 31%. Under the simplifying assumption that the minor increase in rolling reduction *prior to* interannealing will not affect the recrystallization texture upon interannealing, the slight reduction in temper rolling *after* interannealing should lead to lower ±45° ears and, hence, to a further improvement of the overall earing behaviour.

Fig. 5(a) shows the microstructure of the final gauge material after lacquering (i.e. temper H44), produced with the refined processing route. Obviously, in comparison to the original processing route (Fig. 2), the microstructure looks very similar, since the minor reduction in temper rolling from 35% to 31% is not discernible in the optical micrographs.

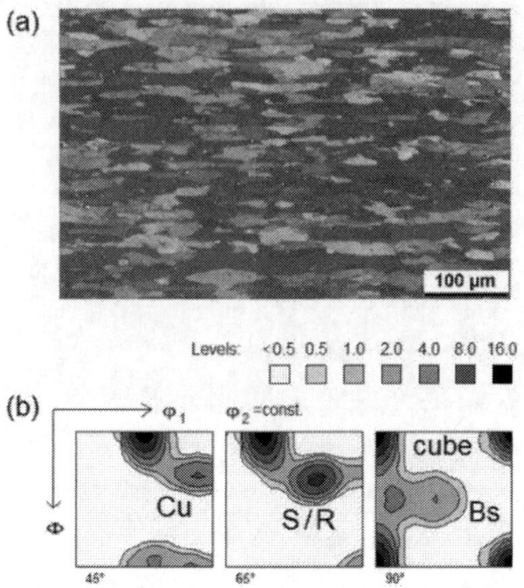

Fig. 5. (a) Microstructure and (b) texture of AA 5050A sheet at final gauge (temper H44) produced with reduced interannealing thickness.

On first sight, the final gauge texture (Fig. 5(b)) resembles the texture of the material with standard processing (Fig. 3(e)), in that both ODFs comprise mixed rolling and recrystallization texture orientations. A detailed comparison by analysing the so-called difference ODF disclosed some significant differences, however (see [16]). More specifically, the standard material revealed slightly stronger orientations of the rolling texture, i.e. Cu, Bs and, to some extent, also the S-orientation. Vice versa, the modified material had higher intensities of the cube orientation and maybe the R-orientation.

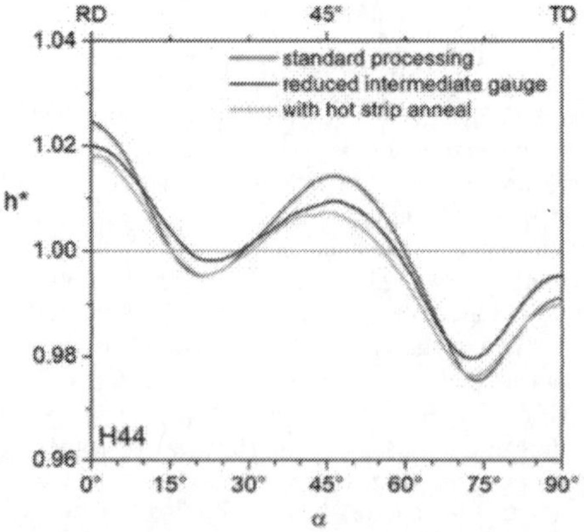

Fig. 6. Comparison of the earing profiles of AA 5050A sheet at final gauge (temper H44) produced with standard and amended processing (symmetrized and normalized).

Thus, analysis of the difference ODF revealed that the material with the modified processing comprised stronger recrystallization texture orientations but weaker rolling texture orientations. Accordingly, the earing behaviour was improved by lowering the earing profile, as apparent from Fig. 6. This diagram shows that the earing profile of the material with the modified processing (blue curve) is in all peaks and valleys closer to unity than the earing profile of the standard material (red curve). This improvement is also reflected in the characteristic earing values Z and ΔZ of 2.3% (rather than 2.7%) and 0.1% (rather than 0.6%), respectively.

In the *second trial* the non-recrystallized hot strip was subjected to an additional batch annealing treatment to enforce recrystallization, which resulted in a coarse-grained recrystallized microstructure (Fig. 7, cf. Fig. 2(a)). The hot strip texture changed from the pronounced β-fibre rolling texture of the non-recrystallized hot strip (Fig. 3(a)) to a typical cube recrystallization texture after hot strip annealing (Fig. 8(a)). Simultaneously, the 0° and 90° earing increased at the expenses of the 45° ears.

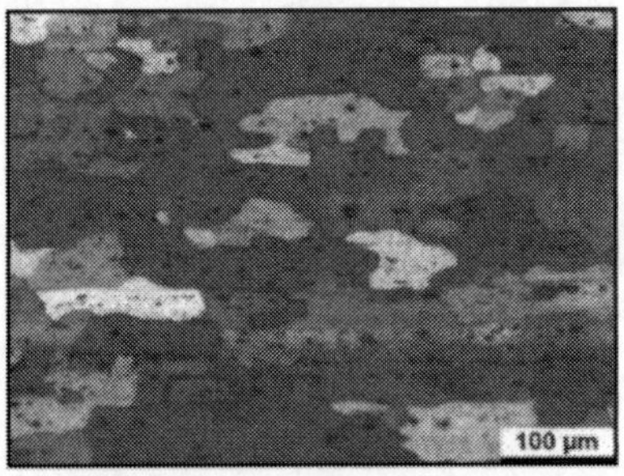

Fig. 7. Microstructure of AA 5050A hot strip after additional recrystallization anneal (optical metallography, longitudinal sections, magn. 200×).

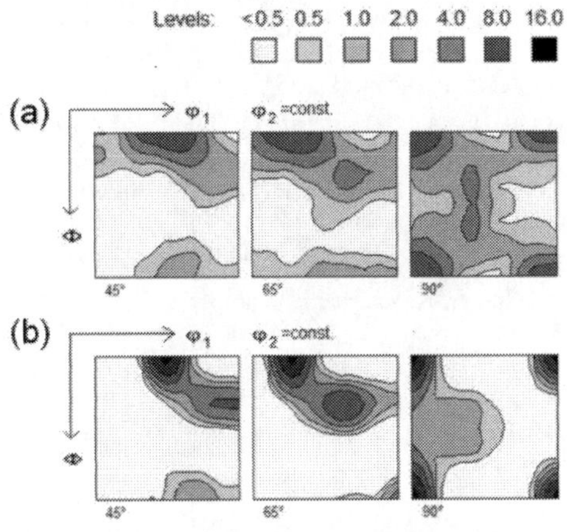

Fig. 8. Evolution of the texture during processing of the AA 5050A sheet; (a) annealed hot strip, (b) lacquered at final gauge, i.e. temper H44 (EBSD, ODF sections φ_2 = 45°, 65° and 90°).

Apparently, the higher level of recrystallization in the annealed hot strip led to larger intensities of the cube orientation throughout the entire downstream processing, although these differences diminished towards the final gauge material (Fig. 8(b)). In turn, the stronger cube texture orientations resulted in a shift of the earing profiles from 45° earing towards 0°/90° earing. Accordingly, the earing behaviour improved with respect to the original processing (Sec. 3.1), as evident from the lowered characteristic earing values of Z = 2.2% and ΔZ = 0.3% (Fig. 6, green curve).

4. Discussion

In the field of aluminium alloys, texture investigations are of great importance since texture affects a number of anisotropic materials properties, which are important for practical applications of Al sheet products. The evolution of texture and texture-related properties is strongly influenced by the various steps of thermo-mechanical processing of Al sheets – homogenization, hot and cold rolling and possible intermediate and/or final annealing (Fig. 1). In the present study it is demonstrated how texture control can be applied during industrial processing of Al alloy AA 5050A sheet in order to minimize earing through the interplay of rolling and recrystallization with the resulting textural effects.

As already mentioned in the introduction, earing in Al alloys is commonly controlled by means of producing mixed textures comprising both rolling and recrystallization texture components. That is to say, the respective 0°/90° and ±45° earing behaviour is superimposed, which results in reduced anisotropy with balanced earing profiles. In the so-called H16 route to produce sheet with minimum earing and medium strength, texture and earing are controlled through an intermediate annealing before a light final temper rolling pass (Fig. 1) [6–9].

It was shown in Sec. 3.1 that cold rolling to intermediate gauge led to the formation of a β-fibre rolling texture together with an appreciable increase of 45° earing (Figs. 2–4). During interannealing the material recrystallized, developing a fine-grained recrystallized grain structure (Fig. 2(c)) and mixed texture comprised of a cube recrystallization texture with medium intensities of a retained rolling, or R, texture (Fig. 3(c)) [11–14]. Accordingly, the interannealed sheet revealed a rather smooth eight-ear profile (Fig. 4).

Final gauge properties were achieved through a final rolling pass to medium strength (H16). This light temper rolling pass of the order of 35% thickness reduction sharpened the rolling texture orientations at the expenses of the recrystallization texture orientations (Fig. 3(d)), yet the rolling reduction was evidently not large enough to significantly degrade the latter. Lacquering of the sheet involving curing of the paint at elevated temperature led to recovery of the materials properties to temper H44, but microstructure and texture remained largely unchanged (Fig. 3(e)). Accordingly, the final gauge H44 material comprised balanced cube and rolling texture orientations and, in consequence, the desired light earing profile suitable for products requiring minimum earing (Figs. 4, 6).

In order to optimize the final gauge earing behaviour two process modifications were probed, (i) lowered intermediate gauge and (ii) hot strip annealing (Sec. 3.2). The above example has demonstrated that texture and earing properties at final gauge depend on the interplay of recrystallization during interannealing plus the deformation applied during final cold rolling. That is to say, the final gauge properties are controlled by the intermediate thickness and the resultant strain imposed upon the final temper rolling pass. In the present case the original processing route led to a weak, balanced earing profile with $Z = 2.7\%$ and $\Delta Z = 0.6\%$. According to Eq. 3 a positive value of ΔZ implies that the 45° earing exceeds the height of the 0°/90° ears, which means that the ratio between rolling and recrystallization texture orientations is shifted towards the former. Accordingly, a reduced interannealing thickness, resulting in a lowered final rolling pass, led to a slight increase in recrystallization texture orientations at the expenses of the rolling texture (Fig. 5(b)) and, in turn, an improved earing profile with $Z = 2.3\%$ and $\Delta Z = 0.1\%$ (Fig. 6, blue curve).

As an alternative, the non-recrystallized hot strip was subjected to a batch annealing treatment to enforce recrystallization. This resulted in a recrystallized microstructure (Fig. 7) with a typical cube recrystallization texture (Fig. 8(a)). Evidently, the cube texture of the annealed hot strip led to larger intensities of the cube orientation and, in consequence, to higher levels of 0°/90° earing throughout the entire process chain, although these differences diminished somewhat towards the final gauge material (Fig. 8(b)). Therefore, at final gauge, i.e. in tempers H16 and, most notably, H44, the earing behaviour was improved with respect to the original processing (see Fig. 6, green curve). With characteristic earing values of $Z = 2.2\%$ and $\Delta Z = 0.3\%$ the material was at a level approximately equal to that of the lowered intermediate thickness (see above). Thus, hot strip annealing offers an alternate processing route to improve earing properties of AA 5050A sheet, albeit at the expenses of higher processing costs.

Summary and conclusions

It is demonstrated how texture and earing can be controlled during industrial processing of sheet of the Al alloy AA 5050A (ISO AlMg1.5(D)). Furthermore, two means to improve earing properties by changing the processing parameters are addressed.

Upon industrial production of low-earing grade Al alloys the 0°/90° earing caused by the recrystallization texture and the ±45° earing due to the rolling texture orientations is superimposed in order to obtain mixed textures with a well-balanced earing profile. In the standard H16 route this is achieved by an intermediate annealing before a light final temper rolling pass.

In the present application the original processing route resulted in sheet with low earing and a positive value of ΔZ, implying that the 45° ears are slightly more pronounced than the 0°/90° ears. Thus, it appeared meaningful to further optimize earing by a strengthening of the recrystallization texture orientations at the expenses of the rolling texture orientations. This was accomplished by two different means, viz. (i) lowering the thickness of the intermediate annealing to weaken the rolling texture at final gauge or (ii) by application of an additional hot strip annealing to strengthen the cube recrystallization texture in the hot strip.

References

1. J.R. Davis (ed.), *ASM Specialty Handbook: Aluminum and Aluminum Alloys*, ASM Int., Materials Park, OH (1993).
2. M.A. Zaidi, T. Sheppard, Mater. Sci. Technol. **1**, 593 (1985).
3. W.B. Hutchinson, H.-E. Ekström, Mater. Sci. Technol. **6**, 1103(1990).
4. B. Ren, Z. Li, J.G. Morris, Scripta metall. mater. **31**, 387 (1994).
5. O. Engler, J. Aegerter, Mater. Sci. Eng. **A618**, 663 (2014).
6. S.E. Naess, Z. Metallkd. **82**, 259 (1991).
7. S. Tsuchida, Y. Baba, in: J.G. Morris, H.D. Merchant, E.J. Westerman, P.L. Morris (eds.), *Aluminum Alloys for Packaging*, TMS, Warrendale, PA, 17 (1993).
8. O. Engler, Mater. Sci. Eng. **A538**, 69 (2012).
9. O. Engler, J. Aegerter, D. Calmer, Mater. Sci. Eng. **A775**, 138965 (2020).
10. O. Engler, V. Randle, *Introduction to Texture Analysis: Macrotexture, Microtexture and Orientation Mapping*, 2nd edition, CRC Press, Boca Raton, FL. (2010).
11. O. Engler, H.E. Vatne, E. Nes, Mater. Sci. Eng. **A 205**, 187 (1996).
12. A. Oscarsson, W.B. Hutchinson, H.E. Ekström, Mater. Sci. Technol. **7**, 554 (1991).
13. O. Sukhopar, G. Gottstein, Int. J. Mater. Res. **107**, 979 (2016).
14. K. Huang, O. Engler, Y.J. Li, K. Marthinsen, Mater. Sci. Eng. **A628**, 216 (2015).
15. O. Engler, Mater. Sci. Eng. **A618**, 654 (2014).
16. O. Engler, K. Knarbakk, in preparation.

 https://doi.org/10.1051/matecconf/202032605003

Influence of homogenization on the microstructure, mechanical properties and texture of the twin-roll cast AA8006 sheet processed by accumulative roll-bonding

Miroslav Karlík[1,2*], *Petr* Homola[2,3], *Peter* Sláma[4], *Jiří* Čapek[5], and *Petr* Harcuba[1]

[1]Charles University, Faculty of Mathematics and Physics, Department of Physics of Materials, Ke Karlovu 5, 121 16 Prague 2, Czech Republic
[2]Czech Technical University in Prague, Faculty of Nuclear Sciences and Physical Engineering, Department of Materials, Trojanova 13, 120 00 Prague 2, Czech Republic
[3]Czech Aerospace Research Centre, Beranových 130, 199 05 Prague 9 – Letňany, Czech Republic
[4]COMTES FHT a.s., Průmyslová 995, 334 41 Dobřany, Czech Republic
[5]Czech Technical University in Prague, Faculty of Nuclear Sciences and Physical Engineering, Department of Solid-State Engineering, Trojanova 13, 120 00 Prague 2, Czech Republic

Abstract. Unhomogenized and homogenized AA8006 (Al-Fe-Mn-Si) alloy sheets twin-roll cast to strip 8.5 mm thick, processed by accumulative roll-bonding (ARB) on 2 mm thick sheets up to 6 cycles were studied. The microstructure was characterized by means of scanning and transmission electron microscopy and electron backscattered diffraction (EBSD), mechanical properties were monitored by hardness measurements and tensile tests. The macroscopic texture was determined by X-ray diffraction. The initial texture of the unhomogenized sheet is rotated cubic {001}<110> combined with recrystallization R texture {011}<211>, while the homogenized sheet has a cubic texture {001}<100> with remaining rolling component C {112}<111>. One ARB cycle leads to the formation of low-angle grain boundaries (LAGB) in original coarser grains and to a low angle rotation of the subgrains bounded by LAGB. After the 3rd and 6th ARB cycle unhomogenized and homogenized sheets show a common rolling texture of cold-rolled aluminium. The thermal stability of sheets processed by 6 ARB cycles was tested by isochronal annealing for 30 min up to 450°C. The homogenized sheet starts to recrystallize at 250°C, while the fine grain structure of the unhomogenized sheet is thermally stable up to 400°C.

1 Introduction

Accumulative Roll-Bonding (ARB) is a technique of grain refinement by severe plastic deformation (SPD) proposed by Saito *et al.* [1]. One ARB cycle consists in sheet cutting, surface treatment, stacking of two pieces, heating (optional), and rolling with 50% reduction in thickness. Since the sheet thickness remains unchanged by this cycle, ARB processing can be repeated as many times as desired, and an extremely high plastic strain is imposed on the material, resulting in structural refinement and increase of strength. ARB does not require any special equipment as other SPD methods such as equal channel angular pressing (ECAP) [2], high pressure torsion (HPT) [3], or cyclic extrusion and compression (CEC) [4], but only a conventional rolling mill. ARB processing has been applied to various metallic materials [5], and it has also been used for creating ultrafine grained metallic composites [6,7]. Ingot cast aluminium alloys are often used as starting materials for ARB [8-10]. An energy efficient technology of twin-roll casting (TRC) is known to produce materials with fine particles and small grain size, which are good starting materials to further grain refinement by ARB processing. Due to a rapid solidification of the metal, TRC sheets have a refined microstructure characterized by a finer dispersion of intermetallic particles and a higher level of matrix supersaturation with respect to the conventional ingot-cast materials [11]. Homogenization annealing reduces the differences between the surface layer and the core of the TRC strip and allows the precipitation of excess solute elements from the matrix [12] resulting in an improvement of the sheet ductility and workability [13,14]. A substantial grain refinement by ARB could be also accompanied by dissolution and reprecipitation of second phase particles [8,15]. Thus, it can have a similar effect as homogenization. A study of the homogenization and ARB processing on the microstructure, texture and mechanical properties of a commercial TRC AA8006 alloy is presented.

2 Experimental details

An AA8006 alloy containing (wt.%) 1.51 Fe, 0.40 Mn, and 0.16 Si was commercially twin-roll cast to an

* Corresponding author: Miroslav.Karlik@fjfi.cvut.cz

8.5 mm thick strip. The unhomogenized material was directly cold rolled down to 2 mm thickness. The second sample was cold rolled to 5.4 mm, homogenized at 610 °C for 18 h and further cold rolled to 2 mm. Unhomogenized and homogenized sheets were then recrystallization annealed for 30 min at 450 °C. One ARB cycle involved five steps: i) degreasing in tetrachloroethylene and brushing with stainless steel 0.3 mm wire brush, ii) stacking of two sheets of 300 mm × 50 mm × 2 mm in dimensions, iii) joining by Al wires to prevent sliding, iv) rolling at ambient temperature with 50% reduction in thickness without lubricant (roll diameter of 340 mm and peripheral speed of 30 m/min), v) trimming and smoothing of specimen edges in order to prevent the propagation of edge cracks. ARB processing was carried out up to 6 cycles. The microstructure of the samples was examined using a FEI Quanta 200 FEG scanning electron microscope (SEM) equipped with EDAX Trident electron backscattered diffraction (EBSD) system. Samples for EBSD were prepared by mechanical grinding and electrolytical polishing (at 12 °C, and 25 V) using Struers Lectropol 5 device filled with the solution of 5% perchloric acid in methanol. Transmission electron microscopy (TEM) samples were prepared by slow diamond blade cutting of slices in long transverse (LT) direction, mechanical grinding to 0.15 mm, punching, and twin-jet electrolytic polishing (at -30 °C, 30 V) in Struers Tenupol 2 unit filled with 6% perchloric acid solution in methanol. A JEOL JEM 2000FX microscope equipped with an energy dispersive spectrometer Bruker and CCD camera Veleta was used for observations of the thin foils at 200 kV accelerating voltage. The macroscopic texture in the sheet centre was determined by X-ray diffraction. The X'Pert PRO MPD diffractometer with cobalt radiation was used to the RX sample surface analysed by X-ray diffraction (XRD). Texture analysis was performed based on orientation distribution function calculation from experimental pole figures (with step size 5×5°), which were obtained on three diffraction lines {220}, {311}, {222} of aluminium phase. The pole figures were corrected for defocusing effect and background intensity. The evolution of mechanical properties after ARB processing was examined by HV10 hardness measurements and tensile test (samples 8 mm wide and 20 mm in gauge length, initial strain rate of 8.3×10^{-4} s^{-1}).

3 Results

3.1 Initial materials

3.1.1 Grains and particles

The grain structure and particle distribution of the sheets in the initial recrystallized condition are presented in Fig. 1. The grains of the unhomogenized sheet are heavily elongated in the rolling direction (Fig. 1a), their mean size in the normal (vertical) direction is 28.5 μm. Furthermore, the grains are not uniform, very coarse ones are present close to the surface and also in the volume. This grain character is accompanied by a higher amount of alloying additions in the solid solution and a non-uniform distribution of dispersoids (Fig. 1c,e). The majority of α-Al(Fe,Mn)Si and Al_6(Fe,Mn) particles is smaller than 500 nm (Fig. 1c), many Al_6(Fe,Mn) precipitates smaller than 100 nm are visible on the TEM micrograph (Fig. 1e). On the other hand, the homogenized sheet presents a quite uniform grain structure with nearly equiaxed grains, having in average 7.8 μm in the normal and 11.6 μm in the rolling direction, respectively (Fig. 1b). Also, the dispersoid particles are coarser (over 500 nm) and quite uniformly distributed (Fig. 1d,f), and the fine precipitate is absent (Fig. 1f).

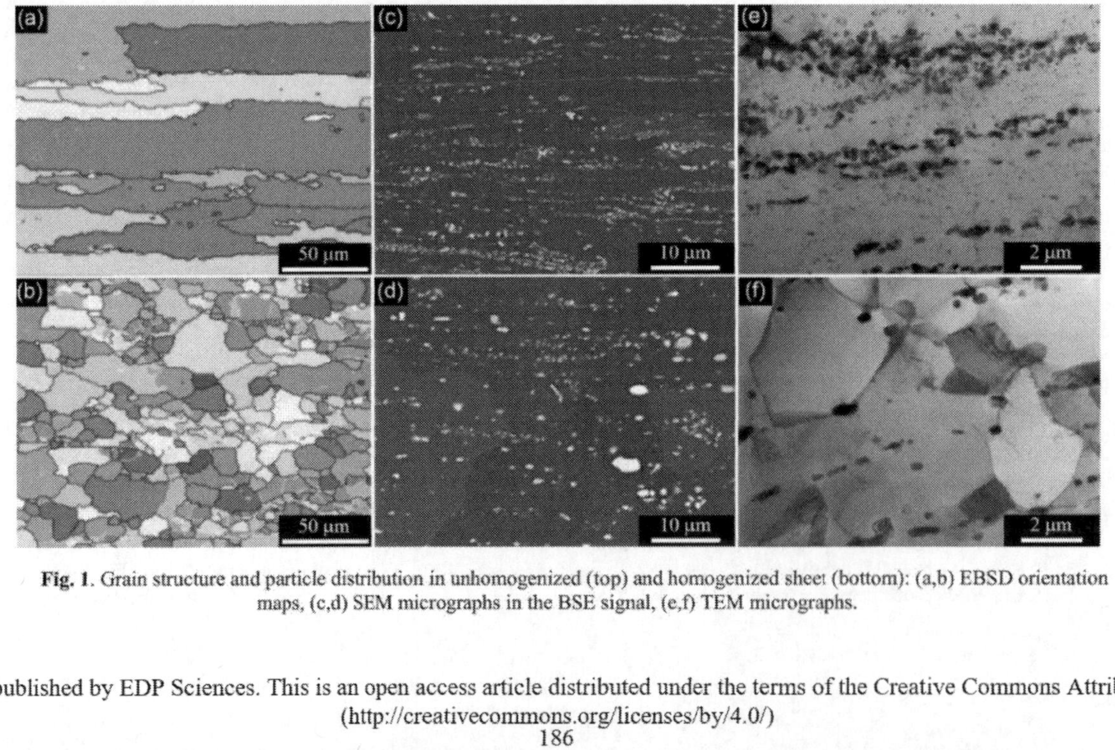

Fig. 1. Grain structure and particle distribution in unhomogenized (top) and homogenized sheet (bottom): (a,b) EBSD orientation maps, (c,d) SEM micrographs in the BSE signal, (e,f) TEM micrographs.

3.1.2 Texture

Fig. 2 shows orientation distribution functions (ODF) in selected Euler's space sections (φ_2 values of 0° (90°), 45°, and 65°). The initial texture of the unhomogenized sheet is rotated cubic {001}<110> combined with recrystallization R texture {123}<634> (Fig. 2a), while the homogenized sheet has a cubic texture {001}<100> with remaining rolling component C {112}<111> (Fig. 2b).

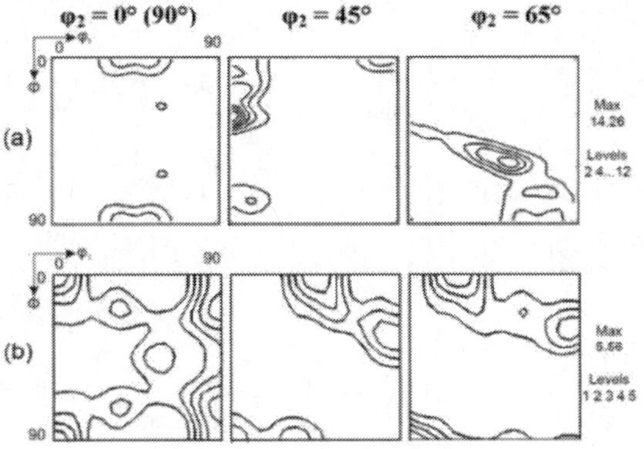

Fig. 2. ODF of the unhomogenized (a) and homogenized (b) sheets in the initial recrystallized condition (Euler's space sections for φ_2 values of 0° (90°), 45°, and 65°).

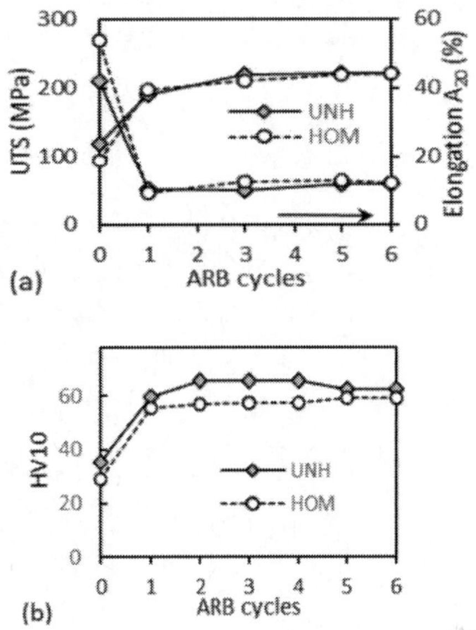

Fig. 3. Evolution of mechanical properties for unhomogenized and homogenized AA8006 sheets ARB processed at ambient temperature: (a) ultimate tensile strength (UTS), elongation A_{20mm}; (b) hardness HV10.

3.2 ARB processed sheets

3.2.1 Mechanical properties

The unhomogenized initial material has higher strength and hardness and lower ductility (Fig. 3) than the homogenized one due to finer dispersion of primary and secondary phase particles (Fig 1) and also somewhat higher level of solute atoms in the matrix (electrical conductivity 29.7 MSm^{-1} and 30.7 MSm^{-1} for unhomogenized and homogenized sheet, respectively, measured by Foerster Sigmatest). For both variants of the initial materials, ARB processing causes the most important increase in strength and decrease of the elongation only during the first cycle (Fig. 3). In the following cycles, the changes are only moderate. Both materials show a small increase of the ultimate tensile strength (UTS) and also a surprising increase of elongation A_{20mm} after the 1st ARB cycle. When comparing the hardness values, there is a noticeable difference after the processing by 2 to 4 ARB cycles, where the unhomogenized sheets exhibit higher values of hardness. After the 5th and 6th ARB cycle the hardness of the unhomogenized sheet slightly decreases and approaches to the value of the homogenized one, while the tensile properties of the unhomogenized and homogenized sheets are practically identical.

3.2.2 Grains and subgrains

EBSD orientation maps showing the evolution of the grain structure of the sheets during ARB processing are in Fig. 4. A gradual refinement of the grain size takes place in both materials. In the samples processed with a lower strain, i.e. after the first two ARB cycles (Fig. 4a,d), the deformed grains are separated by low-angle grain boundaries (LAGB) into subgrains. The LAGB are mostly inclined about 35° with respect to the rolling direction (RD). With increasing number of ARB cycles (increasing strain) the grain boundaries rotate to be parallel with RD and a lamellar grain structure is developed. In the homogenized sheet, the lamellar structure is observed already after the 3rd ARB cycle (Fig. 4e), while in the unhomogenized material there are still inclined shear bands in large areas of the original grains (Fig. 4b). After processing by 6 ARB cycles, both unhomogenized and homogenized sheets present a very fine and homogeneous grain structure (the grain size is about 0.3 μm in the normal direction - ND).

Transformation of LAGB to HAGB with increasing strain is evident from the plots in Fig. 5 indicating quantified results from another representation of EBSD orientation maps in Fig. 4. In this representation of EBSD grain boundary maps (not shown here), the boundaries are marked by different colours according to three ranges of mutual misorientation angles of adjacent grains in three categories: i) LAGB 2 to 5°, LAGB 5 to 15°, and HAGB > 15° and then treated by image analysis. From Fig. 5c it follows that in the initial recrystallized condition, the majority of grain boundaries are HAGB (unhomogenized sheet 90%, homogenized one 70%) During the first two ARB cycles, many LAGB are formed (Fig. 5a,b), and their percentage (addition of values from Fig. 5a and 5b) rises up to 66%. The corresponding fraction of HAGB is low and starting

* Corresponding author: Miroslav.Karlik@fjfi.cvut.cz

https://doi.org/10.1051/matecconf/202032605003

from the third ARB cycle it increases up to 71.5% and 83.8% for unhomogenized and homogenized sheet after 6 cycles, respectively. The curve for homogenized sheet is much steeper than for the unhomogenized one.

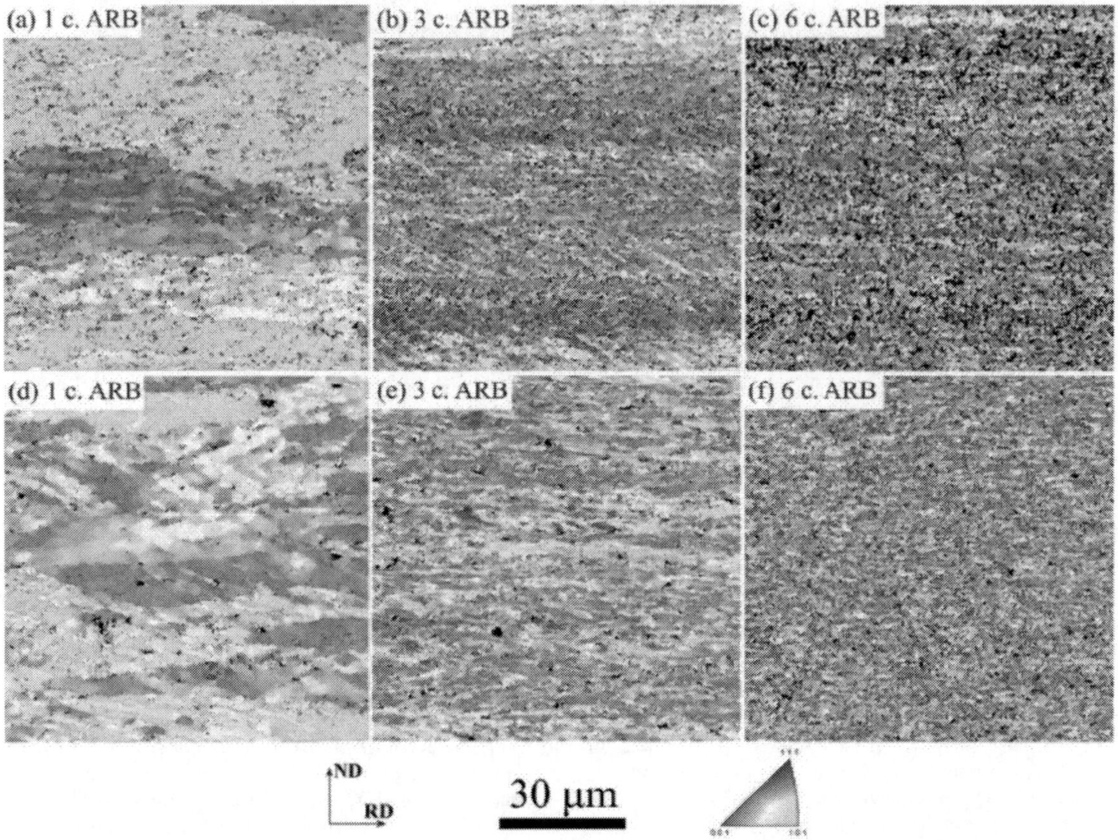

Fig. 4. EBSD orientation maps showing the evolution of the grain structure in the unhomogenized (a,b,c), and homogenized (e,f,g) sheets of the AA8006 alloy.

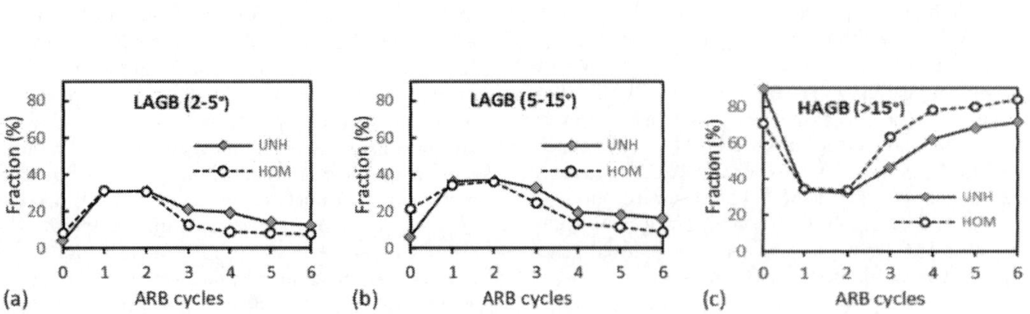

Fig. 5. Evolution of the fraction of low- and high-angle (sub)grain boundaries during ARB processing.

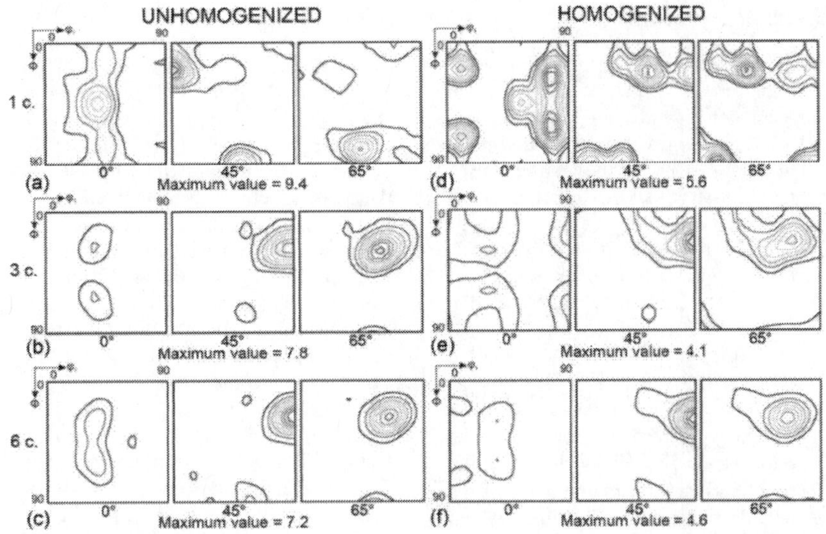

Fig. 6. Comparison of the ODF of the unhomogenized (a,b,c) and homogenized (d,e,f) sheets during ARB processing by 1, 3, and 6 cycles (Euler's space sections for φ_2 values of 0° (90°), 45°, and 65°).

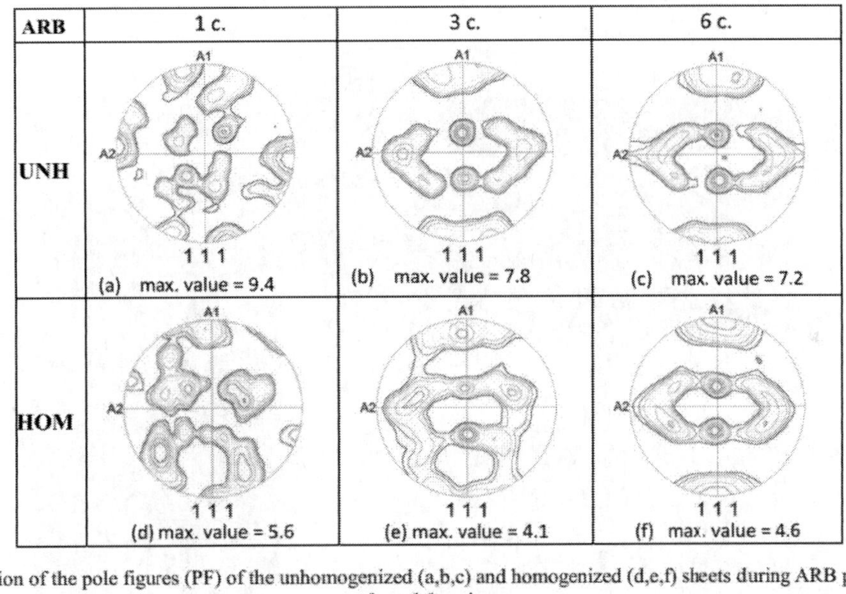

Fig. 7. Evolution of the pole figures (PF) of the unhomogenized (a,b,c) and homogenized (d,e,f) sheets during ARB processing by 1, 3, and 6 cycles.

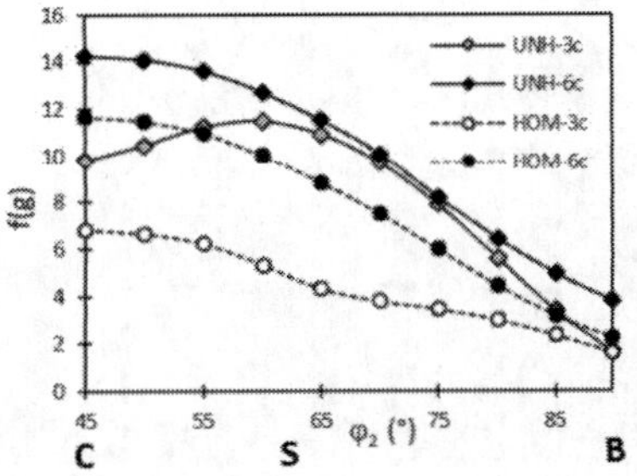

Fig. 8. Evolution of the β-fibre texture components during ARB processing.

3.2.3 Texture during ARB processing

As indicated above, the first ARB processing cycle results in the formation of the low-angle grain boundaries (LAGB) in the original large grains and in small changes in the orientation of subgrains delimited by LAGB. From the point of view of the texture only a rotation of the initial texture takes place in both unhomogenized and homogenized sheets, as it can be seen from the ODF in Fig. 6 and pole figures in Fig. 7. After 3 and 6 ARB cycles the texture corresponds more to a rolling texture of aluminium, with texture components along the β-fibre (Fig. 8). Typical components of the rolling texture, starting from the Cu texture C {112}<111>, through S texture {123}<634> to brass texture B {011}<211>. The component C is the strongest, the S texture prevails only in the unhomogenized sheet after the 3rd ARB cycle. From the β-fibre intensity (Fig. 8) and the maximum values of the ODF (Fig. 6) and the pole figures (Fig. 7) it follows that the unhomogenized sheets present a more pronounced texture that the homogenized ones. After 6 ARB cycles, the texture of the unhomogenized and homogenized sheet is very similar (Fig. 6,c,f, Fig. 7c,f, Fig. 8).

3.2.4 Thermal stability

The thermal stability of unhomogenized and homogenized sheets ARB processed up to 6 cycles was evaluated by an isochronal 30 min (1.8 ks) annealing at temperatures ranging from 50 to 450 °C. The softening behaviour was monitored by hardness measurements and the grain structure was examined by EBSD. From Fig. 9 it follows, that the hardness of both materials has practically the same course up to 200°C. Then at 250°C it drops for the homogenized sheet. From the EBSD maps it is evident that the grain structure of the unhomogenized sheet at 300°C is still deformed, while the homogenized material is already fully recrystallized. The hardness of the unhomogenized sheet decreases only moderately up to the temperature of 400°C. After annealing at 450°C, both materials are fully recrystallized and their hardness values are much closer than in the range from 250 to 400°C. However, the grains of the unhomogenized sheet are very large and not uniform (an abnormal grain growth took place - Fig. 9). On the other hand, the recrystallized homogenized ARB processed sheet shows a much more uniform and finer grain structure.

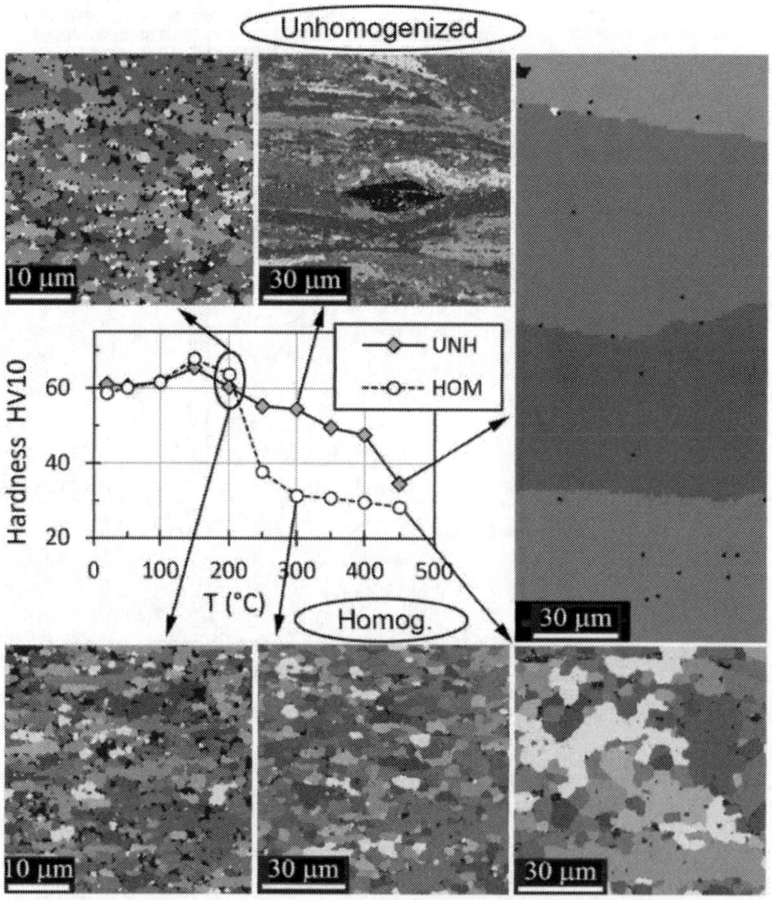

Fig. 9. Thermal stability of the unhomogenized and homogenized AA 8006 alloy sheets processed by 6 ARB cycles (during isochronal annealing for 30 min up to 450°C).

* Corresponding author: Miroslav.Karlik@fjfi.cvut.cz

4 Discussion

4.1 General microstructural features

As expected, ARB processing led to a substantial grain refinement from 28.5 μm and 7.8 μm in normal direction in the case of the unhomogenized and homogenized sheet, respectively (Fig. 1) up to 0.3 μm in the normal direction for both materials (Fig. 4). A similar grain refinement after 6 ARB cycles was achieved also for AA6063 and AA2014 alloy [16], AA5086 alloy [9], or AA3003 alloy [17]. The grain refinement is faster in the homogenized alloy (Fig. 4b,e), accompanied by a higher fraction of HAGB formed after 3 ARB cycles and further (Fig. 5c). After 6 ARB cycles, the grain structure of both materials is very similar (Fig. 4c,f).

4.2 Texture evolution

The textures of the initial materials are different: a rotated cubic {001}<110> combined with recrystallization R texture {123}<634> in the case of the unhomogenized sheet (Fig. 2a), and a cubic texture {001}<100> with remaining rolling component C {112}<111> for the homogenized one (Fig. 2b). However, ARB processing by 3 and more cycles (Figs. 6 and 7) leads to the development of a rolling texture of aluminium, with texture components along the β-fibre (Fig. 8). During ARB processing, the shear texture involving {001}<110> and {111}<110> orientations developed at the surface of the sheet due to the shear deformation induced by a large friction between the rolls and the sheet [18] is progressively introduced to the volume of the sheet. Thus, the final texture in the ARB sheet was mainly composed of Cu texture {112}<111>, through S texture (123)<634> to brass texture B {011}<211>. A similar texture evolution was found also for other Al alloys [8-10, 16-17]. These textural changes during the ARB process contributed to the formation of HAGB. After 6 ARB cycles, the texture of the unhomogenized and homogenized sheet is very similar (Fig. 6,c,f, Fig. 7c,f, Fig. 8). The unhomogenized sheets present only a more pronounced texture that the homogenized ones.

4.3 Mechanical properties

The most important increase in strength and decrease of the elongation during the first ARB cycle was expected. It is typical for ARB processing of many Al based materials [8-10, 16-17]. On the other hand, a slight increase of ductility during next ARB cycles involving higher equivalent strain is somewhat surprising. It was not observed for AA1000 sheet (pure Al) [19], a similar increase was found in the case of the AA8011 (Al-Fe-Si) alloy [8]. Kim et al. [15] explain this behavior by the enhancing of the recovery due to the presence of second phase particles. Our AA8006 alloy contains also numerous particles and so this is probably the reason for the slight increase of ductility after the first ARB cycle.

4.4 Thermal stability

The softening behaviour of the homogenized AA8006 sheet processed by 6 ARB cycles, i.e. the onset of recrystallization at 250°C is similar to the results reported for the same homogenized alloy in [20] after 3 and 5 ARB cycles. Birol [21] also reported that homogenized TRC Al-Fe-Mn sheet starts to recrystallize at much lower temperature (250 – 280°C) than the supersaturated unhomogenized one (400 – 450°C) [14]. Our ARB processed unhomogenized sheets behave in a similar way. However, their recrystallized grains are very large and not uniform due to abnormal grain growth took place (Fig. 9).

5 Conclusions

Unhomogenized and homogenized AAA8006 sheets were processed by accumulative roll-bonding (ARB) at ambient temperature. The main results of their characterization are summarized as follows:

i) ARB processing of both initial materials leads to a significant grain refinement. A homogeneous submicron (0.3 μm) grain size separated by more than 70% fraction of HAGB was achieved after 6 ARB cycles. The grain refinement is faster in the homogenized alloy.

ii) Even if the textures of the initial materials are different: a rotated cubic {001}<110> combined with recrystallization R texture {123}<634> in the case of the unhomogenized sheet, and a cubic texture {001}<100> with remaining rolling component C {112}<111> for the homogenized one, ARB processing by 3 and more cycles leads to the development of a rolling texture of aluminium, with common C,S,B texture components along the β-fibre.

iii) ARB processing causes the most important increase in strength and decrease of the elongation only during the first cycle. In the following cycles, the changes are only moderate. A slight increase of ductility during next ARB cycles is caused by an enhanced recovery due to the presence of second phase particles.

iv) The thermal stability of sheets processed by 6 ARB cycles was tested by isochronal annealing for 30 min up to 450°C. The homogenized sheet starts to recrystallize at 250°C, while the fine grain structure of the unhomogenized sheet is thermally stable up to 400°C.

This research was financially supported by ERDF in the frame of the project "Nanomaterials centre for advanced applications", No. CZ.02.1.01/0.0/0.0/15_003/0000485.

* Corresponding author: Miroslav.Karlik@fjfi.cvut.cz

References

1. Y. Saito, H. Utsunomiya, N. Tsuji, T. Sakai, Acta Mater. **47**, 579-583 (1999)
2. R.Z. Valiev, T.G. Langdon, Prog. Mater. Sci. **51**, 881-981 (2006)
3. R.Z. Valiev, R.K. Islamgaliev, I.V. Alexandrov, Prog. Mater. Sci. **45**, 103-189 (2000)
4. M. Richert, Q. Liu, N. Hansen, Mater. Sci. Eng. A **260**, 275-283 (1999)
5. N. Tsuji, In: Altan, B.S. (Ed.), Severe Plastic Deformation, Toward Bulk Production of Nanostructured Materials, Nova Science Publishers, New York, 545–565 (2005)
6. K. Wu, H. Chang, E. Maawad, W.M. Gan, H.G. Brokmeier, M.Y. Zheng, Mater. Sci. Eng. A **527**, 3073-3078 (2010)
7. P. Qu, L. Zhou, H. Xu, V.L. Acoff, Metall. Mater. Trans. A **45**, 6217-6230 (2014)
8. H.W. Kim, S.B. Kang, N. Tsuji, Y. Minamino, Metall. Mater. Trans. A **36**, 3151-3163 (2005)
9. S. Roy, D. Satyaveer Singh, S. Suwas, S. Kumar, K. Chattopadhyay, Mater. Sci. Eng. A **528**, 8469-8478 (2011)
10. H. Xie, M.P. Wang, W. Chen, Y. Jia, J. Mater. Eng. Perform. **25**, 1199-1210 (2016)
11. S. Sarkar, M.A. Wells, W.J. Poole, Mater. Sci. Eng. A, **421**, 276 (2006)
12. M. Slámová, P. Sláma, M. Cieslar, Mater. Sci. Forum **519–521**, 365 (2006)
13. I. Peter, M. Rosso, In: ICAA13, Ed: H. Weiland, A. D. Rollett, W. A. Cassada, TMS (The Minerals, Metals & Materials Society). 177-182 (2012)
14. Y. Birol, Scr. Mater. **60**, 5-8 (2009)
15. H.W. Kim, S.B. Kang, N. Tsuji, Y. Minamino, Acta Mater.**53**, 1737-1749 (2005)
16. V.G. Arigela, J. Scharnweber, L. Lisenhoeft, P. Chekonin, R. Schaarschuch, S.K. Kolli, N.R. Palukuri, J. Rengaswamy, W. Skrotzki, Kovove Mater. **53**, 245-249 (2015)
17. H. Pirgazi, A. Akbarzadeh, R. Petrov, J. Sidor, L. Kestens, Mater. Sci. Eng. A **492**, 110-117 (2008)
18. C.H. Choi, J.W. Kwon, K.H. Oh, D.N. Lee, Acta Mater. **45**, 5119-5128 (1997)
19. H. Pirgazi, A. Akbarzadeh, R. Petrov, J. Sidor, L. Kestens, Mater. Sci. Eng. A **497**, 132-138 (2008)
20. M. Cieslar, M. Poková, Materials **7**, 8058-8069 (2014).
21. Y. Birol, Scr. Mater. **59**, 611-614 (2008)

 https://doi.org/10.1051/matecconf/202032605004

Thermo-mechanical treatment regulation of microstructure and comprehensive performance in Al-Zn-Mg-Cu alloy

Zhiguo Chen[1,2,*], *Chenghua* Lu[2], *Jing* Peng[2], and *Zhengui* Yuan[2]

[1]Department of Materials Engineering, Hunan University of Humanities, Science and Technology, Loudi 417000, P.R. China
[2]School of Materials Science and Engineering, Central South University, Changsha 410083, P. R. China

Abstract. The comprehensive performance of Al-Zn-Mg-Cu alloy can be significantly improved by a proposed novel thermo-mechanical treatment (NTMT). The influence of the NTMT on the properties and microstructure was investigated by tensile test, corrosion resistance test, X-ray diffraction (XRD), and transmission electron microscopy (TEM). Results show that Al-Zn-Mg-Cu alloy treated by the NTMT can obtain an excellent combination of strength and ductility. The highest yield strength and ultimate tensile strength reached 643 MPa and 664 MPa respectively, and the elongation was 9.7%. Meanwhile, electrochemical corrosion resistance and intergranular corrosion resistance in the aluminum alloy can be improved after the NTMT. The mechanism of the excellent combination of strength and ductility is thought to be the synergistic effect of dislocations substructures, texture configuration, and nanoprecipitates. The improvement of intergranular corrosion resistance of the aluminum alloy is caused by changes in the micro-morphology of grain boundary precipitates after the NTMT, which can block anodic dissolution channels along grain boundaries to reduce the rate of anodic dissolution and avoid hydrogen embrittlement.

1 Introduction

Al-Zn-Mg-Cu alloys can be used in aerospace and construction materials due to their high strength, low density, and good thermal processing performance, and they have a great application prospect [1-4]. However, they are sensitive to intergranular corrosion (IGC), stress corrosion cracking (SCC), and exfoliation corrosion (EXCO) [5, 6], which limit their practical application. In order to meet the application requirements, it is essential to properly select and design processes to improve the strength and corrosion resistance of aluminum alloys. In the early 1960s, Alcoa developed T73 and T76 processes, which improved the stress corrosion resistance of 7075 and 7079 aluminum alloys in the peak ageing process, but at the same time, part of the strength of the alloy was sacrificed [7]. In 1974, the Israeli Cina [8] put forward a multi-step ageing process named the regression and re-ageing (RRA), in which the corrosion resistance of aluminum alloys was improved without sacrificing strength simultaneously. However, it is not convenient for large-scale applications since the alloy plate is required to be kept at a high temperature for a short retrogression time.

Numerous studies have shown that the thermo-mechanical treatment (TMT) has the advantages of high productivity, low cost, and practicality of continuous production [9], and it is a practical method to improve the mechanical properties of aluminum alloys. Huo et al. [10] proposed a TMT combining with warm rolling and continuous rolling treatment, which successfully refined grains of 7075 aluminum alloy to 10 μm, improving the strength and ductility of aluminum alloys. Li et al. [11] used 83% cold deformation and then ageing (100°C, 72 h), and the yield strength of 2024 aluminum alloy reached 711 MPa. Zuo et al. [12] proposed a TMT including pre-deformation, intermediate short-term annealing, and final hot rolling, obtaining

* Corresponding author: hngary@163.com

https://doi.org/10.1051/matecconf/202032605004

superior elongation while maintaining high strength in AA7055 alloy.

In our previous work, a good combination of strength and ductility could be achieved in 2E12 aluminum alloy by a novel thermo-mechanical treatment (NTMT) [13]. Moreover, we achieved an excellent combination of strength and ductility in Al-Mg-Si alloy by the NTMT based on cluster strengthening [14]. Given the results above, we proposed the NTMT based on continuous Solid solution-Hot rolling to solve the preparation of high strength and good ductility Al-Zn-Mg-Cu alloy with good intergranular corrosion resistance, which might provide a new idea for the preparation of high comprehensive properties aluminum alloys to meet the demand for aerospace and other high technology. The effects of the NTMT on the comprehensive properties in Al-Zn-Mg-Cu alloy are discussed. More importantly, the involved mechanism is discussed in depth, which can lay the foundation for the research in other series of aluminum alloys.

2 Materials and methods

The material used in this research was an Al-Zn-Mg-Cu alloy plate with an original thickness of 12 mm. The chemical composition of the studied alloy is given in Table 1.

Table 1. Chemical composition of Al-Zn-Mg-Cu alloy (w.t. %).

Zn	Mg	Cu	Fe	Si	Zr	Mn	Al
6.20	2.35	2.23	0.15	0.12	0.10	0.10	Bal.

The samples were processed by T6 and RRA, respectively. The T6 treatment was solution treatment (475°C/1 h) +water quenching+ageing (120°C/24 h). The RRA treatment was solution treatment (475°C/1 h) +ageing (120°C/24 h) +regression heat treatment (195°C/10 min) +ageing (120°C/24 h). The schematic diagram of NTMT is illustrated in Fig. 1. The steps included continuous Solid solution-Hot rolling (SSHR), deformation treatment, and ageing treatment. The continuous SSHR was the alloys first heated at 475°C for 1 h, and immediately followed by hot rolling with a deformation rate of 30%. The reduction value of cold rolling treatment was 60%. Finally, different ageing treatments were carried out.

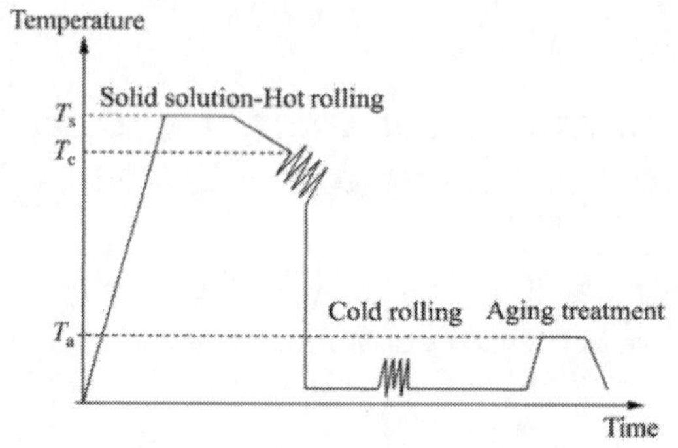

Fig. 1. Schematic diagram of the NTMT in aluminum alloys.

The tensile specimens (the gauge length/width/thickness of 30 mm/8 mm/2.4 mm) were performed on the MTS-858 universal testing machine at room temperature with a constant strain rate of 2.0 mm/min. Three parallel samples were tested for each data point to obtain consistent yield strength, tensile strength, and elongation. The optical microscopy (OM) of the specimens was observed using a Leica DMILM optical microscope model. The susceptibility to intergranular corrosion (IGC) was assessed according to the GB/T7998-2005 standard. The specimens (testing area/ solution volume < 20 mm^2/mL) were immersed in the reagent (57 g/L NaCl+10 mL/L H_2O_2 solution) at a temperature of 35°C and for a time of 6 h. The TEM specimens were mechanically thinned to 80-100 μm, then the mixture of nitric acid and methanol with a volume ratio of 1:3 was used for twin-jet electropolishing below -25°C. TEM observations were performed on TecnaiG^{2}20 transmission electron microscopy with an operating voltage of 200 kV. The electrical conductivity is obtained by the relationship between the resistivity and the electrical conductivity. The resistance of the samples was measured by a double-arm bridge detector, and then the resistivity ρ can be obtained by the Equation (1):

$$R = \rho l/S \tag{1}$$

where ρ is the resistivity, l is the sample length, and S is the sample boundary area. XRD measurements were made on Rigaku D/max 2500 X-ray diffractometer to observe the texture evolution and measure the dislocation density of samples.

3 Results

3.1 Tensile properties

As shown in Table 2, the yield strength (YS) and ultimate tensile strength (UTS) of the Al-Zn-Mg-Cu alloy in the T6 and RRA tempers are very close. However, the elongation (δ) of the aluminum alloy in the T6 is higher than that of RRA. Compared with T6 and RRA, the UTS and YS of NTMT processed samples are significantly improved. The elongation of the aluminum alloy after the NTMT is lower than that of T6, but it is close to the RRA. An outstanding combination of strength and ductility is achieved in the condition of 30% continuous Solid solution-Hot rolling (SSHR) +60% cold rolling (CR) +ageing treatment (120°C/6 h), the highest YS and UTS reached 643 MPa and 664 MPa respectively, meanwhile, the elongation was 9.7%. The elongation value of NTMT processed alloy increased first and then decreased. The elongation of H3C6A18 was 10.6%, which is the best ductility in NTMT processed samples, but the strength of sample H3C6A18 is lower than that of H3C6A6.

3.2 Electrical conductivity and corrosion property test

Table 3 shows the electrical performance parameters of Al-Zn-Mg-Cu alloys after different processes. It can be seen that the conductivity of NTMT processed samples is between T6 (30.66%IACS) and RRA (35.28%IACS), and its conductivity gradually increases with the extension of ageing time. The H3C6A24 sample has the highest conductivity in NTMT processed samples, which is close to that of RRA. The H3C6A6 sample has the lowest conductivity in NTMT processed samples, but significantly higher than that of T6.

As shown in Table 3, after 30%SSHR+60%CR, the longer the ageing time, the smaller the electrochemical corrosion current density (i_{corr}), the slower the corrosion rate, and the better the corrosion resistance. The corrosion current density is lower than that of T6 when the ageing time reaches 12 hours, and it is even less than that of RRA when the ageing time reaches 18 hours.

During the intergranular corrosion test, many white corrosion products could be observed on the surface of T6, RRA and NTMT processed samples. Some samples had the phenomenon of surface warping after corrosion. Due to the different treatments, the intergranular corrosion morphology of each sample is also quite different. The corrosion morphology of the T6 and RRA processed samples shows an obvious corrosion ditch, as shown in Fig. 2(a, b). However, the intergranular corrosion cross-section of the sample exhibited a fibrous morphology due to the rolling deformation (Fig. 2(c-f)). The corrosion depths of samples in the T6 and RRA were 81.67 μm and 52.50 μm, respectively, implying that the intergranular corrosion resistance of the T6 was inferior to that of the RRA. The corrosion depths of samples in the H3C6A6, H3C6A12, H3C6A18, and H3C6A24 were 62.84 μm, 39.86 μm, 29.05 μm, and 27.70 μm, respectively. It indicates that the intergranular corrosion depth of NTMT processed samples gradually decreases, which means that the intergranular corrosion resistance of the samples is improved with the extension of ageing time. It should be noted that the intergranular corrosion resistance in the NTMT processed samples is

Table 2. Tensile properties of Al-Zn-Mg-Cu alloys in different states.

Sample	Treatment	UTS (MPa)	YS (MPa)	δ (%)
T6	475 C/1 h + 12°C/24 h	601±3	525±5	15.0±0.7
RRA	475°C/1 h + 120 C /24 h + 195°C/10 min + 120 °C/24 h	608±4	529±4	11.6±0.5
H3C6A6	30%SSHR+60%CR+120°C/6 h	664±5	643±4	9.7±0.5
H3C6A12	30%SSHR+60%CR+120°C/12 h	658±4	636±3	10.1±0.8
H3C6A18	30%SSHR+60%CR+120°C/18 h	651±5	633±4	10.6±0.7
H3C6A24	30%SSHR+60%CR+120°C/24 h	642±4	620±4	9.4±0.7

greatly improved compared with that of T6. The intergranular corrosion depth of the samples after 12 hours ageing is lower than that of RRA.

Table 3. The electrical performance parameters of Al-Zn-Mg-Cu alloys.

Sample	Conductivity (%IACS)	E_{corr} (V)	I_{corr} ($A \cdot mm^{-2}$)
T6	30.66	-1.364	1.50×10^{-6}
RRA	35.28	-1.289	1.05×10^{-6}
H3C6A6	32.16	-1.317	1.54×10^{-6}
H3C6A12	33.10	-1.347	1.07×10^{-6}
H3C6A18	34.25	-1.324	8.62×10^{-7}
H3C6A24	34.63	-1.319	7.36×10^{-7}

3.3 Metallographic observation

Fig. 3 shows the optical micrographs of Al-Zn-Mg-Cu alloys in different states. It can be seen that most grains of T6 and RRA processed samples have recrystallized and grown up at the solid solution temperature, and their grain sizes are similar, both are equiaxed crystals. However, the grains in NTMT processed samples are elongated due to rolling deformation and formed fibrous structures, which also explains why the intergranular corrosion morphologies of the samples in Fig. 2 are fibrous. What's more, the fibrous grain structure formed in deformation hinders the corrosion crack growth.

3.4 Dislocation density and texture analysis

The dislocation density is calculated using the following Equation (2) [15]:

$$\rho = 2\sqrt{3}\varepsilon /(db) \tag{2}$$

where ρ represents the dislocation density, ε represents the micro-strain, d represents the crystallite size, and b=0.286 nm is the Burgers vector of α-Al. The crystallite size and micro stress are obtained by the half-width of the diffraction peak according to the Williamson-Hall method [16], which can be expressed by the following Equation (3):

$$(B\cos\theta/\lambda)^n = (1/D)^n + (4\ \varepsilon\sin\theta/\lambda)^n \tag{3}$$

where B is half-width of the diffraction peak, λ is the wavelength of X-ray, D is the crystallite size, ε is the micro-strain, θ is the semi-diffraction angle, and n is the deconvolution parameter. Fig. 4 illustrates the dislocation densities of alloys under the process of 30%SSHR+60%CR. It can be concluded that the dislocation density decreases with the increase of ageing time, which indicates that the dislocations have a certain degree of recovery, causing some entangled dislocations to disappear and the dislocation density to decrease.

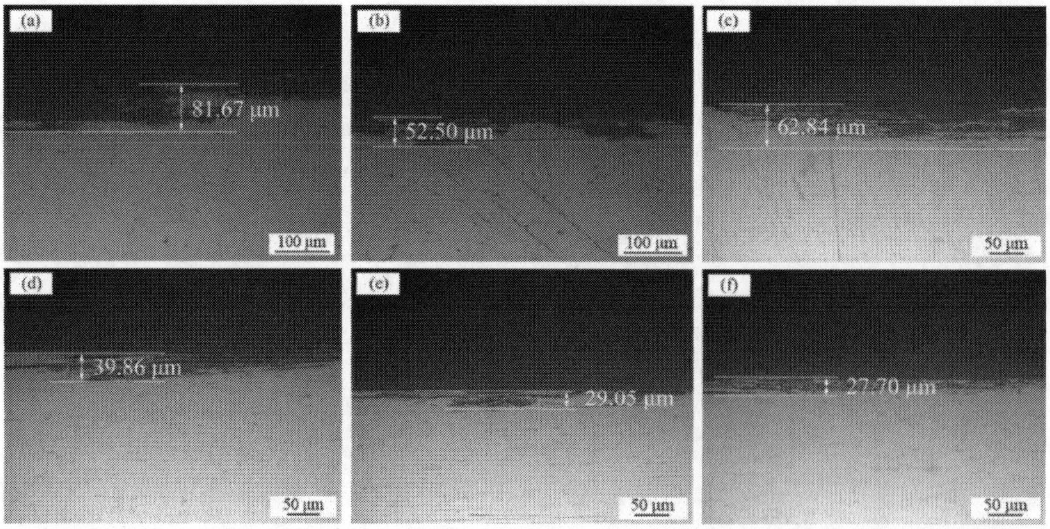

Fig. 2. Intergranular corrosion morphologies of samples in different states: (a) T6, (b) RRA, (c) H3C6A6, (d) H3C6A12, (e) H3C6A18, (f) H3C6A24.

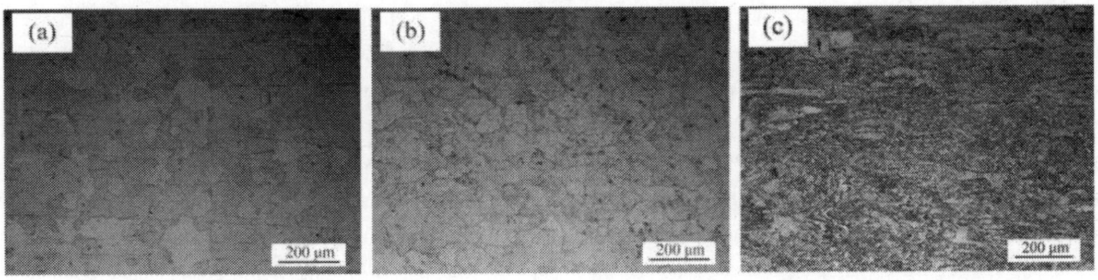

Fig. 3. Optical micrographs of Al-Zn-Mg-Cu alloys in different states: (a) T6, (b) RRA, (c) 30%SSHR+60%CR.

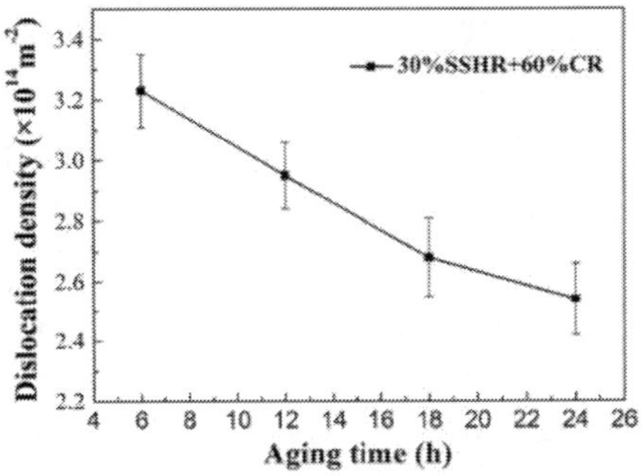

Fig. 4. Dislocation density of NTMT processed Al-Zn-Mg-Cu alloys.

Fig. 5 shows the orientation distribution functions (ODF) of aluminum alloys in T6 and NTMT states. Fig. 5(a) shows the texture components in the T6 sample consist mainly of the recrystallization textures of Cube {001}<100> and strong deformation texture S {123}<634>. Compared with the T6 sample, the rolling textures increase remarkably while the recrystallization textures of Cube {001}<100> and R-Cube {001}<110> decrease. This is attributed to the strong rearrangement of grain orientations that occurred during rolling deformation. So the texture components in the NTMT processed aluminum alloys are mainly deformation textures, such as Copper {112}<111> and S {123}<634> as shown in Fig. 5(b). In addition, it is worth noting that the texture P {011}<111> can be observed in aluminum alloy after the 30%SSHR+60%CR process.

3.5 TEM observation

The transmission electron micrographs of Al-Zn-Mg-Cu alloys in different states are presented in Figs. 6 and 7. From Fig. 6(a, b), it can be found that the sizes and the distribution of intragranular precipitates in T6 and RRA samples are similar, and the fine η′ phases are uniformly

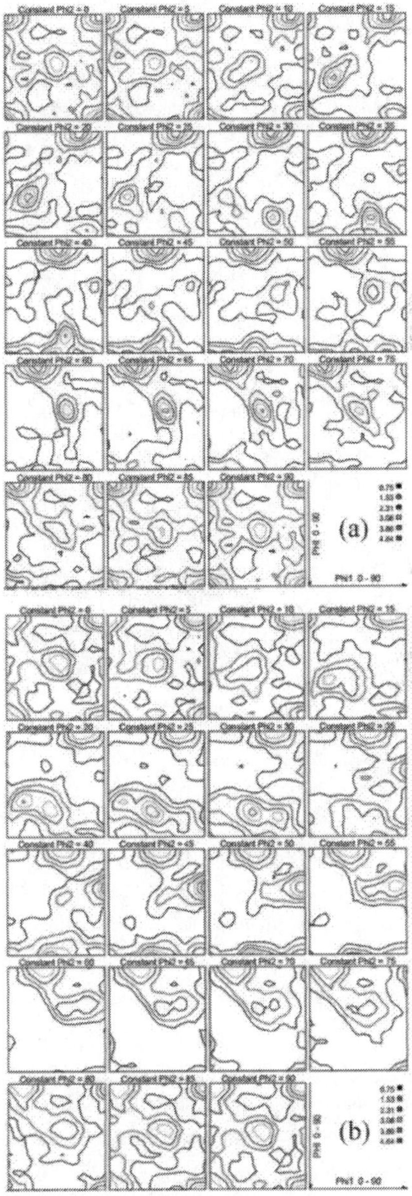

Fig. 5. ODF charts of aluminum alloys in different states: (a) T6, (b) 30%SSHR+60%CR.

distributed in the matrix. The grain boundary precipitates (GBPs) in the T6 and RRA tempers are larger than intragranular precipitates. However, the GBPs in the T6 are continuous, while those in the RRA are discontinuous. In addition, obvious precipitation-free zones (PFZ) can be found by observing the GBPs in the T6 and RRA.

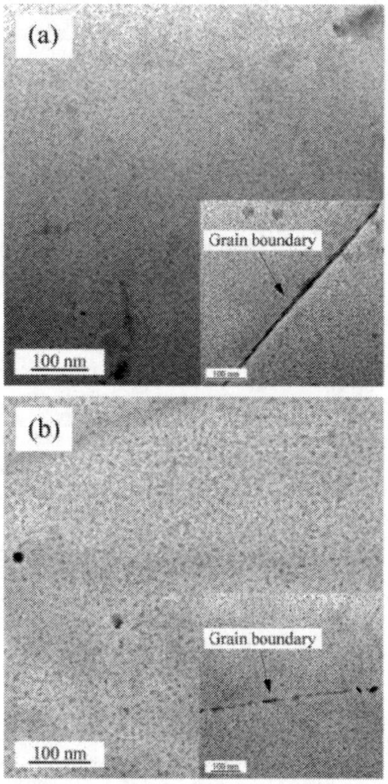

Fig. 6. Transmission electron micrographs of (a) T6 and (b) RRA processed Al-Zn-Mg-Cu alloys.

Due to a large amount of deformation in the NTMT, there are many dislocations in alloys. Since the ageing temperature is insufficient to eliminate dislocations formed during deformation, thus many dislocations are preserved and tangled with each other. And the nano-sized η′ phase gradually grows (Fig. 7), and the dislocation density gradually decreases as the ageing time increases. According to the diffraction pattern in Fig. 7(c, d), it can be concluded that there is a GP zone in the ageing time of 6 to 18 hours, but there is no obvious GP zone diffraction spot after ageing for 24 hours, which indicates the transition of the GP zone to the η′ phase. The precipitation sequence was GP zone→η′→η [17].

Fig. 8 represents the TEM images of grain boundary in NTMT processed samples. The GBPs in NTMT processed samples change from continuous distribution to discontinuous distribution. The GBPs of H3C6A6 and H3C6A12 samples are continuously distributed, while the GBPs of samples processed by H3C6A18 and H3C6A24 are coarse and discontinuously distributed.

4 Discussions

4.1 Effect of NTMT on tensile properties in Al-Zn-Mg-Cu alloy

Al-Zn-Mg-Cu alloy is a typical ageing hardening alloy, and nanoprecipitates formed during ageing treatment are the main factor determining the strength of aluminum alloys. Plastic deformation can introduce dislocation lines in aluminum alloys [18], and the dislocations introduced by deformation in the NTMT also have an impact on the tensile properties of aluminum alloys. Thus, the tensile properties of samples obtained by different treatment processes are quite different. For ageing hardening aluminum alloys, the yield strength can be described by the following Equation (4) [19]:

$$\sigma_y = \Delta\sigma_{gb} + M[\tau_0 + \Delta\tau_c + (\Delta\tau_d^2 + \Delta\tau_p^2)^{1/2}] \quad (4)$$

where σ_y is the yield strength of aluminum alloys, $\Delta\sigma_{gb}$ is the strength change caused by grain boundary strengthening, M is an orientation factor (often termed the Taylor factor), τ_0 is the friction stress of ~7 MPa, $\Delta\tau_c$ is the strength change caused by solution strengthening, $\Delta\tau_d$ is the strength change caused by dislocation, and $\Delta\tau_p$ is the strength change caused by precipitates.

All samples have been processed by overageing and can be regarded as their solid solution atoms precipitated, so the effect of solution strengthening on the strength of aluminum alloys can be ignored. The sizes of the precipitates of T6 and RRA samples are similar, and they are uniformly distributed on the aluminum matrix (Fig. 6). In addition, the dislocations in T6 and RRA samples are considered to be substantially eliminated during solid solution. Therefore, the effect of dislocations on strength can be ignored for the alloys in the T6 and RRA. Therefore, the strength of the T6 and RRA samples is similar due to the similar grain sizes.

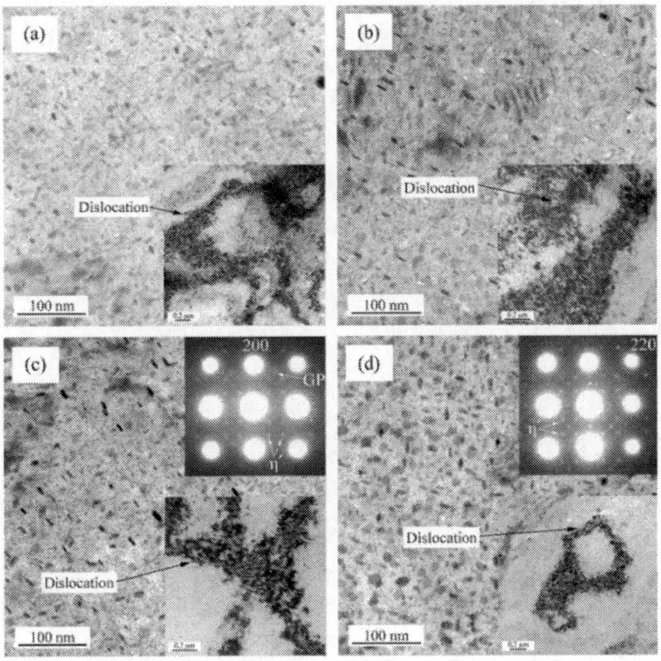

Fig. 7. Transmission electron micrographs of NTMT processed Al-Zn-Mg-Cu alloys: (a) H3C6A6, (b) H3C6A12, (c) H3C6A18, (d) H3C6A24.

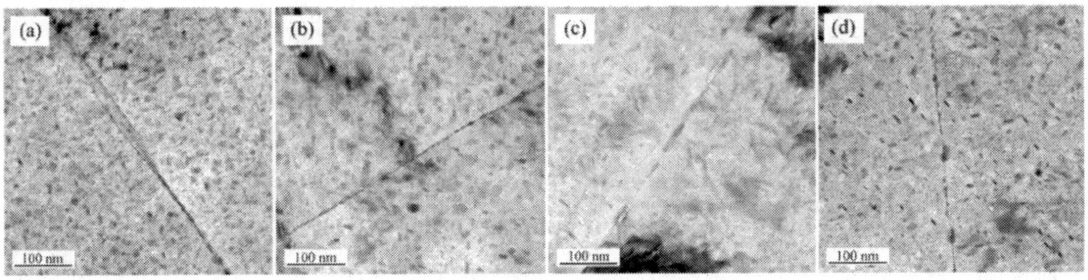

Fig. 8. Transmission electron micrographs of grain boundary in NTMT processed samples: (a) H3C6A6, (b) H3C6A12, (c) H3C6A18, (d) H3C6A24.

With the introduction of dislocations in the NTMT, the contribution due to dislocations ($\Delta\tau_d$) can be described by the following Equation (5) [20],

$$\Delta\tau_d = \alpha G b \rho^{1/2} \qquad (5)$$

where ρ is the dislocation density, α, G and b are constants. Therefore, the greater the dislocation density, the greater its contribution to strength. With the extension of ageing time, the dislocation density of NTMT processed samples gradually decreases (Fig. 4), and the nanoprecipitate grows, resulting in the decrease of strength. In addition, the dislocations create links between nanoprecipitates, improving the efficiency of the transformation of GBPs from small nanoprecipitates to large ones, which can lead to the increase of the size of precipitates after NTMT compared to the T6 and RRA (Fig. 6 and Fig. 7). Thus, the strength of NTMT processed aluminum alloys is obviously higher than that of T6 and RRA because of the ageing precipitation strengthening and dislocations strengthening.

The Taylor factor value of typical deformation textures (Copper, S, Brass) is obviously higher than that of recrystallization Cube textures, and the Taylor factor value of P texture is the highest. P texture only appears in the aluminum alloy under the process conditions of 30%SSHR+60%CR (Fig. 5). Combined with Fig. 5 and

Table 4, it can be concluded that the average Taylor factor of samples after 30%SSHR+60%CR process is much higher than that of T6, which can also explain the improvement of the strength of NTMT processed aluminum alloys.

The ductility of samples is related to dislocations during the tensile test. After the NTMT, the dislocations are introduced into the aluminum alloys due to deformation, but as the ageing time increases, the dislocation tangles gradually decrease, which provides more space for the accumulation of dislocations in the tensile test, so the elongation of samples is improved (Table 2). The elongation of H3C6A6 was 9.7%. The elongation of H3C6A12 and H3C6A18 increased to 10.1% and 10.6%, respectively. However, the elongation of the sample H3C6A24 decreased to 9.4% due to the growth of the precipitates, which was not conducive to the ductility of the aluminum alloy.

Table 4. Taylor factor of typical texture.

Textrue	Talor factor (M)
Copper {112}<111>	3.70
S {123}<634>	3.33
Brass {011}<211>	3.17
Cube {001}<100>	2.45
P {011}<111>	3.97

4.2 Effect of NTMT on electrical conductivity in Al-Zn-Mg-Cu alloy

The electrical conductivity of NTMT processed samples is between T6 and RRA, and the electrical conductivity of NTMT processed samples increases gradually with the extension of ageing time (Table 3). According to Mathiessen rule [21], the electrical conductivity is described by the following Equation (6):

$$\rho = \rho_0 - \Delta\rho_s + \Delta\rho_p - \Delta\rho_v - \Delta\rho_d - \Delta\rho_g \qquad (6)$$

where ρ_0 is the conductivity of pure aluminum, $\Delta\rho_s$ is the effect of solid solution of alloy elements on conductivity, $\Delta\rho_p$ is the effect of precipitates on conductivity, $\Delta\rho_v$ is the effect of the vacancy on conductivity, $\Delta\rho_d$ is the effect of dislocation on conductivity, and $\Delta\rho_g$ is the effect of grain boundary on conductivity. Defects such as vacancies and dislocations lead to the decrease of electrical conductivity of aluminum alloys. Precipitates of alloys have two effects on its electrical conductivity. On the one hand, the lattice distortion of the aluminum matrix is reduced due to the precipitation of solute atoms, which is conducive to the improvement of conductivity. On the other hand, the nanoprecipitates can hinder the movement of electrons and scatter electrons, which is not conducive to electrical conductivity.

The electrical conductivity of RRA processed samples is the highest among all samples (Table 3). The aluminum alloy in the RRA is overageing, so the solute atoms precipitate more completely than T6. It is more helpful to reduce the adverse effect of lattice distortion on electrical conductivity, and the formation of coarse and discontinuous GBPs is also beneficial to the electrical conductivity. Also, NTMT treatment introduces dislocations into the samples. For one thing, it promotes the formation of precipitates in aluminum alloys and improves its electrical conductivity. For another, it can strengthen the lattice distortion and have an adverse effect on the electrical conductivity. By comparing the electrical conductivity of samples with different treatments, it can be found that the promotion of dislocations on the precipitation of aluminum alloys has a greater effect on the conductivity. Therefore, the electrical conductivity of NTMT processed aluminum alloys is higher than that of T6. With the ageing time prolonging, precipitates of aluminum alloys are more sufficient, the density of dislocations gradually decreases due to the recovery, so the conductivity gradually increases. The electrical conductivity of aluminum alloys can reflect the resistance to intergranular corrosion to a certain extent, and there is a positive correlation between corrosion property and electrical conductivity. The electrical conductivity of NTMT processed alloys is significantly increased compared to T6, which indicates that the intergranular corrosion resistance is improved.

4.3 Effect of NTMT on corrosion property in Al-Zn-Mg-Cu alloy

The intergranular corrosion resistance of aluminum alloys is mainly affected by the GBPs [22, 23]. The intergranular corrosion of aluminum alloys is mainly electrochemical corrosion [24]. The potential difference between the aluminum matrix and the grain boundary makes the grain boundary more susceptible to corrosion.

The corrosion potential of η phases along grain boundaries is more negative than that of PFZ and matrix, causing GBPs to undergo anodic dissolution. The anodic dissolution rate of aluminum alloys mainly depends on the dissolution rate of GBPs.

After pre-ageing at 120°C/24 h, the aluminum alloy in the RRA formed fine and uniformly distributed precipitates in the crystals. When the temperature rose to 195°C, the precipitates inside the grain would re-dissolve into the aluminum matrix and GBPs would continue to grow and the intermittent distribution would form. Finally, finely dispersed precipitates re-formed inside the grain after 120°C/24 h, and coarse and intermittently distributed precipitates at the grain boundaries were retained (Fig. 6(b)). The GBPs in the T6 are continuously distributed, which tend to be an anodic corrosion channel in the corrosive medium, accelerating corrosion cracking. Therefore, the intergranular corrosion depth of the aluminum alloy in the RRA is much smaller than that of T6.

Meanwhile, the dislocations introduced by NTMT can create links between nanoprecipitates, improving the transformation efficiency of GBPs from small precipitates to large ones. The GBP of NTMT processed aluminum alloy is mainly the η phase. The coarser and more spaced GBPs can capture the surrounding hydrogen atoms and provide a position for them to form hydrogen molecules to escape, thus reducing the concentration of hydrogen atoms on the grain boundaries [25]. Then hydrogen embrittlement will be avoided and the rate of crack growth will be reduced, thereby improving the intergranular corrosion resistance [26]. In addition, the coarse GBPs can also reduce the corrosion current density, which can improve the corrosion sensitivity of aluminum alloys. As shown in Fig. 8(a, b), the GBPs in H3C6A6 and H3C6A12 (under ageing) are continuously distributed, which lead to a high corrosion sensitivity. However, the H3C6A18 and H3C6A24 (peak ageing) states samples show some discontinuity in the arrangement of precipitates along the grain boundary, which can block anodic dissolution channels and have better corrosion resistance [22]. Therefore, it can be concluded that the intergranular corrosion resistance gradually improves with the extension of ageing time.

Besides, the fibrous grain structure formed in deformation hinders the stress corrosion crack growth.

According to references [25, 27, 28], the Cu content of the GBPs is an important factor affecting the corrosion resistance of Al-Zn-Mg-Cu alloys, followed by the morphology of matrix precipitate (MPT), and other factors such as the segregation of solute atoms at grain boundaries, the width of PFZ, and the number of solid atoms.

5 Conclusions

1) The NTMT can significantly improve the strength of Al-Zn-Mg-Cu alloys while maintaining good ductility. On the treatment of 30% continuous Solid solution-Hot rolling (SSHR) +60% cold rolling (CR) +ageing (120°C/6 h), the highest yield strength and ultimate tensile strength reached 643 MPa and 664 MPa respectively, meanwhile, the elongation can be maintained at 9.7%.
2) The NTMT enables Al-Zn-Mg-Cu alloys to improve electrochemical corrosion resistance and intergranular corrosion resistance. Under 30%SSHR+60%CR, the electrochemical corrosion current density of aluminum alloys after ageing at 120°C for 18 hours is lower than that of RRA, and the intergranular corrosion depth of NTMT after ageing for 12 hours is lower than that of RRA.
3) The mechanism of the excellent combination of strength and ductility is thought to be the synergistic effect of dislocations substructures, texture configuration, and nanoprecipitates. The recrystallization texture P {011}<111> present in NTMT processed aluminum alloys is an important factor affecting the strength of the alloys.
4) The improvement of intergranular corrosion resistance results from changes in the dislocations and micro-morphology of GBPs. The introduction of dislocations into NTMT processed alloys creates links between precipitates and improves the transformation efficiency of GBPs from small precipitates to large ones. The formation of coarse and discontinuous GBPs after ageing can block anodic dissolution channels along grain boundaries to reduce the rate of anodic dissolution and avoid hydrogen embrittlement, thereby improving the corrosion resistance.

The authors gratefully acknowledge the financial support from the National Natural Science Foundation of China (Grant No. 51011120052 and 50871123) and the double first-class discipline construction program of Hunan Province.

References:

[1] T. Tang, G.W. Zhou, Z.H. Li, D.Y. Li, L.M. Peng, Y.H. Peng, P.D. Wu, H.M. Wang and M. Lee, Int. J. Plast. **116**, 159-191 (2019).

[2] M. Tisza and I. Czinege, Int. J. Light Mater. Manuf. **1**, 229-238 (2018).

[3] A. Heinz, A. Haszler, C. Keidel, S. Moldenhauer, R. Benedictus and W. S. Miller, Mater. Sci. Eng. A **280**, 102-107 (2000).

[4] W.S. Miller, L. Zhuang, J. Bottema, A.J. Wittebrood, P. De Smet, A. Haszler and A. Vieregge, Mater. Sci. Eng. A **280**, 37-49 (2000).

[5] X.Y. Peng, Q. Guo, X.P. Liang, Y. Deng, Y. Gu, G.F. Xu and Z.M. Yin, Mater. Sci. Eng. A **688**, 146-154 (2017).

[6] L.L. Wei, Q.L. Pan, L. Feng, Y.L. Wang and H.F. Huang, Mater. Corros. **66**, 54-60 (2015).

[7] R. Woodward, Mater. Des. **10**, 248-254 (1989).

[8] B.M. Cina, *Reducing the Susceptibility of Alloys, Particularly Aluminium Alloys, to Stress Corrosion Cracking*, US Patents (1974).

[9] E.D. Russo, M. Conserva, F. Gatto and H. Markus, Met. Trans. **4**, 1133-1144 (1973).

[10] W.T. Huo, L.G. Hou, H. Cui, L.Z. Zhuang and J.S. Zhang, Mater. Sci. Eng. A **618**, 244-253 (2014).

[11] H. Li, W. Xu, Z.X. Wang, B.J. Fang, R.G. Song and Z.Q. Zheng, Mater. Sci. Eng. A **650**, 254-263 (2016).

[12] J. Zuo, L. Hou, J. Shi, H. Cui, L. Zhuang and J. Zhang, Mater. Charact. **130**, 123-134 (2017).

[13] Z.G. Chen, J.K. Ren, J.S. Zhang, J.Q. Chen, H. Y. Jin and L. Fang, Rare Metal Mat. Eng. **44**, 2341-2346 (2015).

[14] J.K. Ren, Z.G. Chen, J. Peng, W.J. Ma and S.P. Ringer, J. Alloy. Compd. **764**, 679-683 (2018).

[15] Y.H. Zhao, X.Z. Liao, Z. Jin, R.Z. Valiev and Y.T. Zhu, Acta Mater. **52**, 4589-4599 (2004).

[16] T. Ungár, S. Ott, P.G. Sanders, A. Borbély and J.R. Weertman, Acta Mater. **46**, 3693-3699 (1998).

[17] C.H.J. Davies, N. Raghunathan and T. Sheppard, Acta Metall. Mater. **42**, 309-318 (1994).

[18] Z.G. Chen, J.Q. Chen, X.B. Guo, J.K. Ren and Y.L. Deng, Mater. Sci. Eng. A **669**, 33-40 (2016).

[19] M.J. Starink and S.C. Wang, Acta Mater. **51**, 5131-5150 (2003).

[20] P. Luo, D.T. McDonald, W. Xu, S. Palanisamy, M. S. Dargusch and K. Xia, Scr. Mater. **66**, 785-788 (2012).

[21] J.S. Dugdale and Z.S. Basinski, Phys. Rev. **157**, 552-560 (1967).

[22] L.L. Wei, Q.L. Pan, L. Feng, Y.L. Wang and H. F. Huang, Mater. Corros. **66**, 54-60 (2015).

[23] X.Y. Peng, Q. Guo, X.P. Liang, Y. Deng, Y. Gu, G.F. Xu and Z.M. Yin, Mater. Sci. Eng. A **688**, 146-154 (2017).

[24] Y. Luo, X.J. Xu, G.C. Wu, Y.K. Zhang, T. Song, B. Wang, F.B Zhang, C. Cheng and Z.D. Fei, Rare Metal Mat. Eng. **41**, 262-265 (2012).

[25] D.A. Hardwick, A.W. Thompson and I.M. Bernstein, Metall. Trans. A **14**, 2517-2526 (1983).

[26] D. Nguyen, A.W. Thompson and I.M. Bernstein, Acta Metall. **35**, 2417-2425 (1987).

[27] S.P. Knight, N. Birbilis, B.C. Muddle, A.R. Trueman and S.P. Lynch, Corrosion Sci. **52**, 4073-4080 (2010).

[28] J.K. Park and A.J. Ardell, Acta Metall. Mater. **39**, 591-598 (1991).

https://doi.org/10.1051/matecconf/202032605005

Texture formation process of 6063-type aluminium alloy during hot extrusion

Masahiro Araki[1,*], *Kenji* Matsuda[2], *Seungwon* Lee[2], *Taiki* Tsuchiya[2], *Susumu* Ikeno[3]

1 Development Headquaters, YKK AP Inc., Kurobe-shi, Toyama, 938-8613, Japan
2 Graduate School of Science and Engineering for Research, University of Toyama, Toyama-shi, Toyama, 930-8555
3 Professor emeritus, University of Toyama, Toyama-shi, Toyama, 930-8555

Abstract. This study investigated the texture formation process of A6063 alloy hot extruded material using electron backscatter diffraction (EBSD) measurements. The cube component that becomes the main orientation during hot extrusion was formed near the bearing inflow in the extrusion chamber and grew with continuous recrystallization. The Goss component with the same ED // <100> relationship as the cube component was not formed in the chamber or the bearing but was formed by discontinuous recrystallization after passing through the bearing. The TD // <111> orientation component of the surface layer was formed in the bearing and then grew with discontinuous recrystallization. However, because the internal cube and Goss components expanded preferentially, the surface layer TD // <111> components were replaced by the cube and Goss components after passing through the bearing. The cubic texture formation of extruded aluminium alloys is noted to be analogous to the formation of rolling sheet materials, formed by recovery and recrystallization from a plane strain deformation structure. However, for the A6063 alloy extruded under conditions close to industrial production, the cube components were mainly formed by recovery and continuous recrystallization.

1 INTRODUCTION

The texture of aluminium and aluminium alloys has been investigated for a long time. In the 1920s, studies reported the [111] orientation structure developed in the drawing direction of an aluminium wire rod and a double fiber structure of [111] and [100] in a round wire [1]. Although there are fewer studies on this topic compared to research on rolled materials, extensive research has been conducted on the relationship between the texture and mechanical properties of extruded Al-Mg-Si alloys since the 1960s. Research on the recrystallized texture of aluminium alloys gained remarkable attention in the 2000s, [2-3]. Adachi et al. [4] evaluated the effect of zirconium on the texture formation of extruded aluminium and showed that Al_3Zr promotes dynamic recrystallization. VanGeertruyden et al. [5] described the geometric dynamic recrystallization that occurs during the pressing of a Cr-added 6061 alloy. Niisato et al. [6] described the cross-sectional shape of extruded Al-Zn-Mg-Cu alloys and the texture and material strength after extrusion. They also described the recovery and recrystallized structures and their crystal orientations. Wu et al. [7] reported the effect of Si content in extruded Al-Si-Mg alloys on the recovery structure and recrystallization behavior after extrusion.

In this context, we previously investigated the influence of extrusion conditions on the texture of the A6063 alloy under conditions close to industrial production. Our results indicated that the cube component is the main orientation. The abundance ratio of the cube component changes with the extrusion temperature and extrusion speed [8]. Furthermore, the abundance ratio of the cube component changes with the amount of solute elements such as Mg and Si; however, the main orientation (i.e., the cube orientation) does not change [9].

In recent years, attention has been paid to the extrusion and recrystallized texture of the 6000 series alloys, but only a few studies have reported on the texture forming process during hot extrusion, [10-11]. However, as with production of rolled sheet materials, it is important to know the details of texture formation in the hot extrusion process to improve material properties by texture control.

Thus, we herein report on the details of the microstructure and texture formation process from the extrusion chamber to the bearing part of the hot extruded material of the A6063 alloy.

2 Experimental Method

2.1 Materials and processing

Table 1 shows the chemical compositions of the alloys employed herein. The material is a commercial A6063 alloy. Billets with a diameter of 152 mm were prepared by continuous casting using this alloy. The billets were homogenized in an electric furnace and cut further into

* Corresponding author: author@e-mail.org

Table 1 chemical composition of specimens (mass%)

Alloy	Mg	Si	Fe	Cu	Mn	Cr	Ti	Al
A6063	0.45	0.46	0.13	0.005	0.008	0.003	0.014	Bal.

billets of a 42-mm diameter and 100-mm length. Subsequently, a profile with an H-shape (Fig. 1) was extruded at an extrusion mold temperature of 813 K and a ram speed of 10 mm/s. The calculated extrusion ratio was 28.9. After extrusion of the sample for chamber and bearing structure observations, the mold was removed and water-cooled.

The extrusion die has a structure via which the bearing part can be divided, and after the die is cooled, the material remaining in the chamber is taken out. The sample for microstructural observations of the chamber and bearing was prepared based on the position shown in Fig. 2.

2.2 Microstructure Analysis

The crystal orientations were observed and measured using field-emission scanning electron microscopy (FE-SEM, JSM-7001F) coupled with electron backscatter diffraction (EBSD, TSL : Digiview IV OIM detector). OIM 6 was used to analyze the crystal orientation. In this report, a boundary was defined when the angle difference between adjacent measurement points was 2° or more. In addition, the area surrounded by the boundary with an angle difference of 5° or more was considered as a crystal grain, and the average crystal grain size was calculated using the area method. A boundary with an angle difference of 15° or more from an adjacent crystal grain was defined as a high-angle grain boundary, and a boundary with an angle difference between 2° and 15° was defined as a low-angle grain boundary.

The SEM and EBSD samples were prepared by electro-polishing at 273 K and 15 V for 180 s using a polishing solution with a perchloric acid:ethanol ratio of 1:9. The structure of the extruded material was observed in the ND / ED plane (cross section, perpendicular to TD). A texture component having a plane parallel to the ED / TD plane is represented by the index { h k l }, and the direction parallel to the ED is represented by the index 〈 u v w 〉. Table 2 shows the extrusion conditions for sample preparation.

Table 2 Extrusion conditions

Die temperature (K)	Ram velocity (mm/s)	Extrusion ratio
813	10	28.9

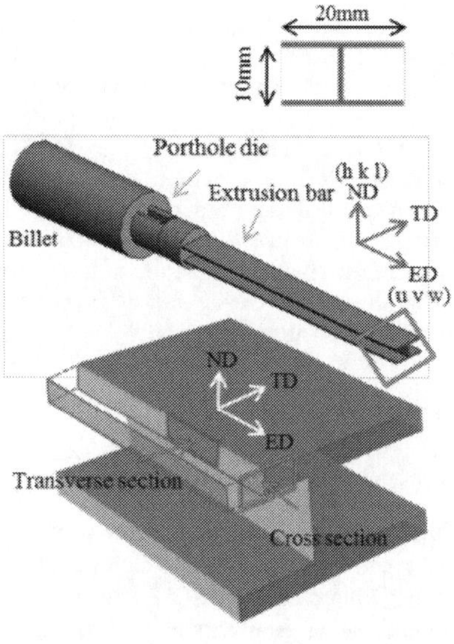

Fig. 1 Schematic of extruded sample.

Table 3 Specific orientations

Orientation name	Orientation indices {h k l}〈u v w〉
Cube	{0 0 1}〈1 0 0〉
Goss	{0 1 1}〈1 0 0〉
Brass	{0 1 1}〈2 1 1〉
TD30Bs	{1 1 2}〈1 1 0〉
Q	{0 1 3}〈3 3 1〉
S	{1 2 3}〈6 3 4〉

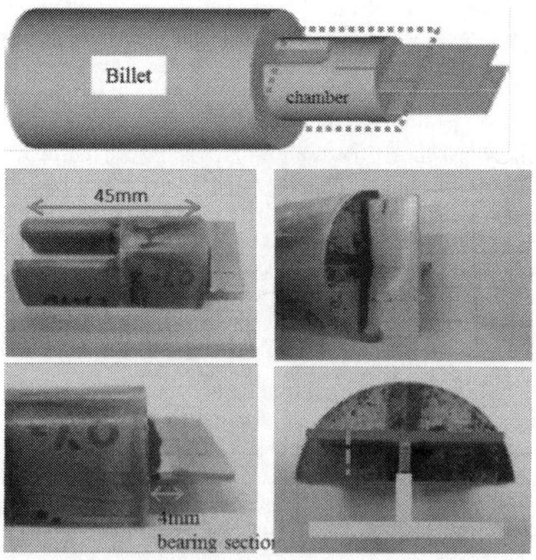

Fig. 2 Sampling locations of extrusion chamber and bearing section

https://doi.org/10.1051/matecconf/202032605005

2.3 FEM simulation

Flow simulation was conducted using the FEM software HyperXtrude (Altair Engineering, Inc.)
Deformation during hot extrusion was calculated by the following model.

$$\sigma = 1 / \alpha \sinh^{-1} (Z / A)^{1/n} \quad (1)$$

$$Z = \dot{\varepsilon} \exp (Q / RT) \quad (2)$$

Formura (2) is the Zener-Hollomon parameter, Q is the activation energy, R is the Universal gas constant. α, A, n, Q are material constants these are shown in Table 4. Die temperature is set to 813K and heat transfer coefficient between aluminium and die is 3000 W/(m^{2}K). Viscoplastic friction model with a friction coefficient of 0.3 were used at the bearing and chamber section.

Table 4 Material constants used for FEM simulation

α (1 / MPa)	ln A	n	Q (J / mol)
0.04	22.5	5.385	141 550

3 Results and discussions

3.1 Microstructure in the extrusion chamber

To observe changes in the structure and crystal orientation during the hot extrusion process of the A6063 alloy, the structure in the extrusion chamber was monitored. Microstructural observations were performed by cutting the billet remaining in the extrusion chamber along the ED/TD plane. Fig. 3 shows the results of microstructural observations by SEM-EBSP. From the inverse pole figure map (IPF-map) in Fig. 3 (a), the same color is observed to be continuously arranged along the metal flow. The crystal orientation map shown in Fig. 3 (b) indicates that specific crystal orientations such as Brass, TD30Bs, and Q components are arranged along the metal flow. Cube-oriented grains that become the main orientation after extrusion were scarcely present in the chamber, until nearly before the bearing inflow. The ratio of cube grain components in the observation area was approximately 1%.

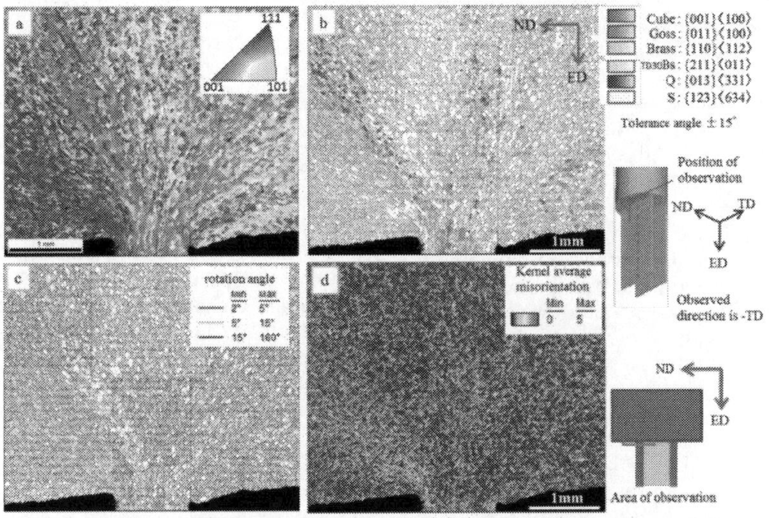

Fig. 3 (a) IPF map, (b) Crystal orientation map, (c) Grain boundary rotation angle image, (d) Kernel average misorientation map on extruded chamber section of A6063 alloy.

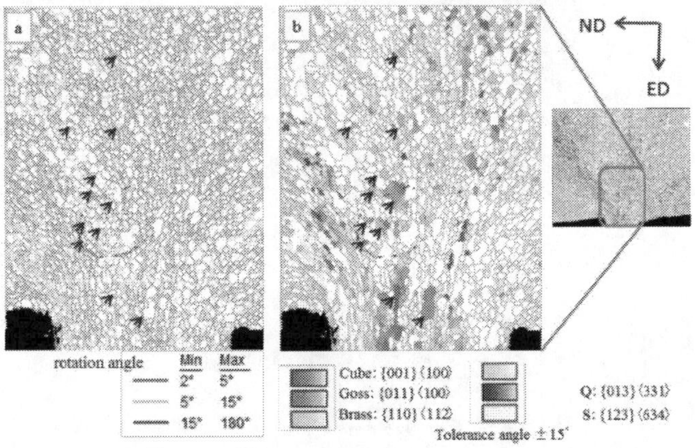

Fig. 4 Magnified image of bearing inlet section on extruded A6063.
(a) Grain boundary rotation angle image and (b) Crystal orientation map.

Analysis of the grain boundary map shown in Figure 3 (c) showed that about 40% of the grain boundaries were low angle grain boundaries (<15 °). Fig. 3 (d) is a Kernel average misorientation map (KAM map) of the same observation area. It can be seen that the green/yellow area is widely distributed in the blue area. The KAM value shows the angle difference from other measurement points in the crystal grain and does not indicate the amount of strain of the material. However it is assumed that the yellow area is the state where distortion remains. As shown in the grain boundary map of Fig. 3(c), many low-angle boundaries existing in the grains were observed. Therefore, the microstructure in the chamber part did not only include recrystallized grains but also consisted of recovered structures. It is noted that the structure formed by the presence of the cube orientation in the chamber was confirmed immediately before the bearing inflow. However, to determine its formation process, the region was expanded and analyzed.

The results are shown in Fig. 4. The crystal orientation map in Fig. 4(b) indicates that many cubic grains (shown in red grains in the figure) exist before and at the bearing inflow. The positions of the cube-oriented grains are also shown. Arrows on the grain boundary map in Fig. 4 (a) indicate the positions where cube-oriented grains exist. These arrows point to grains with low angle grain boundaries with a direction difference of 2 to 15 °. And many of the grains were surrounded by low-angle grain boundaries. Cube-oriented grains were not observed in crystal grains surrounded by high-angle grain boundaries observed in other parts of the chamber or in crystal grains that did not contain low-angle grain boundaries. It is thought that cube components not formed by the discontinuous recrystallization but mainly formed by the recovery.

3.2 Microstructure in the bearing part

The cube orientation component has been reported to become the main orientation after hot extrusion in Al-Mg-Si alloys [12]. As a result of observing a cross-section of the sample prepared in this study, the cube component present in an area ratio of approximately 30% became the main orientation. However, in the chamber shown in Fig. 3, approximately only 1% of the cube orientation component was present. The average crystal grain size was 52 μm in the chamber but increased to 98 μm in the extruded profile. This suggests that the cube component increases and expands from the chamber to the passage through the bearing. Therefore, we monitored the microstructure from the chamber to the bearing part and beyond.

Fig. 5 shows the results of microstructural observations of the bearing part from the chamber of the extruded A6063 alloy. The length of the parallel part of the bearing is 4 mm. In order to examine the relationship between the bearing position and the microstructure, the bearing is shown schematically in Fig. 5. From the crystal orientation map, the cube components shown in red in the figure rapidly increase from the bearing inflow. In the bearing part, TD30Bs components are formed on the surface of the extruded.

Fig. 6 shows an enlarged image of the microstructure from the inflow to the bearing. Fig. 6 (a), (b), and (c) are enlarged views of the bearing outflow and Fig. 6 (A) and (B) are the bearing inflows. A comparison of the microstructure of the surface layer between the inflow and outflow of the bearing confirms that the outflow grains are growth rapidly.

The grain boundary map circled areas in Fig. 6 (b) shows that there are few low-angle boundaries appear in the surface of the outflow component.

Looking at the KAM map in Fig. 6 (c), the area inside the profile is displayed in green or yellow. while the surface area surrounded by a circle is displayed in blue.

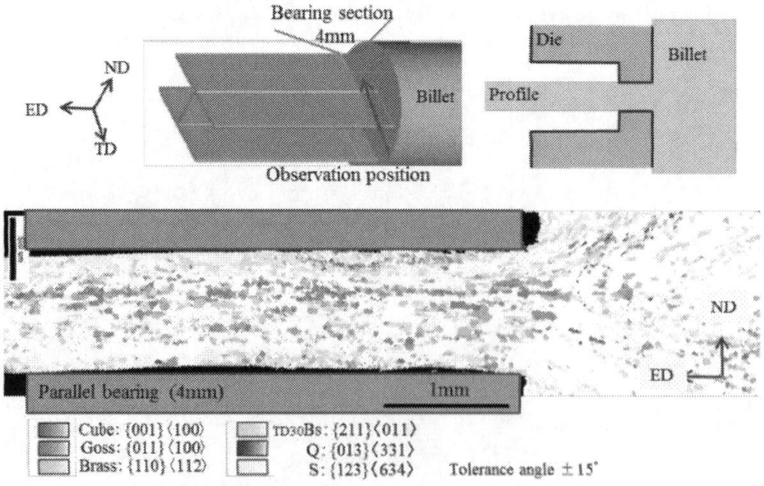

Fig. 5 Crystal orientation map on extrusion bearing part of A6063 alloy.

The KAM value represents the orientation difference within the same crystal grain. And the surface layer is displayed in blue indicate that the orientation difference within the crystal grain is smaller than that inside crystal grain and close to zero. The orientation difference in the crystal grains does not directly represent the amount of strain, but the crystal grains on the surface layer have no misorientation in the crystal grains, so these are considered to be discontinuous recrystallized grains.

The location of cube grains is shown Fig. 6. The positions of cube grains shown in red in Figs. 6 (a) and (A), these are correspond to the arrows on the grain boundary maps in Figs. 6 (b) and (B). The location of the cube grains is shown on the grain boundary map. These locations overlap with the location where low-angle grain boundaries were observed, as indicated by green and red lines in Fig. 6 (b) and (B). These observations indicate that cube grains are formed in front of the bearing and recovery at the extrusion bearing part. On the grain boundary map, there is no change in the crystal grain size and the grain boundary angle.

In the extruded profile on the bearing part, low-angle grain boundaries were continuously connected. Focusing on the bearing from the chamber, cube orientation grains are formed in this region without significant changes in grain size and orientation. In addition, cube components are formed in a region where many low-angle boundaries exist. So it suggested that cube grains generated by crystal rotation and dynamic recovery accompanying metal flow. The location of the cubic grains is indicated by arrows on the KAM map in Fig. 6 (c); however, the cube components do not appear in the recrystallized region shown in blue in the figure. Therefore, it can be considered that the cube orientation component dose not formed by discontinuous recrystallization. Several phenomena were confirmed to occur simultaneously in the bearing part, such as the formation of the cube component and the recrystallization of the profile surface layer.

3.3 Comparison with simulation results

The microstructural observations and crystal orientation analysis in the bearing part indicate that cube components are formed by recovery in the profile. Brass and TD30Bs components are formed at the surface layer and grow in the bearing outflow. To investigate differences in the structure between the interior of the profile and the surface layer and the cause of the microstructure change at the bearing outflow, the metal flow velocity and the shear stress generated in the bearing were calculated by simulation.

Fig. 7 shows a visualized image of the metal flow velocity and shear stress in the bearing and a grain boundary map corresponding to the simulation. Considering the metal flow velocity results in Fig. 7 (a), the material in contact with the bearing at the inflow is a dead metal that hardly moves. The interior material flow velocity is large. The presence of dead metal is inferred

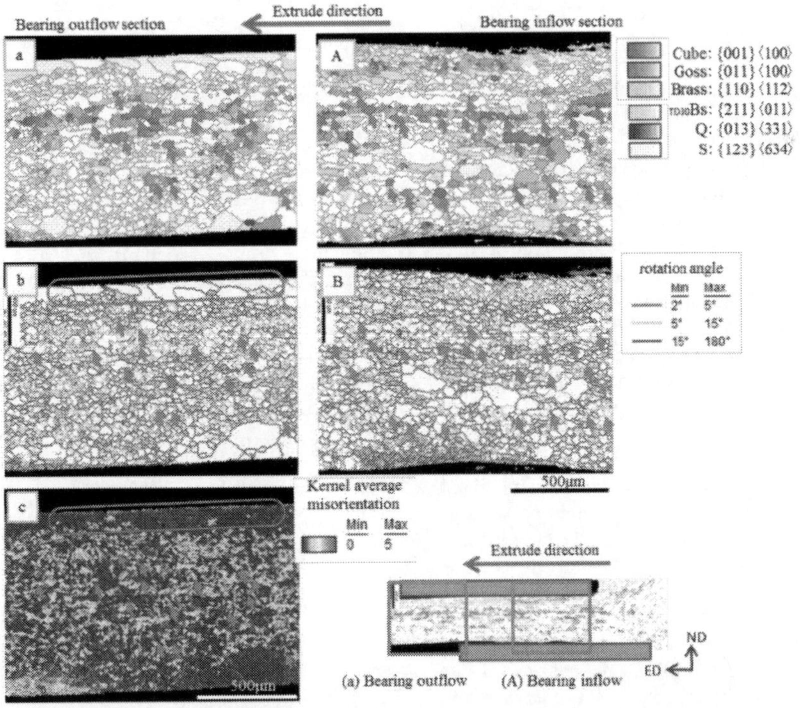

Fig. 6 Crystal orientation map (a), (A). grain boundary rotation angle image (b), (B) and KAM map (c) at bearing section.
(a), (d), and (c) are bearing outlet sections, (A), and (B) are bearing inlet sections

from the black deposits on the profile of the bearing inflow in Fig. 2. The area with large flow velocity difference overlaps with the area with large shear stress. Consequently, the surface layer shear stress is large up to nearly the middle of the bearing, and the shear stress decreases in the latter half of the bearing. Compared to the microstructural change shown in Fig. 7 (c), the region where recrystallized grains began to appear on the surface of the profile and the part where the shear stress decreased almost coincide. This indicates that the formation of the microstructure in the bearing part is affected by the shear stress and the relaxation of the stress owing to metal flow from the chamber to the bearing and the difference in flow velocity between the profile surface and the interior of the bearing.

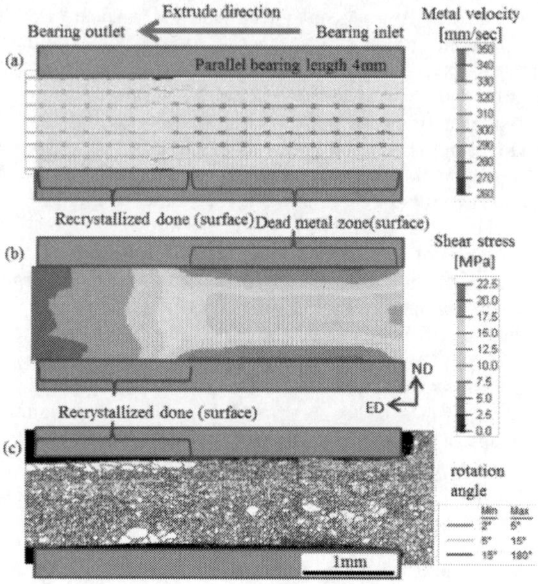

Fig. 7 Extrusion simulation results of (a) metal flow velocity, (b) shear stress at bearing section, and (c) grain boundary rotation angle image at the extruded bearing section

3.4 Microstructure and texture of extruded profile

The microstructural observations of the bearing showed that the cube component was formed at the bearing inflow of the chamber and increased in the bearing. However, the ratio of the cube component was approximately 20% in the bearing, which was smaller than that of the extruded material. Table 5 shows the main component abundance ratios in the chamber, bearing, and extruded material. After passing through the bearing, the profile extruded by 1.5 m had a cube orientation component of 30% and a Goss orientation component of 9.1%; these values were greater than those of the bearing. The ratio of Brass, TD30Bs, and S components decreased. In particular, the Goss orientation component was scarcely present in the chamber and the bearing; however, the abundance ratio was the second largest (after the cube orientation component) in the extruded profile after passing through the bearing. Fig. 8 shows a crystal orientation map and a grain boundary map of the ND/ED plane (transverse section) of the extruded material after passing through the bearing. Show in the crystal orientation map in Fig. (a), the proportions of the cube and Goss components are higher than that in the bearing, and these components extend to the surface layer of the profile. The location of the cube orientation grains in the crystal orientation map is indicated by an arrow, and the location of the Goss orientation grains is indicated by a gray triangle in the crystal grain map in Fig. 8 (a) and (b). Considering the grain boundary map, the region with many low-angle boundaries are represented by red and green lines, and crystal grains surrounded by high-angle grain boundaries are observed. It can be seen that many cube components are present in the crystal grains surrounded by the low-angle boundaries or including the low-angle boundaries. This tendency is similar to the bearing part. In contrast, many Goss-oriented grains were present in grains surrounded by high-angle grain boundaries. Goss components were scarcely present in the chamber and the bearing; however, they are abundant in the extruded profiles. Because Goss grains are crystal grains surrounded by high-angle grain boundaries, they were considered to be newly formed and grown by discontinuous recrystallization after passing through the bearing.

The texture formation process of the aluminium hot extruded material was examined by analyzing the microstructure and crystal orientation of the chamber, the bearing, and the extruded material after passing through the bearing. Fig. 9 shows a schematic of the formation of the cube orientation (i.e., the main orientation after extrusion) and the TD // <111> orientation component of the surface layer.

Table 5 Area fractions of specific direction (%)

Orientation name	chamber	Bearing	profile
Cube	1.3	20.3	30.0
Goss	1.9	2.1	9.1
Brass	7.8	6.5	0.7
TD30Bs	10.8	12.8	2.6
Q	8.9	6.4	5.4
S	6.1	8.1	4.2
Grain size (μm)	52	62	98

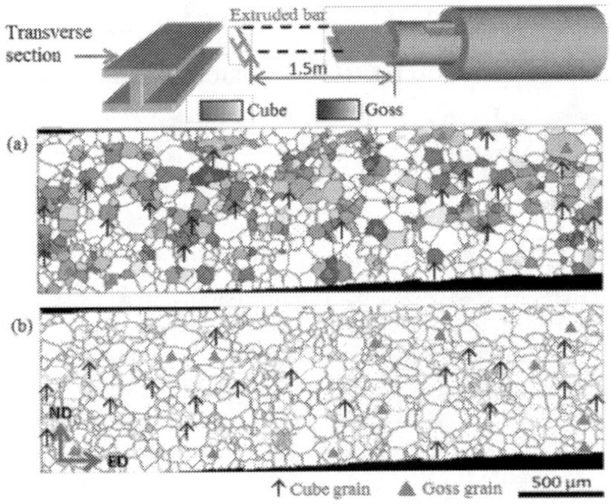

Fig. 8 (a) Crystal orientation map and (b) Grain boundary map of transverse section on extruded profile

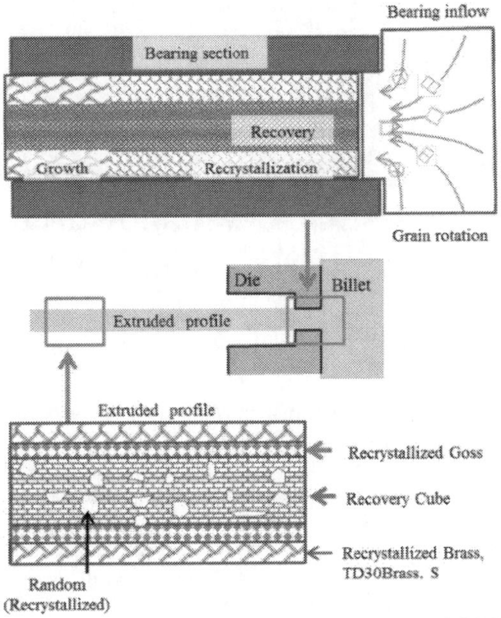

Fig. 9 Schematic of texture formation process of hot extruded A6063.

4 Summary

In order to confirm the formation process of cube-oriented grains in an extruded A6063 alloy, the microstructure in the bearing part from the extrusion chamber was observed. Consequently, cube components were confirmed to be generated before the bearing inflow in the chamber and increased in the bearing part. The cube orientation component existed in structures surrounded by low-angle boundaries and in crystal grains containing low-angle boundaries. From these results, it was concluded that the cube component was primarily formed by recovery.

It has been reported that formation process of cube components of extruded aluminium alloys is similar to that of rolling sheet materials. This process has been explained by recovery and recrystallization from the plane strain deformation structure [13]. In this study, we could not confirm the processed structure due to the plane strain in the chamber and bearing and the cube orientation component recrystallized from the processed structure.

However, cube grains appeared before the bearing inflow, suggesting that they were formed by crystal rotation due to metal flow in the chamber. In the hot extrusion process of the A6063 alloy, recrystallization proceeds from the surface layer of the profile when passing through the bearing part, and crystal grains with a TD // <111> orientation relationship, such as Brass, S, and TD30Bs, are formed. However, most recrystallized grains formed inside the extrusion die showed either Goss or random orientations. This may be due to the differences in shear stress and metal flow velocity between the surface and the interior of the extruded profile.

Acknowledgements : The authors would like to thank Hideaki Nakagawa from YKK Machinery and Engineering Group, support on the extrusion simulation.

References

1. A. Ono. Men. Coll. Engeg.Kyoshu, **3** (1923/25),195.
2. K. Dies and P. Winciers. Z.Metallkd, **57** (1966), 141, 227.
3. H. Inoue. J. JILM **52** (2002) 617-620
4. H. Adachi, K. Osamura, K. Kikuchi and J. Kusui. Mater Trans JIM **46** (2005) 211-214
5. W. H. VanGeertruyden, W. Z. Misiolek, and P. T. Wang. J Matrl Sci **40** (2005) 3861-3863
6. Y. Niisato, T. Minoda and H. Yoshida. J. JILM **63** (2013) 218-222
7. Y. Wu, H. Liao, J. Yang and K. Zhou. J. Mater. Sci. Tehnol. **30** (2014) 1271–1277
8. M. Araki, K. Matsuda, S. Lee, T. Tsuchiya and S. Ikeno. J. JILM **69** (2019) 327–331
9. M. Araki and K. Matsuda. Mater Trans **61** (2020) 104-110
10. C. Zhang, C. Wang, Q. Zhang, G. Zhao, and L. Chen. J. Mater. Proc. Tech. **270** (2019) 323-334
11. Yu, J. Q., Zhao, G. Q., Zhang, C. S., Chen, L., Mater. Sci. Eng. A **682** (2017) 679–690
12. K. Ihara, T. Shikama and K. Morita. Kobe Steel engineering reports **62** (2012) 43-47
13. J. Baumgarten, W. Bunk, and K. Lücke: Z.Metallkd., *72 (1981)*, 162-168

Effects of Si and Zr addition on strength and recrystallization behavior of Al-Mn alloy fin stocks for automotive heat exchanger

Daisuke Shimosaka[1,*], *Mamoru* Ueno[1]

[1]Nippon Light Metal Company, 421-3203 Shizuoka, Japan

Abstract. For an Al-Mn alloy heat exchanger fin stocks which continuous casting was simulated in, the effects of Si and Zr addition on strength and recrystallized grain structures after brazing heat treatment were investigated. As a result of tensile test, pre-brazed tensile strength was increased with increasing Si and Zr contents in the Al-Mn alloy. On the other hands, Zr addition have a small effect on post-brazed tensile strength regardless of Si contents. This was attributed the dissolution of Al_3Zr or $(Al,Si)_3Zr$ precipitates during brazing heat treatment. Post-brazed grain structure were refined with increasing Si contents. Meanwhile, Zr addition led to larger post-brazed grain structures. It was found that Al_3Zr or $(Al,Si)_3Zr$ precipitated during intermediate annealing played to major role in retarding recrystallization during brazing heat treatment.

1 Introduction

Al-Mn alloys, having good brazeability, thermal conductivity, corrosion resistance are widely used for the fin stocks of automotive heat exchanger. Recently, thickness of fin stocks tends to decrease in order to weight reduction and miniaturization of heat exchanger. However, buckling of fin stocks more likely to occur with decreasing thickness of fin stocks. One of the main reasons for buckling is erosion in which molten filler metal diffuse into the fin stocks along grain boundaries. As a countermeasure against erosion, it is widely known that coarsening of recrystallized grain structure after brazing is effective method due to decrease of diffusion pathway. Therefore, fin stocks manufactured by continuous casting, such as twin-roll casting and twin-belt casting, are widely applied to heat exchanger. Coarsely recrystallized grain structure was obtained in twin-roll cast Al-Fe-Ni alloy after brazing heat treatment. It was considered that recrystallization of twin-roll cast Al-Fe-Ni alloys was affected less by particle stimulated nucleation (PSN) than is that of direct-chill cast Al-Fe-Ni alloys [1]. Post-brazed properties of twin-belt cast Al-Mn alloy fin stocks and Al-Fe alloy fin stocks were studied [2]. As a result, it was obvious that twin-roll cast Al-Mn alloy fin stocks have high post-brazed tensile strength and coarse recrystallized grain structure compared to Al-Fe alloy fin stocks. It has been suggested that the twin-belt cast Al-Mn alloy fin stocks have more fine dispersoids and higher amount of solute Mn in the aluminium matrix than that of Al-Fe alloy fin stocks after brazing heat treatment. It is widely known that fine dispersoids can hinder or delay recrystallization by exerting the Zener drag on moving the subgrain and grain boundaries [3]. Thus, it was supported that twin-belt cast Al-Mn alloy fin stocks have coarse recrystallized grain structure. One of the commercial Al-Mn alloys is AA3003. Maximum chemical compositions of this alloy are: 0.6wt%Si, 0.7wt%Fe. Additions of Si and Fe enhance the precipitation of $Al_6(Fe,Mn)$ or $Al_{12}(Mn,Fe)_3Si$. Furthermore, in the Zr-containing alloys, Si and Fe increase the rate of formation of Al_3Zr dispersoids, acting as catalysis [4]. Especially, addition of Si to Al-Zr alloys enhance the formation of the fine spherical Al_3Zr dispersoids and results in an increased precipitation hardening response [5]. Effect of Zr addition on sagging resistance and crystallization behaviour of AA3003 was studied [6]. As a result, coarsely recrystallized grain structure was obtained in AA3003 alloy with Zr addition after brazing heat treatment. As mentioned above, in continuous cast Al-Mn-Si alloy with Zr, recrystallized grain size and tensile strength after brazing heat treatment may be further increased.

The aim of this paper is to investigate the effects of Si and Zr addition on strength and recrystallization behaviour after brazing heat treatment of Al-Mn alloy fin stocks that simulate continuous casting.

2 Experimental

The chemical compositions of the alloys in this study are given in Table 1. In order to consider a continuous strip casting process, the Al-Mn-Si and Al-Mn-Si-Zr alloys were cold rolled to a thickness of 0.15 mm prior to intermediate annealing without homogenization and hot rolling. Intermediate annealing was performed at 400 °C for 2 h to fully annealed on the as cold rolled sheets at 0.15 mm thickness, because of same temper condition being associated with same final cold rolled reduction in

* Corresponding author: author@e-mail.org

 https://doi.org/10.1051/matecconf/202032605006

the fin stocks. Following intermediate annealing, sheets were cold rolled to a final thickness of 0.10 mm as fin stocks. The brazing process of the automotive heat exchanger was simulated by heating the fin stocks at 600 °C for 3 min.

Table 1. Chemical compositions of alloys (mass%).

Alloy	Si	Fe	Cu	Mn	Zr	Al
S15	0.15	0.50	0.02	1.06	–	Bal.
S15Z	0.15	0.50	0.02	1.06	0.10	Bal.
S8	0.80	0.50	0.02	1.06	–	Bal.
S8Z	0.80	0.50	0.02	1.06	0.10	Bal.

Pre- and post-brazed recrystallized grain structure was observed and softening behavior was investigated during brazing heat treatment by Vickers hardness. Furthermore, pre- and post-brazed tensile properties were measured. The deistribution of dispersoids was examined by scanning electron microscopy (SEM) and transmission electron microscopy (TEM) observations. Chemical composition of the dispersoids was identified by energy dispersive X-ray spectroscopy (EDS). In order to evaluate the variation of solute atoms in matrix, the amounts of Mn and Zr solute in the matrix was analyzed by inductively coupled plasma atomic emission spectrometry (ICP-AES) after chemical extraction in phenol.

3 Results

3.1 Tensile properties

Pre- and post-brazed tensile properties of fin stocks are shown in Table 2.

Table 2. Pre- and post-brazed ensile properties of fin stocks.

Alloy	Pre-brazed			Post-brazed		
	UTS (MPa)	0.2% YS (MPa)	EL (%)	UTS (MPa)	0.2% YS (MPa)	EL (%)
S15	167	161	2.4	102	43	8.5
S15Z	177	171	1.9	105	43	9.5
S8	188	177	4.8	125	48	7.4
S8Z	195	184	3.6	128	51	6.8

Pre-brazed tensile strength was increased with increasing Si and Zr contents in Al-Mn alloys. Post-brazed tensile strength was increased by approximately 20MPa with increasing Si addition. Meanwhile, Zr addition have a small effect on post-brazed tensile strength regardless of Si contents.

3.2 Recrystallized grain structure

Fig. 1 and 2 show pre- and post-brazed recrystallized grain structure and grain size after brazing heat treatment of fin stocks.

Alloy	Before brazing heat treatment
S15	
S15Z	
S8	
S8Z	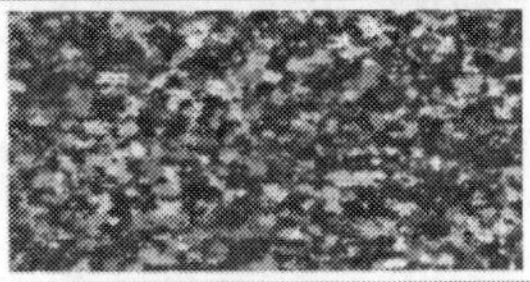

5mm

Fig. 1. Pre-brazed recrystallized grain structure of fin stocks.

Alloy	After brazing heat treatment	Grain size (μm)
S15		1400
S15Z		2200
S8		700
S8Z		1200

Fig. 2. Post-brazed recrystallized grain structure and grain size of fin stocks.

5mm

As revealed in Fig. 1 and 2, in all fin stocks, post-brazed recrystallized grains was larger than that of pre-brazed fin stocks. Post-brazed recrystallized grain size tended to decrease with increasing Si contents. On the other hands, Zr addition led to larger post-brazed grain structures regardless of Si contents. Fig. 3 shows the softening behavior during brazing heat treatment. The hardness of all fin stocks dropped off at 580 °C.

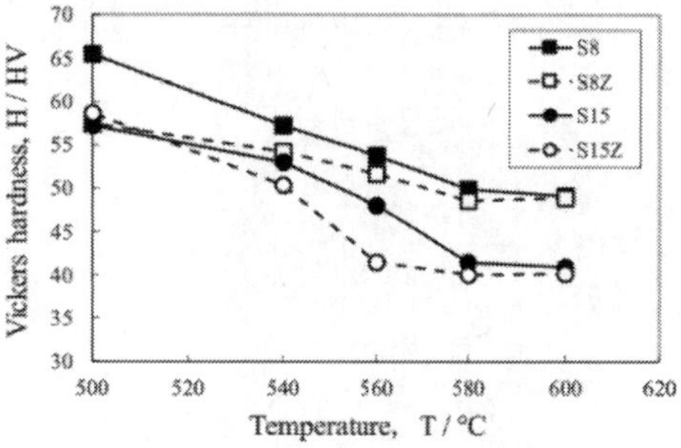

Fig.3. Softening behaviour during brazing heat treatment.

4 Discussion

4.1 Pre-and post-brazed tensile properties and microstructure

Microstructure of fin stocks were observed by SEM and TEM to investigate the effects of dispersoids distribution on tensile properties. Fig. 4 shows the distribution of secondary particles before brazing heat treatment by SEM. Based on the SEM images shown in Fig. 4, the frequency of equivalent diameters for the secondary particles was calculated using the LUZEX analysis software. Fig. 5 shows the frequency of equivalent diameters for the secondary particles in fin stocks.

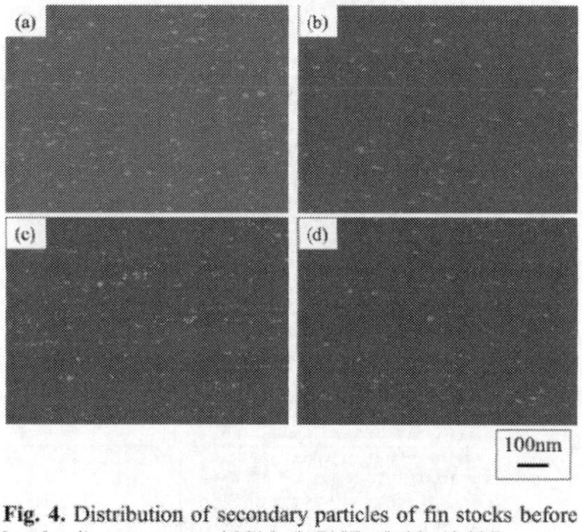

Fig. 4. Distribution of secondary particles of fin stocks before brazing heat treatment. (a)S15, (b)S15Z, (C)S8, (d)S8Z.

As shown in Fig. 5, there are no difference of distribution of secondary particles in fin stocks. It was reported that addition of Si to Al-Mn alloys have small effect on the distribution of coarse secondary particles, meanwhile, Si addition tend to increase the number of fine dispersoids [7].

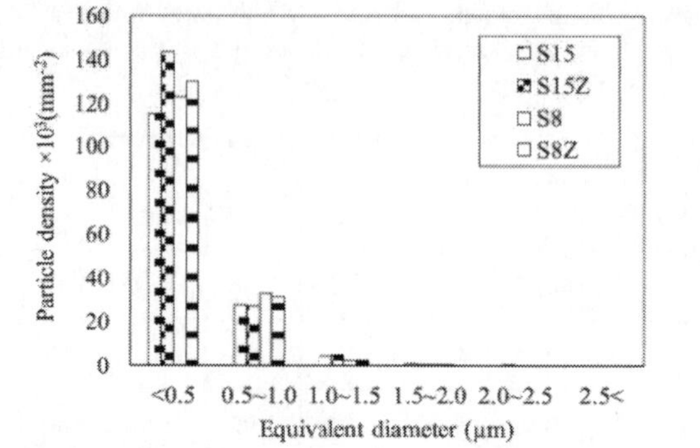

Fig. 5. Particle density of equivalent diameter for secondary particles in fin stocks.

Fig. 6 shows the TEM image of S15Z and S8Z samples before brazing heat treatment. In this study, a number of fine dispersoids were observed in S8Z sample (high Si contents). In addition, Al(Mn,Fe)Si dispersoids were detected by the TEM-EDS.

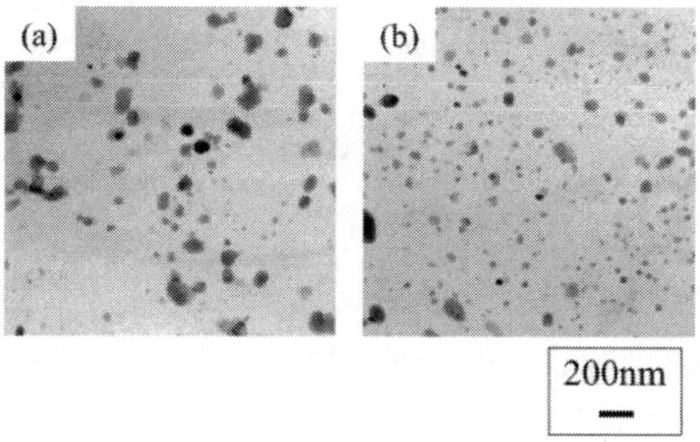

Fig. 6. Distribution of fine dispersoids of fin stocks before brazing heat treatment. (a)S15Z, (b)S8Z,

Zr-containing dispersoids were observed in S15Z and S8Z samples before brazing heat treatment (Fig.7). Fine dispersoids were observed in both samples. Regarding Zr-containing dispersoids, peaks of Al and Zr were detected in S15Z sample, but peaks of Al, Zr and Si were analysed in S8Z sample before brazing heat treatment by TEM-EDS. The EDS analysis results reveal that Si is incorporated inside dispersoids to produce $(Al, Si)_3Zr$ in S8Z sample. The amounts of Mn and Zr solute in fin stocks before and after brazing heat treatment were shown in Table 3. There are less amounts of Mn and Zr solute in matrix before brazing heat treatment. On the other hands, the amounts of Mn and Zr were increased after brazing heat treatment. Fine dispersoids, such as Al_3Zr, $(Al,Si)_3Zr$ and Al(Mn,Fe)Si, likely to be dissolved in brazing heat treatment at 600 °C. Therefore, these results suggest that the reason for the effect of Zr addition on post-brazed tensile strength was small is due to the decrease of the number of fine dispersoids during brazing heat treatment.

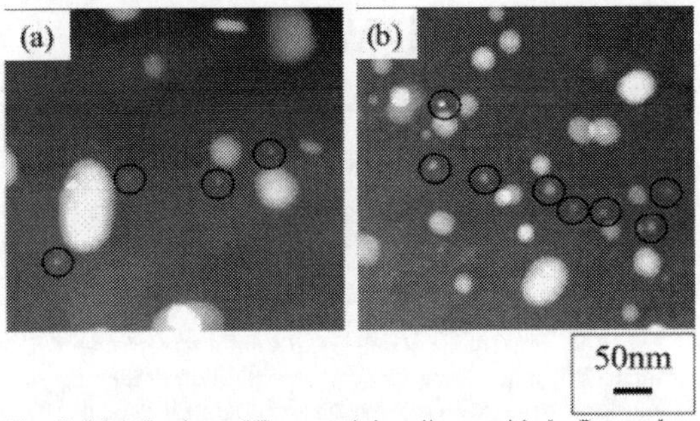

Fig. 7. Distribution of Zr-containing dispersoids in fin stocks before brazing heat treatment. (a) S15Z, (b) S8Z. Black circules indicate Zr-containing dispersoids.

Table 3. Amounts of Mn and Zr solute in matrix before and after brazing heat treatment. (mass%)

	Mn		Zr	
	Before brazing	After brazing	Before brazing	After brazing
S15	0.16	0.34	–	–
S15Z	0.13	0.30	0.02	0.04
S8	0.04	0.13	–	–
S8Z	0.04	0.12	0.03	0.04

4.2 Effects of Si and Zr on Recrystallization behavior

In previous studies, it is widely known that the recrystallization behavior was affected by (1) secondary particles (Particle Simulated Nucleation; PSN), (2) solid solute atoms in the aluminum matrix and (3) fine dispersoids (Zener-Pinning) [3].

It was reported that large secondary particles (> 1 μm) act as nucleation sites of recrystallization. As shown in Fig. 4 and 5, Addition of Si and Zr to Al-Mn alloys had little effect on the frequency of equivalent diameter for large secondary particles, which was more than 1 μm, in addition, the frequency of secondary particles exceeding 1 μm was obviously lower than that of the secondary particles less than 1 μm in all fin stocks. Therefore, this result shows that a similar effect in the larger secondary particles on the recrystallized nucleation sites in all fin stocks.

It is supposed that the solutes in the aluminum matrix would act as obstacles to dislocation migration for recrystallization in the fin stocks during brazing heat treatment. As shown in Table 3, Mn and Zr solutes before brazing heat treatment were insignificant amounts. Therefore, the influence of the solute atoms on recrystallization behavior during brazing heat treatment was considered to be similar.

Fine dispersoids of less than about 100 nm affected the recrystallization behavior because they exert a drag force on the movement of subgrain boundaries. As shown in Fig.7, the size of Al_3Zr and $(Al,Si)_3Zr$ dispersoids were refiner than that of Al(Mn,Fe)Si dispeorsoids. Therefore, it was considered that Al_3Zr and $(Al,Si)_3Zr$ despersoids effectively act as pining the grain boundaries and subgrains. In this studies, chemical composition of dispersoids was changed to $(Al, Si)_3Zr$ from Al_3Zr by addition of Si. However, no change of size, distribution and a number of disersoids were observed. Thus, although addition of Zr led to larger recrystallized grain structure during brazing heat treatment, it was considered that increase of Si have a relatively small effect in continous cast Al-Mn-Zr alloy fin stocks.

4 Discussion

To the application of the continuous cast fin stocks, effects of Si and Zr addition on pre- and post-brazed tensile properties and recrystallization behaviour after brazing heat treatment of Al-Mn alloy fin stocks were studied. The main results can be summarized as follows:

1. Pre-brazed tensile strength was increased with increasing Si and Zr contents in Al-Mn alloys. It was supposed that fine Al(Mn, Fe)Si and Al_3Zr or $(Al,Si)_3Zr$ dispersoids were distributed in matrix. However, Zr addition have relatively small effect on post-brazed tensile strength regardless of Si contents. This is because fine Al_3Zr or $(Al,Si)_3Zr$ dispersoids were dissolved during brazing heat treatment.

2. Coarsely recrystallized grain structure after brazing heat treatment of Al-Mn-Si alloy fin stocks was obtained by addition of 0.1%Zr. It was considered that fine Al_3Zr and $(Al,Si)_3Zr$ were effective for Zener-drag. However, no observation of size, distribution and a number of disersoids were confirmed. Although addition of Zr led to larger recrystallized grain structure during brazing heat treatment, it was considered that increase of Si have a relatively small effect in continous cast Al-Mn-Zr alloy fin stocks.

References

1. A. Niikura, A. Kawahara, G. Kimura and T. Doko, Mater. sci. Fourm, **519-521**, 1635 (2006)
2. T. Kokubo, T. Anami, H. Teramoto, S. Teshima and T. Toyama, Proceeding of VTMS Conference, 149 (2015)
3. F.J. Humphreys and M. Hatherly, *Recrystallization and Related Annealing Phenomena Second Edition* (2004)
4. H. Westengen, O. Reiso and L. Auran, Aluminum, 12, 768 (1980)
5. T. Sato, A. Kamio and Gordon W. Lorimer, Mater. Sci. Fourm, **217-222**, 895 (1996)
6. K. Tohma, M. Asano and Y. takeuchi, J. Japan Inst. Light Metals, **37**,119 (1987)
7. M. Karlik, T. Manik, H. Lauschmann, J. Alloys and compound, **515**, 108 (2012)

Low-Pressure Casting of Aluminium AlSi7Mg03 (A356) in Sand and Permanent Molds

Franco Chiesa[1]*, *Bernard* Duchesne[2], and *Gheorghe* Marin[1]

[1]*Centre de Métallurgie du Québec, 3095 Westinghouse, Trois-Rivières, Qc, Canada G9A 5E1*
[2]*Collège de Trois-Rivières, 3500 de Courval, Trois-Rivières, Qc, Canada, G9A 5E6*

Abstract. Aluminium A356 (AlSi7Mg03) is the most common foundry alloy poured in sand and permanent molds or lost-wax shells. Because of its magnesium content, this alloy responds to a precipitation hardening treatment. The strength and ductility combination of the alloy can be varied at will by changing the temper treatment that follows the solutionizing and quenching of the part. By feeding the mold from the bottom, the low-pressure process provides a tranquil filling of the cavity. A perfect control of the liquid metal stream is provided by programming the pressure rise applied on the melt surface. It compares favorably to the more common gravity casting where a turbulent filling is governed by the geometry of the gating system.

1 The low-pressure permanent mold process

The low-pressure permanent mold casting (LPPM) [1], schematized in Figure 1a and shown in Figure 1b and 1c, is a common process producing high quality castings resulting from two main characteristics:

1) The perfectly controlled quiescent filling from the bottom of the mold when compared to the turbulent flow associated with gravity casting; also, the liquid metal is drawn up the transfer tube from under the melt surface, preventing dross entrainment into the mold cavity.

2) The very efficient feeding achieved via the bottom feeder tube by applying a pressure of up to one atmosphere to the melt; the resulting yield is very high, typically 80-90% versus 50-60% for gravity casting. However, the low-pressure process cannot pour any casting geometries.

Low-pressure sand casting (LPS) is much less common than LPPM; the sand mold rests on top of the pressurized enclosure as shown in Figure 1d. The similarity between LPPM and LPS lies in the controlled quiescent filling of the mold with a dross-free melt; both processes also share the ability to pour thinner castings than gravity pouring would.

Fig.1. Principle of the low-pressure casting (a), with a permanent mold (b, c) and a sand mold (d).

2 Conversion of a gravity poured sand cast bellhousing to a low-pressure permanent mold one

When the order for a bellhousing designed for racecars reached one thousand units a year, it became economical to consider switching from sand casting to permanent mold. Indeed the productivity of permanent mold is higher than that of sand molding in terms of parts produced per day-employee. However, the tools and capital costs are typically 10 times more in permanent mold and the economic break-even point can be easily worked out when the cost of the tooling is known. As we will see, permanent mold provides substantially higher mechanical properties, which is particularly important in the realm of amateur car racing, as most of these parts will be used until failure. Like in passenger car wheels, the shape of the bellhousing lends itself particularly well to low-pressure casting.

Low-pressure is more expensive in capital cost than gravity casting but offers a cleaner metal, as mooth controlled filling and a higher yield and productivity.

*Corresponding author: franco.chiesa@cegeptr.qc.ca

2.1 The original gravity poured sand bellhousing

The belhousing is 400 x 350 x 165 mm in overall size and weighs 7.0 kg. In gravity sand casting, making up for the loss in volume accompanying solidification (i.e. "feeding" the casting) is insured by four 40mm diameter insulated risers and one conical flow-off, 30mm in diameter at its base (Figure 2). The AlSi7Mg (A356) melt was poured at 740°C after refining with Al5Ti1B (15 ppm B) and degassing with argon to a reduced pressure test sample density of 2.60 or 0.12 ml/100g of hydrogen [2] (Mulazimoglu, 1989). The gating system (sprue/runners/gates aeras ratio is 1 : 4 : 4) imposes a filling time of 14 seconds. Two castings were poured from a melt, the composition of which is given in Table 1.

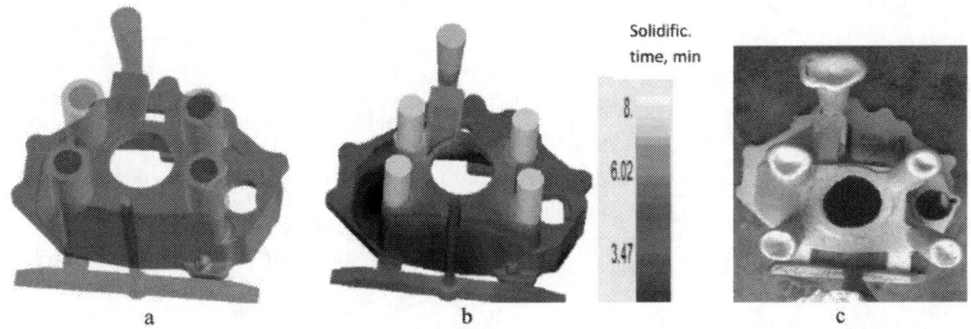

Fig. 2. a) Model of the sand cast bellhousing – b) Solidification sequence (min) – c) Casting at shakeout.

Table 1. Composition of the gravity sand poured AlSi7Mg03 (A356) casting.

Element	Si	Fe	Cu	Mn	Mg	Zn	Cr	Ti	Sr
Mass %	6.56	0.05	0.01	0.00	0.31	0.06	0.00	0.12	0.001

As shown in Figure 2b, the solification time in the casting varies between 1.2and 8.0 minutes. The castings were heat treated to a "soft" T6; i.e solutionized for 8h at 540°C, then quenched in 65°C water and aged for 4 hours at 155°C.

2.2 The low-pressure poured bellhousing

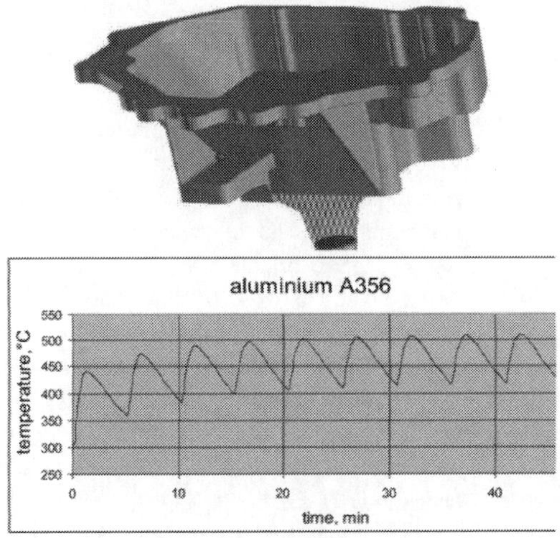

Fig. 3. Model of the low-pressure casting and predicted temperature cycling at the red dot.

Contrary to what takes place in gravity casting, the low-pressure process gives a total freedom in selecting the filling time; it was chosen at 4 seconds by applying a pressure ramp of 10mB/s on the melt. As seen on the model of Figure 3, the ratio of the casting weight/alloy poured is higher than in gravity casting (0.87 versus 0.68), which is typical of low-pressure permanent mold casting. As seen on the graph of Figure 3, the dynnamic thermal equilibrium is reached after 8 castings are poured when the initial temperature of the mold is 300°C. The production rate is close to 12 castings/hour.

The feeding of the casting is insured by applying a pressure of 900mB on the melt, corresponding to an aluminium pressure head of 3600 mm, compared to 100mm in the sand cast set-up. However, for adequate feeding of the casting, directional solidification must take place from top to bottom, in the direction of the transfer tube. The Grade 40 grey iron mold was originally machined at the dimensions of the gravity cast bellhousing. In the early trials, shrinkage appeared at several locations which were easily pinpointed as "hot spots" by solidification modeling (Fig. 4).

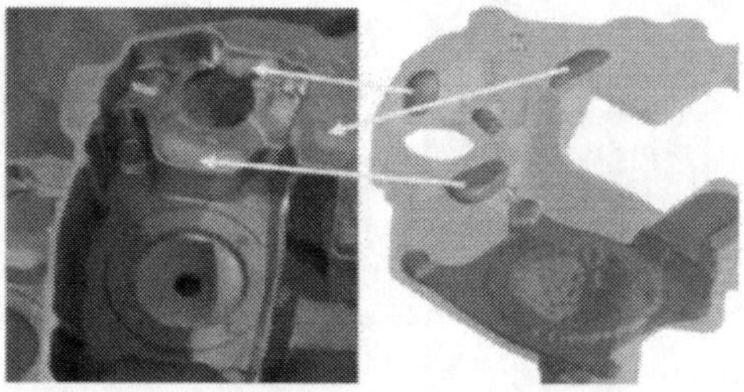

Fig. 4. Shrinkage in the first low-pressure castings and corresponding "hot spots" in the thermal model

In sand casting, chills and risers are used to achieve directional solidification, which is not the case in permanent mold casting. The solutions of the problems highlighted in Figure 4 were as follows:

1) Reducing the thickness of the base flange from 13 to 11 mm, which was possible because of the higher strength of the permanent mold cast part versus the sand cast one.
2) Reducing the mass of alloy at location R by molding a previously drilled 10mm hole.
3) Padding the casting at location X to ensure a feeding path towards the transfer tube.

The melt was argon degassed for 20 min, refined with 15ppm of boron (Al5Ti1B) and modified with 120ppm of strontium (Al10Sr); contrary to the practise in sand casting, where many consider that modification increases the risk of hydrogen pick up, modification is the rule in permanent mold casting. The melt was then poured at 720°C because of the short filling time. The composition of the A356 alloy cast in low-pressure is given in Table 2. As shown in Figure 5, the predicted solidification time in the LPPM bellhousing varies from 0.20 to 3.6 min, expectedly much shorter than in the sand cast part (1.2 to 8.0 min)

Table 2. Composition of the alloy A356 poured by low-pressure

Element	Si	Fe	Cu	Mn	Mg	Zn	Cr	Ti	Sr
Mass%	7.14	0.12	0.03	0.02	0.34	0.04	0.01	0.10	0.012

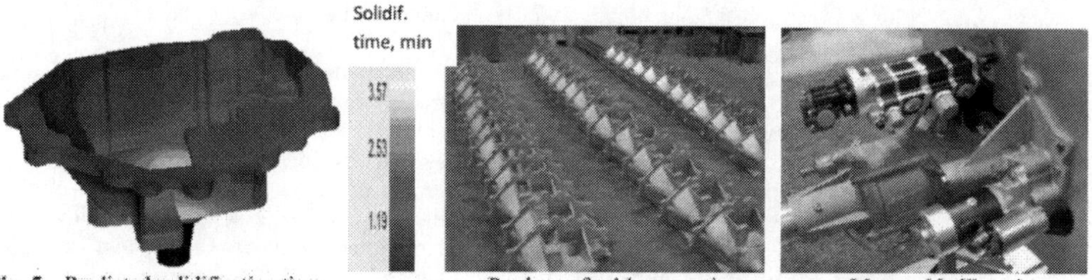

Fig. 5. Predicted solidification time — Produce of a 4 hour casting run — Mounted bellhousing

The LPPM castings were heat treated to a harder T61 temper, i.e. solutionized for 8h at 540°C, then quenched in 65°C water and aged for 8 hours at 155°C instead of 4h for the sand cast bellhousing. This was possible because the permament mold process provides a higher ductility than the sand process for identical heat treating conditions, so that bringing the temper closer to peak aging did not bring about a lower ductility (versus the sand cast part) while allowing to achieve higher yield and ultimate tensile strengths (YS and UTS).

3 Mechanical properties of the gravity sand cast and the low-pressure bellhousings

Mechanical properties may vary considerably within the same casting due to different solidifying conditions [3] (Chiesa, 2018). The properties depend mainly on the local solidification time (time elapsed between the beginning and end of solidification) and to a lesser degree, on the thermal gradient in the direction of the feeders. Note that the thermal gradient is mathematically tied to the solidus velocity; a high thermal gradient corresponds to a low solidus front velocity and either may be used in predicting the metallurgical and mechanical properties at a location id the casting as will be later shown (Equation 1).

Round and flat tensile tests bars were excised at 3 locations from a pair of castings

https://doi.org/10.1051/matecconf/202032606001

produced via each process; this was done at locations T, S and W indicated in Figure 6. The specimens reduced sections were 6 mm in diameter for round specimens and 6 x 12 mm for the flat ones. Since two castings per process were used, the average of the yield strength (YS), ultimate tensile strength (UTS) and elongation (El) were calculated from the results of four samples at locations T and F and of eight samples at location W excised from 2 castings.

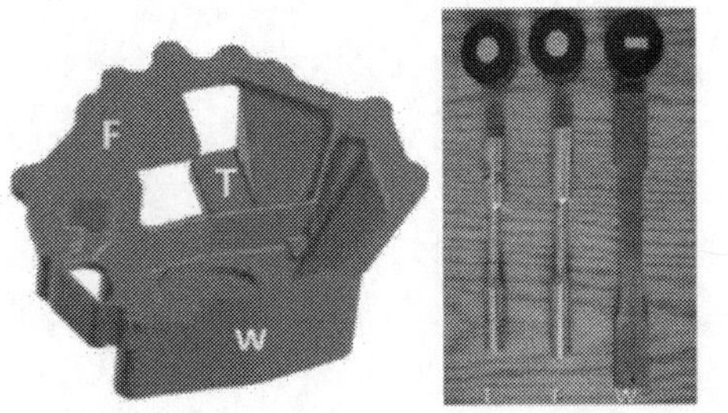

Fig. 6. Location of the tensile test bars T, F, W excised from the sand and low-pressure permanent mold castings

The results of the tensile tests are shown in the graph of Figure 7; the average and the standard deviation are indicated for the three locations and the two processes. The spread in the results is higher in LPPM, which is particularly noticeable on the elongation results. This will be explained later in the present section.

The secondary dendrite arm spacing, DAS, depends only on the local solidification time; it is thus smaller in permanent mold, and smallest in the W test bars. The porosity is also less for short solidification times, but also less with a good local feeding, which requires a strong positive thermal gradient in the direction of the feeder (or a low solidus velocity).

The finer metallurgical structure and the lower porosity of the permanent mold bellhousings explain their superior tensile properties when compared to those of the sand castings. However, the repeatability of the sand cast properties is better than the low-pressure one because the initial temperature or the mold (i.e. room temperature) is much more repeatable than in permanent mold as the thermal régime may be disrupted by changes in mold opening time due to ejection incidents, mold coating retouch, etc., so that the thermal dynamic equilibrium fluctuations are inevitable. The predicted thermal history in Figure 3 represents a perfect situation where filling of the cavity, mold opening and mold closing times would be repeated identical at each cycle, which is an unrealistic expectation in a production context. The 8h aging time of the LPPM castings results in a higher yield strength than that of the sand castings (which was aged for 4 h) while still retaining a higher ductility as represented by the elongation at break of the tensile test.

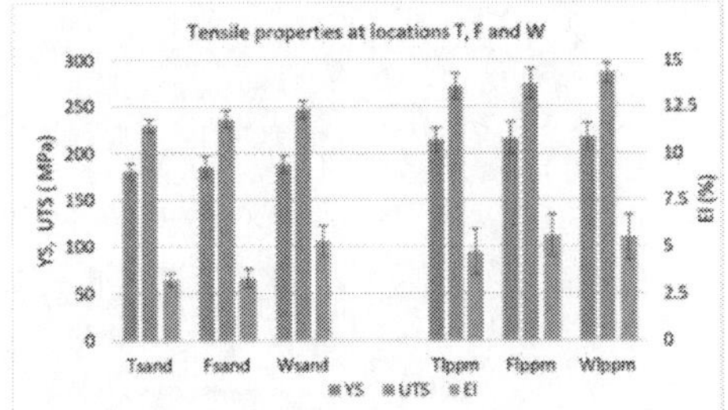

Fig. 7. Tensile properties at three locations in the sand and the LPPM bellhousings

4 Metallographic observations

Sections of the tensile samples were observed under an optical microscope; the level of microposity was determined by image analysis on the mosaics shown in Figure 8. The level of porosity is indicated along with the secondary dendrite arm spacing (DAS).

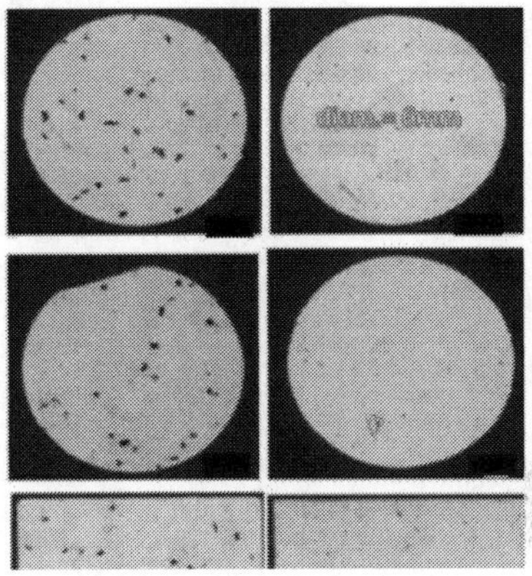

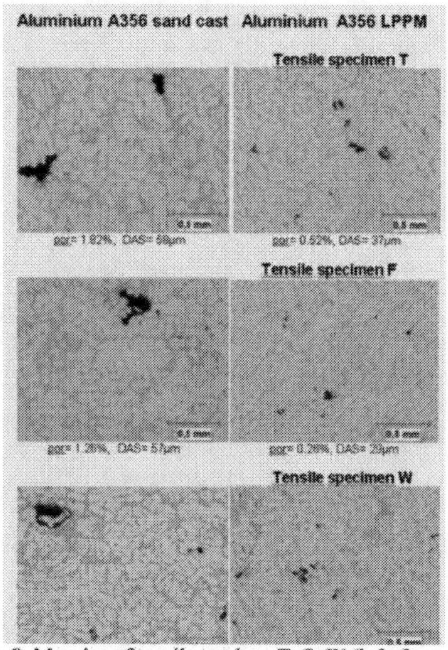

Fig. 8. Mosaics of tensile test bars T, S, W (left, from top to bottom) and corresponding micrographs

https://doi.org/10.1051/matecconf/202032606001

5 Predicting local tensile properties via solidification modeling

The quality index Q is a convenient tool to assess the metallurgical quality of Al-Si-Mg alloys (A356, 357). Q lumps strength and ductility together and gives a value in MPa independent of the alloy temper [4](Drouzy, 1980):

$$Q = UTS + 150 \text{ Log El}$$

In a previous study [5](Chiesa, 2006) showed that the quality index Q could be predicted inside a casting if a solidification modeling was performed. Q depends on the local solidification time t_{sl} in min and the solidus velocity V_s in cm/min via the relationship:

$$Q \text{ (MPa)} = 465 - 10\, t_{sl}^{2/3}\, V_s^{1/5} - 60\, t_{sl}^{1/3}$$

for an alloy A356-T6 degassed to 0.1 ppm [H]

When this equation is applied to the results of the solidification modeling, the maps of Figure 8 are obtained for the sand and LPPM castings respectively.

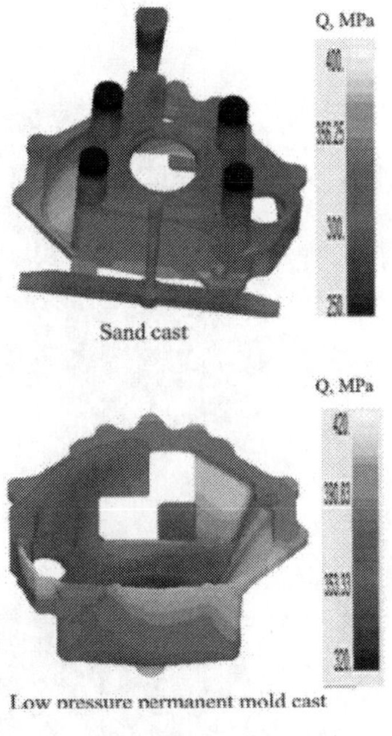

Fig. 9. Predicted distribution of the Quality index Q in the sand and LPPM castings

5.1 Calculation of UTS and El from the values of Q and YS

In A356/357 alloys, YS, UTS and El are related by the relationship [6](Jacob, 1999):

$$YS \text{ (MPa)} = UTS - 60 \text{ Log El\%} - 13$$

Since UTS = Q - 150 Log El, the above equation may be re-written as:

$$YS = Q - 210 \text{ Log El} - 13$$

or

$$\text{Log El} = (Q - YS - 13) / 210$$

Hence, the values of El % and UTS may be calculated from the knowledge of Q and YS:

$$El = 10^{(Q - YS - 13)/210} \quad (1)$$

$$UTS = Q - 150\,(Q - YS - 13) / 210 \quad (2)$$

The yield strength YS has been shown to depend only on the magnesium content of the alloy and the temper treatment; for the present conditions, using the graph from [6] page 222, the results are as follows:

Sand cast: % Mg = 0.31% aged 4 h at 155°C
YS = 187 MPa

LPPM cast: % Mg = 0.34% aged 8 h at 155°C
YS = 222 MPa

Applying equation 1 and 2 with YS =187 MPa and YS = 222 MPa gives the results in Table 3.

Table 3. UTS and Elongation as a function of the Quality index Q, for the conditions of the sand (top) and LPPM castings (bottom).

Mg=0.31%, aged for 4h at 155°C		
	YS= 187 MPa	
Q, MPa	UTS, MPa	El, %
300	229	3.0
320	234	3.7
340	240	4.6
360	246	5.8
380	251	7.2

Mg=0.34%, aged for 8h at 155°C		
	YS= 222 MPa	
Q, MPa	UTS, MPa	El, %
340	265	3.2
360	271	3.9
380	276	4.9
400	282	6.1

When the Q scales in the maps of Figure 9 are narrowed, the values of the Quality index can be accurately determined at location T, F and W for the sand and LPPM castings:

Sand casting: 306, 301 and 365 MPa at locations F, T and W
LPPM casting: 367, 388 and 396 MPa at the same locations

Refering to the results in Table 3, the predicted values of UTS and El can be determined. These predicted values are compared to the experimental ones in Figure 10.

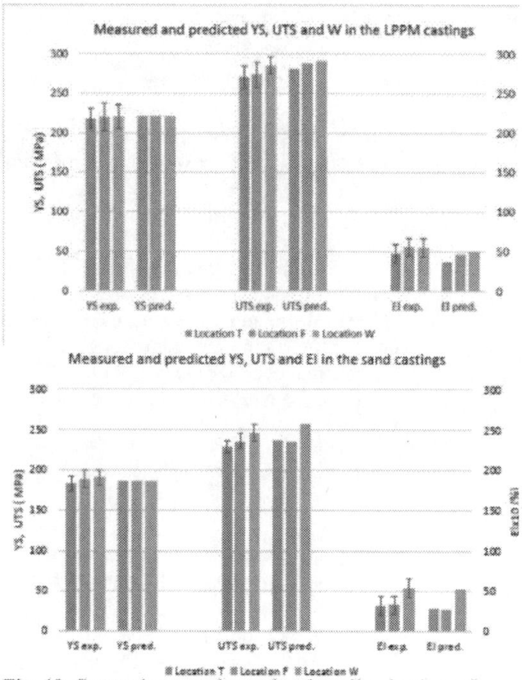

Fig. 10. Comparing experimental and predicted values of YS, UTS and El at 3 locations in the casting.

The agreement between predictions and experiments is reasonable and proves that the tensile properties can be reliably predicted for a casting poured from a clean and degassed AlSiMg melt, as long as the magnesium content and temper conditions are known. Alternatively, it is possible to determine the temper conditions necessary to achieve a set value for the elongation at a designated zone of a casting.

6 Low-pressure filling of a sand cast tank

The mold, 1200 x 1800 x 1800mm in size is shown in its filling position on the low-pressure machine at the top of Figure 11. The tank after shakeout is on the right of the same figure, upside down with respect to its position when poured by low-pressure. The runners are fed by an uncoated 1.6 mm thick steel tube; no noticeable iron pick-up was detected, which is not surprising given that the residence time of the melt in the tube is about one second. A thermocouple was inserted in the runner in order to make sure that it was fully solidified (around 12 min) when the mold was removed from the machine. The filling time was set to 20 seconds, resulting in a liquid metal velocity in the gates of 0.32 m/s, well below the critical velocity above which aluminium oxidation occurs [campbell]. The temperature of the liquid metal front is shown in Figure 12 (a) as it fills the mold cavity; it shows that the coldest metal front temperature is 690°C, i.e. 70°C above the solidus temperature of the alloy.

Fig. 11. The mold on the low-pressure machine (top); The casting at shakeout with its gating system (bottom)

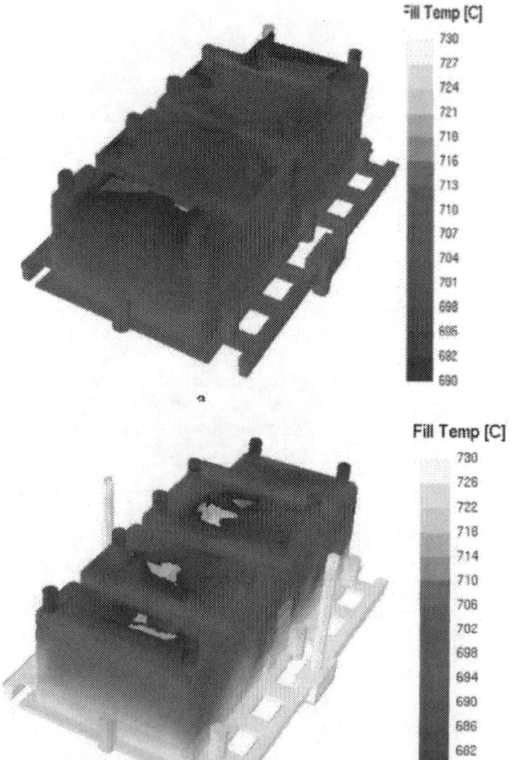

Fig. 12. Temperature of the liquid metal front during the filling: a) by low-pressure; b) by gravity

https://doi.org/10.1051/matecconf/202032606001

If one considers filling the mold by gravity for the same maximum gate velocity of 0.32m/s, since the filling slows down as the liquid metal rises in the cavity (it takes 7.5 s, or 25% of the fill time to fill the last 10% of the mold cavity), the filling time must be increased to 30 s; this is obtained with a tapered sprue bottom diameter of 40mm; the consequent liquid metal front temperature is that of Figure 12b. In gravity casting, the coldest liquid metal front is 674°C, i.e. 16°C lower than in low-pressure. This demonstrates the higher capacity of the low-pressure process to produce thin castings in sand molds; the same has been demonstrated to be the case in permanent mold castings [8].

7 Conclusions

Because of their shorter solidification time, the bellhousings cast by low-pressure permanent mold (LPPM) exhibit better tensile strength and ductility than those cast in sand molds. However, in low-pressure casting, the shape of the casting must be such that directional solidification is insured from the top to the bottom of the casting where the feeding tube is located. In sand casting, non-directional solidification may be tackled with risers, which is not the case in permanent mold. Compared to gravity casting, LPPM involves a cleaner melt because the liquid alloy is drawn from under the melt surface. Low-pressure casting results in a much better yield than gravity casting (i.e. casting weight/poured metal weight of 80-90% versus 50-60% typically for gravity casting).

The concept of the quality index allows predicting the yield strength, ultimate tensile strength and elongation in AlSiMg alloys when the magnesium content, the aging time and temperature are known.

The low-pressure sand casting is a very productive process which should first come to mind when large premium quality aluminium castings must be produced. The process should also be considered when walls are too thin to be obtained by gravity casting (e.g. 3 mm over a 400mm height). However the shape of the casting must lend itself to the process, so that directional solidification is maintained throughout the solidification process.

The authors wish to acknowledge the support of the NSER Council of Canada and the Quebec MESI Ministry. This study was conducted within the framework of the REGAL aluminium research centre.

References

1. D. Neff, S.P. Thomas, S.P., *Aluminum Casting Technology*, Third Edition, American Foundry Society Publications, ISBN-978-087433-456-2, 293 (2018)
2. H. M. Mulazimoglu, N. Handiak and J.E. Gruzleski, J.E. (1989). Some Observations on the Reduced Pressure Test and Hydrogen Concentration of Modified A356 Alloy, *AFS Transactions*, **97**, 225 (1989).
3. F. Chiesa et al., "How Properties Vary Throughout a Casting", *Modern Casting*, August 2018, 39 (2018)
4. M. Drouzy, S. Jacob and M. Richard, M, Interpretation of Tensile Results by Means of the Quality Index, *AFS International Cast Metal Journal,* June 1980, 43 (1980)
5. F. Chiesa, B. Duchesne and G. Morin, *Proceedings of the International Symposium on Aluminium: From Raw Materials to Applications,* CIMM Publications, 115 (2006)
6. S. Jacob, Quality Index in Prediction of Properties of Aluminium Castings - A Review, *AFS Transactions*, **107**, 811 (1999)
7. J. Campbell, *Complete Casting Handbook*, Elsevier Publications. ISBN-13:978-1-85617-809-9, 46 (2011)
8. F. Chiesa, J. Baril and P. Dubé P. (2009), Lights, Camera, Castings! *Engineered Casting Solutions*, March - April, 24 (2009)

Ultrasonic processing of aluminium alloys above the liquidus: the role of Zr

Dmitry Eskin[1,2*]

[1]Brunel Centre for Advanced Solidification Technology, Brunel University London, UB8 3PH Uxbridge, The United Kingdom

[2]Tomsk State University, Tomsk 634050, Russian Federation

Abstract. Ultrasonic melt processing (USP) is gaining quite an interest in recent years due to the benefits of this technology to the melt quality and structure refinement. A number of mechanisms have been identified that govern the effects of USP at different stages of melt processing. Technologically it is advantageous to apply USP to the fluid melt rather than to a mushy solidifying alloy. In this case heterogeneous nucleation on available or activated/multiplied substrates is the main mechanism. Among these substrates, primary crystals of Al_3Zr phase were shown to be potent and effective. This paper gives a review of the own research into the role of Al_3Zr in structure refinement in various groups of Al alloys, from solid-solution type to hypereutectic. This overview includes the evidence of a possible eutectic reaction between Al and Al_3Zr in Al-rich alloys, mechanisms of Al_3Zr formation and refinement under USP (that enables these primary crystals to be active substrates for Al and some other primary phases), the role of USP in facilitating primary solidification of Al_3Zr in the Al-Zr system, and the additional benefits of solute Ti presence. The paper is illustrated with the data obtained over the last 15 years of research led by the author.

1. Interaction of Zr with aluminium

It seems like a strange topic to discuss as it is well known that the Al_3Zr phase reacts with liquid aluminium through a peritectic reaction L + Al_3Zr = (Al) at 661 °C (934 K) [1] with the peritectic point at 0.28 wt% Zr and the maximum liquid solubility of Zr in (Al) 0.11 wt% at 661 °C [2]. This assumption works well for most practical purposes and seems to fit with most of experimental data. Nevertheless, some experimental observations, most notably the occurrence of small Al_3Zr particles in the interdendritic spaces (i.e. grain/dendrite boundaries) after solidification, raise concerns about the solidification reaction in the Al-rich alloys with Zr.

It is worth noting that most of the prior work on the equilibria in the Al corner of the Al-Zr phase diagram has been done by assessment rather than by direct measurements [1]. However, there were some obvious inconsistencies. McPherson and Hansen [3] reported a thermal arrest at 655 °C for a sample containing 6.4 at.% (18.7 wt%) Zr though there had been no discussion of this particular measurement. More recently Janghorban et al. [4] performed careful thermal analysis involving pure Al reference samples and observed thermal arrests for a sample containing 1 at.% (3.3 wt%) Zr at 654 and 656 °C, which indicated a eutectic reaction with the invariant temperature approximately 5 °C below the melting point of Al. They dismissed these results and ascribed them to errors in the measurement of the liquid compositions, i.e. the Zr content of the liquid in equilibrium with Al_3Zr.

Therefore, we performed a series of studies to clarify the phase equilibrium in the Al-Zr system by experiments and thermodynamic calculations [5, 6]. These studies have shown that, at least under certain, close to normal (0.5 to 5 K/s) casting conditions, the formation of fine Al_3Zr particles at dendrite boundaries is a typical phenomenon in Al-Zr alloys containing 0.15 to 1 wt% Zr (Fig. 1). Zr should, of course, segregate to the centre of a grain if it follows a peritectic solidification reaction. Its segregation to the grain boundaries may be a good indicator of a eutectic solidification. The nature of the particles as Al_3Zr has been established using transmission and scanning electron microscopy and energy dispersive X-ray analysis [5, 6]. Accurate differential thermal analysis of as-cast samples identified a temperature arrest at 659 °C [5]. More refined DTA analyses on samples annealed for long times showed the thermal arrest to be close to 659.9 °C [6]. The discrepancy between these results may be due to lack of equilibration of the as-cast samples. This correlates well with the thermodynamically predicted decrease of the invariant reaction temperature using the results of *ab initio* and CALPHAD calculations (0.1 to 0.2 K)

[6]. The experimentally determined temperature is lower than the melting point of pure aluminium, which confirms the presence of a eutectic reaction that may look like shown in Fig. 2. Note that the eutectic point in the new version almost coincides with the peritectic point (maximum liquid solubility) in the currently adopted version of the Al–Zr phase diagram. The question whether this eutectic solidification reaction is equilibrium or nonequilibrium remains open.

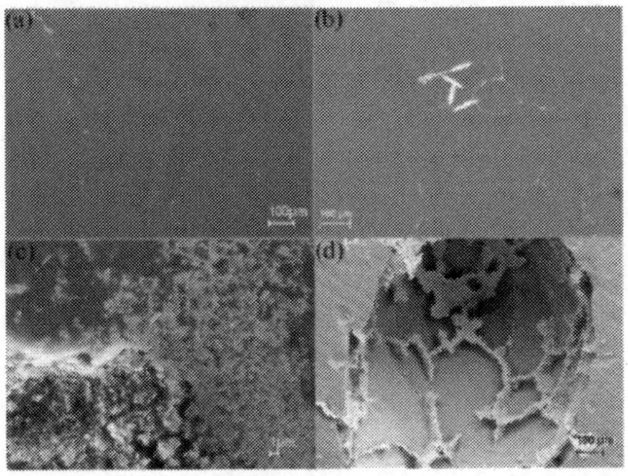

Fig 1. Secondary electron images of grain-boundary particles formed during solidification in an Al–0.4 wt% Zr alloy at cooling rates of 5 K/s (a) and 1 K/s (b–d): a, b, general view; c, higher magnification of grain boundary particles; and d, deep etching of the matrix [5].

It is well known that the formation of Al_3Zr primary phase is rather sluggish, which is a reason of Zr supersaturating the aluminium solid solution during solidification. As the alloys with Sc (eutectic phase diagram) prove, it is not the consequence of the type of the phase diagram but rather the kinetics effect of the primary phase formation. Therefore, the suggested change of the solidification reaction does not change the fact that the Al–Zr alloys are forming supersaturated solid solution during solidification, albeit the amount of Zr available for that will be the function of how much Zr has been consumed by primary and, now, eutectic particles.

In addition to the contribution to the general knowledge on the Al–Zr solidification sequence, the revealed eutectic nature of the Al–Zr phase diagram has implications for the applications that are related to the formation of the primary Al_3Zr phase, e.g. grain refinement that will be considered later in the paper. As only primary particles are useful for the grain refinement, the occurrence of the eutectic reaction renders part of Zr introduced into the alloy useless. Therefore, we need to look at the conditions when the primary solidification is promoted while eutectic reaction is suppressed.

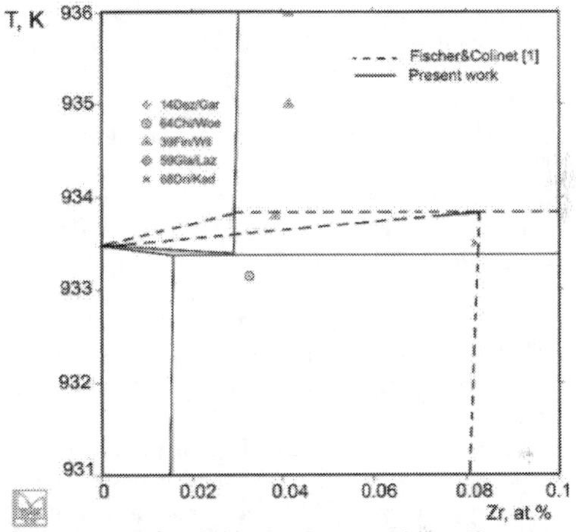

Fig 2. Comparison of the reference phase diagram for Al rich corner of the Al–Zr system with the new version of the phase diagram based on *ab initio* and CALPHAD results [6].

2. Effects of Zr in grain refinement of aluminium alloys

Zirconium is added to aluminium alloys for several purposes: to prevent or control recrystallization [7], to enhance mechanical properties and thermal stability while retaining electrical conductivity [8], and to control grain size during solidification [9]. In the first two cases, supersaturation of the Al solid solution is achieved during solidification and the metastable $L1_2$ Al_3Zr phase precipitates during annealing, pinning dislocations and subgrain boundaries [7]. In the last case, primary crystals of the equilibrium Al_3Zr phase act as nucleating substrates for Al grains [10]. Non-equilibrium accelerated solidification of Al-rich Al-Zr alloys typically results in either formation of a supersaturated solid solution of Zr in Al or the formation of the metastable Al_3Zr phase [11]. So the solidification conditions that promote the formation of the primary particles need to be met to achieve the desired grain refining effect.

Primary Al_3Zr particles should, however, be of rather small size to realize the potential of a grain refiner. Ultrasonic cavitation has long been known to be an efficient means of primary crystal refinement, and also the role of primary Al_3Zr particles in grain refinement upon ultrasonic processing (USP) has been reported already in the 1960s [12]. Later on, the important role of Zr in the improved efficiency of USP for grain refinement of a range of aluminium alloys has been demonstrated on laboratory and industrial scale [13, 14, 15, 16].

In the studies reported here we used a 5-kW water-cooled magnetostrictive transducer coupled with a Nb sonotrode (17.5 kHz, amplitude at least ±20 μm). The alloys were prepared in an electric furnace and the USP has been done in the temperature range above the formation of primary (Al) grains.

Ultrasonic melt processing facilitates the refinement of primary aluminides, including Al_3Zr, through enhanced nucleation on oxides and through fragmentation [17]. To achieve the best results, the processing should be applied above the liquidus of (Al) but within the temperature range of primary solidification of Al_3Zr [15, 17]. The refined to a size of a few μm primary aluminides act as a very efficient substrate for (Al) grains. Figure 3 illustrates this for a number of alloying systems and a range of casting processes. One can see that the significant grain refinement can be achieved not only in hypoeutectic alloys (Fig. 3a-f), but also in such heterogeneous alloys as hypereutectic Al–Ni and Al–Fe (where eutectic grains become refined, Fig. 3i, j)) [16]. Small additions of Ti facilitate the grain refining effect of Al_3Zr upon USP as has been studied in detail elsewhere [15] and thought to be due to the growth restriction effect of solute Ti and a possible change in the surface energy of Al_3Zr doped with Ti [9]. As a result of these studies, an Al–Zr–Ti master alloy has been suggested as a specialised grain refiner for aluminium alloys cast with USP [18] (Fig. 3g, h).

It has been shown that USP of the melt above and through the primary solidification range of Al_3Zr promotes its nucleation and refinement as illustrated in Fig. 4a, b [9, 15, 17]. The number of primary particles increase dramatically after USP while their size decreases. One can notice that the morphology of the particles also changed from a flat dendrite to a plate, which should be more favourable for acting as a nucleating substrate. What is more striking is that the amount of Al_3Zr particles at the dendrite boundaries decreased significantly, which is further illustrated in Fig. 4c, d.

The enhanced nucleation of primary Al_3Zr particles upon USP leaves less free Zr in the system to react by the eutectic reaction, effectively suppressing it. The potential of Al_3Zr (refined during USP) becomes available for grain refinement.

A small addition of Ti seems to assist in promoting the primary phase formation at the expense of the "eutectic" one. Figure 4e shows that there are more primary Al_3Zr particles in the presence of Ti with somewhat decreased amount of grain-boundary crystals. The primary particles are then refined upon USP with virtually no grain-boundary particles formed as shown in Fig. 4f.

These observations suggest one more mechanism of positive effect of Ti on grain refinement induced by primary Al_3Zr, i.e. suppressing eutectic reaction in the Al–Zr system and promoting formation of primary crystals. There may be three reasons for that: (a) shifting the equilibrium towards the peritectic solidification and/or (b) changing the thermodynamic properties of the Al_3Zr phase to facilitate its formation; and/or (c) changing the activity of oxides that act as substrates for Al_3Zr, e.g. via increased wettability.

3. Conclusions

1. The interaction between Al and Zr may be more complicated than it is accepted. An equilibrium or nonequlibriun (preudo) eutectic reaction may take place in Al-rich alloys under conventional casting conditions. This results in the formation of Al_3Zr particles at the dendrite boundaries, effectively decreasing the amount of primary Al_3Zr phase.
2. Ultrasonic melt processing performed above the liquidus of (Al) but in the range of primary formation of Al_3Zr results in significant refinement of primary Al_3Zr particles through enhanced nucleation on oxides and fragmentation. An additional effect of USP is in the decreased amount of Al_3Zr particles formed at the dendrite boundaries.
3. Small additions of Ti seem to suppress the eutectic reaction and promote primary solidification of Al_3Zr. The mechanism of this is still unclear and may be related to the shift in thermodynamic equilibrium or changed surface energy of the phases involved.
4. The enhanced grain refinement of a wide range of Al alloys by USP in the presence of Zr and Ti is due to the favourable combination of factors, i.e. suppression of eutectic reaction, promotion of primary Al_3Zr formation and refinement of its particles through enhanced heterogeneous nucleation and fragmentation.

Acknowledgements

This overview paper is a summation of efforts of many researchers with whom I collaborated over many years. I would like to acknowledge the contribution of T. Atamanenko, L. Zhang, F. Wang, T. Subroto, G. Salloum-Abou-Jaoude, S. Chankitmunkong, A. Khvan, C. Fang, and A. Dinsdale. Also the financial support of M2i (The Netherlands) for projects MC4.05215 and M11.5.11419 and EPSRC (The United Kingdom) for projects Ultra-Cast (Grant EP/L019884/1) and UltraMelt2 (Grant EP/R011095/1) is highly appreciated.

https://doi.org/10.1051/matecconf/202032606002

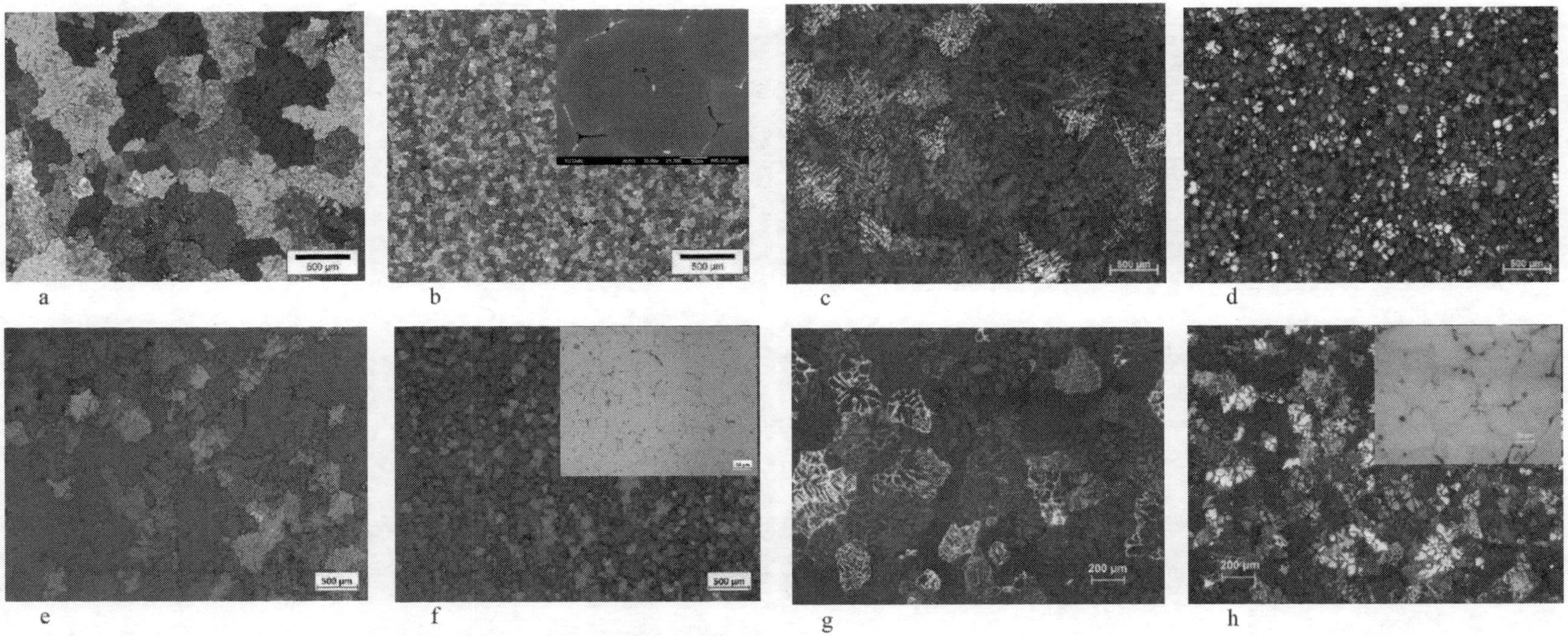

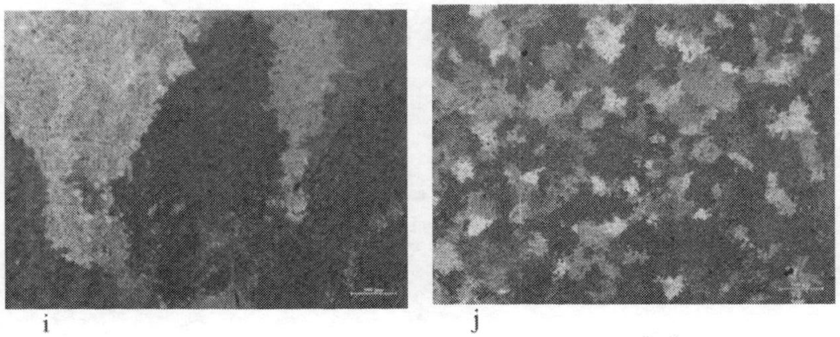

Fig 3. Effects of Zr addition and USP on grain refinement in various Al-based alloys: (a) AA7075; 0.18% Zr+ 0.08% Ti, No USP; gravity casting, grain size 150 µm; (b) AA7075; 0.18% Zr+ 0.08% Ti, USP, gravity casting, grain size 30 µm; (c) AA6XXX, DC casting, grain size 350 µm; (d) AA6XXX+0.2 Zr, DC casting, grain size 120 µm; (e) AA6XXX+0.2 Zr, DC casting, grain size 210 µm; (f) AA6XXX+0.2 Zr, DC casting with USP in a launder, grain size 100 µm; (g) Al–4% Cu, USP; grain size 310 µm; (h) Al–4%Cu + AlZrTi MA, USP; grain size 140 µm; (i) Al–8% Ni–0.3% Zr, gravity casting; and (j) Al–8% Ni–0.3% Zr, USP, gravity casting. Insets show primary Al_3Zr particles acting as substrates.

https://doi.org/10.1051/matecconf/202032606002

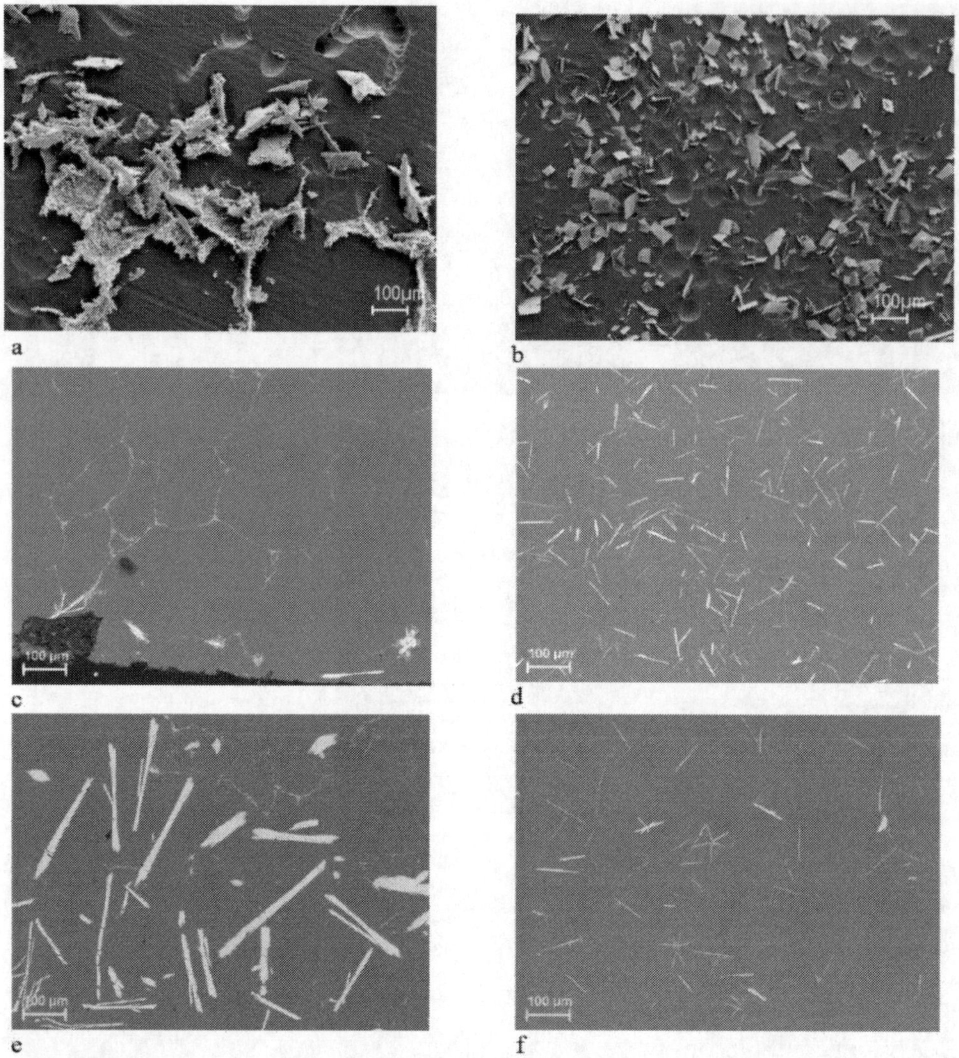

Fig 4. Effect of USP and Ti addition on the formation of Al_3Zr phase during solidification: (a) Al–0.4% Zr, deep etched (note primary Al_3Zr flat dendrites and fine grain-boundary crystals); (b) Al–0.4% Zr Ti, USP (note plate-shaped primary Al_3Zr); (c) Al–0.4% Zr; (d) Al–0.4% Zr, USP; (e) Al–0.4% Zr–0.1% Ti; and (f) Al–0.4% Zr–0.1% Ti, USP.

References

[1] E. Fischer, C. Colinet, J. Phase Equilbr. Diff. **36** (2015) 404–413.

[2] L.F. Mondolfo, *Aluminum Alloys: Structure and Properties*, Butterworths, 1976.

[3] D. J. McPherson, M. Hansen, Trans. ASM **46** (1954) 354–74.

[4] A. Janghorban, A. Antoni-Zdziobek, M. Lomello-Tafin, C. Antion, Th. Mazingue, A. Pisch, J. Therm. Analysis Calorim. **114**(3) (2013) 1015–1020

[5] F. Wang, D.G. Eskin, A.V. Khvan, K.F. Starodub, J.J.H. Lim, M.G. Burke, T. Connolley, J. Mi. Scr. Mater. **133** (2017) 75–78.

[6] A.V. Khvan, D.G. Eskin, K.F. Starodub, A.T. Dinsdale, F. Wang, C. Fang, V.V. Cheverikin, M.V. Gorshenkov, J. Alloys Comp. **743** (2018) 626–638.

[7] N. Ryum, Acta Metall. **17** (1969) 269–278.

[8] N.A. Belov, A.N. Alabin, I.A. Matveeva, D.G. Eskin, Trans. Nonferr. Met. Soc. China **25** (2015) 2817–2826.

[9] T.V. Atamanenko, D.G. Eskin, M. Sluiter, L. Katgerman, J. Alloys Comp. **509** (2011) 57–60.

[10] F. Wang, D. Qiu, Z.-L. Liu, J.A. Taylor, M.A. Easton, M.-X. Zhang, Acta Mater. **61** (2013) 5636–5645.

[11] E. Nes, H. Billdal, Acta Metall. **25** (1977) 1031–1037.

* Corresponding author: dmitry.eskin@brunel.ac.uk

[12] G.I. Eskin. *Ultrasonic processing of molten aluminium* (Metallurgiya, Moscow, 1965).

[13] G.I. Eskin. *Ultrasonic treatment of light alloy melts*, 1st Edn. (Gordon and Breach, Amsterdam,1998).

[14] G.I. Eskin, D.G. Eskin. *Ultrasonic treatment of light alloy melts*, 2nd Edn. (CRC Press, Boca Raton, 2015).

[15] T.V. Atamanenko, D.G. Eskin, L. Zhang, L. Katgerman. Metall. Mater. Trans. A **41A** (2010) 2056–2066.

[16] S. Chankitmunkong, D.G. Eskin, C. Limmaneevichitr. Mater. Sci. Eng. A. 2020 (submitted)

[17] F. Wang, D. Eskin, T. Connolley, J. Mi. Trans. Nonferr. Met. Soc. China **27** (2017) 977–985.

[18] V.M. Sreekumar, D.G. Eskin. JOM **68** (2016) 3088–3093.

Strip casting using a roll caster equipped with a scraper

Toshio Haga[1,*]

[1]Osaka Institute of Technology, Department of Mechanical Engineering, 5-16-1 Omiya Asahi-ku Osaka city, Japan

Abstract. An Al-Mg strip without center segregation could be cast using a single-roll caster equipped with a scraper at speed of 40 m/min. The scraper was useful for flattening a free solidified surface and for cooling the solidification layer by pushing the solidification layer to the roll. Clad strips consisting of 1) an Al-Mn base strip and an Al-Mg overlay strip and 2) an Al base strip and an Al-Sn-Cu overlay strip could be cast using an unequal-diameter twin-roll caster equipped with a scraper at speeds of 30 m/min and 15 m/min, respectively. The base strip and overlay strip were strongly bonded at the interface between the base strip and the overlay strip. The elements of the overlay strip did not diffuse into the base strip. The scraper played two roles in the casting of the clad strip: prevention of the mixture of two kinds of molten metal and making the surface of the base strip a semisolid of high solid fraction.

1 Introduction

Al-Mg alloy strip cast using a twin-roll caster has center segregation [1]. A scraper was proposed and attached to a single-roll caster to cast Al-Mg strip without center segregation [2,3]. The melt drag process uses a single-roll caster to cast a strip without center segregation. However, the thickness distribution is not uniform because the free solidified surface is not flat. In the melt drag process, cooling is insufficient due to poor heat transfer between the roll and the solidification layer because the solidification layer is not pushed into the roll, and the Al-Mg strip is easily broken. As a result, Al-Mg strip cannot be continuously cast using the conventional melt drag process. The thickness distribution and cooling of the strip can be improved by a scraper. The scraper scribes the semisolid layer on the solidification layer, and the thickness distribution and the flatness of the free solidified surface are improved. The heat transfer between the roll surface and the solidification layer becomes greater as the scraper pushes the solidification layer into the roll surface. As the result, the strip cooled enough to be cast continuously.

The scraper was attached to an unequal-diameter twin-roll caster to cast clad strip [4,5]. Two kinds of molten metals can be separated by the scraper and are not mixed. The scribed surface is a semisolid of high solid fraction. The bonding interface does not contact the oxidizing atmosphere due to the effect of the scraper. The solidification layers (two strips) are strongly bonded free from oxide film. In this manner, the scraper is very useful in the casting of both single strip and clad strip. The unequal-diameter twin-roll caster is suitable for scraper attachment and for casting of clad strip depending on the position and size of the rolls. A copper roll is used in order to increase late cooling of the strip in these casters. Therefore, the strips are rapidly solidified, and the casting speed can exceed 15 m/min.

In the present paper, a single-roll caster and a twin-roll caster equipped with a scraper, as well as the properties of the single strip and the clad strip, are shown. The clad strip was cast using a vertical-type tandem twin-roll caster, and the strips cast by the two types of twin-roll casters were compared.

2 Roll caster

The single-roll caster equipped with a scraper and an enlarged view near the scraper are shown in Figs. 1 and 2, respectively [2,3]. The scraper was made of mild steel plate and was covered by an insulator sheet of 2 mm in thickness. The scraper scribed the semisolid layer on the solidification layer by a small load.

Two types of twin-roll casters to cast clad strip are shown in Figs. 3. A vertical-type tandem twin-roll caster (VTTRC) is shown in Fig. 3(a), and the unequal-diameter twin-roll caster equipped with a scraper (UDTRCS) is shown in Fig. 3(b) [4-8].

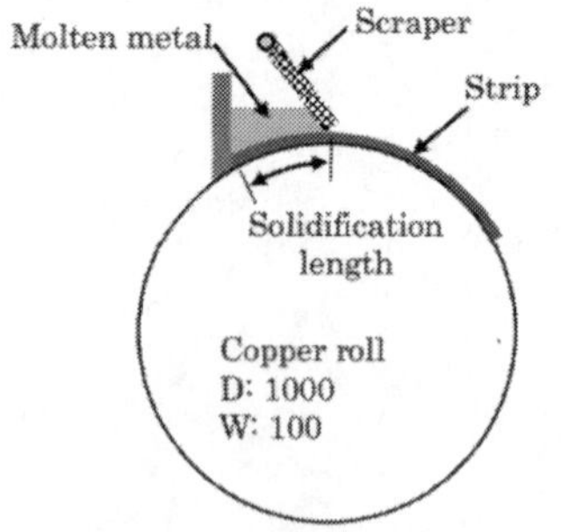

Fig. 1. Schematic illustration of a single roll caster equipped with a scraper

* Corresponding author: toshio.haga@oit.ac.jp

 https://doi.org/10.1051/matecconf/202032606003

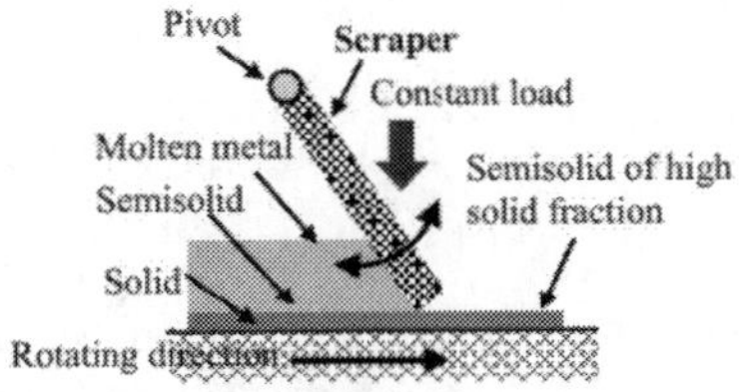

Fig. 2. Schematic illustration near the scraper

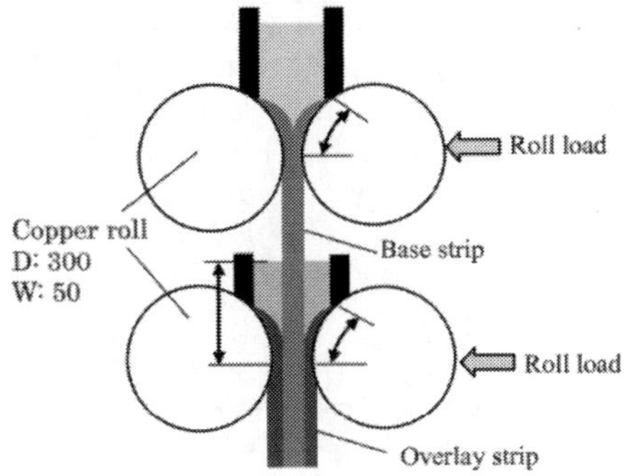

(a) Vertical type tandem twin roll caster

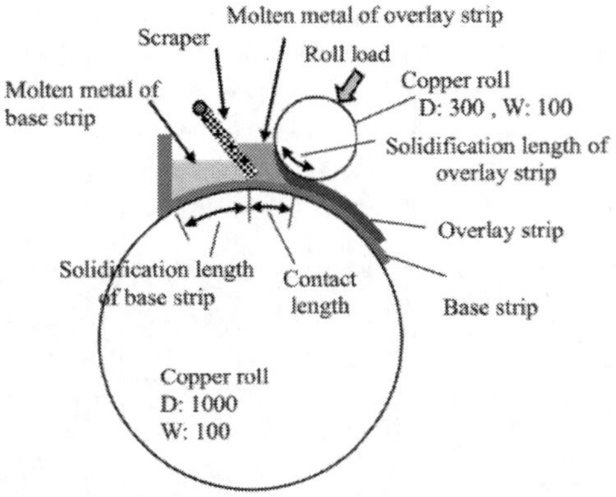

(b) Unequal diameter twin roll caster equipped with a scraper

Fig. 3. Schematic illustration of roll casters for casting clad strip

3 Experimental conditions

A single strip of Al-Mg alloy was cast using a single-roll caster equipped with a scraper, as shown in Fig. 1. The roll speed was 40 m/min, and the scraper load was 100 N. The solidification length was 200 mm. The casing temperature of the Al-Mg alloy was 705°C. A tension test, a cup test, and microscopy of the cross section of the strip were conducted on the strip, which was cold rolled down to 1 mm and annealed at 380°C for 1.8 ks. In the cup test, the diameter of the punch was 32 mm. The gage length and width of the test piece for the tension test were 31 mm and 15 mm, respectively.

Two types of clad strips were cast. One clad strip consisted of an Al-Mg alloy overlay strip and an Al-Mn alloy base strip, and the other clad strip consisted of an Al-Sn-Cu alloy overlay strip and an Al base strip. The clad strip consisting of Al-Mg and Al-Mn was cast at 30 m/min using the casters shown in Figs. 3(a) and 3(b). In Fig. 3(a), the solidification length was 60 mm, and the contact length was 100 mm. In Fig. 3(b), the solidification lengths of the base strip and the overlay strip were 100 mm and 60 mm, respectively. The contact length was 80 mm. The casting temperature of the Al-Mg and Al-Mn were 670°C. The roll load was 5 kN, and the scraper load was 40 N. The clad strip consisting of an overlay strip of Al-Sn-Cu and a base strip of Al was cast using the caster shown in Fig. 3(b). The casting temperatures of Al and Al-Sn-Cu were 684°C and 641°C, respectively. The solidification lengths of the base strip and the overlay strip were 190 mm and 170 mm, respectively. The contact length was 80 mm. The roll load was 5 KN, and the scraper load was 40 N. The roll speed was 15 m/min. A tension shear test was conducted on the clad strip of Al-Sn-Cu and Al.

The chemical compositions of the aluminum alloys are shown in Tables 1 through 4. The solidification temperatures are shown in Table 5.

Table 1. Chemical composition of Al-Mg alloy

Cu	Si	Mg	Zn	Fe	Mn	Ti	Bal.
0.00	0.06	4.81	0.00	0.07	0.45	0.03	Al

Table 2. Chemical composition of Al-Mn alloy

Cu	Si	Mg	Zn	Fe	Mn	Ti	Bal.
0.15	0.28	0.00	0.01	0.61	1.24	0.01	Al

Table 3. Chemical composition of Al

Cu	Si	Mg	Zn	Fe	Mn	Ti	Bal.
0.00	0.04	0.00	0.01	0.09	0.00	0.00	Al

Table 4. Chemical composition of Al-Sn-Cu alloy

Sn	Cu	Si	Mg	Zn	Fe	Mn	Bal.
40.40	0.97	0.01	0.00	0.00	0.03	0.00	Al

Table 5. Solidification temperature

Alloy	Solidus temperature	Liquidus temperature
Al-Mg	577°C	638°C
Al-Mn	643 °C	655 °C
Al	650°C	660°C
Al-Sn-Cu	228°C	610°C

4 Result and discussion

4.1. Single strip

Al-Mg strip could be continuously cast. The surfaces of as-cast strip are shown in Fig. 4. The scribed surface had no metallic luster and was dull. A ripple mark existed at the roll contact surface, but not at the scribed surface. This may indicate that scribing by the scraper was steady state. No specific defect due to scribing occurred at the roll contact surface of the strip.

The cross section of as-cast Al-Mg strip cast by the single-roll caster is shown in Fig. 5. The grains near the roll contact surface were very fine. In other area, the grain size was almost uniform. Center segregation did not occur. A low degree of segregation of Mg occurred near the scribed surface. Cold rolling could be conducted, and the surfaces were indifferentiable after cold rolling.

The results of the tension test are presented below. In the casting direction, the tensile stress, proof stress, and elongation were 304 MPa, 155 MPa, and 26.5%, respectively. In the width direction, the tensile stress, proof stress, and elongation were 272 MPa, 140 MPa, and 16.8%, respectively. The mechanical properties of the casting direction were superior to those of the width direction.

The results of the cup test are shown in Fig. 6. Which surface was facing outward, i.e., the scribed surface or the roll contact surface, did not affect the limiting drawing ratio (LDR), which was 2.0.

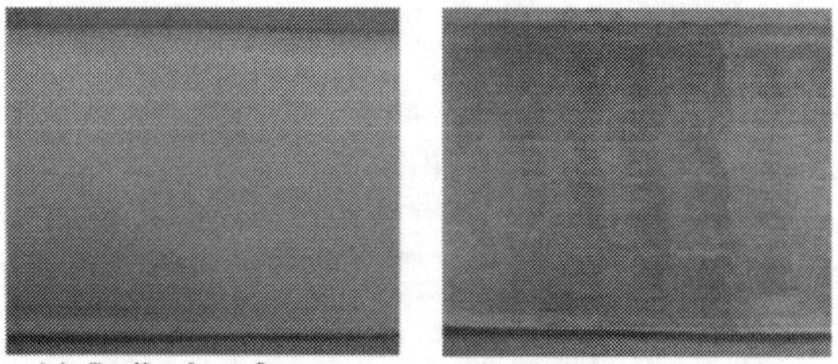

(a) Scribed surface (b) Roll contact surface

Fig. 4. Surface of as-cast Al-Mg strip cast by the single roll caster shown in Fig. 1

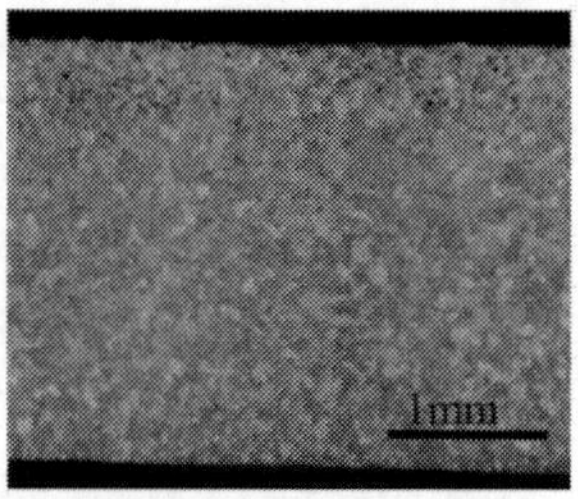

Fig. 5. Cross section of as-cast Al-Mg strip

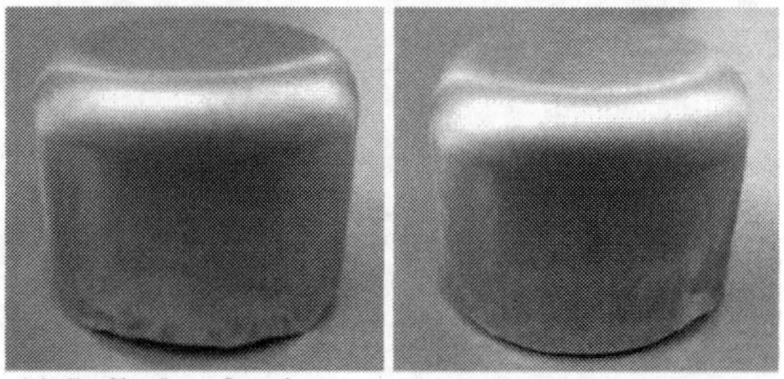

(a) Scribed surface is facing outward (b) Roll contact surface is facing outward

Fig. 6. Results of cup tests of Al-Mg strip

4.1 Clad strip

The cross sections of clad strips consisting of Al-Mg and Al-Mn are shown in Fig. 7. In the clad strip cast by the VTTRC, there were gaps between the Al-Mn base strip and Al-Mg overlay strip. On the other hand, there was no gap between the base strip and the overlay strip of the clad strip cast by the UDTRCS. The surface of the base strip was not heated to the temperature at which bonding with the overlay strip was possible in the VTTR. For example, the bonding temperature is the semisolid temperature. In the UDTRCS, the surface of the base strip is a semisolid of high solid fraction near the scraper. The surface of the base strip did not contact the oxidizing atmosphere. Therefore, the base strip is suitable for cladding. The scraper is useful for clad strip casting from two perspectives. One is that the surface of the base strip becomes a semisolid of high solid fraction. The bonding is easy, and mixture of the two types of aluminum alloy does not occur due to the effect of the high solid fraction. The other is that the surface of the base strip does not contact the oxidizing atmosphere, preventing the generation of oxide film. Clad strips consisting of various combinations of many types of aluminum alloy can be cast using the UDTRC.

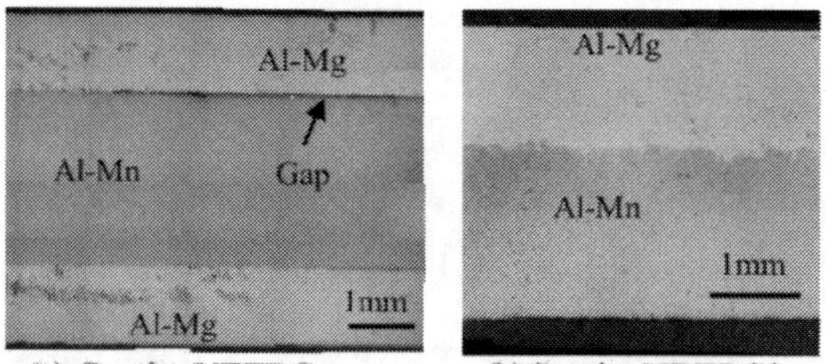

(a) Cast by VTTRC (b)Cast by UDTRCS

Fig. 7. Cross sections of clad strips

Casting of the clad strip consisting of Al-Sn-Cu and Al was attempted using the UDTRCS. The base strip was Al and the overlay strip was Al-Sn-Cu. The difference in solidus lines of the base strip and the overlay strip was more than 400°C. Some problems were predicted for casting. One was that the overlay strip of Al-Sn-Cu was not solidified by the heat from the Al base strip. If the Al-Sn-Cu solidifies, the strip thickness may become thinner. The temperature of the Al base strip after releasing the lower roll was higher than 500°C. This temperature is higher than the solidus line of Al-Sn-Cu. Therefore, the Al-Sn-Cu overlay strip may re-melt. The casting of clad strip is shown in Fig. 8. The forepart of the strip in the figure is the Al base strip. The overlay strip connected to the base strip at the middle part. Meandering and warping did not occur on the base strip.

The bonding of the overlay strip did not influence the base strip. The base strip and overlay strip were not hot rolled, because the roll load was very small. The small roll load was one reason for the non-occurrence of meandering and warping. The solidus line of the Al-Sn-Cu overlay strip is much lower than that of the Al base strip. The Al-Sn-Cu overlay strip did not completely solidify when the clad strip released from the roll. Therefore, the Al base strip was subjected to heat effect by latent heat from the Al-Sn-Cu overlay strip on the conveyer. However, meandering and warping of the clad strip did not occur. The heat effect by the Al-Sn-Cu overlay strip on the Al base strip was not large enough to have a heat effect on the Al base strip. The length of the non-steady area of the forming clad strip was only 100 mm. Re-melting and sweating did not occur at the Al-Mg-Cu overlay strip due to the heat from the Al base strip. Roll coating occurred. The Al-Mg-Cu overlay strip must be semisolid, but the strip did not stick to the roll.

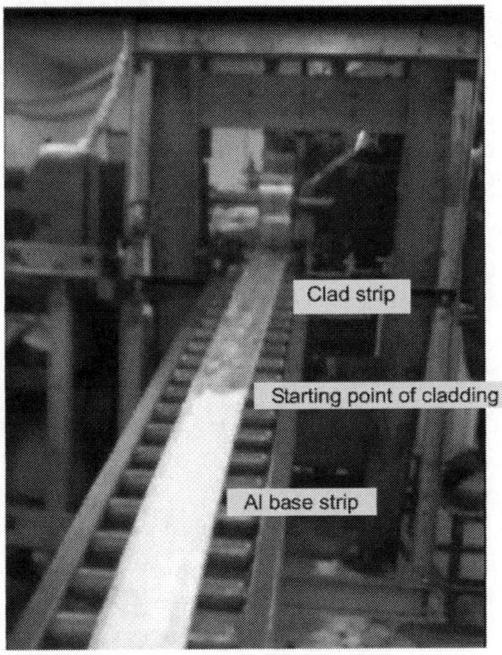

Fig. 8. Casting of clad strip in operation

The as-cast strip and the strip cold rolled down to 2 mm were used for the tension-shear test. The clad strip was not peeled by the cold rolling. The results of the tension-shear test are shown in Fig. 9. Breaking did not occur at the interface for either condition. This means that the two strips were strongly bonded at the interface. The roll load of the caster was very small. The two strips were not bonded by the hot rolling. The surface of Al was a semisolid of high solid fraction, which provided strong bonding. Waving of the interface was not caused by the cold rolling. The Al-Sn-Cu overlay strip did not completely solidify on the Al base strip after cooling by the roll, and the Al base strip solidified. The semisolid condition of the Al-Sn-Cu overlay strip did not influence the bonding ability.

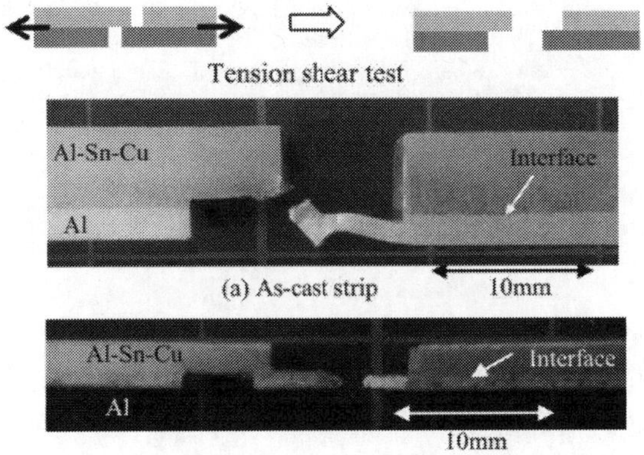

(a) As-cast strip

(b) Cold rolled 2 mm-thick-strip

Fig. 9. Cross section of the clad strip after the tension shear test

The results of the line analysis of Sn near the interface of the as-cast strip and the cold rolled 2-mm-thick strip are shown in Fig. 10.

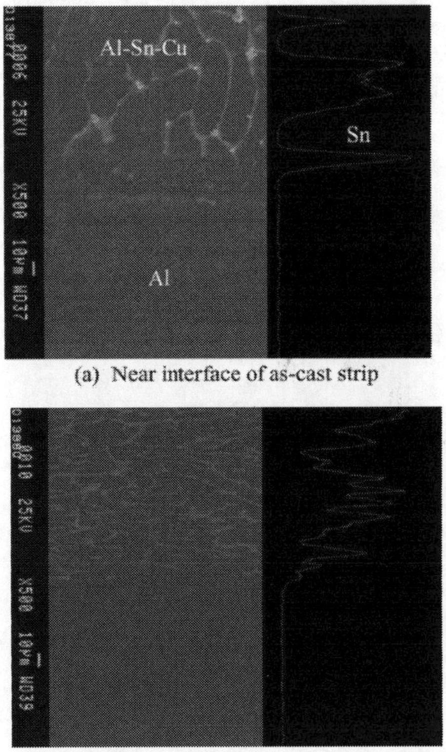

(a) Near interface of as-cast strip

(b) Near interface of cold rolled 2 mm-thick-strip

Fig. 10. Results of line analysis

The Sn did not diffuse into the Al base strip from the Al-Sn-Cu overlay strip. When the molten metal of the Al-Sn-Cu contacts the Al base strip, the surface of the Al base strip was a semisolid of high solid fraction. However, the Sn did not diffuse into the Al base strip. The casting temperature of the Al-Sn-Cu molten metal was lower than the solidus line of the Al base strip. The semisolid layer of the surface of the Al base strip might have been immediately solidified by the Al-Sn-Cu molten metal when contacting the Al base strip. Therefore, the Sn did not diffuse into the Al base strip. However, excellent bonding was attained by the semisolid metal of the Al and the molten metal of the Al-Sn-Cu. The Al-Sn-Cu overlay strip was previously in semisolid condition on the Al base strip, and the Sn did not diffuse into the Al base strip during this time.

5 Conclusion

The Al-Mg strip without the center segregation could be cast at a high speed of 40 m/min using a single-roll caster equipped with a scraper. The scraper was very useful for flattening the free solidified surface and increasing the heat transfer between the roll and the solidification layer.

An unequal-diameter twin-roll caster equipped with a scraper can cast a two-layer clad strip with a clear interface. Two clad strips consisting of 1) an Al-Mg overlay strip and an Al-Mn base strip and 2) an Al-Sn-Cu overlay strip and an Al base strip were able to be cast at high speeds of 30 m/min and 15 m/min, respectively. In the tensile-shear test, the clad strip did not peel at the interface, but rather broke in the strip. In these clad strips, the base strip and the overlay strip were strongly bonded, and the bonding strength at the interface was stronger than that of the base strip or the overlay strip. The elements of the overlay strip did not diffuse into the base strip. The cause of the strong bonding and non-diffusion at the interface might be that the surface of the base strip was a semisolid of high solid fraction.

References

1. M. S. Kim, S. H. Kim, H. W. Kim, Scripta Materialia 152(2018)
2. T. Haga, K. Akitsu, T. Yamabayashi, S. Kumai , H. Watari, Key Engineering Materials **44**(2010)
3. T. Haga , K. Akitsu , S. Kumai , H. Watari, Advanced Materials Re
4. search **445**(2012)
5. T. Haga , R. Kozono , S. Nishida , H. Watari, Advances in Materials and Processing Technologies **3** (2017)
6. T. Haga , K. Okamura, S. Nishida, H. Watari, K. Matsuzaki, Materials Science Forum **879**(2017)
7. T. Haga , R. Nakamura , S. Kumai , H. Watari, Archives of Materials Science and Engineering **37**(2009)
8. R.Nakamura , T.Haga , H.Tsuge , S.Kumai , H.Watari, Materials Science Forum **675-677**(2011)
9. T. Haga , T. Nakamura , S. Kumai , H. Watari, Applied Mechanics and Materials **52-54**(2011)

Microstructure and mechanical properties of high-manganese-containing high-speed twin-roll cast Al-Mn-Si alloy strips and their cold-rolled sheets

T.Ha Nguyen[1,*], *Ram* Song[1], *Yohei* Harada[2] ,*Shinji* Muraishi[2], *Shinji* Kumai[2]
Shun Maruno[3], *Michihide* Yoshino[3], and *Shohei* Iwao[3]

[1]Graduate student, Tokyo Institute of Technology, 2-12-1 Ookayama, Meguro-ku, Tokyo, Japan
[2]Department of Materials Science and Engineering, Tokyo Institute of Technology, 2-12-1 Ookayama, Meguro-ku, Tokyo, Japan
[3]Product Research and Development Dept., Mitsubishi Aluminum Co.,Ltd, 85 Hiramatsu Susono-city, Shizuoka, Japan

Abstract. Al-Mn based alloys with high-manganese content are expected to have improved mechanical properties due to solid solution hardening and/or dispersion hardening. However, the increase of Mn solubility of the alloy is difficult by using the conventional DC casting. In order to solve this problem, in the present study, we focused on the twin-roll casting method which is characterized by high cooling rates. Several kinds of high Mn-containing Al-Mn-Si alloy strips were fabricated by using a vertical-type high-speed twin-roll caster equipped with a pair of water-cooled copper rolls. Direct temperature measurement of the liquid melt during the casting was also performed. The alloy strips of various compositions containing up to 4 Mn and 2 Si (wt%) were successfully obtained. By observing the microstructure of the cross section of the strip, we found the characteristic solidified structure. The solidified structure consisted of three layers. Two solidified shells with a columnar dendrite structure grew from the roll surfaces toward the strip center. In the mid-thickness region, the band structure consisting of equiaxed dendrites and globular grains was observed between the solidified shells. Very fine primary particles were observed in the matrix near the strip surface, while, relatively coarse particles with blocky and needle-like shape were observed in the central band of the as-cast strip. The electric conductivity measurement was performed for the as-cast strips. Mn solubility in Al matrix was estimated from the obtained values. The estimated Mn solubility in the Al-2Mn-xSi strips was between 1.5 ~ 1.8wt% Mn. It was over 1.43wt%Mn for the Al-4Mn-xSi strips. We found that the Mn solubility of the as-cast strips was considerably high. The strips were cold-rolled to the sheets and then annealed at various conditions. They were subjected to the tensile tests, and the effects of solid solution hardening and dispersion hardening are discussed.

1 Introduction

Al-Mn based alloys contain Mn as a principal alloying element, and Si as a secondary alloying element. Fe is usually included as impurities. The alloy possesses good ductility, high corrosion resistance and thermal conductivity. Thin sheets of 3003, 3004 and 3105 alloys have been used widely for architecture and packaging materials and automobile heat-exchangers [1]. In order to improve the fuel efficiency, reduction in the thickness is required for further weight reduction. Therefore, we need to increase the strength of the alloy to achieve this task. Solid solution hardening and dispersion hardening are the promising mechanisms in the alloy [2, 3].

The twin-roll casting (TRC) is one of the near-net-shape casting technique which can produce a metal alloy sheet directly from the molten metal. TRC is classified into two types, horizontal-type and vertical-type, depending on the roll construction. The casing speed of the horizontal-type TRC is a relatively low, therefore, a limited number of commercial alloys can be used for producing the strip by this method. In addition, the alloys with a wide freezing range often suffer from some kinds of casting defects [4]. In contrast, the vertical-type high-speed TRC (HSTRC) has a much higher casting speed compared to the horizontal-type. The strip can be produced at a very high cooling rate (over 10^3C/s) [5, 6]. The high cooling rate is considered to be effective not only for grain refinement but also for increasing the super-saturation of the alloying elements which leads to solid solution hardening and dispersion hardening.

In the present study, Al-Mn-Si alloys with a high Mn and Si content were fabricated by HSTRC. The microstructural characteristics of the as-cast strips was intensively investigated. The cast strips were also cold-rolled to thin sheets and then annealed. The mechanical properties of the sheets were examined. The aim of the present study is to examine the potential of the HSTRC for producing the high-strength Al-Mn-Si alloy cold-rolled sheets by enhancement of the Mn supersaturation.

2 Experimental procedure

Al-Mn-Si-base alloys with a high amount of Mn and Si were used in this study. The nominal composition of the alloys is listed in Table 1. A fixed amount of 0.2 wt% Fe was added into the ingot. The original ingots were fabricated by Mitsubishi Aluminum Co. ltd.

* Corresponding author: nguyenha.aa@m.titech.ac.jp

Table 1. Chemical composition of alloys (wt%)

Alloy	Mn	Si	Fe	Al
A1	2.00	1.00	0.20	Bal.
A2	2.00	1.50	0.20	Bal.
A3	4.00	1.0	0.20	Bal.
A4	4.00	1.50	0.20	Bal.
A5	4.00	2.00	0.20	Bal.

The schematic diagram of HSTRC is shown in Fig. 1. The caster is equipped with a pair of pure copper rolls with a diameter of 300mm and a width of 100mm. No lubricant is applied on the roll surface. The roll is cooled continuously by the running water inside the rolls. The steel plate nozzle is used for obtaining the stable melt head condition which provides the hydrostatic pressure. The nozzle is covered with heat-insulating materials. One of the rolls is fixed firmly to the base, while the other roll is installed with a series of springs for controlling the initial roll separating force.

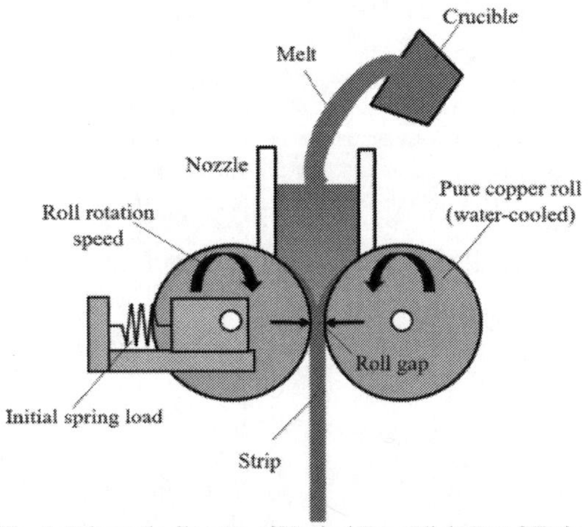

Fig. 1. Schematic diagram of Vertical Type High-Speed Twin-Roll Caster

The alloy strips were fabricated by HSTRC. About 2.2kg of the alloy ingot was melted in a clay-graphite crucible by using an electric resistance furnace. The melt was degassed for 20 min by introducing the Argon gas bubbles into the melt alloy before casting. The melt was solidified on both roll surfaces to form solidification shells. The solidification shells were introduced to the roll gap by the roll rotation and pressed to form the strip. The details of the casting condition are described in Table 2.

In order to estimate the supersaturated Mn content in solid solution, electrical conductivity was measured at room temperature by using SIGMA tester with 500kHz probe (AutoSigma 3000, GE Inspection Technology) The measurement was carried out on the normal-direction plane of the samples.

Table 2. Casting condition for each alloy

Alloy	Pouring temperat-ure (°C)	Roll rotation speed (m/min)	Initial spring load(kN)	Roll gap (mm)
A1	680	60	60	1
A2	670	60	60	1
A3	760	60	60	0.8
A4	760	60	60	0.8
A5	760	60	60	0.8

The cast samples were cold-rolled to 0.1mm-thick sheets. The sheets were annealed at 400°C for 24h, and then subjected to the tensile tests. Fig. 2 shows the schematic diagram of the tensile test specimen. The specimen was machined from the sheet along the casting and the rolling direction.

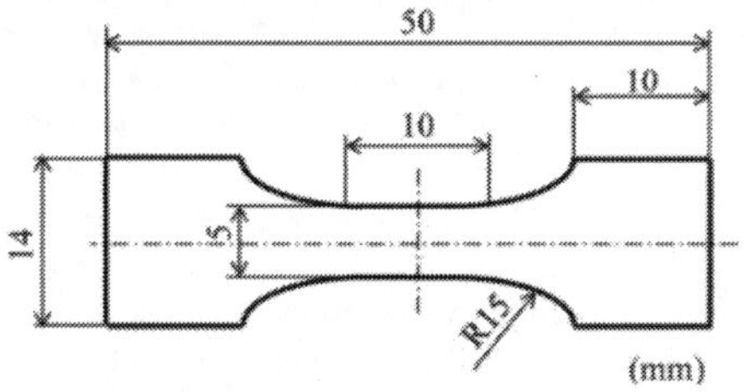

Fig. 2. Tensile test specimen (unit in mm)

The tensile tests were carried out by using Instron-type machine (Shimadzu AX-10kN) at room temperature. The cross-head speed was 0.5mm/min. The change in the gage length was measure by a non-contact type laser displacement meter (Keyence LS-7500). The test was performed 3 times for each condition, and 0.2% proof stress, UTS, and elongation were obtained.

3 Results and discussion

3.1 Thickness profile of as-cast strips

Fig. 3(a) shows the thickness change of the strips, A1 and A2, along the casting direction. 2.6m-long continuous strips were successfully fabricated by HSTRC. The average thickness in the stable region was 2.3mm for A1 and 2.2mm for A2, respectively. Fig. 3(b) shows the thickness change of A3, A4 and A5. About 3m-long continuous strips were obtained. The thickness of three strips was about 1.6mm on average. In total, five types of alloy were successfully cast into thin strips by using HSTRC. The microstructure was carefully examined for the part of the constant thickness, which is marked by the black boxes in Fig. 3(a) and 3(b).

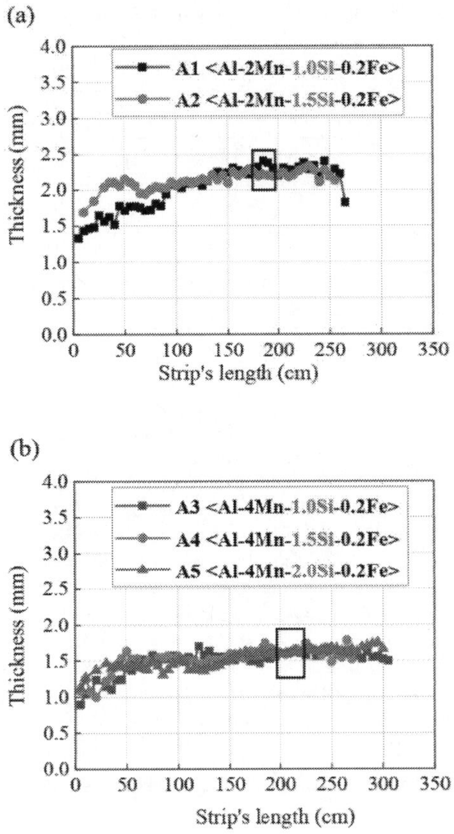

Fig. 3. Strip thickness of five Al-Mn-Si cast strips

3.2 Grain structure of as-cast strips

The microstructure of the cross section of as-cast strips was examined. Fig. 4 shows a representative grain structure of A1 and A3 (2 and 4 wt% Mn-containing) along the casting direction. As shown in Fig. 4(a), the solidified shells grown from the both roll surfaces were observed at the near-surface region. This area consists of very fine chill crystals with random orientation. Dendritic columnar grains and fine equiaxed grains were observed in the sub-surface region. In the mid-thickness region, a band area which consists of fine globular grains and equiaxed grains was observed. Fig. 4(b) shows the macrostructure of A3. This is similar to that of A1, in which the solidified shells are observed in both surfaces. At the near-surface region, some fine chilled crystals were observed. This layer was formed due to the very high cooling rates at the roll surfaces. After the fine chilled layer, long dendritic columnar grains were observed instead of the mixed structure of dendritic columnar and equiaxed grains in A1. The structure change is considered to be due to the fact that the casting temperature of A3 was much higher than that of the A1 (from liquidus temperature +60˚C and +20˚C for A1 and A3, respectively). A central band was also observed at the mid-thickness region of A3.

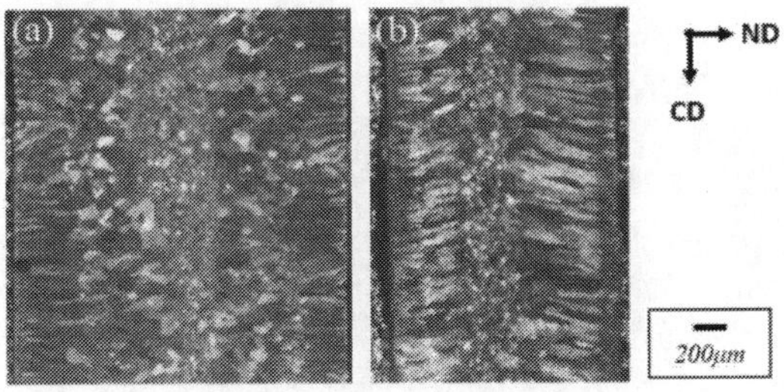

Fig. 4. Representative grain structure of the cross-section of strip. (a) A1 and (b) A3 alloy strips

In the HSTRC process, the central band structure was formed due to the entrapment of the floating crystals or the fragmented dendrite arms grown on the roll surfaces by the solidified shells. When the floating crystals are formed, they can grow to the globular grains under the convection in the nozzle. Breaking of the dendrite branches also possibly occurs at the roll gap [7]. In the A1, both of these two factors could be the origin of the globular grains in the central band. However, for the A3, due to the high casting temperature, the formation of floating crystals in the nozzle can be avoided. Therefore, the growth of fragmented dendrites in the residual liquid should be the main factor for the central band formation. Direct temperature measurement of the liquid melt during the casting was also performed. It was confirmed that the temperature of the mid-central part of the strip just after leaving from the roll nip was a little higher than the solidus temperature of the alloy.

3.3 Microstructure of as-cast strips

Fig. 5 shows a representative microstructure of A1, and A3 at the near-surface, a quarter of the thickness and the central regions. The microstructure was the typical dendritic structure, which was similar to that of the HSTRC 3003 strip [8]. At the near-surface region, cellular and equiaxed dendrite crystals were observed. A large amount of fine constituent particles was distributed inside the grains. The distribution became denser as increasing the Mn concentration. The structure gradually changed into a typical dendritic structure at a quarter of the thickness region. In this region, large constituent particles were frequently observed. In the mid-thickness area, globular grains were observed. Not only fine constituent particles but also a number of blocky/needle-like particles were observed. Both the amount and the size of the particles increased considerably when Mn was 4 wt%. By the EPMA analysis, the large particles in the central region of A1 strip were mainly Al(Mn,Fe)Si. While in A3 strips, large needle-like/or platelet particles were determined as $Al_6(Mn,Fe)$ and Al(Mn,Fe)Si.

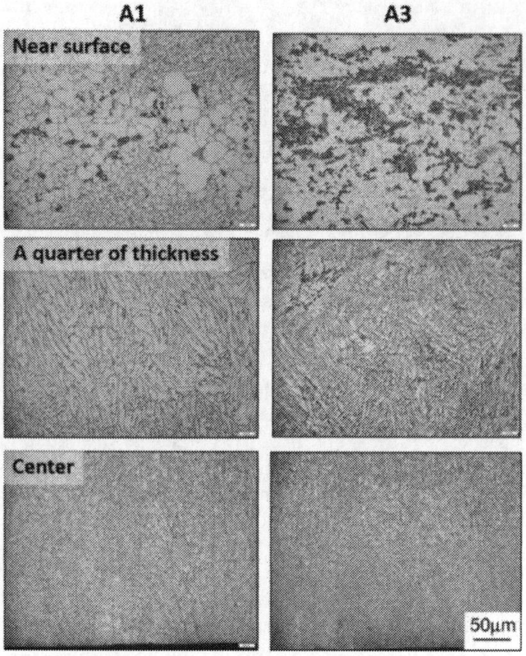

Fig. 5. Representative microstructure of the cross-section of A1, and A3 along thickness direction

3.4 Mn solubility estimation by electrical conductivity (EC) values

The EC of the alloy has a relationship with the concentration of an alloying elements in the solid solution. The present alloys contain Mn, Si and a small amount of Fe. The maximum solid solubility of iron in the Al matrix is as small as 0.005wt% [9]. It is also considered that the concentration of Si in solid solution has a negligible contribution to the EC change due to the very small contribution factor of 0.0068 according to the following equation from [3]:

$$1/\sigma = 0.0267+0.032Fe_{SS}+0.033Mn_{SS}+0.0068Si_{SS} \quad (1)$$

where σ is the electrical conductivity, Fe_{SS} , Mn_{SS} , Si_{SS} are concentration of Fe, Mn, and Si in the solid solution, respectively. Therefore, it is considered that the change of EC is mainly due to the content of Mn in the solid solution. The Mn concentration in the solid solution can be estimated by:

$$Mn\ (\mathrm{wt\%}) = (\rho_{measure}-\rho_{pure\ Al})/\Delta\rho_{Mn\ in\ Al\ alloy} \quad (2)$$

where $\rho_{measure}$ is the electrical resistivity of alloy converted by the measured electrical conductivity, $\rho_{pure\ Al}$ is the electrical resistivity of pure aluminum, which is 26.5 nΩm at room temperature, and $\Delta\rho_{Mn\ in\ Al\ alloy}$ is the electrical resistivity contribution for Mn solute atom in Al matrix. The value of $\Delta\rho_{Mn\ in\ Al\ alloy}$ is 31 nΩm wt%$^{-1}$ at room temperature [10].

Table 3. Estimated Mn concentration in Al-Mn-Si strips

Alloy	Mn concentration (wt%)	
	Surface	Center
A1	1.87	1.79
A2	1.60	1.45
A3	2.26	2.05
A4	1.74	1.46
A5	1.46	1.43

The estimated Mn concentration in the supersaturated solid solution is shown in Table. 3. Compared to the alloy composition for Al - 2 and 4 wt% Mn strip, the estimated Mn contents were decreased. However, all the samples still showed high Mn concentration. The decrease in Mn concentration resulted from the formation of intermetallic phases including Mn. The surface region showed a relatively higher Mn concentration than that of the central region. The cooling rate was significantly higher at the surface than the central region [6]. This might be the main factor of the difference in Mn concentration between them.

3.5 Mechanical properties of cold-rolled sheets

A1, A3 and A5 were cold-rolled to 0.1 mm and annealed at 400°C for 24h. It was possible to apply a large reduction rate (about 95%) to these strips without any homogenized steps. However, in the surface of some rolled sheets, surface defects were observed. The tensile specimens were collected from the sheet without the surface defects. Average values of the 0.2% proof stress, ultimate tensile stress (UTS) and elongation are listed in Table 4.

The sheets obtained from the HSTRC alloy strips show excellent strength. The main factor contributed to the high strength is considered to be the high supersaturated Mn in solid solution due to the high cooling rate of the as-cast strip. Compared to A1, A3 shows higher 0.2% proof stress and UTS due to the higher Mn concentration. However, A5 strip also had a higher strength than A1 in spite of the low Mn concentration. The Mn concentration was 1.87 and 1.46-wt% at the surface of A1 and A5, respectively. However, A5 has a higher Si concentration than that of A1. Therefore, it is considered that the increased amount of Si was also a factor for the improvement of strength. Compared to A1, A3 and A5 showed low elongation with a large variation. Large amount of needle-like intermetallic particles was observed in the central band of both as-cast A3 and A5-strips. $Al_6(Mn,Fe)$ particles are generally brittle. The particles with a high aspect ratio of needle-like/or platelet particles tend to act as stress raisers and reduce the elongation [11].

Table 4. Mechanical properties of cold-rolled sheets

Sheet	0.2% proof stress (MPa)	UTS (MPa)	Elongation (%)
A1	122.3 ± 5.5	169.9 ± 5.3	11.5 ± 1.3
A3	162.7 ± 17.5	190.2 ± 1.8	10.7 ± 5.9
A5	156.4 ± 5.7	193.5 ± 3.5	9.9 ± 8.3

We also observed the precipitated particles in the three sheets by using FE SEM. Besides the constituent particles, a large amount of fine dispersed particles was detected. In Table 4, all the sheets showed high 0.2% proof stress, and the difference between the proof stress and UTS was relatively small (about 48, 27 and 40 MPa in average for A1, A3 and A5, respectively). This character is not common for the annealed solid solution strengthened alloys, but is similar to that of the age-hardened alloys strengthened by the precipitates. This result suggests that the present sheets fabricated from the HSTRC strips are strengthened by the dispersed particles.

4 Conclusion

In total, five kinds of Al-Mn-Si strips containing up to 4Mn and 2Si (wt%) were successfully fabricated by HSTRC. Due to the high cooling rate, the substantial increase of supersaturated Mn concentration in solid solution was successfully obtained. The estimated Mn concentration for all samples were over 1.43wt% and in particular, the highest value was about 2.2wt% at the surface region of Al-4wt%Mn-1wt%Si alloy strip. Higher supersaturated Mn contributed to the high strength of cold-rolled sheets. The dispersion strengthening is considered to be the reason for the improved strength of the present sheets fabricated from the HSTRC strips.

References

1. Kissell, J. & Pantelakis, Syros & Haidemenopoulos, Gregory. *Aluminum and Aluminum Alloys.ch9.* (2004)
2. Yohei Harada, Nan Jiang, and Shinji Kumai: Mater. Trans. **60** (2019) pp. 2435-2441
3. Astrid Marie F. Muggerud, Eva Anne Mørtsell, Yanjun Li, Randi Holmestad, Mater. Sci. Eng. A, **567**, (2013) pp. 21-28
4. M. Yun, S. Lokyer, J.D. Hunt, Mater. Sci. Eng. A, **280**, (2000) pp.116-123
5. Min-Seok Kim, Yoshiyuki Arai, Yasuharu Hori, and Shinji Kumai: Mater. Trans. **51** (2010) pp.1854-1860
6. Ram Song, Yohei Harada, and Shinji Kumai, Mater. Trans. **59** (2018) pp. 110-116
7. Min-Seok Kim and Shinji Kumai: Mater. Trans. **52** (2011) pp.856-861
8. Ram Song and Shinji Kumai: Mater. Trans. **60** (2019) pp.322-329
9. A.L. Kearney: *Properties of Cast Aluminum Alloys, Properties and Selection: Nonferrous Alloys and Special-Purpose Materials*, **vol. 2**, ASM Handbook, ASM International (1990)
10. Shin-ya Komatsu, Shin-Itiro FujiKawa, J. Jpn. Inst. Light Met. **47** (1997) pp.170-181
11. J. R. Davis, *Aluminum and aluminum alloys* (ASM international, 1993)

Interfacial microstructure formation in A356/steel compound castings using metal coating

Aina Opsal Bakke[1,*], *Arne* Nordmark[2], *Lars* Arnberg[1], and *Yanjun* Li[1]

[1]Norwegian University of Science and Technology, Trondheim, Norway
[2]SINTEF Industry, Trondheim, Norway

Abstract. Compound castings between aluminum and steel have great potential for applications in the automotive industry. However, due to large differences in thermal and mechanical properties between steel and aluminum, and the formation of stable aluminum oxides at the interface, it is difficult to form high strength metallic bonding between the two metals. In this work, A356/steel compound castings were produced through a gravity casting process. Various metal coatings, including galvanizing, aluminizing and brass-coating, were applied on the steel inserts to ensure that the A356 aluminum melt could react sufficiently with an oxide-free steel surface, resulting in a high-quality metallurgical bond. The reaction layer formed between the alloys was investigated using Optical Microscopy (OM), Scanning Electron Microscopy (SEM) and Energy Dispersive X-ray Spectroscopy (EDS). In addition, Vickers Micro-hardness was measured across the aluminum-steel interface. Results showed that metallurgical bonding could be achieved with all three coatings. However, for the brass-coated components only local bonding areas were found. In the aluminized and galvanized components, thick reaction layers consisting of binary Al-Fe and ternary Al-Fe-Si phases formed in the aluminum-steel interface. Between the A356 aluminum and aluminized layer, nearly no reaction layer formed. The mechanism for the formation of the various intermetallic phases at the reaction layers are discussed.

1 Introduction

Aluminum alloy A356 is widely used in automotive castings due to its great castability, corrosion resistance and strength-to-weight ratio, especially in the heat-treated condition [1, 2]. However, steel is still superior in terms of strength which makes it difficult to completely replace steel with aluminum alloys. Therefore, the need for a multi-material solution, such as an aluminum-steel compound, arises. To achieve specific property- and functional requirements, new methods must be developed to joint aluminum and steel effectively.

Compound casting is a casting process where two materials, one in solid state and one in liquid state, are joined together. In the process, the liquid material is cast around the solid, thus forming a diffusion zone between the two metals, which results in a metallurgical bond. This joining process is considered economically favorable in large-scale production, due to reduction of process steps [3].

A challenge in compound casting is that a range of brittle Al-Fe intermetallic phases form at the interface, which is detrimental to the overall bonding strength [4]. Compound casting is also difficult due to aluminum oxides forming spontaneously on the metal surface [5]. These oxides are very stable and have high melting points [6]. Therefore, it is difficult to achieve desired wettability between aluminum and steel, which will hinder formation of a metallurgical bond.

Numerous approaches have been considered in order to achieve metallurgical bonding between aluminum and steel. The most common method is to apply a thin coating layer on the steel surface, such as flux coating or coating with a low melting point metal. Zinc coating has been reported to improve bonding, as it prevents the steel from oxidizing and has a lower melting point than aluminum [6]. This allows the zinc coating to melt during casting, leaving a clean steel surface for the aluminum to react.

Springer, Szczepaniak and Raabe tested both hot-dipped Zn coating and electrolytic-coated Zn on low-carbon steel in a dip-test experiment in liquid Al. They found that Zn-coating accelerated the growth of intermetallic phases as well as formation of a more uniform reaction layer. A hot-dip Zn layer appeared to have the largest effect on the results [7]. Kang and Kim joined Al5052 and aluminized steel through an arc braze welding process. The aluminized layer was approximately 30 μm thick and consisted of an AlSi layer and an $Fe_x(AlSi)_y$ layer, where the latter formed during the aluminizing process. It showed that although the aluminized steel showed slightly lower wettability compared to a galvanized steel sample, the layer of intermetallic phases was thinner for the aluminized joints [8].

The influence of various surface treatments of steel inserts in a compound casting process with aluminum alloy ZL114A, has been studied by Jiang et al. [9]. The hot-dip aluminized layer was found to improve the bonding compared to the uncoated steel. By using NH4Cl as

*e-mail: aina.o.bakke@ntnu.no

a surface modifier prior to aluminizing, a reaction layer formed in the interface causing the formation of a metallurgical bond, which significantly improved the overall bonding. Hot-dip galvanizing of cylindrical steel inserts in a Zn melt with 0.1wt% Ni improved wettability between the steel and the ZL114A aluminum melt. Although this helped formation of a metallurgical bond, it resulted in a relatively thick reaction layer (~650 μm thickness) [10]. Combining galvanizing and aluminizing also showed improved bonding compared to joining aluminum ZL114A to a bare grey iron surface. Galvanizing followed by aluminizing caused the aluminized layer to form a reaction layer with the grey iron due to enhanced wettability. This reaction layer then reacted further with the cast aluminum alloy, resulting in a near uniform reaction layer in the final component [11].

In the present research, the interface structure and bonding properties between S235JR steel pipes, coated with three different metallic coating layers, and aluminum alloy A356 were investigated. The different coatings were: galvanizing, aluminizing and brass-coating. Aluminum-steel compound castings were produced in a sand mold through a gravity casting process. Resulting microstructure in the interface was investigated. In addition, Vickers microhardness was measured across the interface of the various castings in order to assess the overall bonding properties of the castings.

2 Experimental Procedure

Compound castings between aluminum and steel were produced through a gravity casting process using a sand mold, shown in Figure 1. S235JR steel pipes with an outer diameter of 20 mm, a wall thickness of 1.5 mm and a length of 155 mm, were used as inserts in the sand mold. Aluminum alloy A356 was used as the casting alloy. Compositions of both materials are given in Table 1.

2.1 Surface Treatments

Three different surface coatings were tested in the experiment. One third of the steel pipes were galvanized through an industrial hot-dipping process. The steel pipes were immersed into the Zn bath in a centrifuge to ensure uniform coating. One third of the steel pipes were brass-coated. The coating was performed by dipping the steel pipes in a cartridge brass (C260) bath at approximately 1200°C. Boric acid is added to the bath to ensure better wetting and prevent oxide formation during coating. The remaining steel pipes were aluminized. First, they were ground with 1200 grit paper, cleaned in ethanol and then immersed in a 10% NH_4Cl solution for five minutes at 75°C followed by air-drying. The salt-coated steel pipes were then preheated to 150°C and then dipped in a liquid aluminum AlSi7 melt at 800°C for five minutes.

2.2 Material Characterization

Samples with a thickness of 1 cm were cut from each casting perpendicular to the aluminum/steel interface. Following cutting, the samples were ground to 4000 grits on

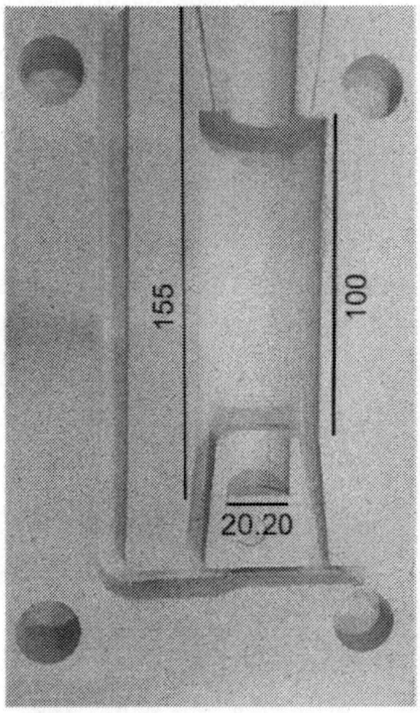

Figure 1: Sand mold used in the casting experiments. All dimensions are given in millimeters.

a Struers LaboPol-21 grinding machine and subsequently polished with 3 μm and 1 μm diamond suspension on a Struers Tegramin-20 polishing machine. To examine the bonding properties and microstructures at the interface, a Zeiss Supra 55VP Field Emission Scanning Electron Microscope (FESEM) with Energy Dispersive X-ray Spectroscopy (EDS) was used. The FESEM was operated with a 10 mm working distance and 15 kV acceleration voltage.

3 Results

3.1 Influence of galvanized coating

In Figure 2a, a macroscopic image of the compound casting sample where a galvanized steel pipe was used. On one side of the interface a large cavity has formed, whereas the remaining interface appears to show good bonding between the steel pipe and cast aluminum. Figure 2b shows an optical micrograph of the aluminum-steel interface in the same casting sample. The irregular steel surface, along with a thin layer at the surface that appears to be slightly darker, suggests formation of intermetallic phases. Within certain areas large intermetallic particles are seen growing from the thin layer towards the cast aluminum. In addition, some platelet shaped particles can be observed in the cast aluminum adjacent to the interface. As these particles are regarded as part of the reaction layer, it shows that there is a slight variation in the reaction layer thickness.

The platelet particles can be better observed in the backscattered SEM images in Figure 3. The contrast clearly shows that intermetallic particles are growing from the steel surface towards the cast aluminum. In this area of the casting, the reaction layer at the interface has a thickness varying from 20-50 μm, whereas platelet particles can

Table 1: Chemical composition in wt.% of S235JR steel and A356 aluminum.

	Mn	C	P	S	Fe				
S235JR [12]	1.4	0.17	0.045	0.045	Bal.				
	Si	Mg	Ti	Fe	Sr	Ga	Zn	Others	Al
A356.0	7.03	0.4052	0.1076	0.0822	0.0131	0.0089	0.0042	0.0029	Bal.

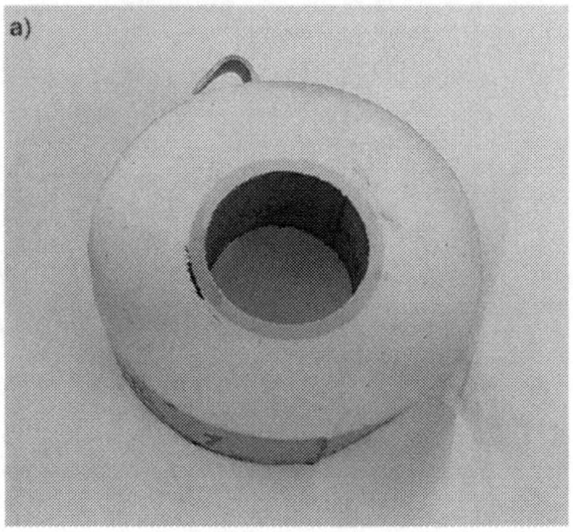

(a) Image of a casting sample with a galvanized steel pipe. A cavity can be observed on one side of the aluminum-steel interface.

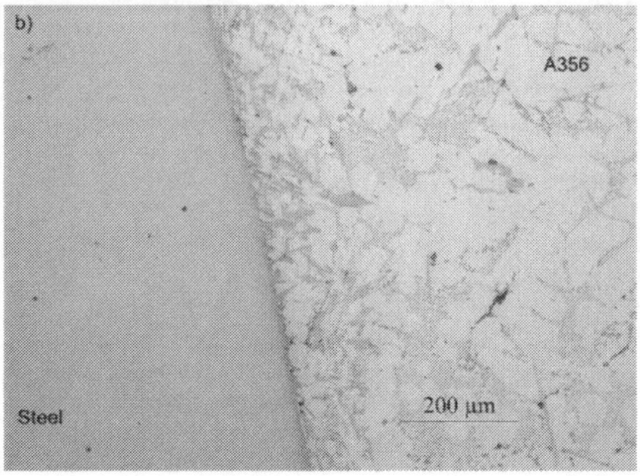

(b) Optical micrograph of the aluminum-steel interface in the galvanized casting. Intermetallic particles with platelet shape can be seen in the interface

Figure 2: Macro- and microscopic images of the compound casting with a galvanized steel insert.

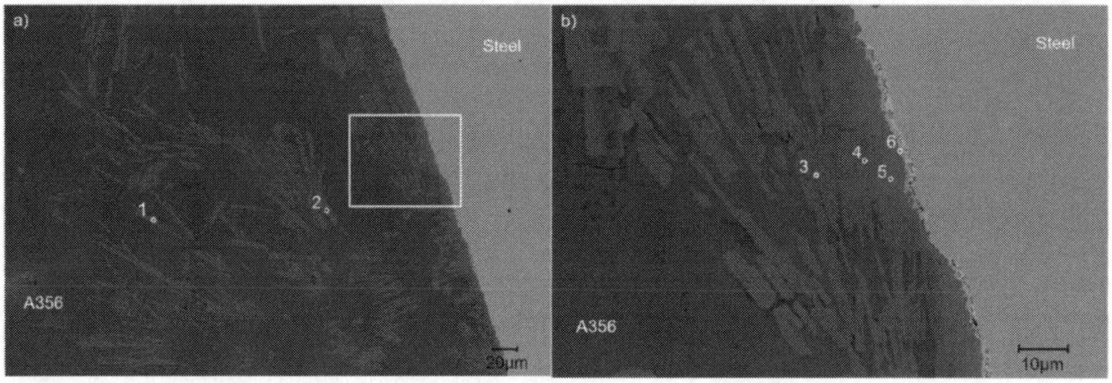

Figure 3: Backscattered micrograph showing the intermetallic phases formed at the aluminum-steel interface in the galvanized casting. b) shows the area marked with a white rectangle in a). Marked areas were analyzed by EDS.

be observed up to 300 μm from the interface. The intermetallic particles framed by the white rectangle in Figure 3a were further investigated at a higher magnification. As seen in Figure 3b, there is a slight contrast difference in the intermetallic layer growing at the steel surface. This suggests that multiple phases have formed in the reaction layer, as confirmed by the EDS analysis shown in Table 2.

From Table 2 it can be seen that the platelet-shaped particles formed outside the interface are the ternary eutectic phase, $Al_{4.5}FeSi$. Due to the sharp edges, such particles can be areas of stress initiation and would therefore decrease the overall bonding strength [13]. Similar platelet particles were observed at the aluminum-steel interface and was again determined to be β-$Al_{4.5}FeSi$, as seen in Area 3 in Table 2. In addition, two other phases were found to form between β-$Al_{4.5}FeSi$ and the steel pipe, as can be observed by the slight contrast variation in Figure 3b. The compositions suggest formation of α-$Al_{7.4}Fe_2Si$

 https://doi.org/10.1051/matecconf/202032606005

Table 2: Chemical composition detected through EDS analysis and possible formed phases in the marked areas in Figure 3.

Area	Composition [at%]				Possible Phase
	Al	Si	Fe	Zn	
1	67.19	16.95	14.99	0.87	β-$Al_{4.5}FeSi$
2	67.76	16.87	15.38	-	β-$Al_{4.5}FeSi$
3	67.08	16.03	15.13	1.76	β-$Al_{4.5}FeSi$
4	69.40	9.94	19.89	0.77	α-$Al_{7.4}Fe_2Si$
5	72.69	2.97	23.49	0.85	Al_3Fe
6	63.43	3.62	30.63	2.31	Al_5Fe_2 or Al_2Fe

and Al_3Fe with the latter closest to the steel pipe. At the immediate steel surface, there is an area with a layer of brighter contrast. The increased Zn concentration in Area 6 suggests that this bright layer is a Zn-enriched layer, showing that some of the galvanized layer remained at the interface. In this area, a higher concentration of Fe was detected, with the aluminum-iron ratio suggesting formation of Al_2Fe. However, as Al_2Fe and Al_5Fe_2 have similar compositions and Al_5Fe_2 often has been reported as the dominating phase forming in the Al-Fe system, it is possible that Al_5Fe_2 has formed adjacent to the steel surface [14]. This shows that the iron concentration increases in the intermetallic phases towards the steel pipe, which coincides with higher iron diffusion at the steel surface. All phases detected in the reaction layer are known phases in the ternary Al-Fe-Si and binary Al-Fe systems [15]. It should also be noted that zinc is barely detected in any of the analyzed areas, showing that upon re-melting of the galvanized layer, zinc has dissolved in the aluminum.

3.2 Influence of aluminized coating

For the compound castings with aluminized steel inserts, no defects could be observed in the aluminum-steel interface macroscopically (Figure 4a), suggesting metallurgical bonding. However, microscopic investigations revealed that the aluminized layer remained in the interface throughout the casting. This can be seen in Figure 4b, where the aluminized layer has a different color compared to the cast aluminum. Good bonding between the cast aluminum and the aluminized layer appears to have formed. The aluminized layer has a relatively uniform thickness of approximately 60 μm. Also observed in the aluminized casting, is the tongue-like intermetallic particles growing from the aluminized layer into the bulk of the steel pipe. Such microstructures are known to form in aluminized steels [16]. This differs from the compound castings with galvanized steel, where the intermetallic phases grew into the cast aluminum.

Figure 5 shows two BSE images of different areas of the reaction layer formed between aluminum and aluminized steel. The interface between the aluminized layer and the cast aluminum can be clearly seen. In Figure 5a the interface can be seen based on the difference in eutectic structure. Bright long platelet-shaped particles are laying along the interface in Figure 5a (Area 1). In Figure 5b, there is a gap between the cast aluminum and aluminized layer, showing that no metallurgical bonding formed in this area. Additionally, a large amount of bright long particles have formed in the aluminized layer.

Table 3: Composition and possible phases formed in the reaction layer of an aluminized casting. Areas coincide with the shown areas in Figure 5.

Area	Composition [at%]			Possible Phase
	Al	Si	Fe	
1	69.07	16.63	14.30	β-$Al_{4.5}FeSi$
2	75.31	0.75	23.94	Al_3Fe
3	72.36	-	27.64	Al_5Fe_2
4	85.99	-	14.01	Al_3Fe
5	90.51	-	9.49	Al_3Fe
6	76.55	-	23.45	Al_3Fe

From Table 3, it appears that the particles observed in the boundary between the aluminized layer and the cast aluminum are ternary $Al_{4.5}FeSi$ (Area 1 in Figure 5a). The particles appear to have a platelet shape, which coincides with the platelet structure observed for the castings with galvanized steel in Figure 3. However, as they are placed between the cast aluminum and the aluminized layer, their placement is likely detrimental to the bonding strength. The intermetallic particles in Area 2 and 3 in Figure 5a appear to be the binary phases Al_3Fe and Al_5Fe_2, although their chemical composition is very similar. Al_5Fe_2 can be recognized by its characteristic tongue-like morphology as it grows towards the steel pipe [15]. As the aluminized layer can still be observed in the interface, it is assumed that the binary Al-Fe phases formed during the aluminizing process. Nonetheless, bonding was observed locally in multiple areas.

In the aluminized area shown in Figure 5b, the particles appear to be binary Al-Fe particles. The compositions detected through EDS (Area 4 and 6 in Table 3) do not coincide with any known Al-Fe phase, but as the particles are very thin it is likely that some of the surrounding Al matrix has been detected in the analysis. The intermetallic tongue-like reaction layer has a slightly different contrast at the interface with the aluminized layer compared to the outer particles. This can also be observed in 5a. Area 2 and Area 6 have similar compositions, suggesting that the same phase has formed in both areas, likely Al_3Fe. Therefore, it is possible that the binary particles formed in the aluminized layer in Figure 5b are also Al_3Fe.

3.3 Influence of brass-coating

Figure 6 shows an image of a casting sample with a brass-coated steel insert. Multiple small cavities can be seen throughout the aluminum-steel interface, suggesting poor bonding.

Further investigation of the compound castings with brass-coated steel showed that the brass layer remained on the steel surface in most areas. Figure 7 shows two BSE

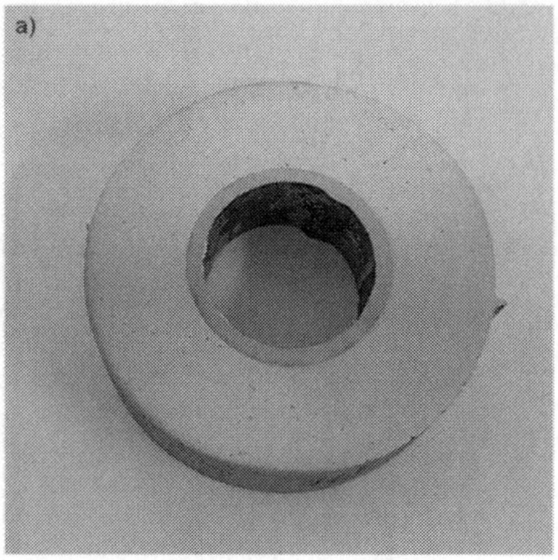

(a) Image of a casting sample with an aluminized steel pipe. No visible defects can be observed in the aluminum-steel interface.

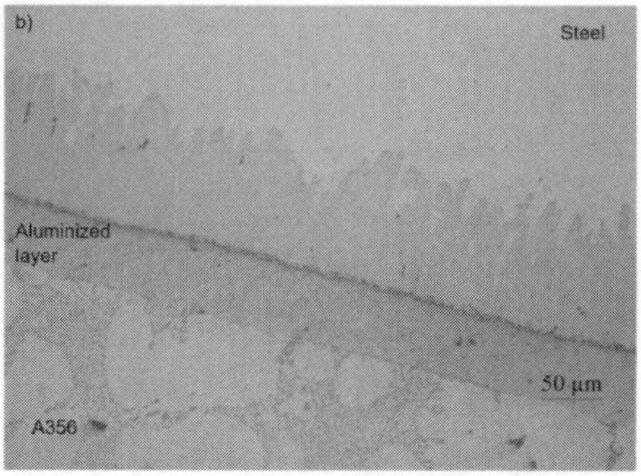

(b) Optical micrograph of the interface in the aluminized casting. An intermetallic phase can be seen growing towards the steel pipe, and the aluminized layer is still intact after casting.

Figure 4: Macro- and microscopic images of the compound casting with an aluminized steel insert.

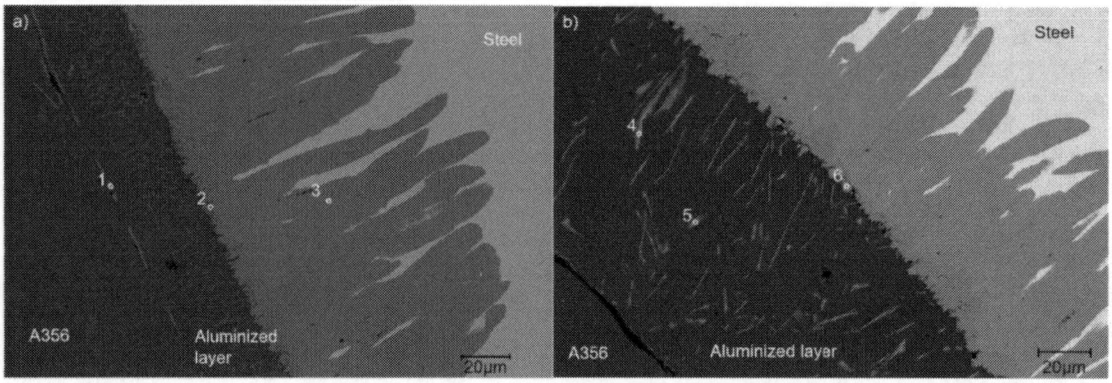

Figure 5: Micrographs of the reaction layer in two different areas in the aluminum-steel interface of an aluminized casting. Marked areas were analysed through EDS.

SEM images of different areas of the aluminum-steel interface. In Figure 7a, most of the brass layer remained on the steel surface, and no metallurgical bonding between this layer and the cast aluminum can be observed. Figure 7b, however, shows an area where a reaction layer has formed. However, several cracks and gaps can be observed on the cast aluminum side outside the reaction layer, indicating that no continuous bond has formed despite the reaction layer.

Although most of the brass layer remain unchanged on the steel surface, small local areas that reacted with the cast aluminum could be found (Areas 2, 3 and 4 in Figure 7a). The reaction area has a droplet-like shape, where the contrast in the figure suggests that multiple phases have formed. From the EDS analysis in Table 4, these phases were determined to be Al_2Cu_3, AlCu and a quaternary phase $Al_5Cu_2Mg_8Si_6$, which is frequently observed in Al-Si-Cu-Mg alloys with high Cu/Mg ratio [17]. The binary Al-Cu phases are suggested to form through solid state phase transformation from the eutectic Al_2Cu phase [18]. As the reaction area is relatively small, it is believed that some local melting of the brass layer has occurred due to local differences in temperature at the interface.

EDS analysis from the reaction layer shown in Figure 7b shows that ternary Al-Fe-Si phases have formed. The intermetallic phase adjacent to the steel surface with a slightly brighter contrast (Area 6 in Table 4) has a composition close to the α-$Al_{7.4}Fe_2Si$ phase. The composition of the intermetallic particles close to the cast aluminum side is close to the ternary eutectic phase β-$Al_{4.5}FeSi$. Interestingly, most of the areas analyzed showed low concentrations of copper, while no zinc was detected. This suggests

https://doi.org/10.1051/matecconf/202032606005

Table 4: Composition and possible phases formed in the reaction layer of a brass-coated casting. Areas coincide with the shown areas in Figure 7.

Area	Composition [at%]							Possible Phase
	Al	Si	Fe	O	Cu	Zn	Mg	
1	1.38	-	3.34	2.96	69.88	22.45	-	Cartridge brass
2	33.31	-	3.26	2.25	54.14	7.04	-	Al2Cu3
3	37.55	-	2.52	2.69	44.15	13.09	-	AlCu
4	23.11	27.87	-	0.34	12.75	-	35.94	Al5Cu2Mg8Si6
5	67.90	15.88	15.44	-	0.78	-	-	Al4.5FeSi
6	67.41	10.14	17.83	-	4.63	-	-	Al7.4Fe2Si

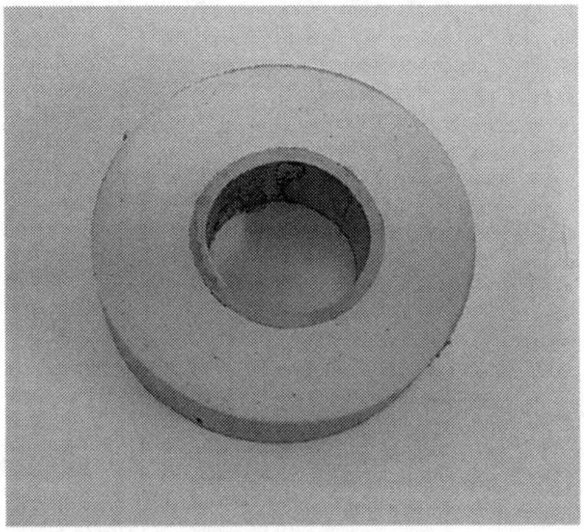

Figure 6: Image of a casting sample with a brass-coated steel pipe. Small cavities can be observed throughout the aluminum-steel interface.

that both copper and zinc have dissolved and diffused into the cast aluminum upon re-melting during casting.

3.4 Vickers microhardness

Vickers micro-hardness was measured across the reaction layer in each casting. The hardness profiles are plotted in Figure 8a for comparison. Microhardness indentations for the aluminum-steel interfaces in the galvanized, aluminized and brass-coated castings are shown in Figures 8b, 8c and 8d, respectively. Origin in the figure is set at the interface between the steel pipe and the cast aluminum in the galvanized and brass-coated castings, and the cast aluminum and the aluminized layer in the aluminized casting.

From the graph, it is clear to see that the intermetallic Al-Fe and Al-Fe-Si phases forming at the reaction layer, have much higher hardness than the steel pipe and cast aluminum. For the aluminized castings, the hardness increases gradually from the steel pipe towards the aluminized layer due to the formation of Al_5Fe_2 and Al_3Fe. The gradual increase is likely due to the tongue-like morphology causing an increased area of Al_5Fe_2 towards the aluminized layer. In addition, the hardness reaches a peak close to the aluminized layer, suggesting that Al_3Fe is harder than Al_5Fe_2. The reaction layer in the galvanized castings also show higher hardness than the steel pipe and cast aluminum, however not to the same extent as the aluminized castings. The hardness peak in the cast aluminum of the galvanized casting is due to the indentation being at an area of intermetallic particles outside the reaction layer at the steel surface. For the brass-coated castings, no hardness peaks were observed. This suggests that the remaining brass layer has a hardness between those of A356 and steel.

4 Discussion

In all three compound castings, binary Al-Fe or ternary Al-Fe-Si phases have formed in the reaction layers. By comparing the three surface treatments, it can be concluded that a more continuous metallurgical bond can be achieved by using galvanized steel inserts, deeming this the best surface treatment out of the three. However, it should be noted that in the castings with galvanized steel inserts, large Al-Fe-Si intermetallic particles are also found in the bulk of the cast aluminium with some distance to the continuous reaction layer at the interface. These particles are likely formed as a result of the galvanized layer remelting upon casting. During the galvanization process, diffusion of iron and zinc atoms lead to the formation of several Fe-Zn phases in the final galvanized layer. Usually the first phases to form during this process are the binary phases $FeZn_{10}$ and $FeZn_{13}$ [19]. The temperature of the liquid A356 is above the peritectic temperatures for both these phases through the reactions $L+\Gamma\text{-}Fe_3Zn_{10} \leftrightarrow \delta\text{-}FeZn_{10}$ at 672°C and $L+\delta\text{-}FeZn_{10} \leftrightarrow \zeta\text{-}FeZn_{13}$ at 530°C [20, 21]. This allows for remelting during casting and thus leave excessive iron atoms in the melt. Therefore, iron atoms diffuse from the steel surface into the aluminum melt and ternary eutectic $\beta\text{-}Al_{4.5}FeSi$ particles can form in the cast A356. This phase is known for its characteristic platelet morphology, which can be detrimental to bonding strength as the sharp edges can induce stress concentrations and therefore fracture can propagate along the interface of the particle-aluminum matrix interface [13]. In addition, the size of the $\beta\text{-}Al_{4.5}FeSi$ particles are important for the overall strength of the casting. It was reported by Ma et al. that there is a significant decrease in the ultimate tensile strength and elongation when the particle length increases to approximately 70 μm [22]. Such particles were also ob-

 https://doi.org/10.1051/matecconf/202032606005

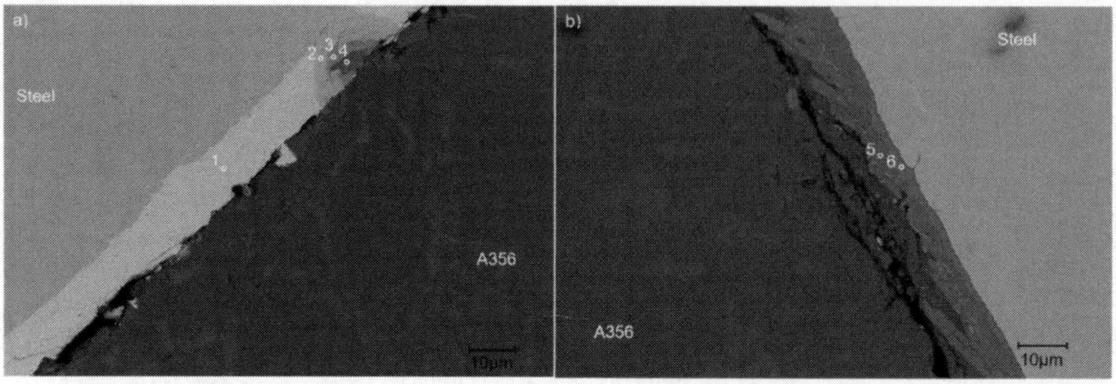

(a) Area where the brass-layer remained at the steel surface. (b) Area of a reaction layer in the aluminum-steel interface.

Figure 7: BSE micrographs of the aluminum-steel interface in a brass-coated casting. The marked areas were analyzed through EDS.

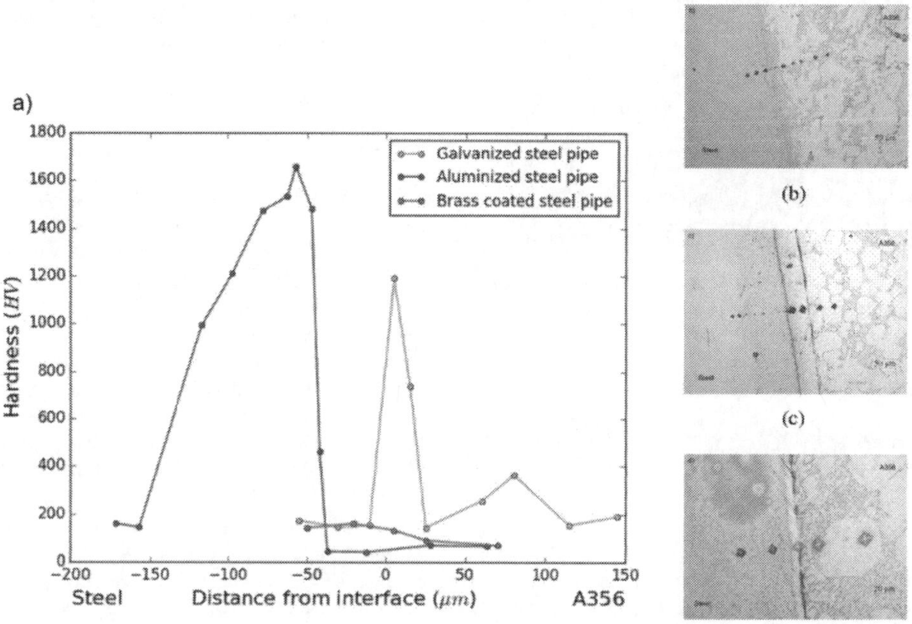

(a) (b) (c) (d)

Figure 8: Micro-hardness measured across the reaction layer for the castings with various surface treatment. Origin is set at the interface between the steel pipe and the cast aluminum in the galvanized and brass-coated castings, and the cast aluminum and the aluminized layer in the aluminized casting.

served at the interface between the A356 aluminum and aluminized layer on the steel inserts. As the aluminized layer did not remelt during casting, less Fe will be available. The number of β-$Al_{4.5}FeSi$ particles is therefore much less in this casting, with the size also being significantly smaller. However, as these particles appear to have formed along the interface between the cast A356 and the A356 from the aluminizing process, it is likely to reduce strength of the overall bond and that fracture can easily occur in this interface.

The eutectic point in the binary Al-Cu system is at 17 at% Cu and 550°C [23], whereas in the Al-Zn system the eutectic point is at 88.7 at% Zn and 381°C [24]. This means that in principle the brass coating-layer could remelt and react with the liquid cast aluminum. However, for the castings with brass-coated steel inserts, the coating-layer remained intact in most areas. This is likely due to the oxide layers forming on the liquid aluminum and brass surfaces. Qiu and Legraf tested the oxidation of commercial brass sheets (20 wt.% Zn) in humidified air and found that a two-phase oxide layer consisting of large areas enriched with Cu_2O with smaller areas of ZnO in between [25]. Such passive oxide layers significantly reduce wettability and reaction with the liquid cast A356. In certain local areas, where the oxide layer is thin enough, the brass layer could remelt and form a reaction layer, which in this case consists of Al-Fe-Si phases. Interestingly, no zinc and barely any copper were detected in this reaction layer, and no Al-Cu phases had formed. While Zn has a relatively high solubility in Al, Cu was expected to form the eutectic binary Al_2Cu phase instead. The low concentration of Cu at the interface would suggest that Cu along with Zn has diffused into the cast A356 upon melting. It should also be noted that compared to the other coatings, the thickness of the reaction layer with intermetallic phases formed in the brass-coated casting is significantly thinner and that such a thickness is preferable in terms of bonding strength.

By comparing the resulting reaction layers from the various metallic coatings, it can be seen that the reaction layers appear to grow into the cast A356 alloy when galvanized and brass-coated steel inserts are used. However, for the castings with aluminized steel inserts, nearly no reaction layer has formed in the aluminum-aluminum interface, which can be attributed to the inert oxide layer on the aluminized coating surface. The characteristic tongue-like morphology of the Al_5Fe_2 phase that grow into the bulk of the steel pipe, was formed during the aluminizing process. It has previously been reported that Al_5Fe_2 is the dominating phase forming between Al and steel in an aluminizing process [4, 26]. Although Fe atoms diffuse faster in liquid Al than Al atoms in solid Fe, Al has a solubility of around 22 at% in iron at 600°C [27]. This can thus contribute to the formation of Al_3Fe at the aluminum steel interface, which then leads to the formation of Al_5Fe_2 through further diffusion [4]. Addition of silicon to the aluminum alloy has been reported to act as a growth inhibitor of the reaction layer, as it promotes the formation of a more uniform Al_3Fe layer that then acts as a diffusion barrier for further growth of Al_5Fe_2 [28]. Despite the silicon present in the A356 alloy, a relatively thick and irregular reaction layer is observed in the aluminized casting. This is believed to be due to a relatively long dipping time used in the aluminizing process, which then promotes further diffusion and phase growth.

For the castings with brass-coated and galvanized steel inserts, the reaction layers form upon casting. In this process, the existing metallic coating is expected to remelt, thus leaving a fresh surface for the aluminum and steel to react. However, the experimental results indicate that there is limited remelting of the brass layer and nearly no remelting of the aluminized layer. This is due to the inert aluminum oxide layers on both the aluminized insert and aluminum melt. Cheng and Wang found that an addition of 10 wt.% Si caused mainly formation of ternary Al-Fe-Si phases and significantly reduced the fraction of binary Al-Fe phases that were observed with lower Si concentration [29]. The A356 alloy contains 7 wt.% silicon, which then diffuses alongside Al at the steel surface and then form ternary Al-Fe-Si phases as detected through the EDS analyses. In the casting process, the diffusion time is much shorter compared to the aluminizing process and therefore further diffusion of Al atoms into the steel and Fe atoms into the cast aluminum, and thus formation of thicker binary Al-Fe phases will be prevented. If also the formation of a ternary phase at the interface is preferred, it can act as a diffusion barrier for growth of the Al_3Fe and Al_5Fe_2 phases, which then would cause preferred growth direction to be towards the cast A356. A thinner reaction layer is believed to have formed in the brass-coated casting compared to the galvanized casting, due to excessive Fe atoms already being present in the cast A356 for the latter.

5 Conclusions

The following conclusions can be made from this research:

It is possible to achieve metallurgical bonding between A356 aluminum and steel using metallic coating, such as galvanizing, aluminizing and brass-coating.

Compound casting with galvanized steel showed the most continuous metallurgical bonding, making it the best coating out of the three. Ternary Al-Fe-Si phases formed in these castings growing towards the cast A356. In addition, ternary Al-Fe-Si particles could be observed in the bulk of the cast aluminum with some distance from the reaction layer at the interface.

In the compound castings with aluminized steel, no reaction layer formed between the aluminized layer and cast A356, but local areas still showed metallurgical bonding. Instead, a reaction layer growing towards the steel pipe with the characteristic tongue-like morphology of the Al_5Fe_2 phase, formed during the aluminizing process.

The brass layer remained in most of the aluminum-steel interface in the compound castings with brass-coated steel. Only local reaction areas could form in the castings, where ternary Al-Fe-Si phases were found to grow towards the cast A356.

Acknowledgements

The authors are grateful for the financial support and contribution from the Norwegian Research Council though the IPN project "AluLean" (90141902), Aludyne Norway for material and FerroZink AS and OS ID for coating of steel pipes.

References

[1] Wang, Q. G., Metall Mater Trans A **34**(12), p. 2887-2899 (2003)

[2] Sebaie, O. E., et al., Mater Sci Eng A **480**, p. 342-355 (2008)

[3] Tayal, R. K., et al., Proc Ntl Conf Trends Advances Mechanical Engineering (2012)

[4] Springer, H., et al., Acta Mater **59**(4), p. 1586-1600 (2011)

[5] Papis, K. J. M., Loeffler, J. F., and Uggowitzer, P. J., Sci China Ser E **52**(1), p. 46-51 (2009)

[6] Aylward, G., and Findlay, T., *SI Chemical Data 5th ed.* (John Wiley Sons Australia, Milton, 2002)

[7] Springer, H., Szczepaniak, A., and Raabe, D., Acta Mater **96**, p. 203-211 (2015)

[8] Kang, M. and Kim, C. Mater Design **81**, p. 95-103 (2015)

[9] Jiang, W., Fan, Z., and Li, C., J Mater Process Tech **226**, p. 25-31 (2015)

[10] Jiang, W., et al., J Alloy Comp **678**, p. 249-257 (2016)

[11] Jiang, W., et al., J Alloy Comp **688**, p. 742-751 (2016)

[12] Smith Stål, *Lagerkatalog-Smith Stål* (2013)

[13] Seifeddine, S., Johansson, S., and Svensson, I. L., Mater Sci Eng A **490**(1-2), p. 385-390 (2008)

[14] Springer, H., et al., Mater Sci Eng A **528**, p. 4630-4642 (2011)

[15] Maitra, T. and Gupta, S. P., Mater Charact **49**(4), p. 293-311 (2002)

[16] Sasaki, S. and Yakou, T., Surf Coat Technol **201**(6), p. 2131-2139 (2006)

[17] Zheng, Y., et al., J Alloy Comp **649**, p. 291-296 (2015)

[18] Tavassoli, S., Abbasi, M., and Tahavvori, R., Mater Design **108**, p. 343-353 (2016)

[19] Pokorny, P., et al., Metalurgija **55**(1), p. 111-114 (2016)

[20] Baker, H. and Okamoto, H., *Binary Alloy Phase Diagrams*, in *ASM Handbook, Volume 03 - Alloy Phase Diagrams* (ASM International, 1992), p. 206

[21] Su, X., Tang, N.-Y., and Toguri, J. M., J Alloy Comp **325**(1-2), p. 129-136 (2001)

[22] Ma, Z., et al., Mater Sci Eng A **490**(1-2), p. 36-51 (2008)

[23] Ponweiser, N., Lengauer, C. L., and Richter, K. W., Intermetallics **19**(11), p. 1737-1746 (2011)

[24] Murray, J. L., B Alloy Phase Diag **4**(1), p. 55-73 (1983)

[25] Qiu, P. and Leygraf, C., Appl Surf Sci **258**(3), p. 1235-1241 (2011)

[26] Bahadur, A. and Mohanty, O. N., Mater Trans **32**(11), p. 1053-1061 (1991)

[27] Kattner, U. and Massalski, T., *Binary Alloy Phase Diagram* (ASM International, Material Park, 1990)

[28] Takata, N., Manamu, N., Kobayashi, S., and Takeyama, M., Intermetallics **54**, p. 136-142 (2014)

[29] Cheng, W. J. and Wang C. J., Intermetallics **19**(10), p. 1455-1460 (2011)

CFD Analysis of Mechanisms Underlying the Porosity-reducing Effect of Atomized Flows in High-pressure Die Cast Products

Eitaro Koya[1],* *Masahiko* Nakagawa[1], *Shinya* Kitagawa[1], *Jun* Ishimoto[2] *Yoshikatsu* Nakano[2] *and Naoya* Ochiai[2]

[1]Honda Motor Co., Ltd, Motorcycle Operations, 3-15-1, Senzui, Asaka, Japan
[2]Tohoku University, Institute of Fluid Science, 2-1-1, Katahira, Sendai, Japan

Abstract. In high-pressure die casting, attention has been paid to the J factor, which is defined by the speed of the metal injected at the gate and the shape of the gate. In casting experiments using a piston die, the porosity of the product can be reduced by increasing the J factor such that the metal flow passing through the gate forms an atomized flow. To clarify the underlying mechanisms, we developed a system for simulating a two-phase flow of air and aluminum by large-scale calculations of turbulent flow. During the development of the system, we injected metal into an open space and performed imaging to confirm the state of the atomized flow. The system was then verified by reproducing the atomized flow. The analysis results visualized the many small turbulent vortices generated in the thick part far from the gate. We demonstrated that the change from small to longitudinal vortices promoted entrainment of air into the aluminum and increased the efficiency of air expulsion outside the die through an increase in the J factor.

1 Introduction

High-pressure die casting (HPDC) is a method of casting at high speed and high pressure that offers the advantages of rapid cooling and production of fine microstructures. For example, the use of HPDC to fabricate pistons takes advantage of these characteristics. In this example, an Al-Si alloy with a Si content above the eutectic composition was used to form a fine network of eutectic silicon and intermetallic compounds surrounding α-phase aluminum, thereby improving high-temperature strength [1]. However, one issue with HPDC is that the strength of manufactured pieces is degraded by porosity arising from gases such as air that are entrained into the metal when it is injected into the die at high speed. Approaches for solving this issue have generally involved use of a strong vacuum or substitution of oxygen in the cavity [1,2]. In addition, a localized high-pressure squeezing method is applied to the thicker areas of manufactured parts to reduce porosity.

Another approach to reducing porosity focuses on the J factor, which is defined by the speed of the metal injected at the gate and the shape of the gate. If the J factor exceeds some threshold value, the metal flow atomizes when it passes through the gate. It has been reported that when the J factor is further increased (the high J factor method), the mechanical properties of HPDC products are enhanced and porosity is reduced [3, 4]. If the quality of HPDC can be enhanced by increasing the J factor, then it would be possible to manufacture high-quality HPDC parts inexpensively by using a general-purpose casting machine without the need for specialized equipment with features such as complex injection mechanisms, dies with complicated seal mechanisms, and auxiliary pressure application devices.

However, the mechanisms underlying the reduction in porosity through the use of a high J factor are not yet clear, and attempts are being made to clarify them [5]. To accomplish this, it is necessary to visualize the flow within the die during HPDC. Attempts to do so have been made using photographic methods but have so far been unsuccessful in capturing high-speed molten flow at microscopic resolutions [6].

In computational fluid dynamics (CFD), large-scale turbulence simulations using high-speed computational processing (e.g., parallel processing) are being applied to understand an increasingly wide variety of phenomena. Against this background, research has been conducted into basic simulations to reproduce the injection condition of the metal at the gate in order to visualize the flow during HPDC [7].

In this study, we modeled an actual casting die design and developed a simulation system to reproduce the experimental results of a casting experiment involving a piston die. The aim of the developed system was to visualize and quantify the detailed state of the air inside the die by using two-phase flow calculations of the air within the molten aluminum.

This paper first presents the results photographic experiments examining various J factors used for the metal interface at the gate and discusses the actual state of the atomized flow. Next, the aluminum filling and air expulsion processes within the cavity of the piston die are investigated using CFD. Finally, the results are

* Corresponding author: eitaro_koya@jp.honda

explained by discussing the mechanisms underlying the reduction in porosity with a higher J factor.

2 Casting experiment on J factor and porosity

2.1. Casting experiment methods

To investigate the relationship between the J factor and the internal quality of HPDC parts, we performed a casting experiment using a piston-shaped die that examined the relationship between the J factor and the porosity of the thicker parts of the piston. The flow state when the metal passes through the gate in HPDC is defined by the J factor in the following equationaccording to Wallace [8]:

$$J = D \times \rho \times V_g^{\ 1.71} \tag{1}$$

Here, V_g is the ingate speed of the metal, ρ is the specific gravity of the metal, and D is the gate shape factor defined by gate area A_g, gate depth a, and gate width b as follows:

$$D = A_g/(a + (a + b) \times 2) \tag{2}$$

As the J factor becomes larger, the flow of metal at the gate changes from a continuous jet to a coarse particle flow and then to an atomized flow. In the case of aluminum, the threshold J factor for atomization is 1064. Figure 1 shows the shape of the cast piston and the gate shape. In the casting experiment, a two-piece die was used, and the J factor was changed by changing the highinjection speed with a constant low injection speed. The die uses chill vents that are open to the atmosphere; no vacuum equipment was used. The piston diameter was 54 mm and its weight was 90 g. A casting machine with a die cramping force of 350 tons was used for casting. The casting pressure was 70 MPa. The material used was an Al-13%Si-3.5%Cu-1.5%Mg alloy.

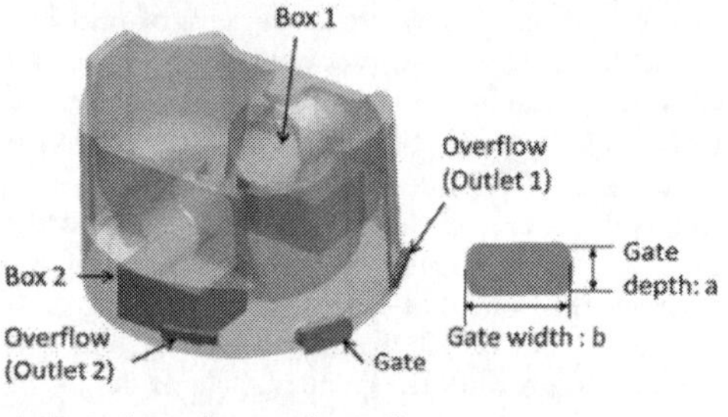

Fig. 1. Test piston shape and gate shape

2.2 Relationship between J factor and porosity

The amount of porosity in the areas labeled Box 1 and Box 2 in Figure 1 was measured with a 225-kV flat-panel micro computed tomography (micro-CT) scanner. Imaging was performed at a resolution of 20 μm, and the porosity was derived by imaging processing. Figure 2 shows the derived porosity images. The porosity measurement points in the thick parts (i.e., the pin boss areas) of the piston are taken as Box 1 and Box 2, and the relationship between the porosity volume fraction and J factor at each of these locations is shown in Figure 3. When the J factor was increased from 1000 to 2000, the porosity volume fraction was greatly reduced. Above 2000, a trend was observed in which the rate of reduction decreased and saturated. This clarified the trend of porosity decreasing as the J factor increased. Furthermore, at the low J factor of 1000, the porosity volume fraction in Box 1 was 1.5 times that in Box 2.

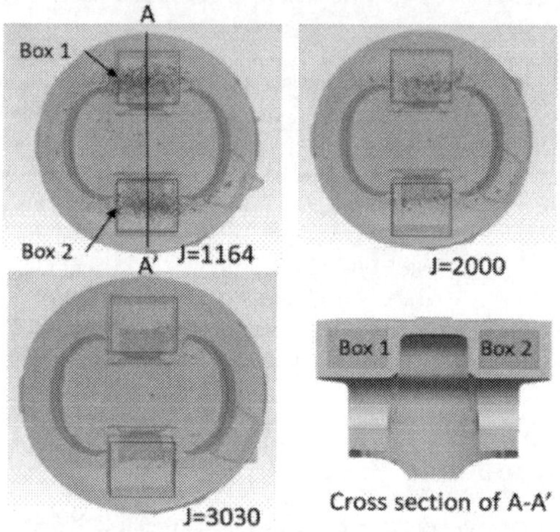

Fig. 2. Porosity images derived from micro-CT

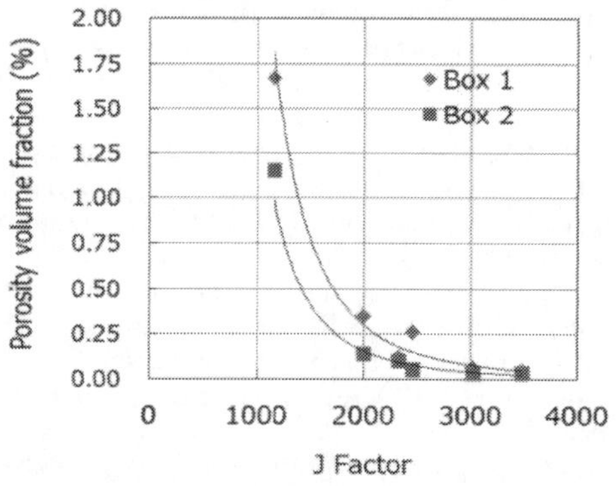

Fig. 3. Relationship between porosity volume fraction and J factor

2.3 Experimental results and research aims

Although a relationship was demonstrated between increased J factor and quantitatively enhanced casting quality, the J factor is an empirical formula that defines the flow of metal as it passes through the gate, whereas the internal quality that we aim to enhance in this experiment is in the thick parts, which are far from the gate. To establish a technique for improving casting quality by increasing the J factor, the following questions must be clarification.

1. What is the microscopic state of the atomized flow as it passes through the gate?
2. What is the state of the molten metal after injection when it flows into the die as an atomized flow at the gate, and what effect does this state have on reducing the porosity of thicker areas?

Question 1 can be investigated by photographic experiments that capture the metal as it is injected into an open space. For question 2, the internal state of the metal flow cannot be determined by observation because aluminum reflects light. We therefore decided to use the latest CFD techniques to create a simulation to visualize the microscopic flow of the metal. In CFD, a microscopic metal flow is simulated as a two-phase flow of aluminum and air. Because the flow in HPDC is turbulent, turbulence calculations need to be performed in the two-phase flow analysis. We developed a system for performing turbulence calculations of this two-phase flow and investigated the mechanisms underlying the reduction in porosity by increasing the J factor. The following gives an overview of the verification process used in this research.

3 Verification process in this research

3.1. Photography experiment on metal injected into an open space

First, we performed a photographic experiment to clarify the changes in the metal flow interface when it passes through the gate; these changes were defined according to the J factor. The method reported by Aida et al. for injecting metal into an open space was used for the experiment [11]. In HPDC, metal is injected such that its speed is at least 30 m/s while passing through the gate. A pulse laser was used to perform transmission imaging to gain a detailed understanding of the microscopic changes in the metal interface as it moves at high speed. Nanosecond-order exposure was made possible by synchronizing the pulse laser with a high-speed camera. By using transmitted light, the metal flow interface can be clearly imaged without interference from light scattering. We used this imaging technique to perform a comparative investigation of the process by which the droplets were produced at the metal flow interface by atomization while varying the J factor. Furthermore, we measured the size distribution of the generated droplets and used the results to set the resolution (i.e., mesh size) required for simulating this phenomenon.

3.2. Computational investigation of two-phase fluid flow of metal inside die

3.2.1 Development of the simulation system

In the flow analysis currently used in computer-aided engineering of regular casting, the macroscopic filling state of single-phase flows of aluminum is simulated using differential methods and finite volume methods. Because of this, the behavior of air cannot be calculated directly.
In this research, model calculations were performed using high-precision finite volume methods to simulate the two-phase flow of aluminum and air at the microscopic level. Numerical calculations of turbulence were performed by large eddy simulation (LES), and interface changes were numerically simulated by the volume of fluid (VOF) method. In other words, the two-phase flow inside the die was simulated by the LES-VOF method through coupling of the LES and VOF methods.

3.2.2 Governing equations for flow analysis

The numerical model represents the simultaneous unsteady flow of two immiscible, incompressible fluids, each having a constant viscosity and surface tension. Surface tension is taken into account through the continuum surface force (CSF) model, in which the surface force is transformed into a body force which is only non-zero in the interface region of limited thickness. [9,10]

The mass conservation equation is

$$\frac{\partial \rho}{\partial t} + \nabla \cdot (\rho \boldsymbol{U}) = 0. \tag{3}$$

The momentum equation is

$$\frac{\partial \rho \boldsymbol{U}}{\partial t} + \nabla \cdot (\rho \boldsymbol{U}\boldsymbol{U}) = -\nabla P + \nabla \cdot \boldsymbol{\tau} + \boldsymbol{F}_S + \rho \boldsymbol{g}. \tag{4}$$

Here, ρ is density, $\boldsymbol{U}$ is a flow velocity vector, $\boldsymbol{\tau}$ is a viscous stress tension, P is absolute pressure, and $\boldsymbol{g}$ is a gravitational acceleration vector. The $\boldsymbol{F_s}$ term in Eq. (4), follows the CFS model. For the surface tension coefficient σ_S, the phase volume fraction α, the curvature of the two-phase interface *k*, and a unit normal vector perpendicular to the interface $\boldsymbol{n_s}$, the body force based on the CSF model is given by

$$\boldsymbol{F_s} = \sigma_S + k\nabla_\alpha, \tag{5}$$

$$\mathrm{K} = -\nabla \cdot n_s, \quad \boldsymbol{n}_S = \frac{\nabla_\alpha}{[\nabla \alpha]} \tag{6}$$

The interface between the phases is simultaneously computed using a surface-capturing methodology that employs the volume fraction of one of the phases (taken here to be liquid-phase) as an indicator function (VOF) to identify the different fluids (surrounding air-phase). The interface is not defined as a sharp boundary, and a transition region exists where the fluid is treated as a mixture of the two fluids on each side of the interface, which in reality is a discontinuous step. When the interface is advected by the flow, the evolution of the VOF advection equation, namely, the transport equation

 https://doi.org/10.1051/matecconf/202032606006

of the volume fraction of liquid α_l, is given by the following conservative form:

$$\frac{\partial \alpha_l}{\partial t} + \nabla \cdot (\alpha_l \boldsymbol{U}) = 0, \quad (7)$$

where α is the volume fraction, the subscript l represents the aluminum phase, and

$$\alpha_l + \alpha_g = 1, \quad (8)$$

where the air phase is indicated by the subscript g. Using the volume fraction of each phase, density ρ and viscosity coefficient μ are

$$\rho = \alpha_l \rho_l + (1 - \alpha_l)\rho_g, \quad (9)$$
$$\mu = \alpha_l \mu_l + (1 - \alpha_l)\mu_g. \quad (10)$$

3.2.3 System verification method

To investigate the fundamental reproducibility of the atomization phenomena when the new system was being developed, we compared the simulation results with the photography results for the injection state of the metal [9]. Then, we modeled and verified the casting die design used in the casting experiment presented in Section 2. Verification consisted of comparing the relationship of the aluminum filling ratio and J factor of the thicker parts at the pin bosses when filling was complete with the trends between porosity volume fraction and J factor as observed by micro-CT.

4 Photography experiment for metal injection

4.1 Photography method

Figure 4 shows the structure of the die used in the injection experiment and the placement of the light source and camera. Photography was performed by using a pulse laser to illuminate the die from the rear horizontal direction and positioning the camera in front of the die. An acrylic plate was attached to the side of the die to transmit the laser light. The injection experiments were performed using a casting machine with a die cramping force of 350 tons. An Al-11%Si-2.5%Cu-0.2%Mg (JIS ADC12) alloy was used as the injected material to allow comparison with the experiments reported by Aida et al. The laser light was illuminated onto the side of the die from a transmitter via optical fiber. Photography was performed using a high-speed camera at 8000 fps. Figure 5 shows an example plot of injection speed in the experiment with J=3030.

4.2 Photography results

Figure 6 shows the photography results for various J factors. At a J factor of 240, the state of the metal flow interface was a continuous jet and the interface was continuous and smooth. In this flow state, almost no

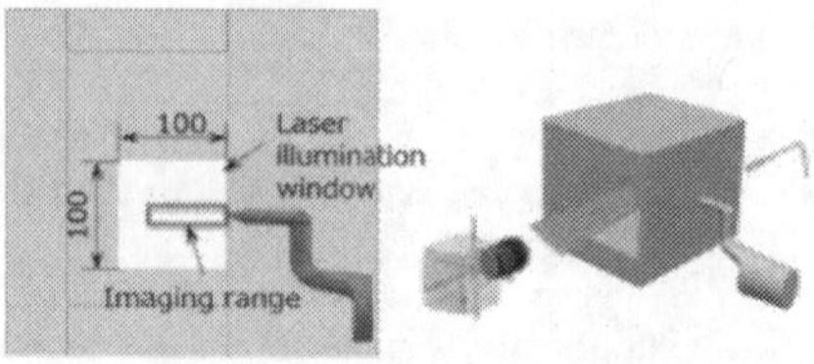

Fig. 4. Layout of the die and photograph equipment for imaging of the injected flow

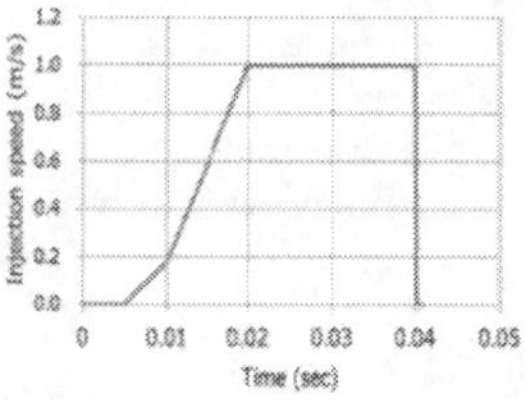

Fig. 5. Injection speed

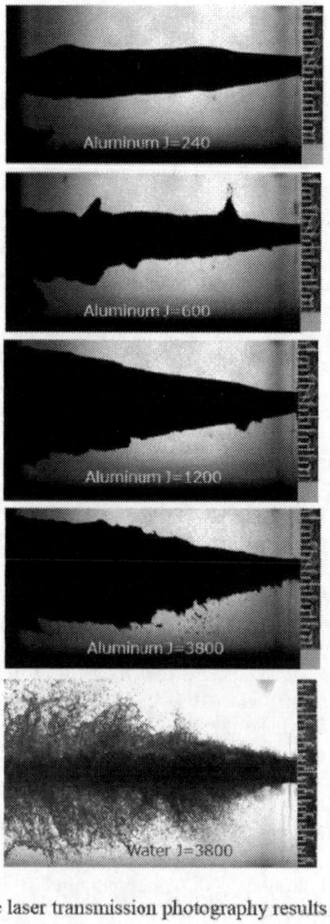

Fig. 6. Pulse laser transmission photography results

vertical spreading of the metal was observed. At a J factor of 600, the metal flow interface began to fluctuate, and the metal flow began to spread in the vertical direction. At a J factor of 1200, which exceeds the threshold for atomized flow, the atomization process was

observed; liquid columns were formed due to K-H instability induced at the metal flow interface and droplets broke off from the liquid columns.
Furthermore, the spread of the metal flow in the vertical direction became larger. At a J factor of 3800, the fluctuations in the metal flow interface further increased and the frequency of droplet generation increased. The photography results for water at a J factor of 3800 are also shown for reference. The internal atomization state could be observed because of the laser light passing through the water. The vertical spreads of water and aluminum were nearly the same. Although the internal state of the aluminum could not be viewed because it is opaque to light, the vertical spread suggests that the atomized flow within the transmission image has a state similar to that of water. Furthermore, whereas the water droplets were spherical, the aluminum droplets were ribbon-shaped.

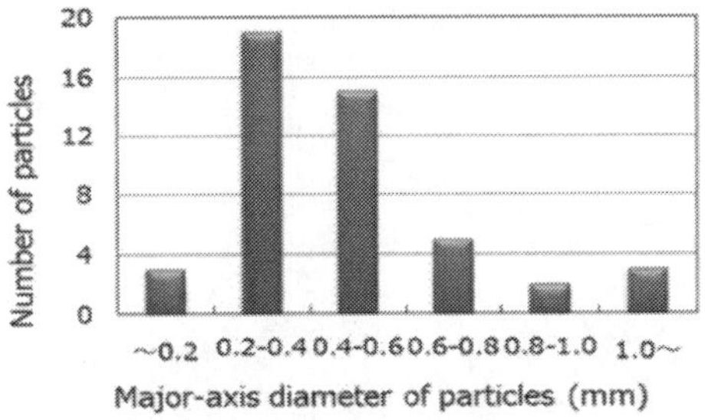

Fig. 7. Major axis diameter distribution of droplets

Figure 7 shows the distribution of the major-axis diameter for the droplets generated from the metal interface. Many droplets of size 0.2 to 0.3 mm were observed. In the atomized flow region defined by the J factor from the above results, a jet flow consisting of a two-phase flow of aluminum and air was formed by atomization upon passing through the gate. Furthermore, the results showed that a fine mesh size with a size of around 0.2 to 0.3 mm was needed for the resolution of the simulation.

5 Simulation system verification

5.1 Computational piston die design model and conditions

Table 1 shows the physical values used in the simulation. LES (Smagorinsky type) was used for the turbulence model. Figure 8 shows the model of the piston part and the boundary conditions. The computational model consisted of the sleeve, runners that branch into two pistons from the sleeve, and two pistons. The sleeve end was taken as the inlet for the moving boundary model.
The measurement results for the plunger speed taken at intervals of 1 ms from the casting experiment were derived for the amount of aluminum fed into the inlet. Furthermore, non-slip boundary conditions with walls were set for the sleeve, runners, and piston component of the inner wall parts. The overflow was not fully modeled but was modeled up to the area where the overflow was modeled from the product up to the minimum cross section of the overflow. This minimum cross-section area was taken as the outlet that opens to the air. The total number of cells in the model mesh was 1 million. The time when the inlet injection speed began to increase from low to high when the metal tip reached the gate was set as 0 s.
The calculations were performed until the plunger speed suddenly decreased after the injection was finished and the flow of the metal inside the die had completely stopped. Note that although the plunger begins to apply pressure after it decelerates in an actual casting machine, the pressurization calculation was not performed in the current simulation in order to clarify the changes in the two-phase flow accompanying the change in metal speed.

Table 1. Physical quantities used in the calculations.

	Aluminum	Air
Kinematic viscosity (m^2/s)	1.77E-06	1.60E-05
Density (kg/m^3)	2550	1.165
Surface tension (kg/s^2)	0.878	–

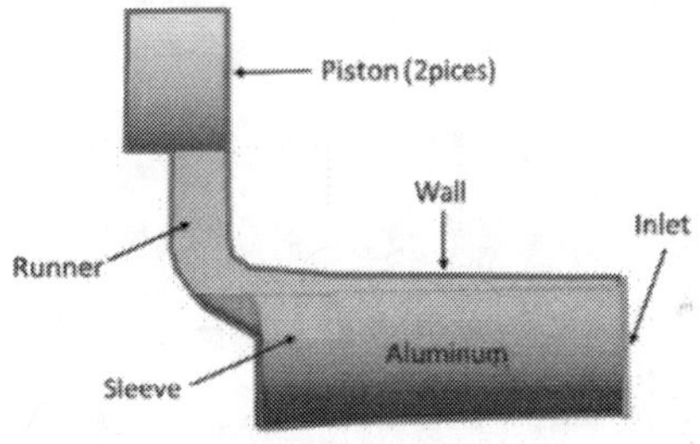

a) Boundary condition

b) Piston calculation model and names of each part

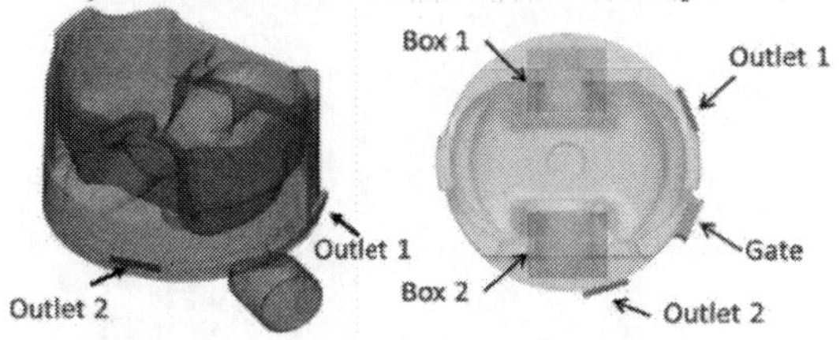

Fig. 8. Casting experiment model, boundary conditions

5.2 Verification of computational results

The validity of the computational model was verified according to whether the relationship between porosity volume fraction and J factor in the cast product was reproduced. Figures 9 and 10 show the filling ratio at each injection time of the parts (Box 1 and Box 2) where the porosity volume fraction was measured by micro-CT.

https://doi.org/10.1051/matecconf/202032606006

Figure 11 shows porosity images derived from micro-CT for Box 1 and Box 2. At both Box 1 and Box 2, the rapid increase in the filling ratio and the enhanced filling ratio with increasing J factor were reproduced. Furthermore, the trend of a better filling ratio at Box 2 compared with Box 1 was also reproduced. Note that the difference in maximum filling ratio between Box 1 (90%) and Box 2 (99.5%) was clearly observed in the calculations. These results are discussed in detail in the next section; briefly, the trends in the phenomena investigated here were fully reproduced.

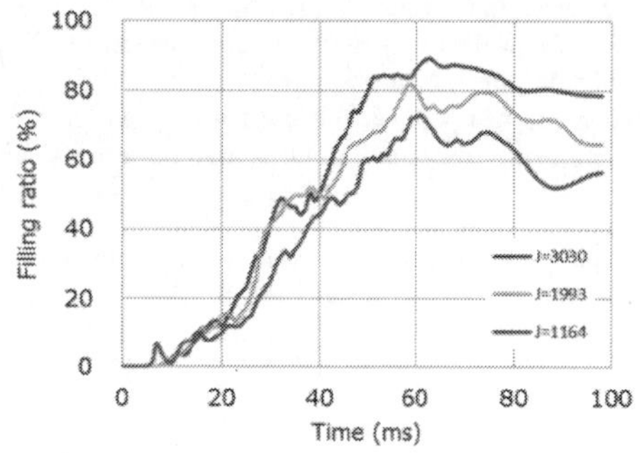

Fig. 9. Changes in the filling ratio at Box 1

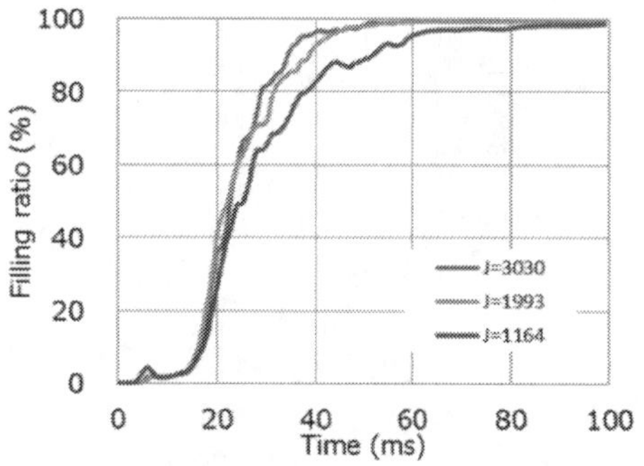

Fig. 10. Changes in the filling ratio at Box 2

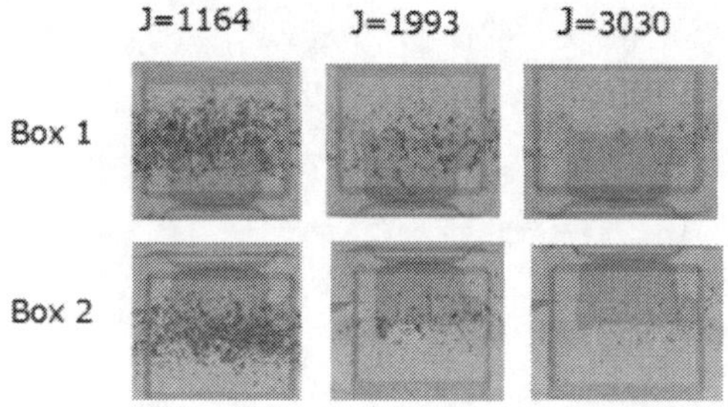

Fig. 11. Porosity images of micro-CT in each Box

6 Discussion of atomized flow by higher J factor

We analyzed the state within the mold when the J factor becomes high enough to produce the atomized flow observed in the experiment involving injection into an open space (see Section 4). The analysis was performed by comparing the computational results for the two injection conditions of J=1164 and J=3030.

6.1 Visualization of flow inside piston by using flow lines

Figure 12 shows the flow lines at 30 ms after acceleration to high speed. The high-speed flow of metal from the gate collides with the piston wall, splits to the left and right, and forms a fast flow from the upper part of the walls of the piston boss part to the outlet (section A). However, the flow speed decreases and the flow toward the outlet weakens in the thick part (section B). Even though the metal is injected at high speed with a high gate speed, there are areas where the flow speed is low. At J=1164 in particular, the speed in the thick part at the pin boss decreases to less than 10 m/s and the flow is observed to stagnate.

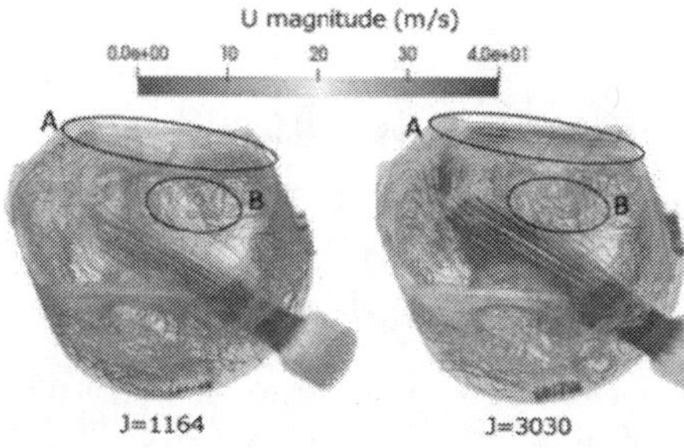

Fig. 12. Flow lines inside the piston (30 ms after acceleration to high speed)

6.2 Changes in the aluminum volume fraction isosurface during the filling process

Figure 13 shows the change in the isosurface of the volume fraction of aluminum $\alpha_{AL} = 0.5$ from 15 ms after accretion to 30 ms after filling.

In the figure, the scalar value of the velocity is color mapped onto the isosurface. The displayed region shows an area in which there is a substantial amount of entrained air with $\alpha_{AL} = 0.5$ or less. The changes in the $\alpha_{AL} = 0.5$ isosurface visualize the void formation process.

Figure 13(a) shows the isosurface at 15 ms after acceleration while the injection is still accelerating. Here, the metal from the gate spreads in the horizontal direction through the entire piston and flows toward the outlet in a mixed state of aluminum and air.

Figure 13(b) shows the isosurface at 30 ms after acceleration when the rate of high-speed injection is constant. In areas where the flow speed from the gate is high, the flow contains little air. Air voids have formed in Box 1 and Box 2 in the thick parts of the piston. These voids are expelled by the gas phase extrusion effect

caused by the negative pressure gradient of the liquid phase in the outlet direction.
Figure 13(c) shows the isosurface at 50 ms after acceleration when the high-speed injection has finished. Because the void on the Box 2 side is near Outlet 2, the air from the void is expelled and the occupancy of voids in Box 2 decreases. In Box 1, the flow of air from the void to Outlet 1 remains.
Figure 13(d) shows the isosurface at 30 ms after the high-speed injection has finished. The filling is complete, the speed of the metal is zero, and the voids inside the product are spherical. These spherical voids form a spherical pore shape called blowholes in HPDC. In actual casting, point D corresponds to pressurization by the casting machine at 30 ms after the injection is finished, and the voids are compressed at a casting pressure of around 60 MPa.

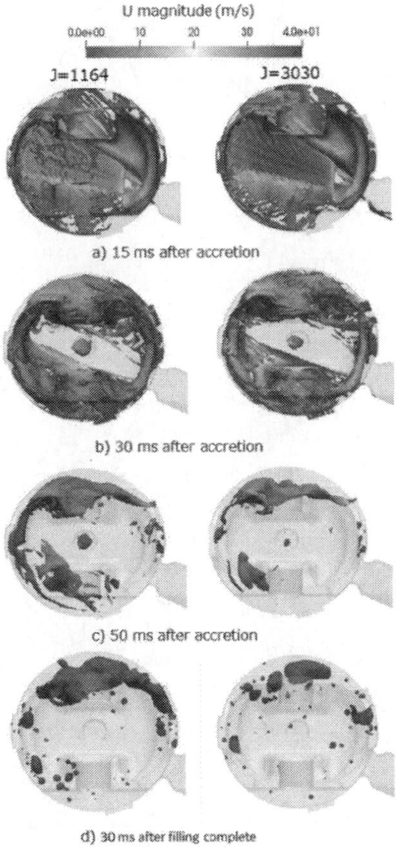

Fig. 13. Void visualization by $\alpha_{Al} = 0.5$ isosurface

6.3 Changes in vortex structure

The HPDC flow is turbulent, and the vortex structure of the turbulent flow is thought to affect the transport efficiency of the air in the two-phase flow of aluminum and air. The vortex structure is difficult to determine through photography experiments but can be identified through numerical calculations. We therefore attempted to visualize the changes in the vortex structure of the metal within the die.
A method that uses the second invariant Q of the velocity gradient tensor has been proposed as a method for identifying vortex structure [12]. The definition of Q is given by Eq. (11).

$$Q \equiv -\frac{1}{2}\frac{\partial u_i}{\partial x_j}\frac{\partial u_j}{\partial x_i} = \frac{1}{2}\left(\|\boldsymbol{\Omega}\|^{2} - \|\boldsymbol{S}\|^{2}\right) \quad (11)$$

Here, $\boldsymbol{\Omega}$ is the vorticity vector and $\boldsymbol{S}$ is the shear velocity vector. The vortex structure can be identified by extracting the isosurface of the region where Q exhibits a positive threshold value. This means that the regions where the rotational motion is larger than the shear motion are extracted as vortex cores. The developed system has the resolution required for calculating a small turbulence vortex using the LES model, which reproduces the atomization phenomena. Because of this, the vortex structure could be visualized using the Q isosurface
Figure 14 shows the changes in the isosurface from 15 ms to 30 ms after accretion. The scalar value of vortex strength, called enstrophy, is color mapped onto the Q isosurface.
At 15 ms after acceleration during injection acceleration (Figure 14a), many small vortices with low vorticity are formed at the position of void formation.
At 30 ms after acceleration when the injection speed is constant (Figure 14b), the small vortices change into longitudinal vortices toward the outlet under the J=3030 condition.

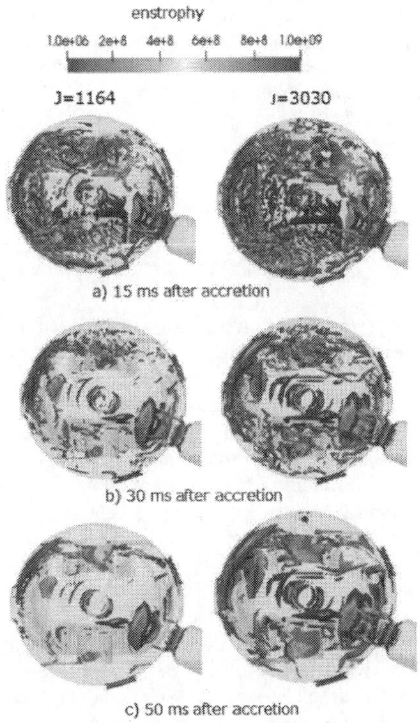

Fig. 14. Visualization of the vortex structure by Q isosurface

https://doi.org/10.1051/matecconf/202032606006

On the other hand, the small vortex structure is maintained at the position of void formation under the J=1164 condition.

The formation of a turbulence vortex was observed in the entire metal flow until 50 ms after acceleration when the high-speed injection finished (Figure 14c). For J=1164, the filling is completed with the small vortices remaining in the part of Box 1 with void formation.

The results of the above visualization of the vortex structure clarified the following two points regarding the effect of the J factor on the filling flow characteristics.

1. Under the atomized flow condition, small vortices are generated by turbulent vortices during filling acceleration. Microbubbles are captured near the center of the turbulent vortex due to the pressure gradient in the vortex. The small vortices are maintained in the thick part until filling is complete, and a strong correlation was observed with the presence of porosity.
2. As the J factor is increased, the vortices change from small vortices to those having a structure with a pressure gradient to the outlet. It is thought that the porosity of the thick part is reduced through an effect whereby the microbubbles are expelled to the outlet.

6.4 Quantifying the expulsion of air from the outlet

We next investigate the effect of the changes in vortex structure described in Section 6.3 on the amount of air expelled from the outlet.

The amount of aluminum expelled V_{AL} and the amount of air expelled V_{AIR} were found from Eq. (12) below using the cross-sectional area of the outlet A_{OUTLET}, the speed of the aluminum U_{MAG} at each time, and the aluminum volume fraction α_{AL}.

$$U_{mag} = \sqrt{Ux^2 + Uy^2 + Uz^2}$$

$$V_{AL} = A_{OUTLET} \times U_{mag} \times \alpha_{AL}$$

$$V_{AIR} = A_{Outlet} \times U_{mag} \times (1 - \alpha_{AL}) \qquad (12)$$

Figure 15 shows the amount of air expelled from the outlet at each time (timestep = 1 ms). The expulsion of air reaches a maximum at point A, 15 ms after acceleration. This corresponds to when small air vortices form to the greatest extent as described in Section 6.3. When aluminum flows into Outlet 1 at this time and blocks the expulsion of air, the amount of air expelled from Outlet 2 suddenly increases. After this, the air is expelled from the outlet in a multiphase flow with aluminum. The amount of air expelled at point B when the maximum speed is reached corresponds closely to the changes in vortex structure.

Figure 16 shows the changes in the proportion of air expelled from each outlet. The ratio of the amount expelled from each outlet as the J factor changes is expressed with the maximum amount expelled from Outlet 1 at J=3030 normalized to 100. At point A when the aluminum reaches Outlet 1, 30% to 40% of the air is expelled. After this, at the high J factor of J=3030 at point B when the speed is constant, a multiphase flow with 80% to 90% air is expelled. For effective expulsion of air, it appears important to lengthen the acceleration time and to actively utilize the vortex structure from the turbulence.

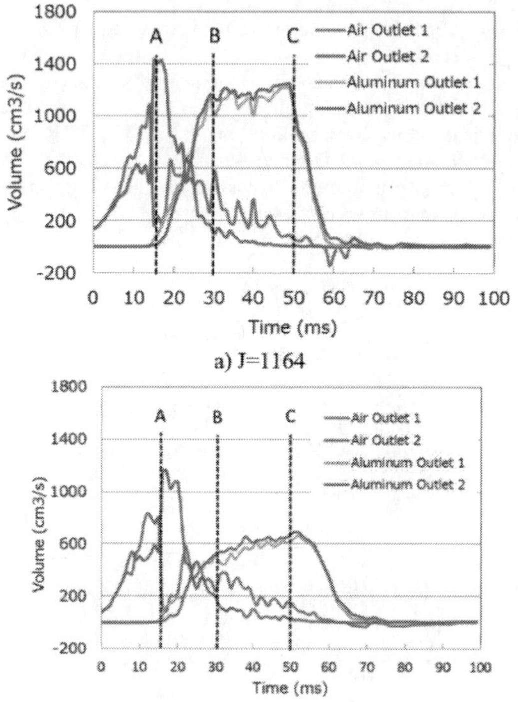

a) J=1164

b) J=3030

Fig. 15. Amount of air and aluminum expelled from the outlet.

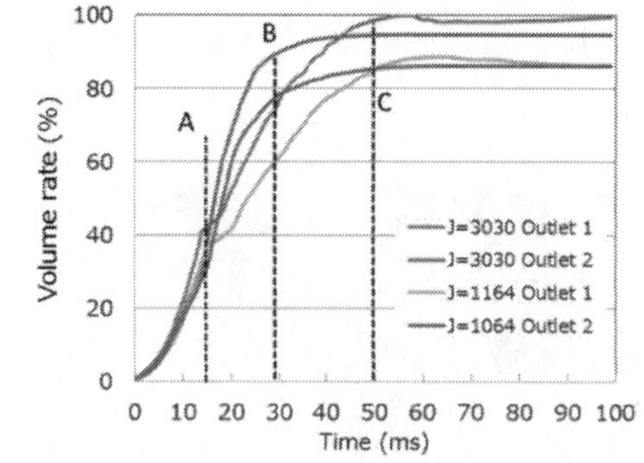

Fig. 16. Proportion of air expelled from the outlet

6.5 Changing the filling pressure

Figure 17 shows the time course of metal pressure in Box 1 and Figure 18 shows the relationship between the maximum metal pressure and J factor.

As the metal flow speed is increased and the J factor is increased, the metal pressure inside the product part increases because the metal flow is constricted at the outlet. As shown in Figure 18, the maximum metal pressure increases linearly with J factor. Although the actual casting pressure is pressurized up to 70 MPa, around 50 ms is required to reach this pressure. Thus, increasing the metal pressure increases the pressurization effect during the initial pressurization stage.

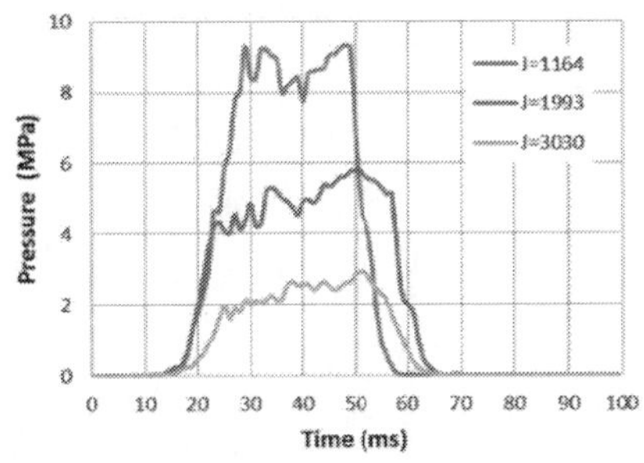

Fig. 17. Time course of metal pressure for various J factors

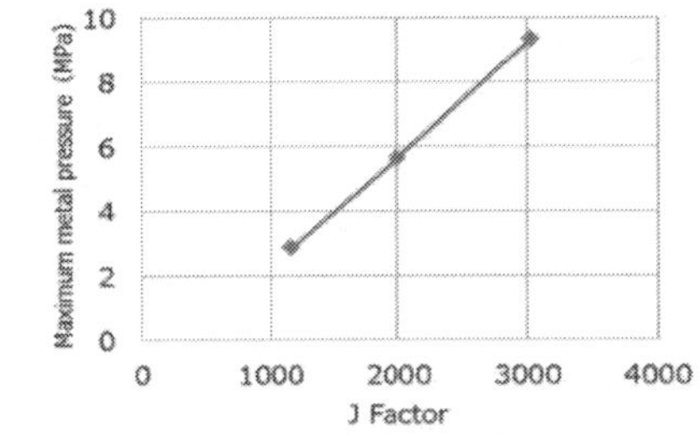

Fig. 18. Maximum metal pressure and J factors

6.6 Sources of porosity

We now discuss the results showing that the differences in filling ratio between Box 1 and Box 2 in the calculations were larger than the differences in porosity volume fraction in the cast products, as described in Section 5.2.

Oxygen was not detected by gas analysis of the cast test pieces, but around 60% nitrogen and 40% hydrogen were detected. Oxygen, which constitutes around 20% of air, undergoes an oxidation reaction with the molten aluminum metal during the filling process and does not remain as a gas in the product. The calculations in the developed system are based on a fixed volume of gas without any changes in composition. Because the air and aluminum form a multiphase flow in the initial filling stage, the decrease in the volume of oxygen gas is thought to decrease at an earlier stage with increasing J factor.

Considering that the volume in Box 1 when the void forms decreases to 20%, the large voids disappear at the maximum filling ratio of 90% for J=3030 and 80% for J=1993, which matches the experimental results.

Figure 19 shows SEM micrographs of the porosity of a piston fabricated in the casting experiment. An SEM micrograph of a 0.25 mm blowhole from air incorporated in regular HPDC is shown for comparison.

The porosity in the test piece was characterized by a small pore size of around 0.1 mm, and clear dendrites were observed. For HPDC, a model has been proposed for the formation of porosity by hydrogen in solid solution emitted between the dendrites during solidification [13].

In HPDC, hydrogen is produced during the filling process by reactions such as those between the mold release agent and the molten metal. Although the filling ratio in Box 2 was 99.5% in the high J factor state and no large voids were observed, fine porosity originating from hydrogen was actually observed. This is attributed to hydrogen or nitrogen being finely dispersed into sizes smaller than the mesh size of the calculation and forming nuclei for porosity. Then, fine porosity is formed when the hydrogen is emitted during the solidification process.

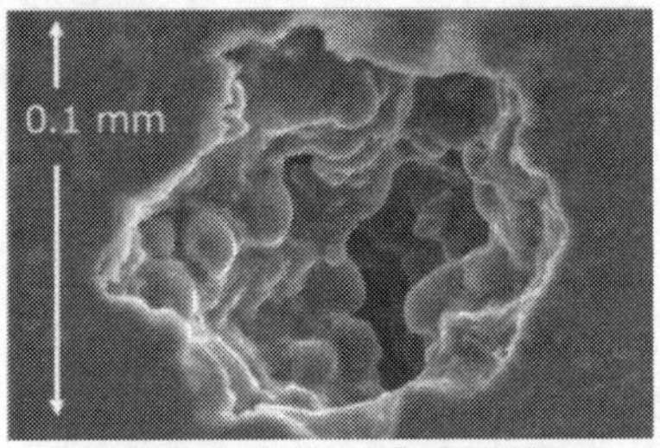

a) Porosity in Box 1 for J = 3030

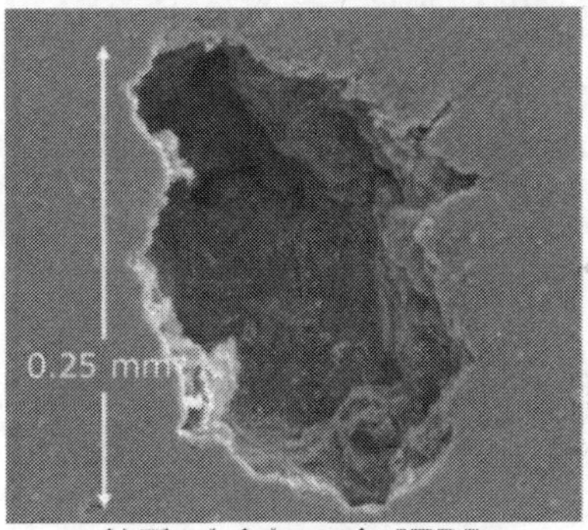

b) Blowhole in regular HPDC

Fig. 19. SEM micrographs of high J factor porosity and a blowhole

7 Conclusion

Experimental results were obtained in which a metal flow passing through the gate formed an atomized flow and the porosity in a HPDC piston was reduced by increasing the J factor. The underlying mechanisms were clarified by analysis of the two-phase flow of aluminum and air using the LES-VOF method for turbulence calculations.

1. In the die cavity, an atomized flow generates small vortices via turbulent vortices during filling acceleration, and these were accompanied by small bubbles near the vortex centers. A strong correlation was observed between the locations of these vortices

and the generation of high porosity in the thick parts.

2. It is thought that the porosity of thick parts can be decreased by changing the vortex structure to longitudinal vortices toward the overflow and decreasing voids by increasing the J factor.
3. We showed the feasibility and utility of visualizing states that cannot be understood from experiments by performing turbulence calculations and accurately solving the state of the multiphase flow in HPDC. This system can be applied to various types of HPDC depending on the definitions of the physical properties of materials.

Acknowledgements: We thank Mr. Youhei Sekiguchi, Mr. Yuki Fujisawa, Mr. Yuji Okada, and Mr. Tomohiro Fujimoto of Honda Foundry Co., Ltd. for performing the casting experiments and collecting the casting data. We also thank Mr. Masanori Tsukamoto of Softflow Co., Ltd. for support with post-processing of the calculation results by POSTFLOW.

References

1. A. Tanihata, N. Sato, K. Katsumata., T. Shiraishi, SAE 2006-01-098 (2006)
2. M.Suzuki, SETC 2006-32-0031
3. T. Kaneuchi, H. Nakano, J-DE JD08-20 (2008)
4. Y. Yamada, H. Yoshi, S. Mochizuki, et al.,SETC 2011-32-0504 (2011)
5. S. Aida, I. Kuboki, Y. Hayashi, M. Fujimoto, et al., J-DE Transactions, JD12-8 (2012).
6. I. Takahashi, K. Anzai, M. Itamura, Reports of 170th J.JFS Meeting, p 21 (2017).
7. E.Koya, M. Nakagawa, S. Kitagawa, J. Ishimoto ,Y. Nakano, N Ochiai, SAE, 2018-01-1393, (2018)
8. E.A.Herman, NADCA, "GATING MANUAL" , p35-36, Publication #5
9. J. Ishimoto, T. Sato, A. Combescure, IJHE, Vol42, Issue 15 (2017)
10. J. Ishimoto, F. Sato, G. Sato, JEGTP Vol.132, 082801 (2010)
11. E.Koya, M. Nakagawa, S. Kitagawa, J. Ishimoto ,Y. Nakano, N Ochiai, SAE, 2018-01-1392, (2018)
12. J.C.R.Hunt, A.A. Wray, P.Morin, Center for Turbulence Research CTR-88, p193, (1988)
13. K.Kubo, J-JFES, 78, p620 (2006)

INDUSTRIAL SEMISOLID CASTING PROCESS FOR SECONDARY ALUMINIUM ALLOYS FOR DECARBONISING LIGHTWEIGHT PARTS IN AUTOMOTIVE SECTOR

Fabrizio D'Errico[1,*], *Daniele* Casari[2], *Mattia* Alemani[2], *Guido* Perricone[2] , *Mauro* Tosto[2]

[1]Politecnico di Milano, Department of Mechanical Engineering, 20156 Milan, Italy
[2]Brembo SpA, 24040, Stezzano, Italy

Abstract. The life cycle holistic approach for the automotive sector highlighted how much important is working on decarbonisation of Al casting processes to produce vehicle components. Broadening the use of recycled aluminium alloys, instead of high energy intensive primary aluminium alloys, is key for environment preservation. While primary aluminium alloys are preferred by automotive original equipment manufacturers (OEMs) because impurities (mainly Fe) present in secondary aluminium alloys might cause fatigue properties decay, a new semisolid state process route has been developed at Brembo to drastically reduce the sensitivity of cast aluminium to Fe impurities. Based on these premises, during the CRAL European project in the framework of the LIFE Programme, an industrial prototype machinery - a vertical high pressure die casting machine - has been designed and constructed to inject secondary aluminium in the semisolid state. A number of A357 Fe-enriched brake callipers manufactured via the new process route have been fully investigated by fatigue tests, SEM and OM analyses in order to validate the new eco-sustainable product compared to conventional ones manufactured with primary aluminium by gravity casting.

1 Introduction

It is common knowledge today that lightweight vehicles are fundamental for decreasing traffic-related pollutant emissions by conventional internal combustion engine powered vehicles. The EU legislation is continuously evolving for more restrictive emission targets for new vehicles. Different studies show potential weight reduction scenarios going from 15% to 35% weight reduction [1, 2], estimating a potential greenhouse gas emissions reduction in the range 9-18 Gtons CO2eq between 2010 and 2050 [5]. Reducing unsprung weight onboard vehicles is the key to improving handling, reducing inertia adverse effect, improving powertrain efficiency. The lower the unsprung weight, the less work the shocks and springs have to do to keep the tires in contact with the road over bumpy surfaces. Lot of problems, if not all of them, is caused in fact by inertia. Heavier unsprung mass (such as brake assemblies, tires, hubs, knuckles) means higher inertia; the higher the inertial dynamic forces, the more is the energy consumed by suspension spring and dampers for accommodating shocks to keep tyres on the ground. Similarly, if unsprung components have a high mass, they are harder to accelerate/decelerate and thus it is more difficult for the suspension to maintain a consistent tyre load.
Reducing mass in these locations improves the vehicle handling and performance, and provides fuel efficiency [3]; it is estimated that cutting one kilogram of unsprung weight is comparable, due to vehicle dynamics, to reducing the sprung mass weight by 3-6 kg (depending on inertial energy absorbed by dampers, i.e. on road irregularities). The effect of less unsprung weight means that the vehicle's suspension can control the wheel and, in turn, the tyres' contact patch more effectively, resulting in better traction and vehicle control. Reducing the weight of the braking system has been a key area of focus for automobile manufacturers.
The manufacturers are reducing the unsprung weight of today's modern vehicles everywhere they can and these lighter assemblies give the driver more effective cornering, handling, braking, and improved road feel.
A key role in this challenge is played by lightweight materials for structural and massive components such as aluminium alloys. Old cast-iron brake calipers have been replaced by ones made of aluminium.
Nevertheless, aluminium alloys currently used in conventional casting processes for manufacturing lightweight unsprung subsystem components such as knuckles and brake system assemblies suffer from environmental drawbacks over the whole cradle-to-grave cycle. Specifically, high-purity primary aluminium is required to produce the final casting alloys and the extraction process of primary aluminium is an energy consuming process with a high environmental impact [4-6].
If on one hand, weigthsaving by putting onboard lighter aluminium components instead of heavier ones can lower the fuel consumption and pollutant emissions per

* Corresponding author: fabrizio.derrico@polimi.it

https://doi.org/10.1051/matecconf/202032606007

kilometre for all internal combustion engine vehicles, on the other hand such benefit could not be fully achieved if a negative net balance for the total CO2 emitted over the manufacturing phases occurred.. More specifically, the use of secondary aluminium alloys derived from recycling patterns instead of primary aluminium alloys would allow a 95% net energy saving and a 80% reduction in total greenhouse gases emissions, avoiding any type of chemicals [7,8]. Therefore, it is clear that a wider use of secondary alloys in casting components for the automotive sector is an environmentally conscious choice for a significant cleaning-up of the vehicles' manufacturing stage. Unfortunately, original equipment manufacturer (hereinafter OEMs for brevity) use mostly primary Al-based alloys. However, this choice is due to the fact that the use of recycled aluminium could introduce some metal impurities such as iron and manganese, residual of recycling patterns. Such impurities are detrimental for the toughness of final cast components because they cause the precipitation, during the solidification phase, of hazardous needle-like or blocky phases [9-12]. Conventional casting technologies such as gravity casting or low and high pressure die casting produce coarse intermetallic β-Al5FeSi platelets, resulting in poorer mechanical properties.

1.1 Effect of shear on Fe intermetallic formation

The Fe tolerance of Al–Si based alloys is closely related to the morphology of the primary Fe-containing intermetallic compounds, and the detrimental effect of Fe can be reduced or even eliminated by the removal of the large needle-shaped primary β-Al5FeSi compounds. Despite being just a few, research studies exist in which it is shown that shearing improves the Fe tolerance of Al–Si-based casting alloys [13]. Fan et al. claim that Rheo die cast processes (hereinafter RDC for brevity) can increases the tolerance to Fe contents [14]. RDC samples can tolerate a Fe concentration up to 1.5 wt.% without any severe damages to the mechanical properties. This is because intensive shearing changes the morphology of iron-containing compounds from long needles or plates to much smaller particles.

Fang et al [13] investigated the effects of intensive forced-melt convection, generated by RDC processing, on the mechanical properties of Fe-containing Al–Si alloys, noticing some interesting aspects. Their activity on the LM25 commercial alloy, to which different amounts of iron were added, showed that RDC is effective in modifying the morphology of the primary Fe-bearing phases. Such processing route appears to be effective in spheroidising both the α-Al and the monoclinic β-Al5FeSi phases. An intensive melt shearing in the semisolid state is effective in turning the long needle morphology into less detrimental short bars or discs with a substantial reduction in the aspect ratio.

An explanation for the mentioned effect is also given [14]. They believe that the formation of a needle-shaped morphology is owed to a significantly different solid–liquid (S–L) interfacial energy and to the associated disparity in growth rates along different crystallographic directions. Consequently, the modification of aspect ratios for the β-Al5FeSi phase suggests that melt shearing could reduce the differences in solid-liquid interfacial energies and in turn slow down preferential growth in the direction of the longitudinal needle axis.

1.2 The CRAL project

The 3and ½ years European Commission co-financed LIFE CRAL project, coordinated by Brembo Braking System Spa, ended in December 2019. It was aimed at designing and validating a novel industrial semisolid process plant that would have used low-purity (i.e. Fe-enriched) Al alloys to manufacture lightweight high-performance brake callipers, reducing the total cradle-to-gate carbon footprint compared to primary aluminium brake callipers manufactured by the conventional gravity casting process.

Based on previous literature [13,14], the improvement of the semisolid casting technology was considered as a key aspect to reach such objectives.

The semisolid process adopted for the industrial pilot plant is based on the rheocasting process route, which consists in stirring a metallic alloy in a temperature range where solid and liquid particles coexist, i.e. during the first stage of solidification.

The stirring and controlled cooling actions change the initial dendritic solid particles into globular ones, leading to a more homogeneous microstructure.

The pilot semi-solid method (hereinafter SSM for brevity) line is shown in Fig.1. It is a completely automated die casting area equipped with a stationary holding furnace, two anthropomorphic robotised arms and a 10.000 kN 4-columns vertical die-casting press. The 3D render of the press is shown in Fig.2. The press is equipped with a rotating plate having three containers, each for a specific process stage. When the container is in the first position, namely out of the press moving plate, the container works as a pouring chamber: the pouring robot picks up the aluminium alloy from the holding furnace and pours it into the container. Then, the stirring device enters the container and starts mixing the molten metal while the cooling system allows to control the temperature into the solidification interval.

Subsequently, the container rotates under the moving plate, mechanically coupling with the injection piston and preparing for the injection of the semisolid metal.

The injection pressure nominally available for the CRAL press machinery is 3.000 kN, capable of injecting the liquid metal with a 1.800 bar metal pressure. Fig. 3 displays the rheo-cast component manufactured via the CRAL pilot SSM line, i.e. a specifically developed monoblock brake calliper.

Microstructural investigations as well as bench and road tests were then performed in order to assess the performance of the component. The interactions of the brake calliper with the wheel corner and the suspensions were taken into account during these tests as well as the vehicle dynamics and vibration modes that could lead to unwanted noise generation (Fig.4).

https://doi.org/10.1051/matecconf/202032606007

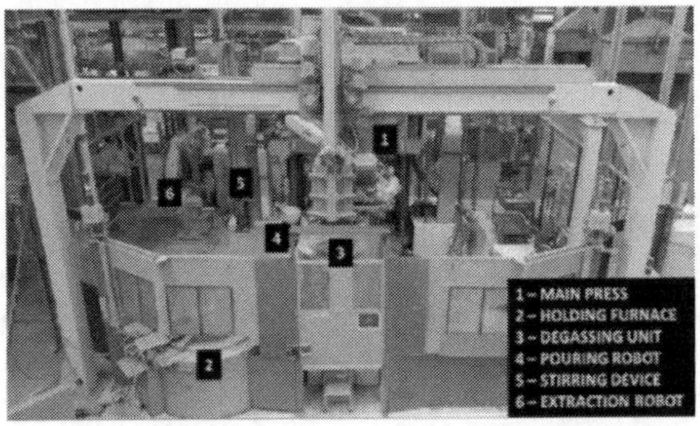

Figure 1. The CRAL semisolid rheo-casting plant.

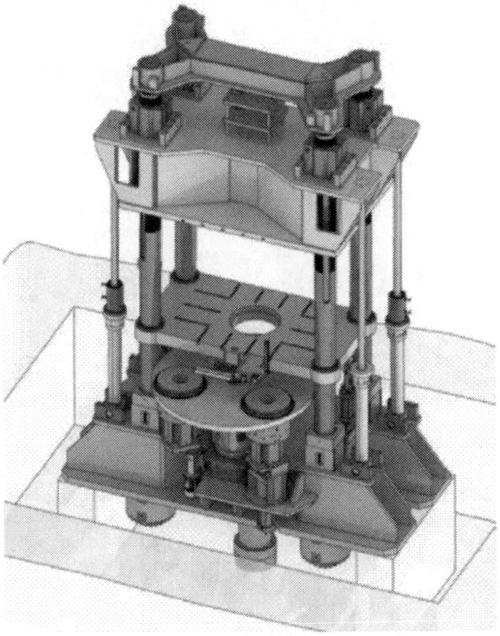

Figure 2. The CRAL vertical rheo-casting press machinery.

Figure 3. Demonstrator brake calliper made of secondary aa356 alloy by semisolid high pressure die casting process.

Figure 4. The CRAL secondary Al calliper mounted on wheel hub for road tests.

2 Experimental part

The material used for the test campaign is an A357 Brembo-type alloy with a 0.80 - 0.95 wt.% Fe concentration. In order to fulfil OEMs specifications for new products, the Brembo team focused on a number of process parameters considered as crucial for improving the metallurgical quality of the parts. It was found by several tests that the main parameters to control during the CRAL process can be divided into two groups, the out-of-press and in-press parameters, such as:

- Out-of-press parameters: pouring temperature and stirring time of the semisolid slurry. Several trial tests were carried out to evaluate heat losses and temperature decrease occurring when the liquid metal is transferred from the holding furnace to the thermo-regulated containers. Three different starting temperatures were investigated. Experiments aimed at clarifying the evolution of the solid fraction as a function of stirring time were performed..
- In-press parameters: temperature of the mould, plunger speed and intensification pressure.: Casting trials were performed in order to determine the optimal injection parameters After the process parameters were fine-tuned, a small batch of 100 brake callipers was produced.

During casting trials, the defects observed in the cast parts were summarised in a preliminary list in order to have an inventory of metallurgical defects that could correlate with process parameters.

Table 1. Major metallurgical defects observed in preliminary casting trials.

Issue	Parameters influencing defects	Undertaken action
Shrinkage porosities	geometry	Modified geometry
	solidification kinetics	Optimisation of injection curves
Oxides and Blisters	Metallurgical cleanliness	Modified lubricant
	Air entrapment	Study on specific heat treatments and/or post heat treatments
	Lubrication system/lubricant type	

A number of samples were extracted from specific regions of the CRAL secondary Al callipers (Figure 5) and subsequently machined in conformity with internal drawings to obtain tensile specimens.

The outcome of tensile tests is reported in Table 2. Results from CRAL primary Al specimens and samples extracted from gravity cast callipers are also included. It can be observed that the CRAL process allows for: (i) enhanced mechanical properties (CRAL primary Al vs. gravity casting); (ii) adequate performances even if secondary Al alloys are employed.

Figure 6 shows the representative microstructures of primary Al gravity cast and secondary Al semi-solid samples, respectively. Primary α-Al and Al-Si eutectic phases can be observed.

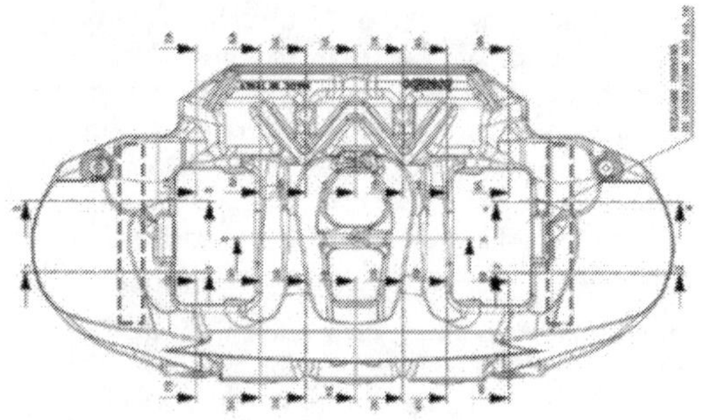

Figure 5. Technical drawing of the CRAL secondary Al calliper; the tensile specimens are extracted from the regions marked with dashed red lines, while the black arrows indicate the sections undergoing standard metallographic investigations in Brembo aluminium foundry Quality Lab.

Table 2. Comparison between the tensile properties of specimens extracted, respectively, from gravity cast, CRAL primary Al and CRAL secondary Al callipers.

Alloy	$Rp_{0.2}$ [MPa]	Rm [MPa]	A%
Gravity casting	321 ± 22	285 ± 17	2.95 ± 1.68
CRAL primary Al	≥ 320	≥ 280	≥ 3
CRAL secondary Al	≥ 290	≥ 250	≥ 1

All the specimens exhibit a very homogeneous microstructure, globular and rosette-type finely distributed grains. Fe-bearing intermetallic platelets can also be noted on the fracture surface of broken samples (Fig. 7). These particles are well known in literature to be highly detrimental to the mechanical properties of aluminium alloys.

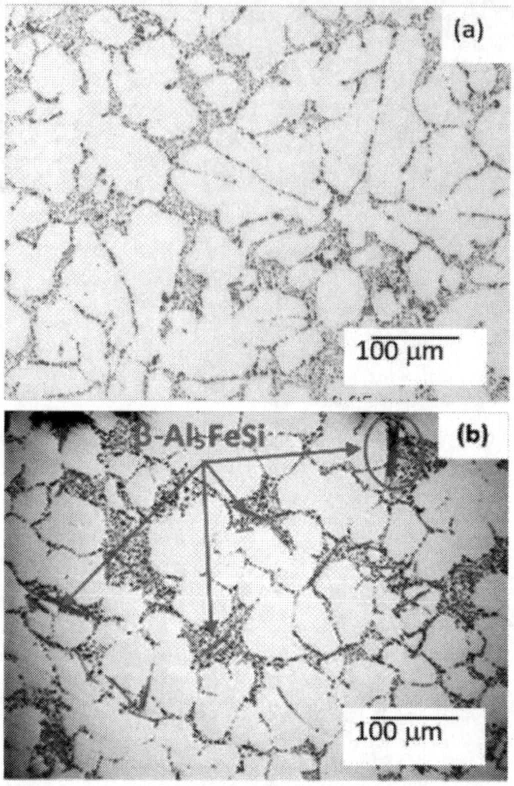

Figure 6. Microstructures at low and high magnification: a) gravity casting, conventional process; b) CRAL-secondary Al casting. Arrows identify the β-Al5FeSi phase. Such particles have a crucial role in the mechanical behaviour of the alloy, prompting fracture initiation and subsequent propagation even at low applied loads (cracked particle marked with the red oval).

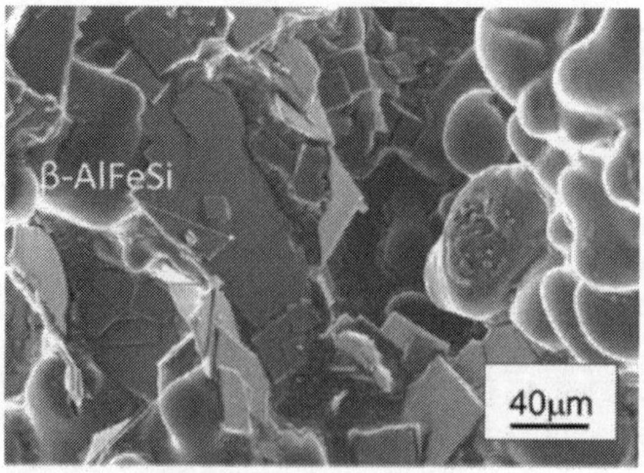

Figure 7. SEM image of the fracture surface of a tensile specimen revealing the presence of Fe-bearing intermetallic phases.

Eventually, the Design and Verification Plan (DVP) ensured that the prototype callipers responded to usual operating conditions, especially in case of emergency braking. Table 3 shows the different kinds of bench-tests carried out on the CRAL secondary Al callipers, all successfully passed (Fig.8). The DVP also included road tests: about 1000 km were covered on Italian roads in order to validate the brake calliper for the use on vehicles.

Table 3. CRAL secondary Al calliper Design and Verification Plan.

Basic Characteristics	
BTS 61.01	Air Pressure Seal Test
BTS 61.02	Hydraulic Pressure Tightness
BTS 61.03	Vacuum Tightness
BTS 61.21	Bleedability
Functionality	
BTS 61.06	Fluid Displacement
BTS 61.07	Luft - Roll-Back
BTS 61.20	Dragging Torque
BTS 61.07	Knock-back
BTS 61.24	Squeak Test
BTS 61.07	Piston sliding pressure
BTS 61.07	Pistons return load
BTS 61.04	Fluid Pressure Stiffness
Strength	
BTS 61.05	Burst pressure
BTS 61.08	Torque Strength
BTS 61.22	Strength of coupling and screw-bleeder
Durability	
BTS 61.19	Vibration Test
BTS 61.11	Pulsating Pressure Endurance
BTS 61.11	Pulsating pressure endurance - Temp.
BTS 61.12	Pressure and Torque Durability - Passenger Car

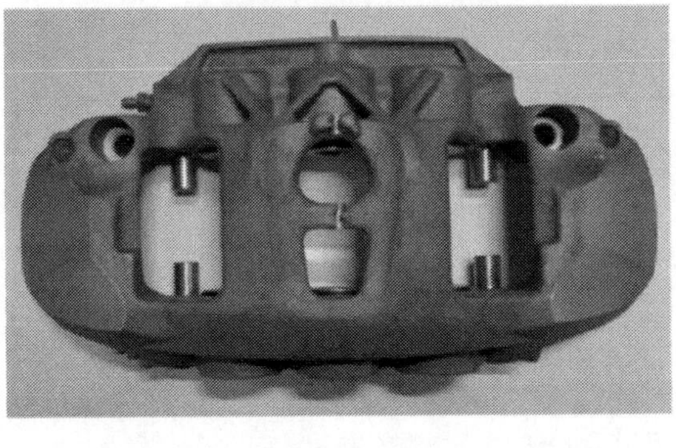

Figure 8. CRAL secondary Al calliper after pressure and torque durability test.

3 Discussion

In the recent years, many processes were investigated for modifying the harmful Fe-bearing intermetallic compounds by rapid solidification, addition of neutralising elements and melt superheating. Nevertheless, low efficiency, complicated processes or high costs accompany the above approaches. Few studies can be found in which the role of slurry stirring on the size and distribution of iron-based intermetallic compounds is investigated. The use of neutralising elements is by far the most common response to iron in the casting industry, but not the proper one. The Fe tolerance of Al–Si alloys is closely related to the morphology of the primary Fe-containing intermetallic compounds. It has been demonstrated that the detrimental effect of Fe can be reduced or even eliminated by modifying the large needle-shaped primary β-Al5FeSi compounds via shearing [13].

Furthermore, the globular structure achieved by means of the CRAL process can counterbalance the harmful effect of Fe-containing platelets, allowing for mechanical properties in compliance with Brembo and OEMs standards. Such platelets were observed to be much smaller than those usually found in gravity cast components (~ 50 μm vs. 200 μm), hence highlighting the advantage of semisolid stirring of the metal. Intensive melt shearing in the semisolid state is considered very effective in turning the long needle morphology into less detrimental short bars or discs with a substantial reduction in the aspect ratio. Further positive results over the intensive shearing is obtained by combined effect of shearing action and the intensive forced convection and the much more homogenous temperature in the mass during steering that help stabilisation of the α-AlFeMnSi intermetallic compounds with respect to the β-AlFeSi.

In order to assess the efficiency of RDC processing in hindering the detrimental effect of Fe on mechanical properties of cast Al alloys, the concept of Fe tolerance was used in this manuscript; since the elongation of die cast components is the most significant property, being severely affected by impurities such as iron, this parameter was linked to the concept of Fe tolerance.

4 Conclusions

The high-pressure rheo-casting industrial process developed in the CRAL project appears to be consistent with the metallurgy phenomena indicating that the higher is the refinement of the primary globular cells, the lower is the growth rate of the intermetallic phases forming in the eutectic spaces (such as Fe-containing intermetallics in secondary Al-Si alloys). Also, the stirring produces a more homogeneous microstructure, which can counteract the detrimental effects of Fe-bearing intermetallic compounds, leading to reasonably good mechanical properties as well as performances in line with internal and OEMs standards.

The Brembo team equipped a real vehicle with recycled aluminium brake callipers. The DVP results were very promising, since a reasonably good mechanical response was obtained for recycled pressure-die casting aluminium brake callipers. This represents a breakthrough for the whole automotive industry, which will be also extended to further Brembo core products.

Acknowledgements

This work makes use of results produced by the CRAL project (http://www.cralproject.eu/en), co-funded by the European Commission within the 2015 LIFE+ Programme. Authors thank support provided by European Commission for supporting project dissemination activities. This work reflects only the authors' views. The Community is not liable for any use that may be made of the information contained therein.

References

1. McKinsey & Company, Lightweight, Heavy Impact (2012).
2. P. Martino, 4th Int. Conf. Advanced Materials and Tech. for Transport, Turin (2014)
3. A.K. Sachdev, R.K. Mishra, A. Mahato, A. Alpas, ICAA13, Pittsburgh (2012)
4. D.R. Gunasegaram , A. Tharumarajah, Met. and Mat. Trans. B **40** (4) (2009)
5. R.B.H. Tan,, H.H. Khoo, J. of Cleaner Prod., **13** (6) (2005)
6. B.J. Welch, M.M. Hyland, B.J. James, JOM, **53** (2) (2001)
7. T.E. Norgate, W.J. Rankin, Int. Symposium on Greenhouse Gases in the Metallurgical Industries: Policies, Abatement and Treatment (2001)
8. R.B.H. Tan, H.H. Khoo, J. of Cleaner Prod. **13** (6) (2005)
9. S.G. Irilzap and N. Saklakoglu, Eng. Sc. and Tech. **17** (2014)
10. Q.G Wang, D. Apelian, D.A Lados, J. Light Metals, **1** (1) (2001)
11. A. Fabrizi, G.Timelli, S. Ferraro, F. Bonollo, Int. J. of Materials Research, 106 (7) (2015)
12. L.A. Narayanan, F.H. Samuel, J.E. Gruzleski, Met. and Mat. Trans. A, 25 (8) (1994), 1761-1773
13. X. Fang, G. Shao, Y. Liu and Z. Fan, Mat. Sc. and Eng. A (2007)
14. Z. Fan, X. Fang and S. Ji, Mat. Sc. and Eng. A 412 (2005)

Review, Role of $L1_2$ Modified $(Al_{1-x}Me_x)_3Ti$ Intermetallic Compounds on Heterogeneous Nucleation of Alpha Aluminum Grains

Yoshimi Watanabe*

Department of Physical Science and Engineering, Nagoya Institute of Technology, Gokiso-cho, Showa-ku, Nagoya, Japan

Abstract. Al_3Ti intermetallic compound with the tetragonal $D0_{22}$ structure undergoes a phase transformation to the high-symmetry $L1_2$ cubic structure by addition of third elements, *Me*. The lattice constants of some $L1_2$ modified $(Al_{1-x}Me_x)_3Ti$ intermetallic compounds are closed to that of alpha aluminum. Therefore, it is expected that the addition of $L1_2$ modified $(Al_{1-x}Me_x)_3Ti$ intermetallic compound particles show good grain refining performance of cast aluminum. In this paper, our recent results on novel refines containing heterogeneous nucleation site particles of $L1_2$ modified $(Al_{1-x}Me_x)_3Ti$ intermetallic compounds have been reviewed.

1 INTRODUCTION

Equiaxed grain structure ensures uniform mechanical properties, reduced ingot cracking, improved feeding to eliminate shrinkage porosity, distribution of secondary phases and microporosity on a fine scale, improved machinability and cosmetic features. The grain refining innoculants commonly used in the aluminium industry are usually master alloys of Al-Ti or Al-Ti-B [1]. Grain refinement of aluminum and its alloys by the addition of Al-Ti-B refiners to liquid melt prior to casting is a common practice in order to achieve a fine equiaxed grained microstructure in a casting which otherwise solidifies usually with coarse columnar grain structure [2]. Figure 1 shows microstructural feature of Al-Ti-B refiner. As can be seen, the Al-Ti-B refiner contains TiB_2 heterogeneous nucleation site particles and platelet-shaped Al_3Ti heterogeneous nucleation site particles [3-8] with $D0_{22}$ structure (a = 0.3851 nm, c = 0.8608 nm). It is well accepted that the effective heterogeneous nucleation site particles have better lattice matching with the solidified alpha aluminium (a = 0.4049 nm).

It is also known that the Al_3Ti intermetallic compound with the tetragonal $D0_{22}$ structure can undergo a phase transformation into $(Al_{1-x}Me_x)_3Ti$ intermetallic compounds with $L1_2$ cubic structure by addition of third elements of *Me* [9-11]. Moreover, the lattice constant of some $L1_2$ modified $(Al_{1-x}Me_x)_3Ti$ intermetallic compounds is closed to that of alpha aluminum, *e.x.*, the lattice constants of $L1_2$ modified $Al_{2.7}Fe_{0.3}Ti$, $Al_{2.75}Ni_{0.25}Ti$ and $Al_{2.5}Cu_{0.5}Ti$ intermetallic compounds are a = 0.393 nm, a =0.394 nm and a = 0.3927 nm, respectively. An $L1_2$ modified $Al_{2.5}Cu_{0.5}Ti$ intermetallic compound can be formed at the Cu/Al_3Ti interface in the Cu/Al_3Ti diffusion couple specimen heated at elevated temperature as shown in Fig. 2 [12].

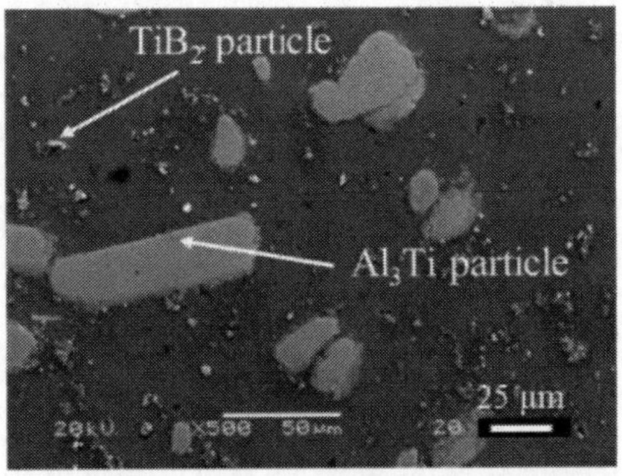

Fig. 1. SEM micrograph of Al-Ti-B refiner.

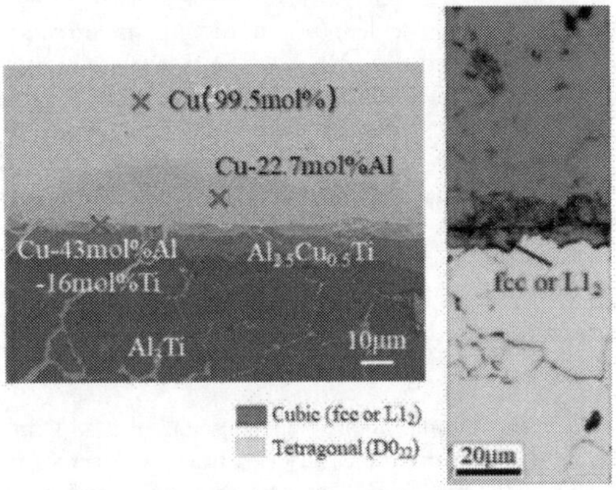

Fig. 2. Micrographs showing the interface in the Cu/Al_3Ti diffusion couple specimen heated at 600 °C for 4 h [12].

* Corresponding author: yoshimi@nitech.ac.jp

Therefore, it is expected that these intermetallic compounds show good grain refining performance. In our studies [13-17], refining performance of novel refines with $L1_2$ modified $Al_{2.5}Cu_{0.5}Ti$, $Al_{2.7}Fe_{0.3}Ti$ and $Al_{2.7}Ni_{0.3}Ti$ particles are studied. In this paper, our recent results on novel refines containing heterogeneous nucleation site particles of $L1_2$ modified $(Al_{1-x}Me_x)_3Ti$ intermetallic compounds have been reviewed.

2 LATTICE MATCHING

It is important to describe the lattice matching between heterogeneous nucleation site particles and alpha aluminum, because an effective refiner contains heterogeneous nucleation site particles with a better lattice matching with the solidified phase. One of the most famous equations is eq. (1) proposed by Turnbull and Vonnegut [18].

$$\delta = (|a_S - a_C| / a_C) \times 100(\%) \qquad (1)$$

where δ is called as disregistry between the heterogeneous nucleation site and the solidified phase, a_S and a_C are the lattice parameters of the heterogeneous nucleation site particles and solidified phase without deformation, respectively. However, Al_3Ti has the low-symmetry tetragonal $D0_{22}$ structure. Therefore, eq. (1) must be modified for application to two-dimensional lattices. The plane disregistry was proposed by Bramfitt [19], and calculated by

$$\delta^{(hk\lambda)_s}_{(hk\lambda)_n} = \frac{1}{3}\sum_{i=1}^{3}\left\{ \frac{\left| d[uvw]^i_s \cos\theta - d[uvw]^i_n \right|}{d[uvw]^i_n} \right\} \times 100\,\% \qquad (2)$$

where $(hk\lambda)_s$ and $(hk\lambda)_n$ are the low-index planes of the heterogeneous nucleation site and solidified phase, and $[uvw]_s$ and $[uvw]_n$ are the low-index orientations on $(hk\lambda)_s$ and $(hk\lambda)_n$, respectively. Moreover, $d[uvw]_s$ and $d[uvw]_n$ are the interatomic spacing distances along $[uvw]$, and θ is the angle between $[uvw]_s$ and $[uvw]_n$.

However, it is unclear why three in-plane directions are considered when evaluating the plane disregistry. Moreover, there are some possibilities to select the third axis [15]. Kato *et al.* have proposed parameter M [20], which is approximately proportional to the specific misfit strain energy. In this study, parameter M defined by eq. (3) will be adopted to discuss the effective heterogeneous nucleation site [21,22].

$$M = \varepsilon_1^{\,2} + \varepsilon_2^{\,2} + (2/3)\varepsilon_1\varepsilon_2, \qquad (3)$$

In eq. (3), ε_1 and ε_2 are the principal misfit strains calculated from the principal distortions. The parameter M values of some crystallographic orientation relationships between Al_3Ti and aluminum are listed in Table 1. As can be seen, different planes of Al_3Ti show different parameter M values. In case of $L1_2$ modified $(Al_{1-x}Me_x)_3Ti$ intermetallic compounds, better lattice matching between the heterogeneous nucleation sites and the solidified phase can be achieved, as shown in Table 1.

Table 1. Parameter M values between heterogeneous nucleation site particles and aluminum.

Orientation Relationship	M value
$(100)_{Al_3Ti} // (100)_{Al}$	4.31×10^{-3}
$(001)_{Al_3Ti} // (100)_{Al}$	6.38×10^{-3}
$(110)_{Al_3Ti} // (110)_{Al}$	4.31×10^{-3}
$(112)_{Al_3Ti} // (111)_{Al}$	2.24×10^{-3}
$(0001)_{TiB_2} // (111)_{Al}$	9.95×10^{-3}
$(hk\lambda)_{Al_{2.5}Cu_{0.5}Ti} // (hk\lambda)_{Al}$	2.57×10^{-3}
$(hk\lambda)_{Al_{2.7}Fe_{0.3}Ti} // (hk\lambda)_{Al}$	2.50×10^{-3}
$(hk\lambda)_{Al_{2.7}Ni_{0.3}Ti} // (hk\lambda)_{Al}$	1.93×10^{-3}

3 EXPERIMENTAL METHODS

3.1 Fabrication of Refines

Two types of particles, *i.e.*, prepared by mechanical crushing of bulk $L1_2$ modified $(Al_{1-x}Me_x)_3Ti$ intermetallic compounds and directly prepared by gas atomization, were used [13-17]. Hereafter, the particles prepared by the crushing of an arc melted bulk sample and prepared by the gas atomization will be denoted as "crushed particles" and "gas-atomized particles", respectively. The bulk $L1_2$ modified $(Al_{1-x}Me_x)_3Ti$ intermetallic compound was prepared by an arc melting method under an argon atmosphere. The obtained bulk $(Al_{1-x}Me_x)_3Ti$ intermetallic compound was homogenized at elevated temperature, and then crushed into fine particles by a hammer. Both crushed and gas-atomized particles were sieved into the size of 75-150μm. The sieved $(Al_{1-x}Me_x)_3Ti$ particles are mixed with pure aluminum particles (99.9%, 106-180μm), where the volume fraction of $(Al_{1-x}Me_x)_3Ti$ particles was 10 vol%, 20 vol%, 30 vol% and 40 vol%. Sintering of the mixed powder using a spark plasma sintering (SPS) [23-25] was performed at 500 °C for 5 min under an applied stress of 45 MPa. Flow diagram for fabrication process of the refiners with $L1_2$ modified $(Al_{1-x}Me_x)_3Ti$ particles is shown in Fig. 3.

3.2 Refining Performance

To evaluate the grain refinement performance of the fabricated refiners containing $L1_2$ modified $(Al_{1-x}Me_x)_3Ti$ heterogeneous nucleation site particles, following casting experiments were conducted. Commercially pure aluminum ingot (99.99%, 148.8 g for 10 vol% refiner, 149.4 g for 20 vol% refiner, 149.6 g for 30 vol% refiner, and 149.7 g for 40 vol% refiner) was melted in an alumina crucible using the electrical resistance furnace at 750 °C in argon gas atmosphere. The mass fractions of the refiners as additives were 0.8%, 0.4%, 0.27% and 0.2% for the 10, 20, 30, and 40 vol% refiners, respectively, so that the total amount of

$L1_2$ modified $(Al_{1-x}Me_x)_3Ti$ particles within the melt was the same. After the addition of the refiner into the melt, the melt was stirred with a rod for 30 s, after which no further stirring was carried out. The melt was cast into a cylindrical steel mold of 45mm inner diameter, 70mm outer diameter, and 70mm height after a holding time. Here, the holding time is the time interval from when the stirring was finished to when the melt was cast into the mold. The classical metallographic route was used to prepare the samples for microstructural analysis. The specimen for grain structure examination was horizontally cut from the bottom part (5 mm from the bottom) of each cast sample. The macrostructural and microstructural features of the casts were studied by optical microscopy or scanning electron microscopy after etching the polished surface with a 10% hydrofluoric acid aqueous solution. Flow diagram of grain refining performance test is summarized in Fig. 4.

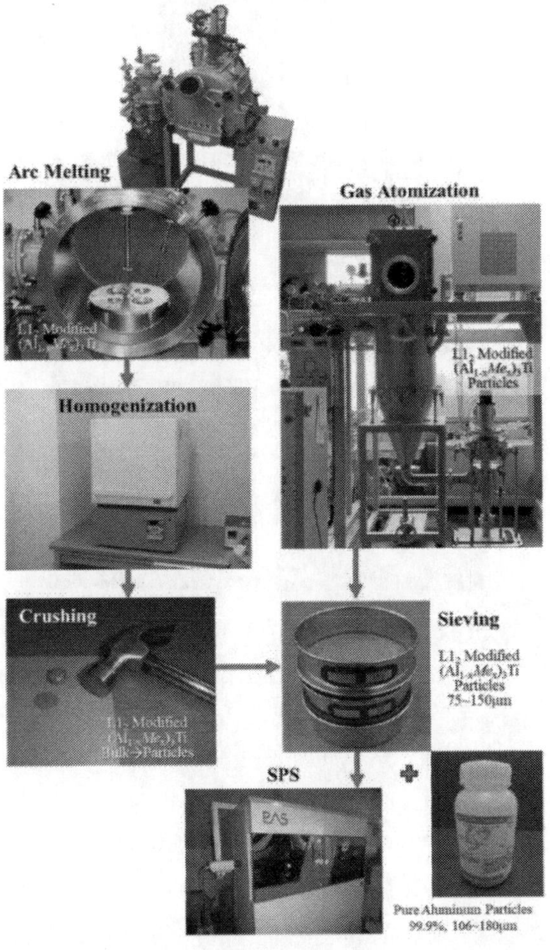

Fig. 3. Flow diagram for fabrication process of the refiners with $L1_2$ modified $(Al_{1-x}Me_x)_3Ti$ particles.

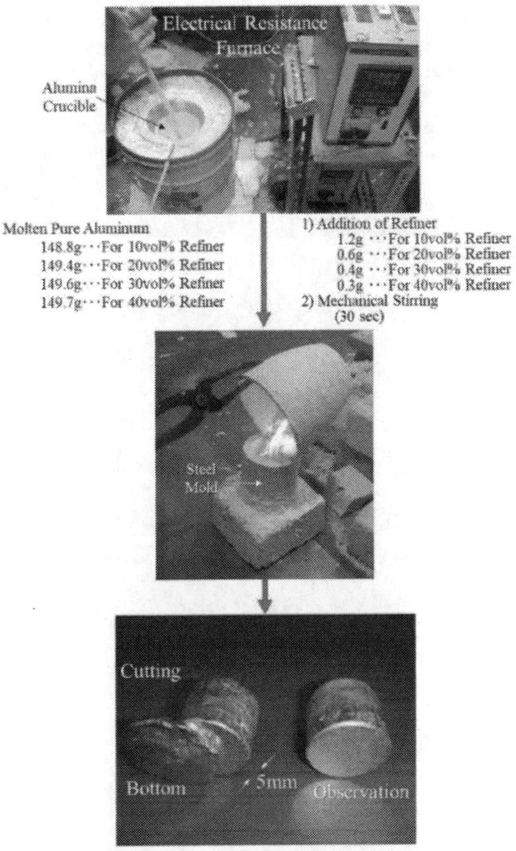

Fig. 4. Flow diagram for casting experiment of the refiners with $L1_2$ modified $(Al_{1-x}Me_x)_3Ti$ particles.

4 RESULTS AND DISCUSSION

4.1 Refined by Al-10 vol%$Al_{2.7}Fe_{0.3}Ti$

The SEM photomicrographs of Al-10 vol% $Al_{2.7}Fe_{0.3}Ti$ refiners with homogenized crushed particles and gas-atomized particles are shown in Figs. 5 (a) and (b), respectively [13]. It is seen that reaction between the $Al_{2.7}Fe_{0.3}Ti$ intermetallic compound and the aluminum matrix is not found at the interface, even though the refiners were fabricated by the sintering route. However, heating of Al-10 vol% $Al_{2.7}Fe_{0.3}Ti$ refiners with gas-atomized particles at 750°C for 300s causes microstructural evolution, as shown in Fig. 5 (c) [14]. The particles in the heated Al-10 vol% $Al_{2.7}Fe_{0.3}Ti$ refiners with gas-atomized particles have a core and mantle microstructure, and the chemical compositions of the core region and mantle region are a titanium rich $Al_{2.7}Fe_{0.3}Ti$ phase and $Al_3(Ti,Fe)$, respectively.

It is also visible from cross-sectional observations of gas-atomized particles (Fig. 5 (b)) that the gas-atomized particles contain a mesh-shaped secondary phase. However, such secondary phase is not found in the inside of the crushed particle as shown in Fig. 5 (a). Instead, secondary phase was found on the surface of crushed particles. This may be evidence of the fragmentation behavior of the arc-melted intermetallic compound sample, caused by crushing with a hammer. Namely, fragmentation occurs across the secondary phase, and as a result, the secondary phase exists not in the inside of the crushed particle but on its surface. Anyhow, it can be concluded that the $L1_2$ modified $Al_{2.7}Fe_{0.3}Ti$ intermetallic compound particles can be successfully embedded in the aluminum matrix by SPS without any reaction.

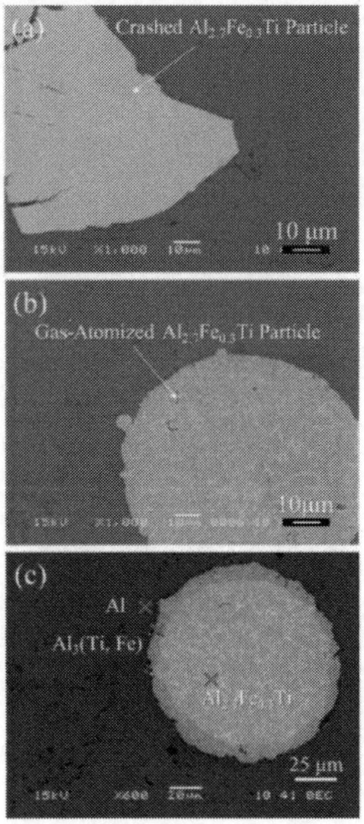

Fig. 5. SEM photomicrographs showing Al-10 vol% $Al_{2.7}Fe_{0.3}Ti$ refiners with (a) crushed particles and (b) gas-atomized particles before heating [13], and (c) after heating at 750°C for 300 s [14].

By using the Al-10 vol% $Al_{2.7}Fe_{0.3}Ti$ refiners with homogenized crushed particles and gas-atomized particles, the casting experiments are carried out, and the results are shown in Figs. 6 (b) and (c), respectively [13]. A photomacrograph taken before addition is shown in Fig. 6 (a). It is seen from Figs. 6 (a) to (c) that the higher grain refining efficiency is found for the refiner with gas-atomized particles compared with the refiner with homogenized crushed particles. This is because the secondary phase in the gas-atomized particles is mainly located inside the particles, while that in the crushed particles located on the surface of the particles, which may lead to large parameter M value of heterogeneous nucleation site particles.

The lattice matching between the particles and the solidified phase is critical to the effectiveness of the heterogeneous nucleation site. To discuss the effects of the lattice matching on the grain refining performance, the grain refining performance of the unhomogenized crushed refiner with a certain amount of the secondary phase on the surface of particles and heated gas-atomized refiner with partially decomposed $Al_{2.7}Fe_{0.3}Ti$ particles are studied. The results are also shown in Figs. 6 (d) and (e), respectively [14]. As can be seen, the grain-refining efficiency of the unhomogenized crushed refiner or heated gas-atomized refiner is inferior to that of the homogenized crushed refine or unheated refiner, respectively. This is because the surfaces of the unhomogenized crushed particles and heated particles are covered with secondary phase and mantle phases, which may have a lower lattice matching. Different refining performance characteristics appeared for refiners with different phases, even though the chemical compositions of the refiners themselves were the same.

Fig. 6. Grain refining performance test results of the Al-10 vol% $Al_{2.7}Fe_{0.3}Ti$ refiner with (b) crushed particles and (c) gas-atomized particles [13]. A photomacrograph taken before addition is also shown in (a). Results of (d) unhomogenized Al-10 vol% $Al_{2.7}Fe_{0.3}Ti$ refiner and (e) heated Al-10 vol% $Al_{2.7}Fe_{0.3}Ti$ refiner [14]. Holding time of the grain refining performance test is 600 s.

4.2 Refined by Al-(10-40)vol%$Al_{2.5}Cu_{0.5}Ti$

SEM micrographs of the Al-10vol%, 20 vol%, 30 vol% and 40 vol% $Al_{2.5}Cu_{0.5}Ti$ refiners with gas-atomized particles are shown in Figs. 7 (a) to (d), respectively [16]. Spherical shaped gas-atomized particles are homogeneously distributed in the aluminum matrix, even when there is a high volume fraction of refiner. The SPS process thus successfully produced fully dense refiners consisting of a high volume fraction of $Al_{2.5}Cu_{0.5}Ti$ particles embedded in the aluminum matrix.

 https://doi.org/10.1051/matecconf/202032606008

The refiners containing different amounts of heterogeneous nucleation site particles showed differences in the grain refining performance with the holding time, even though the total amount of $Al_{2.5}Cu_{0.5}Ti$ particles within the melt was the same. The best holding times for the 10 vol%, 20 vol%, 30 vol%, and 40 vol% atomization refiners were determined to be 300 s, 420 s, 540 s and 600 s, respectively. The results of grain refinement performance tests for the 10 vol%, 20 vol%, 30vol% and 40 vol% atomization refiners with holding times of 300 s, 420 s, 540 s and 600 s are shown in Figs. 8 (a) to (d), respectively [16]. Longer holding time is required to achieve a uniform dispersion of $Al_{2.5}Cu_{0.5}Ti$ particles in the melt for the refiners with high volume fraction of heterogeneous nucleation site particles. In this way, fading of the grain refinement effect occurs slower when the volume fraction of heterogeneous nucleation site particles increased.

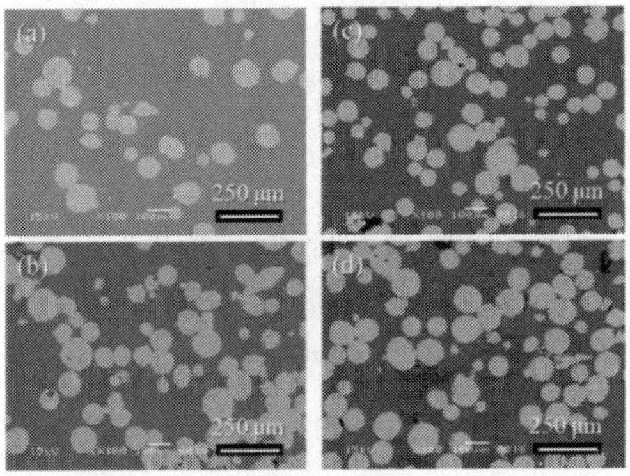

Fig. 7. SEM micrographs of fabricated (a) Al-10vol%, (b) 20 vol%, (c) 30 vol% and (d) 40 vol% $Al_{2.5}Cu_{0.5}Ti$ refiners with gas-atomized particles [16].

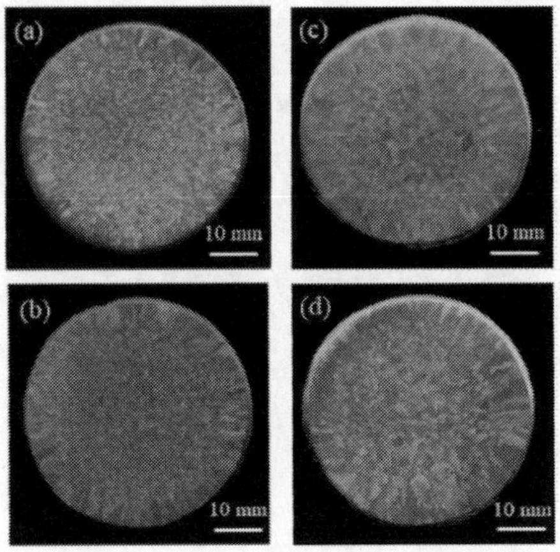

Fig. 8. Macrostructures of aluminum casts refined with (a) Al-10vol%, (b) 20 vol%, (c) 30 vol% and (d) 40 vol% $Al_{2.5}Cu_{0.5}Ti$ refiners with gas-atomized particles by hold times of 300 s, 420 s, 540 s and 600 s, respectively [16]. The levels of addition were 0.8%, 0.4%, 0.27% and 0.2% for the 0, 20, 30, and 40 vol% atomization refiners, respectively, so that the total amount of $Al_{2.5}Cu_{0.5}Ti$ particles within the melt was the same.

4.3 Refined by Al-10 vol%$Al_{2.7}Ni_{0.3}Ti$

The minimum grain size can be achieved for the cast refined by the Al-10 vol% $Al_{2.7}Ni_{0.3}Ti$ refiner after the holding time of 690 s, as shown in Fig. 9 (a) [15]. This size is about one ninth that of the unrefined cast. However, a lower grain refinement efficiency is found for the Al-10 vol% $Al_{2.7}Ni_{0.3}Ti$ refiner than for the Al-10 vol% $Al_{2.7}Fe_{0.3}Ti$ refiner shown in Fig. 6 (c). This phenomenon cannot be explained by simple lattice matching between the heterogeneous nucleation site and the solidified phase, since the M value between $Al_{2.7}Ni_{0.3}Ti$ and aluminum, $M = 1.93 \times 10^{-3}$, is smaller than that between $Al_{2.7}Fe_{0.3}Ti$ and aluminum, 2.50×10^{-3}.

Next, the thermal stability of the $Al_{2.7}Ni_{0.3}Ti$ phase in the Al-10 vol% $Al_{2.7}Ni_{0.3}Ti$ refiner was investigated. For this purpose, the Al-10 vol% $Al_{2.7}Ni_{0.3}Ti$ refiner was heated at 750 °C for 300 s, and result is shown in Fig. 9 (b) [15]. It is seen that heating for a short time causes the decomposition of the $Al_{2.7}Ni_{0.3}Ti$ phase into Al_3Ti and Al_3Ni phases. On the other hand, in the case of the heated Al-10 vol% $Al_{2.7}Ni_{0.3}Ti$ refiner, the particles have a core and mantle microstructure as shown in Fig. 5 (c), and the chemical composition of the core region is still $Al_{2.7}Fe_{0.3}Ti$ phase. By comparing the decomposition phenomena in the Al-10 vol% $Al_{2.7}Ni_{0.3}Ti$ and Al-10 vol% $Al_{2.7}Fe_{0.3}Ti$ refiners, it is found that the thermal stability of the $Al_{2.7}Ni_{0.3}Ti$ phase in the refiner is lower than that of the $Al_{2.7}Fe_{0.3}Ti$ phase. This may be the reason for the relatively lower grain refinement performance of the Al-10 vol% $Al_{2.7}Ni_{0.3}Ti$ refiner. In this way, the thermal stability of the heterogeneous nucleation site also affects the grain refinement performance.

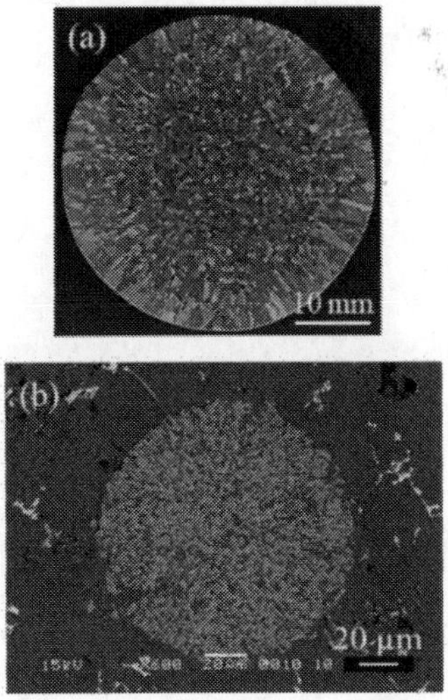

Fig. 9. (a) Grain refinement performance test result of the Al-10 vol% $Al_{2.7}Ni_{0.3}Ti$ refiner after holding time of 690 s. (b) Microstructure of Al-10 vol% $Al_{2.7}Ni_{0.3}Ti$ refiner heated at 750 °C for 300 s [15].

5 CONCLUSIONS

In this paper, the role of $L1_2$ modified $(Al_{1-x}Me_x)_3Ti$ phases on the heterogeneous nucleation of alpha aluminum primary crystal was reviewed. The refiners with $L1_2$ modified $(Al_{1-x}Me_x)_3Ti$ particles prepared by gas atomization or crushing were fabricated by SPS. The results are summarized as follows.

1) $L1_2$ modified $(Al_{1-x}Me_x)_3Ti$ particles can be prepared by mechanical crushing of bulk $L1_2$ modified $(Al_{1-x}Me_x)_3Ti$ intermetallic compounds and directly prepared by gas atomization. The $L1_2$ modified $(Al_{1-x}Me_x)_3Ti$ particles embedded in an aluminum matrix do not react with aluminium nor transformation into stable phases during sintering.

2) The aluminum cast without the refiner has coarse and inhomogeneous grains. On the other hand, the grain size is smaller in the aluminum cast with the Al-$(Al_{1-x}Me_x)_3Ti$ refiners. Therefore, $L1_2$ modified $(Al_{1-x}Me_x)_3Ti$ particles can act as effective heterogeneous nucleation sites for primary aluminum in the solidification process.

3) Different refining performance characteristics appeared for refiners with different phases, even though the chemical compositions of the refiners themselves were the same. This is because the lattice matching between heterogeneous nucleation site particles and solidified aluminum is critical to obtain finer grain structures for aluminum casts.

4) Although $L1_2$ modified $Al_{2.7}Ni_{0.3}Ti$ phase has small M value, the grain refinement performance of Al-10 vol% $Al_{2.7}Ni_{0.3}Ti$ refiner is relatively weaker than expected. This is because the Al-10 vol% $Al_{2.7}Ni_{0.3}Ti$ refiner does not have a high thermal stability. The thermal stability of the heterogeneous nucleation site also affects the grain refinement performance.

This work was supported by The Light Metal Educational Foundation Inc. of Japan, the New Energy and Industrial Technology Development Organization (NEDO) of Japan and Japan Science and Technology Agency (JST) "Collaborative Research Based on Industrial Demand" Grand Number JPMJSK1614.

References

1. P. S. Mohanty, J. E. Gruzleski, Acta Mater., **43**, No. 5, 2001 (1995).
2. B. S. Murty, S. A. Kori, K. Venkateswarlu, R. R. Bhat, M. Chakraborty, J. Mater. Proc. Tech., **89-90**, 152 (1999).
3. Y. Watanabe, N. Yamanaka, Y. Fukui, Z. Metallkd., **88**, No. 9, 717 (1997).
4. Y. Watanabe, N. Yamanaka, Y. Fukui, Metall. Mater. Trans. A, **30A**, No. 12, 3253 (1999).
5. Y. Watanabe, H. Eryu, K. Matsuura, Acta Mater., **49**, No. 5, 775 (2001).
6. P. D. Sequeira, Y. Watanabe, Y. Fukui, Scrip. Mater., **53**, No. 6, 687 (2005).
7. P. D. Sequeira, Y. Watanabe, H. Eryu, T. Yamamoto, K. Matsuura, Trans. ASME, J. Eng. Mater.Tech., **129**, No. 2, 304 (2007).H. Sato, Y. Noda, Y. Watanabe, Mater. Trans., **54**, No. 8, 1274, (2013).
8. Y. Watanabe, P. D. Sequeira, H. Sato, T. Inamura, H. Hosoda, Jpn J. Appl. Phys., **55**, No. 1, 01AG03 (2016).
9. J. Tarnacki, Y.-W. Kim, Scr. Metall., **22**, No. 3, 329 (1988).
10. H. Mabuchi, K. Hirukawa, H. Tsuda, Y. Nakayama, Scr. Metall. Mater., **24**, No. 3, 505 (1990).
11. Y. Nakayama, H. Mabuchi, Intermetallics, **1**, No. 1, 41 (1993).
12. Y. Watanabe, K. Hirata, H. Sato, Jpn J. Appl. Phys., **57**, No. 1S, 01AF04 (2018).
13. Y. Watanabe, T. Hamada, H. Sato, Jpn J. Appl. Phys., **55**, No. 1, 01AG01 (2016).
14. Y. Watanabe, T. Hamada, H. Sato, Jpn J. Appl. Phys., **56**, No. 1S, 01AG02 (2017).
15. Y. Watanabe, K. Maekawa, H. Sato, Jpn J. Appl. Phys., **57**, No. 1S, 01AF08 (2018).
16. Y. Watanabe, R. Yamazaki, K. Yamanaka, H. Sato, J. Mater. Proc. Tech., **255**, 400 (2018) .
17. Y. Watanabe, S. Taniai, H. Sato, Jpn J. Appl. Phys., **58**, No. SA, SAAC04 (2019).
18. D. Turnbull, B. Vonnegut, Ind. Eng. Chem., **44**, No. 6, 1292 (1952).
19. B. L. Bramfitt, Metall. Trans., **A 1**, No. 7, 1987 (1970).
20. M. Kato, M. Wada, A. Sato, T. Mori, Acta Metal., **37**, No. 3, 749 (1989).
21. Y. Watanabe, Q. Zhou, H. Sato, T. Fujii, T. Inamura, Jpn J. Appl. Phys., **56**, No. 1S, 01AG01 (2017).
22. Y. Watanabe, M. Sato, T. Chiba, H. Sato, N. Sato, S. Nakano, Metall. Mater. Trans. A, **51A**, No. 3, 1345 (2020).
23. U. Anselmi-Tamburini, J.E. Garaya, Z.A. Munira, Mater. Sci. Eng. A, **407**, 24 (2005).
24. M. Omori, Mater. Sci. Eng. A, **287**, 183 (2000).
25. Y. Watanabe, Y. Iwasa, H. Sato, A. Teramoto, K. Abe, E.Miura-Fujiwara, J. Mater. Proc. Tech., **211**, 1919 (2011).

https://doi.org/10.1051/matecconf/202032607001

Single track scanning experiment on the hypereutectic aluminium alloy Al-8%Zn-7%Ni-3%Mg

Pavel Shurkin[1,*], *Torgom* Akopyan[1, 2], *Alexey* Prosviryakov[1], *Alexander* Komissarov[1], *Natalia* Korotkova[1]

[1]National University of Science and Technology "MISIS", 119049, Leninskii pr., 4, Moscow, Russia
[2]Baikov Institute of Metallurgy and Materials Science, 119991, Leninskii pr., 49, Moscow, Russia

Abstract. Solidification path, hot tearing susceptibility and processability by L-PBF of the hypereutectic aluminium alloy Al-8%Zn-7%Ni-3%Mg were investigated. It was shown that the alloy has an advantageous hot tearing resistance due to formation of an adequate amount of the [(Al)+Al_3Ni] eutectic and a reduced effective solidification range. In that sense, it may be valid for L-PBF. To prepare a feedstock for L-PBF, we produced chips which were subsequently grinded by high-energy ball milling into a 100 μm fine powder batch. The powder contained irregular shape particles that may have led to failure in fabrication of regular single tracks at a laser power of 170-250 W and scanning speed of 200-350 mm/s. Whilst from on-top view all the tracks had defects like balling and irregularity, their cross-sections also revealed a high degree of penetration into 5052 alloy substrate. Most structures showed a level of mixing of 5/1 between the substrate and the experimental alloy. The most successful track obtained at 250 W and 250 mm/s showed an inhomogeneity in distribution of the alloy-rich regions. Meanwhile, no defects and no primary Al_3Ni phase were observed. The present study serves as an initial one for further research which may be focused on producing of a finer powder, assessment of different layer thicknesses and fabrication of bulk specimens.

1 Introduction

Metal-based laser additive manufacturing is a wide field of material science requiring for careful selection of composition that is considered to be the most critical for synthesis of minimal-defect products. Among a great variety of methods the laser powder bed fusion (hereafter referred to as L-PBF) is an advance process based on laser beam travel along the metallic powder batch presenting an opportunity to produce complex functional parts for various purposes [1-3].

Among numerous compositions for L-PBF investigated through years, the aluminium based materials present some complications. In comparison to steels, titanium or magnesium, aluminium is a good heat conductor, but its laser absorption is lower. In this case a higher laser power has to be applied [4]. Nevertheless they are believed to be the most attractive for light-weight application due to their low density (~2.7 g/cm^3), easy conventional manufacturing and possibility of secondary stock usage [5].

It is certainly known that the most studied aluminium alloys for L-PBF are Al-Si alloys, perfect for producing cast products as well [5, 6-9]. A good fusion ability of powder particles is caused by specific solidification path which contains eutectic reaction [L→(Al)+(Si)] at 577 °C. Non-eutectic alloys (Al-Zn-Mg, Al-Cu, etc.) are not appropriate for L-PBF due to wide solidification range [4]. Attempts to produce L-PBF parts from the high-strength Al-Zn-Mg alloys leaded to extensive formation of hot cracks [9]. Therefore several works are devoted to preparing silicon-added mixture based on Al-Zn-Mg matrix which provides enough appropriate cracks resistance [10, 11].

However, we suppose that there are some risks related to loss of strength may occur due to possible Mg_2Si phase formation. There are some eutectic forming elements which are believes to be promising for replacing silicon [12, 13]. Our attitude toward the issue of L-PBF of Al-Zn-Mg alloys is slightly similar, but we decided to use nickel instead of silicon. Al-Zn-Mg-Ni alloys based on the [(Al) +Al_3Ni] are proved to have excellent fluidity and hot tearing resistance [5, 14]. Moreover, since there is no interaction between pair Zn/Mg and Ni it is possible to fabricate a natural particulate-reinforced composite material by an increase in Ni content up to hypereutectic concentration. In favour to this approach, Al-20%Si hypereutectic alloy was fabricated in [15] and after a pseudo-eutectic cellular microstructure was observed as a result. This is due to high cooling rates (more than 10^3-10^7 K/s [16]) provided by rapid laser spot movement.

In this initial work we do not consider the cooper addition because of its harmful effect on the fluidity and hot tearing resistance [17]. We chose the strongest Al-Zn-Mg matrix related to commercial alloys (e.g. 7001, 7090, 7055, etc.) [18]. Research on rapid solidification of the Al-Ni hypereutectic alloys reported that the

* Corresponding author: pa.shurkin@gmail.com

pseudo-eutectic ultrafine structure can be achieved at 7-8%Ni [19, 20]. Thus the main experimental alloy in the present work is a Al-8%Zn-7%Ni-3%Mg. Starting from the first principles the experiment was conducted through powder fabrication by high-energy ball milling (hereafter referred to as HEBM). This approach is an advantageous due to lack of necessity to consider expensive gas atomization. Some successful experiences on L-PBF of Al-based and Fe-based alternative powder feedstock are observed in [21, 22]. The core of the present work is to investigate the molten pool behaviour during L-PBF and the microstructure evolution in order to substantiate the prospects of the alloy and tailor a route for further works.

2 Materials and methods

The work was focused on producing of samples and its characterization. Firstly, we conducted a hot tearing test of three alloys Al-8%Zn-3%Mg, Al-8%Zn-4%Ni-3%Mg and Al-8%Zn-7%Ni-3%Mg in order to substantiate the choice of the alloy composition. Secondly, the analysis of the non-equilibrium solidification was carried out by Scheil-Gulliver simulation in Thermo-Calc software (TTAL5 Al-alloys database) [23] for prediction of the microstructure. Finally, the powder was produced, which was used as a feedstock in a single track experiment.

2.1 Preparation of samples

The alloys were produced via melting of pure aluminium (99.9%Al), zinc (99.9%Zn), magnesium (99.9%Mg) and Al-20%Ni master alloy in an electrical laboratory furnace Nabertherm K 1/13. Hot tearing test was performed via pouring the molten metal into a backbone mould with variation of diameter from 16 mm to 10 mm. The experimental alloy Al-8%Zn-7%Ni-3%Mg was also poured into a graphite cylindrical mould of 40x280 mm in size. The raw material for the single track experiment was irregular shaped powder produced from the alloy chips (fig. 1). The 150 g batch was milled into powder in air using a 4-reel Retsch PM 400 planetary ball mill at 300 rpm. The weight ratio between the stainless-steel balls and the experimental alloy chips was 10:1. After 10 min of dry grinding, the powder was sieved so to achieve a less than 100 μm size, following which the remaining coarse fraction was milled again. The total milling time was 30 min.

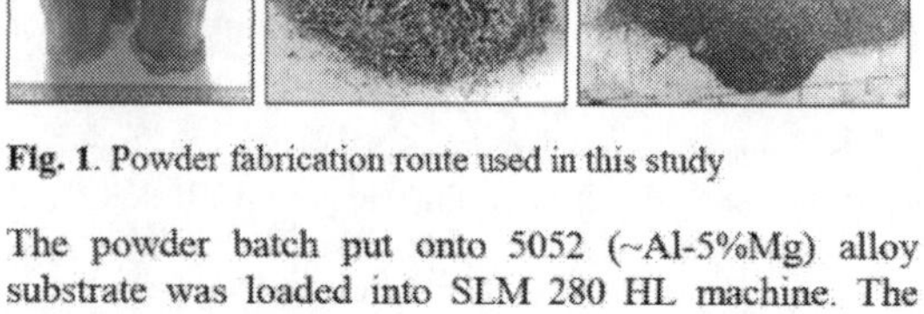

Fig. 1. Powder fabrication route used in this study

The powder batch put onto 5052 (~Al-5%Mg) alloy substrate was loaded into SLM 280 HL machine. The laser was travelled just one time along the lines of 120 mm length. The powder was melted at a laser power of 170, 200, 250 W and a scanning speed of 200, 250, 300, 350 mm/s (12 tracks in total). The thickness of layer was about 150 μm.

2.2 Characterization

The microstructure, width and height of tracks was examined by optical microscopy (OM, Axio Observer MAT) and scanning electron microscopy (SEM, TESCAN VEGA3) equipped with an electron microprobe analysis system (EMPA, Oxford Instruments). The metallographic samples were ground using a SiC abrasive paper and polished with a 1 μm diamond suspension. A 1% hydrogen fluoride (HF) water solution was used for etching. The particle-size distribution of the powder was studied using an Analysette 22 laser particle analyser. The parameters D_{10}, D_{50}, D_{90}, which are the cumulative particle diameters for 10%, 50%, and 90 % of the entire powder volume, were determined. The SPAN parameter was calculated using the equation [24]:

$$SPAN = \frac{D_{90} - D_{10}}{D_{50}} \tag{1}$$

3 Results and discussion

3.1. Hot tearing susceptibility

Hot tearing susceptibility mainly depends on solidification path, which may be described by thermodynamic calculation. A comprehensive description of phase equilibria in Al-Zn-Mg-Ni system is described in our recent paper [25]. It is observed that a Ni content over 3.6% promotes hypereutectic solidification manner including the further reactions [L→Al_3Ni], [L+Al_3Ni+(Al)] and [L+Al_3Ni+(Al)+T]. When considering the Al-8%Zn-7%Ni-3%Mg alloy, a fractions of primary and eutectic Al_3Ni phase are nearly equal and their total amount composes 12.8 vol.%. This phase is reliably indicated by elemental mapping and presented in SEM microstructure (fig.2) as bright constituents of spheroidal agglomerated and rod-like shapes surrounded by grey (Al) dendrites. In the vicinity, dark T phase veins are revealed appeared as a result of

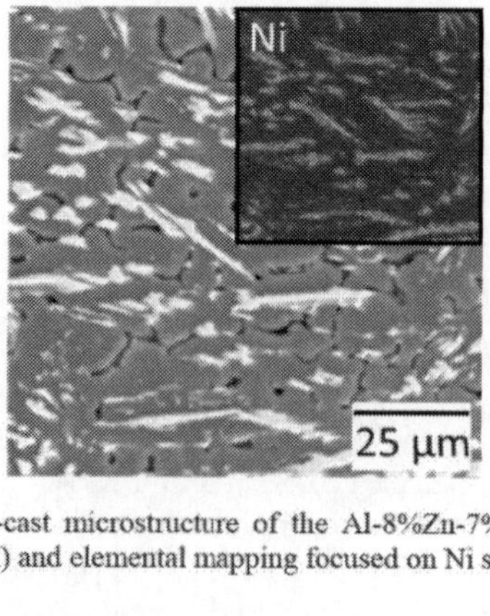

Fig. 2. As-cast microstructure of the Al-8%Zn-7%Ni-3%Mg alloy (SEM) and elemental mapping focused on Ni spectre

non-equilibrium solidification. Generally, the phases retained after solidification agree the calculation data.

According to Scheil-Gulliver curves (fig. 3), in Al-8%Zn-7%Ni-3%Mg the Al_3Ni primary phase occurs prior to (Al) solid solution at the temperature of about 700 °C. Then, after formation of (Al) dendritic network and Al_3Ni eutectic-origin phase, the non-equilibrium solidification T phase appears at 478 °C. The latter temperature is mainly constant and does not depend on Ni content. The results of the hot tearing test showed that HTS index (defined as the least diameter of a crack-free sample) for Al-8%Zn-3%Mg is 16 mm and addition of Ni contributes to improvement of hot tearing tendency down to 12 mm for Al-8%Zn-7%Ni-3%Mg alloy. This improvement can be described by slightly dwindled effective solidification range (ESR). According to [26] the ESR value for Al-Zn-Mg-Ni alloys can be described using Scheil-Gulliver curves as a difference between the temperature at which 85% solid phases were formed (start of the linear shrinkage) and the non-equilibrium solidus. The ESR value descends with an increase in Ni content, but the change is negligible (~ 10 °C). The influence of Ni on the improvement of the HTS index of Al-Zn-Mg-Cu alloys was described in [14]. The key aspect used was the Cracking Susceptibility Coefficient (CSC) developed by Clyne and Davies in [27]. Making the assumption that a time when a solid fraction is at 0.9, 0.99 and 0.4 can be replaced by the temperature, the formula looks as follows:

$$CSC = \frac{T_{0.9} - T_{0.99}}{T_{0.4} - T_{0.9}} \quad (2)$$

Thus it can be seen that the CSC value is significantly reduced with addition of Ni (from 0.87 for Al-8%Zn-3%Mg to 0.62 for Al-8%Zn-7%Ni-3%Mg) even at hypereutectic concentrations.

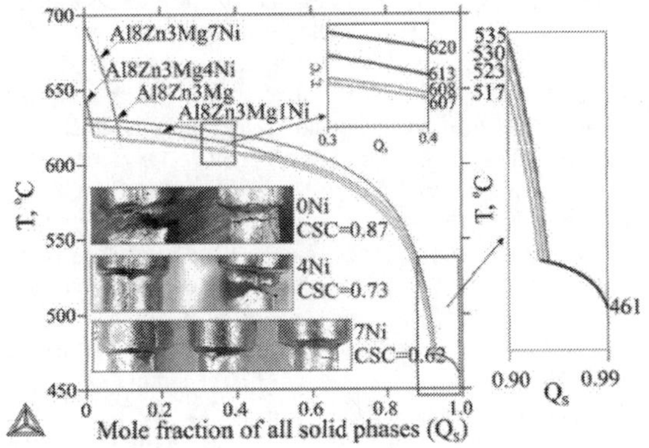

Fig. 3. Scheil-Gulliver curves and hot tearing susceptibility of the experimental alloys

Another factor contributed to the improvement of hot tearing resistance is an adequate amount of eutectic liquid promoting healing of incipient cracks during contraction at the latter stages of solidification. In our test the fewer diameters of sample the higher tensile stress is induced accompanied by an increase in amount of interdendritic channels appeared after rupture of liquid film. In that sense, whilst in the base alloy no eutectic melt can refill these channels, in case of Ni-rich alloy presence of the sufficient eutectic volume [(Al)+Al_3Ni)] shows an observable effect in the test samples of up to 12 mm. It is believed that the achievement can be critical for producing quality L-PBF parts at industrial processing parameters.

3.2 Characterization of powder batch

The initial chips were enough brittle for easily milling within a low duration. A ubiquitous issue in HEBM processing is a contamination of the powder from the milling media. In our case the enhanced oxidation may occur and iron may be captured from a grinding bowl and balls. However, we assume that the milling time is relatively low and these factors may be ignored.

On the other hand it is obvious that the quality of powder has a crucial impact on the defects appearing in the L-PBF parts. It is rather necessary to use spherical powder produced by atomization methods. However, we consider these methods as very expensive ones. In order to preliminary estimate the processability of the experimental alloys by L-PBF, we decided to try out to use milled powder. Thus, the flowability is not supposed to be the key parameter in the present study.

Figure 4 represents a particle-size distribution of the powder from the experimental Al-8%Zn-7%Ni-3%Mg alloy. It can be seen that some occasional particles exceeding 100 µm in size are observed and therefore the experiment promises to be slightly tough. In this case we chose a bit lower scanning speed to provide appropriate melting and fusion of the particles. Nevertheless, the experimental powder has a SPAN parameter of 1.1 which can provide an excellent flowability. Thus, the usage of such powder can be promising in assessment of new alloys for L-PBF.

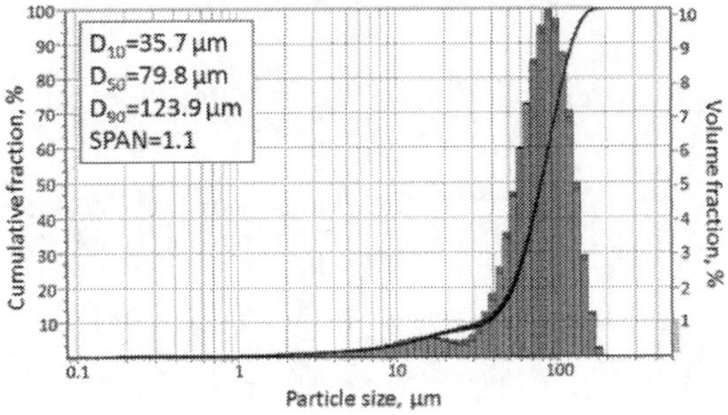

Fig. 4. Particle-size distribution of the experimental powder

The powder consisted of irregular and coarse particles with rough surfaces (fig. 5a). Each coarse plate of Al_3Ni phase, presented in the as-cast state, was split into fine unevenly distributed particles (fig. 5b). This structure looks like a composite in which the particles play a reinforcing role for aluminum-based matrix.

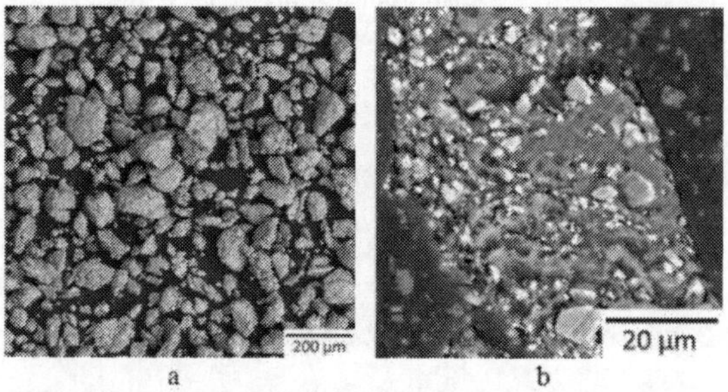

a b

Fig. 5. Shape of the powder particles (a) and the structure the individual particle (b)

Since the SLM 280 HL industrial machine has its own filter and hopper system (without transfer powder from one to another platform) which require for smooth and spherical powder, we were forced to place the powder manually with its sequent levelling by carriage.

3.3 Single track scanning experiment

Twelve single tracks were studied by OM and SEM methods. From on-top IMAGES made at the 5 mm distance from the start of laser travelling, it is clearly seen that all tracks have defects such as irregularity and balling (fig. 6a). It is likely that the prolonged light surface is caused by sufficient fusion and consolidation of the melt pool. Since conditions applied are greatly far from common ones (e.g. scanning speed is about 4 times less than that ones for branded Al10SiMg alloy), we suppose that the laser melting of the ball milled powder was accompanied with a significant wetting of the substrate (fig. 6b), associated with low scanning speed.

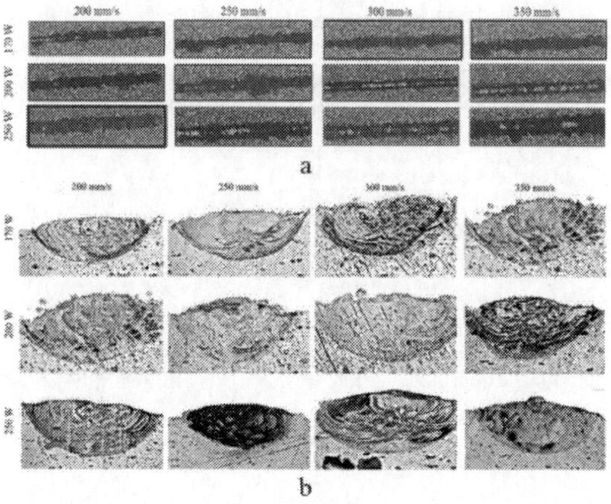

Fig. 6. On-top view (a) and cross-section images (b) of the single tracks

Analysis of the cross-section images (fig. 7) showed that the width of the molten pool increases with an increase in laser power from 140 µm (for 170 W) to 220 µm (for 250 W). Relatively deep penetration of about 50-100 µm was observed for all samples. It is complicated to reveal any associations between processing parameters and molten pool geometry due to a high degree of randomness. For example, the track obtained at 250 W and 350 mm/s has a width of 330 µm (appeared to be the largest one), that is not reasonable because it must have become slender due to an increase in the scanning speed.

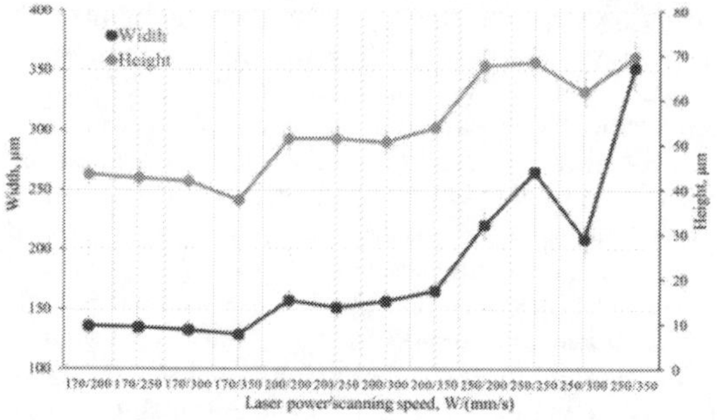

Fig. 7. Width and height measured on the cross-sections of the single tracks

Nevertheless, there is a great difference among the structures of the tracks and some regularity has been found. The traits of the cross-section structures are the evidential inhomogeneity caused by turbulence and Marangoni flow occurred after laser power irradiation. Temperature gradients generated coupled heat- and mass-transfer processes and that's why the concentration fluctuation is well visible in optical images as dark coloured vortexes. They are most likely indicating the level of interaction between the Al-8%Zn-7%Ni-3%Mg alloy powder and the 5052 grade alloy substrate.

We chose the most representative types of structures appeared in individual tracks fabricated at 250 W and studied them by SEM. The track obtained at 250 W and 200 mm/s has a structure (fig. 8a) in which the amount of the experimental alloy decreased approximately fivefold due to lower scanning speed and intense mixing with substrate. Some cracks and porosity are also observed. The most successful track was built at 250 W and 250 mm/s. The OM revealed that it has the darkest surface and from the SEM structure (fig.8b) it is clear that the experimental alloy-rich area (designated as Ni-rich area) is concentrated on the right sight of the track. In comparison to the previous track, there is more than 4%Ni, but just about 2%Zn that is probably due to some evaporation. The left side map shows the Substrate/alloy mixing about 5/1 likewise as in the previous track. The widest track fabricated at 250 W and 350 W has a structure without dark vortexes. Only some segregated parts of the cross-sections are indicated to be the experimental alloy (fig. 8c). It is interesting that the displaced region revealed at the up-left side is enough convex and slightly penetrated into the substrate that confirm its remelting during laser travelling. Ultimately, the structure of the Ni-rich area is much finer than in the powder sample, so that it is believed to have a pseudo-eutectic structure.

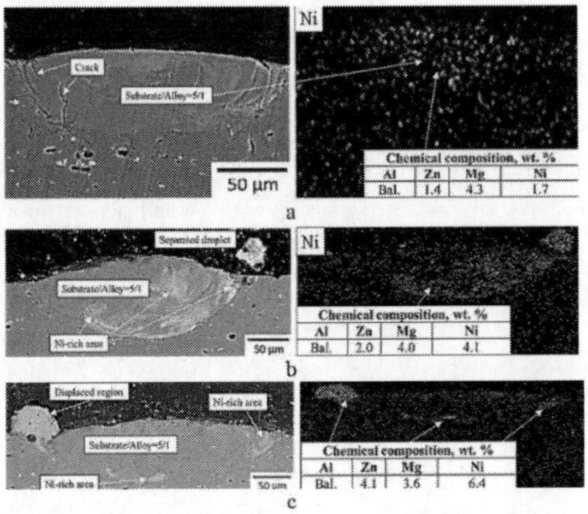

Fig. 8. Cross-section microstructures of the single tracks, fabricated at a laser power and a scanning speed: a – 250W and 200 mm/s; b - 250W and 250 mm/s; c - 250W and 350 mm/s

4 Conclusions

In summary, whilst the experiment was relatively tough for arguing that the new Al-8%Zn-7%Ni-3%Mg alloy is excellent for L-PBF, some important conclusions for further research were drawn:

1) The Al-8%Zn-7%Ni-3%Mg has an advantageous hot tearing resistance due to lower effective solidification range and adequate amount of [(Al)+Al_3Ni] eutectic in comparison to the base Al-8%Zn-3%Mg alloy and near-eutectic Al-8%Zn-4%Ni-3%Mg alloy.

2) The feedstock powder material of 100 µm size for L-PBF was prepared via machining of initial cast ingot into chips and further HEBM. The SPAN and D_{50} values were 1.1 and 79.8 µm respectively. While the powder had mostly irregular shape, the structure contained grinded fine unevenly distributed Al_3Ni intermetallics.

3) A comparative evaluation of the twelve single tracks fabricated at a laser power of 170-250 W and scanning speed of 200-350 mm/s was carried out. The low quality of powder likely caused the irregularity of tracks. Moreover, it was shown that a relatively low scanning speed leaded to significant wetting of the substrate and turbulence. The latter represented as vortexes and mixing between the experimental alloy and the substrate.

4) Three types of the single tracks fabricated at 250 W and different scanning speed were studied in detail. The structure of the first track fabricated at 200 mm/s showed cracks and porosity as well as a substrate/alloy mixing of 5/1. Increase of scanning speed up to 250 mm/s promoted the least level of mixing in the second track. It contained no porosity and cracks, but some amount of zinc was evaporated. The third track fabricated at 350 mm/s showed a random displaced region, which contained more than 6%Ni indicating on the initial alloy. It has an ultrafine structure provided by rapid solidification.

5) Further research will follow the path toward softening of the experimental conditions. The key approach will be focused on more sufficient grinding of the powder and on trying out higher laser velocity and its power. It is believed that the proposed feedstock is promising for assessment of the new alloys for L-PBF.

This research was funded by the Russian Science Foundation (project no. 19–79– 30025)

References

1. T.S. Srivatsan, T.S. Sudarshan, *Additive Manufacturing Innovations. Advances and Applications* (CRC PRESS. Taylor & Francis Group, 2015)
2. R. Deb, T. Wei, H.L. Zuback, J.S. Mukherjee, T. Elmer, J.W. Milewski, J.O. Beese, A.M. Wilson-Heid, A. De, W. Zhang, Progress in Materials Science, **92**, 112-224 (2018)
3. D. Herzog, V. Seyda, E. Wycisk, C. Emmelmann, Acta Mater. **117**, 371–392 (2016)
4. C. Galy, E. Le Guen; E. Lacoste; C. Arvieu, Add. Manuf. **22**, 165-175 (2018)
5. M. Glazoff, A. Khvan; V. Zolotorevsky, N. Belov; A. Dinsdale, *Casting Aluminum Alloys. 2nd Edition: Their Physical and Mechanical Metallurgy* (Elsevier, 2018)
6. J. Liu, Z. Liu, Y. Jiang, G.W. Wang, Y. Yang, L.C. Zhang, JALCOM, **735**, 1414-1421 (2018)
7. W. Pei, W. Zhengying, C. Zhen; D. Jun, H. Yuyang, L. Junfeng, Z. Yatong, Appl. Surf. Sci. **408**, 38-50 (2017)
8. N. Read, W. Wang, K. Essa, M. Attallah Moataz, Mater. Des. **65**, 417-424 (2015)
9. P. Wang, H.C. Li, K.G. Prashanth, J. Echert, S. Scudino, JALCOM, **707**, 287–290 (2017)
10. R. Casati, M. Coduri, M. Riccio, A. Rizzi, M. Vedani, JALCOM, **801**, 243-253 (2019)
11. M.L. Montere-Sistiaga, R. Mertnens, B. Vrancken, X. Wang, B. van Hooreweder, J.-P. Kruth, J. van Humbeeck, J. Mater. Process. Technol. **238**, 437-445 (2016)
12. N.A. Belov. Russian Journal of Non-Ferrous Metals, **3**, 243-249 (2010)
13. N.A. Belov, A.V. Khvan, A.N. Alabin, Mater. Sci. Forum, **519-521**, 395-400 (2006)
14. F. Liu, X. Zhu, S. Ji, JALCOM, **821**, 153458 (2020)
15. P. Ma; K.G. Prashanth; S. Scudino, Y. Jia, H. Wang, C. Zou, Z. Wei, J. Eckert, Metals, **4**, 28-36 (2014)
16. J. Zhang, B. Song, Q. Wei, D. Bourell, Y. Shi, J. Mater. Process. Technol. **35**, 270-284 (2019)
17. K.T. Kim, J.H. Yang, Y.S. Lim, J. Kor. Foundry Soc. **32**, 38-43 (2012)
18. J.E. Hatch, *Aluminum: Properties and Physical Metallurgy* (American Society for Metals, 1984)
19. R. S. Barclay, H. W. Kerr, P. Niessen, Journal of Materials Science, **6**, 1168-1173 (1971)

20. G. Gonzalez, G.A. Lara-Rodriguez, A. Sandoval-Jiménez, W. Saikaly, A. Charai, Materials Characterization, **59**, 1607-1612 (2008)
21. B. Fullenwider, P. Kiani, J. M. Schoenung, K. Ma, Powder Technology, **342**, 562-571 (2019)
22. Q. Han, R. Setchi, S. L. Evans, Powder Technology, **297**, 183-192 (2016)
23. Thermo-Calc Software TTAL5 Al-Alloys. URL: www.thermocalc.com (access date: 17.01.2020)
24. R. Baitimerov, P. Lykov, D. Zherebtsov, L. Shultc, A. Radionova, K.G. Prashanth, Materials, **11**, 742 (2018)
25. P. Shurkin, T. Akopyan, N.Korotkova, A. Prosviryakov, A. Bazlov, A. Komissarov, D. Moskovskikh, Metals, **10**, 762 (2020)
26. V. S. Zolotorevskii, A. V. Pozdnyakov, A. Yu. Churyumov, The Physics of Metals and Metallography, **115**, 286-294 (2014)
27. T.W. Clyne, G.J. Davies, Br. Foundryman, **74**, 65–73 (1981)

Wire and Arc Additive Manufacturing with TiC-Nanoparticle Reinforced AA5183 Alloy

Geir Langelandsvik[1,2,*], *Olav* Ragnvaldsen[2], *Jan E.* Flåm[2], *Odd M.* Akselsen[1], and *Hans J.* Roven[2]

[1]SINTEF Industry, 7491 Trondheim, Norway
[2]NTNU Norwegian University of Science and Technology, Department of Materials Science and Engineering, 7491 Trondheim, Norway

Abstract. Wire and arc additive manufacturing of aluminium-ceramic composites shows great potential to produce high strength materials. By incorporation of nanoparticles in the feedstock wire, fine-grained material with low susceptibility for solidification cracking and enhanced strength can be obtained. In fact, this study utilised the novel screw extrusion method to prepare an aluminium alloy containing TiC nanoparticles. The commercial aluminium alloy AA5183 was selected for WAAM to assess and benchmark the effects of screw extrusion and TiC. The materials have been assessed in terms of microstructure, porosity content and mechanical properties. The presence of TiC reduced the average grain diameter by 70%, while Vickers hardness increased with 13%. However, number of pores per unit volume increased by one order of magnitude. The porosity is believed to stem from hydrogen introduced in the AA5183-material through screw extrusion processing, in addition to hydrogen trapping and pore nucleation on TiC nanoparticles.

1 Introduction

Research and engineering related to additive manufacturing (AM) of metallic materials has accelerated the last decade. The major driving forces for the extensive exploration of AM are the possibilities of manufacturing with shorter lead times, higher material utilisation and novel designs. The modified welding method wire and arc additive manufacturing (WAAM) is among the most intensively researched AM principles. WAAM builds components in a layer-by-layer fashion by fusion of a metallic wire. The material is strategically deposited by preprogrammed robotic or CNC manipulation on a build substrate. WAAM is considered a low-cost, high efficiency method due to a low hardware investment cost (~\$120k), cheap feedstock (~\$30/kg for aluminium), high material- and energy utilisation (close to 100%) and high deposition rates (up to 1 kg/h for aluminium) [1].

Additive manufacturing of aluminium alloys have been defiant for several reasons [2]. Trapped gas porosity from supersaturated hydrogen limit mechanical properties, especially fatigue [3]. The porosity content can be eliminated by appropriate use of wire feedstock, decreasing the arc heat input and application of suitable post-processing [4–6]. Several aluminium alloys have broad solidification ranges and large contraction upon solidification. The columnar dendritic growth during solidification combined with thermal contraction increases the risk of hot tearing and intergranular cracking. WAAM is therefore restricted to a limited number of aluminium alloys. WAAM materials solidify with lower cooling rate compared to powder bed fusion (PBF) technologies, and consequently exhibit a coarser microstructure. The mechanical properties of WAAM materials are thus inferior to PBF counterparts [7].

Introduction of grain refining agents in the wire feedstock has been demonstrated for arc welding to eliminate crack susceptibility, refine the microstructure and enhance mechanical properties. Janaki Ram et al. [9] incorporated combinations of titanium, boron and zirconium to a heat-treatable AA7020 Al-Zn-Mg filler wire in order to induce grain refinement in the weld fusion zone. A slight, although significant, reduction in crack susceptibility after welding was shown. The enhanced mechanical properties of the as-welded structure was attributed to the Hall-Petch effect due to extensive grain refinement. Sokoluk et al. [10] practically eliminated hot cracking of an AA7075 Al-Zn-Mg-Cu alloy weld by introduction of titanium carbide nanoparticles. TiC served as a heterogeneous nucleation agent which altered the dendritic grain structure from columnar to equiaxed. Sales and Ricketts [11] performed WAAM with a scandium-reinforced AA5183 Al-Mg wire. Scandium is well known to promote grain refinement and increase the strength by precipitation of the Al_3Sc phase [12].

WAAM with a ceramic nanoparticle reinforced aluminium feedstock has to the authors knowledge never been demonstrated. This study will examine the effects of material preparation through metal screw extrusion and addition of TiC nanoparticles on the microstructure and mechanical properties. The Al-Mg alloy AA5183 was selected as base material to have a direct benchmark of the WAAM materials, as this alloy is readily available as

*e-mail: geir.langelandsvik@sintef.no

Table 1: Chemical composition of AA5183 Al-Mg WAAM feedstock in wt% [8].

Mg	Mn	Fe	Si	Cu	Ti	Al
4.65	0.66	0.12	0.15	0.01	0.11	Balance

Table 2: CMT input parameters for WAAM of AA5183.

Parameter	AA5183	AA5183-TiC
Current (A)	85	85
Voltage (V)	11.9	16.6
Travel speed (m/min)	0.48	0.48
Wire feed speed (m/min)	5.8	5.0

WAAM feedstock. Although AA5183 has low susceptibility for solidfication cracking [13], it was selected as a proof-of-concept material for microstructural refinement.

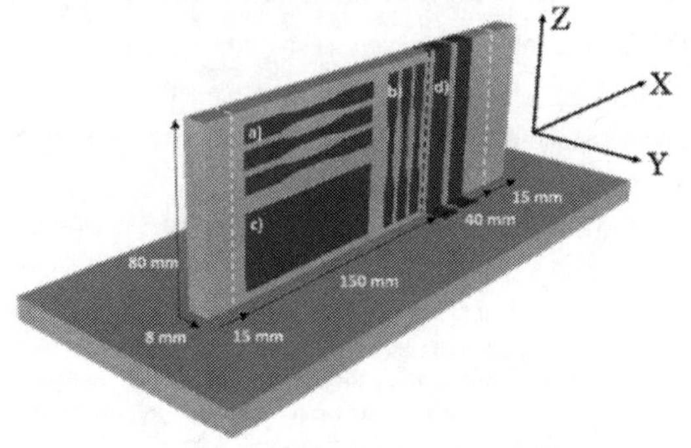

Figure 1: Reference coordinate system, dimensions and sample locations of the WAAM thin wall. a-b) Tensile specimens in deposition direction *X* and through-thickness direction *Z*, respectively c) section for metallography and microhardness d) section for CT.

2 Experimental procedure

The WAAM feedstock of interest was based on the AA5183 Al-Mg alloy. The feedstock was in form of $\phi 1.2$ mm wire manufactured by Safra SpA (Brescia, Italy). The alloy composition is given in Table 1.

Two versions of the AA5183 wire were used as feedstock for WAAM; one monolithic AA5183 wire as delivered from supplier; and one modified AA5183 wire with 1 vol% TiC nanoparticle addition (fcc crystal structure, diameter 40–60 nm). The TiC-modified wire was prepared by mixing monolithic AA5183 wire and TiC nanopowders in a screw extrusion process as described in [14]. The screw extrusion product was further processed to welding wire quality by drawing and wire shaving at the Technische Universität Bergakademie Freiberg, Germany.

WAAM was performed with a Fronius TPS 440i Cold Metal Transfer (CMT) power source. CMT short circuits the arc by dipping the filler material into the weld pool, which provides better arc stability. CMT supplies less energy to the workpiece than conventional electric arc welding, which has been shown desirable for microstructure refinement and porosity reduction [5]. The input welding parameters for the two different WAAM thin walls are given in Table 2.

The CMT torch was mounted to a six degrees of freedom robot arm (ABB IRB2400). The robot arm was programmed by the software RobotStudio® to build single bead thin walls with dimensions as shown in Figure 1. The material was deposited along the *X*-axis in a zigzag pattern on an AA6082-T6 plate substrate with thickness 10 mm.

Mechanical properties were examined by isostatic tensile testing along two perpendicular directions (deposition direction *X* and through-thickness direction *Z*) in agreement with Figure 1. Vickers microhardness measurements with 100 grams force ($HV_{0.1}$) was performed in the frontal plane, i.e. the *YZ*-plane. Grain morphology and size was examined in cross-polarised optical microscopy on samples anodised in Baker's reagent for 180 s. The grain size was determined by the grain intercept method. The presence of TiC was revealed by energy dispersive X-ray spectroscopy (EDS) in field emission scanning electron microscopy (SEM).

Internal porosity was inspected by X-ray computed tomography (CT). The CT data was acquired by a Nikon XT H225 ST instrument. A tungsten reflection target was used, with an acceleration voltage of 135 kV and a current of 195 μA. The X-rays were not filtered. The imaging was done with an integration time of 0.708 second, amplification of 18 dB, with 3142 projections per 360°. The distance from source to sample was 56.18 mm, distance from source to detector was 1124.2 mm, resulting in a voxel size of 10 μm. The images were exported in 16-bit TIFF format and processed in the public domain software ImageJ using scripts developed at SINTEF.

3 Results and discussion

An equiaxed grain morphology was observed in all investigated WAAM materials as shown in Figure 2. The morphology was independent of the addition of TiC, but was rather governed by the reduced heat input from the CMT power source. A low heat input decreased the temperature gradient *G* in the liquid weld pool, which allowed nucleation in the constitutional undercooled zone ahead of the solidification front. This observation is in accordance with earlier studies on WAAM of AA5183 [13, 15].

In fact, a reduced grain size was obtained by the addition of TiC nanoparticles. While the monolithic AA5183

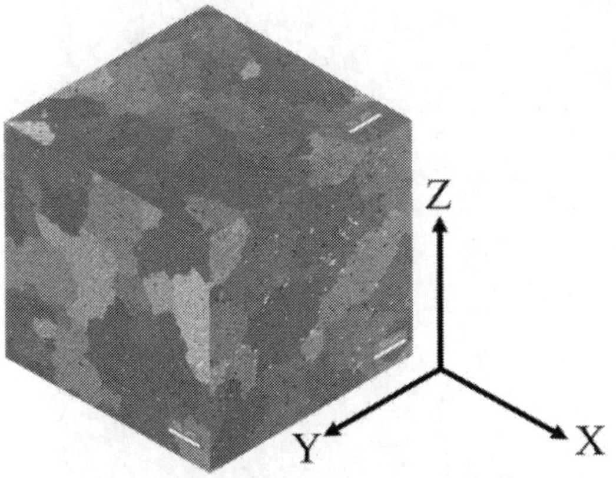

(a) AA5183

(b) AA5183-TiC

Figure 2: Microstructure in three perpendicular planes of the WAAM materials. Grain refinement was achieved in the AA5183-TiC parallel. Scale bars are 50 μm.

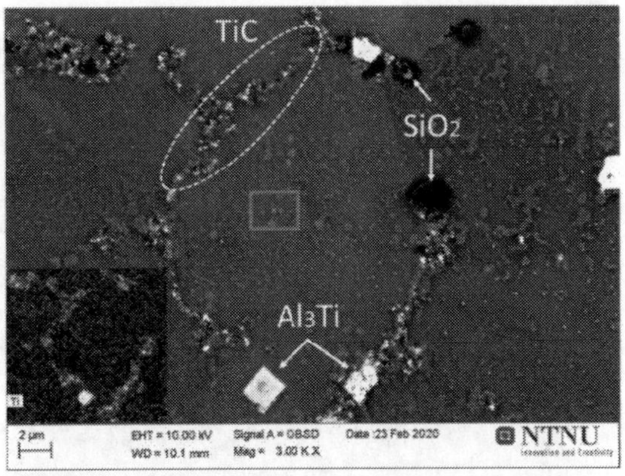

(a) Grain boundaries decorated with TiC. The red rectangle area is shown at higher magnification in b).

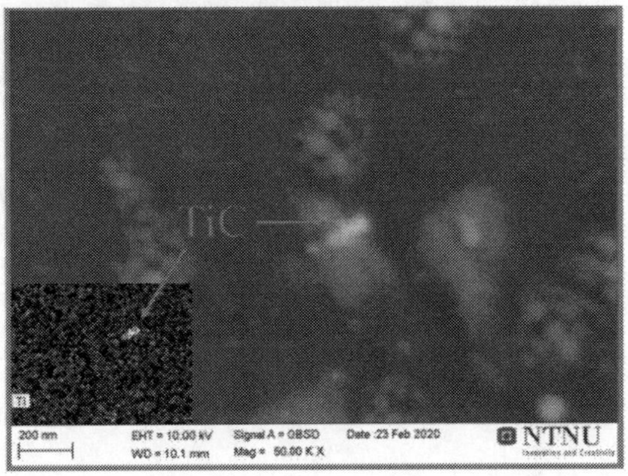

(b) Marked area from a). Possible grain nucleation site on TiC nanoparticle.

Figure 3: SEM micrographs of AA5183-TiC WAAM wall. The presence of titanium is proved by EDS (inserted images at left bottom).

WAAM structure exhibited an average grain diameter of 60±5 μm, the average grain diameter was reduced to 17±1 μm in the AA5183-TiC material. The grain refining effect can be related to heterogeneous nucleation on TiC particles upon solidification in WAAM. The constitutional supercooling required to activate a nuclei decreased with the presence of TiC nanoparticles, due to the low interface energy between aluminium and TiC [16]. The presence of a TiC cluster in the grain center in Figure 3 indicate that TiC may act as the heterogeneous nucleation site. The Al_3Ti intermetallic was also observed in the material. Several authors have proposed this constituent responsible for grain refinement of aluminium alloys [17]. The current study has not been able to determine the grain refining phase.

Two- and three-dimensional porosity maps of the WAAM thin walls are shown in Figure 4. All cavities were solely spherical in both samples, indicating the presence of entrapped gas. The effect of screw extrusion on the AA5183 material is shown in the lower part of Figure 4a). A mixture of micro- and macroporosity can be seen, which is believed to stem from oxidation and further hydroxation of magnesium during screw extrusion. The upper part of the WAAM material in Figure 4a) consists of screw extruded AA5183 with TiC. The porosity formation was extensive in this area. It is believed that incorporation of TiC accounted for the increased pore content. The commercial WAAM feedstock in Figure 4c) exhibited a repeated pore structure along the fusion line of every bead layer. Remelting and rapid solidification upon deposition of a new layer restrict hydrogen gas escape close to the fusion line, and account for the layered structure. A similar effect can also be seen in Figure 4b).

It is well-known that dissolved hydrogen in aluminium create porosity upon solidification, due to the large difference in hydrogen liquid and solid state solubility. Hydrogen can be introduced to WAAM materials in a number

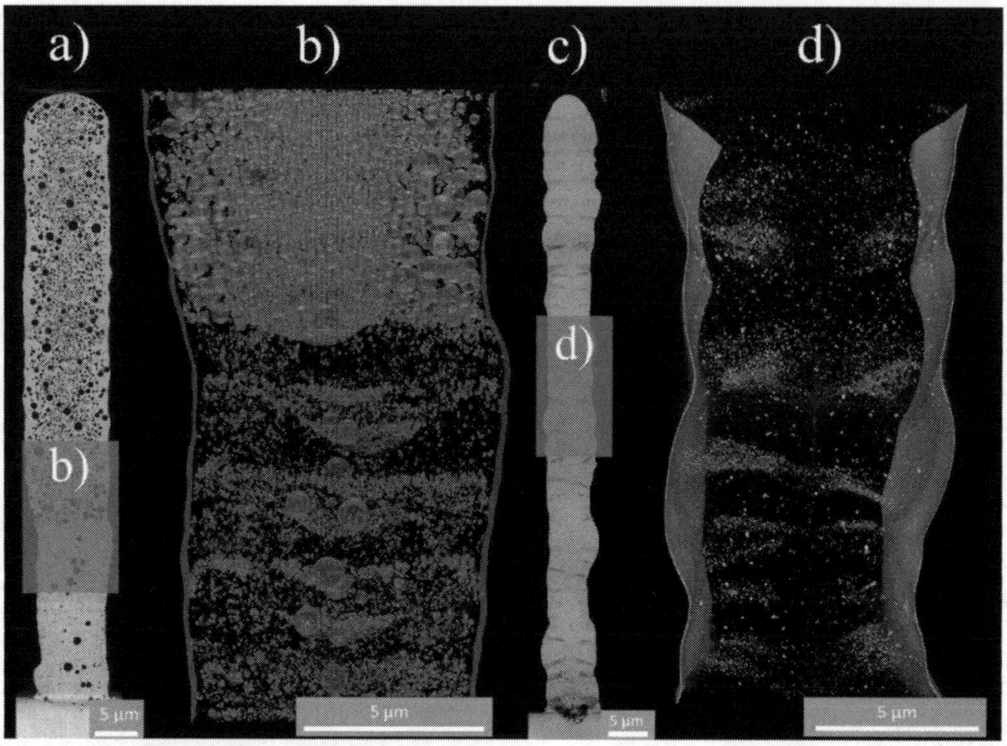

Figure 4: Computed tomograpghy scans of WAAM thin walls. a) 2D section of AA5183-TiC b) 3D reconstruction of AA5183-TiC section (50 layers) c) 2D section of AA5183 b) 3D reconstruction of AA5183 section (900 layers).

of ways. A wire feedstock with poor surface quality can cause arc instability during WAAM, which can introduce moisture from air into the weld pool. Moisture and grit on the wire surface itself is another common source of hydrogen porosity [4]. A higher heat input of the AA5183-TiC material increases the weld pool size and elevate hydrogen pick-up.

All pores detected by CT in a representative volume of the two WAAM materials are presented in Figure 5. Compared to the commercial AA5183 deposit, the AA5183-TiC material possessed a higher pore density. The increase in porosity can be explained by the contamination of TiC nanoparticles in air and the feedstock preparation through screw extrusion. Screw extrusion is a solid-state processing method where metal fragments (e.g. pellets and chips) are compacted together and extruded to a solid rod [18]. In fact, metal fragment surfaces can contain (hydr)oxides, which are broken up and baked into the aluminium matrix during screw extrusion. Thus, the screw extruded AA5183-TiC wire is believed to contain a higher hydrogen content than the commercial AA5183 wire. Measures to significantly reduce the hydrogen content is still under development.

Supersaturated hydrogen is unstable in the liquid weld during solidification in WAAM, and will precipitate and form bubbles. Hydrogen trapping and subsequent precipitation is facilitated by the presence of solid interfaces, like inclusions and particles [19]. With TiC nanoparticles evenly dispersed in the weld pool, porosity formation is enhanced and resulting in a pore structure as seen in Figure 4a-b).

Further, isostatic tensile results are shown in Figure 6. The tensile properties of the commercial AA5183 material were comparable to reported literature data, i.e. ultimate tensile strength 290–300 MPa [13, 20]. Anisotropy in strain between deposition direction X and through-thickness direction Z can be related to the layer-wise accumulation of cavities as shown in Figure 4d). The through-thickness direction Z probably experienced crack growth by pore coalescence as this direction crossed several pore layers. The serrations observed in the stress-strain curves are related to the Portevin-Le Chatelier effect commonly observed in Al-Mg alloys [21].

The strength and ductility of the AA5183-TiC WAAM material were inferior due to the high porosity level. A denser material would provide a rightful comparison to the AA5183 WAAM material. However, the strength can also be assessed through microhardness. An average of minimum 15 indents revealed that incorporation of TiC nanoparticles increased the microhardness by 13% (from 78±5 $HV_{0.1}$ to 88±5 $HV_{0.1}$). Such hardness (and corresponding strength) increase can be related to a number of strengthening mechanisms. Five of these contributions from TiC, i.e. from grain refinement (Hall-Petch), load

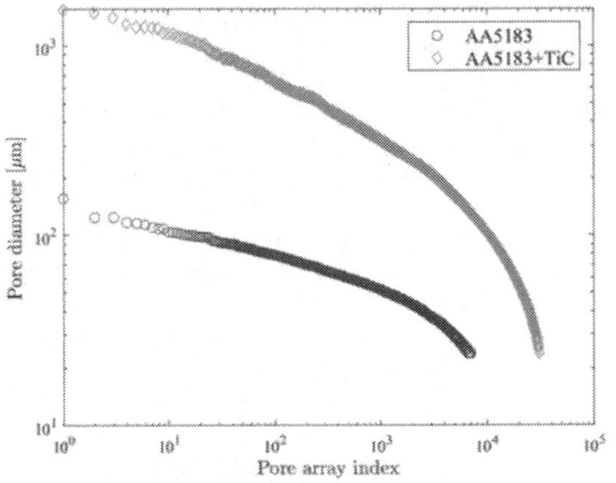

Figure 5: Pore count diagram with pore diameter sorted from largest to smallest in the WAAM materials. Measurement volume was 6×6×1 mm^3 and one index point represent one pore. The AA5183-TiC WAAM material has a higher pore density compared to the monolithic AA5183 WAAM material.

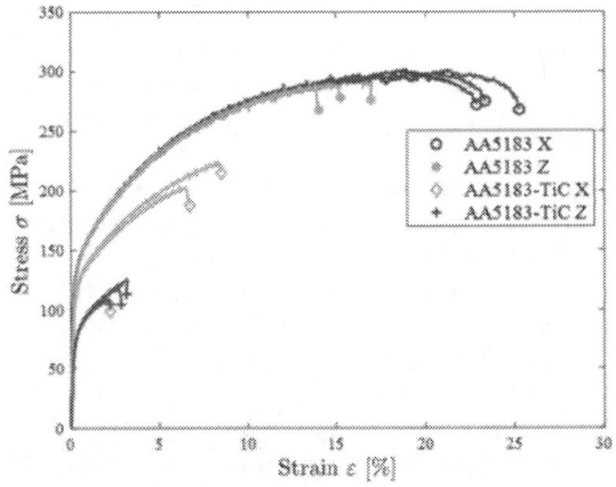

Figure 6: Isostatic tensile results for investigated WAAM materials. TiC-reinforced parallels exhibited lower strength and ductility due to a high porosity content.

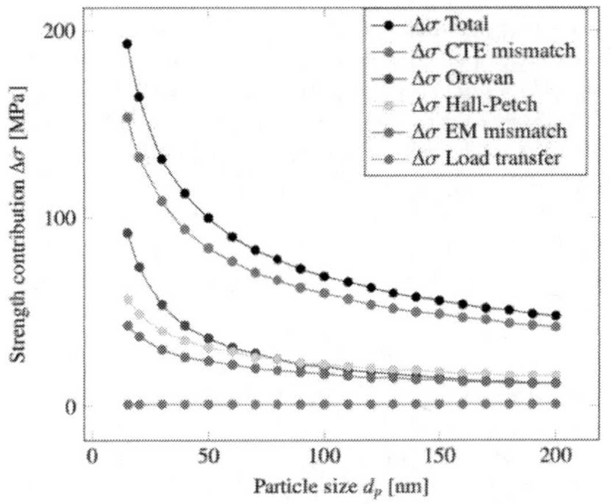

(a) Strength contribution of TiC nanoparticles with respect to particle size d_p. Volume fraction = 1 vol%.

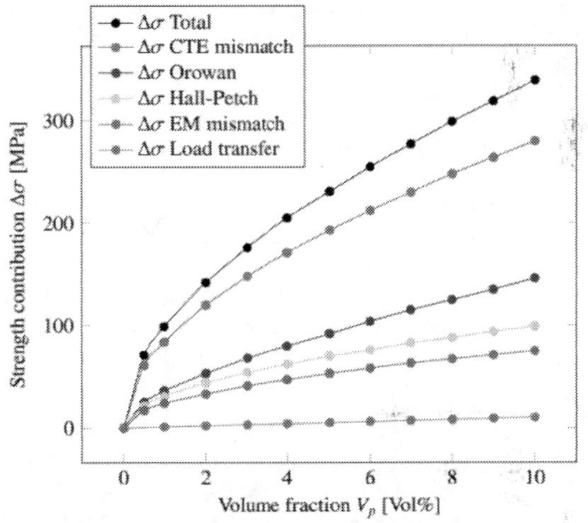

(b) Strength contribution of TiC nanoparticles with respect to volume fraction V_p. Particle diameter = 50 nm.

Figure 7: Theoretical enhancement of ultimate tensile strength ($\Delta\sigma$) from addition of TiC nanoparticles.

transfer, elastic modulus (EM) mismatch, coefficient of thermal expansion (CTE) mismatch and Orowan strengthening, are summarised in Figure 7. Related equations and input parameters can be found in the Appendix. Recall that 1 vol% of TiC was added to the nanoreinforced material. It is evident that for a low volume fraction of TiC (<1 vol%, Figure 7b), the strength and hardness contribution is governed by local strain fields from CTE mismatch.

Incorporation of potent grain refiners in the feedstock wire can expand the aluminium alloy selection for WAAM. Several alloys from, e.g., the AA6XXX Al-Mg-Si and AA7XXX Al-Zn(-Mg-Cu) systems are considered unsuitable for WAAM in monolithic state due to solidification cracking. The cracking mechanism arise due to an unfavourable dendritic microstructure and residual stresses from thermal contraction. By inducing a dendritic to equiaxed grain transition through addition of grain refiners, the residual stresses are distributed over a larger grain boundary area which prevent cracking. This study has worked as a proof-of-concept with a benchmark towards a commercial alloy. Future research will employ screw extruded material to produce WAAM feedstock of crack susceptible alloys from the AA6XXX and AA7XXX series.

Aluminium WAAM components with superior mechanical properties can be achieved by tailored nanoparticle additions in the near future. If high-strength aluminium alloys like AA7075 can be deposited by WAAM, components can reach a tensile strength of ~550 MPa after a post weld heat treatment [10]. In addition, the nanoparticles *themselves* can provide a further strength enhancement, Figure 7. The tensile strength contribution $\Delta\sigma$ increases with increasing particle volume fraction V_p and decreasing particle size d_p. Thus, keeping numerous of nanosized particles evenly dispersed in the wire feedstock and deposited WAAM material is of utmost importance. The authors propose that the solid-state processing of aluminium through screw extrusion is well-suited to reach this objective.

4 Conclusions

The applicability of nanoparticle reinforced feedstock wires for Wire and arc additive manufacturing (WAAM) has been explored. Deposited WAAM thin walls with a TiC-modified AA5183 Al-Mg alloy exhibited grain refinement compared to a commercial AA5183 benchmark. The hardness was also increased by 13% by the addition of TiC. Tensile properties were poor due to extensive pore formation in the TiC-modified alloy.

Acknowledgements

The authors would like to thank the Research Council of Norway for financial support through Contract No. 272402. Further, the authors acknowledge Mr. Morten H. Danielsen, Mr. Nikolai Marhaug, Mr. Pål C. Skaret and Mr. Stein Rørvik for their experimental support.

References

[1] S.W. Williams, F. Martina, A.C. Addison, J. Ding, G. Pardal, P. Colegrove, Mater. Sci. Technol. **32**, 641 (2016), https://doi.org/10.1179/1743284715Y.0000000073

[2] B. Wu, Z. Pan, D. Ding, D. Cuiuri, H. Li, J. Xu, J. Norrish, Journal of Manufacturing Processes **35**, 127 (2018), https://doi.org/10.1016/j.jmapro.2018.08.001

[3] H. Toda, H. Oogo, K. Uesugi, M. Kobayashi, Materials transactions **50**, 2285 (2009), https://doi.org/10.2320/matertrans.M2009123

[4] E.M. Ryan, T.J. Sabin, J.F. Watts, M.J. Whiting, Journal of Materials Processing Technology **262**, 577 (2018), https://doi.org/10.1016/j.jmatprotec.2018.07.030

[5] B. Cong, J. Ding, S. Williams, The International Journal of Advanced Manufacturing Technology **76**, 1593 (2015), https://doi.org/10.1007/s00170-014-6346-x

[6] J. Gu, J. Ding, S.W. Williams, H. Gu, P. Ma, Y. Zhai, Journal of Materials Processing Technology **230**, 26 (2016), https://doi.org/10.1016/j.jmatprotec.2015.11.006

[7] J.J. Lewandowski, M. Seifi, Annual review of materials research **46**, 151 (2016), https://doi.org/10.1146/annurev-matsci-070115-032024

[8] SAFRA, Safety data sheet, SAFRA, Brescia, Italy (2014), available at: http://www.safraspa.it/en/msds/. Last visited 19.02.2020

[9] G.J. Ram, T. Mitra, V. Shankar, S. Sundaresan, Journal of Materials Processing Technology **142**, 174 (2003), https://doi.org/10.1016/S0924-0136(03)00574-0

[10] M. Sokoluk, C. Cao, S. Pan, X. Li, Nature communications **10**, 98 (2019), https://doi.org/10.1038/s41467-018-07989-y

[11] A. Sales, N.J. Ricketts, in *Light Metals 2019* (Springer, 2019), pp. 1455–1461, https://doi.org/10.1007/978-3-030-05864-7_182

[12] J. Røyset, N. Ryum, International Materials Reviews **50**, 19 (2005), https://doi.org/10.1179/174328005X14311

[13] A. Horgar, H. Fostervoll, B. Nyhus, X. Ren, M. Eriksson, O. Akselsen, Journal of Materials Processing Technology **259**, 68 (2018), https://doi.org/10.1016/j.jmatprotec.2018.04.014

[14] O. Ragnvaldsen, Master's thesis, NTNU (2019), http://hdl.handle.net/11250/2624249

[15] B. Zhang, C. Wang, Z. Wang, L. Zhang, Q. Gao, Journal of Materials Processing Technology **267**, 167 (2019), https://doi.org/10.1016/j.jmatprotec.2018.12.011

[16] M.X. Zhang, P.M. Kelly, M.A. Easton, J.A. Taylor, Acta Materialia **53**, 1427 (2005), https://doi.org/10.1016/j.actamat.2004.11.037

[17] T. Quested, Materials Science and Technology **20**, 1357 (2004), https://doi.org/10.1179/026708304225022359

[18] K.G. Skorpen, E. Mauland, O. Reiso, H.J. Roven, Transactions of Nonferrous Metals Society of China **24**, 3886 (2014), https://doi.org/10.1016/S1003-6326(14)63547-0

[19] C. Tian, J. Law, J. van der Touw, M. Murray, J.Y. Yao, D. Graham, D.S. John, Journal of Materials Processing Technology **122**, 82 (2002), https://doi.org/10.1016/S0924-0136(01)01229-8

[20] X. Fang, L. Zhang, G. Chen, X. Dang, K. Huang, L. Wang, B. Lu, Materials **11**, 2075 (2018), https://doi.org/10.3390/ma11112075

[21] A. Benallal, T. Berstad, T. Børvik, O. Hopperstad, I. Koutiri, R.N. De Codes, International Journal of Plasticity **24**, 1916 (2008), https://doi.org/10.1016/j.ijplas.2008.03.008

Appendix

All equations to estimate the tensile strength contribution from TiC nanoparticles in Figure 7 are included in Equation 1–9. Input parameters are provided in Table 3. Please note that the grain size calculation in Equation 2 is based upon Zener pinning and not heterogeneous nucleation.

$$\Delta\sigma_{HP} = \frac{k_y}{d} \tag{1}$$

$$d = \frac{4\alpha d_p}{3V_p} \tag{2}$$

$$\Delta\sigma_{LT} = 0.5V_p\sigma_m \tag{3}$$

$$\Delta\sigma_{Orowan} = \frac{0.81MGb\ln(d_p/b)}{2\pi(1-\nu)^{0.5}d_p\left(\frac{1}{2}\sqrt{\frac{3\pi}{2V_p}}-1\right)} \tag{4}$$

$$\rho_{EM} = \frac{6V_p}{bd_p}\epsilon \tag{5}$$

$$\Delta\sigma_{EM} = Gb\lambda\sqrt{\rho_{EM}} \tag{6}$$

$$\rho_{CTE} = \frac{A\Delta C\Delta TV_p}{bd_p(1-V_p)} \tag{7}$$

$$\Delta\sigma_{CTE} = \kappa Gb\sqrt{\rho_{CTE}} \tag{8}$$

$$\Delta\sigma = \sqrt{\sum_i \Delta\sigma_i^2} \tag{9}$$

Table 3: Input parameters for Equations 1–9 to estimate strength contribution from TiC nanoparticles.

Parameter	Symbol	Unit	Al	TiC
Volume fraction particles	V_p	-		0.01 – 0.10
Particle diameter	d_p	nm		30– 200
Grain size	d	µm		
Yield strength Al	σ_m	MPa	200	
Hall Petch strengthening coefficient	k_y	MPa $mm^{-1/2}$	1.8	
Proportionality constant	α	nm	0.5	
Taylor factor	M	-	3	
Shear modulus	G	GPa	25.5	
Burgers' vector	b	nm	0.286	
Poissons' ratio	ν	-	0.33	
Geometrical factor	A	-	12	
Coefficient of thermal expansion	C	µm/m K	24	7.7
CTE-difference matrix/particle	ΔC	µm/m K		16.3
Temperature difference melt/test	ΔT	K	620	
Uniform deformation	ϵ	-	0.01	
CTE dislocation strengthening coefficient	κ	-	1.25	
EM dislocation strengthening coefficient	λ	-	0.5	
Strength contribution from mechanism i	$\Delta\sigma_i$	MPa		
Dislocation density	ρ_i	m^{-2}		

Lifetime assessment of the process-dependent material properties of additive manufactured AlSi10Mg under low-cycle fatigue loading

Carl Fischer[1,2,*] and *Christoph* Schweizer[1,**]

[1]Fraunhofer Institute for Mechanics of Materials IWM, Woehlerstrasse 11, 79108 Freiburg, Germany
[2]Offenburg University of Applied Sciences, Badstrasse 24, 77652 Offenburg, Germany

Abstract. Systematic low-cycle fatigue (LCF) experiments are carried out on additive manufactured AlSi10Mg specimens for several material conditions with varying layer thickness, heat treatment, building direction and surface quality. The deformation behaviour depends significantly on the heat treatment. It is outlined that the process control and heat treatment can produce fatigue properties comparable with the cast material, whereby an as-built specimen surface leads to a lifetime reduction in all cases. The experiments are accompanied with detailed metallo- and fractographic investigations. For all tested LCF specimens, the defect type and the failure origin defect size are characterized in terms of the $\sqrt{area}$ parameter by using scanning electron microscopy. The failure of the specimen is mostly caused by lack of fusion surface or near-surface defects, whereby the defect size is determined by the SLM process parameters, such as building direction, surface quality and layer thickness. On the basis of the experimental data and the observed defects, a mechanism-based, deterministic lifetime model is developed and adapted to the specific damage mechanisms of the additive manufactured AlSi10Mg alloy.

1 Introduction

The selective laser melting (SLM) process stands at the beginning of a major development for many medical and aerospace applications and also in many other sectors of industry [1]. Although some additive manufacturing systems are technologically similar, parameter sets for laser power, laser focus, scanning strategy and layer thickness are very different and can be freely chosen by the operator [2]. Especially for precipitation hardening materials such as the aluminium alloy AlSi10Mg, the mechanical material and component properties vary locally [1–3]. To design and merchandise highly stressed additive manufactured components, manufacturers must understand the relationships between the process and the resulting microstructure-dependent material properties [1].

The primary objective in the qualification of additive manufactured materials is to achieve the optimum in component density [4, 5]. For this purpose, a wide matrix of parameter sets are systematically studied to obtain the lowest-possible porosity [4, 6]. In order to characterize the mechanical properties, usually micro-hardness and tensile tests are carried out [6]. An overview over the tensile properties of additive manufactured aluminium alloys is given, e.g. in [1, 3, 6]. For dynamically loaded applications, high-cycle fatigue (HCF) tests are performed to determine the fatigue strength [7–10]. Current research is dealing with the material characterization for a wide spectrum of loading conditions, so that in addition to HCF, the low-cycle fatigue (LCF) properties of additive materials are investigated more intensively [11–13].

For additive manufactured aluminium materials it is reported in literature, that the fatigue properties are very sensitive to defects or the defect structure [3, 6, 7], the rough surface quality [6, 11] and residual stresses [6]. The manufacturing process yields preferentially to gas pores and lack of fusion inclusions, which have a major influence on the mechanical properties and therefore have to be avoided or at least reduced [1, 2]. In [3], the fatigue strength of additive manufactured and cast aluminium alloys is correlated with an initial defect size in terms of the $\sqrt{area}$ parameter from Murakami [14]. The parameter $\sqrt{area}$ itself denotes the size of an internal arbitrarily shaped crack [14]. The large material scatter in HCF results is assessed with $\sqrt{area}$ and extreme value statistics, where the defect population and defect size distribution of whole specimens are applied by computer tomography scan measurements in order to predict lifetimes and also the failure probability of additive manufactured AlSi10Mg [15, 16]. The $\sqrt{area}$ concept is also adapted in [9, 11] for the LCF lifetime prediction of additive manufactured AlSi10Mg by using elastic-plastic fracture mechanics based on the cyclic J-integral ΔJ. The identical fracture mechanics approach is applied for the fatigue life prediction of additive manufactured Ti-6Al-4V in [12] and 316L stainless steel in [12, 13].

In this work, systematic LCF tests are carried out for a selective laser melted (SLM) AlSi10Mg alloy. The aim of the investigations is to be able to understand and evaluate the connections between the process-dependent mi-

*e-mail: carl.fischer@iwm.fraunhofer.de
**e-mail: christoph.schweizer@iwm.fraunhofer.de

crostructure, the defect structure and the LCF lifetime properties. The defect type and the failure origin defect size are examined by means of $\sqrt{area}$ from high resolution scanning electron microscopy fracture surface images. With the obtained material parameters from the LCF tests and the initial defect sizes, the lifetimes are assessed using a mechanism-based deterministic lifetime model.

2 Investigated material conditions and experimental test set-up

An EOS M400 system at the Fraunhofer EMI institute was used for additive manufacturing of the AlSi10Mg aluminium samples. The system is equipped with a 1 kW YLR series CW laser with a wavelength of 1070 nm. The chemical composition of the EOS AlSi10Mg powder is given in [17]. The samples are manufactured under argon atmosphere with a maximum oxygen content of 1.2 % on a substrate plate, which was preheated to 165 °C. Five sets of process parameters given in table 1 are used to produce the specimens. The specimens with a layer thickness of 90 µm are produced in a 67° rotating stripe strategy using the parameter sets provided from the system manufacturer. For the reduced layer thickness of 30 µm the parameter set was developed by the Fraunhofer EMI and cannot be given in detail due to confidentiality reasons.

Table 1: Characterized material conditions.

Process	Layer thickness	Heat treatment
1	90 µm	none
2	90 µm	2 h at 300 °C
3	90 µm	Annealing: 1 h at 525 °C Quenching: 60 °C water Ageing: 6 h at 165 °C
4	30 µm	none
5	30 µm	2 h at 300 °C

From the first manufacturing process, the material with a powder layer thickness of 90 µm is characterised without additional heat treatment (referred to as material condition 90 µm). The material from the second manufacturing process (named as 90 µm + T5) is subjected to a stress relief heat treatment for two hours at 300 °C, which is intended to reduce anisotropies caused by the manufacturing process [17]. The stress relief heat treatment reduces the hardness from 121HV10 to 84HV10. The influence of the reduced layer thickness of 30 µm is examined for identical heat treatment conditions for the material conditions 30 µm and 30 µm + T5. Since the casting counterpart is usually exposed to a strength-enhancing T6 heat treatment [18], the influence of a T6 heat treatment on the additive manufactured AlSi10Mg alloy is investigated in this work (named as 90 µm + T6).

After the SLM process, mechanical properties similar to the T6 condition of cast components are reported in [17]. For this purpose, a T6 heat treatment is not recommended for laser sintering components and instead a stress relief heat treatment for two hours at 300 °C is suggested by the system manufacturer to reduce or eliminate feasible anisotropic properties due to the layer structure by suitable thermal post heat treatments [17]. However, in order to investigate the extent to which positive effects on the mechanical properties can be achieved by adjusting the material's microstructure, a batch of additive manufactured samples was subjected to an additional strength-enhancing T6 heat treatment. In literature, strongly varying periods for the solution heat treatment are specified for SLM AlSi10Mg alloys (cf. [7, 8, 19–22]). Therefore, the influence of the solution heat treatment time on the microstructure was investigated first. Figure 1 shows the evolution of the microstructure after solution heat treatment compared to the initial state. In the initial state, the eutectic silicon (Si) regions are existent in fine, coherent cell structures. By applying a solution heat treatment, the silicon phase first becomes spheroidal in shape and then coarsens with increasing time, which is also observed in [22]. Furthermore, the needle-like ferrous β-Al_5FeSi phase is formed [20–24]. According to [21, 22], the solution heat treatment also forms beneficial Mg_2Si precipitations, which are largely responsible for the increase in strength during the T6 heat treatment. On the basis of the preliminary investigations, a short solution heat treatment time of one hour was chosen, as a fine microstructure with spherical phases was achieved. In addition, the formation of the detrimental β-Al_5FeSi is impeded, which could lead to high stress concentrations due its platelet-like shape [25] and is thus prone to fracture [26] as well as crack nucleation and propagation reducing the fatigue lifetime [25, 27].

Figure 2 shows the microstructure of the additive manufactured AlSi10Mg alloy from the first manufacturing process in the initial 90 µm material condition. The metallographic section is prepared with Murakami etching for 15 s in order to visualise the individual layers and the heat-affected zones. The microstructure contains isolated gas pores as well as some lack of fusion areas due to incompletely melted powder during the manufacturing process.

In order to investigate the influence of the building direction, the specimens are manufactured in three orientations of 0° (specimen axis parallel to manufacturing layer), 45° (specimen axis diagonal to manufacturing layer) and 90° (specimen axis perpendicular to manufacturing layer). Due to rapid cooling rates during the manufacturing process, residual stresses are introduced in the material, so that especially large-volume, horizontally oriented specimens are prone to premature crack formation [2, 28]. In order to minimize this problem, additional support structures acting as thermal bridges for improved heat dissipation and also short specimen geometries are applied [28]. 80 mm long, smooth cylindrical specimens with 9 mm gauge length and a diameter of 8 mm were sampled and polished from the raw material (see figure 3a). Some of the 45° oriented specimens were also manufactured with as-built surface (see figure 3b).

Strain-controlled LCF tests were performed to experimentally characterize the process-dependent mechanical properties of the additive manufactured aluminium alloy. The LCF specimens were tested in ambient air with a

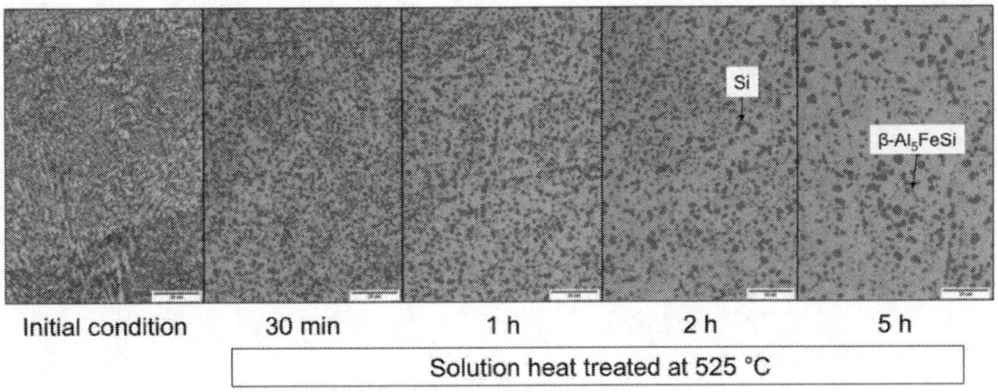

Figure 1: Evolution of the microstructure with solution heat treatment time at 525 °C compared to the initial condition (Murakami etched light microscopy images).

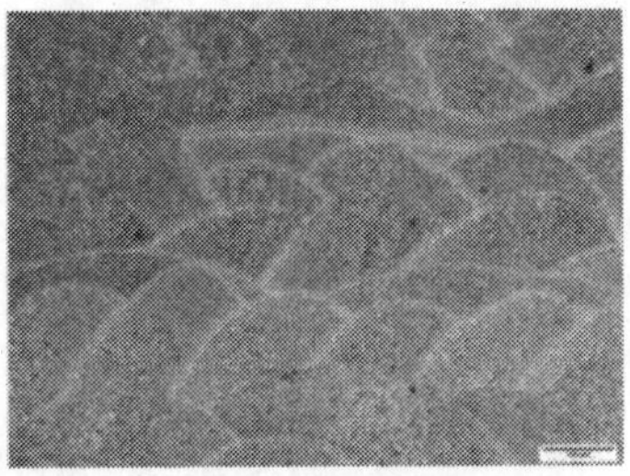

Figure 2: Light microscopy image of the 90 µm SLM AlSi10Mg microstructure after 15 s Murakami etching.

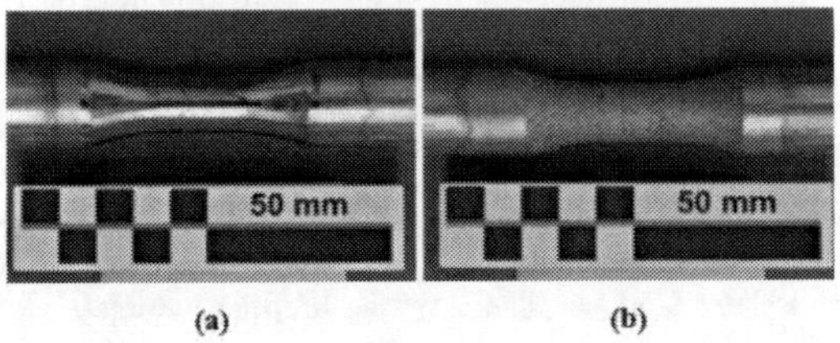

Figure 3: (a) Polished and (b) as-built surface specimen conditions (tested specimens).

100 kN electromechanical Instron 8862 universal testing machine. A capacitive extensometer with a calibrated gauge length of 8 mm from Epsilon Technology Corp is employed for measuring and controlling the axial strain. The temperature is measured with a type K ribbon thermocouple placed in the middle of the gauge length. A LabView software application is used to enable the test conditions and for data collection. All LCF tests are carried out in a triangular cycle with a constant strain rate of $d\varepsilon/dt = 10^{-3}$ 1/s and fully reversed loading ($R_\varepsilon = -1$) and the mechanical strain amplitudes are chosen to ε_a^{mech} = 0.3 %, 0.4 % and 0.5 %.

3 Results of the low-cycle fatigue tests

3.1 Deformation behaviour at half lifetime

The influence of the building direction, layer thickness and heat treatment on the deformation behaviour is explained in Figure 4 by means of stabilized stress-strain hysteresis loops at half lifetime $N_A/2$ for a mechanical strain amplitude of ε_a^{mech} = 0.5 %. The hysteresis loops of the 90 µm material condition in figure 4a show a slight tension-compression asymmetry, which is most pronounced for the building direction of 90° and the 45° oriented specimen in as-built condition. The stress response as well as the elastic material stiffness of the 45° as-built specimen are significantly lower than for the machined and polished specimens. The stress relief heat treatment diminishes the observed stress asymmetry and leads to a deformation behaviour which is independent of the building direction and the surface condition for the 90 µm + T5 material (see figure 4b). However, the heat treatment significantly reduces the overall mechanical strength, whereby the cyclic hardening behaviour is also decreased.

Although the T6 heat treatment has a strong influence on the microstructure morphology of the investigated SLM AlSi10Mg alloy (see figure 1), the heat treatment shows a negligible effect on the hysteresis loops in comparison to the 90 µm condition as depicted in figure 4c. Consequently, the 90 µm + T6 material condition is comparable to the initial 90 µm material condition, which received no further heat treatment after manufacturing. It has to be mentioned, that this conclusion only holds for small strains, since tensile tests on the 90 µm + T6 condition revealed a decrease in hardening for T6 compared to the initial 90 µm condition. The cyclic hardening behaviour as well as the ductility are gently reduced by the peak enhancing T6 heat treatment. Unfortunately, no specimens of the T6 material condition were available with as-built surface.

Figure 4d illustrates for the 30 µm material condition, that a reduced powder layer thickness does not affect the hysteresis loops considerably (compare with figure 4a).

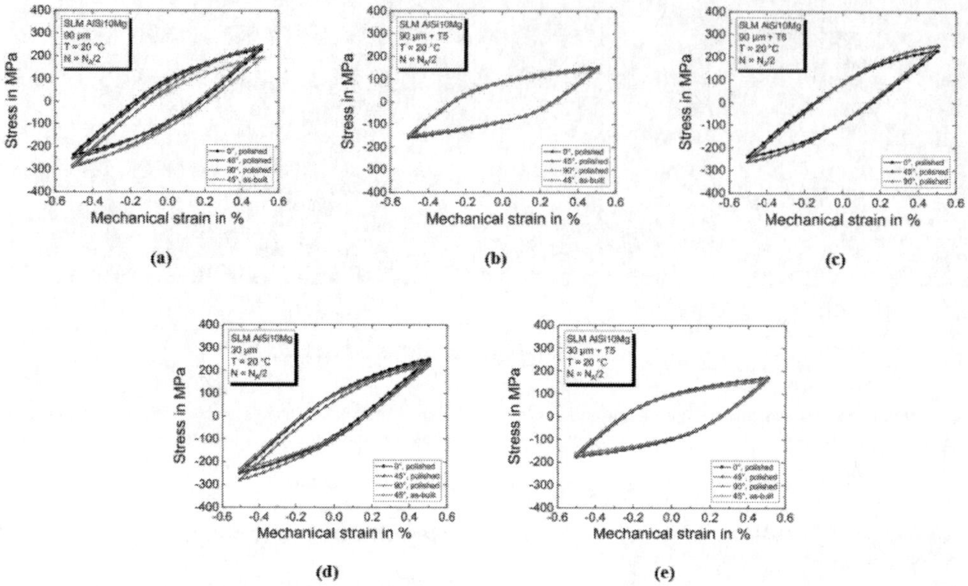

Figure 4: Deformation behaviour of half life hysteresis loops for a mechanical strain amplitude of ε_a^{mech} = 0.5 % for the investigated material conditions: (a) 90 µm, (b) 90 µm + T5, (c) 90 µm + T6, (d) 30 µm and (e) 30 µm + T5.

The tension-compression asymmetry is also apparent and most pronounced for the 90° oriented specimen. Moreover, the resulting stresses are somewhat reduced for the as-built specimen surface. The reduced layer thickness of 30 µm in combination with the stress relief heat treatment (see figure 4e) has hardly any effect on the mechanical strength in comparison to the 90 µm + T5 material condition.

3.2 Lifetime behaviour

In addition to the results shown in figure 4, further LCF tests with mechanical strain amplitudes of ε_a^{mech} = 0.3 % and 0.4 % are performed. The lifetimes of all LCF tests for the different material conditions are presented in figure 5. If possible, the lifetimes are determined by a 5 % drop of the maximum stress as recommended in [29]. For classification and comparability reasons, the obtained test results are plotted together with LCF lifetimes from [11], which correspond to a not further heat-treated SLM AlSi10Mg alloy with two different layer thicknesses and two building directions (horizontal and vertical), which were manufactured in three processes (process 1 to process 3). For all considered material conditions, independent of heat treatment and the used powder layer thickness, there is no markable influence of the building direction on the lifetime. The 90 µm condition in figure 5a exhibits the shortest lifetimes. Compared to the literature data from [11], the lifetimes are found at the lower scatter band. The as-built surface condition reduces the lifetime significantly, whereby this trend seems to be more pronounced at lower mechanical strains. Although the stress relief heat treatment reduces the strength of the material (cf. figure 4b) noticeably, the lifetime behaviour in figure 5b is enhanced in comparison to the initial material condition. The detrimental influence on the lifetime due to as-built surface conditions is also significantly attenuated by the subsequent stress relief heat treatment. The T6 heat treatment does not have a significant effect on the lifetime (cf. figures 5a and 5c). Please note, that no as-built specimens were available for this material condition. The reduced powder layer thickness of 30 µm has a positive effect on the lifetime (see figures 5d and 5e), especially at high mechanical strains. The as-built surface condition has a detrimental influence on the lifetime, which is also reduced by the heat treatment. The obtained results correspond very well with the experimental tests of process 3 from [11], where the layer thickness is also 30 µm. The longest lifetimes are achieved with a powder layer thickness of 30 µm and subsequent stress relief heat treatment (see figure 5e).

In figure 5f, the SLM lifetimes of the polished specimens are compared to similar cast aluminium alloys from the literature [18, 30–35]. The scatter of the cast aluminium materials is very pronounced. The lifetimes of the SLM AlSi10Mg alloy are found at the upper scatter band of the cast aluminium alloys for the mechanical strain amplitudes of ε_a^{mech} = 0.4 % and 0.5 %. At a mechanical strain amplitude of ε_a^{mech} = 0.3 %, the lifetimes tend to be shorter, since the level of the mechanical load has a more pronounced influence on the lifetime behaviour of the SLM alloy than for the cast aluminium materials. A T5 heat treatment results in the shortest lifetimes for

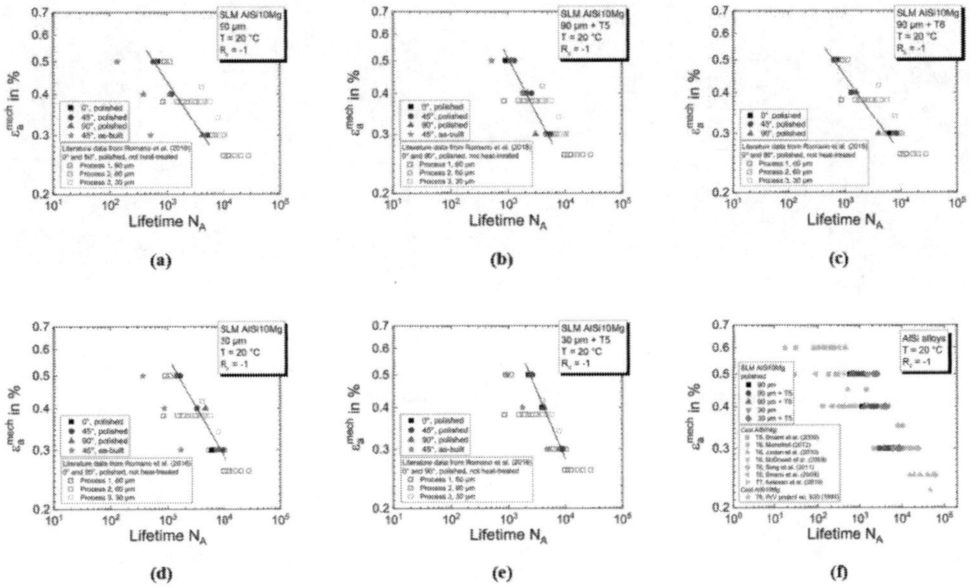

Figure 5: Lifetime behaviour of the investigated material conditions in comparison to LCF tests from [11]: (a) 90 µm, (b) 90 µm and T5, (c) 90 µm and T6, (d) 30 µm and (e) 30 µm and T5. (f) Comparison of SLM lifetimes of polished specimens with literature data from [18, 30–35] for similar cast aluminium alloys.

the AlSi7Mg cast material. The lifetime behaviour is enhanced with a T6 heat treatment, whereby further artificially over ageing to T7 has no significant impact. In contrast to the observations for the cast materials, a T5 heat treatment of the SLM AlSi10Mg alloy leads to prolonged lifetimes. The T6 heat treatment does not improve the lifetime behaviour in case of the additive manufactured AlSi10Mg alloy. To sum up, the lifetimes of the SLM AlSi10Mg alloy are comparable to those of the cast aluminium alloys.

4 Defect structure analysis

By means of high-resolution scanning electron microscopy images of the fracture surfaces, the defect structure and the failure origin defect size of all LCF specimens is characterised in terms of the $\sqrt{area}$ parameter from Murakami [14]. Figure 6 shows exemplary defect sizes and defect types at the failure origin. The failure of the specimens is caused by single lack of fusion surface or near-surface defects. Although all specimens contained gas porosities, gas pores were not found to act as detrimental crack initiation sites. Only three specimens failed due to internal defects, which were also induced by lack of fusion. In some cases, multiple defects or defect clusters are observed. Especially the as-built surface specimens are very sensitive to multiple crack initiation. A considerable number of defects are found at surface and the existing roughness leads to complex surface profiles and exposed defect locations, which are very favourable for crack initiation.

The defect size is determined using the image processing software Fiji [36]. Only the defect size that led to the failure of the specimen is evaluated. However, one or more secondary cracks are present on almost all fracture surfaces. The defect size distribution by means of $\sqrt{area}$ is presented in figure 7 for different building directions and surface conditions. The building direction of the samples has a noticeable influence on the defect size, as the horizontal (0°) and vertical (90°) oriented specimens tend to have larger initial defects. The defect size is generally reduced by the lower layer thickness of 30 µm. For the horizontally oriented specimens a large scatter in defect size is observed. The smallest defects are found for the 45° specimens independent of the powder layer thickness. The overall defect size is increased again for the 90° specimens, whereas two outliers are found for the 90 µm + T5 condition. Also one outlier is observed for the 30 µm material condition, which is caused by a large internal defect. The defect size is significantly larger for the as-built specimens, whereby a smaller layer thickness also has a positive effect on the defect size.

5 Mechanism-based deterministic lifetime modelling

In this section, the obtained lifetimes from section 3.2 are assessed by using a mechanism-based lifetime model. In [37], two cast aluminium alloys used for cylinder heads and pistons in combustion engines were extensively characterised and the fracture mechanics based

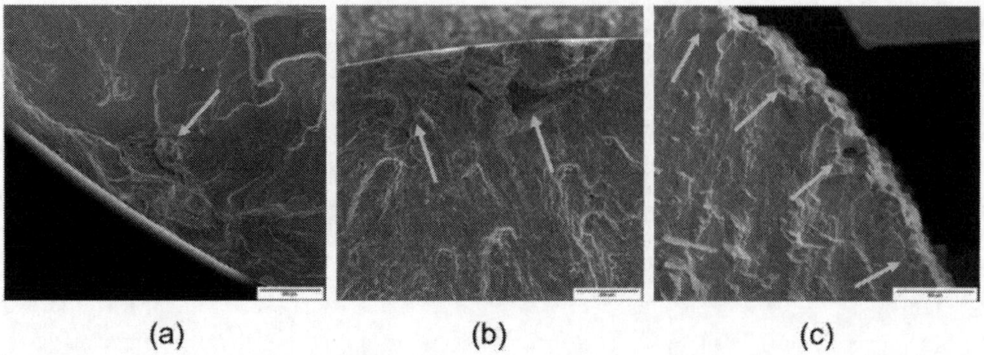

Figure 6: Examples for defect sizes and defect types at failure origin: (a) single lack of fusion surface defect (90 µm specimen), (b) lack of fusion cluster (90 µm specimen) and (c) typical multiple crack initiation at as-built surfaces (30 µm specimen).

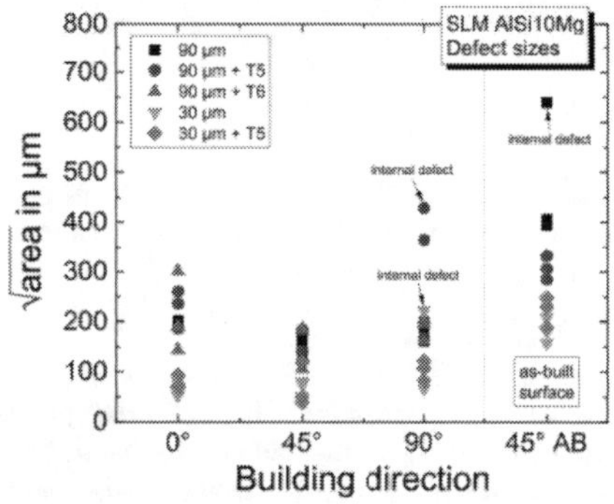

Figure 7: Defect sizes in dependence of building direction, layer thickness and surface condition.

damage parameter $D_{TMF,brittle}$ was developed, which accounts the temperature-dependent influence of the mechanical strength (i.e. cyclic yield stress σ_{CY}) on the damage mechanism more accurately:

$$D_{TMF,brittle} = d_{n'} Z_D \sigma_{CY}^{q} F(t, \sigma, T). \quad (1)$$

The pre-factor $d_{n'}$ depends on the hardening behaviour of the material by means of the cyclic Ramberg-Osgood [38] hardening exponent n′. Both for plane strain and plane stress states, $d_{n'}$ values are tabulated in [39], whereby a third order polynomial function for plane strain is published in [40]. Furthermore in equation (1), Z_D denotes the damage parameter from [41], σ_{CY} the cyclic yield stress at a plastic strain offset of 0.2 % and setting the exponent to $q = 0.2$ has turned out to be a good choice for cast aluminium materials. The function $F(t, \sigma, T)$ (see e.g. in [42]) accounts for time-dependent effects and arbitrary stress-temperature histories. Since this work only considers room temperature LCF tests, $F = 1$ holds for all tests. The damage parameter $D_{TMF,brittle}$ is based on the D_{TMF} parameter from [43], which takes the form:

$$D_{TMF} = \frac{Z_D}{\sigma_{CY}} F(t, \sigma, T). \quad (2)$$

D_{TMF} is modified accounting for the brittle fracture of hard inclusions (e.g. primary silicon particles or hard intermetallic phases) in a ductile matrix, which promotes sudden crack growth e.g. in aluminium-silicon alloys. Since brittle fracture is usually a stress-controlled process, the cyclic yield stress σ_{CY} appears in the numerator of $D_{TMF,brittle}$.

A power-law correlation between the crack growth increment per cycle da/dN and the damage parameter $D_{TMF,brittle}$ is assumed:

$$\frac{da}{dN} = \beta (D_{TMF,brittle} a)^{B}. \quad (3)$$

In $D_{TMF,brittle}a$ from equation (3), $Z_D a$ provides an estimate of the effective cyclic J-integral ΔJ_{eff} for semicircular surface cracks:

$$\Delta J_{eff} = Z_D a = \left(1.45 \frac{\Delta\sigma_{eff}^2}{E} + 2.4 \frac{\Delta\sigma \Delta\varepsilon^{pl}}{\sqrt{1 + 3n'}}\right) a. \quad (4)$$

The stress range $\Delta\sigma$, plastic strain range $\Delta\varepsilon^{pl}$ and the cyclic hardening Ramberg-Osgood exponent n′ are usually determined from stabilized hysteresis loops at half lifetime. For the effective stress range follows $\Delta\sigma_{eff} = \sigma_{max} - \sigma_{op}$ with the maximum stress σ_{max} and the crack opening stress σ_{op}. Crack closure effects are taken into account by means of the analytical system of equations published in [44]. In order to assess internal defects, the solution for a penny-shaped crack in an infinite medium under mode I loading with the stress intensity factor given e.g. in [45] and the plastic solution from [46] is applied to cyclic loading:

$$\Delta J_{eff} = \left(\frac{4}{\pi} \frac{\Delta\sigma_{eff}^2 (1 - \nu^2)}{E} + \frac{6}{\pi} \frac{\Delta\sigma \Delta\varepsilon^{pl}}{\sqrt{1 + 3n'}}\right) a. \quad (5)$$

According to [47], equation (4) is obtained from equation (5) by the multiplication of the pre-factors $4/\pi$ and

https://doi.org/10.1051/matecconf/202032607003

$6/\pi$ with a squared correction factor of 1.12^2 and inserting $\nu = 0.3$ for the Poisson's ratio.

In LCF tests, the lifetime N_A is determined by crack initiation, subsequent short crack growth and the formation of a technical, ca. 1 mm long macroscopic crack. Especially in the LCF regime with high strain amplitudes, the crack initiation phase has no significant influence on the lifetime [37, 48–50]. The lifetime N_A is integrated from equation (3) by separation of variables from an initial crack length a_0 to a final crack length a_f. Assuming that $D_{TMF,brittle}$ does not change significantly over the number of cycles one obtains:

$$N_A = \frac{A}{D^B_{TMF,brittle}}, \tag{6}$$

where

$$A = \ln\left(\frac{a_f}{a_0}\right) \quad \text{for } B = 1, \tag{7}$$

and:

$$A = \frac{a_f^{1-B} - a_0^{1-B}}{(1-B)\beta} \quad \text{for } B > 1. \tag{8}$$

In order to assess the lifetimes in figure 5, the material properties E, σ_{CY} and n′ are determined first from the stabilised hysteresis loops at half lifetime from figure 4. The pre-factor $d_{n'}$ is calculated by applying the third order polynomial function given in [40]. Surface or near-surface defects are considered as semicircular surface cracks, therefore equation 4 is used for the calculation of Z_D. For specimens where an internal defect resulted in failure, a penny-shaped crack geometry is assumed and equation 5 is applied.

The mechanism-based lifetime assessment in terms of the $D_{TMF,brittle}$ parameter is presented in figure 8. The impact of the initial defect size on the lifetime is indicated by the colour scale, where larger initial defects mostly lead to reduced lifetimes. This trend is especially evident in the case of as-built and not heat-treated 90 µm specimens, where initial defect sizes larger than 600 µm occur. In addition, the influence of varying sizes for $\sqrt{area}$ on the predicted lifetime curves is shown in figure 8. To this end, $\sqrt{area}$ is converted either to a semicircular surface $a_0 = \sqrt{area}\sqrt{2/\pi}$ or to a penny-shaped initial crack length $a_0 = \sqrt{area}/\sqrt{\pi}$. The exponent B = 1.34, i.e. the slope of the lifetime model, is fitted to the experiments with the smallest initial defect size (i.e. polished 30 µm specimens), because a minimum lifetime scatter is assumed and logically the slope of the lifetime curve should not be affected much by outliers. Afterwards, the factor β is fitted with the defect sizes of the polished 30 µm specimens from figure 7 according to equation (8) with the previous determined factor B and a final crack length of $a_f = a_0 + 1$ mm. With the obtained result for $\beta = 2.4 \times 10^{-4} m^{1-B}/(\text{cycle} \cdot MPa^{1.2B})$, equation (8) is applied with the exemplary $\sqrt{area}$ sizes shown in figure 8. Most of the LCF tests with defect sizes of $\sqrt{area}$ between 50 µm and 400 µm are well described with the corresponding curves. The lifetimes show a strong dependency on $\sqrt{area}$. However, lifetimes of the as-built specimens are overestimated despite of the considered initial defect size.

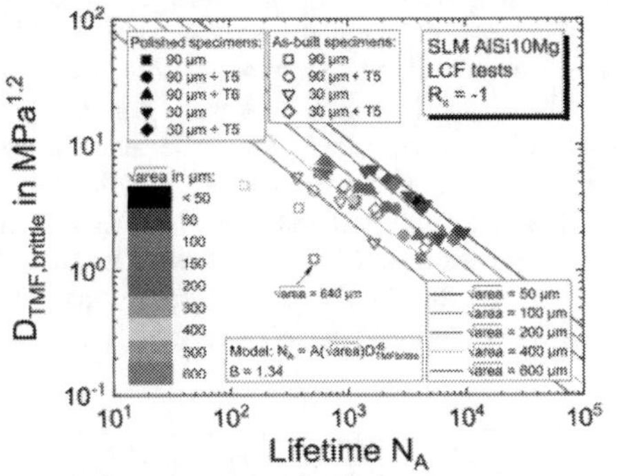

Figure 8: Lifetime assessment by means of $D_{TMF,brittle}$ and influence of $\sqrt{area}$ on the lifetime curves.

The deterministic lifetime prediction is performed in figure 9 considering the initial defect size according to equation (8) in the pre-factor $A(\sqrt{area})$ with the previous determined parameters B, β and a_f. A probability density function is adjusted to the difference between the logarithmic experimental and predicted lifetimes. The given scatter band factor is calculated with the standard deviation σ to $10^{2\sigma}$ and corresponds to a 95 % confidence interval. By

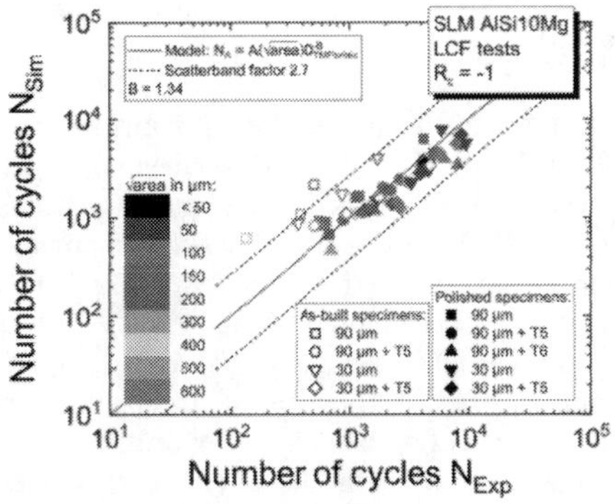

Figure 9: Deterministic lifetime prediction by means of $D_{TMF,brittle}$ and the initial defect size $\sqrt{area}$.

considering the individual initial defect size $\sqrt{area}$ of each specimen in the mechanism-based, deterministic lifetime model, the determined scatter band factor is reduced from 4.4 to 2.7. However, the predicted lifetimes of the non-heat treated and as-built specimens are still overestimated. Thus, the effective defect size $\sqrt{area_{eff}}$ [14] is taken into account, which allows an estimate for the effect defect size of irregular shaped defects by considering a smooth enveloping contour, for the interaction between a near-

https://doi.org/10.1051/matecconf/202032607003

surface defect and the specimen's surface or for the interaction effect between two adjacent defects. The lifetime prediction in terms of the effective defect size $\sqrt{\text{area}_{\text{eff}}}$ is presented in figure 10. Please note, that the identical

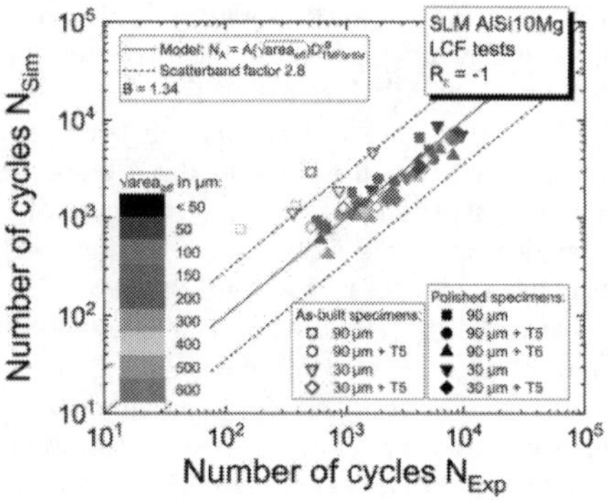

Figure 10: Deterministic lifetime prediction by means of $D_{\text{TMF,brittle}}$ and the initial effective defect size $\sqrt{\text{area}_{\text{eff}}}$.

colour scale from figure 9 is used and due to larger effective defect sizes the symbol colour might change. The expectation to obtain an enhanced description of the experimental results is not fulfilled. The achieved scatter band by a factor of 2.8 is even slightly larger than in figure 9. Especially the as-built specimens are found more closely at the upper scatter band or even further outside. The scatter band factor is slightly increased also for the other experiments.

6 Discussion

The cyclic hysteresis loops in figure 4 demonstrate, that the mechanical strength is slightly dependent on the building direction and show a tension-compression asymmetry in case of the 90° specimens from the material conditions 90 µm and 30 µm (see figures 4a and 4d). In [11], the tension-compression asymmetry in the 90° specimens is attributed to the manufacturing process. This is supported by the results of this work, since the asymmetry visible for the 90° specimens vanishes by introducing a stress relief heat treatment for 2 h at 300 °C (compare figures 4a with 4b and 4d with 4e). If this has anything to do with residual stresses, which usually are known to vanish during LCF loading, remains an open issue. By using finite element crystal plasticity simulations and considering the orientation dependent texture of the same AlSi10Mg alloy, the anisotropy in the elastic stiffness and the yielding behaviour could be explained for tensile tests [51], which show the same tendency as the hysteresis loops in figure 4.

The pronounced lifetime reduction in the cyclic LCF tests for the as-built specimens is attributed to the rough surface quality and increased defect density close to the surface. Defects with a $\sqrt{\text{area}}$ size of more than 600 µm were detected. The proposed fracture mechanics model using $D_{\text{TMF,brittle}}$ and $\sqrt{\text{area}}$ does not yet account for the detrimental surface effect even if an effective area size is considered (see figures 9 and 10). In [1], the component performance is attributed to surface roughness. The surface roughness is affected by so-called 'staircase effects' [52, 53], whereby the curvature radius of the specimen is approximated stepwise layer by layer. The surface roughness is decreased by a reduced layer thickness [54]. A second effect is addressed to the inhomogeneous melting of individual powder particles or balling networks [1]. As observed in figure 6, the fracture surfaces indicate clearly, that a rough and rugged surface preferably leads to multiple crack initiation and consequently crack coalescence, which can potentially accelerate the damage evolution. According to [8], the lifetime is also decreased in the HCF range for as-built specimens in comparison to machined and heat-treated specimens. An enhanced lifetime behaviour is obtained by a subsequent heat treatment for both layer thicknesses of 30 µm and 90 µm (see figure 5), which is in accordance with the findings in [8]. A smaller layer thickness also leads to increased LCF lifetimes, which is confirmed in [11] for LCF tests and in [16] for HCF tests.

The LCF lifetimes of various additively manufactured materials are assessed in [9, 11–13] with the effective cyclic J-integral ΔJ_{eff} expression from Dowling [55]. The approaches with similar formulations for ΔJ_{eff} have also demonstrated to be successful for lifetime assessment. However, it is not expected that the Dowling formulation with the modifications for internal and surface cracks would automatically lead to a better lifetime prediction for as-built surface specimens, since multiple crack initiation is not taken into account.

The last point, which is discussed, is the final crack length. In the lifetime predictions, the final crack length which determines the end of the fatigue life is chosen to $a_f = a_0 + 1$ mm, which is longer than a 1 mm technical fatigue crack. Since fatigue lives were evaluated according to a load drop criterion, a certain crack growth increment should be required to influence the maximum stress evolution. As in some cases very large initial defect sizes occur, the chosen formulation seems to be appropriate. In [9] the fatigue crack length is set to 3 mm for the lifetime predictions, as fatigue crack lengths between 2 mm and 5 mm were measured on the fracture surfaces. Because the crack growth is quite accelerated at these crack lengths (see e.g. in [11, 37]) only a negligible number of cycles are accumulated. It is mentioned, that the final crack length could also be further limited by the fracture toughness, which was not considered for reasons of simplicity.

7 Conclusions

Low-cycle fatigue tests on additive manufactured AlSi10Mg with different layer thicknesses, heat treatments and surface conditions were performed at room temperature and accompanied by detailed microstructure investigations. The introduction of a solution heat treatment (e.g. within a T6 heat treatment) coarsens the eutectic regions, which form fine, coherent cell structures in the initial state. The mechanical properties and the deformation behaviour of the characterized aluminium alloy AlSi10Mg is strongly affected by a subsequent heat treatment. A T5 stress relief heat treatment reduces the mechanical strength significantly. An as-built specimen surface in the initial material condition leads to slightly lower mechanical strength and considerably reduced lifetimes. Subsequent machining and polishing of the specimens is therefore recommended. A reduced powder layer thickness enhances strength and also the lifetime behaviour. The negative surface influence is significantly reduced by a heat treatment. The achieved lifetimes are in good accordance with available SLM and also comparable cast aluminium data from the literature.

High-resolution scanning electron microscopy images were taken from the fracture surface. The defect structure was analysed by means of the $\sqrt{\text{area}}$ parameter. Only the defects, that were responsible for the specimen failure were evaluated. In nearly all cases, manufacturing-related defects such as incompletely melted powder (lack of fusion) were found. Apart from three internal defects, surface or near-surface defects act as crack initiation sites. Especially for the as-built specimens, the fatigue crack is found to initiate at defect clusters. The defect size depends on the building direction, layer thickness and also the surface quality. Compared to the polished specimens, significantly more and larger defects are observed in the as-built specimens, which are prone for multiple crack initiation. For machined and polished specimens, the smallest defects are found for the 45° orientation. The reduction of the powder layer thickness from 90 μm to 30 μm also reduces the defect size.

The lifetimes are assessed by means of a mechanism-based lifetime model and the $D_{TMF,brittle}$ parameter. The lifetimes show a strong dependence on the initial defect size, since larger defects usually lead to shorter lifetimes. Consequently, the initial defect size is taken into account in a deterministic lifetime model, which allows a significantly improved lifetime prediction. However, the negative surface influence cannot be explained solely by the initial defect size. Therefore, further mechanisms such as multiple crack initiation and the crack coalescence have to be considered in the lifetime model.

Acknowledgements

The authors greatly acknowledge the financial support of the Fraunhofer-Gesellschaft. Special thanks go to Aron Pfaff, Klaus Hoschke and Martin Jaecklein from the Fraunhofer Ernst-Mach-Institut EMI for manufacturing and preparation of the specimens.

References

[1] T. DebRoy, H. Wei, J. Zuback, T. Mukherjee, J. Elmer, J. Milewski, A. Beese, A. Wilson–Heid, A. De, W. Zhang, Progress in Materials Science **92**, 112 (2018)

[2] E. Olakami, R. Cochrane, K. Dalgarno, Progress in Materials Science **74**, 401 (2015)

[3] S. Beretta, S. Romano, International Journal of Fatigue **94**, 178 (2017)

[4] D. Buchbinder, Tech. Rep. 01RIO639A-D, Bundesministerium für Bildung und Forschung (2010)

[5] N. Aboulkhair, N. Everitt, I. Ashcroft, C. Tuck, Additive Manufacturing **1–4**, 77 (2014)

[6] N. Aboulkhair, M. Simonelli, L. Parry, I. Ashcroft, C. Tuck, R. Hague, Progress in Materials Science **106**, 100578 (2019)

[7] E. Brandl, U. Heckenberger, V. Holzinger, D. Buchbinder, Materials and Design **34**, 159 (2012)

[8] N. Aboulkhair, I. Maskery, C. Tuck, I. Ashcroft, N. Everitt, Materials and Design **104**, 174 (2016)

[9] S. Romano, S. Beretta, S. Foletti, LCF response of AlSi10Mg obtained by Additive Manufacturing, in Eighth International Conference on Low Cycle Fatigue (LCF8), Dresden (2017), pp. 53–58

[10] N. Uzan, S. Ramati, R. Shneck, N. Frage, O. Yeheskel, Additive Manufacturing **21**, 458 (2018)

[11] S. Romano, L. Patriarca, S. Foletti, S. Beretta, International Journal of Fatigue **117**, 47 (2018)

[12] B. Torries, R. Shrestha, A. Imandoust, N. Shamsaei, Fatigue Life Prediction of Additively Manufactured Metallic Materials Using a Fracture Mechanics Approach, in Solid Freeform Fabrication 2018: Proceedings of the 29th Annual International Solid Freeform Fabrication Symposium (2018), pp. 1181–1190

[13] R. Shrestha, J. Simsiriwong, N. Shamsaei, Additive Manufacturing **28**, 23 (2019)

[14] Y. Murakami, Metal fatigue: *Effects* of Small Defects and Nonmetallic Inclusions, 1st edn. (Academic Press, 2002)

[15] S. Romano, A. Brandão, J. Gumpinger, M. Gschweitl, S. Beretta, Materials & Design **131**, 32 (2017)

[16] S. Romano, A. Brückner-Foit, A. Brandão, J. Gumpinger, T. Ghidini, S. Beretta, Engineering Fracture Mechanics **187**, 165 (2018)

[17] EOS GmbH - Electro Optical Systems, Materialdatenblatt EOS Aluminium AlSi10Mg (2014), downloaded 2017-02-08

[18] B. Flaig, K. Lang, Tech. Rep. 567, Forschungsvereinigung Verbrennungskraftmaschinen e.V. (1994)

[19] I. Maskery, N. Aboulkhair, M. Corfield, C. Tuck, A. Clare, R. Leach, R. Wildman, I. Ashcroft, R. Hague, Materials Characterization **111**, 193 (2016)

[20] U. Tradowsky, J. White, R. Ward, N. Read, W. Reimers, M. Attallah, Materials and Design **105**, 212 (2016)

[21] N. Takata, H. Kodaira, K. Sekizawa, A. Suzuki, M. Kobashi, Materials Science & Engineering A **704**, 218 (2017)

[22] A. Maamoun, M. Elbestawi, G. Dosbaeva, S.C. Veldhuis, Additive Manufacturing **21**, 234 (2018)

[23] X. Cao, J. Campbell, Metallurgical and Materials Transactions A **35**, 1425 (2004)

[24] S. Belmares-Perales, A. Zaldivar-Cadena, Materials Science and Engineering B 174, 191 (2010)
[25] H. Ammar, A. Samuel, F. Samuel, Materials Science and Engineering A 473, 65 (2008)
[26] A. Samuel, F. Samuel, Metallurgical and Materials Transactions A 26, 2359 (1995)
[27] P. Huter, S. Oberfrank, F. Grün, B. Stauder, International Journal of Fatigue 88, 142 (2016)
[28] D. Zhang, Ph.D. thesis, RWTH Aachen (2004)
[29] International Standard ISO 12106, Metallic materials - fatigue testing - Axial-strain-controlled method (2003)
[30] A. Emami, S. Begum, D. Chen, T. Skszek, X. Niu, Y. Zhang, F. Gabbianelli, Materials Science and Engineering A 516, 31 (2009)
[31] S. Michelfeit, Ph.D. thesis, Technische Universität Darmstadt, Fachbereich Maschinenbau (2012)
[32] J. Jordon, M. Horstemeyer, N. Yang, J. Major, K. Gall, J. Fan, D. McDowell, Metallurgical and Materials Transactions A 41, 356 (2010)
[33] D. McDowell, K. Gall, M. Horstemeyer, J. Fan, Engineering Fracture Mechanics 70, 49 (2003)
[34] M. Song, Y. Kong, M. Ran, Y. She, International Journal of Fatigue 33, 1600 (2011)
[35] E. Natesan, S. Eriksson, J. Ahlström, C. Persson, Materials 12, 3033 (2019)
[36] J. Schindelin, I. Arganda-Carreras, E. Frise, V. Kaynig, M. Longair, Nature methods 9, 676 (2012)
[37] C. Fischer, C. Schweizer, S. Mittag, T. Seifert, K. Kuhlen, Tech. Rep. 1188, Forschungsvereinigung Verbrennungskraftmaschinen e.V. (2019)
[38] W. Ramberg, W. Osgood, Tech. Rep. Technical Note No. 902, National Advisory Committee For Aeronautics, Washington DC (1943)
[39] C. Shih, Tech. Rep. MRL E–147, Providence, R.I.: Division of Engineering, Brown University (1983)
[40] C. Schweizer, Ph.D. thesis, Karlsruher Institut für Technologie (KIT), Fakultät für Maschinenbau (2013)
[41] H. Heitmann, H. Vehoff, P. Neumann, in Fracture 84, edited by S. Valluri, D. Taplin, P.R. Rao, J. Knott, R. Dubey (Pergamon, 1984), pp. 3599 – 3606
[42] M. Metzger, B. Nieweg, C. Schweizer, T. Seifert, International Journal of Fatigue 53, 58 (2012)
[43] W. Schmitt, R. Mohrmann, H. Riedel, A. Dietsche, A. Fischersworring-Bunk, in Fatigue 2002 – Proceedings of the Eighth International Fatigue Congress, edited by A. Blom (Warley, West Midlands: EMAS, 2002), pp. 781–788
[44] J.C. Newman Jr., International Journal of Fracture 24, 131 (1984)
[45] H. Tada, P. Paris, G. Irwin, The Stress Analysis of Cracks Handbook, 3rd edn. (ASME Press, 2000)
[46] M. He, J. Hutchinson, J. Appl. Mech 48, 830 (1981)
[47] H. Riedel, Fracture at high temperatures, 1st edn. (Springer–Verlag Berlin Heidelberg, 1987)
[48] C. Schweizer, T. Seifert, B. Nieweg, P. von Hartrott, H. Riedel, International Journal of Fatigue 33, 194 (2011)
[49] C. Chauvot, M. Sester, Computational Materials Science 19, 87 (2000)
[50] G. Maier, H. Riedel, T. Seifert, J. Klöwer, R. Mohrmann, Time and temperature dependent cyclic plasticity and fatigue crack growth of the nickel–base Alloy617B – experiments and models, in Euro Superalloys 2010 (Trans Tech Publications Ltd, 2011), Vol. 278 of Advanced Materials Research, pp. 369–374
[51] T. Hanemann, Analyse der Mikrostruktur und Simulation der mechanischen Eigenschaften von generativ gefertigtem AlSi10Mg (2017), Seminar paper, Karlsruher Institut für Technologie (KIT)
[52] S. Rahmati, E. Vahabli, Int J Adv Manuf Technol 79, 823 (2015)
[53] A. Boschetto, L. Bottini, F. Veniali, Journal of Materials Processing Technology 241, 154 (2017)
[54] C. Qiu, C. Panwisawas, M. Ward, H. Basoalto, J. Brooks, M. Attallah, Acta Materialia 96, 72 (2015)
[55] N. Dowling, in Cyclic stress-strain and plastic deformation aspects of fatigue crack growth (ASTM International, West Conshohocken, 1977), Vol. STP637-EB, pp. 97–121

Formation of bulk-laminated materials by localized deformation during ECAP of an AA6060 aluminum alloy

Martin Franz-Xaver Wagner[1,*] and *Philipp* Frint[1]

[1]Materials Science Group, TU Chemnitz, 09107 Chemnitz, Germany

Abstract. This paper addresses localized plastic flow during equal-channel angular pressing (ECAP) of an AA6060 aluminum alloy. We observe an alternating formation of shear bands and matrix bands during ECAP that leads to pronounced strain partitioning without cracking. Local deformation is analyzed by considering the distortion of indents along flow lines in the center of a split billet. We estimate equivalent strains of about 3.6 inside the shear bands, whereas plastic deformation in the adjacent matrix bands is almost negligible. Microstructural analysis by SEM and STEM confirms that the shear bands exhibit typical features of severely plastically deformed microstructures at the onset of forming an ultrafine-grained microstructure. We further present statistics of band widths, and we discuss the roles material hardening as well as ECAP die geometry (in terms of the inner die radius) in facilitating the recurrent localized deformation that, in the absence of crack nucleation, leads to the production of an interesting and novel type of bulk-laminated materials by ECAP.

1 Introduction

Localization of deformation is a common phenomenon in materials science; examples include Lüders bands (in mild steels) [1], Portévin-Le Châtelier bands (e.g. in aluminum alloys) [2,3], or martensite bands related to the stress-induced phase transformation (e.g. in shape memory alloys [4–6]). In many cases, localized deformation is directly related to material failure. This is, for instance obvious, for adiabatic shear bands formed at high strain rates, where a locally increased amount of plastic deformation results in (locally) more pronounced energy dissipation, a (local) increase of temperature, and thus in (locally confined) thermal softening that in turn leads to an even more pronounced concentration of further deformation inside the forming shear band [7,8]. ECAP – one of the most promising severe plastic deformation methods that allows to produce bulk ultrafine-grained (UFG) materials [9–11] – is also often limited by shear band formation accompanied by crack nucleation and growth that ultimately leads to billet fragmentation [12–14].

In the light of these predominantly detrimental effects, here we consider a somewhat surprising and special situation where repeated shear localization actually results in the generation of a new type of UFG material with bulk lamellar structure. We have recently studied ECAP of an AA6060 aluminum alloy that has been cold-worked by extrusion at room temperature prior to ECAP, [15]. Interestingly, the material exhibits a pronounced and recurrent formation of shear bands, separated by much less deformed regions (which we refer to as matrix bands in the remainder of this paper). No damage occurs under the processing conditions used in our ECAP experiments. The resulting billets consist of alternating layers of shear and matrix bands that exhibit very different microstructural and mechanical properties, respectively. In this paper, we briefly present information on the initial material and on the ECA-pressing procedure. We analyze local deformation by considering the distortion of indents in a split billet, and we discuss the resulting microstructural features of shear vs. matrix bands. We furthermore study band widths and their distributions in the billets. Finally, we briefly discuss two potential factors that influence the recurrent formation of shear and matrix bands, material hardening capacity and ECAP die geometry.

2 Material and experiments

An AA6060 aluminum alloy of commercial purity was investigated. The material with the chemical composition 0.49Mg-0.45Si-0.16Fe-0.026Mn-0.003Cu-0.003Zn-(bal.)Al (in wt.-%) was obtained in a cast condition. It was first homogenized at 530 °C for 12 hours followed by water quenching. Prior to ECA-processing, the material was processed by extrusion at room temperature, introducing an equivalent strain of approximately 1.3. In contrast to commercial extrusion, this processing step can be considered as a first part of the severe plastic deformat-

* Corresponding author: martin.wagner@mb.tu-chemnitz.de

https://doi.org/10.1051/matecconf/202032608001

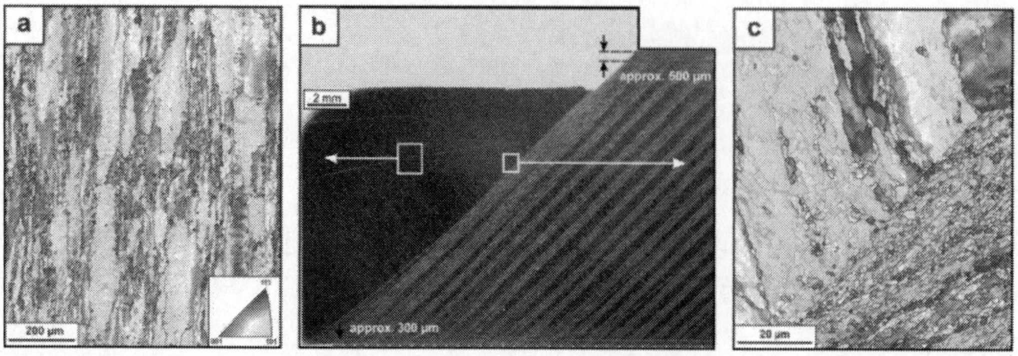

Fig. 1. Overview of the repeated formation of shear and matrix bands during ECAP of the cold-worked AA6060 aluminum alloy. (a) EBSD micrograph of the undeformed initial microstructure. (b) Low-magnification side-view on the end-piece of a billet after ECAP; shear bands (bright) and matrix bands (dark) form an alternating pattern. Band thickness is approximately constant from top to bottom of the billet, but in the very top/bottom layers (indicated by red dashed lines), the deformation appears to be more homogeneous. (c) EBSD micrograph of a region encompassing both undeformed material (upper left) and material that has passed the ECAP die's plastic deformation zone (lower right), consequently exhibiting pronounced grain refinement.

ion process (which is then followed by the main step during ECAP itself) since the pressing temperature is below the recrystallization temperature of the material.

ECAP billets with the dimensions of 15x15x50 mm³ were cut from the cold extruded material parallel to the extrusion direction. ECAP was conducted in a laboratory scale 90°-die (details of its design, which features movable walls to reduce friction between die and billet, are given in [16]) at room temperature and a pressing speed of 20 mm/min. Assuming an idealized shear zone geometry, as proposed by Iwahashi et al. [17], a single ECAP pass in a 90°-die is often estimated to result in an equivalent strain of 1.15. In order to further reduce friction during processing, a suitable wax-based lubricant (Bechem Beruforge 150D) was used [18]. A back-pressure of 150 MPa was applied to ensure that the deformation zone remained relatively thin and that no corner gap (i.e., an empty region near the outer corner of the die) occurred.

Microstructural analysis was performed by optical microscopy, electron backscatter diffraction (EBSD) and low voltage scanning transmission electron microscopy (STEM). The samples were prepared using a standard grinding and polishing procedure followed by either etching in a 4 % sodium hydroxide solution (optical microscopy) or in colloidal silica (EBSD). STEM foils were prepared by electrolytic and ion polishing. Electron microscopy was conducted using a Zeiss Neon40EsB field emission microscope. During EBSD analysis, a suitable post-processing of the raw data was performed, including a slight clean-up, neighbor confidence index (CI) correlation and a grain CI standardization with a minimum CI of 0.1.

3 Strain partitioning during ECAP

Fig. 1a shows an EBSD micrograph of the microstructure of the investigated aluminum alloy, taken from the end piece of a billet (shown as a macro scale micrograph in fig. 1b). This region has not passed through the plastic deformation zone in the ECAP die and therefore it is representative for the initial cold extruded material. The macroscopic view in fig. 1b reveals an alternating arrangement of two different types of bands (both with approximately similar volume fractions), oriented parallel to the main shear plane in the ECAP die: shear bands that appear much brighter in the micrograph and dark matrix bands. Both types of bands exhibit an almost constant thickness from the top to the bottom of the billet; because of the geometrical constraints posed by the rigid exit channel of the ECAP die, the very top (width ≈ 500 µm) and bottom (width ≈ 300 µm) regions appear to be deformed more homogeneously as indicated in fig. 1b (indeed, as discussed in greater detail in section 4, these regions are characterized by band branching). Figure 1c presents a magnified EBSD micrograph taken from the region surrounding the ECAP die's plastic deformation region (as highlighted in fig. 1b). An effective grain refinement is observed in the lower right hand-side of the plot (black lines indicate high angle grain boundaries), which represents a material volume that has been passed through the plastic deformation region. In contrast, coarse grains are observed on the upper left hand-side of the micrograph, corresponding to material that has not passed the plastic deformation region.

One focus of our previous study [15] was a detailed analysis of the local deformation path of material points passing through the plastic deformation region. We used a simple visio-plastic mapping technique by tracking indents that were prepared on a polished inner surface of a lengthwise split billet. An equidistant grid of hardness indents (Rockwell cone indenter) with a spacing of 400 µm was prepared on the interior surface (see fig. 2). To prevent a plastic collapse of the indents during ECAP, a filling with an epoxy resin was applied prior to reassembling the billet. In order to ensure the characterization of the "steady-state" deformation condition, ECA-pressing was interrupted and the L-shaped billet (as shown schematically in fig. 2a) was extracted without further deformation by removing two walls of the die.

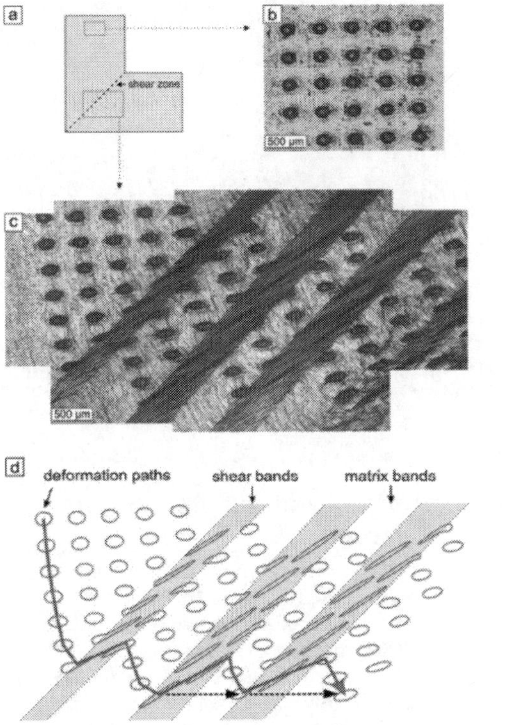

Fig. 2. (a) Schematic view of the L-shaped billet produced by interrupting an ECAP pass, showing the undeformed and deformed regions studied regions studied in greater detail. (b) Magnified view of the grid of indents in the undeformed region. (c) Grid of indents adjacent to the thin shear zone. Shear bands, characterized by pronounced plastic deformation, appear darker than the neighboring and much less deformed matrix bands. (d) Extracted positions and shapes of individual indents, determined from (c). By following the displacement of one line of indents, the deformation path through the ECAP die can be documented (indicated by the red line).

Fig. 2b shows a low magnification optical micrograph of the square grid of indents located in the entrance channel of the die (i.e. in the region that had not been subjected to plastic deformation in the shear zone of the die prior to interrupting the test). Despite the split billet having been put back together and subjected to a partial ECAP pass, the grid is undistorted, as are individual indents. This further indicates that our approach of analyzing grids in a split billet primarily records plastic deformation once it happens in the shear zone. Once material points are pushed close to the shear zone of the ECAP die, plastic deformation obviously commences, figs. 2c and d. This is obvious both from the shapes of single indents, and from the representative deformation path highlighted by a red line in fig. 2c, located on the left of figs. 2c and d. The deformation path already deviates significantly from the vertical orientation in the entrance channel of the die prior to reaching the shear zone itself (which is located close to the position of the left-most shear band in figs. 2c and d). In good agreement with analytical models of plastic flow in ECAP dies, like Beyerlein and Tomés fan-shape model [19,20] or Tóth et al.'s super-elliptical flow line function approach [21,22], some initial plastic deformation is already accumulated along this part of the deformation path.

As discussed in detail in [15], a careful reconstruction of the shapes of the distorted indents (by fitting ellipses, fig. 2d) allows to analyze the deformation history along representative paths. From the considerably elongated indents inside the shear bands it is obvious that these regions have been severely deformed far exceeding the nominal shear strain that is typically introduced by a single pass of ECAP using a 90° die. A simple analysis comparing indents in matrix and shear bands yields an estimated equivalent plastic strain due to this massive shear of about 3.6, [15]. In contrast, the matrix bands next to those severely sheared regions experience much less strain. Indeed, a comparison of the shapes of the corresponding indents with those in front of the shear zone reveals no differences, which is a strong indication that the matrix band regions have not been sheared further at all while passing the ECAP die's deformation zone.

It is particularly worth noting that the macroscopic deformation introduced into the billet must of course conform to the boundary conditions imposed by the ECAP die's geometry. The severe shear deformation inside the shear bands thus must be counterbalanced by the considerably reduced amount of plastic deformation in the matrix bands – a process that can be aptly referred to as recurrent strain partitioning. Considering the effective displacement vectors associated with deformation in a shear band an in an adjacent matrix band, this is obviously the case: the overall deformation results in horizontal material flow, i.e., parallel to the die's top and bottom in the exit channel (see black arrows in fig. 2d).

The low-resolution optical micrograph in fig. 3a, taken from the same perspective as in figs. 1 and 2, was further used to evaluate band widths, and to document band branching that occurs near the top and bottom surfaces of the billet. Fig. 3a confirms our estimate that shear and matrix bands exhibit similar width distributions; since they are arranged in an alternating pattern, their volume fractions are approximately 50%. We highlight that band thickness does not vary significantly across the entire thickness of the billet, as shown by additional analyses from different cutting and viewing directions, [15]). By placing lines perpendicular to the interfaces between the bands, and employing a simple line intersection analysis, the shear band width distribution plot shown in fig. 3b was determined. A closer inspection of the band structures in fig. 3a reveals that band branching occurs primarily near the bottom of the billet in a relatively wide zone, providing a gradual transition from the predominant reg-

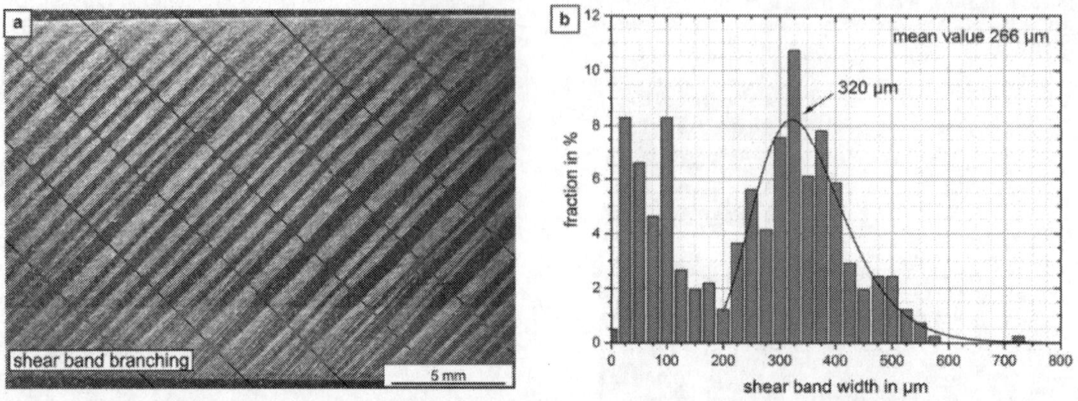

Fig. 3. (a) Shear and matrix bands, observed from the side of the billet exhibit a mostly constant thickness throughout the billet, but towards the top and (more pronounced) the bottom surface, band branching occurs (this transition region is highlighted in red). (b) Histogram of band width values determined from (a) using the line-intercept method. Shear band width exhibits a bimodal distribution with two peaks that can be directly related to the different regions in the billet (band widths in transition region: red; band width values in the main volume of the billet: grey).

ion of constant shear band width towards the outer material layers at the very bottom of the billet, which (due to the geometric constraints of the die) exhibits an apparently homogeneous deformation. Consequently, the shear band width distribution is bimodal with two distinct peaks that can be directly related to the different regions in the billet. In the region of band branching (also highlighted in red in fig. 3a), thin bands (red bars in the histogram, fig. 3b) with widths below 200 µm are observed. Shear bands in the center part of the billet (grey bars in fig. 3b) approximately follow a normal distribution with a peak at a width of about 320 µm. Simple averaging of all band widths over the entire billet results in a mean value of 266 µm. To summarize, the statistical analysis demonstrates that the shear bands in this billet exhibit quite constant widths, typically on the order of several hundreds of microns.

4 Microstructural features of shear and matrix bands

Representative results of our detailed microstructural analysis are presented in fig. 4. EBSD was performed in the two different orientations shown schematically in Fig. 4a. The EBSD micrographs shown in figs. 3b and c highlight representative microstructures for both, shear and matrix bands. Both images show a considerable grain refinement in the shear bands originating from a large amount of plastic deformation. In contrast, relatively large and much less deformed grains are clearly visible in the matrix bands. Even the elongated morphology of the grains formed by the previous room temperature extrusion can still be observed. It is moreover obvious that there still remains a pronounced crystallographic texture, especially in the matrix band regions. These findings agree well with

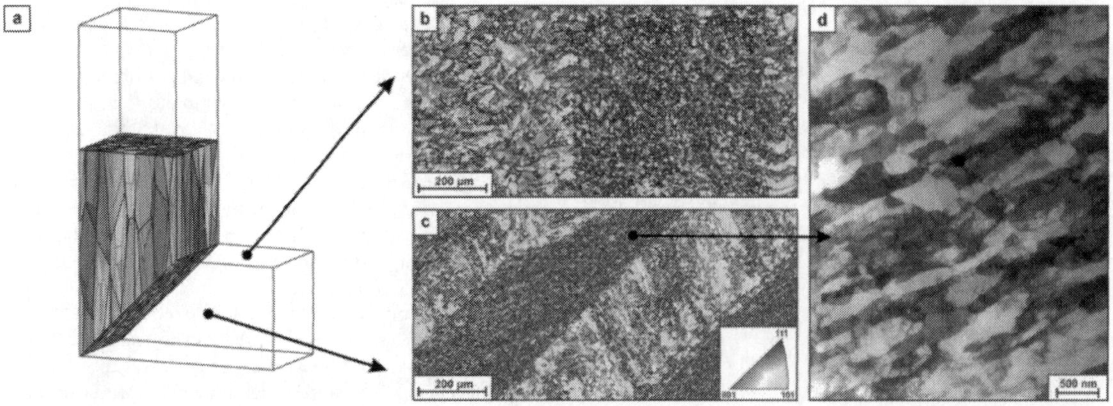

Fig. 4. Microstructural analysis of shear and matrix bands: (a) Schematic representation indicating the orientation of elongated grains (resulting from prior cold extrusion) and highlighting the viewing directions corresponding to the EBSD micrographs in (b) and (c). (d) The STEM image at higher magnification of the microstructure inside a single shear band clearly shows the formation of an UFG microstructure.

the results of our strain analysis described above. Because of the very small grain size inside the shear bands and the physical limitations of EBSD measurements, additional low voltage scanning transmission electron microscopy (STEM) was performed, fig. 4d. These regions clearly are characterized by a fully developed UFG microstructure: The small, elongated grains have a length of approximately 500 nm and a width of about 200 nm. The sharply defined boundaries between individual grains in the STEM image indicate substantial misorientations between grains – i.e., the boundaries have the characteristics of high angle grain boundaries. We emphasize that this observation is uncommon for 6xxx-series aluminum alloys deformed by a single ECAP pass; usually, multiple-pass ECAP, with a number of passes of about 4 (which corresponds to an equivalent strain of 4.4) is needed to generate the microstructural characteristics found in our shear bands.

5 Effect of strain hardening and ECAP die geometry on recurrent strain partitioning

The key scientific question raised by our experimental observations is why, and how, recurrent strain partitioning by shear banding during ECAP of the cold-worked AA6060 alloy occurs. Quite generally, unstable plastic flow may be expected when material hardening is insufficient to counterbalance softening effects (such as thermal softening, strain rate softening in materials with negative strain rate sensitivity, transformation softening, texture softening, or additional microstructural softening mechanisms), [23–26]. Indeed, shear band formation in aluminum subjected to high strains (but with shear banding on a much smaller length scale) has been investigated in detail, and discussed in relation to hardening behavior by Korbel and co-workers in the 1980s [27–30].

The role of different types and amounts of material hardening can also be evaluated in a straight-forward manner from a phenomenological, continuum mechanical point of view. Horn et al. recently analyzed our ECAP process by dedicated finite element simulations, carefully modeling the geometry of our die and varying the material behavior of the billet, [31]. The results of this study agree remarkably well with our experiments, particularly in terms of the resulting band widths. The numerical study demonstrated that the assumption of kinematic hardening is required to properly model shear band formation, whereas isotropic hardening only leads to homogeneous deformation in the simulations. Interestingly, when one numerically considers materials with different hardening rates and strain hardening capacities, distinct regions (associated with different types of homogeneous or inhomogeneous deformation during ECAP) can be identified in a representative stress-strain plot, fig. 5. One might expect from our general discussion of unstable plastic flow that a material with an overall low hardening rate and a low strain hardening capacity might be prone to localized deformation. However, such a material behavior (represented by flow curves in the lower region in fig. 5) is shown in [31] to only result in minor strain fluctuations; no distinct shear or matrix bands are formed and plastic strains only exhibit small gradients inside the billet. A more pronounced degree of strain hardening (i.e., stress-strain curves typically located inside the center region of the diagram in fig. 5), as expected, leads to stable plastic flow; no bands are formed and the deformation is mostly homogeneous (with some deviations, of course, in regions at the very top or bottom of the billet, where the ECAP die geometry affects the local stress states). Most importantly, the simulations reveal that materials with a pronounced initial strain hardening rate that quickly levels out after using up a certain amount of strain hardening capacity are prone to recurrent strain partitioning, forming well-defined shear and matrix bands as observed in our ECAP experiments.

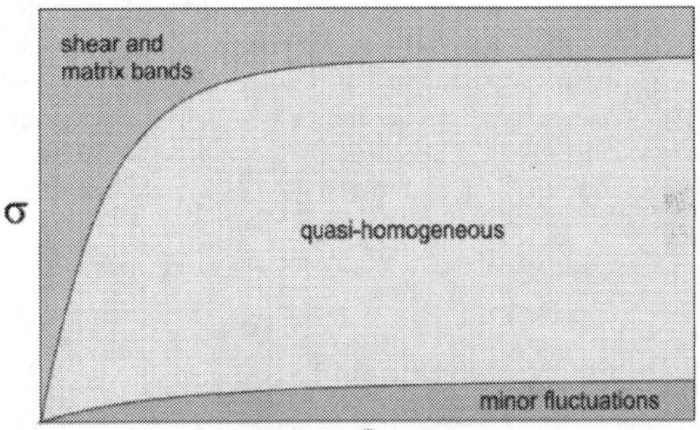

Fig. 5. Schematic comparison of different uniaxial flow curves as function of effective plastic strain, highlighting different material responses to ECAP, [31]: materials with a low hardening capacity exhibit only minor fluctuations in terms of equivalent plastic strain in only weakly defined bands. An increased hardening capacity suppresses this effect and leads to an apparently homogeneous deformation of the billet. In contrast, materials with a high initial hardening rate and a sufficiently large hardening capacity exhibit fully developed shear and matrix bands as observed in our experiments.

From a materials science point of view, it is furthermore critical to understand which microstructural conditions promote strain partitioning. Clearly, the cold-worked microstructure prior to ECAP leads to the kind of mechanical behavior identified by the phenomenological finite element simulations, [31]. Moreover, we have shown in [15] that the material behavior during ECAP exhibits a pronounced orientation dependence: billets cut with different orientations with respect to the cold extrusion direction, when subjected to similar ECAP processing conditions, can exhibit either strain partitioning or conventional, homogeneous shear deformation. This obviously demonstrates that texture plays an important role. While a first tentative analysis of texture effects is discussed in [15], we acknowledge that further work is required to fully understand the effects of texture on shear band formation and unstable plastic flow in our AA6060 alloy during ECAP.

We also recorded the temporal evolution of the pressing force during ECA-pressing, analyzing distinct stress drops, [15]. Because of the optimized lubrication conditions (and because billets of the same alloy, but with different orientations, can exhibit homogeneous deformation under similar processing conditions), these stress drops cannot be related to stick-slip phenomena. Indeed, we clearly document that the stress drops can be directly associated with individual shearing events. The following qualitative scenario can therefore be used to describe recurrent strain partitioning mechanistically: as the billet is pressed through the die, stresses build up locally in the ECAP die's shear zone region (associated with strain hardening as discussed above) until they reach a critical value that leads to nucleation and rapid growth of a shear band through the billet's cross section. This dynamic process, and severe plastic deformation inside the shear band, result in mechanical unloading of the adjacent material, which follows the leading shear band through the deformation zone without additional massive shearing – until stresses gradually increase again and the entire process is repeated. This scenario implies a crucial role of local stress states, and this further suggests that specific details of the ECAP die geometry may have a significant effect on unstable plastic flow and shear band formation.

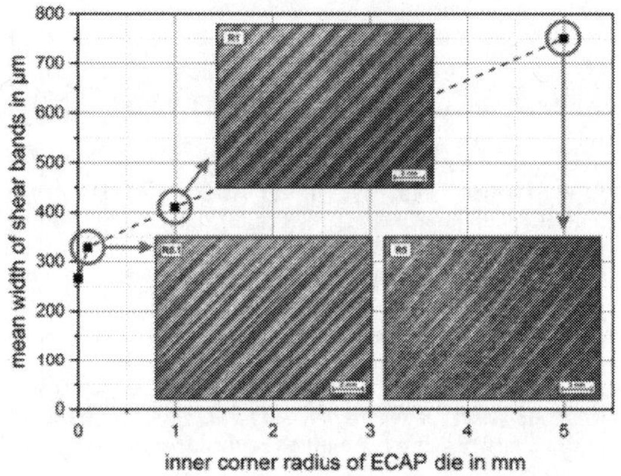

Fig. 6. Variation of mean shear band width as a function of the inner corner radius of the ECAP die. Insets show optical micrographs of shear and matrix bands (similar viewing direction as in figs. 1-3). With increasing corner radii, shear bands become wider and less well defined; for R = 5 mm, a transition towards an almost homogeneous type of deformation is observed.

To test this hypothesis, we modified a special detail of our ECAP die, its inner corner radius, where stress peaks are likely to occur during ECA-pressing, [31,32]. Fig. 6, where the average shear band width is plotted as a function of the inner corner radius, summarizes the results of this experimental parameter study. The micrographs presented as in-sets in fig. 6 further give an impression of the resulting shear and matrix band structures in the billets that were otherwise subjected to similar pressing conditions. Our earlier ECAP experiment was performed in the die with a nominally sharp inner corner radius (R ≈ 0 mm), which acts as a local stress raiser and leads to pronounced stress gradients in its vicinity. Increasing corner radii lead to reduced local stress peaks and less pronounced stress gradients. This obviously correlates with increasing (mean) shear band widths. An inner corner radius of 5 mm still results in recurrent strain partitioning; however, the shear bands are much less well-defined than in the cases involving smaller radii. While an in-depth analysis of local deformation in the different shear band populations produced using different die geometries is on-going work, it appears that the case of R = 5 mm represents a transition state towards an almost homogeneous shear deformation. Clearly, our novel experimental results confirm a pronounced effect of ECAP die geometry, most likely related to the local stress state and its interaction with the nucleating shear bands, on the characteristic widths of shear and matrix bands. This observation opens an additional pathway that potentially allows to modify the structure of bulk laminated billets produced by ECAP of AA6060 by relatively simple engineering approaches, particularly ECAP die design.

6 Summary and conclusions

The experimental and numerical results presented in this paper address a novel phenomenon during ECAP of a cold-worked AA6060 alloy: strain partitioning by recurrent formation of shear bands and matrix bands. The resulting materials are characterized by alternating layers of these bands, with mean bands widths of several hundreds of microns, and can be considered as a type of bulk laminated material with partially UFG layers. Plastic deformation in the shear bands is significantly larger than expected for a single ECAP pass in a 90° die, and this excessive straining is counter-balanced by considerably smaller amounts of straining in the matrix bands. Microstructural analysis by electron microscopy confirms this observation; even after a single pass, microstructural features typically associated with UFG materials are identified inside the shear bands.

Additional numerical analysis and novel experimental work was performed in order to understand which factors promote unstable plastic flow and shear banding during ECAP of this alloy. From a mechanical point of view, kinematic hardening and a high initial hardening rate combined with a low hardening rate at higher strains leads to the formation of distinct shear and matrix bands that agree well with our experimental observations. This kind of mechanical behavior is of course intricately linked to texture of the initial material. Finally, it was shown that the local stress state (and the stress gradient) at the inner die corner, which can be manipulated by modifying the corner radius geometry, has a pronounced effect on shear band widths.

We note in closing that the bulk laminated AA6060 material produced by ECAP, in addition to posing some scientifically challenging questions, exhibits interesting mechanical properties, with the UFG microstructures inside the shear bands significantly deviating from the mechanical behavior of the less deformed matrix bands. A more detailed analysis of both the local and global mechanical behavior of the bulk laminated materials will be addressed in a companion paper.

The authors would like to acknowledge fruitful discussions and a highly enjoyable scientific collaboration related to the simulation of strain partitioning with Tobias Horn, Christian Silbermann, and Jörn Ihlemann, all from the Solid Mechanics Group in Chemnitz. Early stages of this work have been funded by the German Research Foundation (Deutsche Forschungsgemeinschaft, DFG) through the Collaborative Research Center SFB 692.

References

1. A.H. Cottrell, B.A. Bilby, Proc. Phys. Soc. Sect. A. **62**, 49-62 (1949)
2. A.H. Cottrell, Philos. Mag. J. Sci. **44**, 829-832 (1953)
3. S.D. Antolovich, R.W. Armstrong, Prog. Mater. Sci. **59**, 1–160 (2014)
4. J. Khalil-Allafi, B. Hasse, M. Klönne, M. Wagner, T. Pirling, W. Predki, W.W. Schmahl, Materwiss. Werksttech. **35**, 280–283 (2004)
5. C. Elibol, M.F.-X. Wagner, Mater. Sci. Eng. A **643**, 194-202 (2015)
6. M.F.-X. Wagner, G. Eggeler, Mech. Mater. **38**, 1012–1025 (2006)
7. M.A. Meyers, Y.B. Xu, Q. Xue, M.T. Pérez-Prado, T.R. McNelley, Acta Mater. **51**, 1307–1325 (2003)
8. M. Dolinski, D. Rittel, A. Dorogoy, J. Mech. Phys. Solids. **58**, 1759–1775 (2010)
9. R.Z. Valiev, Y. Estrin, Z. Horita, T.G. Langdon, M.J. Zehetbauer, Y.T. Zhu, JOM **58**, 33–39 (2006)
10. R. Valiev, Nat. Mater. **3**, 511–516 (2004)
11. Y. Iwahashi, Z. Horita, M. Nemoto, T.G. Langdon, Acta Mater. **46**, 3317–3331 (1998)
12. M. Gaspérini, C. Pinna, W. Swiatnicki, Acta Mater. **44**, 4195–4208 (1996)
13. R. Lapovok, L.S. Tóth, A. Molinari, Y. Estrin, J. Mech. Phys. Solids **57**, 122–136 (2009)
14. K. Hockauf, M.F.-X. Wagner, T. Halle, T. Niendorf, M. Hockauf, T. Lampke, Acta Mater. **80**, 250–263 (2014)
15. P. Frint, M.F.-X. Wagner, Acta Mater. **176**, 306–317 (2019)
16. S. Frint, M. Hockauf, P. Frint, M.F.-X. Wagner, Mater. Des. **97**, 502–511 (2016)
17. Y. Iwahashi, J. Wang, Z. Horita, M. Nemoto, T.G. Langdon, Scr. Mater. **35**, 143–146 (1996)
18. P. Frint, M.F.-X. Wagner, S. Weber, S. Seipp, S. Frint, T. Lampke, J. Mater. Process. Technol. **239**, 222–229 (2017)
19. I.J. Beyerlein, C.N. Tomé, Mater. Sci. Eng. A **380**, 171–190 (2004)
20. I.J. Beyerlein, L.S. Tóth, Prog. Mater. Sci. **54**, 427–510 (2009)
21. L.S. Tóth, Adv. Eng. Mater. **5**, 308–316 (2003)
22. L.S. Tóth, R. Arrufat Massion, L. Germain, S.C. Baik, S. Suwas, Acta Mater. **52**, 1885–1898 (2004)
23. A. Vinogradov, Y. Estrin, Prog. Mater. Sci. **95**, 172–242 (2018)
24. T.A. Lebedkina, M.A. Lebyodkin, J.-P. Chateau, A. Jacques, S. Allain, Mater. Sci. Eng. A **519**, 147–154 (2009)
25. H. Ovri, E.T. Lilleodden, Acta Mater. **89**, 88–97 (2015)
26. H. Ovri, E.T. Lilleodden, Mater. Des. **125**, 69–75 (2017)
27. A. Korbel, F. Dobrzanski, M. Richert, Acta Metall. **31,** 293–298 (1983)
28. A. Korbel, V.S. Raghunathan, D. Teirlinck, W. Spitzig, O. Richmond, J.D. Embury, Acta Metall. **32**, 511–519 (1984)
29. A. Korbel, M. Richert, Acta Metall. **33**, 1971–1978 (1985)
30. A. Korbel, P. Martin, Acta Metall. **34**, 1905–1909 (1986)
31. T.D. Horn, C.B. Silbermann, P. Frint, M.F.-X. Wagner, J. Ihlemann, Metals **8**, 55 (2018)

[32] R.B. Figueiredo, I.P. Pinheiro, M.T.P. Aguilar, P.J. Modenesi, P.R. Cetlin, J. Mater. Process. Technol. **180,** 30–36 (2006)

Investigation on electrical conductivity and hardness of 6xxx aluminum conductor alloys with different Si levels

Siamak Nikzad Khangholi[1,*], *Mousa Javidani*[1], *Alexandre Maltais*[2] and *X.-Grant Chen*[1]

[1]Department of Applied Sciences, University of Quebec at Chicoutimi, Saguenay, QC, Canada, G7H 2B1
[2]Arvida Research and Development Centre, Rio Tinto Aluminum, Saguenay, QC, Canada, G7S 4K8

Abstract. The impact of Si levels on electrical conductivity (EC) and hardness in Al-Mg-Si 6xxx aluminum conductor alloys was studied at different aging times. Three experimental alloys with the Si levels of 0.4, 0.66, 0.89 wt.% containing a constant Mg content of 0.65 wt.%, designated as Si4, Si6, and Si9, were selected. It is found that the average peak hardness of Si4, Si6, and Si9 alloys were 114, 127 and 140 HV, respectively. The Si9 alloy with the highest Si level exhibited the highest peak hardness, which was consistent with the number density of strengthening precipitates. However, the EC of all studied alloys at peak aging was lower than the minimum required one (52.5 %IACS). The higher Si level in the alloy, the longer aging time required to enhance EC. The Si4, Si6, and Si9 alloys reached the minimum required EC after 3, 5 and 34 h aging time, of which the corresponding hardness was 103, 119 and 111 HV, respectively. Consequently, the Si6 alloy represented a better trade-off between EC and hardness among three studied alloys. The quantitative analyses of precipitates at peak aging and overaging states were carried out by a transmission electron microscope (TEM) investigation.

1. Introduction

In recent years, the application of Al-Mg-Si alloys (such as AA6101 and AA6201) in the electric transmission line are increasingly grown since they give a wide variety of the strength due to precipitation hardening along with the desired electrical conductivity [1, 2]. Aluminum conductors can be generally regrouped into two categories: aluminum conductor steel reinforced (ACSR) and all-aluminum alloy conductor (AAAC). In the ACSR, commercial pure aluminum alloys such as AA1350 alloy are widely used, of which the core consists of the steel to reinforce the structural function in power lines. To meet the requirement of high mechanical properties and good corrosion resistance, the AAAC, especially Al-Mg-Si 6xxx alloys, are mostly used in high voltage transmission lines because of their high strength [1-5].

Magnesium and silicon are known as the major alloying elements in Al-Mg-Si 6xxx alloys, making them age-hardenable alloys [6]. It was found that the various Mg/Si ratios could affect the strength and electrical conductivity of Al-Mg-Si conductor alloys to a great extent [6-8]. In the coarse-grained structure, the excessive Si (1<Mg/Si<2) led to the higher hardness and strength with a slight decrease in EC [6, 7], while in the fine-grained structure it resulted in a higher EC owing to the grain boundary precipitates [7]. In general, the Mg excessive may not have a significant impact on strength, while affecting negatively EC [6].

In general, precipitation hardening, work hardening, grain boundary strengthening, and solute solution strengthening are the main strengthening mechanisms in aluminum conductor alloys. However, in the metallurgical aspect, most parameters enhancing the strength of aluminum alloys tend to reduce their electrical conductivity. Therefore, the mechanical properties and electrical conductivity should be simultaneously taken into account in development of Al-Mg-Si conductor alloys [9]. To ensure the high performance of aluminum conductor alloys, the norms and regulations describing the design requirement have been established. For example, the minimum requirement of AA6201 alloy wires with the diameter between 3.25 mm and 4.75 mm for UTS and EC are 305MPa and 52.5%IACS, respectively [5, 10, 11].

Considering that the Mg/Si ratio of the β" precipitates as a strengthening phase is one [12, 13], the effect of different Si levels relative to the Mg/Si ratio of β" on EC and hardness has been rarely studied. The aim of this study is to investigate the effect of various Si levels on EC and hardness. The underlying microstructures for balancing the electrical conductivity and hardness were explored by the quantitative analysis of strengthening precipitates.

2. Experimental

Three experimental alloys with the Si contents of 0.4, 0.66, 0.89 wt.% and a constant Mg content of 0.65 wt.%,

* Corresponding author: siamak.nikzad-khangholi1@uqac.ca

designated as Si4, Si6, and Si9, were investigated in the present work, and their chemical compositions are listed in Table 1. The general wire fabrication process consists of casting, homogenization (560 °C for 6 h), followed by hot rolling at 350-480 °C (74% area reduction). The hot-rolled sheets were machined into the square bars, after which a solution treatment at 540 °C for 2 h was applied. Afterwards, the cold wire drawing (area reduction of 50%) for the final diameter of wires of 4.7 mm was conducted, followed by the artificial aging at 180 °C for different aging times. It was reported that 180 °C is an appropriate aging temperature to get a better optimization of strength and EC in Al-Mg-Si conductor alloys [14].

The electrical conductivity based on ASTM E1004 was measured via Sigmascope method with frequency of 480 kHz at different aging times. Fifteen measurements were recorded and calculated as an average value. The microhardness tests were carried out with the force of 25 g and dwell time 20 s, and the average value of eight measurements was reported. The intermetallic phases in three alloys were investigated using a scanning electron microscope equipped with energy dispersive spectroscopy (EDS). To characterize the strengthening precipitates, the TEM samples were taken from the drawn samples and prepared using the electropolishing technique. The characteristics of precipitates, such as their number density, were quantified using the method in Ref. [15].

3. Results and Discussion

Fig. 1a and 1b show the microhardness and electrical conductivity (EC) of experimental alloys as a function of the aging times. The microhardness of the alloys first increased sharply to reach their peak values and then gradually decreased with increasing aging time. In general, the higher Si content, the higher microhardness is during entire aging period. The peak microhardness of Si4, Si6, and Si9 alloys are 114, 127 and 140 HV, respectively. However, the EC values of Si4, Si6, and Si9 alloys at the peak aging are 52.3, 49, and 47.5 %IACS, respectively, which are less than the minimum required EC (52.5 %IACS). Therefore, the aging time needs to be prolonged to enhance EC. It was found that the EC of the Si4, Si6, and Si9 alloys started to exceed 52.5% IACS after 3, 5, 34 h, respectively. The corresponding microhardness values for Si4, Si6, and Si9 alloys are 103, 119 and 111 HV, respectively.

Fig. 1c shows the microhardness versus EC for the entire aging times, and Fig. 1d magnifies Fig. 1c to represent the microhardness versus EC above the required 52.5 %IACS. Results of Fig. 1c show that in the low EC range (44-51 %IACS), the high Si alloy (Si9) exhibits the highest microhardness. However, with increasing EC, the microhardness of the Si6 alloy exceeds the microhardness of the Si9 alloy. It can be seen in Fig. 1d, the Si6 alloy shows the highest hardness above the minimum required EC among three alloys, suggesting as an appropriate candidate for conductor applications provided the strength is taken into account in the priority. However, with a further increase in aging time, the microhardness of the Si6 alloy sharply drops with an insignificant growth in EC, exhibiting a small window of high microhardness above the minimum required EC. Similarly, the Si9 alloy has the second highest hardness but shows a limited window of microhardness above 52.5 %IACS. On the other hand, in the high EC range (54.5-55.5 %IACS), the Si4 alloy still shows a reasonably high strength (90-100 HV). Therefore, the low Si alloy (Si4) could be considered if the EC becomes the first priority in the design.

Fig. 2 shows the type and distribution of Fe-rich intermetallics after homogenization in the three alloys. In general, the type of intermetallics can be distinguished according to the Fe/Si ratio and morphology [16, 17]. The predominant Fe-rich intermetallics in the Si4 alloy are spherical/Chinese-script shaped α-Al_8Fe_2Si (Fig. 2a), while rod-like β-Al_5FeSi is the dominant Fe-rich intermetallics in the Si6 and Si9 alloys, as displayed in Fig. 2b and c. Fig. 2d shows the image analysis results of the distribution of Fe-rich intermetallics in the three alloys. The type and distribution of the intermetallics in those alloys could be correlated to the Fe/Si ratio of the alloys and homogenization temperature [18]. Since the β-Al_5FeSi intermetallics consume more Si compared to α-Al_8Fe_2Si intermetallics [19], it could be generally expected that the presence of β intermetallics slightly enhances the electrical conductivity and reduces the precipitation degree due to the Si solute extraction from the matrix. In addition, the β intermetallics have more stress concentration due to their plate-like morphologies relative to α intermettalics [20], affecting the strength. In reality, the volume fractions of the Fe-rich intermetallics in all three alloys are quite low (less than 0.5%). Therefore, it is reasonable to expect that the Fe-rich intermetallics have insignificant effects on both strength and EC of experimental alloys.

Fig. 3 displays bright field TEM images of the precipitate distribution at the peak aging for three alloys. As mentioned above, the peak aging conditions are the Si4 alloy at 2 h (Fig. 3a), the Si6 alloy at 1 h (Fig. 3b), and the Si9 alloy at 1 h (Fig. 3c). The density numbers of precipitates are shown in Fig. 3d, indicating that the average density number of precipitates increased with the Si level. Fig.3e displays the precipitate length and inter-precipitate spacing for three alloys at peak-aged states. Based on Fig.3e, the average precipitate lengths for Si4-2h, Si6-1h, and Si9-1h alloys were approximately 50 nm. It was reported that the majority of precipitates were identified as β' precipitates since the cold drawing could suppress the formation of β" phases [21, 22]. It is found that the Si9 alloy exhibited the highest number density of precipitates, which corresponds well with the highest microhardness at peak aging. On the other hand, the Si4 alloy possesses the lowest number density, resulting in the lowest microhardness at peak aging (Fig. 1a). It stems from the fact that Si addition increases the number of the clusters in the early aging, promoting the precipitation of the strengthening phase [6, 23]. Moreover, the Si_2 columns were observed in the core of GP zones and the subsequent precipitates [24], showing the key role of Si in precipitation. Based on Fig.3e and Fig.1a, the inter-precipitate spacing is inversely proportional to peak-hardness, meaning that Si-excessive alloys (Si6-1h and Si9-1h) with a smaller inter-precipitate spacing are more resistant against dislocation movement. On the other hand, a shorter inter-precipitate spacing causes more electrical resistivity [25], especially in Si6-1h and Si9-1h alloys compared to Si4-2h alloy. Therefore, lower EC in peak-aged condition for Si6 and Si9 alloys could mainly result from both solute (especially Si) and precipitates.

Table 1 Chemical compositions of Al-Mg-Si conductor alloys with different Si levels

Alloys	Compositions (wt.%)				Mg/Si
	Mg	Si	Fe	Al	(at.%)
Si 4	0.67	**0.4**	0.09	Bal.	1.94
Si 6	0.66	**0.66**	0.13	Bal.	1.15
Si 9	0.68	**0.89**	0.10	Bal.	0.86

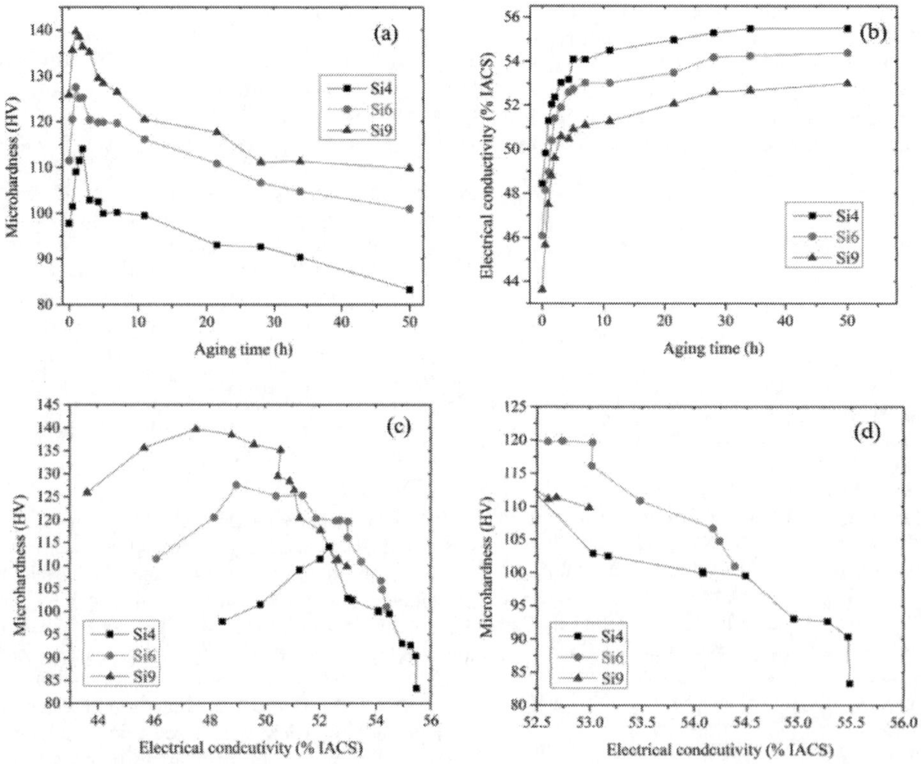

Figure 1 (a) Microhardness and (b) electrical conductivity as a function of the aging time at 180 °C; (c) microhardness versus EC over entire aging times and (d) microhardness versus EC above the minimum required 52.5 %IACS

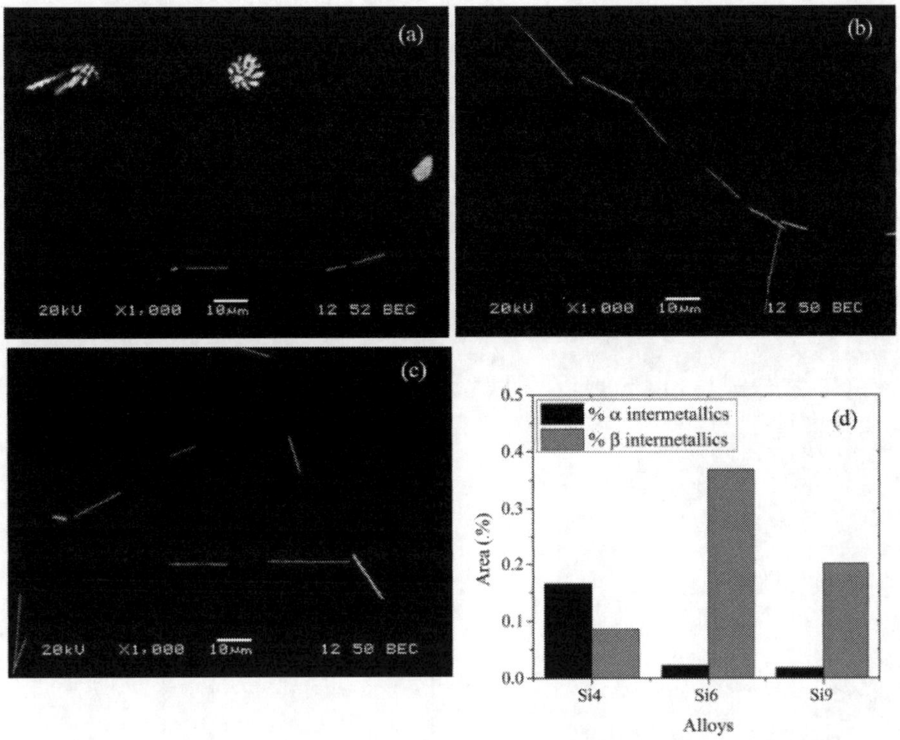

Figure 2 Distribution of Fe-rich intermetallics after homogenization in (a) Si4 (b) Si6 (c) Si9 alloys; (d) the area percentage of Fe-rich intermetallics in three alloys

As mentioned earlier, the Si4, Si6, and Si9 alloys met the minimum required EC after 3, 5, and 34 h aging, respectively. Obviously, with an increase in Si levels, the aging time has to be largely increased to enhance the EC, which causes the microstructure and hardness changes. The different aging times to reach the required EC in the experimental alloys could be attributed to the different Si solute levels in the aluminum matrix. Fig 4 displays the precipitates of Si4 alloy at 3 h, Si6 alloy at 5 h, and Si9 alloy at 34 h aging. Fig. 4d shows the number density of precipitates, and Fig. 4e displays the precipitate length and inter-precipitate spacing at the corresponding aging times for three alloys. Measurements on the TEM images (Fig. 4e) reveal that the average precipitate lengths for Si4-3h, Si6-5h, and Si9-34h alloys were around 60, 80, and 170 nm, respectively. Therefore, it can be identified that the Si9-34h alloys were subjected to more precipitate coarsening to reach the required EC in relation to Si4-3h and Si6-5h. Compared Fig. 3d with Fig. 4d, it could be seen that the number density of the precipitates in the Si9-34h alloy significantly drops after prolonged aging to meet the required EC, resulting in a remarkable reduction of the strengthening effect (Fig. 1d). The Si6-5h alloy exhibited the highest number density of precipitate among three alloys, thereby showing the highest hardness. The number density of precipitates decreases from Si6-5h to Si4-3h and further to Si9-34h alloys after reaching the minimum required EC. Based on Fig. 4e, the Si9-34h alloy has the largest inter-precipitate spacing, suggesting that the electrical resistivity from precipitates greatly decreased. Therefore, electrical resistivity in Si9-34h alloys mainly comes from solutes (especially Si).

It is worth mentioning that the Si9 alloy after 34 h aging showed the higher hardness relative to the Si4 alloy after 3 h aging (Fig. 1d), although its number density of precipitates is less than the Si4 alloy (Fig. 4d). In general, the precipitation hardening plays a primary role in controlling the strength for the heat-treatable aluminum alloys. However, the other strengthening mechanisms such as strain hardening and solid solution strengthening during wire drawing process can also make a remarkably contribution to the alloy strength. It could be deduced that both strain hardening and solid solution strengthening in the Si9 alloy are presumably higher than those in the Si4 alloy owing to its high Si level.

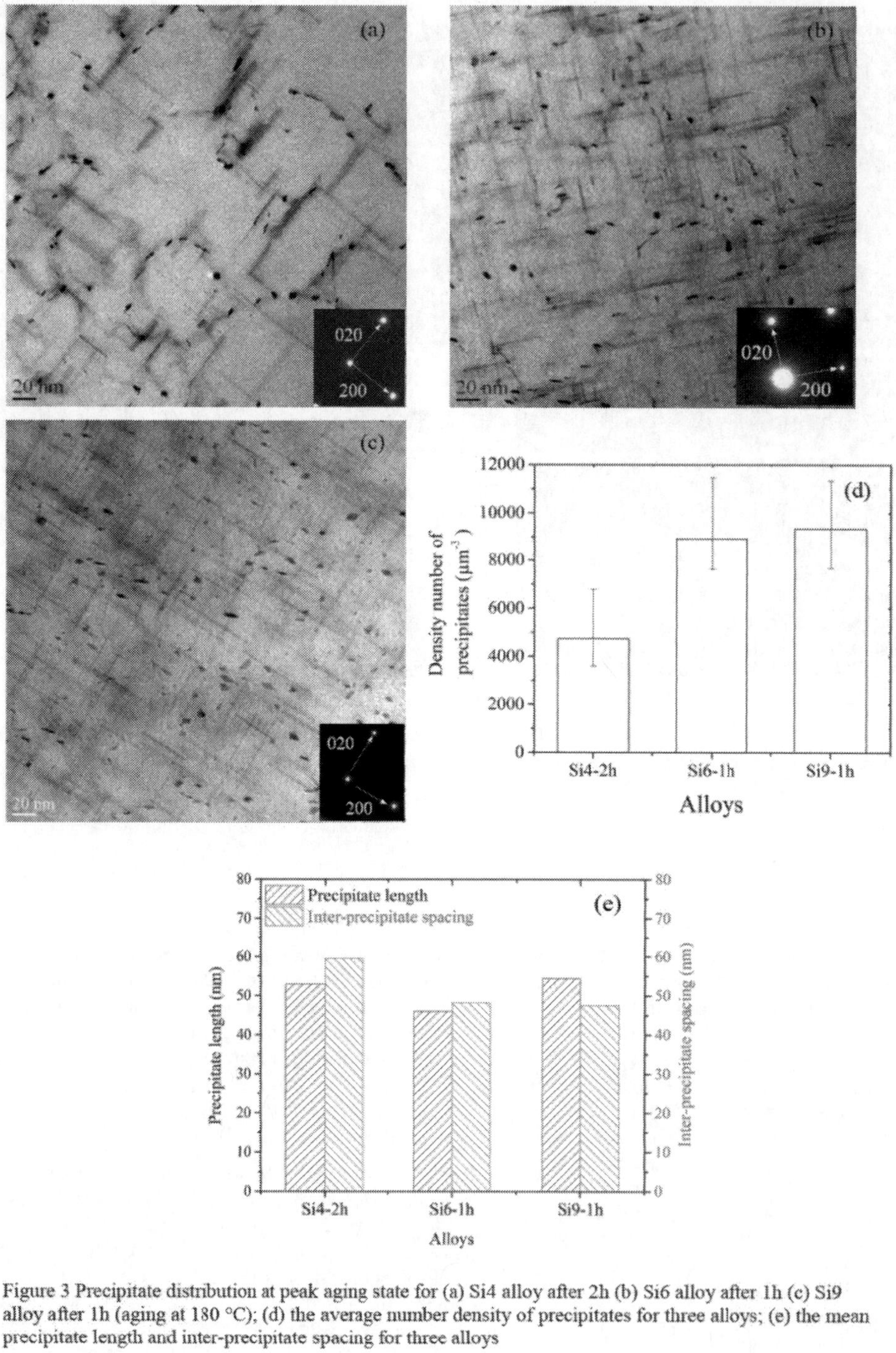

Figure 3 Precipitate distribution at peak aging state for (a) Si4 alloy after 2h (b) Si6 alloy after 1h (c) Si9 alloy after 1h (aging at 180 °C); (d) the average number density of precipitates for three alloys; (e) the mean precipitate length and inter-precipitate spacing for three alloys

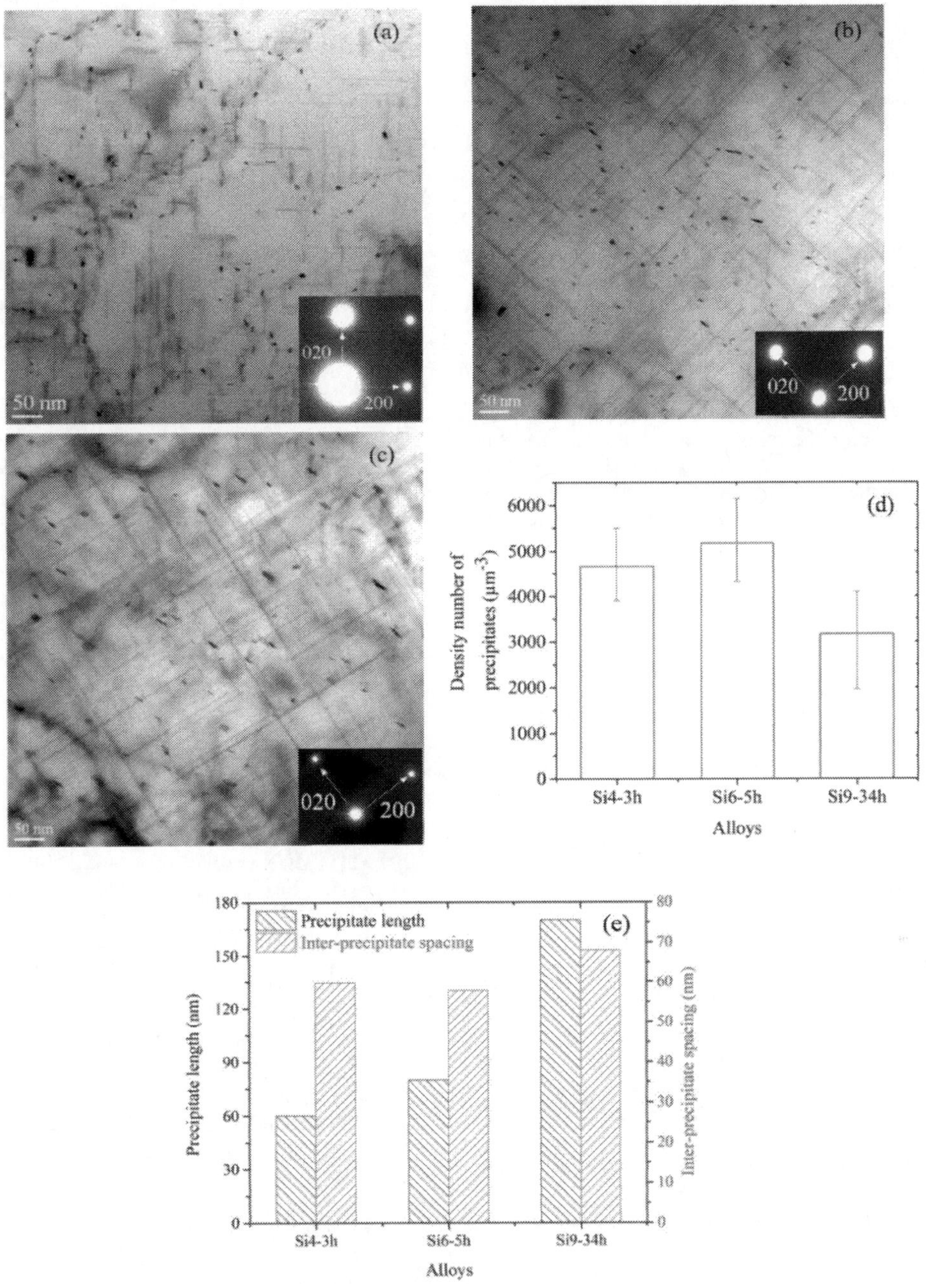

Figure 4 Precipitate distribution for (a) Si4 alloy after 3h (b) Si6 alloy after 5h (c) Si9 alloy after 34 h aging at 180 °C; (d) the corresponding number density of precipitates for three alloys; (e) the mean precipitate length and inter-precipitate spacing for three alloys

4. Conclusions

1. The peak hardness of the Al-Mg-Si conductor alloys increased with Si levels irrespective of the electrical conductivity, mainly resulting from precipitation hardening of nanoscale precipitates.
2. Longer aging times required with increasing Si levels in Al-Mg-Si conductor alloys, especially in the Si-excessive alloy (Si9) to reach the minimum

required EC. The hardness of the Si9 alloy sharply decreased from the peak aging to the prolonged overaging at which the minimum required EC reaches, showing a limited window of hardness above 52.5 %IACS.

3. The Si6 alloy after 5 h aging showed the highest hardness above the minimum required EC. Its corresponding number density of precipitates was considerably higher than other two alloys above 52.5 %IACS. Owing to its better trade-off between EC and hardness, the Si6 alloy can be considered as an appropriate candidate for aluminum conductor applications.
4. The Si9 alloy at 34 h aging exhibited the lowest number density of precipitates but still has higher hardness than the Si4 alloy at 3 h aging, suggesting the other strengthening mechanisms such as strain hardening and solid solution strengthening during wire drawing process can also make a remarkable contribution to the alloy strength.

Acknowledgments

The authors would like to acknowledge the financial support of the Natural Sciences and Engineering Research Council of Canada (NSERC) and Rio Tinto Aluminum through the Research Chair in the Metallurgy of Aluminum Transformation at University of Quebec at Chicoutimi.

References

[1] S. Karabay, Materials & Design, **27**, 821-832, (2006).

[2] S. Karabay, Materials & Design, **29**, 1364-1375, (2008).

[3] M. Iraizoz, N. Rossello, M. Amado, Light Metals 2015, Springer International Publishing, Cham, 183-187, (2016).

[4] F.U. Flores, D.N. Seidman, D.C. Dunand, N.Q. Vo, , Light Metals 2018, 247-251, (2018).

[5] M.G. E. Cervantes, J. A. Ramos, S.A. Montes, Mater. Res. Soc. Symp. Proc, **1275**, 03-09, (2010).

[6] X. Xu, Z. Yang, Y. Ye, G. Wang, X. He, Materials Characterization, **119** , 114-119, (2016).

[7] S. Jiang, R. Wang, Journal of Materials Science & Technology, **35**, 1354-1363, (2019).

[8] Y. Han, D. Shao, B.A. Chen, Z. Peng, Z.X. Zhu, Q. Zhang, X. Chen, G. Liu, X.M. Li, Journal of Materials Science, **52**, 4445-4459, (2016).

[9] Q. Zhao, Z. Qian, X. Cui, Y. Wu, X. Liu, Journal of Alloys and Compounds, **666**, 50-57, (2016).

[10] E. Standard, EN 50183. Conductors for overhead lines, aluminium magnesium silicon alloy wires, 2000.

[11] ASTM, Annual Book of ASTM Standards, Electrical Conductors, 2002.

[12] S.J.A. H.W. Zandbergen, J. Jansen, Science, **277**, 1221-1225, (1997).

[13] H.W.Z. S.J. Andersen, J. Jansen, C. TrÆholt, U. Tundal, O. Reiso, Acta Materialia, **46**, 3283-3298, (1998).

[14] G. Lin, Z. Zhang, H. Wang, K. Zhou, Y. Wei, Materials Science and Engineering: A, **650**, 210-217, (2016).

[15] J. Kim, C. Daniel Marioara, R. Holmestad, E. Kobayashi, T. Sato, Materials Science and Engineering: A, **560**, 154-162, (2013).

[16] Y. Wu, J. Xiong, R. Lai, X. Zhang, Z. Guo, Journal of Alloys and Compounds, **475**, 332-338, (2009).

[17] Y. Birol, Journal of Materials Processing Technology, **148**, 250-258, (2004).

[18] L.-z. Yan, Y.-a. Zhang, X.-w. Li, Z.-h. Li, F. Wang, H.-w. Liu, B.-q. Xiong, Transactions of Nonferrous Metals Society of China, **24**, 939-945, (2014).

[19] M.H. Mulazimoglu, A. Zaluska, F. Paray, J.E. Gruzleski, Metallurgical and Materials Transactions A, **28**, 1289-1295, (1997).

[20] M.H. Mulazimoglu, A. Zaluska, J.E. Gruzleski, F. Paray, Metallurgical and Materials Transactions A, **27**, 929-936, (1996).

[21] H. Nemour, D. Mourad Ibrahim, A. Triki, Journal of Thermal Analysis and Calorimetry, **123**, 19-26, (2015).

[22] K. Teichmann, C.D. Marioara, S.J. Andersen, K.O. Pedersen, S. Gulbrandsen-Dahl, M. Kolar, R. Holmestad, K. Marthinsen, Philosophical Magazine, **91**, 3744-3754, (2011).

[23] C.D. Marioara, S.J. Andersen, H.W. Zandbergen, R. Holmestad, Metallurgical and Materials Transactions A, **36**, 691-702, (2005).

[24] J.H. Chen, E. Costan, M.A. van Huis, Q. Xu, H.W. Zandbergen, Science, **312**, 416-419, (2006).

[25] B. Raeisinia, W.J. Poole, D.J. Lloyd, Materials Science and Engineering: A, **420**, 245-249, (2006).

Influence of the Mg and Bi content on brazeability of MONOBRAZE material under flux-free brazing conditions

Kanokploy Phumprasop[1,*], Dai Yamamoto[1], Noriyuki Yamada[1,] and Hirokazu Tanaka[1]

[1]Research & Development Division, UACJ Corporation, Japan

Abstract. In the conventional brazing, the clad materials with the filler layer and the flux coating have been used to get good brazeability. However, an innovative Al-Si based single-layer brazing material (MONOBRAZE® material) has been developed recently which does not need filler layer by supplying molten filler from the inside of the material during brazing. In addition, a new flux-free brazing technique has also been developed. In the flux-free brazing, Mg and Bi are added to break the oxide film and to improve the wettability of molten filler, respectively. However, it is not clarified how these element works on the brazeability of MONOBRAZE® material in flux-free brazing. Therefore, in this study the effect of Mg and Bi content on the brazeability of MONOBRAZE® material was investigated. From the result, it is found that MONOBRAZE® material is applicable for flux-free brazing by adding Mg. Mg is considered to break the oxide film into fine particles and Bi is considered to assist the destruction of oxide film. From this cause, a new aluminum substrate surface was exposed which allow molten filler to mcve on the surface and contribute to the formation of the fillet.

1 Introduction

1.1 Conventional brazing technology

Aluminum brazing has been applied to connect a lot of components of automobile heat exchanger efficiently. Basically, filler alloy (Al-Si alloy; 9-13mass%Si) is cladded on core alloy (Al-Mn alloy) and it works as molten filler at temperature over 577℃ to connect with other components, while core alloy does not melt[1)]. In order to achieve the good bonding, flux is sprayed on the filler alloy surface to remove the oxide film during brazing.

There are several methods of aluminum brazing such as vacuum brazing, inert gas brazing, VAW brazing and CAB brazing. Currently, NOCOLOK® brazing is a mainstream of aluminum brazing[2)] which use non-corrosive flux to remove the oxide film.

1.2 MONOBRAZE® material

In recent years, an innovative Al-Si based single layer brazing material (MONOBRAZE® material) has been developed which does not need a filler layer by supplying a molten filler that consists of Si around 2.5mass% from the inside of the material during brazing[3)]. From the thermodynamics calculation by JMatPro® as shown in Fig.1, it is clear that the fraction of liquid phase of 2.5mass% Si is 17mass% at the temperature 600℃, while that of 10mass% Si (4045 aluminum alloy) is 100%. Due to the less amount of liquid phase, the alloy with 2.5mass% Si maintains its structural strength, while the liquid phase, generated at mainly grain boundary and Si particles, works as a connecting molten filler sufficiently.

1.3 Flux-free brazing technology

To minimize the manufacturing cost, a new brazing technique without using flux has been developed. Under a vacuum or inert gas atmosphere, it is known that Mg works to destroy the aluminum oxide film during brazing when using a Mg-containing aluminum alloy[4,5)]. Also, adding Bi is known to improve the brazeability by increasing the wettability of the molten filler[4)]. There are reports about brazeability of clad materials using flux-free brazing[4,5)], but it has not ever surveyed whether flux-free brazing can be applicable to MONOBRAZE® material. Therefore, in this study, the effect of the Mg and Bi contents on the brazeability of the MONOBRAZE® material during flux-free brazing was investigated.

* Corresponding author: kanokploy-phumprasop@uacj.co.jp

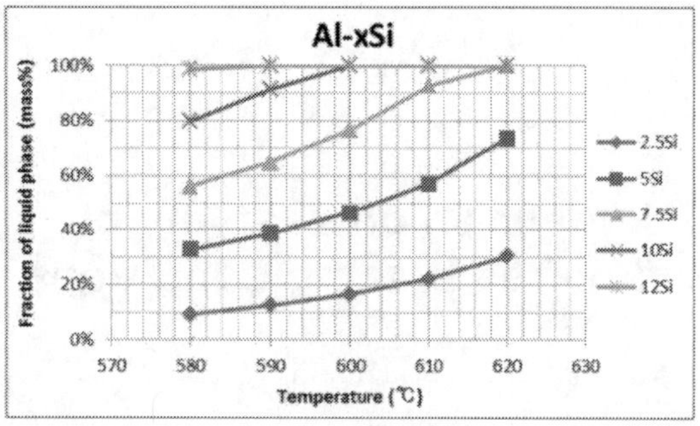

Fig. 1. The fraction of the liquid phase at different Si contents calculated by JMatPro®

2 Experiment procedures

2.1 Material

2.1.1 Chemical composition

The chemical composition of the MONOBRAZE® material is shown in Table 1. In this paper, the concentration of elements means mass concentration (denoted "%" as abbreviation). As the base composition, the MONOBRAZE® material contained 2.5% Si.

For No.2, 0.05% Mg was added with no added Bi. For No.3 and No.4, 0.025% and 0.05% Mg was added with 0.02% Bi, respectively.

Table 1. The chemical composition of MONOBRAZE® material

No.	Chemical composition (mass %)					
	Si	Fe	Mn	Zn	Mg	Bi
No.1	2.5	0.2	1.1	1.5	0	0
No.2	2.5	0.2	1.1	1.5	0.05	0
No.3	2.5	0.2	1.1	1.5	0.025	0.02
No.4	2.5	0.2	1.1	1.5	0.05	0.02

Table 2. The chemical composition of alloy 3003+1.5Zn

Alloy name	Chemical composition (mass %)				
	Si	Fe	Mn	Zn	Cu
3003 + 1.5 Zn	0.25	0.3	1.25	1.5	0.1

2.1.2 Manufacturing condition

The manufacturing condition of the MONOBRAZE® materials is shown in Table 3. Each sample was casted by a continuous casting process into a plate with the thickness of 6 mm. The plate was annealed at 420℃ for 2 hours, rolled down to the thickness of 0.2 mm, and finally annealed again at 370 ℃ for 2 hours. Grain recrytallization of the entire plate was confirmed. Moreover, in order to remove the oxide film that formed during the manufacturing processes, the material was dipped in 2% nitric acid / 1% hydrofluoric acid at room temperature for 90 sec.

Table 3. The manufacturing condition of MONOBRAZE® material under flux-free brazing condition

Process	Condition
Casting	Continuous casting (CC)
Intermediate annealing	420℃ for 2h
Cold rolling	0.20 mm
Final annealing	370℃ for 2h
Etching	2% nitric acid and 1% hydrofluoric acid at room temperature for 90 sec

2.2 Brazing process

In this study, the sample temperature was raised to the target temperature of 600℃ for 5 min., then cooled to room temperature as shown in Fig.2. Brazing was performed under a nitrogen gas atmosphere.

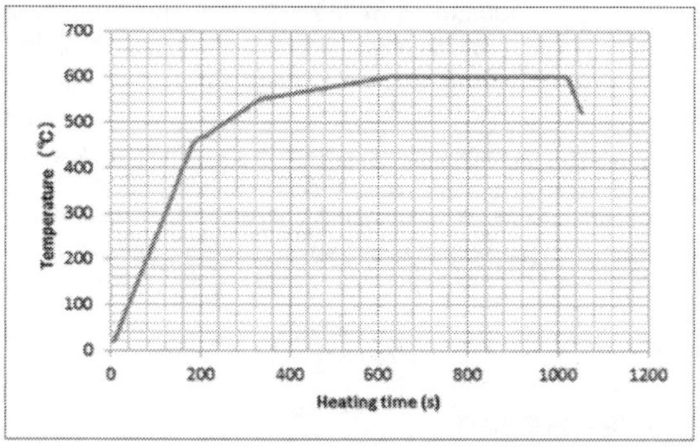

Fig. 2. Brazing heat profile

2.3 Evaluation of brazeability

2.3.1 In situ observation of surface during brazing

In order to investigate how the fillet would be formed during brazing, the T-joint sample was brazed up to 600℃ in a chamber, then cooled to room temperature while continuously taking a video of the surface of horizontal plate (No.1 and No.2) from the top side over the vertical plate (3003, t1.0 mm) as shown in Fig.3. Since sample No.1 was used as a reference of conventional brazing, NOCOLOK® flux was sprayed on the surface of No.1 before assembling the T-joint specimen. After brazing, the cross section of the bonding part between vertical and horizontal plates was observed using a microscope.

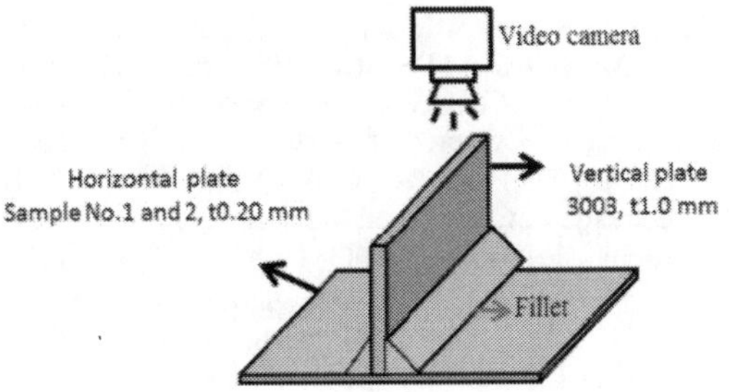

Fig. 3. T-joint sample set up for observation of molten filler alloy

2.3.2 Evaluation of bonding ratio of miniature-core sample

To evaluate the brazeability of the MONOBRAZE® material under flux-free brazing condition, the bonding ratio between the fin and tube of a miniature-core sample was measured. Figure 4 shows a schematic diagram of the miniature-core sample. Samples No.1 to No.4 were used as plates to sandwich the corrugated fin stock (t0.08 mm, H14 temper). The chemical composition of the fin stock is shown in Table 2. The fin pitch was adjusted to 3 mm, and the number of connecting points between fin and tube was 12 per a miniature-core.

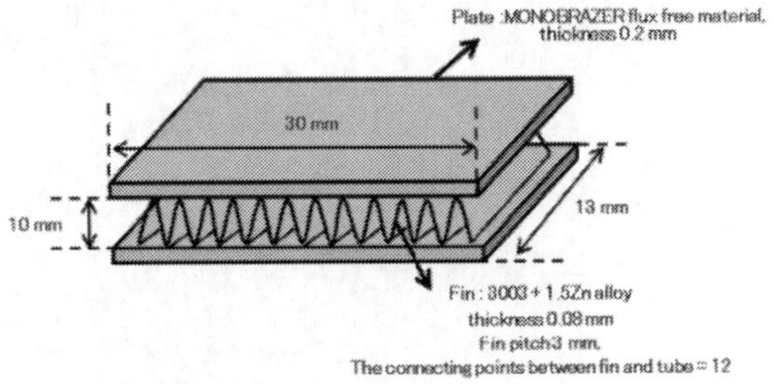

Fig. 4. The schematic diagram of the miniature-core sample

After brazing the miniature-core sample, the fin was peeled off and the lengths of the remaining fillets were measured as shown in Fig.5. The bonding ratio was then calculated by equation (1).

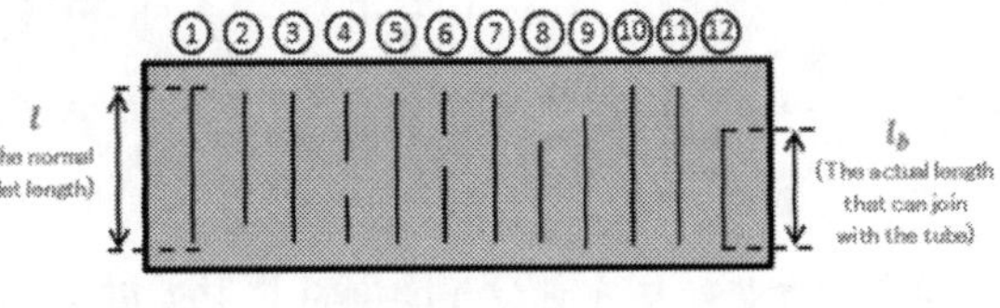

Fig. 5. The schematic of the miniature-core sample

$$\text{Bonding ratio} = \sum_{b=1}^{12} \frac{l_b}{l} x100 \qquad (1)$$

3 Results

3.1 T-joint sample

Figure 6 shows images of the surface of T-joint samples that were taken by a video camera. At a temperature <570 ℃, molten flux was observed on sample No.1 (0Mg-0Bi). Stripe patterns formed by rolling can be seen on the surface of No.2 (0.05Mg-0Bi). At a temperature over 600℃, web-like patterns and small dots, which did not appear at the temperature <570℃, could be seen on the surface of No.2. These patterns were also seen on the surface of No.1 although less visible probably due to coverage of the molten flux. Web-like patterns and small dots are considered to correspond to melt sites of the grain boundary and precipitated Si particles.

Figure 7 shows the cross section of a T-joint specimen. It is clear that the vertical plate and horizontal plate were joined with filler. From the result, there was no difference in the size of the fillet between the conventional brazing and MONOBRAZE® flux free brazing. Moreover, no void was seen for the specimen using MONOBRAZE® flux free brazing, while some voids were found in the specimen with conventional brazing using flux.

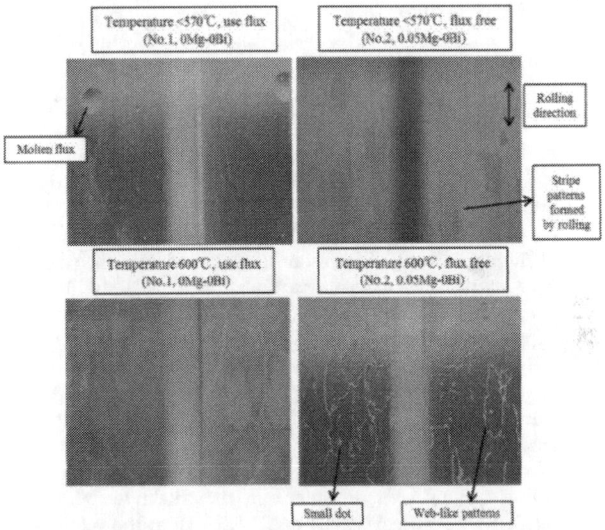

Fig. 6. Surface of T-joint sample during brazing at temperature <570℃ and 600℃

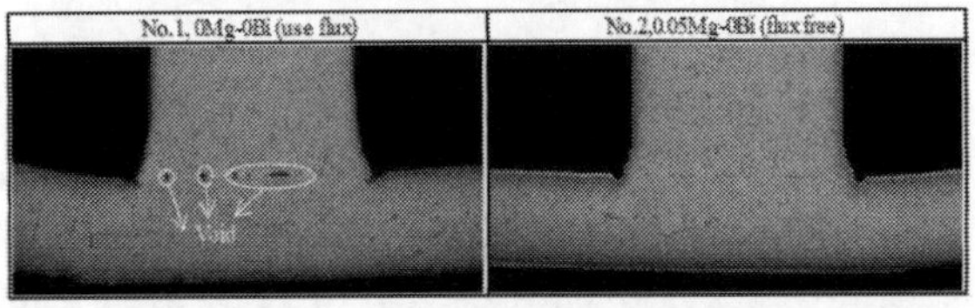

Fig. 7. Cross section photomicrograph at bonding part after brazing

3.2 Bonding ratio of miniature-core sample

Figure 8 shows the result of the bonding ratio for different Mg and Bi contents. When Mg was not added, there was no bonding without flux (No.1). However, when 0.05% Mg was added (No.2), the bonding ratio increased from 0% to 86%. From comparing No.2 and No.4, the addition of 0.02% Bi was considered to increase bonding ratio. When comparing No.3 and No.4, it is clear that the increase of Mg content has positive effect on the increase of bonding ratio.

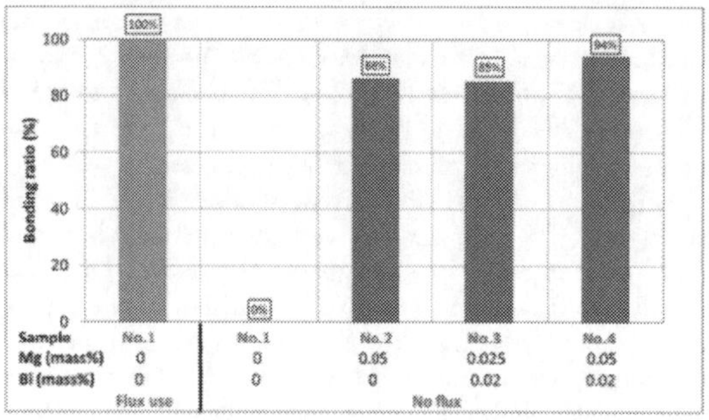

Fig. 8. The result of bonding ratio

4 Discussion

4.1 The effect of Mg and Bi on the destruction of oxide film of plates (MONOBRAZE material itself)

Since the breazeability of a material during flux-free brazing is considered to be closely related to the condition of the oxide film during brazing, the thickness of the oxide film during brazing was investigated by an XPS analysis. To estimate the thickness of the oxide film, the width at half maximum (WHM) of oxygen from the depth profile was measured. Figure 9 shows the WHM of oxygen obtained from samples with different Bi contents which were heated to the target temperature Based on the result of the XPS analysis, there was a tendency that the thickness of the oxide film gradually increased as the brazing temperature increased, and there was no significant difference in the thickness of the oxide film between 0% Bi and 0.02% Bi at the melting temperature 580℃.

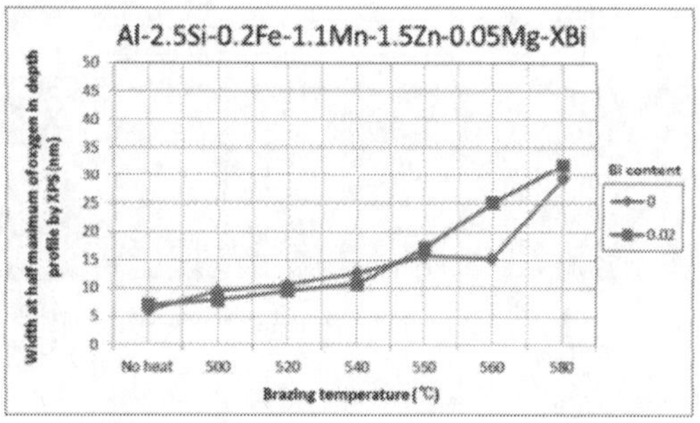

Fig. 9. The relation between the thickness of the oxide film and Bi content during brazing

In addition, the transformation of the oxide film on the surface of MONOBRAZE® material during brazing was also observed by SEM. Figure 10 shows the secondary electron images of surface of samples No.1 (0Mg-0Bi), No.2 (0.05Mg-0Bi) and No.4 (0.05Mg-0.02Bi) which were suddenly cooled at each temperature (550 and 580℃) during heating to 600℃ When focusing on the surface of No.2 at 580℃, fine white particles can be seen on the entire surface, whereas no particles seen on No.1. This tendency became more remarkable with the addition of 0.02%Bi, clearly showing separated areas of fine white particles and black region.

According to Yamayoshi *et al.*[3)], this is the result of the oxide film broken into complex oxide particles by the following reduction reaction;

$$Mg + 4/3Al_2O_3 \rightarrow MgAl_2O_4 + 2/3Al$$

When this reaction happens, the molten filler is considered to move on the surface of the MONOBRAZE® material and allowing flux free brazing.

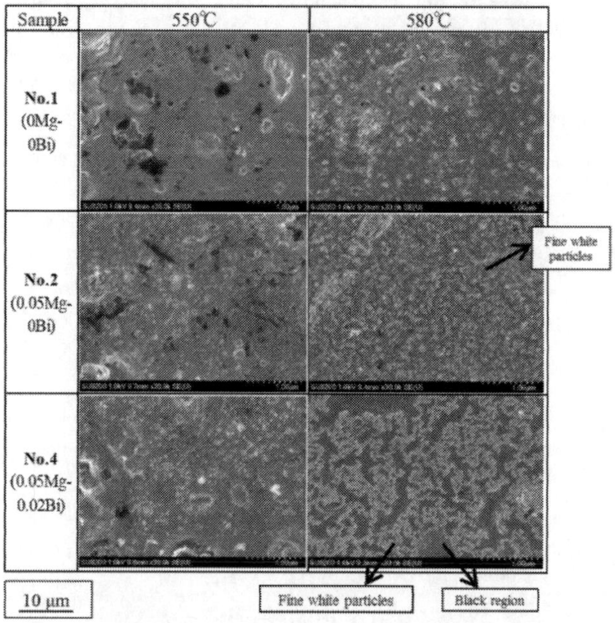

Fig. 10. The secondary election images of MONOBRAZE® material under flux-free brazing condition surface during brazing process

4.2 The effect of Mg condensed in the fillet on the destruction of oxide film of fin stocks

To achieve a good bonding between a plate and a fin stock, the oxide film of fin stock is necessary to be destructed as well. Therefore, in order to clarify the effect of Mg content on the destruction of the oxide film of fin stock, the concentration of Mg in the liquid phase of Al-2.5Si-0.2Fe-1.1Mn-1.5Zn-0.05Mg was calculated by JMatPro® as shown in Fig.11. At the temperature of 590℃, the material is considered to be partially melted and the molten filler is supposed to have a contact with the fin stock. From Fig.11, the Mg concentration in the molten filler is estimated to be about 0.20%. This is 4 times higher than added Mg content (0.05%). This highly concentrated Mg is considered to attack the oxide film of fin stock, resulting in formation of the fillet. This theory is in good accordance with our experiment result that the addition of 0.05% Mg increased the bonding ratio.

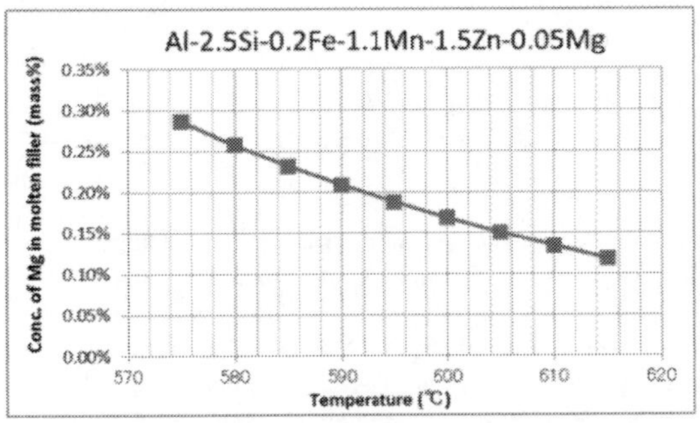

Fig. 11. Concentration of Mg in molten filler of Al-2.5Si -0.2Fe-1.1Mn-1.5Zn-0.05Mg calculated by JMatPro®

However, the result in this study was considered not sufficient to explain the characteristics of the oxide film changes with the Bi content. Therefore, the relation of the oxide film and the characteristics of Bi in terms of the chemical composition and crystal structure will be investigated in a future study.

5 Conclusions

In this research, the influence of the Mg and Bi contents on the brazeability of the MONOBRAZE® material under flux-free brazing conditions was investigated. Followings are conclusions in this study.

1. The MONOBRAZE® material is applicable for flux-free brazing by adding Mg.
2. Mg is considered to break the oxide film into fine particles.
3. Bi is considered to assist the destruction of oxide film.

Base on the above result, a new aluminium substrate surface was exposed which allows molten filler to move on the surface and contribute to the formation of the fillet.

6 References

1. R.S. TIMSIT and B.J. JANEWAY, The Welding Journal, **73**, 119-121 (1994)
2. A.Witterbrood, A. Burgur, S. Kirkham, **10**, 35 (2011)
3. T. Kurosaki, T. Murase, K. Terayama, M. Seki, Y. Betsuki, J. Ninomiya and A. Niikura, The Journal of Japan Institute of Light Metals, **68**, 125-132 (2018)
4. T. Yamayoshi, Y. Itoh and A. Fukumoto, The VTMS13: Vehicle Thermal Management Systems, **4**, 53-57 (2017)
5. DR. W. SCHULTZE and H. SCHOER, The Welding Journal of International Brazing Conference , **4**, 645 (1973)

MATEC Web of Conferences **326**, 08004 (2020) https://doi.org/10.1051/matecconf/202032608004
ICAA17

Similarity and dissimilarity of joint interface morphology in magnetic pulse welded Al/Cu and Al/Ni plates

Shingo Kimura[1*], *Shinji* Muraishi[2], and *Shinji* Kumai[2]

[1]Graduated student, Tokyo Institute of Technology, 2-12-1, Ookayama, Meguro-ku, Tokyo, Japan
[2]Tokyo Institute of Technology, Department of Materials Science and Engineering, 2-12-1, Ookayama, Meguro-ku, Tokyo, Japan

Abstract. A characteristic wavy morphology often appears at the joint interface of magnetic pulse welding (MPW), and an intermediate layer is formed in some metal combinations. It has been known that the wavy morphology changes mainly depend on the density difference between the metals. A sinusoidal wavy interface is formed for the combination of similar metals (Al/Al, Cu/Cu) and that of dissimilar metals having almost the same density (Cu/Ni). In contrast, a trigger-like wavy interface is formed for the combination having a large density difference (Al/Cu, Al/Fe). The difference in strength (hardness) of the solid metal is also assumed to affect the wavy interface morphology. In the present study, two metal combinations (Al/Cu and Al/Ni) were subjected to the MPW to elucidate the effect of hardness difference, since Cu and Ni have almost the same density, but different hardness. Both the MPWed Al/Cu and Al/Ni joints showed a trigger-like wave interface. The wave size (wave-height and wavelength) of Al/Ni was smaller than that of Al/Cu. In Al/Ni, the distribution of intermediate phase was more continuous tracing the outline of the wave. The numerical simulation of the wave formation process was performed using the Smoothed Particle Hydrodynamics (SPH) method. It was revealed that the extent of metal jet penetration into the metal in the process of joining behind the collision point was weaker in Ni than in Cu. This is considered to be due to the larger deformation resistivity of Ni, which is harder than that of Cu.

1 Introduction

Resent years, demand for dissimilar metal joints is increasing in many fields like automobile or electronic devices. Solid-state welding, which does not include melting phenomenon at the joint interface, has been gaining more attention recently.

Impact welding is one of the high-speed solid-state welding methods which can establish strong bonding even in the dissimilar metal combination. In impact welding, a characteristic wavy morphology often appears at the joint interface. An intermediate layer forms in some metal combinations. It has been known that the wavy morphology changes mainly depending on the density difference between the metals. A sinusoidal wavy interface is formed for the combination of similar metals (Al/Al, Cu/Cu) and that of dissimilar metals having almost the same density (Cu/Ni) (Fig. 1 (a), (b)). In contrast, a trigger-like wavy interface is formed for the combination having a large density difference (Al/Cu (Fig. 1 (c)), Al/Fe). In impact welding, it is said that the solid metal around the collision point acts like a fluid due to the extremely high pressure caused by impact. Therefore, the wavy interface morphology is said to be affected by mainly the density which is determined by the fluid's characteristics. However, the difference in strength (hardness) of the solid metal is also assumed to affect the

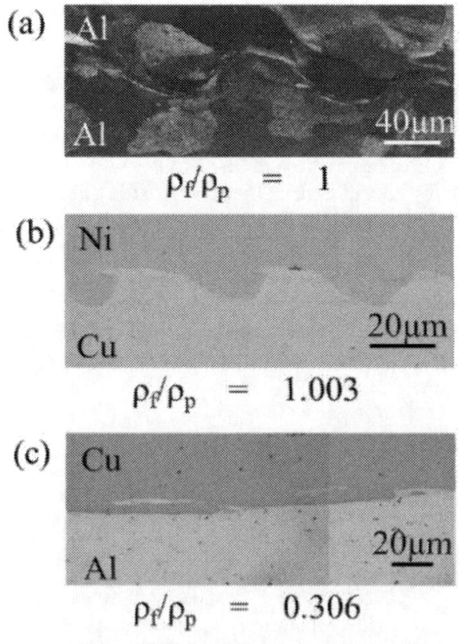

Fig. 1. The joint interface morphology in different density ratio (a) Al/Al, (b) Cu/Ni and (c) Al/Cu fabricated by MPW. (ρ_f: density of flyer plate, ρ_p: density of parent plate.)

* Corresponding author: kimura.s.aj@m.titech.ac.jp

wavy interface morphology.

Magnetic pulse welding (MPW) is one of the impact welding methods which uses electromagnetic force to accelerate and collide two metals together for joining. MPW is used for joining of thin plates or pipes [1]. In the present study, two metal combinations (Al/Cu and Al/Ni) were subjected to the MPW to elucidate the effect of hardness difference on the joint interface morphology, since Cu and Ni have almost the same density, but different hardness.

2 Experimental procedure

Bmax MP12.5/25 with capacitance 40μF was used for MPW. Fig.2 shows a schematic diagram of the experimental setup. A pure Al plate (hereafter Al, 200mm × 70mm × 0.8mm) was used as a flyer plate and a pure Cu and Ni plate (hereafter Cu and Ni, 200mm × 70mm × 0.4mm) were used as a parent plate to fabricate Al/Cu and Al/Ni joint. The initial surface condition was as-received and quite smooth, which underwent cold rolling, so no surface treatment was added. The flyer plate's overlap length over the coil has an important role in MPW [2]. In this study, the overlap length is fixed to 3mm which is equal to the upper side width of coil. The experimental conditions performed in the present study are shown in Table 1. In both combinations, the surface of metal plate was cleaned by acetone before MPW in order to eliminate the effect of surface contamination.

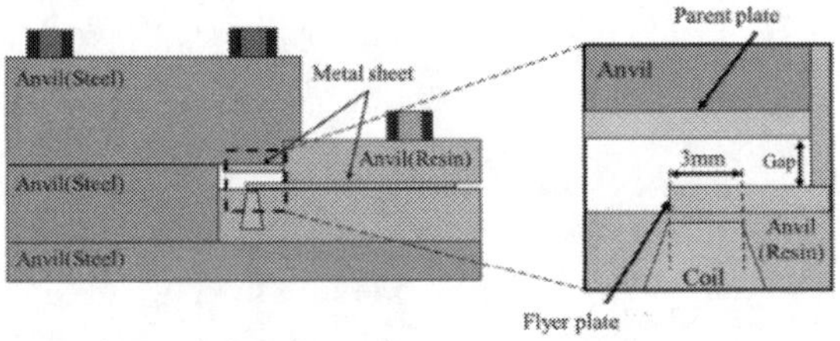

Fig. 2. Schematic diagram of experimental setup.

Table 1. Experimental Condition

Parent plate	Hardness of parent plate (HV)	Charging energy (kJ)	Gap (mm)
Cu	76.9	4	2.4
Ni	181		

After MPW, the joint interface was observed by using an optical microscope and SEM.

3 Numerical analysis method

A series of numerical analyses were performed in order to investigate the interface forming behaviour of both combinations. The analysis was divided into two steps including (1) high-speed oblique collision process caused by electromagnetic force and (2) wavy interface formation process. The numerical analysis models described below are mainly based on the model reported by Kambe et al. [3].

3.1 Deformation manner of the flyer plate and impact angle and velocity analysed by ANSYS Emag-mechanical

In the first step, the magnetic field and deformation manner of the Al plate by electromagnetic force were reproduced by ANSYS Emag-mechanical, which uses the finite element method (FEM) to simulate a situation. Fig. 3 shows the FEM model used in this step. The left side shows a FEM circuit which calculates the current runs through the coil. After the calculation of the current, magnetic flux generated around the coil was calculated with the FEM model shown in the right side of Fig.3. Then, the induced electromagnetic force was calculated, and the deformation manner of the metal plates was reproduced. Finally, the impact velocity and impact angle were obtained from the simulation result.

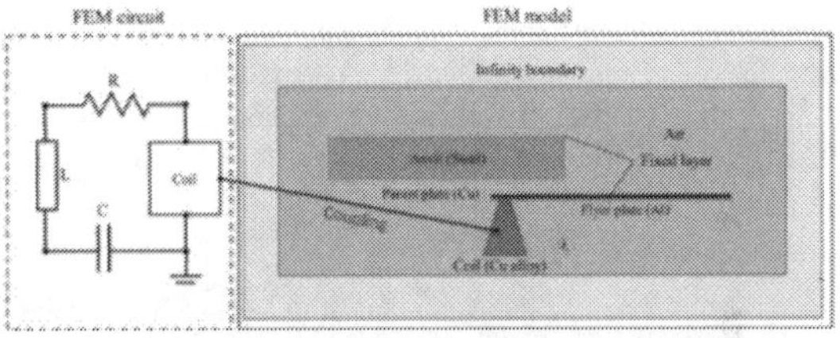

Fig. 3. The FEM model used for ANSYS Emag-mechanical.

3.2 The metal jet emission and the wavy interface formation analyzed by ANSYS Autodyn

In the next step, the collision process including the metal jet emission and wavy interface formation behaviour was reproduced using the smoothed particle hydrodynamics (SPH) method. In this method, materials are discretized into particles which defined as interpolated points of physical quantities. The particles are distributed at a specific distance defined as the smoothing length, h. Since there is no limitation of mesh shape deformation, this method is suitable for large deformation problems like MPW. Fig.4 shows the model used for this step. The smoothing length was set to 1μm near the surface and gradually increased as going away from the surface in order to reduce the calculating time. This analysis aims at reproducing the metal jet emission and the wavy interface formation. The impact velocity and the impact angle obtained by the first step simulation were used for the analysis.

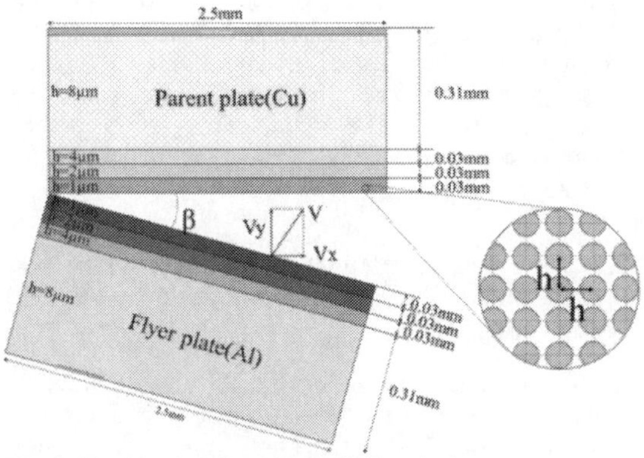

Fig. 4. The Simulation model for SPH analysis.

4 Results and discussion

Fig. 5 shows the obtained interface morphology of each combination. In MPW, the impact velocity and impact angle change through the collision process and these conditions dominate the wavy morphology [4]. To compare the wavy interface formed at the same impact velocity and angle conditions, these pictures were taken from the same distance from the initial impact point. Both joint interfaces showed a trigger-like wavy morphology and contained the intermediate layer. The wave size of Al/Ni was smaller than that of Al/Cu, and the shape of the wave was more clearly observed in Al/Cu than Al/Ni. In addition, the distribution of the intermediate layer was formed more continuously tracing the outline of the wave.

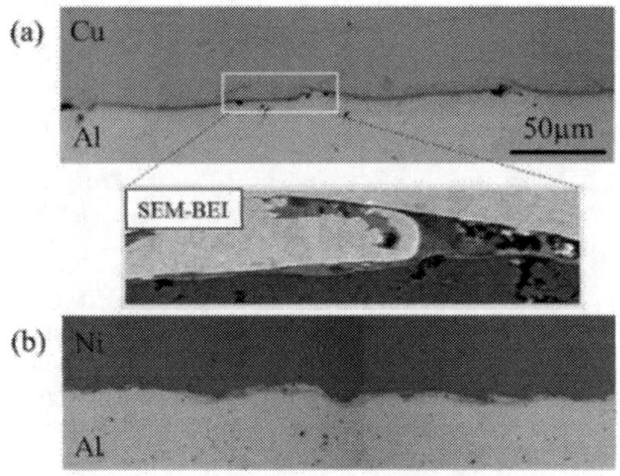

Fig. 5. The joint interface morphology of (a) Al/Cu, (b)Al/Ni

Fig. 6 shows the reproduced wavy interface morphology by using the SPH method. Both interfaces showed a trigger-like wavy morphology and the wave size was smaller in the Al/Ni interface. These results show a good correspondence with the experimental results. Fig. 7 shows the wavy interface formation behaviour reproduced by the SPH method. In this figure, the particles with different colors were used to locate the particles position from the surface in order to detect the material movement at the joint interface in detail. The metal jet emission was reproduced as the SPH particle emission from the collision point. The metal jet was emitted mainly from the Al surface in both combinations. As the collision proceeded, a part of the metal jet penetrated

Fig. 6. The reproduced joint interface morphology of (a) Al/Cu, (b)Al/Ni

into the parent material (indicated by a red arrow) and the rest of metal jet continued emitting forward. In the case of Al/Cu, the penetration became deeper and the wave continued growing after the collision point passed. On the other hand, the penetration was weakened, and the wave did not grow much in the case of Al/Ni. Due to the growth of the wave, a swirling motion occurred in Al/Cu. This is considered to be due to the hardness difference of the parent material. The metal jet could not penetrate much into the Al/Ni because the deformation resistivity of Ni is larger than that of Cu.

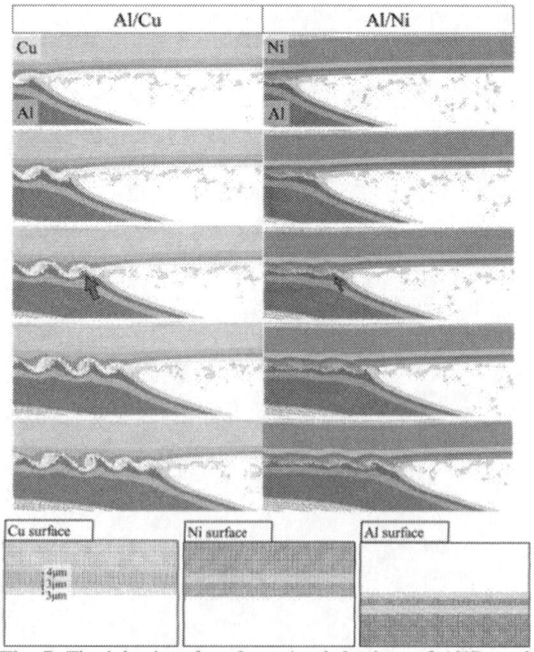

Fig. 7. The joint interface formation behaviour of Al/Cu and Al/Ni reproduced by SPH method.

Fig. 8 shows the temperature distribution and the area above the melting temperature, in which the local melting is assumed, reproduced by using the SPH method. From Fig. 8 (a), a significant temperature rise occurred at the joint interface, especially at the location where the metal jet penetration occurred. In Fig. 8 (b), the red area consisting of red particles indicate that exceeding the melting point, which may result in the formation of intermediate layer in the successive cooling process. In Al/Cu, the distribution of the red area was not continuous and separated from wave to wave. On the other hand, the distribution of the red area was continuous along the interface in Al/Ni. The trace of the red area (Fig. 8) showed good agreement with the trace of the intermediate layer formed at the MPWed Al/Cu and Al/Ni joints (Fig. 5).

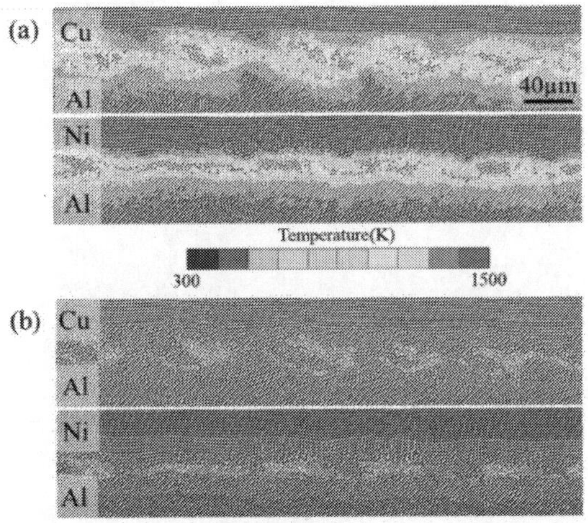

Fig. 8. The temperature distribution (a) and possible local melting zone distribution (b) of reproduced joint interface.

Conclusions

Magnetic pulse welding for two combination, Al/Cu and Al/Ni, were performed. Both joint interfaces showed trigger-like wavy morphology, and the wave size was smaller in Al/Ni combination. The distribution of intermediate layer was formed more continuously tracing the outline of the wave. Numerical analyses were performed to reproduce the wavy interface formation behaviour by using SPH method. The analysis successfully reproduced the wavy interface formation behaviour. The metal jet emission reproduced the emission of SPH particles, and the metal jet was mainly composed of Al particles in both combinations. The wavy interface was formed due to the penetration of the metal jet into the parent material. In the case of Al/Ni, the penetration of the metal jet into the parent material was weaker than in the case of Al/Cu, so a smaller wavy interface was formed for Al/Ni. This is considered to be due to the difference in hardness of the parent material.

Acknowledgement

A part of the present work was supported by JSPS-KAKENHI (JP19K05028) and the scholarship from the Light Metal Educational Foundation. Their supports are deeply acknowledged.

Appendices

The parameters used in this study is shown in the table below. The Mie-Grüneisen EOS based on the shock Hugoniot was applied for the materials in this study. The constitutive model used for this study was the Steinberg-Guinan constitutive model, which represents the mechanical behavior of metals. [3]

Table.2 parameters used for numerical analysis

Symbol	Material property	Al	Cu	Ni
ρ	Density /g·cm^3	2.707	8.930	8.900
Γ	Guruneisen coefficient	1.97	2.02	1.93
c_0	Bulk sound speed /m·s^{-1})	5386	3940	4650
s	Parameter S1	1.339	1.489	1.445
C_v	Specific heat /J·kg^{-1}·K^{-1}	884	383	401
G_0	Initial shear modulus /GPa	27.1	47.7	85.5
Y_0	Initial yield stress /GPa	0.04	0.12	0.14
Y_{max}	Maximum yield stress /GPa	0.48	0.64	1.2
β	Hardening constant	400	36	46
n	Hardening exponent	0.27	0.45	0.53
G'_P	Derivative dG/dP	1.767	1.350	1.393
G'_T	Derivative dG/dT /GPa·K^{-1}	-0.01 669	-0.01 798	-0.02 783
Y'_P	Derivative dY/dP	0.002 608	0.0033 96	0.002 282

References

1. T. Aizawa, J. Japan Inst. Light Met. **54**, (2004), pp153-158
2. Z. Lu, W. Gong, S. Chen, T. Yuan, C. Kan, X. Jiang, J. Manu. Proc. **46**, (2019), pp59-66
3. T. Kambe, Y. Kedo, S. Muraishi, S. Kumai, Mater. Trans. **61** (2020), pp346-354
4. M. Watanabe, S. Kumai, Mater. Trans. **50**, (2008), pp286-292

Development of laser welding of high strength aluminium alloy 2024-T4 with controlled thermal cycle

Mete Demirorer[1], *Wojciech* Suder[1,*], *Supriyo* Ganguly[1,*], *Simon* Hogg[2,*], and *Hassam* Naeem[2]

[1]Welding Engineering and Laser Processing Centre, Cranfield University – United Kingdom
[2]Department of Materials, Loughborough University – United Kingdom

Abstract. An innovative process design, to avoid thermal degradation during autogenous fusion welding of high strength AA 2024-T4 alloy, based on laser beam welding, is being developed. A series of instrumented laser welds in 2 mm thick AA 2024-T4 alloys were made with different processing conditions resulting in different thermal profiles and cooling rates. The welds were examined under SEM, TEM and LOM, and subjected to micro-hardness examination. This allowed us to understand the influence of cooling rate, peak temperature, and thermal cycle on the growth of precipitates, and related degradation in the weld and heat affected area, evident as softening. Although laser beam welding allows significant reduction of heat input, and higher cooling rates, as compared to other high heat input welding processes, this was found insufficient to completely supress coarsening of precipitate in HAZ. To understand the required range of thermal cycles, additional dilatometry tests were carried out using the same base material to understand the time-temperature relationship of precipitate formation. The results were used to design a novel laser welding process with enhanced cooling, such as with copper backing bar and cryogenic cooling.

1 Introduction

Aluminium (Al) is most commonly preferred due to its superior features, such as strength-to-weight ratio, high formability, high durability, corrosion resistance with remarkable cost-efficiency, manufacturability and abundance (Miller *et al.*, 2000; Fridlyander *et al.*, 2002; Sakurai, 2008; Hong and Shin, 2017). There is a growing body of interest that recognises the importance of aluminium processing and its strengthening procedures. The standard features of aluminium can be advanced through some heat treatments, such as age-hardening (Oguz, 1990; Jordan, 2016; Andersen *et al.*, 2018). On the other hand, aluminium alloys are very sensitive to their processing conditions (Nascente *et al.*, 2002) in spite of the fact that these hardening processes provide advantageous properties through generating secondary phases, i.e. precipitates, in the matrix, which create internal stresses by inhibiting dislocation motion which is required for the plastic deformation. Obstacles to the deformation provides to increase the hardness and strength rates crucially (Ambriz and Jaramillo, 2014).

In accordance with results from literature (Sánchez-Amaya *et al.*, 2012; Alshaer, Li and Mistry, 2017; Niu *et al.*, 2017; Ahn *et al.*, 2018) about aluminium welding, previous studies generally focus on problems which occur in fusion zone (FZ), and research to date has not yet determined how the coarsening of precipitates can be inhibited in the over-aged zone, i.e. heat-affected zone (HAZ). As mentioned above, these secondary phases are susceptible to the heat which has a primary role during welding (Zhang *et al.*, 2015). In addition to other problems like hot cracking, porosity etc., softening related to negative precipitate evolution is inevitable upon welding of aluminium alloys. Generally, welding of age hardenable aluminium alloys is avoided because of the sensitivity of these secondary phases to the heat of welding, which results in coarsening of the precipitates in HAZ (Zervaki and Haidemenopoulos, 2007; Zhang *et al.*, 2016). Arora *et al.* (2010) also give essential information about the peak temperature interval of HAZ of welded 2219 aluminium alloy and HAZ usually occurred between 400 °C and 150 °C. Mishra and Sidhar (2017) identified that 1 s - 2 s is enough to make the precipitates in 2xxx aluminium alloys coarsened in higher temperature rates. When we consider this information, and examine the TTP diagrams of 2xxx aluminium alloys, it seems that critical average cooling rate should be higher than 250 °C/s to prevent harmful precipitation coarsening in HAZ, between the temperature intervals from 400 °C to 150 °C.

It makes the aluminium welding process challenging. Although these problems - which occur because of heat retention - can be reduced through generating novel design, considering the process characteristics and dynamics. This novel design should aim to mitigate HAZ softening.

* Corresponding author: w.j.suder@cranfield.ac.uk
* Corresponding author: s.ganguly@cranfield.ac.uk
* Corresponding author: S.C.Hogg@lboro.ac.uk

https://doi.org/10.1051/matecconf/202032608005

2 Experimental Setup

To understand what thermal cycle is needed to mitigate/minimise HAZ softening, a range of experiments were performed, altering the parameters to vary thermal welding cycles. The effect of thermal cycling on the microstructural evolution and related mechanical properties was assessed and it was supported by light optical microscope (LOM) and secondary electron microscope (SEM) analyses.

2.1 Welding Equipment

Diode-pumped continuous wave (1.070 μm) IPG YLR ytterbium fibre laser with 8 kW of maximum power was used in these experiments, as seen by the setup found in Figure 1. To achieve a 0.6 mm beam diameter, 24.3 cm focal length was set between the baseplate and the lens. The angle between the laser beam and the base plate was set to 5° to protect the laser equipment against reflection.

Fig. 1. Experimental setup.

To assess the effect of passive cooling during laser welding, a copper backing bar was added to the jig as a heat sink, which is shown schematically in Figure 2, and it was tried to increase the cooling rate by passive cooling. Thermocouple measurements recorded change in the thermal cycle. Average cooling rates in HAZ were calculated using Equation 1 between the temperature rates of 400 °C and 150 °C, which are dangerous for harmful precipitate evolution, which results in HAZ softening. The active cooling rate to mitigate HAZ softening, was decided.

$$Average\ Cooling\ Rate\ [°C/s]=(400\ °C-150\ °C]/\Delta Time\ [s] \quad (1)$$

Table 1. Composition of material.

Substrate (AA 2024-T4), %					
Al	Ref	**Cu**	4.30	**Cr**	0.03
Si	0.06	**Mn**	0.52	**Zn**	0.18
Fe	0.10	**Mg**	1.50	**Ti**	0.004

$$Heat\ Input\ [kJ/m] = Power\ [kW]\ /\ Speed\ [m/s] \quad (2)$$

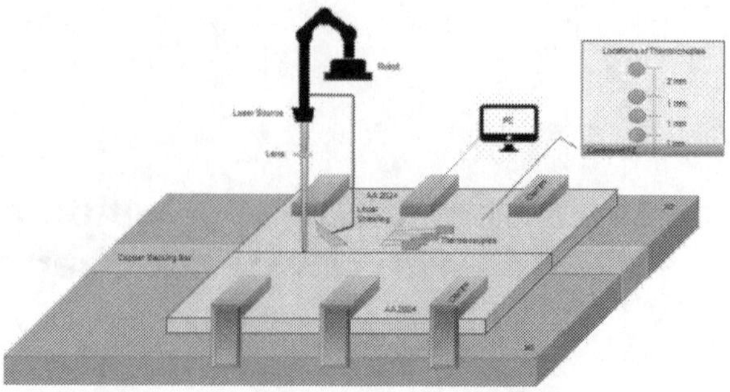

Fig. 2. Schema of the copper bar added laser fixture.

2.2 Welded Materials and Parameters

During the experiments, 2 mm thick AA 2024 alloys in T4 condition were used, and the composition of AA 2024-T4 alloys used can be found in Table 1. Parameters which were used in the experiments can be found in Table 2. Heat input rates were calculated by using Equation 2.

Table 2. Parameters used in laser welding experiments.

	Power P, [W]	Power P,[kW]	Speed V, [cm/min]	Speed V,[m/s]	Heat Input [kJ/m]
1	2500	2.5	100	0.017	150.00
2	8000	8.0	1000	0.167	48.00

2.3 Analysing Data

2.3.1 Macro- and Microstructural Analysis

After performing the laser welding experiments, samples were cut and mounted by casting the pieces of samples into the resin. When the cold mounting was completed, samples were ground by using silicon carbide abrasive papers from coarse P240g, fine P1200g and finer P2500g. After grinding, samples were polished with the chemo-textile clothes with using 6 μm and 9 μm diamond abrasives suspensions, and at the final stage, samples were polished with 0.05 μm OP-S colloidal silica suspension. Finally, samples were etched with Keller's reagent. Afterwards, macro- and micro-images were examined by using LOM and SEM to show and compare the precipitate coarsening with varying the various heat input rates. SEM of Oxford Instruments was used to understand the characteristics of the selected areas of the specimen and to figure out the precipitate coarsening as microstructural analysis in this study.

2.3.2 Hardness Testing

To investigate the change in the mechanical properties, hardness mapping was conducted using the Zwick/Roell ZHV hardness machine in the facilities of the Cranfield University. The load of the indentations was 100 g and dwell time was 15 s. Hardness mapping was performed 0.05 mm below the top surface of the weldment, and the distance between two indentations was 0.5 mm. The first indentation was made into the centre of fusion zone, and 39 indentations were made during the mapping of the weld profile.

2.4 Finite Elements Modelling (FEM)

FEM model was developed on ABAQUS software to simulate the laser welding process. The domain was meshed as seen in Figure 3. Figure 4 shows that the region which was affected by the heat of welding. It was meshed densely, but unaffected areas were meshed coarsely. This model contains 3D volume elements with eight nodes (C3D8T) to make coupled temperature-displacement numerical analysis of heat distribution, and 8800 elements were generated on the domain.

Transient heat transfer analysis was conducted to obtain thermal cycle and to calculate the cooling rate. Moving Gaussian heat source was used by using a DFLUX subroutine, which was programmed using FORTRAN (Tsirkas, Papanikos and Kermanidis, 2003), and it was used with the generated model while loading the conditions.

User-defined surface heat flux which was managed by the DFLUX subroutine - which was used to simulate the laser welding process, and power-travel speed variables were changed in this DFLUX subroutine.

Material properties were taken from the studies of Zhao *et al.* (2012). 5.67 x 10-8 Wm^2/K^4 as Stefan-Boltzmann constant and -273.15 as absolute zero temperature to obtain results in degrees Celsius were set in the model creation stage.

Experimental data were used to validate this FEM model. The increased cooling rate provided by active cooling was simulated, as seen in Figure 5. The simulated cooling rate was enough to exhibit HAZ softening and provide a remarkable clue for developing novel design by including the active cooling method.

Thermal profile and cooling rate of FEM were presented and compared with the results which were obtained from laser only welding simulations. Boundary condition was implemented into this numerical model.

The surface of the plate, except the region where the laser was applied, was fixed to 5 °C. This boundary condition represents the cryogenically cooled surface. It was supposed that the temperature at the top surface of the substrate will drop into the pre-defined lower temperature (5 °C) through using of cryogenic gas.

The boundary conditions were initialised during the welding step to prevent the substrate from pre-cooling. A pre-cooled substrate will not melt.

Fig 3. FEM model generated on ABAQUS software

Fig 4. Coarse and dense mesh on the modelled part.

Fig 5. Model representing cryogenic cooling

3 Methodology

A systematic literature review was conducted of studies regarding how age-hardenable aluminium alloys lose their advanced properties during welding processes, and how much cooling rate is required to mitigate HAZ softening in welded structures. It was found from Arora *et al.* (2010) and Mishra and Sidhar (2017) that processing temperatures between 400 °C and 150 °C are very critical, and the exposure to these temperatures for more than 1 s or 2 s causes disadvantageous precipitate coarsening. Therefore, it must be noted that the average cooling rate must be more than 250 °C/s. A series of welds was achieved with varying thermal cycles and its effect on resulting microstructure and properties assessed. Comparison between the data from literature and experiments was made. The outcome was used to derive a range of possible cooling rates to mitigate HAZ softening offered by standard laser processing, which includes beneficial instrumentations to reduce side effects of the heat of welding process. FEM on ABAQUS software was used to simulate laser only welding process, and to establish if the experimentally obtained cooling rate was enough to inhibit HAZ softening. At the last stage, it was checked which cooling rates are feasible with standard laser welding process and with enhanced cooled ones.

4 Results and Discussion

4.1 Thermal Analysis

Heat input rate changes the cooling rate and peak temperature exposure. It was revealed in Figure 6, which was obtained at the distance of 1 mm from the centre of FZ that using the higher heat input rate (150 kJ/m) during laser processing causes decreased cooling rate. Still, lower heat input (48 kJ/m) provides a higher cooling rate. It means that temperature drops in shorter times in the lower heat input laser welding processing. Hence, narrower weld profiles can be obtained by achieving small HAZ.

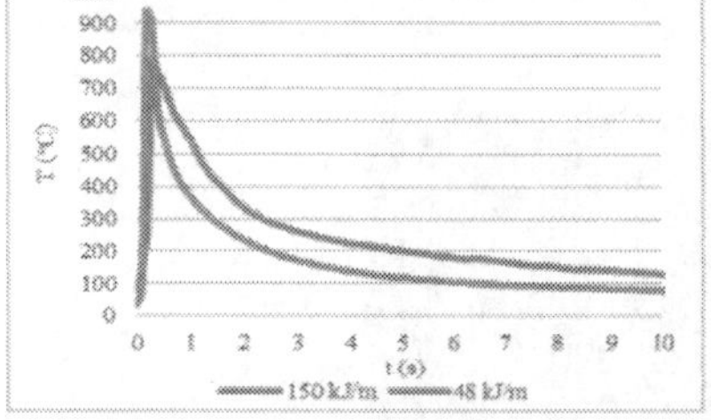

Fig 6. Thermal cycle of lower and higher heat input rates.

As the cooling rate between processing temperatures of 400 °C and 150 °C is very critical, possible cooling rates in HAZ of standard laser welding and copper cooled one were measured and illustrated in Figure 7.

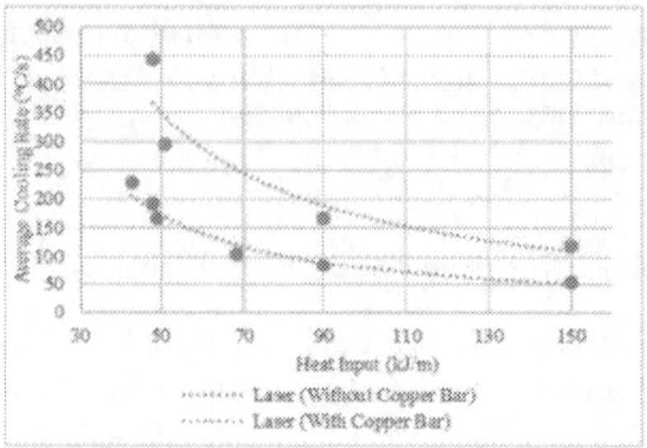

Fig 7. Changing in the average cooling rate in HAZ with alteration of heat input of various processes.

As seen in Figure 7, the application of higher heat input rate of laser only welding causes 55 °C/s as the average cooling rate in HAZ, but lower heat input rate provides 192 °C/s average cooling rates in HAZ. However, it is crucial to say that using a copper bar as heat sink increases the cooling rate remarkably in the heat-affected zone. Higher heat input rate with using the copper bar as heat sink gives us 119 °C/s as the average cooling rate in the heat-affected zone. Similarly, if the lower heat input rate when using the copper bar is applied, it gives us 444 °C/s as the cooling rate in HAZ. Although using heat sink makes the weld profile narrower, it is not enough to prevent HAZ occurrence, so there is a need to use an external cooling device to get rid of heat-affected in the welded structures.

4.2 Macro- and Microstructural Analysis

The SEM image of the parent metal and the HAZ of the laser-welded sample can be found in Figure 8. Some dark, dark-bright and bright phases can be seen in various sizes, and a big difference in the size of the precipitates can be remarkably seen. It was revealed in Figure 8 a) that parent metal consisted of many small-sized and uniformly distributed precipitates, approximately 3 – 4 μm. These should be S' and S" particles. Presence of the fine and evenly distributed metastable precipitates enhances higher strength rates because of several reasons.

Firstly, they are coherent with the matrix and the interface between fine precipitates and other phases is sufficient to limit dislocation movement in each slipping plane through a decreased rate of interfacial free energy rate. Secondly, fine precipitates are distributed in the matrix uniformly, so the structure is more homogenous. It cannot be seen in any precipitate-depleted zone, which causes to reduce strength rate of the material based on the increased rate of inter-particle spacing.

Coarsened constituents, which were illustrated in Figure 8 b), were coarsened from coherent particles due to the heat exposure during the laser welding process. Laser welding with 150 kJ/m causes the cooling rate of 55 °C/s, which is enough to coarsen these fine-sized precipitates and these became 15 – 20 μm and elongated. It means that the amount of the precipitates was decreased due to ease of migration of the atoms due to excessive heat exposure and coagulation of more solute atoms to form large-sized precipitates, which are equilibrium S phases in this case. It causes precipitate-depleted areas to occur and this decreases the mechanical properties. Coarsening and elongation of these precipitates cause to reduce the coherency, and they become incoherent with the aluminium matrix (α-phase).

The coarsening phenomenon can be clearly understood from Figure 8 that HAZ occurs because of coarsening of these main strengthening constituents (precipitates).

Achieving excellent mechanical properties by controlling the cooling rate is the primary purpose of this study. Because of this, parameters were varied to achieve higher cooling rate, which provides sufficient properties in HAZ. Application of heat input rate of 48 kJ/m offers to reduce the duration of heat exposure and it can be figured out that less number of solute atoms collide to coagulate. The distribution of precipitates was also affected by lowering the heat input rate.

Distribution of precipitates is relatively uniform after application of heat input of 48 kJ/m. Still, higher heat input rate (150 kJ/m) creates more precipitate-depleted areas than the application of heat input of 48 kJ/m. Figure 8 c) reveals that the size of precipitates is less (10 – 12 µm) than the alloy which was laser welded with 150 kJ/m. It means that using a lower heat input lowers the thermal degradation, but 192 °C/s average cooling rate is not enough to suppress the harmful precipitate transition.

Changes in the weld characteristics in terms of appearance can be identified in Table 3. Lowering the heat input makes the weld zone smaller. Underfill was decreased and weld became hourglass shape by reducing the heat input and providing stable melt during the laser welding process. Furthermore, it can be easily recognised that using a copper bar increases the cooling rate and HAZ of laser-welded structures which were additionally cooled as the cooper bar became narrower. However, solidification shrinkage and macro-sized porosity occurred because of the higher cooling rate.

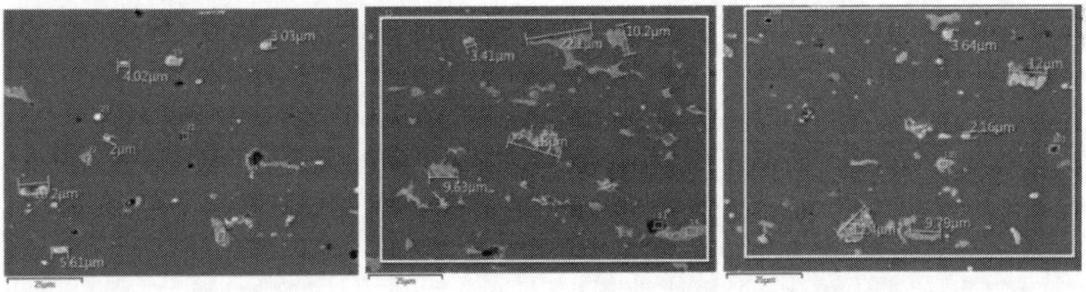

Fig 8. SEM images of AA 2024-T4 alloy, **a)** as received, **b)** laser welded with 150 kJ/m and **c)** laser welded with 48 kJ/m.

4.3 Hardness Measurements

The difference between the copper backing bar in the jig and no copper bar in jig can be understood very well by checking the mechanical properties. It was revealed in Figure 9 that using the copper bar with lower heat input rates increases the material properties and makes the HAZ narrower. It was illustrated that FZ cooled faster through using the copper backing bar. Centre of FZ showed 109.1 HV with using the copper backing bar, but if no copper bar used, hardness in this area becomes 108.4 HV.

Notably, significant differences can be recognised in the area, which consists of columnar grains. As mentioned before, columnar grains occur because of slower cooling rate in this area. However, using a copper bar increases the cooling rate and, they become much smaller, so hardness rates become higher. The hardness of the areas which were not cooled by the copper backing bar are 102.4 HV and 101.3 HV. On the other hand, the hardness of the regions which were cooled by copper backing bar 107.2 HV and 104.4 HV.

Table 3. Macro-graphs of laser-welded structures.

Macro Images of Laser Welded AA 2024 Alloy		
Without Using Copper Bar	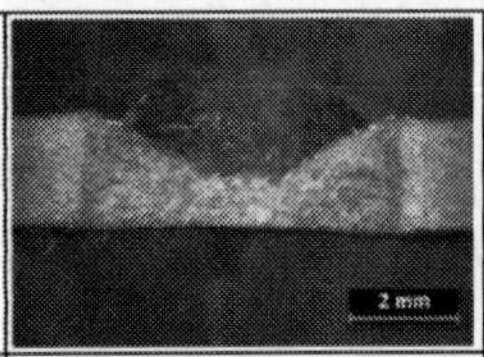	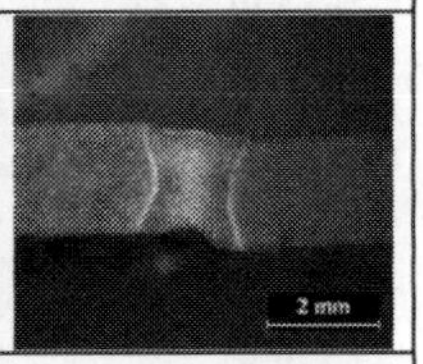
Heat Input	150 kJ/m	48 kJ/m
With Using Copper Bar	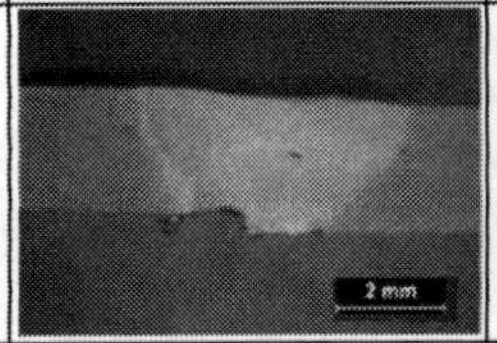	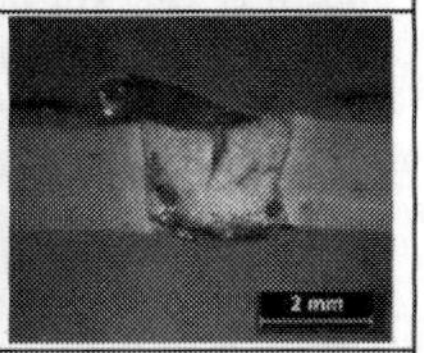
Heat Input	150 kJ/m	48 kJ/m

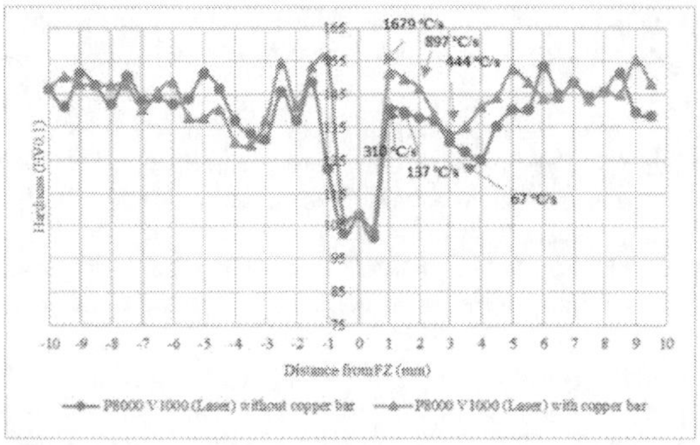

Fig 9. Hardness map of cooled/non-cooled laser welded samples.

The real difference can be seen in the solution strengthened zone. The hardness rates in this zone which were not cooled by the copper backing bar are 140.5 HV in 1 mm, 139.6 HV in 1.5 mm and 137.9 HV in 2 mm, but the hardness rates in this zone which were cooled by the copper backing bar are 151.2 HV in 1 mm, 149.7 HV in 1.5 mm and 146.7 HV in 2 mm. In this area, precipitations were dissolved in the matrix because of the exposed higher temperatures rate which can be seen as an orange line in Figure 6. This temperature rate is not enough to melt this zone. However, it is enough to dissolve the precipitates. Then, a higher thermal gradient in this zone causes to re-precipitate in the grain boundaries in this area. As can be understood from the graph in Figure 9, HAZ locates next to the solution strengthened zone. Hardness in HAZ (3 mm) of laser-welded structure which was not cooled by a copper backing bar was around 130.7 HV, although hardness in HAZ of the copper cooled structure was around 133.2 HV. Increasing the cooling rate from 192 to 444 °C/s by using copper heat sink provides approximately 3 HV increase in HAZ. The reason for this improvement can be clarified by that copper backing bar provides more rapid cooling, which has a positive effect on the evolution and distribution of precipitates in the aluminium matrix. More rapid cooling reduces the number of solute atoms migrating to collide and coagulate, as it can be seen in Figure 8 c). Finer and relatively uniformly distributed precipitates in the aluminium matrix, which are provided by increased cooling rate through using the copper backing bar, contribute to improving the weld profile in terms of micro-hardness rates. It is clearly recognised that using a copper backing bar increased the hardness in HAZ while decreasing the width of this area. However, it is not enough to mitigate HAZ by using copper heat sink as passive cooling.

Figure 9 also shows that the width of HAZ of laser-welded structure, which was not cooled by the copper backing bar, was approximately 4 mm, although the width of HAZ of the copper cooled structure was approximately 3 mm. It means that using the copper backing bar increases the cooling rate from 192 °C/s to 444 °C/s in this case and improves the weld profile due to decreasing the HAZ width.

4.4 FEM Analysis

If the data is established before is examined, it should be noted that using heat sink is not enough to prevent HAZ occurrence, although it seems that it makes the weld profile narrower. However, as the highest hardness rate locates at the distance of 1 mm from the centre of FZ, the amount of average cooling rate in this zone should be enough to create sufficient welded profile. Therefore, there is a need to use an external cooling device to mitigate HAZ in the welded structures.

FEM was used to simulate active cooling and to demonstrate what type of results active cooling can give. Realistically, a cooling rate of 1679 °C/s can be achieved with the backing bar, but a sufficient and homogenous cooling rate cannot be obtained in the distant locations from FZ. It means that some areas suffer from the lower cooling rate, which results in softening in HAZ, which is caused by precipitate coarsening. For this reason, FEM is a handy tool to simulate active cooling and experienced thermal cycle which can be found in Figure 10. Peak temperature rates are decreased in the cooled structure and temperature rates dropped very rapidly, while laser welded and the non-cooled sample was experiencing 900 °C. As it was shown in Figure 10, 0.35 s is enough to cool the laser-welded sample from peak temperature (700 °C) to room temperature.

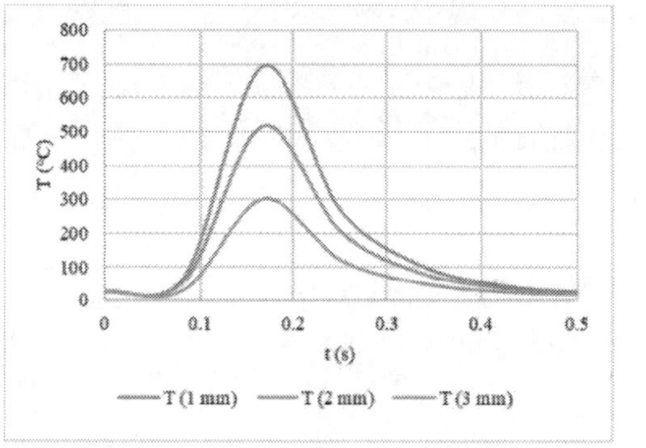

Fig 10. Thermal cycle of the actively cooled model.

Figure 11 shows that decreasing the surface temperature from room temperature to 5 °C during laser welding provides an increase in the cooling rate, and to mitigate HAZ through average cooling rate above 2000 °C/s - which is above the experimentally measured average cooling rate of 1679 °C/s - which provided by passive cooling through using a copper heat sink. It means that lowering the surface temperature through active cooling provides a higher cooling rate and prevents HAZ softening. This type of external trailing cooling device enhances to suppress HAZ occurrence in high strength aluminium alloys after the laser welding process. It provides to reduce the need for post welding heat treatment (PWHT) which can provide to return pre-weld and as-received properties with the additional processes and extra costs.

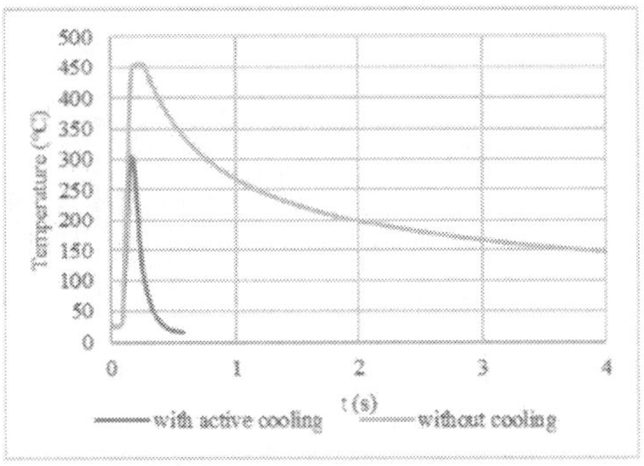

Fig 11. Comparison of the thermal cycle in the distance of 3 mm from the centre of FZ of the actively cooled model with the non-cooled model.

To summarise up, it must be worth bearing in mind that it is not possible to achieve enough cooling rate across HAZ of laser only processed structures. Therefore, some innovative design should be made for laser welding of high strength aluminium alloys, which consists of improved features in HAZ. FEM analysis of active cooling will likely be the right candidate for the prevention of HAZ softening.

5 Conclusion

In this work, an initial correlation between welding conditions, thermal cycle and the resulting microstructural response has been studied.

- Higher heat input causes a decreased rate of cooling rate.
- Heat release from weld zone causes precipitates to get coarsened in HAZ and this coarsening of precipitates decreases mechanical properties. Thermal cycle and cooling rate of welding have a critical influence on the mechanical properties due to negative precipitate evolution.
- Passive cooling through using copper bar increases the cooling rate. Experiments showed that in HAZ copper bar increased cooling rate from 192 °C/s to 444 °C/s, but this improvement is not enough to prevent HAZ occurrence.
- Solid solution strengthened area of the laser-welded sample shows higher hardness rate then, because this zone experienced a higher cooling rate (1679 °C/s) which suppressed the precipitate coarsening.
- It has been proven by FEM modelling that using additional external active cooling device/equipment will be an excellent solution to provide rapid cooling with the cooling rate above 2000 °C/s for prevention of softening in initial zones. It can also increase the productivity of this process by improving the process characteristics with a novel design.

References

Ahn, J., Chen, L., He, E., Dear, J. P. and Davies, C. M. (2018) 'Optimisation of Process Parameters and Weld Shape of High Power Yb-Fibre Laser Welded 2024-T3 Aluminium Alloy', *Journal of Manufacturing Processes*, 34(April), pp. 70–85. DOI: 10.1016/j.jmapro.2018.05.028.

Alshaer, A. W., Li, L. and Mistry, A. (2017) 'Effect of filler wire properties on porosity formation in laser welding of AC-170PX aluminium alloy for lightweight automotive component manufacture', *Proceedings of the Institution of Mechanical Engineers, Part B: Journal of Engineering Manufacture*, 231(6), pp. 994–1006. DOI: 10.1177/0954405415578584.

Ambriz, R. R. and Jaramillo, D. (2014) 'Mechanical Behavior of Precipitation Hardened Aluminum Alloys Welds', in Monteiro, W. A. (ed.) *Light Metal Alloys Applications*. Intech Open Access, pp. 35–59. doi: 10.5772/58418.

Andersen, S. J., Marioara, C. D., Friis, J., Wenner, S. and Holmestad, R. (2018) 'Precipitates in Aluminium Alloys', *Advances in Physics: X*, 3(1), pp. 790–813. DOI: 10.1080/23746149.2018.1479984.

Arora, K. S., Pandey, S., Schaper, M. and Kumar, R. (2010) 'Microstructure Evolution during Friction Stir Welding of Aluminum Alloy AA2219', *Journal of Materials Science and Technology*, 26(8), pp. 747–753. DOI: 10.1016/S1005-0302(10)60118-1.

Fridlyander, I. N., Sister, V. G., Grushko, O. E., Berstenev, V. V., Sheveleva, L. M. and Ivanova, L. A. (2002) 'Aluminum alloys: Promising materials in the automotive industry', *Metal Science and Heat Treatment*, 44(9–10), pp. 365–370. DOI: 10.1023/A:1021901715578.

Hong, K. M. and Shin, Y. C. (2017) 'Prospects of laser welding technology in the automotive industry: A review', *Journal of Materials Processing Technology*. Elsevier B.V., 245, pp. 46–69. DOI: 10.1016/j.jmatprotec.2017.02.008.

Jordan, A. (2016) *Microstructure Characterisation and Corrosion Properties of Two Recycled Aluminium Alloys AA5050 and AA5011*. PhD Thesis. The University of Manchester.

Miller, W. S., Zhuang, L., Bottema, J., Wittebrood, A. J., De Smet, P., Haszler, A. and Vieregge, A. (2000) 'Recent development in aluminium alloys for the automotive industry', *Materials Science and Engineering: A*, 280(1), pp. 37–49. DOI: 10.1016/S0921-5093(99)00653-X.

Mishra, R. S. and Sidhar, H. (2017) 'FSW of Al – Cu and Al – Cu – Mg Alloys', in Gifford, C. (ed.) *Friction Stir Welding of 2XXX Aluminum Alloys Including Al - Li Alloys*. Oxford, UK: Butterworth-Heinemann, pp. 47–77. DOI: 10.1016/b978-0-12-805368-3.00004-2.

Nascente, P. A. P., Bolfarini, C., Benassi, C. L., Alcântara, N. G. and Santos, J. F. (2002) 'Surface and Microstructural Characterisation of Laser Beam Welds in an Aluminum Alloy', *Journal of Vacuum Science & Technology A: Vacuum, Surfaces, and Films*, 20(4), pp. 1416–1419. DOI: 10.1116/1.1487868.

Niu, L. Q., Li, X.-Y., Zhang, L., Liang, X.-B. and Li, M. (2017) 'Correlation between Microstructure and Mechanical Properties of 2219-T8 Aluminum Alloy Joints by VPTIG Welding', *Acta Metallurgica Sinica (English Letters)*, 30(5), pp. 438–446. DOI: 10.1007/s40195-016-0516-9.

Oguz, B. (1990) 'Aluminium Alloys', in *Welding of the Non-Ferrous Metals*. Oerlikon.

Sakurai, T. (2008) 'The Latest Trends in Aluminum Alloy Sheets for Automotive Body Panels', *KOBELCO Technology Review*, 28, pp. 22–28.

Sánchez-Amaya, J. M., Boukha, Z., Amaya-Vázquez, M. R. and Botana, F. J. (2012) 'Weldability of Aluminum Alloys with High-Power Diode Laser', *Welding Journal*, pp. 155s-161s. Available at: http://files.aws.org/wj/supplement/WJ_2012_05_s155.pdf (Accessed: 2 May 2019).

Tsirkas, S. A., Papanikos, P. and Kermanidis, T. (2003) 'Numerical simulation of the laser welding process in butt-joint specimens', *Journal of Materials Processing Technology*, 134(1), pp. 59–69. DOI: 10.1016/S0924-0136(02)00921-4.

Zervaki, A. D. and Haidemenopoulos, G. N. (2007) 'Computational kinetics simulation of the dissolution and coarsening in the HAZ during laser welding of 6061-T6 Al-alloy', *Welding Journal*, 86(8), pp. 211s-221s.

Zhang, L., Li, X., Nie, Z., Huang, H. and Niu, L. (2016) 'Comparison of microstructure and mechanical properties of TIG and laser welding joints of a new Al-Zn-Mg-Cu alloy', *Materials and Design*. Elsevier Ltd, 92, pp. 880–887. DOI: 10.1016/j.matdes.2015.12.117.

Zhang, L., Li, X., Nie, Z., Huang, H. and Sun, J. (2015) 'Microstructure and Mechanical Properties of a New Al-Zn-Mg-Cu Alloy Joints Welded by Laser Beam', *Materials & Design*, 83, pp. 451–458. DOI: 10.1016/j.matdes.2015.06.070.

Zhao, N., Yang, Y., Han, M., Luo, X., Feng, G. and Zhang, R. (2012) 'Finite element analysis of pressure on 2024 aluminum alloy created during restricting expansion-deformation heat-treatment', *Transactions of Nonferrous Metals Society of China (English Edition)*, 22(9), pp. 2226–2232. DOI: 10.1016/S1003-6326(11)61453-2.

Correlation between TEM, SAXS and DSC to investigate the influence of SPD on precipitation mechanisms of an Al-Zn-Mg-Cu alloy

Amandine Duchaussoy[1,*], *Alexis* Deschamps[2], *Xavier* Sauvage[1], *Frédéric* De Geuser[2], *Zenji* Horita[3]

[1]Normandie Univ, UNIROUEN, INSA Rouen, CNRS, Groupe de Physique des Matériaux, 76000, Rouen, France

[2]Univ. Grenoble Alpes, CNRS, Grenoble INP, SIMaP, F-38000, Grenoble, France

[3]School of Engineering, Kyushu Institute of Technology, Kitakyushu, 804-8550, Japan

Abstract. This study allowed to understand the influence of severe plastic deformation by High Pressure Torsion (HPT) on an Al-Zn-Mg-Cu alloy and more especially the influence on precipitation mechanisms thanks to the correlation between DSC, SAXS and STEM data. A shear strain of $\gamma \approx 200$ was first applied to the alloy. This deformation leads to nanostructuration, creation of high density of defects but also solute segregation to boundaries and dynamic precipitation. These specific microstructural features give rise to an acceleration of precipitation kinetics during heat treatments but also to a lower precipitation temperature and a modification of final precipitates size distribution as compared to the un-deformed alloy.

1 Introduction

7### aluminium alloys are high strength alloys mainly used for aerospace applications due to their high specific mechanical properties [1]. However, in order to further increase these properties, we propose to combine precipitation hardening with grain boundary strengthening thanks to ultrafine-grained (UFG) structures. It has been shown that severe plastic deformation (SPD) achieved for example by equal-channel angular pressing (ECAP) or high-pressure torsion (HPT) may efficiently refine coarse-grained microstructures down to the sub-micrometer scale in a large variety of aluminum alloys [2–5]. However, the extreme level of strain applied during these processes leads to a very high level of stored energy that usually gives a poor microstructure thermal stability and a faster precipitation kinetics [6,7]. This is attributed to an increased atomic mobility promoted by crystalline defects and to a lowering of the nucleation barrier.

In this study, results obtained by small angle X-Ray scattering (SAXS), differential scanning calorimetry (DSC) and transmission electron microscopy have been correlated to understand the deformation influence on the precipitation of an Al-Zn-Mg-Cu alloy.

2 Experimental

2.1 Material

The material investigated in the present study has the following composition (wt.%) of: Al 87.19%, Zn 9.46%, Mg 1.87%, Cu 1.33%, Fe 0.8% and %<0.2 Zr+Sc. This alloy has been cast with an ultra-fast solidification rate to promote a fine initial intermetallics distribution. It was subsequently solution heat treated at 474°C during 1h.

2.2 High Pressure Torsion (HPT)

The HPT process (**Fig. 1**) has been used to deform the alloy. Disc shape sample of 10mm diameter and 1mm thickness was deformed under a pressure of 6 GPa at RT with a rotation speed of 1 rpm. The shear deformation is a function of the number of revolutions *N*, the distance from the disc centre *r* and its thickness after HPT *h* [8]:

$$\gamma = \frac{2\pi N.\, r}{h} \quad (1)$$

* Corresponding author: amandine.duchaussoy@uclouvain.be

https://doi.org/10.1051/matecconf/202032608006

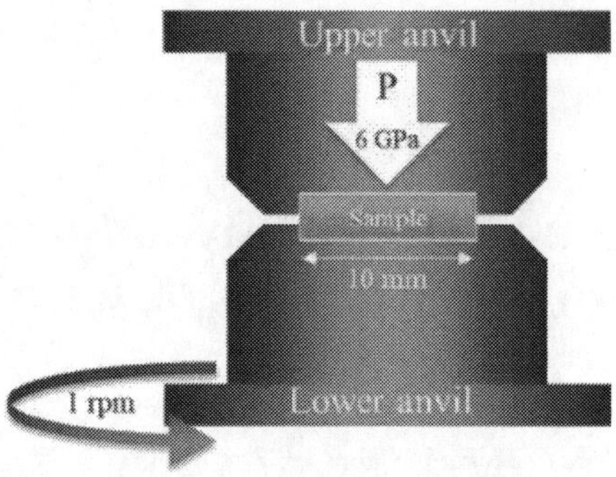

Fig. 1 : Schematic representation of HPT process

In this study, the level of deformation has been achieved by applying 10 revolutions. The anvil temperature, monitored during processing with an infra-red temperature measurement device, never exceeded 50 °C.

2.3 Characterization methods

Microstructure and more specifically precipitates evolution were observed using several complementary techniques.

First, scanning electron microscopy (SEM) was performed with a Gemini Leo1530 FEG-SEM microscope using back-scattered electron (BSE) detector.

Scanning Transmission Electron Microscopy (STEM) was carried out using a JEOL ARM-200F operated at 200 kV. Dark Field (DF - collection angles 20–80 mrad) and high-angle annular dark field (HAADF – collection angles 80–180 mrad) images were recorded with a probe size of 0.2 nm and a convergence angle of 34 mrad. TEM samples were prepared as 3 mm diameter discs whose center was halfway between the center and the edge of the HPT disc. At this location, the corresponding shear strain can be estimated thanks to eq.(1) to $\gamma \approx 200$. Then TEM foils were prepared by conventional mechanical polishing down to 100 μm thickness, followed by double-jet electropolishing (performed on a Stuers TenuPol-5) with a mixture of 30% HNO3 + 70% CH3OH at a temperature of −30 °C (~20 V).

Differential Scanning Calorimetry (DSC) measurements were performed using a Mettler Toledo DSC 3+ at a heating rate of 20 °C/min up to 500 °C. Samples were placed in pure Al crucibles with a capacity of 40 µl able to receive samples with an average size of 3*3 mm and a weight between 30 and 50 mg.

X-ray diffraction (XRD) analysis was carried out with diffractometer operating with CuKα irradiation with an accelerating voltage of 40 kV and a current of 40 mA. The Kα2 radiation was filtered out from the XRD profile.

SAXS was performed on a custom built laboratory setup with a rotating anode source at the Kα energy of Cu (about 8 keV) and a Pilatus 2D detector. CCD camera data were corrected for read-out noise, distortion, flat-field and background noise. *In situ* experiments were carried out using a dedicated furnace in which the samples, thinned to 70–100 µm, were placed. The diameter of the X-ray beam was 1 mm, and the signal was recorded from a location placed at the mid-distance between the center and the edge of the HPT-processed disc in order to probe a reproducible microstructure from one sample to the other and to be able to correlate this data with STEM observations. The SAXS 2D-data was radially averaged and normalized into the 1D-profile of scattered intensity I versus the scattering vector amplitude q; detailed procedures for data interpretation and used model can be found in [9–11]. This model needs some hypothesis which can be given by other experimental methods like

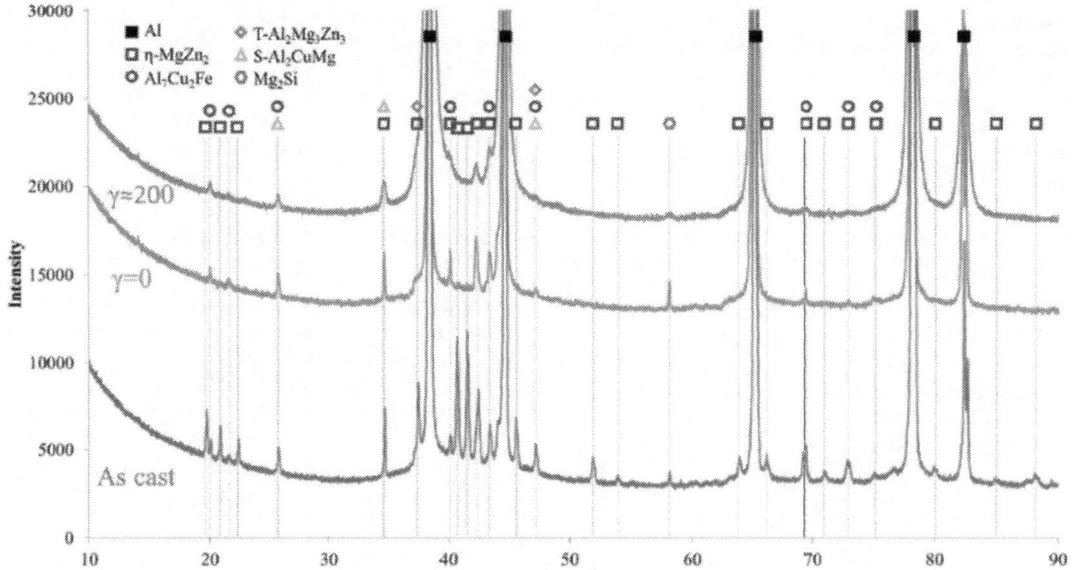

Fig. 2 : XRD patterns of an Al-Zn-Mg-Cu alloy after casting, after solution treatment ($\gamma = 0$) and after deformation by HPT ($\gamma \approx 200$).

https://doi.org/10.1051/matecconf/202032608006

precipitates composition in Zn (around 53%) and their shape (assumed to be spherical).

3 Results

3.1 Influence of deformation on microstructure

3.1.1 Influence on present phases

First, XRD analyses (**Fig. 2**) show the presence of different phases after casting such as η, T, S which are mainly composed of Zn, Mg and Cu. But also the presence of Al_7Cu_2Fe intermetallic phase. After solution treatment at 474 °C during 1 h, most of the η peaks disappear which means a large dissolution of this phase. Some enriched Zn-Mg-Cu remain after solution treatment but with a small amount. And insoluble Al_7Cu_2Fe intermetallic particles are still present. After HPT deformation, all peaks are still exhibited and no other appears, which means that dissolution or precipitation of a new phase induced by deformation probably did not occur.

As expected, the aluminium peaks of the deformed ultra-fine grain material are broader than the ones of the un-deformed coarse grain material. This broadening is due to the increase of defects introduced during the HPT intense deformation, namely grain boundaries, sub-grain boundaries and dislocations [12]. The diminution of some peaks intensity could the results of fragmentation of particles down to the nanoscale.

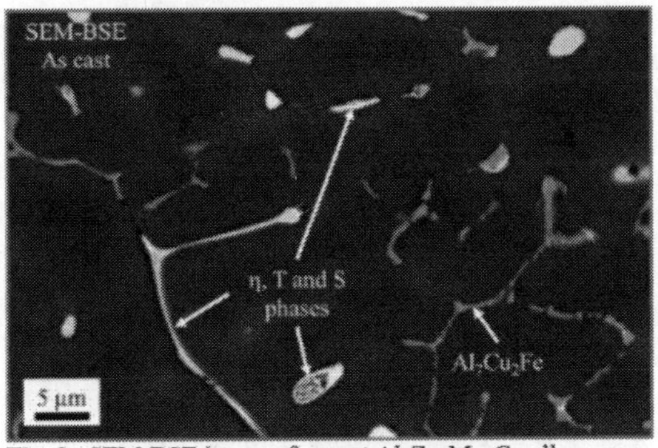

Fig. 3 : SEM-BSE image of as cast Al-Zn-Mg-Cu alloy

SEM-BSE images of as cast alloy (**Fig. 3**) confirm the presence of the different phases revealed by the XRD pattern. Image on **Fig. 4**(a), after solution treatment shows the un-dissolved intermetallic particles. SEM-BSE images after deformation (**Fig. 4**(b)) shows only a limited fragmentation of these intermetallic particles. Quantitative measurements reveal that the average intermetallic particles size evolved from 0.76±0.01 µm to 0.55±0.02 µm. And that the surface fraction is stable.

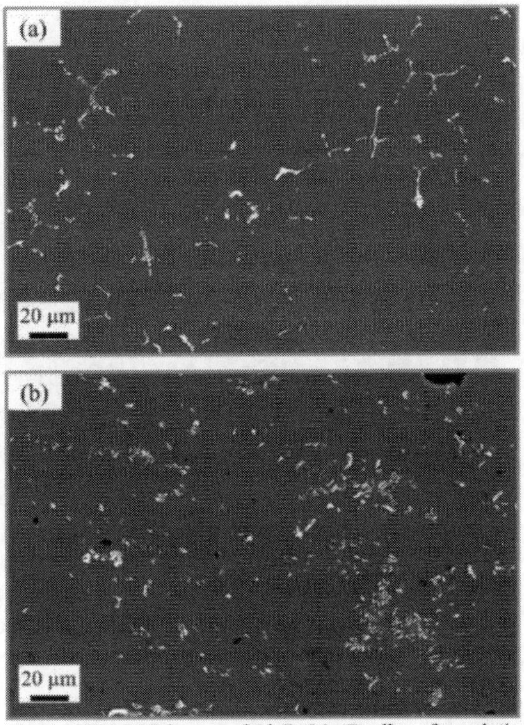

Fig. 4 : SEM-BSE images of Al-Zn-Mg-Cu alloy after solution treatment (a) and after deformation by HPT (b)

3.1.2 Nanostructure formation by intense deformation

The STEM-HAADF images of **Fig. 5**(a) and (b) recorded in the un-deformed material (after quenching from solution treatment and natural ageing) shows a sub-grain boundary on which some heterogeneous precipitation occurred. A higher magnification image (**Fig. 5**(c)) allows to visualize Al_3(Sc, Zr) dispersoïds and to distinguish some very small precipitates which can be identified as GP-zones thanks to previous works [13–15].

After deformation (**Fig. 5**(d,e,f)) the microstructure is significantly different. The DF image (**Fig. 5**(d)) exhibits lots of disorientations, which can be grain or sub-grain boundaries or even dislocations. The HAADF Z contrast images (**Fig. 5**(e,f)) highlight segregation of heavy elements on these defects but also locally precipitation. EDS measurements (shown in Table 1) reveal that segregations and precipitations are enriched in Zn, Mg and Cu.

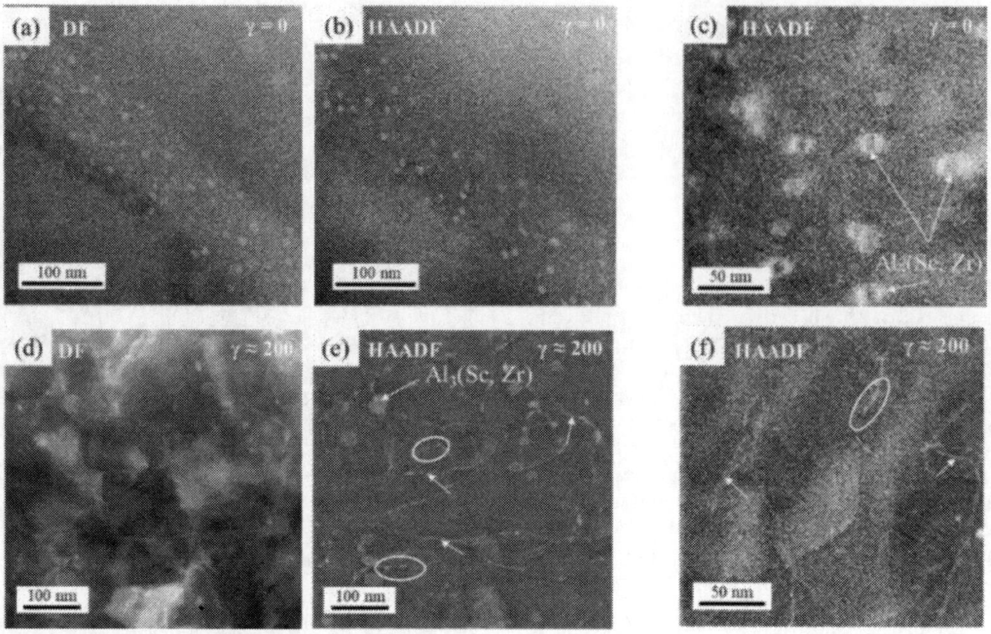

Fig. 5 : Dark field (a,d) and HAADF (b, c, e, f) images of un-deformed sample (a,b,c) and HPT deformed sample (d, e, f) of an Al-Zn-Mg-Cu alloy.

wt.%	Matrix	σ	Precipitates	σ
Al	71.4	0.5	61.8	0.5
Zn	21.3	0.4	28.9	0.4
Mg	1.8	0.2	2.6	0.2
Cu	5.0	0.2	6.2	0.3
Sc	0.0	0.0	0.0	0.0
Zr	0.2	0.3	0.1	0.3
Fe	0.4	0.1	0.4	0.1

Table. 1 : EDS measurements in the matrix and in precipitates (heterogeneous precipitation) for the un-deformed sample.

These results show that HPT deformation has a huge influence on the microstructure, and this can clearly influence precipitation during subsequent heat treatments. Before investigating such heat treatments, SAXS analyses were performed at 25°C to correlate the STEM data with the size distribution found by SAXS.

3.1.3 Correlation between STEM and SAXS results

The precipitates size distributions were estimated from SAXS data fitting with a Monte Carlo procedure similar to that of the McSAS algorithm [10,11]. It has been performed for both the deformed and undeformed alloys, and they are presented in **Fig. 6**. First, these curves show that both samples contain a population of precipitates with a size between 0.5 and 1.5 nm, consistent with GP zones [9].

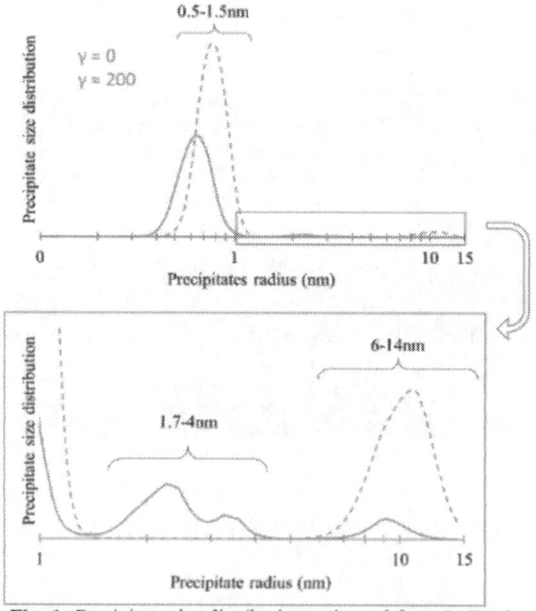

Fig. 6 : Precipitate size distribution estimated from SAXS data fitting for un-deformed (dashed line) and deformed (full line) Al-Zn-Mg-Cu alloy.

However, the areas under the curves suggest that there is a higher volume fraction of these precipitates for the un-deformed sample. Besides, for this latter material, another population of precipitates is found with size between 6 and 14 nm. This size corresponds to the inter-granular precipitates observed by STEM. For the deformed sample, this population is also exhibited but with a much lower volume fraction as compared to un-

https://doi.org/10.1051/matecconf/202032608006

deformed material. Nevertheless, a third population, with a size distribution between 1.7 and 4 nm, is also exhibited but only in the deformed sample. This is consistent with the precipitates observed on defects by STEM.

These first results show that deformation by HPT leads to a nano-structuration of the structure with high level of defects like grain boundaries, sub-grain boundaries and dislocations. The deformation leads to precipitation of Zn, Mg and Cu closely related to their segregation on the defects. A different precipitate size distribution as compared to un-deformed alloy is then achieved. This change could modify the alloy behavior during heat treatment. Indeed, it is well known that high density of defects usually leads to faster diffusion and decreased nucleation barrier. In previous works, precipitation sequence was changed or accelerated during heat treatment after severe plastic deformation [9,16]. The observed segregations could also have an influence during heat treatment.(ZH: I do not see description for Fig.7 before Fig.8.)

3.2 Alloys behaviour during heating ramp

DSC curves (**Fig. 8**) show clearly a different behaviour during annealing in continuous heating conditions between deformed and undeformed samples.

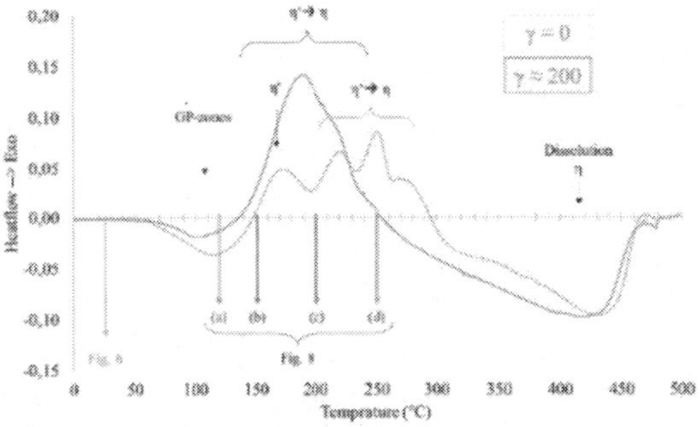

Fig. 8 : DSC curves of un-deformed (dashed line) and deformed (continue line) of Al-Zn-Mg-Cu alloy during a temperature ramp of 20°C/min

Indeed, for the un-deformed sample, the precipitation sequence during 20°C/min ramp is decomposed in several steps: Dissolution of GP-zones (~100°C), then, formation of η' precipitates (~170°C) and evolution of η' in stable η precipitates (~200 to 300°C). Finally, there is dissolution of precipitates between 320 and 450°C [15,17,18]. In the deformed sample, the GP-zones dissolution appears at lower temperature, the precipitation of η' also occurs at lower temperature (~80°C) and all the following transformations are gathered within the same exothermic peak (excepted for dissolution of precipitates). This shows a diminution of precipitation temperature followed by an acceleration of the precipitation kinetics. Indeed, the precipitation

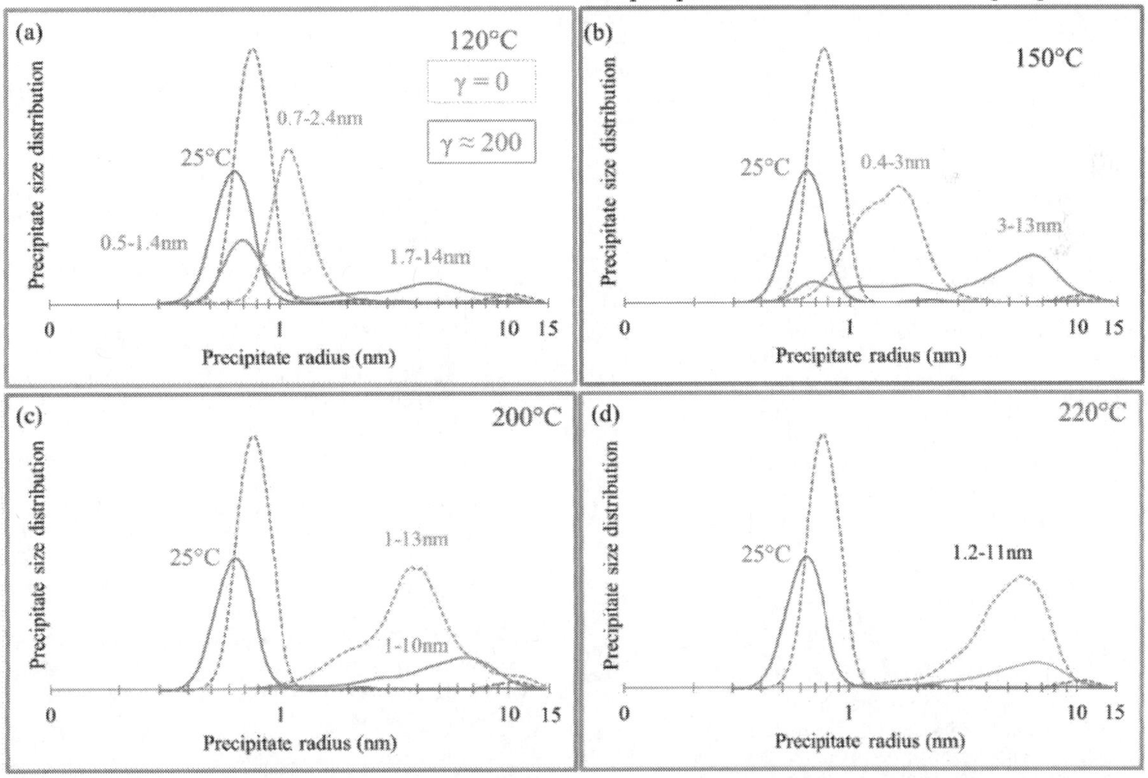

Fig. 7 : Precipitate size distribution estimated from fitting SAXS data for un-deformed (dashed line) and deformed (full line) Al-Zn-Mg-Cu alloy at different temperatures during a ramp of 5°C/min

completion is around 300°C for the un-deformed sample and around 250°C for the deformed sample.

It can be noted that the area of the first endothermic peak, corresponding to the GP-zones dissolution, decreases with the deformation. It also seems to shift towards lower temperatures (from 120 ° C to 100 ° C). According to [15], this would indicate the presence of smaller GP zones before annealing. Furthermore, the decrease in the peak area indicates a decrease in the GP-zone volume fraction present in the samples deformed by HPT. Thus, the deformation induces a decrease in the size but also in the volume fraction of GP-zones. The segregations observed after deformation (**Fig. 5**) could reduce the quantity of elements available for homogeneous precipitation and may contribute to this difference. These results are in agreement with the SAXS curves at 25°C (**Fig. 6**) which show a lower volume fraction of GP-zones for the deformed sample with a smaller average size.

Then, different temperatures were chosen, and are indicated in **Fig. 8** by arrows, to display SAXS precipitate distributions. It should be noted that the ramp is not exactly the same during DSC and SAXS experiments: 20°C/min and 5°C/min respectively.

Precipitate size distribution at 120°C (**Fig. 7(a)**) for the un-deformed sample shows a monomodal distribution with precipitate size between 0.7 and 2.4 nm. So there is an increase of precipitate size and a decrease in volume fraction as compared to the distribution at 25°C (grey curve).

For the deformed sample, there is also a decrease in volume fraction and in the average size of this smaller population of precipitates. But the growth rate is lower and the decrease in volume fraction is more important (~50% compared to ~35%) than for the un-deformed sample. Moreover, the volume fraction of bigger precipitates population (2-14 nm) increases.

At 150°C (**Fig. 7(b)**), the area under the curve of the monomodal distribution for the un-deformed sample is 40% lower as compared to 120°C. This shows an increase in volume fraction. A small increase in average size (~1.5 nm to ~2 nm) is also observed. For the deformed sample, the behavior is completely different. Indeed, the volume fraction of smaller population continues to decrease and that of the bigger precipitates continues to increase.

At 200°C (**Fig. 7(c)**), the volume fraction and the average precipitate size continue to increase for the precipitates of un-deformed sample. For the deformed sample, the small population of precipitates totally disappears. The average size of the larger precipitates population continues to increase but the volume fraction is now decreasing.

Finally, at 220°C, the precipitate population of un-deformed sample simply grows (increases in size but not in volume fraction). Contrarily to the precipitates population of the deformed sample which don't grow and exhibits a decrease in volume fraction.

These results show that in the undeformed sample, a progressive growth of precipitates occurs during the temperature ramp with first a decrease in volume fraction and then a kind of stabilization. For the deformed sample, there is a rapid loss of the small precipitates balanced by the formation of precipitates with a wide size distribution which rapidly decreases in volume fraction certainly because of the bigger precipitate population (>15nm). The rapid precipitation kinetics and the wide range of size precipitates is in agreement with DSC curves, and also with the lower precipitation temperature.

3.3 Alloys behaviour during isothermal treatment at 100°C.

Isothermal heat treatment was carried out on both samples and the microstructure evolution was followed both by SAXS and STEM.

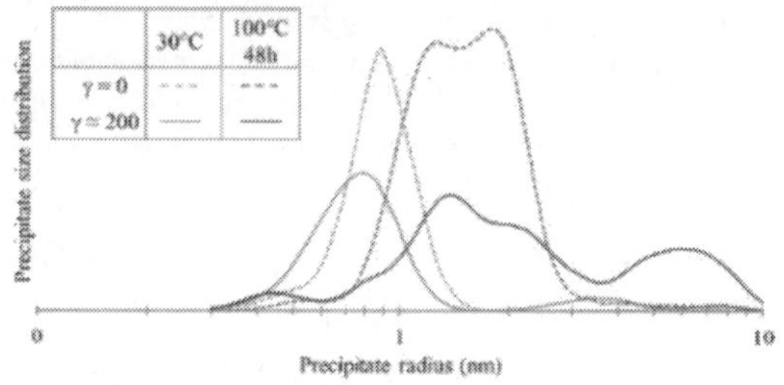

Fig. 9 : Precipitate size distribution estimated from fitting SAXS data for un-deformed and deformed sample, before and after thermal treatment at 100°C during 48h

Precipitates size distributions (Fig. 9) show a progressive growth of precipitates after 48h at 100°C for the un-deformed sample. For the deformed sample, there are less GP-zones before the thermal treatment, (as discussed in previous section), and there is a simultaneous growth of these GP-zones and some precipitates which can be identified as η' or η. A large increase in volume fraction is also observed for these precipitates contrarily to the GP-zones volume fraction which is stable during this treatment.

The STEM images for un-deformed sample (**Fig. 10 (a, b, c)**) after this annealing treatment show homogeneous precipitation with precipitates size in a range of 0.5-3 nm and some heterogeneous precipitation (size 4-10 nm).

For the deformed sample, dark field images (**Fig. 10 (d)**) show that the grain size is still below the micron size. HAADF images (**Fig. 10 (e, f)**) show two populations of precipitates, one on the defects with a size between 3 and 15 nm, and one inside grains with a size between 0.7 and 3 nm.

These observations are in good agreement with the SAXS measurements that revealed one population of precipitates for the un-deformed sample with a size between 0.4 and 1.5 nm (0.5-3nm for STEM measurements) and two populations for deformed sample: 0.3 - 3nm for smallest (0.7-3nm in STEM) and 3-10 nm (3-15nm in STEM). The population of heterogeneous precipitates observed in STEM for the un-deformed sample was not seen in SAXS certainly because of its very low volume fraction.

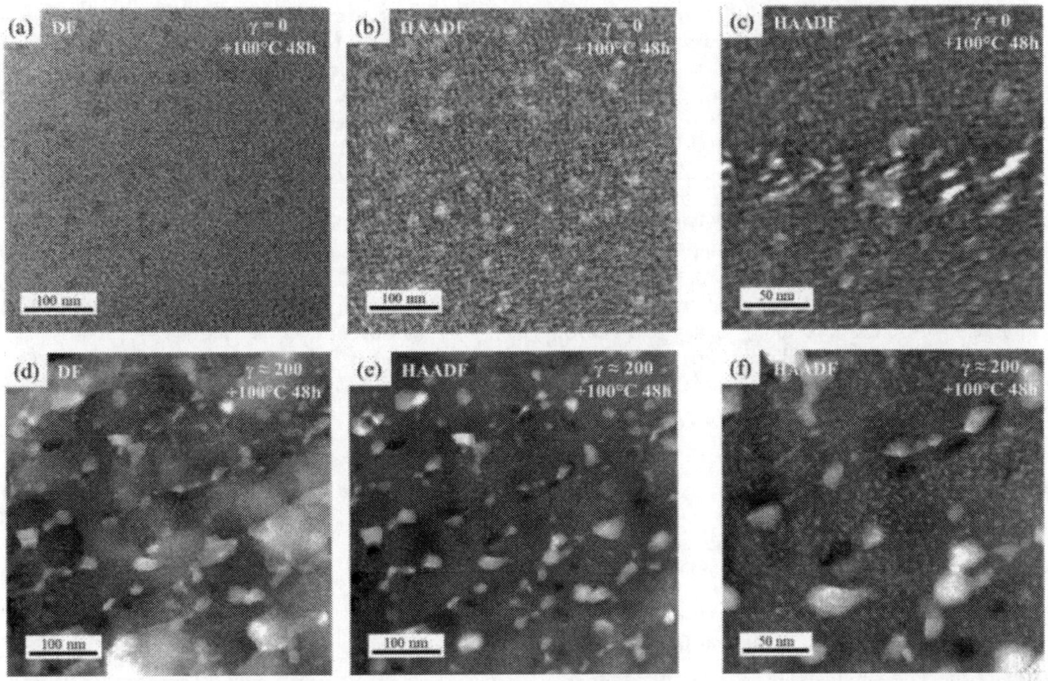

Fig. 10 : Dark field (a,d) and HAADF (b, c, e, f) images of un-deformed sample (a,b,c) and HPT deformed sample (d, e, f) of an Al-Zn-Mg-Cu alloy heat treated by 100°C during 48h

4 Discussion

First, this study shows that deformation by HPT does not influence the phases present in the alloy, meaning that there is no deformation induced phase transformation or total dissolution of another. However, many microstructural changes are caused by this severe plastic deformation. Observations by SEM show a fragmentation of Al_7Cu_2Fe intermetallic particles, but it is rather, limited since fragment sizes are still in the micrometer range.

Deformation-induced segregations on defects was observed after HPT, mostly on grain and sub-grain boundaries and probably along some dislocations. It has been proposed that this kind of segregation could be formed by several mechanisms. Dislocations can bring some solute to grain or sub-grain boundaries when annihilating. Segregations can also be formed by solute drag during the migration of a grain boundary [19]. Vacancies are created constantly during deformation and thus they may give rise as well to non-equilibrium segregations. Indeed, vacancies annihilate at grain boundaries which are defect sinks, a flow of vacancies towards the grain boundaries is then established during the deformation. If solute-vacancies complexes form, then vacancies bring solutes to boundaries creating segregation [20]. In the present case, STEM-EDS revealed such Zn, Mg, Cu segregations.

STEM images show also some precipitation of Zn, Mg, Cu localized on defects. It can be the consequence of solute/defects interaction which increase the solute diffusion combined with high density of crystalline defects which decrease the nucleation barrier.

The distribution of solute atoms is therefore largely influenced by severe plastic deformation, and this can also have an influence on the following heat treatment or the thermal stability of the alloy as observed in this study.

Correlation between SAXS, DSC and STEM measurements allowed to follow the precipitation mechanisms in this deformed structure and to compare with the undeformed material subjected to the same annealing treatments. DSC shows a precipitation that starts at lower temperature than in the un-deformed alloy (100°C against 120°C). It also shows that the stages of precipitation are not anymore sequential but become simultaneous, leading to a faster precipitation kinetics. SAXS results confirm this by evidencing a bimodal precipitates distribution compared to the monomodal one for the un-deformed sample and also by a larger size of precipitates and a faster decrease in volume fraction of precipitates smaller than 15 nm.

The isothermal treatment made at 100 °C during 48 h by *in-situ* SAXS allows to further evidence this accelerated precipitation kinetics. In order to understand why two populations of precipitates are present in the deformed sample, observations with STEM were necessary. They show that the population of large precipitates is mostly localized on grain boundaries after annealing and the population of smaller particles is inside the grain. Thus, deformation leads to higher density of inter-granular precipitation compared to un-deformed alloy. Intra-granular precipitation still occurs in the deformed material but in a limited way (lower volume fraction of small precipitates observed by SAXS and STEM analyses). This balance can result from the high density of grain boundaries which generate high density of bigger precipitates, reducing the amount of solute for the intra-granular precipitation. Furthermore, the smaller size of intra-granular precipitates in the deformed sample (shown by SAXS and STEM) can also be a consequence of this lack of available solute.

4 Conclusions

This study allowed to understand first the influence of deformation on solute distribution in solid solution and more generally on the microstructure of an Al-Zn-Mg-Cu alloy. But also the influence of this deformation on precipitation mechanisms by following the precipitate evolution using different observation methods. By DSC, and more specifically to followed precipitates size distribution by SAXS and the spatial arrangement of these precipitates thanks to STEM.

Acknowledgements

Both the Agence Nationale de la Recherche and the joint CNRS-JSPS 2017 summer program for the support to the trans-national collaboration are gratefully acknowledged for financial support (PRASA project-ANR-15-CE08-0029). This work was also supported in part by Grant-in-Aid for Scientific Research (S) from the MEXT, Japan (No. 26220909). HPT was carried out in the International Research Center on Giant Straining for Advanced Materials (IRC-GSAM) at Kyushu University.

Thomas Dorin at Deakin university is thanked for casting the alloy.

References

[1] J.C. Williams, E.A. Starke Jr., Acta Materialia 51 (2003) 5775–5799.
[2] O. Andreau, J. Gubicza, N. Xian Zhang, Y. Huang, P. Jenei, T.G. Langdon, Materials Science and Engineering: A 615 (2014) 231–239.
[3] C. Xu, M. Furukawa, Z. Horita, T.G. Langdon, Acta Materialia 51 (2003) 6139–6149.
[4] Y. Estrin, A. Vinogradov, Acta Materialia 61 (2013) 782–817.
[5] D.C. Patil, K. Venkateswarlu, S.A. Kori, G. Das, M. Das, S.N. Alhajeri, T.G. Langdon, IOP Conf. Ser.: Mater. Sci. Eng. 63 (2014) 012085.
[6] S.-S. Wang, J.-T. Jiang, G.-H. Fan, A.M. Panindre, G.S. Frankel, L. Zhen, Acta Materialia 131 (2017) 233–245.
[7] A. Deschamps, G. Fribourg, Y. Bréchet, J.L. Chemin, C.R. Hutchinson, Acta Materialia 60 (2012) 1905–1916.
[8] K. Edalati, Z. Horita, Materials Science and Engineering: A 652 (2016) 325–352.
[9] A. Deschamps, F. De Geuser, Z. Horita, S. Lee, G. Renou, Acta Materialia 66 (2014) 105–117.
[10] I. Bressler, B.R. Pauw, A.F. Thünemann, J Appl Cryst 48 (2015) 962–969.
[11] B.R. Pauw, J.S. Pedersen, S. Tardif, M. Takata, B.B. Iversen, J Appl Cryst 46 (2013) 365–371.
[12] H. Queudet, S. Lemonnier, E. Barraud, J. Guyon, J. Ghanbaja, N. Allain, E. Gaffet, Materials Science and Engineering: A 685 (2017) 227–234.
[13] G. Sha, A. Cerezo, Acta Materialia 52 (2004) 4503–4516.
[14] H. Zhao, B. Gault, D. Ponge, D. Raabe, F. De Geuser, Scripta Materialia 154 (2018) 106–110.
[15] S. Liu, C. Li, S. Han, Y. Deng, X. Zhang, Journal of Alloys and Compounds 625 (2015) 34–43.
[16] M. Murayama, Z. Horita, K. Hono, Acta Materialia 49 (2001) 21–29.
[17] Y.H. Zhao, X.Z. Liao, Z. Jin, R.Z. Valiev, Y.T. Zhu, Acta Materialia 52 (2004) 4589–4599.
[18] J. Gubicza, I. Schiller, N.Q. Chinh, J. Illy, Z. Horita, T.G. Langdon, Materials Science and Engineering: A 460–461 (2007) 77–85.
[19] L. Priester, Les Joints de Grains: De La Théorie à l'ingénierie, L'Editeur: EDP Sciences, 2006.
[20] X. Sauvage, N. Enikeev, R. Valiev, Y. Nasedkina, M. Murashkin, Acta Materialia 72 (2014) 125–136.

Mechanical and Spectroscopic Characterization of Solute Elements in Aluminum

Seiichiro Ii*, and *Toru* Hara

Research Center for Structural Materials, National Institute for Materials Science, Tsukuba, 305-0047 Japan

Abstract. We characterized the distribution of minor elements such as Si and Fe in Al utilizing a nanoindentation and electron microscopy with an energy dispersed X-ray spectroscopy (EDS) system. Nanoindentation can detect the dislocation nucleation known as "pop-in" event, the critical load (Pc) depends on the solute amount of Fe. However, that in Si-doped Al is rarely changed up to 1.0 at% of Si. That independent Pc in Al-Si is caused by the inhomogeneity of the Si, which is the grain boundary segregation, in Al. The grain boundary segregation of Si was clearly detected by using a newly developed microcalorimeter type EDS system, even at the 0.1 at% Si.

1 INTRODUCTION

The mechanical properties of Al alloys are mainly achieved by the microstructure control, especially the controlling of precipitates [1]. In order to control the shape, size, and volume fraction of the precipitates, the distribution of the supersaturated alloying elements is one of the important factors. The quantitative analysis for alloying elements has been analyzed by several techniques based on chemical and physical properties of alloying elements in Al. As for chemical analyses, an inductively coupled plasma atomic emission spectrometry (ICP-AES) and an atomic emission spectrometry (AES) are often used for the quantitative measurement of the chemical composition [2, 3]. On the other hand, from a physical standpoint, the internal friction and electrical resistivity are also used for the analysis of impurities [4, 5]. However, above techniques measure the only average composition of the specimen or the wide region in the alloys. Recently, the importance of the heterogeneity of the materials in the microstructure is also pointed out [6], the quantitative analysis of the alloying elements at local region becomes more crucial.

As for the local analysis in the chemical and physical methods, several techniques combined with the microscopy are often used. The representative technique is energy dispersed X-ray spectroscopy (EDS) combined with electron microscopy. In a decade, Tanaka et al. have developed the new type of analytical scanning electron microscope (SEM) attached with a microcalorimeter-type EDS [7]. By this analytical SEM, the chemical analysis in the special resolution of SEM can be conducted with an energy resolution of around 10 eV, which is one order higher than the conventional EDS system. It means that the elements in materials are identified with accuracy, even for a multicomponent system. In addition to the improvement of the energy resolution, the sharpness of each peak in the spectrum assists the improvement of the detection limit for the minor elements because of high peak/background (P/B) ratio.

Besides, the alloying elements also act as the obstacle, that is, as solute strengthening [8]. It depends on the misfit strain between alloying elements and Al matrix and the distribution of the alloying elements. Especially, the misfit strain of Fe, which is known as a typical impurity in Al, is remarkable larger than the other elements and it is expected as a solute strengthening element in recent. To investigate the local mechanical responses, the nanoindentation technique is widely used [9]. In the load (P) – displacement (h) curve obtained by the nanoindentation test, the discontinuous change in strain or stress is frequently seen during the loading. This phenomenon is known as "pop-in" event, and it is recognized as the dislocation nucleation and multiplication [10]. Since the solute strengthening is related to a dislocation interaction with the solute atoms, we, therefore, expect the possibility of nanoindentation technique for the mechanical measurement of the element distribution.

In this study, we here report that the mechanical and chemical effects of minor elements such as Fe and Si are characterized by spectroscopic of microcalorimeter-type EDS installed to scanning electron microscopy (SEM) and nanoindentation technique and discuss the possibility as the characterization for the minor elements in Al alloys.

2 Experimental procedure

The pure Al, Al-Si and Al-Fe alloys are fabricated by the high-frequency induction melting technique in an argon atmosphere. Chemical composition of ingots is analyzed

* Corresponding author: II.Seiichiro@nims.go.jp

Table 1. Chemical composition in Al alloys (at%)

	Fe	Si	Cu	Al
Pure Al	< 0.001	0.001	< 0.001	bal.
Al-0.02Fe	0.0251	< 0.001	< 0.001	bal.
Al-0.04Fe	0.0459	< 0.001	< 0.001	bal.
Al-0.1Si	< 0.001	0.1	< 0.001	bal.
Al-1.0Si	< 0.001	1.0	< 0.001	bal.

by an ICP-AES and listed in table 1. The ingots were homogenized at 873 K for 10.8 ksec in air and then hot and cold rolled. The plate was cut into a square with 10 mm one side and arbitral thickness and solutionized in air. The specimens were mechanically and electrochemically polished with a mirror surface. The local mechanical response was evaluated by the nanoindentation (TriboIndentator: TI950 Hysitron Co.) was used. An indentation rate and a maximum load are 0.1μN/sec and 1mN, respectively. We measured the mechanical responses at several ten points in grain interior for a few grains. The chemical analysis at the grain interior and grain boundary was performed by the microcalorimeter EDS equipped onto SEM (Ultra55, Carl Zeiss AG.) installed in Kyushu university. Observation was performed at an acceleration voltage of 5kV. As a reference, the EDS spectra by conventional silicon drift detector were also obtained simultaneously. The grain boundary analyzed by EDS was the random grain boundary with the misorientation angle of 35.5° around [25 -3 16] measured by OIM.

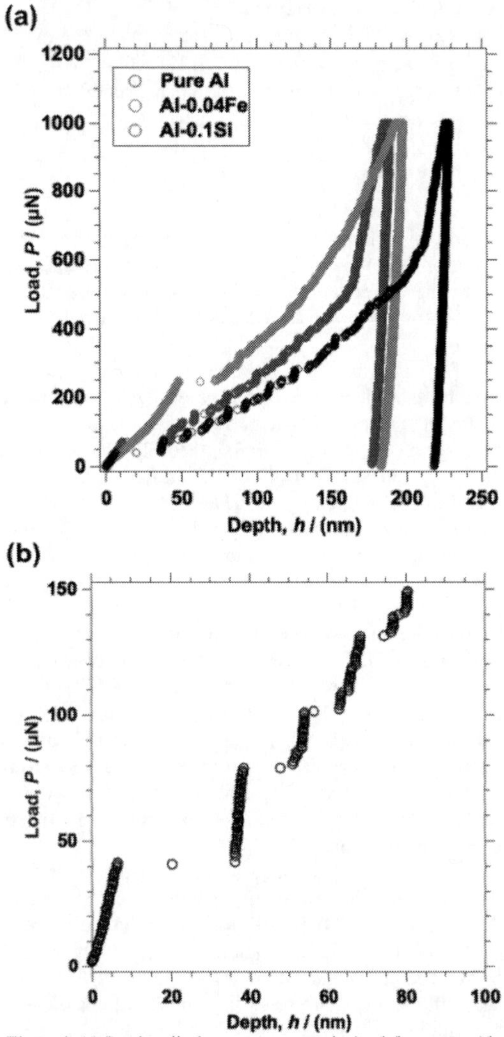

Figure 1 (a) Load – displacement curves obtained from pure Al, Al-0.04 at% Fe and Al- 0.1 at% Si alloy, which are plotted as circle (pure Al), square (Al- 0.04 at% Fe) and diamond (Al- 0.1 at% Si), respectively. (b) Enlarged P – h curve of pure Al showing the beginning of the nanoindentation. The discontinuous strain phenomena are indicated by arrows. This discontinuity is known as “pop-in” event, which is recognized as the nucleation and multiplication dislocation.

3 Results and discussion

3.1 Measurement of local mechanical response by nanoindentation and the effect of solute elements of Si and Fe in Al

Figure 1 (a) shows load (P) – depth (h) curves obtained from pure Al, Al-Fe and Al-Si alloys, respectively. All indentation tests are examined in the grain interior, and the effect of grain boundaries on the mechanical response in the alloys is eliminated. In all P–h curves, one can see the discontinuous strain jump at the early stage of the deformation as shown in fig.1 (b) which is enlarged P-h curve correspoinding to the beginning of the nanoindentation in pure Al. That discontinuous event is known as “pop-in” and this is considered to be a strain burst due to the nucleation and/or multiplication of the dislocations. Furthermore, the load P at the first pop-in event is recognized as a critical load (Pc). This is confirmed as the first event of the dislocation nucleation

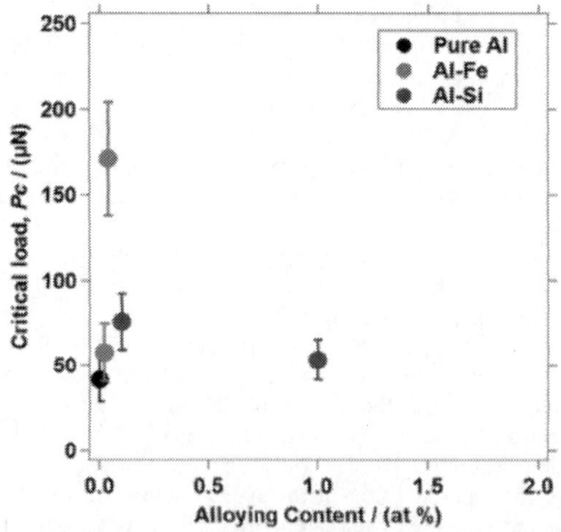

Figure 2 The effect of alloying contents of Fe and Si on the critical load, Pc in load-displacement curve in nanoindentation measurement for Al alloys. Open circle and solid circle are showing the results of Fe and Si, respectively.

and growth at the defect-free region [11]. Figure 2 shows the effect of Fe and Si addition on the Pc in Al. Pc in every specimen fluctuates indicated as an error bar in fig.2. Thereby, we expect that the pop-in event is presumed as a thermal activation process. Fe addition in Al remarkably increases the Pc of Al alloy, and the Pc is almost linearly increased with increasing Fe contents, even if alloying contents of Fe is small, which is approximately 0.046 at%. On the other hand, Si addition up to 1.0 at % fewer influences rather than Fe addition. The relationship between solutions and Pc has been reported in steel [12] and in Mg [13]. Those can be recognized as the solute strengthening, in which the solutions act as an obstacle for dislocation motion. Uesugi and Higashi have studied the solute strengthening in Al from the standpoint of *ab-initio* calculation [14, 15]. They reported that the misfit strain of Fe (-3.9 %) in Al matrix provides much higher misfit strain than the other elements, which are often included in Al such as Si, Mg, and Cu, in their report. Their report is well-consistent with our results obtained by nanoindentation. Therefore, the reason for the increment of Pc is owing to strengthening by solutions and larger misfit strain of solute Fe in Al matrix significantly affects to Pc in comparison with Si. On the other hand, Pc in Al-Si is not change very much, even at 1.0 at% Si. We expect that it is related to chemical inhomogeneity, such as grain boundary segregation. Uesugi et al. also reported that the segregation tendency is likely to be inverse to that of the misfit strain [16, 17]. In order to clarify experimentally, we investigated the distribution of minor solution is investigated by a newly developed EDS system.

3.2 Chemical analysis of Si by microcalorimeter EDS with SEM

Figure 3 shows a backscattered electron image of the Al-0.1 at% Si alloy including the grain boundary. The grain boundary is geometrically characterized by OIM based on the EBSD and that is distinguished as the random boundary. Generally, at the random boundary, much more solutions tend to be segregated at the grain

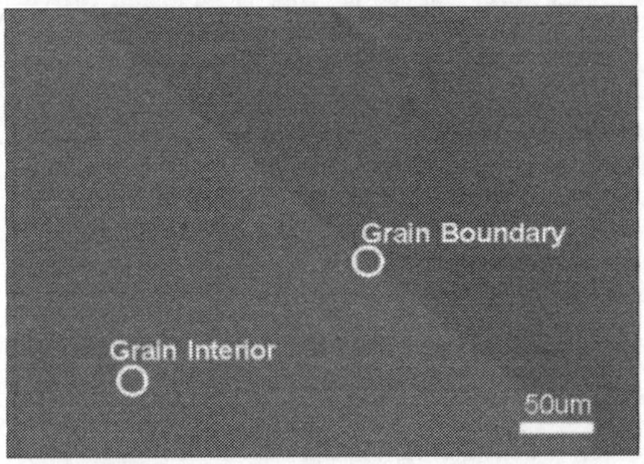

Figure 3 SEM image showing the grain boudanry in Al-0.04 at % Si alloy. The geometrical relation of the both grains is the random boundary. The EDS analyses have conducted both of the grain boundary and grain interior indicated by white circle in the figure.

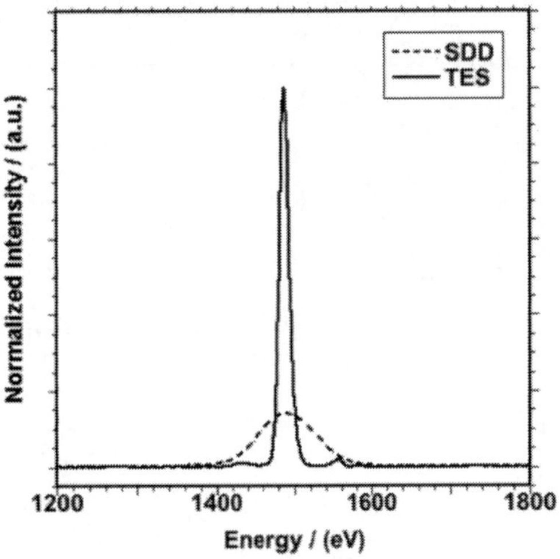

Figure 4 EDS profiles of Al-K_α taken from the grain interior in Al-0.04 at% Si alloy by newly developed microcalorimeter type EDS (solid line) and the conventional SDD-EDS system (dotted line), respectively. Both profiles are normalized by the background. The profile taken by the microcalorimeter type EDS is much sharp than that by SDD-EDS, because of the high energy resolution of microcalorimeter type EDS [6]. Also, the small Al-K_β peak can be seen in the profile of microcalorimeter type EDS.

boundary compared with the low angle and coincidence boundary as low Σ boundary (Σ value of less than 29). In order to clarify the grain boundary segregation of Si in the alloy, we analyzed it by the microcalorimeter type EDS analysis developed by Tanaka et al. [7]. In the same region, we also simultaneously analyzed by the conventional SDD-EDS system. Figure 4 shows the EDS spectra taken from the grain interior by TES and SDD-EDS system, respectively. Both spectra are normalized by their background. The Al-K_α peak taken by the TES-EDS is sharper than that by the conventional SDD-EDS. Both full width of half maximum (FWHM) were 130 eV in SDD-EDS and 15 eV in TES-EDS, respectively. Besides, the small peaks are also seen at both sides of the Al-K_α peak in the spectrum taken by TES-EDS. The small peaks detected at the higher energy side of Al-K_α is recognized as the Al-K_β peak. The Al-K_β peak is usually not seen in case of the SDD-EDS measurement because of its low energy resolution and P/B. The EDS spectrum becomes much broad and the Al-K_β peak is masked by the main Al-K_α peak as shown in the dotted line in figure 4. Figure 5 (a) and (b) are EDS spectra taken from the grain boundary and grain interior by SDD-EDS (a) and TES-EDS (b), respectively. The solid and dotted lines correspond to the spectrum taken from the grain boundary and grain interior, respectively in both figs.5 (a) and (b). All EDS spectra are only normalized with the main peak height of Al-K_α in each SDD and TES-EDS, respectively. In fig. 5 (a), the Si peak is hardly seen in EDS spectrum taken in both of the grain interior and the grain boundary because of the tailed Al-K_α peak. On the other hand, many peaks can be seen in the spectra taken by the TES-EDS system as shown in fig.5 (b). In those spectra, not only Al-K_α and

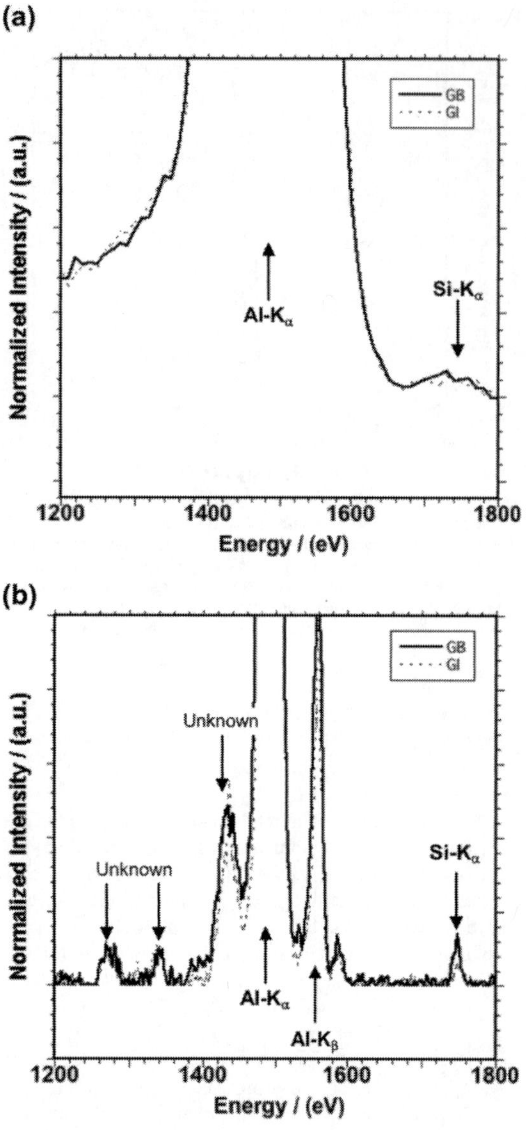

Figure 5 EDS spectrum taken by the conventional SDD-EDS (a) and newly developed microcalorimeter type EDS (b) system, respectively. All spectrum are only normalized by the main paek height of Al-K$_\alpha$. In (a), main Al-K$_\alpha$ and Si-K$_\alpha$ at 1740 eV can be detected, but the peak of the Si-K$_\alpha$ is broad and small. Therefore, the EDS spectra taken at the grain boundary (solid line) and the grain interior (dotted line) are overlapped each other. On the other hand, the EDS spectra of the grain boundary and grain interior taken by microcalorimeter type EDS system in (b), shape Si-K$_\alpha$ peak can be detected. The height of both peaks is different each other and it suggests the grain boundary segregation.

Si-K$_\alpha$ but also the other peak can be detected, because of the significant improvement of Peak / Background (P/B) ratio. The peak detected at slightly higher energy is identified as Al-K$_\beta$, however, unfortunately the other peaks cannot identified. In addition to the detection of Si-K$_\alpha$ peak in the TES-EDS measurement, the height of peaks taken at the grain boundary is higher than that of the grain interior. It suggests that the much Si is in the grain boundary, that is, grain boundary segregation is experimentally detected by the TES-EDS analysis, even if it is the less Si contents of 0.1 at%. Concerning the grain boundary segregation, there are many reports about the impurities such as Sulphur and phosphorous in steel [18]. In the case of steel, the heat treatment at a relatively higher temperature and it is recognized as the equilibrium segregation. Moreover, surface analysis as auger spectroscopy can be easily conducted to the segregated grain boundaries in the steel than the Al. In the case of Al, the quantitative analysis of grain boundary segregation was hard to be conducted, because of the low solubility and high ductility. The obtained result suggests that the quantitative analysis of TES-EDS enable us to analyze the local chemical composition even at low contents.

In our mechanical and analytical experiments, we examined the effect of Fe and Si on the local mechanical response in Al. Furthermore, we succeeded in the detection of the grain boundary segregation of minor Si in Al by the new EDS system. Consequently, we can consider that Fe act as a solute strengthening element in the Al matrix. On the other hand, since small misfit strain and the strong tendency of grain boundary segregation of Si, it less influences their mechanical response in comparison Fe. We found that the new techniques described above provide new findings for the relationship between the distribution of minor solutions and the strengthening. Unfortunately, Fe in Al was not detected simultaneously with Al by TES-EDS, through our experiments in the current situation. For it, the following reasons are expected. One is a detectability, and another is the energy range for the measurement. For the first one, although we have to clarify the lowest detection limit in this system, the modification of EDS system is also progressing. Concerning the energy range, the peak energies of Al-K$_\alpha$ of 1486 eV and Si-K$_\alpha$ of 1740 eV are close because the atomic number is next to each other, while Fe-K$_\alpha$ is 6398eV, and it is far from that of Al-K$_\alpha$. If they are improved, even if Fe content is minimal, it would be detected in the Al matrix. Moreover, to clarify the inhomogeneous distribution, we also need to accumulate both mechanical and analytical data because of the orientation dependence of the mechanical property and the grain boundary character dependence of the segregation. Although we have a few issues, we will be at the stage to overcome the aforementioned problems, and the minor dopants being important in Al alloy can be detected.

4 Conclusions

We investigated the methodology to evaluate the distribution of the solute elements in Al from the standpoint of the mechanical response and chemical analysis. The following results are obtained.

1. In the nanoindentation, the pop-in event related to the dislocation nucleation and multiplication can be seen in Al alloys. The critical load, which is the load at the first pop-in event, increases with increasing of

the Fe contents. However, less tendency of the increase of Pc is found in the Al-Si alloys. This is considered to be owing to their misfit strain and distribution of solute Fe and Si, respectively.

2. Even for the small amount of 0.1 at% Si in Al, we succeeded in the apparent detection of Si peak by the newly developed microcalorimeter type EDS analysis. Besides, the peak height of Si-K_{α} at the grain boundary is higher than that at the grain interior, and it suggests that the grain boundary segregation occurs in Al-Si alloy. The first principle calculations for the tendency of the grain boundary segregation reported previously supports our results.

Acknowledgments

This work is partly supported by the grant-in-aid for scientific research (S) (16H06366). The microcalorimeter type EDS analysis are particularly supported by Advanced Characterization Platform of the Nanotechnology Platform Japan in Kyushu university sponsored by the Ministry of Education, Culture, Sports, Science and Technology (MEXT), Japan. We are also grateful to Dr. K. Tanaka for the fruitful discussion.

References

1. J. R. Davis ed. *Aluminum and Aluminum Alloys* (ASM International, Ohio, 1993)
2. ASTM E3061-17, Standard Test Method for Analysis of Aluminum and Aluminum Alloys by Inductively Coupled Plasma Atomic Emission Spectrometry (Performance Based Method), ASTM International, West Conshohocken, PA, 2017
3. ASTM E716-16, Standard Practices for Sampling and Sample Preparation of Aluminum and Aluminum Alloys for Determination of Chemical Composition by Spark Atomic Emission Spectrometry, ASTM International, West Conshohocken, PA, 2016
4. R. B. Schwarz, L. L. Funk, Acta Metall. **31**, 299 (1983)
5. P. L. Rossiter, *The electrical resistivity iof metals and alloys* (Cambridge University Press, Cambridge, 1991)
6. D. Raabe, M. Herbig, S. Sandlöbes, Y. Li, D. Tytko, M. Kuzmina, D. Ponge, P. -P. Choi, Corr. Opin. Solid St. Mater. Sci. **18**, 253 (2014)
7. K.Tanaka, A. Odawara, A. Nagata, M. Ikeda, Y. Baba and S. Nakayama, Surf. Interface Anal. 38, 1646 (2006)
8. P. M. Anderson, J. P. Hirth, J. Lothe, *Theory of Dislocations* (Cambridge University Press, Cambridge, 2017)
9. C. A. Schuh, Mater. Today. **9, 32** (2006)
10. C. A. Schuh, A. C. Lund, J. Mater. Res. **19**, 2152 (2004)
11. L. Zhang, T. Ohmura, Phys. Rev. Lett. **112**, 145504 (2014)
12. K.Sekido, T. Ohmura, L. Zhang, T. Hara, K. Tsuzaki, Mater. Sci. Eng. A. **530** 396 (2011)
13. H. Somekawa and C. Schuh, Acta Mater. **59**, 7554 (2011)
14. T. Uesugi and K. Higashi, J. Jap. Inst. Light Metal, **54**, 82 (2004) (In Japanese)
15. T. Uesugi, K. Higashi, Mater. Sci. Forum. **561-565**, 977 (2007)
16. T. Uesugi, Y. Inoue, Y. Takigawa, K. Higashi, Mater. Sci. Forum. **551-552**, 331 (2007)
17. Y. Inoue, T. Uesugi, Y. Takigawa, K. Higashi, Mater. Sci. Forum. **561-565**, 1837 (2007)
18. P. Lajček, *Grain boundary segregation in metals* (Springer-Verlag, Berlin, 2010)

https://doi.org/10.1051/matecconf/202032609002

Effect of solution treatment on microstructures and mechanical properties of industrial 7050 thick plate

Zhiqiang Xie [1], Zhihong Jia [1,2,4*], Lin Lin [3], Shunyan Lin [3], and Qing Liu [1,4]

[1] International Joint Laboratory for Light Alloys (Ministry of Education), College of Materials Science and Engineering, Chongqing University, Chongqing 400044, China
[2] Electron Microscopy Center of Chongqing University, Chongqing 400044, China
[3] Southwest Aluminum (Group) Co., Ltd, Chongqing 401326, China
[4] Key Laboratory for Light-weight Materials, Nanjing Tech University, Nanjing 210009, China

Abstract. In this work, effect of solution treatments on the microstructures and mechanical properties of industrial 7050 aluminum alloy thick plate was investigated by tensile test, scanning electron microscopy (SEM), energy dispersive spectrometry (EDS) and transmission electron microscopy (TEM). The results show that the area fraction of residual phases decreases with increasing solution temperature from 475°C to 500°C. The overheating phenomenon identified by compound melting ball and crystal separating phase was observed at a temperature of 485°C or above. The compound melting balls are composed of $S(Al_2CuMg)$ phase dissolved with Zn atoms and Al matrix. The crystal separating phase is consisted of Al_2Cu phase and the amorphous structure containing solute atom Mg. The maximum tensile strength of 529 MPa was obtained for the alloy solution treated at 475°C for 3h followed by peak aging at 120°C for 24h.

1 Introduction

The 7050 aluminum alloy thick plates have been widely used as structural components in aerospace industries due to good combination of the strength and toughness [1]. Therefore, a number of investigations have been carried out on the processes and microstructures of the 7050 aluminum alloy [2-5]. In order to obtain improved mechanical properties, the production of 7050 thick plates involves several heat treatment stages including homogenization, solutionizing, quenching and aging. Precipitation and/or dissolution of second phase particles occur during these heat treatment steps which have great influence on the microstructural evolution in subsequent stages. Generally, the solution treatment, an important step, is to re-dissolve the soluble phases formed during solidification into matrix as much as possible [6-8]. Thus, it is necessary to understand the relation between solution temperature and dissolution of second phases. This understanding is also critical in designing optimized process.

Typical soluble constituent phases in Al-Zn-Mg-Cu 7xxx series aluminum alloys include $Mg(Zn,Mg,Cu)_2$, $S(Al_2CuMg)$ and Al_2Cu phases. After homogenization treatment and thermomechanical processing, the resolvable residual phases are $Mg(Zn,Mg,Cu)_2$, $S(Al_2CuMg)$ [9, 10]. And these residual phases can degrade the effect of age-hardening and aid crack initiation and propagation. Additionally, these remnant phases can lead to micro-galvanic corrosion when the alloys are exposed to aqueous corrosive environments [11-13]. Therefore, an appropriate solution process is required to resolve as many residual phases as possible. The main factors for the solution treatment are temperature, holding time, and cooling rate. The effect of solution temperature is the most obvious. Increasing the solution temperature within a certain range can increase the solid solubility of solute atoms in the matrix, thereby the effect of solution treatment could be improved. The choice of solution temperature depends on the melting temperature of the low melting point residual phases. It has been reported that the melting temperatures associated with $Mg(Zn,Mg,Cu)_2$ and $S(Al_2CuMg)$ are about 475-478°C and 490-501°C , respectively. It can be accepted that the soluble phase can re-dissolve more sufficiently at a higher solution temperature. But, excessive solution temperature will cause overheating of the alloy, which will seriously damage the microstructure and properties [14].

A large number of studies have been carried out on the effect of solution temperature on the microstructure and mechanical properties of 7050 aluminum alloys [15, 16]. The phenomenon of overheating has also been observed when a solution temperature is beyond a limited temperature in 7050 aluminum alloy. Generally, it has been widely accepted that the solution temperature for overheating is about 480°C for 7050 alloy. And the overheated characteristic structure takes on the compound melting ball, crystal separating phase and overheating triangle section. In addition, characteristic structure of overheating is changed with variant degree of overheating, and the mechanical properties also fail to change much when the alloy just started to overheat,

* Corresponding author: zhihongjia@cqu.edu.cn.org

while decrease fast with increasing a solution temperature causing overheating [14, 17, 18].

Therefore, a suitable solution temperature can maximize to re-dissolve the residual phases, which has an important effect on obtaining good comprehensive mechanical properties. At the same time, understanding the effects of different degrees of overheating on the microstructure and mechanical properties of alloys can provide a reference for optimizing solid solution parameters. However, there are little researches focusing on basic characteristics of overheated structure and its influence on alloy properties. In this work, the effect of different solution temperatures on the microstructure and mechanical properties of 7050 alloy is investigated, including statistics of residual phases, phase evolution, and tensile properties testing. In particular, a variety of microanalysis methods are used to systematically study the microstructure of overheating.

2 Experimental

The current study was carried out on the 130 mm thick plate of the hot rolled 7050 aluminum alloy with chemical composition of Al-5.90% Zn-2.10% Mg-2.10% Cu-0.11% Zr-0.07% Fe-0.03% Si (wt.%), which was provided by Southwest Aluminum (Group) Co., Ltd Chongqing, China. For the experiments, the 10 × 10 × 10-mm^3 specimens and the standard tensile specimens were cut from a quarter of the thick plate along the thickness direction. The dimension of tensile specimens is shown in Fig. 1. Samples were solution treated at 475°C/3h, 485°C/3h, 500°C/3h, respectively. The solution treatments were carried out in air furnace, quenched in room temperature water and then aged at 120°C for 24 h.

The microstructure of the alloy was characterized by scanning electronic microscopy (SEM) (AURIGA, Zeiss) and transmission electron microscopy (TECNAI F20, ThermoFisher) operated at 200 keV. Samples for microstructural observations by SEM were prepared by standard mechanical grinding and followed by suspension polishing. The composition and distribution of the residual phases were analyzed by SEM equipped with an energy-dispersive spectrometer detector (EDS) and back scatter electron detector (BSE). The TEM specimens of as-quenched samples were prepared by focused ion beam (FIB) thinning. The ambient temperature tensile tests were performed on a SHIMADZU AG-X 10KN universal testing machine at a tensile velocity of 3 mm/min, and the fractured surfaces were observed by SEM.

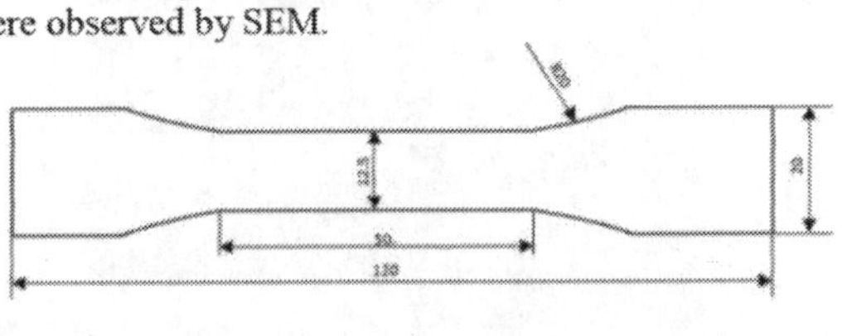

Fig. 1. The size of standard tensile specimen (unit: mm)

3 Results and discussion

Fig. 2 shows SEM micrographs of the hot rolled 7050 aluminum alloy solution treated at different conditions. residual phase particles are reduced graudally with increasing the solution temperature from 475°C to 500°C, but they are still observed after solution treatment at 500°C. In addition, the compound melting balls and overheating triangle secticns are also observed in Fig. 2b and Fig. 2c (inside the red ellipse), which belong to overheated characteristic structure. Ten SEM micrographs taken from the samples treated at each different solution temperature were analyzed by Image J to provide the area fraction and number density of the residual phases, and the statistical results are presented in Fig. 3. The number of residual phases gradually decrease with the increase of solution temperature as shown in Fig. 3a. And it can be seen from Fig. 3b that with the solution temperature increasing, the number density of the residual phases with the particle diameter less than 12 μm gradually decreases, while increases when the particle diameter is greater than 12 μm.

It is well known that the more of the residual phase particles are re-dissolved, the better of the solid solution strengthening and subsequent aging strengthening are. However, the alloy sample has been obviously overheated when a solution temperature reached 485°C or above, indicating the applied temperature exceeding the melting temperature of some phases. According to the reported melting points of the constituent phases in the alloy, $Mg(Zn,Mg,Cu)_2$ phase should be re-dissolved firstly when a solution temperature reaches 475°C. With further increasing a solution temperature up to 485°C or above, the alloy has been overheated because the temperature has surpassed the melting point of the $Mg(Zn,Mg,Cu)_2$ phase. The area fraction of residual phase will be even smaller at higher solution temperature, because in addition to complete dissolution of $Mg(Zn,Mg,Cu)_2$ phase, some phase S phase particles will also start to re-dissolve [14]. At the same time, the compound melting balls produced by overheating have larger diameter, so that the number density of residual phase increases with increasing of solution temperature when the particle diameter is larger than 12μm.

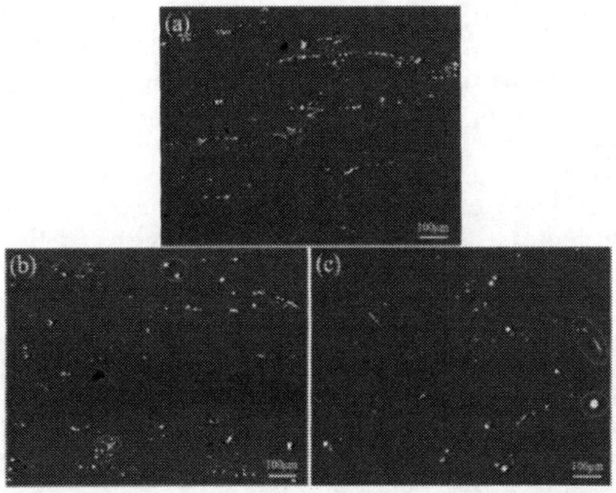

Fig. 2. SEM micrographs of the 7050 aluminum alloy solution treated at different conditions (a) 475°C/3h, (b) 485°C/3h, (c) 500°C/3h.

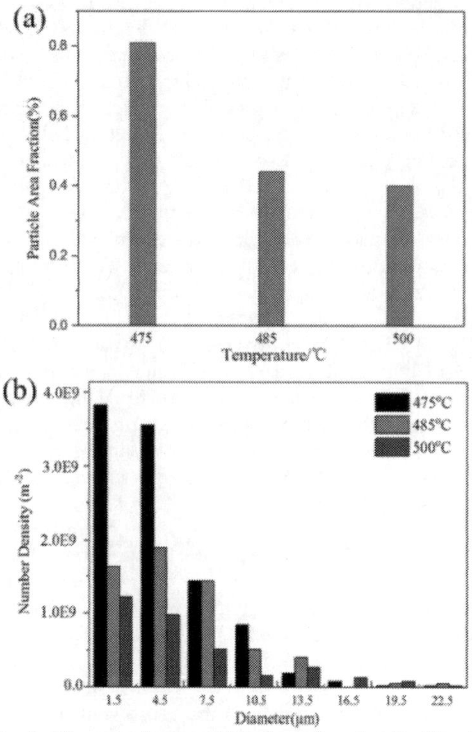

Fig. 3. The area fraction (a) and number density (b) of the residual phases in 7050 aluminum alloy solution treated at different conditions.

In order to systematically analyze the effect of solution temperature on microstructure, the evolution of residual phases were investigated in more detail. Fig. 4 shows SEM micrographs of the typical residual phases and the corresponding EDS mappings of the samples solution treated at different conditions. It can be seen that the main S(Al_2CuMg) phase is left in the alloy after solution treated at 475°C/3h and no overheating phenomenon is found. However, the compound melting balls are observed in the remained residual phases after solution treated at 485°C/3h or 500°C/3h, indicating the occurrence of overheating. The compound melting ball has complex structure as revealed by different contrast through the ball. The EDS mappings in Fig. 4d and Fig. 4f reveal the Zn distribution besides Al, Mg, Cu compositions in this kind of compound melting ball. To understand the microstructure and structure of the ball, TEM investigations were done. The TEM sample of the interesting compound melting ball was prepared from the red-square marked area in Fig. 5a by focused ion beam (FIB). Two kinds of phases with different contrasts are observed clearly in Fig. 5c. The diffraction patterns of two phases in Fig. 5d and e reveal the compound melting ball composed of the S phase and the Al matrix, respectively. The EDS analysis from Fig. 5f and Fig. 5g shows that the Zn atoms in the compound melting ball mainly exist in the S phase.

Fig. 6 presents the crystal separating phase also belonging to the compound melting structure due to overheating. It can be further confirmed that the 7050 alloy has been overheated after solution treated at 485°C/3h or 500°C/3h. Moreover, the TEM sample, prepared by FIB, was used to analyze the composition of crystal separating phase from the alloy after solution treated at 500°C/3h (Fig. 6b yellow box area), and the results are shown in Fig. 7. It can be easily found from Fig. 7a that the crystal separating phase is not made up of one phase. The microstructure at position 1 is shown in Fig. 7b, indicating an amorphous structure. The diffraction spot at position 2 in Fig. 7a is shown in Fig. 7c. Combined with the energy spectrum information of Fig. 7d, it can be determined that it is the Al_2Cu phase at position 2 and the amorphous structure at position 1 containing the Mg solute atom.

Through the above micro-analysis, the compound melting ball and the crystal separating phase are typical overheating microstructure of the 7055 aluminum alloy. The main reason for forming the compound melting structure during overheating is that the solution temperature exceeds the melting temperature of the residual phase. When an overheating temperature is low, the low melting point phases are likely to be enriched at the grain boundary to form small compound melting pellets because of higher energy at the grain boundary. As an overheating temperature continues to rise, sufficient energy is obtained inside the grain and the compound melting balls also form in the gain besides at grain boundaries. With further increasing the overheating temperature, the compound melting balls evolve gradually and form a continuous compound structure at the grain boundaries. It is well known that overheating is detrimental to the comprehensive mechanical properties of the alloy. The compound melting balls with a large particle diameter and the crystal separating phase will cause stress concentration in the 7055 aluminum alloy. On the other hand, it can be seen from Fig. 5g and Fig. 7d that the aging strengthening effect will be weakened because part of solute atoms Zn and Mg are used by the compound melting balls and crystal separating phase at overheating condition.

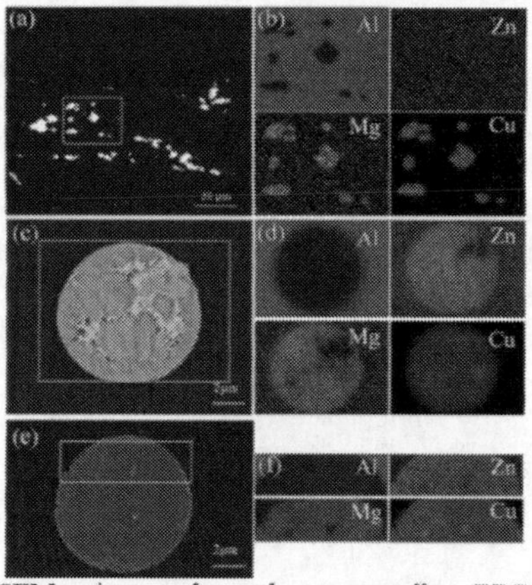

Fig. 4. SEM micrographs and corresponding EDS derived composition maps for Al, Zn, Mg and Cu of 7050 alloy solution treated at different conditions: (a) (b)475°C/3h, (c) (d) 485°C/3h, (e) (f) 500°C/3h.

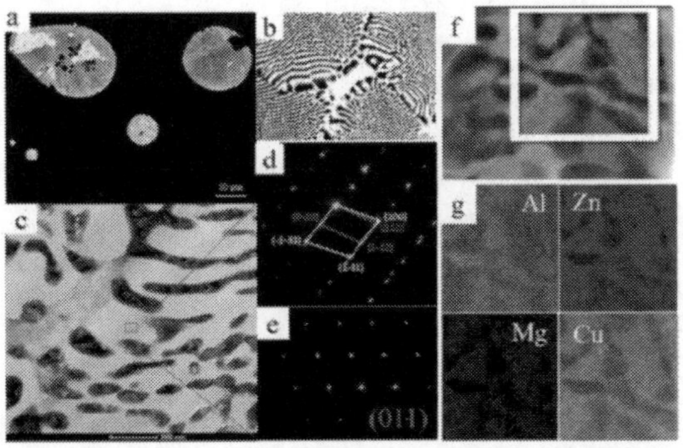

Fig. 5. Microstructure of the compound melting ball of 7050 alloy after solution treated at 500°C/3h: (a) SEM micrograph, (b) high magnification SEM micrograph in (a), (c) TEM bright field image, (d) diffraction pattern of the S (Al_2CuMg) phase projected along [-201], (e) diffraction pattern of the Al matrix taken from the [011] zone axis, (f) STEM image and (g) the corresponding EDS maps.

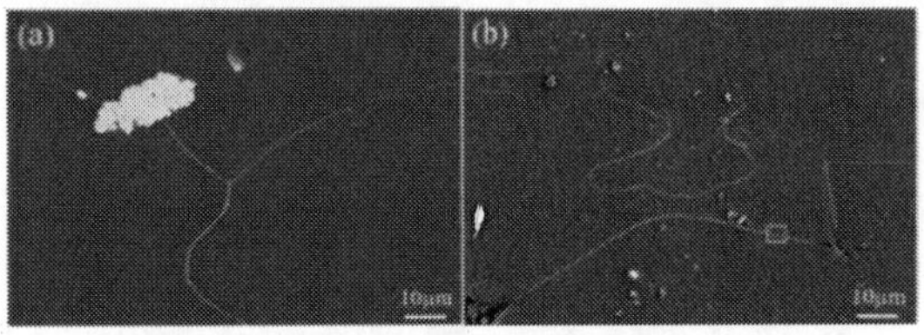

Fig. 6. SEM micrographs of crystal separating phases of 7050 alloy solution treated at different conditions: (a) 485°C/3h, (b) 500°C/3h.

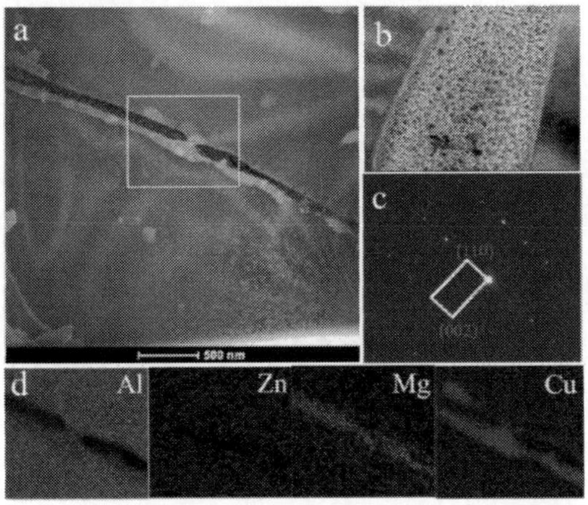

Fig. 7. Microstructure of the crystal separating phases of 7050 alloy after solution treated at 500°C/3h: (a) STEM image, (b) TEM bright field image at the position 1 in (a), (c) diffraction pattern of the Al_2Cu phase taken from the position 2 marked in (a), (d) the EDS maps of the yellow marked area in (a).

The tensile properties of the peaking aged 7050 aluminum alloy with different solution treatments were carried out and the results are listed in Table 1. It clearly shows that the strength and elongation decrease with the increase of solution temperature. The better ultimate tensile strength (σ_b), yield strength ($\sigma_{0.2}$) and elongation of the alloy solution treated at 475°C are obtained as 529.0±5 MPa, 479.4±6 MPa and 8.1±0.3, respectively. Fracture surfaces of the peak aging samples with different solution treatments of 7050 aluminum alloy are presented in Fig. 8. For the sample solution treated at 475°C/3h, it can be seen from Fig. 8a there are many dimples on the fracture surface, which indicates a ductile fracture. For the 485°C/3h and 500°C/3h samples, a large number of holes can be clearly observed in Fig. 8b and Fig. 8c, which are caused by the overheating of constituent particles such as compound melting balls with a large diameter and highlighted with arrows on the fracture surfaces. Generally, the fracture resistance of the 7050 aluminum alloy is very sensitive to the existence of any coarse constituent particles because these sites can act as crack initiation [19]. Additionally, when overheating is significantly worsened, the grain boundaries are also severely weakened because of the crystal separating phase. Thus, the crystal separating phase also adversely affects mechanical properties. It can be concluded from Fig. 5g and Fig. 7d that the aging strengthening effect will be weakened to some extent because less solute atoms Zn and Mg are available at an overheating condition. Hence, compared with solution treatment at 475°C/3h, the tensile properties of the peaking aged 7050 aluminum alloy solution treated at 485°C/3h and 500°C/3h are depressed.

Table 1. The tensile properties of the peaking aged 7050 aluminum alloy with different solution treatments.

Solution temperature	$\sigma_{0.2}$ (MPa)	σ_b (MPa)	δ (%)
475 °C,3 h	479.4±6	529.0±5	8.1±0. 3
485 °C,3 h	447.3±4	503.2±5	6.6±0. 6
500 °C,3 h	432.8±5	485.6±6	6.1±0. 5

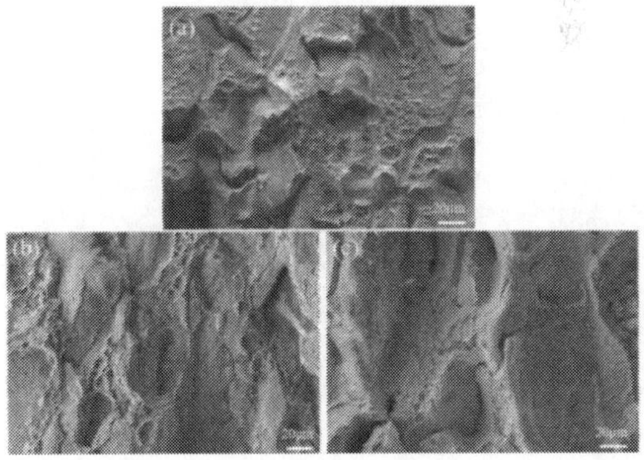

Fig. 8. SEM micrographs of fracture surfaces of the peak aged samples with different solution treatments: (a) 475°C/3h, (b) 485°C/3h, (c) 500°C/3h.

4 Conclusions

This study has been performed on the effect of solution temperature on the microstructure and mechanical properties of hot rolled 7050 aluminum alloy plate after subsequent peak aging. The following conclusions may be drawn:

(1) The area fraction of residual phases gradually decreases with the increase of the solution temperature from 475°C to 500°C. However, the overheating phenomenon identified by compound melting ball, crystal separating phase and overheating triangle section was observed at a temperature at 485°C or above.

(2) The compound melting ball is composed of $S(Al_2CuMg)$ phase dissolved with Zn atoms and Al matrix. The crystal separating phase is consisted of Al_2Cu phase and the amorphous structure containing solute atom Mg.

(3) The maximum tensile strength of 529 MPa was obtained for the alloy solution treated at 475°C for 3h followed by peak aging at 120°C for 24h. Overheating will seriously weaken the comprehensive mechanical properties of the alloy.

This work was supported by the Fundamental Research Funds for the Central Universities of China (Grant No. 2018CDGFCL0002 and 2019CDQYWL029), the National Natural Science Foundation of China (Grant No. 51871035), the Foundation for Innovative Research Groups of the National Natural Science Foundation of China (Grant No. 51421001) and the "111" project (B16007) by the Ministry of Education.

References

1. J.C. Williams, E.A. Starke, Acta Mater. **51** (2003) 5775-5799.
2. F.Y. Xie, X.Y Yan, L. Ding, F. Zhang, S.L. Chen, M.G. Chu, Y.A. Chang, Mater. Sci. and Eng.: A **355**(1-2) (2003) 144-153.
3. J. D. Robson, Mater. Sci. and Eng.: A **382**(1-2) (2004) 112-121.
4. L.L. Wei, Q.L. Pan, Y.L. Wang, L. Feng, H.F. Huang, J. Mater Eng and Perf **22**(9) (2013) 2665-2672.
5. X.M. Zhang, W.J. Liu, S.D. Liu, M.Z. Zhou, Mater. Sci. and Eng.: A **528**(3) (2011) 795-802.
6. N. M. Han, X. M. Zhang, S. D. Liu, D. G. He, R. Zhang, J. Alloys Compd. **509**(10) (2011) 4138-4145.
7. B. Li, Q.L. Pan, C.P. Chen, H.H. Wu, Z.M. Yin, J. Alloys Compd. **664** (2016) 553-564.
8. Y.L. Deng, L. Wan, Y. Zhang, X.M. Zhang, J. Alloys Compd. **498**(1) (2010) 88-94.
9. G.S. Peng, K.H. Chen, S.Y. Chen, H.C. Fang, Mater. Sci. and Eng.: A **641** (2015) 237-241.
10. P.F. Jia, Y.H. Cao, Y.D. Geng, L.Z. He, N. Xiao, J.Z. Cui, Mater. Sci. and Eng.: A **612** (2014) 335-342.
11. L. P. Huang, K. H. Chen, S. Li, M. Song, Scripta Mater. **56**(4) (2007) 305-308.
12. S.P. Knight, N.Birbilis, B.C. Muddle, A.R. Trueman, S.P. Lynch, Corrosion Sci. **52**(12) (2010) 4073-4080.
13. D.K. Xu, N.Birbilis, D.Lashansky, P.A. Rometsch, B.C. Muddle, Corrosion Sci. **53**(1) (2011) 217-225.
14. D.K. Xu, P.A. Rometsch, N. Birbilis, Mater. Sci. and Eng.: A **534** (2012) 234-243.
15. D.K. Xu, P.A. Rometsch, N. Birbilis, Mater. Sci. and Eng.: A **534** (2012) 244-252.
16. W.T. Huo, J.T. Shi, L.G. Hou, J.S. Zhang, J. Mater. Process. Technol. **239** (2017) 303-314.
17. F.H. Gao, N.K. Li, N. Tian, Q. Sun, X.D. Liu, G. Zhao, Trans. Nonferr. Met. **18**(2) (2008) 321-326.
18. G.S. Wang, Q.Q. Chen, K. Tao, Q.C. Chen, Z.H. Zhao, Mater. Sci. For. **877** (2016) 587-592.
19. P.S. Pao, S.J. Gill, C.R. Feng, Scripta Mater. **43** (2000) 391-396.

Advanced imaging and simulations of precipitate interfaces in aluminium alloys and their role in phase transformations

Zezhong Zhang[1,*], *Laure* Bourgeois[2,3,**], *Yong* Zhang[3], *Julian M.* Rosalie[4], and *Nikhil* Medhekar[3,***]

[1]Department of Materials Science and Engineering, Monash University, Victoria, Australia. Present address: Electron Microscopy for Materials Research (EMAT), University of Antwerp, Groenenborgerlaan 171, 2020 Antwerp, Belgium; and Department of Materials, University of Oxford, 16 Parks Road, Oxford OX1 3PH, United Kingdom.
[2]Monash Centre for Electron Microscopy, Monash University, Victoria, Australia.
[3]Department of Materials Science and Engineering, Monash University, Victoria, Australia.
[4]Erich Schmid Institute of Materials Science, Austria. Present address: University of Leoben, Leoben, Australia.

Abstract. Precipitation is accompanied by the formation and migration of heterophase interfaces. Using the combined approach of advanced imaging and atomistic simulations, we studied the precipitate-matrix interfaces in various aluminium alloy systems, aiming to resolve their detailed atomic structures and illuminate their role in phase transformations.

1 Introduction

Being light and strong, aluminium has pinned its special position in aerospace history. From the Wright brothers' first powered flight to the current spacecraft in service, aluminium alloys are dependent on precipitation hardening for their improved strength. Once a precipitate nucleates, the heterophase interface forms as the critical boundary separating the secondary phase from the matrix; while the growth of the precipitate will be accompanied by the migration of the precipitate-matrix interfaces. Classical phase transformation theory can rationalise the critical role of interfaces in terms of thermodynamics and kinetics [1]. For instance, the interfacial energy plays an important role in the precipitate orientation relationship and the equilibrium shape, the nucleation and growth energy barrier, which ultimately governs the microstructure evolution and hardening response.

Classical analytical models assume the interface to be sharp (e.g. Becker model [2]) or diffused (Cahn-Hilliard model [3]). However, recent microscopy studies [4, 5] found that precipitate interfaces may have distinct chemistry and complex intermediate structures different from the bulk phases. The associated atomic mechanisms in phase transformations remain poorly understood. First-principles methods such as density functional theory (DFT) are often used to address the atomic behaviours of precipitation, for example, Ref.[6, 7]. However, the lack of precise experimentally-based interfacial structures prevents accurate predictions even for the simplest cases. The innovation of characterisation techniques fundamentally reshaped the field of materials science. In particular, aberration correction of electron optics significantly pushed the resolution limit of electron microscopy such that we can probe the structure, chemistry and bonding state of materials at atomic scale.

To investigate the detailed structure of precipitate interfaces, we studied the precipitate interfaces in various aluminium alloys systems using aberration-corrected scanning transmission electron microscopy. Based on the atomic structure resolved experimentally, atomistic simulations were performed to explore the underlying mechanisms. This paper will be focused on our recent progress in the textbook Al-Ag system. Together with our studies in other aluminium alloy systems, we will discuss the common observations of precipitate interfaces and their role in phase transformations.

2 Experimental and computational methods

2.1 Alloy processing and sample preparation

The Al-1.68 at.% Ag alloy used in this work was cast from high-purity aluminium and silver. The cast ingots were homogenised at 525°C, then hot- and cold-rolled to 0.5 mm alloys sheets. The samples were in the form of disks 3 mm in diameter and 0.5 mm in thickness, punched from an alloy sheet after rolling. They were solutionised at 525°C in a nitrate salt bath and quenched in water or oil to room temperature. This process was designed to manipulate as-quenched lattice defect concentration (i.e. vacancies and dislocations), which changes the Ag clustering and following precipitation behaviours during ageing [8]. Then the samples were aged at 200°C in an oil bath for a range of times to characterise different stages of precipitation. The TEM specimens were made by mechanically grinding the

*e-mail: zezhong.zhang@uantwerpen.be
**e-mail: laure.bourgoeis@monash.edu
***e-mail: nikhil.medhekar@monash.edu

disks and electro-polishing using nitric acid/methanol at -25°C and 13 V with the current around 200 mA [8, 9]. We prepared samples in other aluminium alloy systems under similar conditions.

2.2 Electron microscopy characterisation and analysis

The alloy microstructures and precipitate atomic structures were characterised by high-angle annular dark-field scanning transmission electron microscopy (HAADF-STEM) to exploit the large difference in the atomic numbers between the solute elements (e.g. Ag) and the Al matrix. High-resolution HAADF-STEM imaging was conducted in a dual-aberration-corrected FEI Titan^3 80-300 field-emission gun transmission electron microscope (FEGTEM) at 300 kV with a 1.2Å probe and 55-200 mrad HAADF detector. Preliminary investigations of the microstructures were carried out on a JEOL JEM 2100F FEGTEM and an FEI Tecnai G^2 F20 Super-Twin lens FEGTEM operated at 200 kV. The *in situ* annealing was performed on a JEOL 2100F using a Gatan 625 heating holder.

To interpret the contrast of HAADF-STEM, electron microscopy images were simulated with the experimental parameters using the μSTEM software [10], implementing the multislice method with quantum excitation of phonons to incorporate elastic and inelastic scattering.

We applied statistical parameter estimation theory to locate the atomic positions at the interface, as implemented in the software StatSTEM [11]. The 3D atomic models were built by applying the known periodicity of the composing phases to the refined 2D positions in the viewing direction.

2.3 Atomistic simulations

First-principles density functional theory calculations were performed using the Vienna *Ab initio* Simulation Package (VASP) [12] with the potential constructed under the generalised gradient approximation with the treatment of Perdew, Burke, and Ernzerhof (GGA-PBE) [13]. Internal and external structural parameters were fully optimised until Hellmann-Feynman forces were less than 0.01 eV/Å. The convergence of the relevant energy differences concerning energy cut-off, k-point sampling and supercell size was better than 1 meV/atom. The formation energies of different phases are given relative to the energy of FCC Al and Ag in the ground state. The atomic structures and energetics of the computed phases are given in table 1, showing a good agreement ($< 5meV/atom$) with results from Materials Genome Project [14].

The defect energy of Ag in solid solution was calculated by an isolated Ag substitutional point defect in an Al supercell containing 108 atoms, giving a substitutional defect energy of 0.09 eV, in reasonable agreement with previous calculations of 0.02 eV using the local density approximation (LDA) [15]. The monolayer of Ag on $\{111\}_{Al}$ plane was modelled using trigonal supercells in which the Ag monolayer was surrounded above and below by Al (representing the infinitely wide two-dimensionally coherent Ag plane(s) surrounded by Al matrix), containing the equivalent of 24 atomic planes of $\{111\}_{Al}$. For coherent interfaces, we calculated structures with and without Ag interfacial segregation to deduce the energy difference per Ag segregated as the strain is negligible. For semi-coherent interfaces, we separated the contributions of elastic strain energy and interfacial energy by measuring the slope of formation energy against 1/n, where n is the number of atomic planes normal to the interface.

The embedded atom method (EAM) was used to estimate the energetics of interfaces in the large systems via the Large-scale Atomic/Molecular Massively Parallel Simulator (LAMMPS) [19]. Energy and force relaxations were performed for each structure.

Understanding the atomic mechanisms of phase transformations requires identification of the local, atomic-level structures and lattice defects. The crystal structures were recognised using adaptive common neighbour analysis [20] and dislocation analysis was performed using the dislocation extraction algorithm [21] as implemented in the Ovito software [22].

3 Results

The γ' (Ag_2Al) phase in the Al-Ag alloy system has served as a textbook example for understanding phase transformations, precipitating hexagonal close packed (HCP, space group: $P6_3/mmc$) crystals in the face-centred cubic (FCC, space group: $Fm\bar{3}m$) aluminium matrix. The broad facets of γ' precipitates are fully coherent. Distinct Ag segregation was found at the coherent interfaces of γ' precipitate in the early stage of aging [9, 23], as shown in Fig. 1 (a-b). In Fig. 1 (c), DFT calculations were performed to calculate the interfacial Ag segregation from the energetics perspective as a function of precipitate thickness [9]. For the semi-coherent interface, the traditionally considered atomic sharp $\{1\bar{1}00\}_{\gamma'} \parallel \{112\}_{Al}$ interface was observed to be reconstructed into a four-layer step of $\{10\bar{1}1\}_{\gamma'} \parallel \{111\}_{Al}$ and a two-layer step of $\{10\bar{1}1\}_{\gamma'} \parallel \{002\}_{Al}$ interfaces. We explained the interfacial reconstruction using the DFT-deduced interfacial energy for each orientation relationship [9].

The edges of γ' precipitates are semi-coherent. According to phase transformation textbooks [1], Shockley partial dislocations are present at the semi-coherent interface to achieve the FCC-HCP structural transformation. We observed that the $\{1\bar{1}00\}_{\gamma'} \parallel \{112\}_{Al}$ interface is periodically reconstructed by nano-steps parallel to low-index $\{002\}_{Al}$ and $\{111\}_{Al}$ planes, which was also seen in previous studies [23, 24]. Here we performed first-principles calculations to deduce the interfacial energies associated with different orientation relationships. For the atomically sharp $\{112\}_{Al}$ interface, the interfacial energy was calculated to be $331mJ/m^2$, in good agreement with the values obtained from previous experimental estimation ($350mJ/m^2$) [25] and DFT calculations ($325mJ/m^2$) [18]. The interfacial energies for the $\{111\}_{Al}$ and $\{002\}_{Al}$ interfaces are $175mJ/m^2$ and $183mJ/m^2$ respectively. The

Table 1. Atomic structures and energetics of face-centred cubic (FCC) and hexagonal close packed (HCP) phases in the Al-Ag alloy system. The schematic digram shows the stacking differences of close-packed planes between different phases.

Phase	Lattice Parameters (Å)			Cohesive Energy (eV/atom)		Formation Energy (eV/atom)	
	Experiment	Previous DFT	This work	Previous DFT	This work	Previous DFT	This work
Al	4.046[a]	4.039[c], 3.98[e]	4.050	−3.748[c]	-3.745	0	0
Ag	4.079[a]	4.161[c], 4.02[e]	4.161	−2.832[c]	-2.827	0	0
AgAl	a=2.88[f] c=27.35[f]	N/A	a=2.975[f] c=26.884[f]	N/A	−3.331[f]	N/A	−0.044[f]
Ag_2Al	a=2.858[b] c=4.607[b]	a=2.922[c], 2.850[d] c=4.618[c], 4.546[d]	a=2.890[g] c=4.600[g]	−3.208[c]	−3.209[g]	−0.071[c], −0.121[d,e]	−0.075[g]

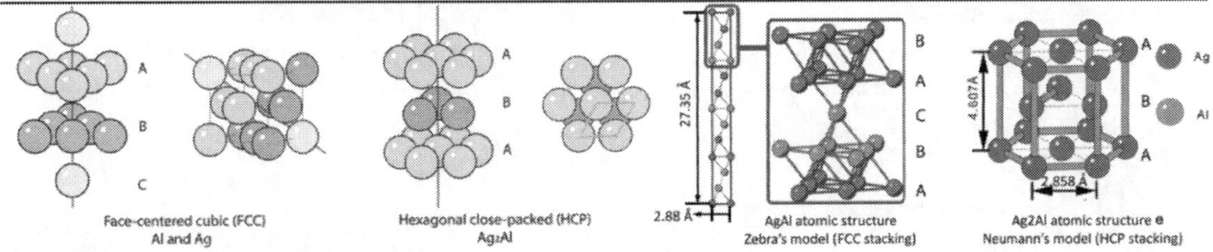

a. X-ray measurements, taken from Ref. [16].
b. X-ray measurements, taken from Ref. [17].
c. PBE-GGA calculations, taken from Materials Genome Project [14].
d. Ultrasoft-LDA calculations, taken from Ref. [6].
e. Ultrasoft-LDA calculations, taken from Ref. [18].
f. Our STEM measurements and PBE-GGA calculations, taken from Ref. [8].
g. Our PBE-GGA calculations, taken from Ref. [9].

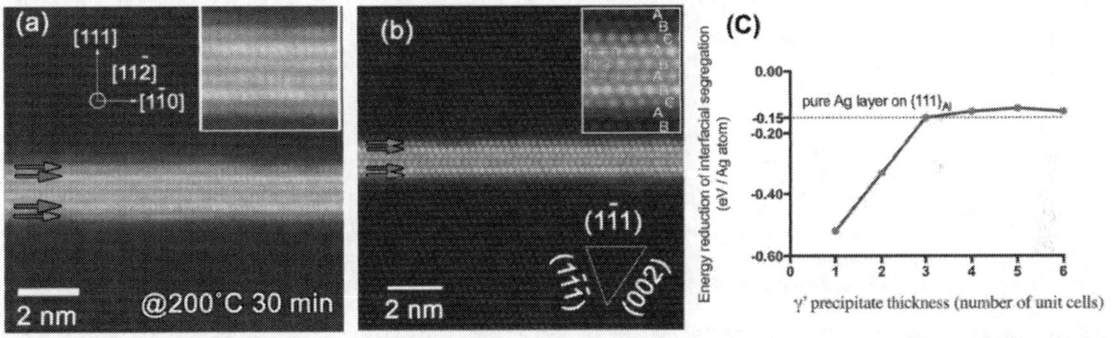

Figure 1. HAADF-STEM images of γ' coherent precipitate-matrix interface viewed in (a) a $\langle 112 \rangle_{Al}$ direction and (b) a $\langle 110 \rangle_{Al}$ direction, where blue arrows indicate the Ag segregation as reported in our previous study [23] while red arrows indicate the Ag segregation found in this study [9]. (c) Energy reduction associated with interfacial Ag segregation as a function of precipitate thickness relative to the energy state of Ag in solid solution. The energy of an Ag mono-layer on $\{111\}_{Al}$ planes (-0.15 eV/Ag atom) is displayed as a dashed line for comparison. Adapted from [9] with permission.

reconstruction can be expected to have an energetic origin: it is energetically favourable if the sum of the interfacial energies of the steps is lower than that of the sharp $\{1\bar{1}00\}_{\gamma'} \parallel \{112\}_{Al}$ interface.

Determining the exact types and locations of dislocations at the FCC-HCP interfaces is experimentally and analytically challenging. Here we combined advanced imaging with atomistic structural analysis that revealed two types of semi-coherent interfaces with unique chemical distribution and dislocation, denoted as type-I interface and Type-II interface in Fig. 2. Fig. 2(a-b) shows the HAADF-STEM image with the refined column positions as highlighted by the green and red plus signs. In the enlarged views, it is clear that the FCC/HCP transition region has a shift normal to the close-packed planes as indicated by a yellow line. By comparing the stacking of FCC (ABCABC) with that of HCP (ABABAB), we identified two layers (highlighted in red letters) that have their stacking exchanged when crossing the FCC/HCP interface. The exchanged stacking results in the atomic column in one close-packed layer being aligned on top of the next close-packed layer in the transition zone, as indicated by

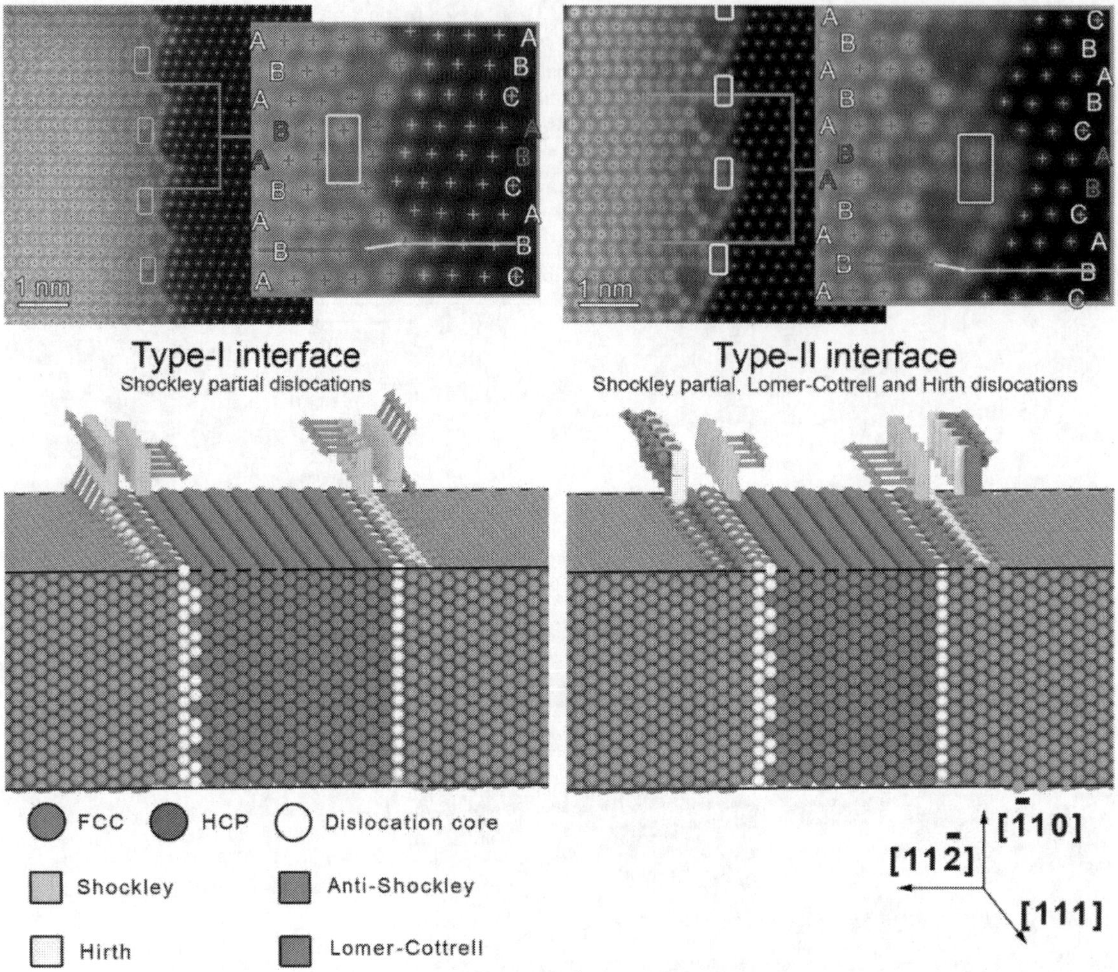

Figure 2. The coupling between chemical distributions and dislocations at semi-coherent Al-γ' FCC/HCP interfaces. The green and red plus signs indicate the refined atomic position via the parameter estimation theory. The yellow rectangles highlight the transition zones of the exchange of stacking from FCC (ABCABC) to HCP (ABABAB). The 3D models were built by applying the periodicity of FCC and HCP to the experimentally measured atomic positions in a $\langle 110 \rangle_{Al}$ direction. The adaptive common neighbour analysis was applied to determine local structure: where red atoms belong to an HCP structure, green atoms belong to an FCC structure and white atoms are inferred to be the dislocation cores. Dislocation extraction algorithm was used to determine the exact types and locations of dislocations confined in both interfaces as labelled. Reprinted from [9] with permission.

the yellow rectangular frames on HAADF-STEM images. With the precise column location obtained from imaging and known crystal periodicities of FCC/HCP in the viewing direction, we built the interfacial models for structural (using common neighbour analysis [20]) and dislocation (using dislocation extraction algorithm [21]) analysis. We should note that this 2D-to-3D model construction is only possible with interfacial edge dislocations parallel to the viewing direction. For a screw dislocation or a dislocation tilted away from the electron beam, we can no longer simply apply the crystal periodicity, which is disturbed by its strain filed. Type-I interface has three pairs of Shockley partial dislocations as expected in the textbook. Specifically, a 90° Shockley partial dislocation is located in between the "exchanged" layers as described above, while two 30° Shockley partial dislocations are located above and below the ridges with single column Ag depletion region as shown in Fig. 2(a). Each Shockley partial dislocation is separated by two layers of close-packed planes as expected. In contrast, the Type-II interface has periodically spaced a 90° Shockley partial, a Lomer-Cottrell and a Hirth dislocation. Specifically, a 90° Shockley partial dislocation is still located in between the "exchanged" layers as described, while a Lomer-Cottrell dislocation is next to the three-column Ag depletion and a Hirth dislocation is next to the one-column Ag depletion as shown in Fig. 2(b). Our EAM simulations using the experimentally-informed interfacial models suggested the dislocation reaction from the Type-I interface to at the Type-II interface as:

$$\underbrace{1/6[1\bar{2}1] + 1/6[\bar{2}11]}_{\text{30° Shockley partials}} \rightarrow \underbrace{1/6[\bar{1}\bar{1}2]}_{\text{90° anti-Shockley partial}} \rightarrow \underbrace{1/6[\bar{1}\bar{1}0]}_{\text{Lomer-Cottrell}} + \underbrace{1/3[001]}_{\text{Hirth}}. \quad (1)$$

As a first-order approximation, assuming the strain energy associated with dislocations is proportional to the sum of the square of their Burgers vectors, the second part of this reaction reduces the strain energy by 50%. Of course, the exact energy involves the coupling between the adjacent Shockley partial, Lomer-Cottrell and Hirth dislocations at the heterophase interface, instead of a simple superposition of their associated strain energy. We also ignored other electronic interactions due to the chemical segregation at the dislocation core, which should be included for a future study.

Despite its textbook status, the Al-Ag system has not revealed all its phase transformation secrets. For instance, we recently discovered a new phase ζ (AgAl, space group: $R\bar{3}m$) in the binary system. As shown in Fig. 3(a), the ζ phase has a characteristic structure composed of alternating Ag and Al-enriched bi-layers on close-packed planes, while its stacking follows that of FCC, see table 1. Fig. 3(b-e) show the time-resolved evolution of GP zone ϵ-ζ-γ' phase transformations as imaged during *in situ* annealing. This phase transformation pathway is triggered by lattice defects as we discovered recently for alloys heating at the nanoscale (e.g. heating the thin TEM foil), irradiation and bulk deformation [8, 26].

4 Discussion

Precipitate interfaces often have chemical segregations to lower the associated interfacial energy. In binary systems, extra Cu and Ag were found at the coherent interfaces of θ' [4] and γ' [9, 23] precipitates respectively. Individual Sn atoms were also observed at the θ' semi-coherent interface [27]. Also, In-Sb nanoparticles can be attached to the coherent or semicoherent interfaces of θ' precipitate depending on thermal history [28] and consequently change the hardening behaviour. The strength of segregation varies with precipitate thickness, which can be explained reasonably well with DFT calculations from energetics perspective. According to classical nucleation theory, interfacial energy is the critical factor that dominates the nucleation energy barrier. Thus, interfacial solute segregation is likely to be strongly favoured for precipitate nucleation.

Though Al-Ag alloys are soft and have limited engineering application, a trace amount of Ag can significantly improve the ageing response of various aluminium alloys by promoting precipitation [29, 30]. Recent studies showed Ag segregates at the precipitate matrix interfaces [31, 32] or participates within the precipitate phases [33, 34]. Interestingly, the ζ phase shares structural similarities with the layered Ag segregation at several precipitate interfaces, including the θ' phase in Al-Cu-Ag [35], the Ω phase in the Al-Cu-Mg-Ag [32], the T_1 phase in Al-Cu-Li-Mg-Ag [36] and the β''/β' phases in Al-Mg-Si-Ag [37]. The micro-alloying effect of Ag is expected to be deeply related to the clustering behaviour of Ag in the aluminium matrix and its interaction with defects and other solute elements.

Precipitate interfaces may develop crystallographic reconstructions and unique intermediate structures that are distinctly different from the composing bulk phases. To lower the associated interfacial energy, the high-index $\{1\bar{1}00\}_{\gamma'} \parallel \{112\}_{Al}$ semi-coherent γ'-Al interfaces is reconstructed by $\{10\bar{1}1\}_{\gamma'} \parallel \{111\}_{Al}$ and $\{10\bar{1}1\}_{\gamma'} \parallel \{002\}_{Al}$ low-index interfaces [9]. The reconstructed γ' semi-coherent interfaces have periodically spaced dislocations to achieve the FCC-HCP structural transformation. The application of atomic position refinement and atomistic analysis techniques successfully determined the types and locations of dislocations at the interfaces unambiguously. We found two types of FCC-HCP interfaces with different combinations of dislocations with unique chemical distribution at the dislocation cores as shown in Fig. 2. Further EAM simulations of the interface models showed the transformation from Type-I interface into the Type-II interface via a dislocation reaction. Previous experiments fitted the growth kinetics of γ' precipitate and found the thickening is inhibited [38]. Based on the *in situ* TEM observations, Laird and Aaronson [39] proposed a ledge mechanism in which both the coherent and the semi-coherent/incoherent interfaces of precipitates advance by ledges with different mobility. The chemical segregation and structural dislocations confined at the interfaces are expected to influence the growth kinetics of ledges, given that the Hirth and Lomer-Cottrell dislocations are know [40]. However, until now it is not clear how to calculate the interfacial mobility from the first principles. Similarly, the θ' precipitate phase in Al-Cu alloys has an intermediate structure composed by θ'' phase and another so-called θ'_t structure. Such observation may serve as the key to address the disagreement between experimental measurement and theoretical prediction of precipitate aspect ratio and growth kinetics. Future studies are needed to predict interfacial mobility based on the experimental derived atomic models.

The crystallography of the composing phases does not fully determine the structure of interfaces. For example, the η (Al_2Au) phase and the θ' (Al_2Cu) phase share similar crystallography but the interfacial structure is remarkably different: the coherent interface does not have extra Au and the semi-coherent interface is a simple combination of η phase and the Al matrix [41]. Our DFT calculations showed that this is attributed to the much lower defect energy in aluminium of solute Au compared with solute Cu. We found that micro-alloying of Au in Al-Cu alloys can significantly improve the ageing response [42]. EDS mapping showed that Au segregates inside of the θ' precipitates and hence promotes the θ' nucleation [42, 43].

5 Conclusion

Since the first engineering application of aluminium alloys more than 100 years ago, their thermal and mechanical

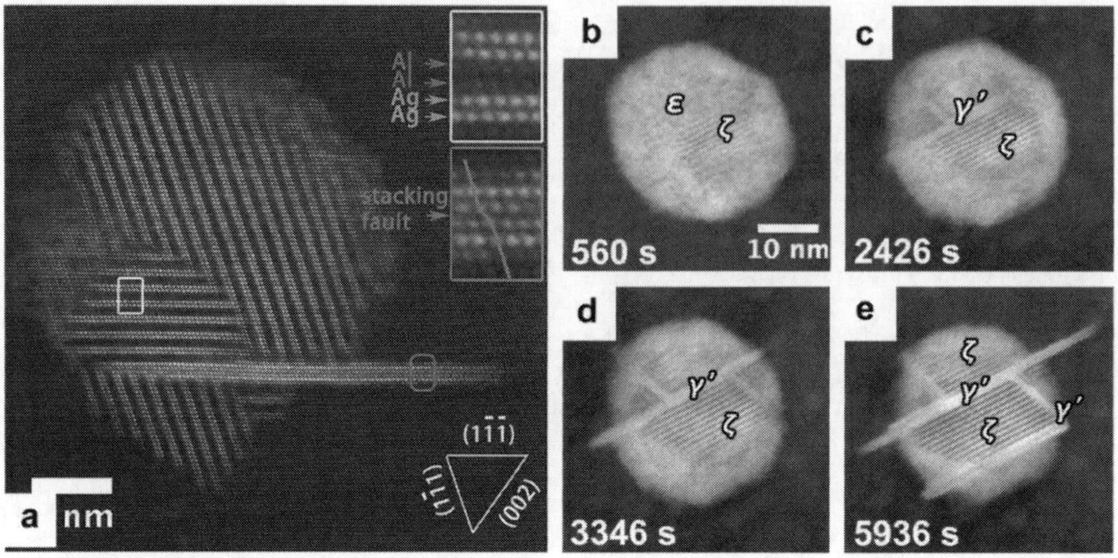

Figure 3. (a) The atomic structure of the ζ (AgAl) phase. The close-packed plane stacking of ζ phase follows that of FCC but with alternating Al- and Ag-enriched bilayers. (b-e) The time-resolved evolution of GP zone ϵ-ζ-γ' phase transformation as imaged during *in situ* annealing at 200°C. Adapted from [8] with permission.

properties have been much improved with our knowledge of precipitation hardening. Here we inherit the insights of pioneer work by revisiting simple alloy systems using advanced imaging and atomistic simulations. A deep understanding of phase transformations requires a careful examination of the precipitate interfacial structures with aberration-corrected electron microscopy. Atomistic calculations can explain the interfacial structures reasonably well and will be increasingly important for precipitation mechanisms study at the atomic scale. Our findings show that the transformations may not be as simple as was believed, due to the intimate interplay between structure and composition confined at interfaces. With an ultimate goal of rational alloy design, further studies are needed to integrate the detailed knowledge of interfaces into the prediction of precipitate nucleation and growth kinetics.

The authors acknowledge funding from the Australian Research Council (LE0454166, LE110100223, DP150100558), computational support from Monash Advanced Research Computing Hybrid, National Computing Infrastructure and Pawsey Supercomputing Centre funded by the Australian Government, and the use of facilities within the Monash Centre for Electron Microscopy.

References

[1] H. Aaronson, M. Enomoto, J. Lee, *Mechanisms of Diffusional Phase Transformations in Metals and Alloys* (CRC Press, 2010), pp. 433–491, ISBN 9781439882535

[2] R. Becker, Ann. Physik **32**, 128 (1938)

[3] J.W. Cahn, J.E. Hilliard, The Journal of chemical physics **28**, 258 (1958)

[4] L. Bourgeois, C. Dwyer, M. Weyland, J.F. Nie, B.C. Muddle, Acta Materialia **59**, 7043 (2011)

[5] L. Bourgeois, N.V. Medhekar, A.E. Smith, M. Weyland, J.F. Nie, C. Dwyer, Physical Review Letters **111**, 046102 (2013)

[6] N.A. Zarkevich, D.D. Johnson, A.V. Smirnov, Acta Materialia **50**, 2443 (2002)

[7] K. Kim, A. Roy, M. Gururajan, C. Wolverton, P.W. Voorhees, Acta Materialia **140**, 344 (2017)

[8] Z. Zhang, L. Bourgeois, J.M. Rosalie, N.V. Medhekar, Acta Materialia **132**, 525 (2017)

[9] Z. Zhang, J.M. Rosalie, N.V. Medhekar, L. Bourgeois, Acta Materialia **174**, 116 (2019)

[10] L. Allen, A. D'Alfonso, S. Findlay, Ultramicroscopy **151**, 11 (2015)

[11] A. De Backer, K. van den Bos, W. Van den Broek, J. Sijbers, S. Van Aert, Ultramicroscopy **171**, 104 (2016)

[12] G. Kresse, J. Furthmüller, Physical Review B **54**, 11169 (1996)

[13] J.P. Perdew, K. Burke, M. Ernzerhof, Physical Review Letters **77**, 3865 (1996)

[14] A. Jain, S.P. Ong, G. Hautier, W. Chen, W.D. Richards, S. Dacek, S. Cholia, D. Gunter, D. Skinner, G. Ceder et al., APL Materials **1**, 011002 (2013)

[15] C. Wolverton, V. Ozoliņš, Physical Review B **73**, 144104 (2006)

[16] W.P. Davey, Physical Review **25**, 753 (1925)

[17] J.P. Neumann, Acta Metallurgica **14**, 505 (1966)

[18] D. Finkenstadt, D.D. Johnson, Physical Review B **81**, 014113 (2010)

[19] S. Plimpton, Journal of computational physics **117**, 1 (1995)
[20] J.D. Honeycutt, H.C. Andersen, Journal of Physical Chemistry **91**, 4950 (1987)
[21] A. Stukowski, A. Arsenlis, Modelling and Simulation in Materials Science and Engineering **20**, 035012 (2012)
[22] A. Stukowski, Modelling and Simulation in Materials Science and Engineering **18**, 015012 (2010)
[23] J.M. Rosalie, C. Dwyer, L. Bourgeois, Acta Materialia **69**, 224 (2014)
[24] J.M. Howe, U. Dahmen, R. Gronsky, Philosophical Magazine A **56**, 31 (1987)
[25] R.D. Doherty, K.E. Rajab, Acta Metallurgica **37**, 2723 (1989)
[26] L. Bourgeois, Y. Zhang, Z. Zhang, Y. Chen, N.V. Medhekar, Nature communications **11**, 1 (2020)
[27] L. Bourgeois, C. Dwyer, M. Weyland, J.F. Nie, B.C. Muddle, Acta materialia **60**, 633 (2012)
[28] Y. Zhang, Z. Zhang, N.V. Medhekar, L. Bourgeois, Acta Materialia **141**, 341 (2017)
[29] I.J. Polmear, Nature **186**, 303 (1960)
[30] J.H. Auld, J.T. Vietz, I.J. Polmear, Nature **209**, 703 (1966)
[31] C. Hutchinson, X. Fan, S. Pennycook, G. Shiflet, Acta Materialia **49**, 2827 (2001)
[32] S.J. Kang, Y.W. Kim, J.M. Zuo, Acta Materialia **81**, 501 (2014)
[33] C.D. Marioara, J. Nakamura, K. Matsuda, S.J. Andersen, R. Holmestad, T. Sato, T. Kawabata, S. Ikeno, Philosophical Magazine **92**, 1149 (2012)
[34] S. Wenner, C.D. Marioara, Q.M. Ramasse, D.M. Kepaptsoglou, F.S. Hage, R. Holmestad, Scripta Materialia **74**, 92 (2014)
[35] J.M. Rosalie, L. Bourgeois, Acta Materialia **60**, 6033 (2012)
[36] S.J. Kang, T.H. Kim, C.W. Yang, J.I. Lee, E.S. Park, T.W. Noh, M. Kim, Scripta Materialia **109**, 68 (2015)
[37] Y. Weng, L. Ding, Z. Zhang, Z. Jia, B. Wen, Y. Liu, S. Muraishi, Y. Li, Q. Liu, Acta Materialia **180**, 301 (2019)
[38] M. Ferrante, R.D. Doherty, Acta Metallurgica **27**, 1603 (1979)
[39] C. Laird, H. Aaronson, Acta Metallurgica Materialia **17**, 505 (1969)
[40] W. Cai, W. Nix, *Imperfections in Crystalline Solids*, MRS-Cambridge Materials Fundamentals (Cambridge University Press, 2016), ISBN 9781107123137
[41] L. Bourgeois, Z. Zhang, J. Li, N.V. Medhekar, Acta Materialia **105**, 284 (2016)
[42] Y. Chen, Z. Zhang, Z. Chen, A. Tsalanidis, M. Weyland, S. Findlay, L.J. Allen, J. Li, N.V. Medhekar, L. Bourgeois, Acta Materialia **125**, 340 (2017)
[43] Y. Zheng, Y. Liu, N. Wilson, S. Liu, X. Zhao, H. Chen, J. Li, Z. Zheng, L. Bourgeois, J.F. Nie, Acta Materialia **184**, 17 (2020)

AUTHOR INDEX